COURSE NOTES ON THE INTERPRETATION OF INFRARED AND RAMAN SPECTRA

COURSE NOTES ON THE INTERPRETATION OF INFRARED AND RAMAN SPECTRA

Dana W. Mayo
Foil A. Miller
Robert W. Hannah

WILEY-INTERSCIENCE
A JOHN WILEY & SONS PUBLICATION

Published by John Wiley & Sons, Inc., Hoboken, New Jersey.
Published simultaneously in Canada.

Library of Congress Cataloging-in-Publication Data:

Miller, Foil A.
Course notes on the interpretation of infrared and Raman spectra / Foil A. Miller, Dana W. Mayo, Robert W. Hannah.
p. cm.
Includes bibliographical references and index.
ISBN 0-471-24823-1 (cloth)
1. Raman spectroscopy. 2. Infrared spectra. I. Mayo, Dana W. II. Hannah, R. W. (Robert Wesley), 1931- III. Title.
QD96. R34M55 2004
543′. 57–dc22 2004000637

Printed in the United States of America

10 9 8 7 6 5 4 3 2

It is with great pleasure and affection that we dedicate this book to the memory of four persons who played enormously important roles in developing and teaching the course on which it is based (the MIT/Bowdoin College course on *The Interpretation of Infrared and Raman Spectra*), but who are no longer with us.

Professor Richard C. Lord of Massachusetts Institute of Technology founded the course in 1950, taught in it for 32 years, and maintained an active interest until his death in 1989.

Professor Ellis R. Lippincott of the University of Maryland participated from 1952 to 1974. His colorful personality and unique lecturing style will long be remembered.

Dr. Lionel J. Bellamy of the Explosives Research and Development Establishment, Waltham Abbey, England was a stalwart of the staff for 22 years. He is well known as the author of several pioneering, widely-used, and influential books on infrared group frequencies. He was as colorful lecturer with a tremendous amount of information on the subject.

Dr. Harry Willis of Imperial Chemical Industries, England brought an extensive knowledge of polymer spectroscopy to his lectures, which extended from 1978 to 1990 (13 years).

All of these individuals contributed to the notes contained herein, and are all greatly missed.

THE AUTHORS

CONTENTS

Foreword ix

Preface xv

1 Introduction 1
Foil A. Miller

2 Characteristic Frequencies of Alkanes 33
Dana W. Mayo

3 Characteristic Frequencies of Alkenes (Olefins) 73
Foil A. Miller

4 Characteristic Frequencies of Molecules with Triple Bonds and Cumulated Double Bonds 85
Robert W. Hannah and Foil A. Miller

5 Characteristic Frequencies of Aromatic Compounds (Group Frequencies of Arenes) 101
Dana W. Mayo

Introduction to Exercises 141

Exercise Section I 145

6 Spectra of X–H Systems (With Emphasis on O—H and N—H Groups) 163
Foil A. Miller

7 Spectra of Carbonyl Compounds of All Kinds (Factors Affecting Carbonyl Group Frequencies) 179
Dana W. Mayo

8 Amides, Carboxylate Ion, and C—O Single Bonds 205
Foil A. Miller

9 Groups Containing N=O Bonds, or Si, P, S, or Halogen Atoms 217
Robert W. Hannah

Exercise Section II 247

10 Infrared Spectra of Polymers: Introduction **261**
Robert W. Hannah and Dana W. Mayo

11 Infrared Spectra of Inorganic Materials **297**
Foil A. Miller

12 Survey of Infrared and Raman Group Frequencies **355**
Dana W. Mayo

Exercise Section III **399**

13 Sample-Handling Techniques **425**
Robert W. Hannah

14 Infrared Spectra of Mixtures **461**
Robert W. Hannah

Answers to Chapter 5 Figure 5.30 **505**

Answers to Exercises **509**

Bibliography **549**

Index **559**

FOREWORD

HISTORY OF THE MIT–BOWDOIN COLLEGE SUMMER INFRARED COURSE: FIRST 51 YEARS—1950–2000

Foil A. Miller

The MIT–Bowdoin College summer course on infrared spectroscopy is the world's longest running short course on this subject. During its first 51 years of operation, over 6900 student-weeks of training have been offered. Consequently it has had a large impact on the use of this technique.

Apparatus for infrared spectroscopy became commercially available at the end of World War II. After a few years, instrument manufacturers became concerned that their sales would be limited by a scarcity of users who were knowledgeable in the measurement and interpretation of infrared spectra. In 1949 Walter Baird and Bruce Billings of Baird Associates and Van Zandt Williams of the Perkin-Elmer Corporation came separately to Professor Richard C. Lord of Massachusetts Institute of Technology (MIT), a leading academic researcher in the field. They explained their concern and asked him to present a short course to provide rapid training in the subject. Lord was interested but did not want to undertake the project by himself, so he invited the author to join him in the venture.

For the first two years the course consisted of two identical five-day sessions held in successive weeks. The enrollment was restricted to 28 students each week because of equipment and manpower needs for the laboratory. There were 15 hours of lectures in the morning, all given by Professor Lord and the author. They were divided about equally between the basic optics of infrared spectrometers and the theory and applications of infrared spectroscopy. Four afternoons were devoted to hands-on laboratory experiments. Three of the experiments were devoted to the properties and use of double-beam instruments and the fourth to single-beam optics and sample handling. Students were divided into groups of seven and rotated among the four experiments. In addition to obtaining various spectra, they did such fundamental operations as cleaving and polishing rocksalt, assembling sealed liquid cells, and focusing a parabola with the Foucault knife-edge test.

The first year tuition was $90, a dormitory room was $2 per night, and meals were paid for by students at the MIT cafeterias. Lectures were held in a room which was not air conditioned and was uncomfortably hot. There was construction work outside, and the lecturers had to compete with the din of jack hammers.

However, it was a very stimulating experience. A number of knowledgeable people from the instrument companies were present, including Van Zandt Williams of Perkin-Elmer, Bruce Billings and David Z. Robinson from Baird Associates, and William S. Gallaway of Beckman Instruments. They commented freely, and the ensuing class discussion was lively and instructive.

It soon became apparent that there were two types of students with different needs. One group had almost no experience in infrared spectroscopy and wanted fundamental information on apparatus, experimental techniques, and applications. The second group had substantial laboratory experience in infrared spectroscopy and wanted much more emphasis on the interpretation of spectra. Therefore, starting with the third year (1952), two separate courses were given in successive weeks. The first was devoted to experimental aspects. In addition to morning lectures, each student had 10 hours of laboratory in the afternoons (2 hours per day for five days). The second course concentrated on the theory and applications of infrared spectra with heavy emphasis on characteristic group frequencies. An important feature was 10 hours devoted to solving problems in the interpretation of unknown spectra.

A laboratory course is labor intensive, requiring instructors and equipment for small groups of students. Attendance in the laboratory course (the first week) was therefore limited to 60. The students were divided into 10 groups of six each. Five of these groups were in the laboratory from 1:00 until 3:00 pm, and the other five from 3:00 until 5:00 pm. The groups rotated among five experiments, one for each afternoon. Instrument manufacturers provided their most recent instruments and also sent skilled personnel to supervise the experiments in which the instruments were used. In addition Professor Lord's graduate students were each assigned to an experiment.

Also in 1952 guest lecturers were added. Some of them were invited to return and ultimately became full-fledged members of the lecture staff. In this way Drs. Ellis Lippincott, Dana Mayo, and Lionel Bellamy became regular lecturers while the course was still at MIT.

The course was well received. By 1955 the demand for the first week had exceeded the limit of 60. (A few excess students took only the lectures but not the laboratory.) Attendance for the second week was over 100. The course was held at MIT each summer for 22 years (1950–1971 inclusive), with annual total attendance during the first 20 years varying between 53 and 207 per year.

In 1970 two new experiments were added, one on Fourier transform infrared spectroscopy and the other on Raman spectroscopy with laser excitation. This was the last year of the laboratory offering. In 1971 there was a precipitous drop in attendance, perhaps related to the sharp economic downturn, and the first week (containing the laboratory) had to be canceled. Only 29 persons attended the second week. MIT informed Professor Lord that it no longer wanted to sponsor the course, so he asked Lippincott, Miller, and Mayo whether any of them wanted to offer the course at their institution. Lippincott and Miller could not do so, but Mayo, who by then was at Bowdoin College in Brunswick, Maine, was enthusiastic. Hence after 22 years at MIT, the course was moved to Bowdoin College, where the 1972 and all subsequent courses have been offered under the direction of Professor Dana W. Mayo. The course has now been at Bowdoin considerably longer than it was at MIT and has had many more students. Professor Mayo continued as Director through 2000, the period covered by this history. He was succeeded in 2001 by Professor Peter Griffiths of the University of Idaho.

Several changes were made when the course was moved to Bowdoin. First, the laboratory portion was abandoned because neither the physical facilities nor the necessary manpower was available. The course therefore consisted of lectures and problem sessions. From 1972 through 1976 only one week was presented.

Second, the daily schedule was changed. Monday morning and afternoon were devoted to lectures, followed by a lobster bake in the evening. Tuesday, Wednesday, and Thursday mornings contained lectures, the afternoons were free, and the evenings were devoted to problem sessions. The course ended Friday at noon. This gave the participants an opportunity to explore the area during three afternoons, a very popular feature. Many attendees brought their families and made the stay part of their vacations.

In 1977 a second week of lectures on advanced topics was added at Bowdoin. The content of the two weeks varied somewhat over the years, but gradually the first week became devoted mainly to infrared and Raman characteristic group frequencies plus a few lectures on instrumentation and sample handling. An important part of the course was the evening exercise sessions, when each student interpreted about 50 unknown spectra. The second week had lectures on more advanced topics. It always had a heavy component on polymers, sampling techniques, instrumentation, and Raman spectroscopy. Other subjects at various times included forensic applications of infrared, quantitative analysis, small samples and microspectroscopy, biological applications, and near-infrared spectroscopy. There were two evenings of problems on polymer spectra, and another evening was devoted to Jeanette Grasselli's famous lecture on the use of combined techniques.

In 1989 a third week was added. Peter Griffiths and James de Haseth had presented a workshop on Fourier transform infrared spectroscopy at the University of Georgia for three years (1986–1988). They proposed moving it to Bowdoin and appending it to the two one-week courses that were already operating there. This was done. The third week differs from the first two in containing a large amount of hands-on work with instruments. Up to eight manufacturers send their instruments and provide personnel to supervise experiments using them. The attendance has been limited to 40 (eight groups of 5 students each) to give every student adequate time on the instruments.

Registrants could take the three weeks in any combination they wished. Most took a week in each of several summers, but a few have taken two weeks and even all three weeks the same year.

During the period that the course has been offered at Bowdoin, the staff has been saddened by the deaths of Ellis Lippincott, Lionel Bellamy, Professor Lord, and Harry Willis. Fortunately Drs. Robert Hannah, Jeanette Grasselli, Peter Griffiths, James de Haseth, and Bruce Chase have been excellent replacements.

SOME DETAILS

A. *Attendance.* At MIT, about 2965 people took the course. (A person attending for two weeks was counted twice. The 1958 attendance figures are missing.) At Bowdoin the number through 2000 was 3938. Thus over 6900 student-weeks of training have been provided.

B. Regular lecturers have been (in the order in which they joined the course):
   1. Richard C. Lord, 1950
   2. Foil A. Miller, 1950
   3. Ellis R. Lippincott, 1956
   4. Dana W. Mayo, 1960
   5. Lionel J. Bellamy, 1962
   6. Jeanette G. Grasselli, 1977
   7. Robert W. Hannah, 1977
   8. Harry Willis, 1978
   9. Peter R. Griffiths, 1983
   10. James A. de Haseth, 1989
   11. Bruce Chase, 2000

C. There were also many guest lecturers over the years. In 1959 they were Norman Colthup, Norman Jones, Norman Sheppard, and Norman Wright. Professor Lord referred to this as "the Norman invasion of infrared."

D. The 40th year of the course in 1989 was marked with a symposium and a gala celebration held during the weekend between weeks 1 and 2.

E. Some traditions
   1. The lecturers greatly enjoy working together to present the course. There is a long tradition of humor, joviality, kidding, and remarks from the back row. Unexpected slides occasionally appear on the screen with comments such as "Baloney," "Hogwash," "Don't believe a word of what follows," or "You have exceeded your time limit." There is a famous slide of Bob Hannah impudently sticking out his tongue. Jeanette Grasselli is notorious for introducing her featured lecture (always an outstanding hit) with a series of slides embarrassing the other lecturers (which, ruefully, is also a hit).
   2. A much-appreciated tradition is the weekly Monday night lobster bake.
   3. The end of the midmorning coffee break is signaled by the ringing of one of two bells. One is a cowbell from India—loud but unmusical. The other is a small melodious ship's bell presented by the students one year because they couldn't stand the tone of the Indian bell.
   4. Another tradition is Foil Miller's presentation concerning the course neckties that the staff has acquired over the years and their analogies to infrared bands.

F. *Courses given abroad.* The course has been presented abroad 13 times as a one-week offering. The material was taken mainly from the first week of the U.S. course, with some omissions to make room for a little material from the second week.

Year	Date	Place
1982	May 10–14	Kristiansand, Norway
1983	May 16–27	Shanghai, Peoples' Republic of China
(One week of material presented over two weeks because of the need to use interpreters.)		
1984	Mar. 26–30	Royal Holloway College near Windsor, England
1985	May 20–24	Strömstad, Sweden
1985	May 28–31	Utrecht, Netherlands
1987	Jan. 26–30	Mexico City, Mexico
1988	May 16–20	Near Malmo, Sweden
1988	May 24–27	Breda, Netherlands
1990	Mar. 19–23	Queretaro, Mexico
1992	Apr. 26–30	Budapest, Hungary
1992	May 4–8	Strömstad, Sweden
1995	May 28–June 2	Lerum, near Gothenburg, Sweden
1999	May 3–7	Halmstad, Sweden

G. A more detailed history of the course has been prepared and can be obtained from any of the regular lecturers.

PREFACE

The notes that form the basis of this text are the distillation of over a half-century of teaching the best strategies for obtaining the maximum amount of molecular structure information from the infrared and Raman spectra of both organic and inorganic materials. Vibrational spectra can reveal significant structural features of complex organic materials. The mission of this text is to develop the logic to extract this information and to transfer the skill to the reader. For this to be successful it is essential that the reader have a working knowledge of modern organic chemistry, particularly of the theory of bonding that has been developed for organic molecules.

The notes are mainly in outline form for brevity, to show the logical organization of the material, and to assist the reader in finding an earlier discussion that becomes important in later arguments.

The Foreword presents a brief historical description of the Summer Course on which the notes are based. The course was founded at MIT in 1950 by Professors Richard C. Lord and Foil A. Miller and was held there each summer for twenty-two years. In 1972 it was moved to Bowdoin College where it has been held each summer up to the present time (2003). In addition the Summer Course has been offered fourteen times outside the United States.

The historical account is followed in Chapter 1 by a general overview of many topics which will be used throughout the book. This includes the origin of infrared and Raman spectra and the definition of terms employed (IX–XIII). There is a section on how vibrations in molecules may be divided into the useful categories of group and fingerprint vibrations (I–IV, XIV). Also introduced at this point is the 1500 cm^{-1} rule (II) and an introduction to the approaches utilized in the interpretation of the spectra along with considerable advice about these latter points (VI, VII). Another topic is an important discussion of vibrational coupling (XV) that will be referred to a number of times in later chapters. Chapter 1 also contains two sections that draw the reader into direct interaction with the text. First, in section V the reader is asked to develop a list of the properties of good group frequencies. Secondly, the reader is taken through the interpretation of the spectrum of a specific fairly complex substance (VIII). While the reader has not yet acquired the background information to make the individual band assignments, the example provides a good demonstration of the *approach* to dealing with an unknown spectrum. This example will be referred to a number of times at the points in the ensuing chapters where the modes used to establish the structure are identified as good group frequencies. The important thing to recognize at this point is the logic of integrating this information to arrive at a potential molecular structure.

Chapters 2–5 contain discussions focused primarily on the spectra related to the vibrations of the hydrocarbon skeletal framework of organic molecules. In Chapter 2 the characteristic *group* frequencies of *alkanes* (the saturated aliphatic hydrocarbons) are considered in detail. These discussions are necessarily limited to modes arising from C—H and C—C stretching and bending vibrations. The displacement of the atoms involved and the frequencies of the modes are described (II). When mode coupling occurs it is noted (II, D, F). The vibrations associated with the normal (or straight chain) hydrocarbons are considered initially (II), followed by coverage of the structurally and vibrationally more complex branched chain and cyclic systems (III). The use of isotopic substitution to identify modes (V), and the influence of hetero-atoms on these group frequencies are also considered toward the end of this chapter (VI). Arguments are developed during these discussions that establish why C—H stretching and bending modes exhibit the characteristics of good group frequencies (II) and why the similar C—C modes give rise to poor group frequencies.

Chapter 3 extends the discussion of hydrocarbon frequencies to *alkenes (olefins)*, that is, to molecules containing carbon-carbon double bonds (>C=C<). The 3000 cm^{-1} rule is introduced (I). The displacement of the atoms leading to alkene group frequencies are described, and in a few cases where coupling occurs it is noted. The discussion covers both open chain and cyclic alkenes (II, A5b), and demonstrates the powerful ability of vibrational spectra to establish the substitution pattern of the unsaturated section of these molecules. The influence of conjugation is noted (II, A5a), followed by the influence of polarization on alkene group frequencies. This latter effect involves the influence of hetero-atoms on the unsaturated C—H and C=C modes (III, C) and can significantly perturb them. The chapter ends with some final advice for dealing with the interpretation of the alkenes (IV). Again, as in Chapter 2, in Chapter 3 the modes discussed here are essentially C—H and C—C vibrations (I, III) with the addition, in this case, of those modes associated with the unsaturated section of the molecule (II).

Chapter 4 examines the group frequencies of triple bonds and cumulated double bonds. The *alkyne* hydrocarbons (acetylenes, —C≡C—) are treated in I and the important nitrile group, —C≡N, in detail in II. In the case of the alkynes the amount and type of substitution can often be identified from vibrational data (I, D, E). Systems containing cumulative double bonds (so-called "back-to-back" double bonds, >C=C=C<) such as the allenes in the hydrocarbon series are considered in III. A number of them contain hetero-atoms as for example, the ketenes (>C=C=O, III, B). Section F presents a problem that allows the reader to determine which of two possible structures is the correct one based on infrared and Raman data.

Chapter 5 completes the discussion of the hydrocarbon frameworks of organic molecules by surveying the group frequencies of *aromatic (arene)* ring systems. The chapter deals primarily with the benzene ring (a carbocyclic aromatic ring system) and its derivatives. The extension of these arguments to aromatic ring systems in general is established and there is a short discussion of heterocyclic aromatic compounds at the end of the chapter (V). A distinct difference between aromatic and aliphatic (saturated) spectra is pointed out and explained (I, A, B). The group frequencies associated with the hydrocarbon section of the benzene derivatives are based primarily on arguments derived from an extension of the vibrational assignments for benzene (II, III). A case where the Rule of Mutual Exclusion, which was developed in Chapter 1, XIII, C, applies is shown in the case of symmetrically *para*-substituted benzene rings (III, A8). The amount of information contained in the vibrational spectra that can be applied to establish the substitution pattern of these rings is large and powerful, and in the case of limited sample size is often the only spectroscopic route to these structural assignments.

The chapter closes with two examples of the application of the aromatic group frequencies to establish molecular structure including the substitution pattern (VI). First, a detailed interpretation of the infrared spectra of styrene and its polymerization product polystyrene is given (VI, A). The reader is then challenged to interpret the infrared and Raman spectra of an unknown sample that contains an aromatic ring system (VI, B). Readers may check their assignments for the "unknown" by turning to the answer section (Chapter 15). This unknown (a hydrocarbon) acts as a bridge to lead into Exercise Section I that follows Chapter 5.

Next comes an introduction to three exercise sections. The introduction explains how the exercises are organized and gives a number of important details about them. It also includes information about individual exercises. It is important that the reader carefully read this material before starting the exercises. It is also profitable to review pertinent sections of the introduction when undertaking a new exercise section later in the text.

Exercise Section I contains four exercises and is designed to accomplish a number of points for the reader. First and most importantly, in Exercises 1–3 the reader is immersed in the interpretation of ten pairs of infrared and Raman spectra, all of which are hydrocarbons. These exercises are arranged to solidly reinforce the discussions of Chapters 1–5 and are a direct extension of the example introduced in Chapter 5 (VI, B). *Secondly, it requires that the reader review in particular detail the discussions of Chapter 1 that involve strategies employed in the interpretation of the spectra.* Finally, Exercise 4 is made up of two parts that involve the interpretation of three infrared spectra of molecules containing hetero-atom functional groups that were discussed in Chapter 3.

This first set of exercises signal that this is a point where the reader is asked to go back and review not just the protocol of interpretation but all of the details of the material discussed in Chapters 1–5.

In approaching the exercises the reader should be very careful to completely read the description of the sample being investigated because a successful solution generally involves utilizing not just the spectrum but all of the information available. For example in Exercise 1, the five unknown samples are described as olefinic hydrocarbons possessing boiling points in the six-carbon range and that they are pure liquids. By adding this information to that obtained from the spectra, it is possible in these examples to narrow each sample to one or two possibilities.

In real life cases we likely will not be able to identify the exact compound but only the general class of materials to which the unknown belongs. An exact match, however, can often be obtained by comparison with spectral libraries. The large majority of the spectra utilized in these unknowns can be found in the Coblentz Desk Book or the Aldrich Handbook. Once the reader has arrived at a solution, the answer may be checked by reference to Chapter 15 that contains detailed solutions to all the problems.

Following the first set of exercises, the discussion shifts to hetero-atom group frequencies starting with Chapter 6. These discussions are rather complete by the time we get to the end of Chapter 9. In Chapter 6 the spectra of two types of X—H systems are examined. After a number of introductory comments about the properties of X—H groups in general and a Summary Table (I), the significant effects of hydrogen bonding on vibrational spectra are considered in some detail (II). The discussion then focuses first on the modes associated with O—H groups (III, IV) and secondly on N—H systems (V). The chapter ends with references to other less important X—H systems that are discussed elsewhere in the text, mainly in Chapter 9 (VI).

Chapter 7 is entirely devoted to discussions of the vibrational frequencies of the carbonyl functional group (>C=O). This group is without question the most important functional one in organic chemistry because it is so pervasive. Fortunately, the vibrations of carbonyls possess all of the characteristics that give rise to excellent group frequencies (I). Perhaps most importantly, those factors that perturb the location of the carbonyl in the spectrum are now well understood in terms of theoretical organic chemistry (I, D). Thus infrared and Raman spectra have become powerful tools that can provide considerable structural information about the local environment of the carbonyl group in an organic molecule. These factors are discussed in detail in part II of the chapter. They are identified as mass (II, A), geometric (II, B), electronic (II, C), and interaction (II, D) effects. In section II both first-order (II, D1a) and second-order (II, D1f) coupling of carbonyls are considered. Also described are examples of intramolecular (II, D1e) and intermolecular (II, D2a) H-bonding of these systems. The discussion covers both field effects (II, D1g) and transannular interactions (II, D1h). The chapter ends with the consideration of amide carbonyl systems (III) and a summary table of specific carbonyl frequencies plus a summary outline of the major factors which effect carbonyls (IV).

Chapter 8 picks up on the discussions of amide carbonyls at the end of Chapter 7 and extends the coverage to all the group frequencies derived from amide systems (I). This includes the N—H stretching and bending modes (which are all given in a Summary Table at the start of the chapter). Of particular interest is the case of secondary amides which undergo coupling of the C—N stretch and the N—H in-plane bend to give two bands (I, D2a). The upper one is found near 1550 cm^{-1} and is strong and isolated while the lower one occurs near or below 1300 cm^{-1} and has both variable intensity and position (I, D2a, 1) and is much less useful than its higher wavenumber companion. Lactam systems are also considered at this point. The carboxylate ion has two modes in the same general region as the amides and is examined next (II).

Finally, C—O single bond systems including ethers (III, B), esters (III, C), anhydrides (III, D), and peroxides (III, E) are examined toward the end of the chapter. Alcohol and carboxylic acid systems are considered in Chapter 6 where the focus is on the stretching vibrations.

Chapter 9 is concerned with functional groups that may contain —N=O bonds or Si, P, S or the halogen atoms. With the exception of the X—H stretches derived from these elements and the nitroso (—N=O) and nitrite groups (—O—N=O), the large majority of the group frequencies associated with these elements fall below 1500 cm^{-1} in the fingerprint region of the spectrum (I).

The chapter opens with a number of general comments (I, II) and then moves to the section on nitroso compounds (II, B). These latter groups can lead to fairly complex spectra because isomerization (enolization) to oximes can occur (II, B4), and in the case of tertiary and aromatic systems dimerization is possible (II, B5). Nitrosoamines also tend to dimerize to give spectra that are consistent with the above assignments (II, D). Finally, there are detailed discussions of nitro (—NO_2 (III, A, B)) and nitrate (—O—NO_2 (III, C)) groups. Here the stretching frequencies are enormously intense and are often the most intense bands in the spectrum.

Next silicon derivatives are considered (I). The more important of the group frequencies associated with this element are the Si—H stretches (II, A, B), the Si—CH_3 methyl deformations (III, A), a mixed Si—CH_3 stretch (III, B) and the Si—O—Si antisymmetric stretch (IV). This section ends with a discussion of halosilanes (V).

Phosphorous compounds are examined in much the same fashion as the silicon systems. A general introduction (I) is followed by discussions of the P—H stretch (II), PH, PH_2 and P—CH_3 systems (III, A, B, C). The P=O stretch (IV) is discussed in some detail. It can provide considerable information about the environment of the P=O group. This section closes with an examination of P—O—C stretches (V) which can be correlated with both aliphatic and aromatic substituents.

Sulfur derivatives behave in a manner similar to phosphorous compounds. This section considers the S—H stretch (I), the C—S (II, A) and S—S (II, B) stretches, and the S=O stretch (III). The latter system (S=O stretch) is examined in detail because considerable information about the local environment of the S=O system can be extracted from the vibrational spectra. The data cover sulfoxides, sulfinic esters, and sulfites (III, C) plus sulfones, sulfonates, and sulfates and the influence of halogen and nitrogen substitution on these systems. (III, D). Finally C=S, C—S and S—S stretches are briefly considered.

The chapter closes with a survey of the halogen group frequencies. The section begins with a look at the C—F stretches which, while intense, do not have any particularly useful correlations except in highly substituted systems (I, II). Next C—Cl, C—Br and C—I systems are examined (III). As in the case of the fluoro derivatives, the other halogens also do not possess good group frequencies because their natural modes lie deep in the fingerprint region and are often and unpredictably mixed with other vibrations. Raman spectra will give the most reliable data because of the intense scattering by these systems. The discussion ends with an interesting example of how conformational isomerism in aliphatic chlorides can be recognized from the vibrational spectra.

This chapter completes the discussions of hetero-atom group frequencies which started in Chapter 6. Exercise Section II which follows will require a review of these four chapters with perhaps a quick glance at the earlier discussions in Chapters 1–5.

Exercise Section II is focused on developing experience at interpreting the spectra of organic systems possessing hetero-atom functional groups. It consists of six exercises (Exercises 5–10) which are composed of six pairs of infrared and Raman spectra, two single infrared spectra plus a very old infrared prism reference spectrum (fifteen spectra total). Exercises 5 and 6 are made up of single pairs of infrared and Raman spectra and form a bridge between Exercise Section I and Exercise Section II because one of the unknown materials is a simple hydrocarbon and the other has a hetero-atom functional group discussed in the earlier section.

Exercise 8 consists of the infrared and Raman spectral pair of a highly moisture-sensitive substance. Once the substance has been identified it will be clear what the differences are between the unknown spectra and the earlier infrared reference spectrum which is included in the problem.

Exercise 10 is the last exercise in this section and is designed to demonstrate the power of infrared data to help establish the course of organic reactions in the research laboratory. These two examples were taken from graduate student research problems of Professor A. C. Cope's laboratory. Once the reader has arrived at a solution the answer may be checked by reference to Chapter 15 which is the most direct route in this latter case because research samples are unlikely to appear in spectral reference collections.

Chapters 10–12 are three chapters that address special areas of interpretation. Chapter 10 is focused on the interpretation of polymer spectra. Exercise Sections I and II have three exercises that involve the identification of relatively simple polymer spectra. These spectra were introduced to demonstrate to the reader that the extension of the group frequencies approach to the interpretation of polymer spectra is, in general, straightforward. However because of the importance of polymer spectra, we now consider this area in some detail in Chapter 10. Section I of the chapter builds on the interpretation of the spectra of hydrocarbon polymers started in exercise sections I and II. In section II the problem of the presence of plasticizers is examined and in addition the polymerization of hetero-atom monomers is explored. The sampling of polymers to acquire infrared and Raman spectra often requires specialized techniques. A short introduction to a few of these techniques is given in Section III. The chemistry involved in the formation of polymers is reviewed in part IV with examples of condensation (nylon) and addition (polyethylene) polymerization presented. Copolymers are examined next (V) with methylmethacrylate-stryene used as an example. The effects on the spectra of block and random copolymerization are also noted. Next crosslinked polymerization is studied (VI) with phenol-formaldehyde. Tacticity (VII) is then explored with evidence for its presence in the spectra of polypropylene. This discussion leads to a concise examination of conformational isomerism (VIII) and the impact of this type of structural modification on polymer spectra. The chapter closes with a discussion and examples of the use of Flow Diagrams in polymer identification.

Chapter 11 is devoted to an examination of the infrared spectra of a large number of inorganic materials. Raman data are valuable but will not be considered in this treatment. To possess internal vibrations a system must be covalently bound. Many inorganic substances do exhibit covalent bonding. Thus, inorganic molecules and polyatomic ions can have normal vibrational spectra.

Following some general introductory remarks (I–III) the chapter surveys the spectra of a large number of well known inorganic substances (IV). There is a summary table of ten inorganic sulfates (IV, B) and a chart of characteristic infrared frequencies covering most common inorganic ions (IV, C). A number of interesting examples are considered. For example, polymorphism is shown to be present in calcium carbonate, and can give rise to two crystalline forms known as calcite and aragonite (IV, E9b). These two different forms give rise to different infrared spectra and it can be easily shown that crab shells have the calcite polymorph while oyster shells have the aragonite form. Sampling techniques for inorganic materials are also considered throughout the chapter (II, III, VIII, & IX). There is a section (XI) containing a brief survey of the infrared spectra of minerals. This section also includes a spectrum of a Type IIa diamond. The chapter ends with a brief introduction to external or lattice vibrations (phonons). These vibrations require the sample to be in the crystalline state. Almost all are below 300 cm^{-1} and therefore out of the range of most infrared instruments. Obviously by their origin they do not contribute to the group frequency information derived from inorganic materials.

Chapter 12 is designed with two purposes in mind. First, it underscores the power of being able to utilize the complementary nature of the infrared and Raman effects simultaneously while viewing these spectra presented together on the same scale. Second, it carries out a complete review of the majority of the essential group frequencies discussed in Chapters 1–9 and prepares the reader for the third and final exercise section. The chapter commences with some general comments about instrumentation and sampling specific to the acquisition of these spectra. A Summary Table of the major group frequencies and their relative intensity in each effect follows. The survey starts with the hydrocarbons (III) and continues on with systems substituted by polar atoms (O, N and the halogens), plus X-H groups (IV). Sulfur systems (V), nitro-group substitution (VI), aromatic systems (VII), nitrile groups (VIII), carbonyl groups (IX), and olefinic systems (X) follow with a number of examples in each section.

The survey ends with three examples where the information from one or both effects solved the problem. First, Raman spectra are shown to be helpful in identifying the presence of carbocyclic ring systems in terpene samples (XI, 1). Second, the application of both infrared and Raman spectra of a series of dimer reaction products unambiguously identified the stereochemistry of the product (XI, 2). Third, an examination of the infrared and Raman spectra of methyl substituted olefins can identify the presence of methyl substituents directly substituted on the double bond carbon atoms (XI, 3).

Exercise Section III consists of nine exercises (Exercises 11–19) that are designed to cover a number of points. These exercises gradually become more challenging compared to Exercise Sections I and II.

Exercise 11 contains two pairs of infrared and Raman spectra of samples that are in the gas phase. These systems exhibit quite different spectra from condensed phase samples. In this case the band contours contain a multitude of sharp spikes, and some bands are missing intensity in the center of the band. The band shapes are governed by the superposition of rotational energy levels on the vibational levels. The spacing of the close-lying rotational levels is controlled inversely by the moments of interia of the molecule. If at least one of the moments is small, the level separation increases. Hence in light molecules these individual transitions may be resolved around a vibrational line, and these bands become very complex. As the molecular weight of the molecule increases the rotational spacing collapses and broad smooth band contours are observed in the gas phase. This is similar to the band shapes observed in condensed phases where free rotation no longer occurs. The spectra observed in Exercise 11 clearly require that the unknown substances are of low molecular weight. Exercise 12 has two parts with two pairs of infrared and Raman spectra in each part. The samples are pure materials. It is possible to make a confident identification of these unknowns by utilizing all of the information supplied with problem. Exercise 13 has two parts that are illustrative of practical problems that depend on just infrared data to reach a solution. Exercise 14 also involves two parts with two pairs of infrared and Raman spectra. In this case the identification is aided by having additional spectra obtained at different sample path lengths available. Exercises 15 and 16 relate to the discussions in Chapter 9. Exercise 15 is a single infrared and Raman pair. It is a pure substance that contains sulfur as the most important clue to the identification. Exercise 16 has two parts and depends only on infrared data for the solutions, however the unknowns have been identified as esters of inorganic acids which greatly reduces the possible solutions. Exercise 17 draws on the discussions of Chapter 11 to correctly identify two inorganic materials from their infrared spectra. Exercises 18 and 19 expand the reader's experience at interpreting polymer spectra and draw on the material presented in Chapter 10. Exercise 18 has only infrared data while Exercise 19 has both infrared and Raman. As with the earlier sections, once the reader has arrived at a solution the answer may be checked by reference to Chapter 15 where there are detailed solutions to all the problems.

This completes the exercise sections. The reader should now have considerable confidence to tackle the real life interpretation of unknown spectra in their own laboratories.

Our discussions conclude with two chapters that are focused primarily on an introduction to sampling procedures for obtaining infrared spectra. Chapter 13 considers the sampling techniques themselves while Chapter 14 examines the more challenging problems involved in preparing mixtures of materials for infrared analysis.

Chapter 13 covers sampling of liquids and solids. Liquid sampling by solution, capillary film, sealed cells, and internal reflection are considered. Solid sampling by solution, thin film, mull, KBr disk, pyrolyzate, specular reflection and diffuse reflection are touched on. Sealed cells are described, along with the solvents generally employed and procedures for accurately determining the cavity thickness. Particular attention is paid to the mulling and pressed KBr disk techniques for handling solids. Reflection techniques are described in some detail, with a number of examples mentioned.

Chapter 14 starts with a general discussion of the approach to examining mixtures, including some valuable advice (I). Because the separation of the components of a mixture is the most powerful approach to unraveling the identification of the substances involved, the techniques of choice are carefully reviewed in the next section (II). A series of examples employing these techniques helps to demonstrate how effective this approach can be (III). These examples cover a wide range of mixtures and separation techniques such as: lubricating grease, lipstick, barbiturates, adhesives, fire retardant, phenolic resin, binder in a grinding wheel, polybutadiene and polyvinylchloride, carbon filled materials such as butyl rubber, and a gasoline additive (III).

Finally, a number of techniques that eliminate the need to carry out the actual separation of the components have been developed. For example, computer subtraction has proven to be valuable and an example of this technique involving gasoline components is given. Secondly infrared microscopes have been able to isolate the spectra of components in solid mixtures such as powders and laminates. GC-IR systems are not covered in this treatment.

Chapter 15 contains detailed solutions of all of the Exercise Sections plus the Unknown Spectrum at the end of Chapter 6 (VI, B). Chapter 15 will be most effective when the reader does not refer to it until they have carefully arrived at their own solution to a particular problem.

The text concludes with a comprehensive bibliography of the infrared and Raman literature. It consists of the following sections: 1. The Basics. A core collection on infrared spectroscopy. 2. General Texts. 3. Libraries of Infrared Reference Spectra, 4. Infrared Group Frequencies, 5. Gases and Vapors, 6. Polymers and Coatings, 7. Inorganics and Organo-Metallics, 8. Biochemical, Drug, and Forensic Applications, 9. Other Classes of Compounds, 9A. Essential Oils, 9B. Isotopically-Labeled Compounds, 9C. Silicones, 9D. Organo-phosphorus Compounds, 9E. Pesticides, 9F. Propellants and Explosives, 9G. Simpler Molecules, Fundamental Frequencies for, 10. Surface Studies, 11. Near Infrared Spectra, 12. Far Infrared Spectra, 13. Instrumentation and Techniques, Miscellaneous Infrared Topics, 14. Raman Spectroscopy.

DANA W. MAYO
FOIL A. MILLER
ROBERT W. HANNAH

ACKNOWLEDGMENTS AND AN ANNOUNCEMENT

There are over five hundred spectra in this book. Although the majority of them were generated in our laboratories, many were obtained from other sources and we wish to gratefully acknowledge this valuable help. All of the spectra, both ours and others, have been modified by computer manipulation to improve them in various cosmetic ways such as (1) by changing from an original linear micrometer presentation to one linear in cm^{-1}, (2) to change the intensity scale, e.g. from absorption to transmission, (3) to darken the spectral curve, or (4) to remove dark background. The very large majority have been placed on new higher resolution grids for easy reading. They are teaching tools, not reference spectra as these manipulations have created small background and band shape distortions.

Jeanette Grasselli, a long-time and valued lecturer in the Week II course, kindly supplied many of the polymer spectra from her laboratory, especially spectra contained in several of the exercises. Two of the unknown polymer spectra used in Exercise Section III and one spectrum in the Chapter 10 are based on work from Siesler and Holland-Moritz.

Data from Norman Jones' group at the National Research Council of Canada were used for two figures on cyclopentanone spectra in Chapter 1, for several of the spectra of alkanes in Chapter 2, and for the band patterns of substituted benzenes in Chapter 5. In Chapter 5 the combination band patterns for pyridine systems were derived from the work of Katritsky at Cambridge University.

Chapter 7 contains a modified version of a Lionel Bellamy figure showing the shift of the carbonyl frequency with variation of the internal bond angle and the mass of the C-X group. A number of the inorganic spectra in Chapter 11 are from the excellent book by Nyquist and Kagel of the Dow Chemical Co.

The normal modes for the benzene ring shown in two figures in Chapter 5 can trace their origin to a modification of those modes given by Clothup, Daly and Wiberley.

The authors would also like to express deep appreciation for the patient encouragement and deep understanding shown by the editorial staff at John Wiley. Darla Henderson, Amy Romano, and Christine Punzo were absolutely essential ingredients in the birth of the text. We also would like to mention an old friend, Barbara Goldman, who really was the catalyst that got the computers into action.

Finally, the authors would like to mention that, following the year 2000 course, Professor Peter Griffiths of the University of Idaho became the new (and 3rd) Director of the course. We are confident that under his guidance this educational program will continue to play a significant role in training individuals to optimize their measurement and interpretation of infrared and Raman spectra.

1 Introduction

FOIL A. MILLER

I. Vibrational frequencies can be divided into two broad categories: group frequencies and fingerprint frequencies.
   A. Group frequencies
      1. These are characteristic of groups of atoms: $-OH$, $-C{\equiv}N$, $-CH_3$, $-C_6H_5$, $-COOH$, $-CONH_2$, $-NO_2$, etc.
      2. The vibrations are largely localized within the group.
   B. Fingerprint frequencies
      1. These are highly characteristic of the specific molecule.
      2. They are due to vibrations of the molecule as a whole rather than being localized within a group.
      3. The numerical values usually cannot be predicted except in a very general way; e.g., the frequency will be between 1300 and 1000 cm^{-1}.
      4. Fingerprint frequencies are valuable for characterizing a molecule.
         a. The infrared (IR) spectrum is the most unique, characteristic, and widely applicable physical property known.
         b. It is therefore termed the "fingerprint" of a molecule. The unique, characterizing vibrations are "fingerprint vibrations."
            1) But the IR spectrum is really better than a fingerprint.
            2) The fingerprint of a person tells nothing about appearance—height, build, weight, color of skin or hair or eyes.
            3) The IR spectrum not only identifies the compound but also tells something about its make-up through the group frequencies. For example, the sample does or does not have a C=O or an O–H or a phenyl group.

Course Notes on the Interpretation of Infrared and Raman Spectra, by Dana W. Mayo, Foil A. Miller, and Robert W. Hannah.
ISBN 0-471-24823-1

C. Historical note

1. Sir William Herschel, an English astronomer, discovered the IR region of the spectrum in 1800.[1]
2. His grandson, William James Herschel, realized that human fingerprints are unique and was the first person to understand their utility.[2] (J. E. Purkyne noted their uniqueness in 1823 but did not understand their usefulness.[3])
3. Thus the grandfather's discovery laid the basis for fingerprinting molecules, and the grandson's discovery laid the basis for fingerprinting humans.

Some abbreviations that will be used:

s, m, w = strong, medium, weak (band intensity)
sp, b = sharp, broad (band width)
v = very

II. Importance of 1500 cm^{-1}, a useful dividing point

A. Above 1500 cm^{-1}

1. If a band has reasonable intensity, it is certainly a group frequency. (Very weak bands above 1500 cm^{-1} may be sum tones or overtones.) The interpretation is usually reliable and free of ambiguities. One can be confident of the deductions.
2. Therefore we start at the high-wavenumber end of the spectrum and work downward.

B. Below 1500 cm^{-1} (the fingerprint region)

1. The band may be *either* a group frequency or a fingerprint frequency.
2. The lower the frequency, the more likely that a band is due to a fingerprint mode.
3. Even if a band in this region has the proper frequency for a group, it does not necessarily follow that that group is present. The band may be there just by chance.
4. For this reason it is desirable for a group frequency below 1500 cm^{-1} to be characterized by something in addition to its frequency— e.g., to be very intense, unusually broad, unusually sharp, or a doublet.
5. In the fingerprint region, the *absence* of a group frequency is more reliable evidence than its presence.
6. The best way to use this region is to raise specific questions in the high-frequency region, and come here to seek the answers. *Examples*:
   a. There is saturated C—H present (bands at 2970–2850 cm^{-1}). Is there evidence for C—CH_3? Look at 1378 $\pm$ 5 cm^{-1} for a weak-to-medium (w–m) band.
   b. There is an O—H ($\sim$3350 cm^{-1}, s, vb). Is it an alcohol? If so, is it aliphatic or aromatic? Look at 1250–1050 cm^{-1} region for a very strong band.

c. There is C=C (1680–1630 cm^{-1}). How is it substituted? Look for the strongest band in the 1000–700 cm^{-1} region.
d. There is a phenyl group (~1600 and ~1500 cm^{-1}). How is it substituted? Look at the strongest bands in the 900–700 cm^{-1} region.

7. In each case we have a reason to look in a narrow part of the fingerprint region to see whether or not there is a band there.

Now we return to group frequencies.

III. Definitions of a group frequency

A. First, ideally, it is a frequency which is *always* found in the spectrum of a molecule containing that group and *always* occurs in the same narrow wavenumber range.
1. We shall have to relax the two "always" in some cases.
2. There are *infrared* group frequencies and *Raman* group frequencies. They are often complementary—one intense, the other weak.
3. This is a practical working definition. There is another less useful one based on theory.

B. Second, it is the frequency of a vibration for which the *form* of the vibration—the pattern of the displacements—is nearly the same in every molecule containing that group.

IV. Group frequencies are determined empirically by studying the spectra of many related molecules.

V. Desirable qualities of a good group frequency. (To be completed by the reader. Answers are at the end of this chapter—but please don't peek until you make your own list.)

A.
B.
C.
D.
E.
F.
G.

No group satisfies all these qualities. Groups coming the closest are >C=O, –O–H, and –C≡N.

VI. The *absence* of a group frequency is often as useful information as its presence. Therefore it is helpful to use cross-hatched chart paper with a wavenumber scale to see precisely where there are *not* certain bands.

VII. Procedure for using group frequencies

Suppose you are presented with an IR spectrum and asked what it tells about the sample. How should you proceed?

A. Note how the sample was run: physical state, thickness, mulling agent or solvent. Mark the positions of the strong bands of any mulling agent or solvent because the instrument may be dead there and give no information.
B. *Start at the high-frequency end* of the spectrum and work downward.
C. Concentrate on the intense bands. Really intense bands are usually due to polar groups.
D. *Do not try to account for all the bands*. Many of them—sometimes most of them—are fingerprint bands.
E. Ask questions of the spectrum as described earlier.
F. Know the reliability of the assignments—whether "certain" or "maybe."
G. Use other information.
   1. There is other information within the spectrum: bandwidths and intensities; in a Raman spectrum, depolarization ratios.
   2. Likely possibilities: One often knows *something* about the sample—that it is a detergent, a polymer, a street drug, etc. This narrows the possibilities dramatically.
   3. Elemental analysis: The sample cannot contain an $-SO_2-$ group if it has no sulfur!
   4. Molecular weight: This may be available from the mass spectrum.
   5. Physical properties: Melting point (mp), boiling point (bp), solubility, color, odor, refractive index.
   6. Other kinds of spectra: nuclear magnetic resonance (NMR), ultraviolet (UV), mass. *These are invaluable aids.*
H. It is helpful to make neat penciled notes right on the spectrum.

VIII. *Example*: See Figure 1.1.
A. Note the format. The spectrum is linear in reciprocal centimeters with no scale change.
B. The intensities are excellent.
C. The sample was run as a KBr disk. Therefore it is a solid, and there are no interfering bands. (Sometimes with KBr disks a weak, very broad water band appears at 3300–3000 cm^{-1}, but there is none in this spectrum.)
D. A preliminary glance at the full spectrum shows many sharp bands. This indicates an aromatic ring or some other relatively rigid structure.
E. Start at the high-frequency end of the spectrum and work downward.
   1. *~3370 cm^{-1}*. Ignore this initially because at this value it is too weak to be a fundamental.
   2. *~3094 and ~3053 cm^{-1}*. Certainly due to C—H stretches of an sp^2 C. Suggests an aromatic or olefinic compound.
   3. *3000–2900 cm^{-1}*. There is nothing significant here, which itself is significant. There is *no* aliphatic C—H. This is an example of the value of negative information.

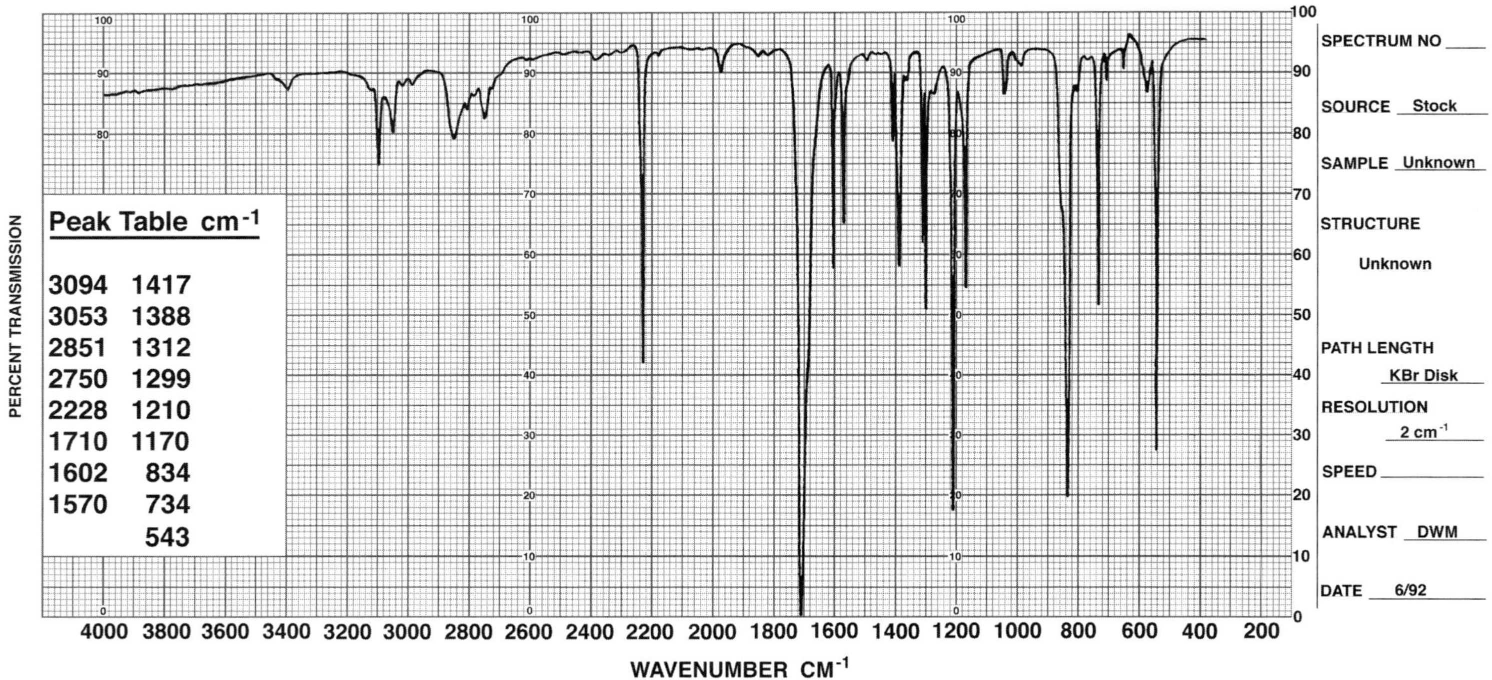

Fig. 1.1 Infrared spectrum of unknown sample run as KBr disk. Linear wavenumber presentation with no scale change.

4. *2851 and 2750 cm^{-1}*. Certainly due to an aldehyde C—H stretch.
5. *2228 cm^{-1}*. Certainly due to a —C≡N group, and from the exact wavenumber it must be conjugated. (Unconjugated —C≡N is at 2250 ±10 whereas conjugated —C≡N is 2225 ±15.)
6. *1710 cm^{-1}*. Certainly due to a carbonyl group. Furthermore it is slightly lower than a normal aldehyde (~1725 cm^{-1}). The two most common effects that lower carbonyl frequencies are hydrogen bonding and conjugation. There is no hydrogen bonding here because there is no O—H or N—H stretch. Therefore the carbonyl *may* be conjugated.
7. *1602 and 1570 cm^{-1}*. Good evidence for a phenyl ring. Because 1570 cm^{-1} is relatively strong (rather than being a shoulder or a weak band), the ring is either conjugated or has an O or an N atom directly attached to it.
8. How is the ring substituted? There are two places to look:
   a. *2000–1650 cm^{-1}*. Unfortunately this region cannot be used because the intense carbonyl band obscures part of it.
   b. Go to the region below 900 cm^{-1} and look for strong band(s). There is one at 834 cm^{-1}, which suggests para-substitution but does not prove it.

F. Now put all this information together and try to deduce some possible structures. One then looks up their spectra and compares them with the unknown spectrum.
   1. The sample has:
      a. A phenyl ring, probably conjugated and possibly para-substituted
      b. An aldehyde
      c. A carbonyl, possibly conjugated (It may well be in the aldehyde.)
      d. A conjugated nitrile group
   2. It does *not* have aliphatic C—H. Nor does it have a number of other groups, such as —COOH, C—O, O—H, or —N—H.
   3. This suggests *para*-cyanobenzaldehyde. It is the correct structure.

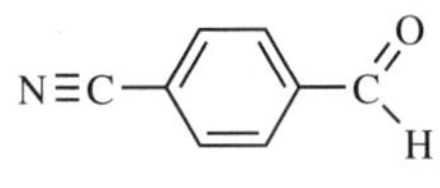

   4. We have obtained the complete structure from the IR spectrum alone. That is unusual. One tries to narrow it down to three or four structures and compare their IR reference spectra against that of the unknown.

G. Note that most of the bands have not been explained. They are fingerprint bands, not group frequencies.

H. Additional comments
   1. *~3370-cm^{-1} band*. This is probably the overtone of the carbonyl stretch. Calculated as $2 \times 1705 = 3410$ cm^{-1}. Observed,

~3370 cm^{-1}. It is typical to find the observed band about 1% lower than the calculated value, which in this case would be 3410 − 34 = 3376 cm^{-1}.

2. Where are the 1500 and 1450 cm^{-1} phenyl bands? These also are characteristic of phenyl rings and are good group frequencies. It is unusual for them to be missing, but this shows that group frequencies are sometimes fallible. It is known that for some aromatic rings conjugated with carbonyl the IR intensity of 1500 cm^{-1} may be quite weak, and that is probably the case here.[4]
3. *1388 cm^{-1}*. This is a sharp, moderately intense band. The C—CH_3 group has a useful frequency near here, at 1378 ± 5 cm^{-1}. It cannot be due to that, however, because there is not enough aliphatic C—H stretch intensity near 2950 cm^{-1}, and because 1388 cm^{-1} is slightly out of the very narrow range of 1378 ± 5. This is an example of an interference near this position. It is not common, but it does occasionally occur.

I. Analyzing a spectrum is fun. Regard it as a game. The famous astronomer and spectroscopist Henry Norris Russell said, "Analyzing a spectrum is exactly like doing a crossword puzzle, but when you get through with it you call the answer research."[5]

IX. Criteria for band intensity and bandwidth

A. Intensity

1. Correct way: Measure the extinction coefficient ε (synonymous with absorptivity *a*) using the *Lambert–Beer law*:

$$\log_{10}(I_0/I) = \log_{10}(1/T) = A = \varepsilon cl \text{ (or } = abc)$$

where I_0 = intensity of light incident on sample
I = intensity of light emerging from sample
T = transmission, I/I_0
A = absorbance, $=\log_{10}(I_0/I)$
ε = extinction coefficient, $\equiv a$ (absorptivity)
c = sample concentration
$l \equiv b$ = sample thickness

a. Using ε removes the effect of concentration and path length. The value of ε at any wavenumber is a property of the substance.
b. But to do this is quite a bit of work; one has to determine the concentration and path length. It is not necessary for our use with group frequencies.

2. Qualitatively:
   a. Adjust the sample thickness or concentration so that the strongest band is 5–10% transmitting. Then rate on an arbitrary, subjective scale as vs, s, m, w, or vw.

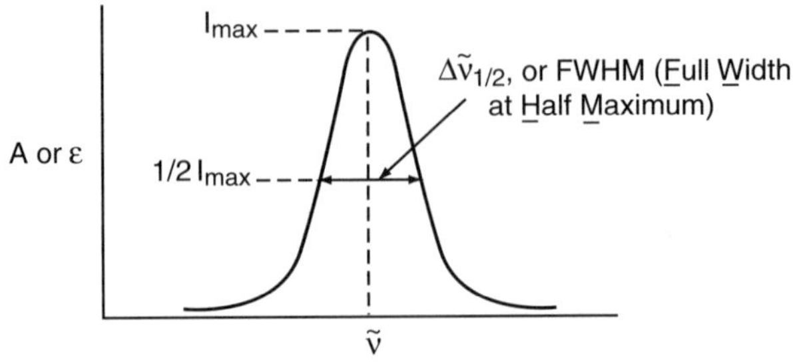

Fig. 1.2 Measurement of half bandwidth.

b. Make crude comparisons between spectra by making allowances for differences in sample thickness and concentration. It is much simpler to do this if the intensity scale is linear in absorbance (A) because $A \propto cl$. If the scale is linear in $\%T$, remember that T depends exponentially on c and l: $T = 10^{-\varepsilon cl}$.

B. Bandwidths

1. Correct way: Plot absorbance (A) or absorptivity (ε) vs. reciprocal centimeters, and measure the width of the band at half intensity (Figure 1.2). This is called the "half band width," symbol $\Delta\tilde{\nu}_{1/2}$.
2. Some typical values of half bandwidths

a. Light gas, individual vibration–rotation lines	$<0.1\ \text{cm}^{-1}$
b. Liquid or solution	5–15
c. Hydrogen-bonded liquid	20–400
d. Solid	5–50

This too is a lot of work, and more than we need to do.

3. Qualitatively, for use with group frequencies:
   a. Look at the bands on either a $\%T$ or an absorbance plot and classify them as sharp, normal, or broad.
   b. *Example*: Hexane vs. hexanoic acid (Figures 1.3a, b).
   c. For comparing bandwidths, a linear wavenumber presentation is desirable. In the early days of IR spectroscopy many excellent reference spectra were plotted linearly in micrometers (previously called microns). A linear micrometer plot distorts bandwidths. It squeezes bands at low micrometers (high wavenumbers) and widens bands at large micrometers. These spectra are still useful, but one must remember this distortion when making comparisons with spectra that are linear in wavenumber.

C. Good reference on band intensities and bandwidths: R. N. Jones and C. Sandorfy, in Weissberger, *Technique of Organic Chemistry*, Vol. 9, Interscience, New York, 1956, Chapter 4.

X. How many bands are expected in an IR or Raman spectrum?

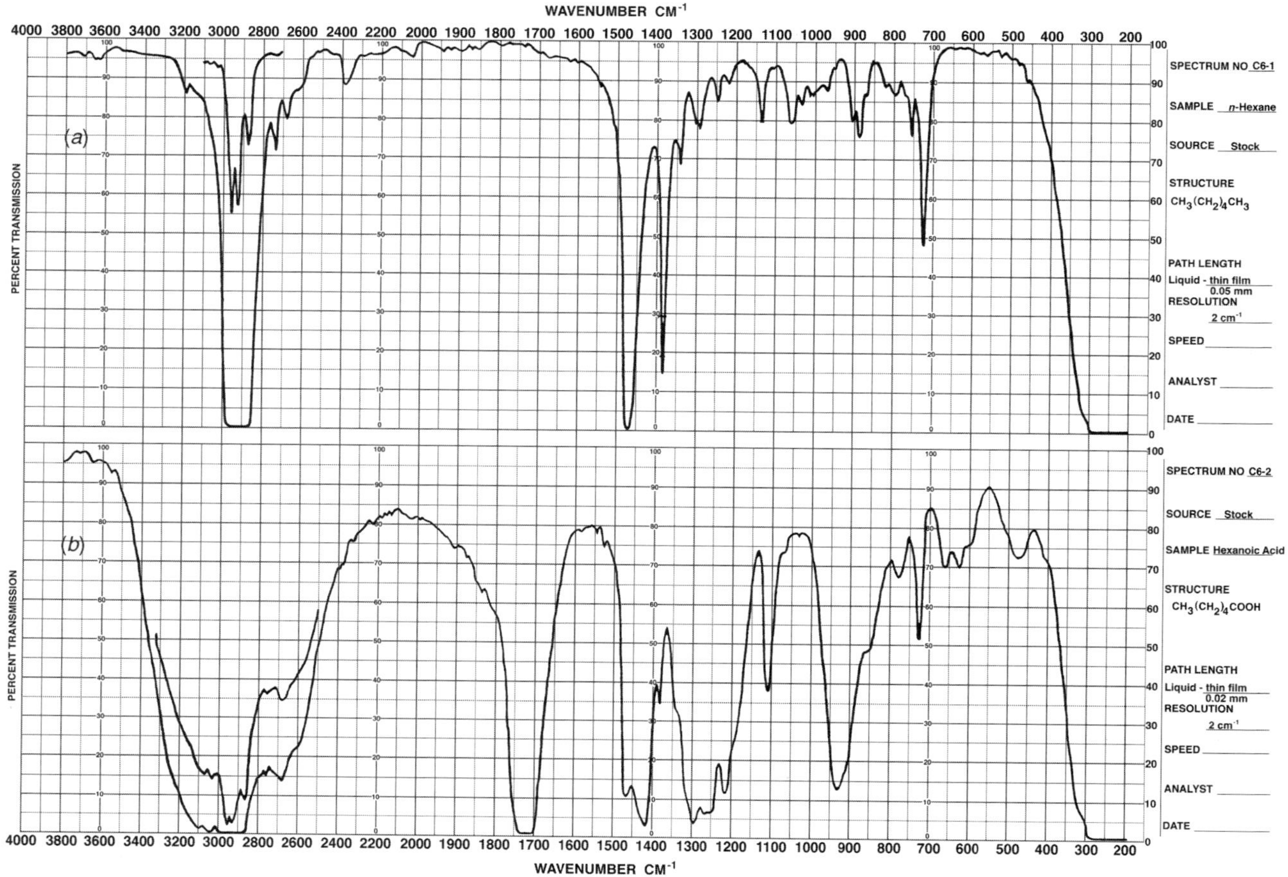

Fig. 1.3 Infrared spectra of (*a*) *n*-hexane and (*b*) *n*-hexanoic acid. Note that there is a 2× scale change at 2000 cm^{-1}. Also note that some of the bands in (*b*) are much broader than any in (*a*).

Answer: Approximately $3N - 6$, where N = number of atoms in the molecule.

A. We are concerned with IR and Raman *vibrational* spectra.
   1. A molecular vibration is a motion of the nuclei *relative to each other.* A vibration has no translation (no net linear momentum) and no rotation (no net angular momentum).
   2. There are $3N - 6$ (or $3N - 5$) fundamental vibrations.
      a. The terms *fundamental vibration*, *normal vibration*, and *normal mode* are synonymous.
      b. In principle, each fundamental vibration can appear in the spectrum, although in practice not all do.

B. Derivation of $3N - 6$ rule
   1. Degrees of freedom (d.f.)
      a. A single particle has 3 d.f. of motion because it can move independently along the three Cartesian directions X, Y, and Z. Alternatively, three coordinates are needed to specify its position in space, and each of these can be varied independently. These are *translational* d.f.
      b. For two particles, there are 6 d.f. Now tie the two together with a stiff spring (a chemical bond) to represent a diatomic molecule. Then there are only three degrees of translational freedom, given by the motion of the center of gravity. What has happened to the other three? They have become one vibration and two rotations, as shown in Figure 1.4.
      c. For a molecule with N atoms:
         1) The total number of d.f. = $3N$. Of these:
            a) Three are translational (i.e., the coordinates of the center of gravity).
            b) If the molecule is *nonlinear*, three more are rotational because there are three principal axes of rotation.
            c) If the molecule is *linear*, only two are rotational because now there are only two principal axes of rotation. (Rotation around the axis of the molecule does not count because there is so little mass off of it.) Examples of

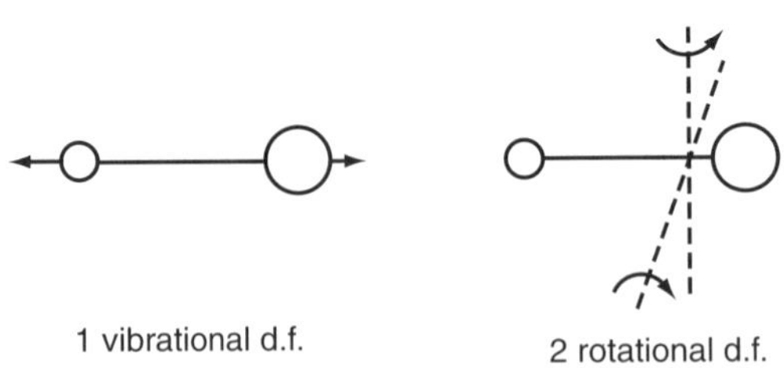

Fig. 1.4 One vibration and two rotations of a diatomic molecule.

linear molecules are all diatomic molecules plus O=C=O, H–C≡C–H, and H–C≡C–C≡N.

d) All the other d.f. must describe motions where the nuclei move *relative to each other*, i.e., must be vibrational. (In translation and rotation, the nuclear arrangement is rigid.)

2) Therefore *for a nonlinear molecule* (by far the most common case):

$$\text{No. of vibrational d.f.} = 3N - 3(\text{translational}) - 3(\text{rotational}) = 3N - 6$$

For a linear molecule:

$$\text{No. of vibrational d.f.} = 3N - 3(\text{translational}) - 2(\text{rotational}) = 3N - 5$$

2. There are as many fundamental vibrations as there are vibrational degrees of freedom: $3N - 6$ or $3N - 5$.
3. Each fundamental vibration has a different pattern of displacements. The displacements are shown by arrows, as in Figure 1.5*a*. Reversing all the arrows, as in Figure 1.5*b*, is an equally good representation because it is just the opposite phase of the same vibration.
4. It is the frequencies of these fundamental vibrations that are observed in IR and Raman spectra. Note that the number of them is finite.

C. But we do not expect to see exactly $3N - 6$ (or $3N - 5$) bands because there are reasons for observing both fewer and more.

1. Reasons for observing fewer bands
   a. In molecules with symmetry, selection rules may forbid some from being observed. The molecule still carries out these vibrations, but they cannot be excited by absorbing IR radiation or by Raman scattering.
   b. Bands may be formally allowed but are too weak to be observed at the sample thickness used.
   c. Bands may be outside the range of the instrument, usually at lower frequency.

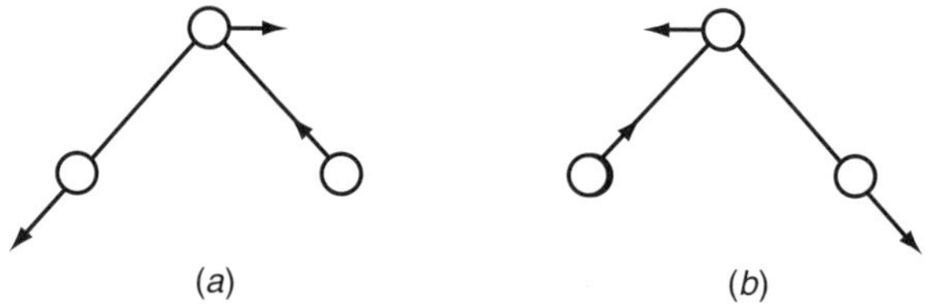

(*a*) (*b*)

Fig. 1.5 Two equivalent representations of the same vibration. (It could be the antisymmetric stretch of H_2O or SO_2.)

d. Degeneracy may cause some to be coincident. Degeneracy is the occurrence of two or more bands at the same frequency. There are two types of degeneracy:
   1) *Required degeneracy.* When a molecule has a three-fold or higher axis of symmetry, some of its vibrations *must* be degenerate. They occur in pairs or triples with identically the same frequency. An example is the doubly degenerate vibration of CO_2:

      (+ means motion out from the plane of the paper; − means motion back from the plane of the paper. They are not electrical charges.) These two bending vibrations have identically the same energy because they differ only in being rotated 90° in space. They are doubly degenerate.
   2) *Accidental degeneracy.* Two or more vibrations have frequencies so close together that the instrument does not resolve them. A good example is the C—H stretches of saturated hydrocarbons. Although *n*-hexane has 14 C—H groups and therefore 14 C—H stretching vibrations, at most only four separate bands are observed at 3000–2800 cm^{-1}. (See Figure 1.3a.)
   3) One cannot identify degenerate bands from their intensities. A doubly degenerate band will be twice as intense as it would be if it were singly degenerate, but twice as intense as what?

2. Reasons for finding more bands than expected
   a. One may observe sum tones. These are classified as:
      Binary: $\tilde{\nu}_a + \tilde{\nu}_b$
      Ternary: $\tilde{\nu}_a + \tilde{\nu}_b + \tilde{\nu}_c$ or $2\tilde{\nu}_a + \tilde{\nu}_b$
      Quaternary, etc.
      In absorbing one photon of light, two or more vibrations are excited simultaneously. This is improbable, so these bands are weak.
   b. Overtones may be seen. They are classified as:.
      First overtone, or second harmonic $2\tilde{\nu}_a$
      Second overtone, or third harmonic $3\tilde{\nu}_a$
      etc.
      Thus overtones are a special case of sum tones where all the frequencies are the same. Like sum tones, they are weak.
   c. Difference tone $\tilde{\nu}_a - \tilde{\nu}_b$ Uncommon and weak

3. The net result is that we do not expect to see exactly $3N - 6$ bands in the spectrum. On the other hand, the number of bands will be finite and often modest.

a. If the symmetry is high, selection rules will forbid the observation of some of the fundamentals in the IR. Other vibrations will be doubly or triply degenerate. Therefore far fewer than $3N - 6$ will be observed.
b. As the symmetry is lowered, more bands are seen.
c. *Example*: Figures 1.6*a*, *b*. Infrared spectrum of CCl_4 vs. CH_2Cl_2. Both have the same numberof atoms, five, and $3N - 6 = 9$ for both. However, CCl_4 is much more symmetrical and has a much simpler spectrum.

XI. Mechanism for the absorption of IR radiation

A. How do molecular vibrations interact with light? A simple classical model can be given which is very useful, although it is not correct in every detail. We shall use a diatomic molecule for simplicity.

B. A diatomic molecule has a natural frequency of vibration given by Hooke's law:

$$\nu(\text{sec}^{-1}) = \frac{1}{2\pi}\sqrt{\frac{k}{\mu}} \qquad \text{or} \qquad \tilde{\nu}(\text{cm}^{-1}) = \frac{1}{2\pi c}\sqrt{\frac{k}{\mu}}$$

where k = force constant.
μ = reduced mass, $= (m_1 \times m_2)/(m_1 + m_2)$
c = speed of light

For a given diatomic molecule, m_1 and m_2 are fixed, and therefore μ is also. So too is the force constant k. Therefore each kind of molecule has a definite frequency of vibration which is unique to it.

C. Imagine an HCl molecule between the plates of an electrical capacitor, as in Figure 1.7.

1. HCl has a dipole moment—i.e., a separation of electrical charges, as shown The charge is not a full electron or the molecule would be ionic, so it is represented as a partial charge.
2. If the capacitor plates are charged as shown and the molecule is oriented as shown, the bond is stretched because of the attraction of dissimilar charges. If the electrical field is reversed, the bond is compressed.
3. If the field could be reversed at a rate equal to the natural frequency of vibration of the molecule, there would be resonance absorption. Energy would be taken from the electrical field and converted into energy of vibration. This in turn would be dissipated by collisions and converted into heat.
4. If the field alternates too rapidly or too slowly, there is no net absorption of energy.
5. This is analogous to pushing a child in a swing. If one pushes too rapidly or too slowly, there is no net transfer of energy. However, if

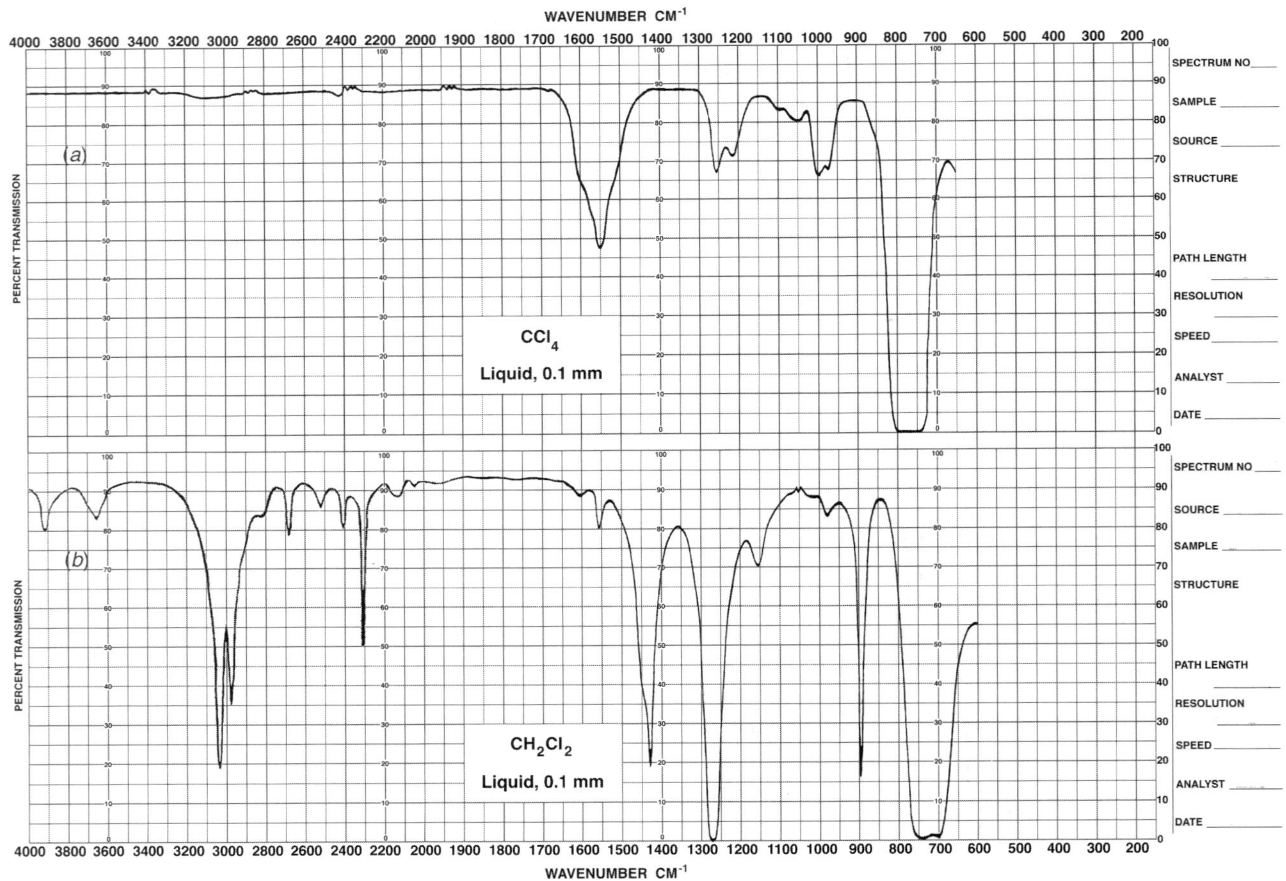

Fig. 1.6 Infrared spectra of (*a*) CCl_4 and (*b*) CH_2Cl_2. Note that CCl_4 has a much simpler spectrum, a consequence of its higher symmetry.

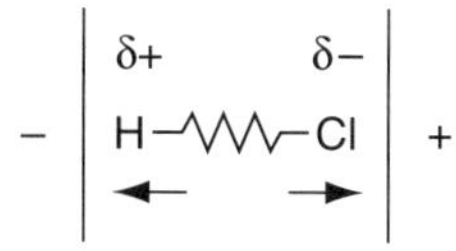

Fig. 1.7 HCl molecule between plates of an electrical capacitor.

one pushes at the natural frequency of the swing, more and more energy is transferred to the swing.

D. No electrical capacitor can be made to alternate its charge with sufficiently high frequency to match the frequency of a molecular vibration. However, a beam of light can provide both the electrical field and the proper frequency. In Figure 1.8 the situation for plane-polarized light is shown for simplicity.
   1. Light has electrical and magnetic fields which vary sinusoidally perpendicular to the direction of propagation. For plane-polarized light the electrical field vibrates in one plane and the magnetic field in the plane 90° to the former. Infrared spectroscopy does not depend on the magnetic field, so no more will be said about it.
   2. The frequency needed is in the IR region, 200–4000 cm^{-1}.
   3. In this region the wavelength is much longer than the size of the molecule.
      a. At 1000 cm^{-1}, $\lambda = 0.001$ cm [because λ (cm) $= 1/cm^{-1} = 1/1000\ cm^{-1}$].
      b. Molecular dimensions are roughly 10 Å or 1 nm or 10^{-7} cm.
      c. Therefore the wavelength is about 10,000 times longer than the molecular diameter.
   4. Hence the molecule is bathed in an electrical field of nearly linear gradient. It acts as though it were in an electrical capacitor.

E. We have deduced *the first requirement for the absorption of IR radiation: The frequency of the light must be equal to the (classical) frequency of vibration of the molecule.*

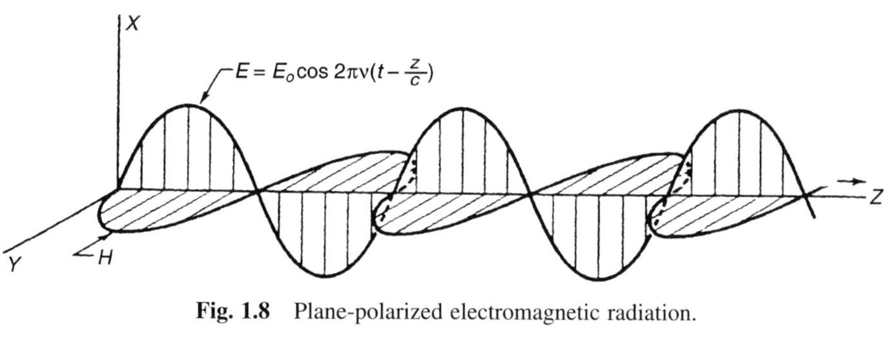

Fig. 1.8 Plane-polarized electromagnetic radiation.

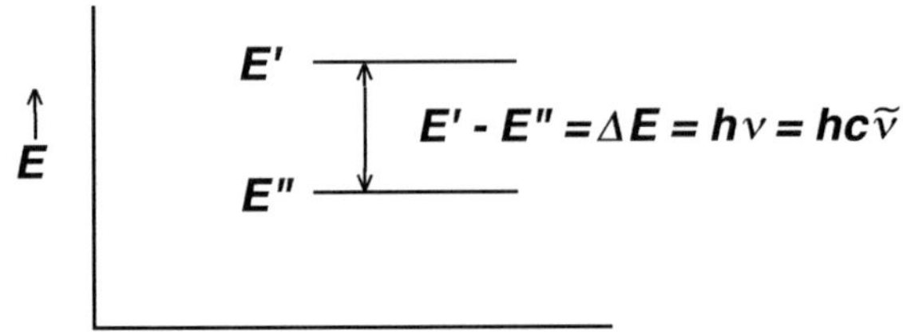

Fig. 1.9 Bohr resonance relation.

1. In modern terms it is stated somewhat differently; the Bohr resonance relation must be obeyed. The energy of the absorbed photon must exactly equal the difference in energy of the two states involved, $h\nu = hc\,\tilde{\nu} = E' - E''$. See Figure 1.9.
2. But this is not enough, as we see from another example.

F. H_2 in a capacitor. See Figure 1.10.

1. Now the external field neither stretches nor compresses the bond because of the symmetry of the molecule.
2. Why the difference from HCl? For one thing, HCl has a permanent dipole moment because of the charge separation. This allows the field to act differently on the two ends of the molecule. H_2 does not have a dipole moment because of its symmetry.
3. Must a molecule have a permanent dipole moment to absorb IR radiation? No, as we see from another example.

G. CO_2 in a capacitor. See Figure 1.11.

Fig. 1.10 H_2 between plates of capacitor.

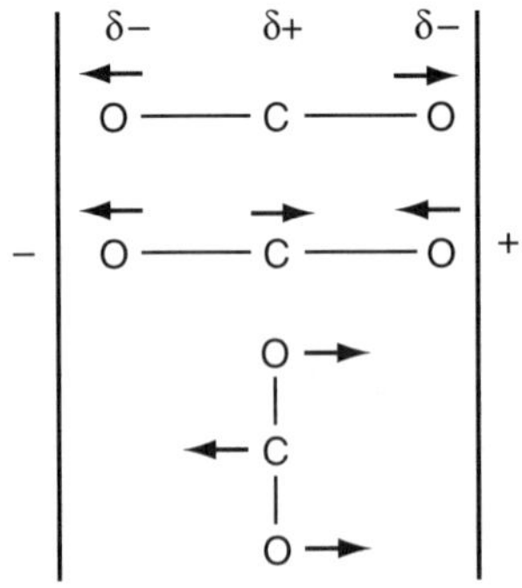

No.	Cm^{-1}	IR active?	Raman active?
1	(1337)	No	Yes
2	2349	Yes	No
3 & 4	667	Yes	No

Fig. 1.11 CO_2 in capacitor.

1. CO_2 has no permanent dipole moment because of its symmetry: O=C=O.
2. However, experimentally it is found to have two very intense IR absorption bands which occur at 2349 and 667 cm^{-1}. These are due to vibrations 2 and 3/4 in Figure 1.11.
3. CO_2 has another vibration, 1, which does *not* absorb in the IR. What is the difference between these?
4. Note that:
   a. Vibrations 2 and 3/4 produce a *change* in the dipole moment during the vibration, and they are the ones which are IR active.
   b. Vibration 1 does *not* produce a change in the dipole moment; the moment remains zero throughout the vibration by symmetry. This vibration is IR inactive.

H. This is a general result, so we have deduced a *second requirement for IR absorption: A vibration must produce a change in the dipole moment.*

XII. Raman spectroscopy

A. Raman and IR spectra are complementary and supplement one another. It is now usually easier to obtain a Raman spectrum than an IR one, so it is a very useful and helpful technique. Because we shall often use Raman spectra in this text, a short description of the phenomenon will be given here.

B. Raman spectra arise from the inelastic scattering of photons by molecules.

C. Schematic experiment. See Figure 1.12.
1. The sample is irradiated with intense monochromatic light from a laser.
2. Assume that the sample is transparent—i.e., that it does not absorb at this frequency. Then nearly all the light passes through the sample.

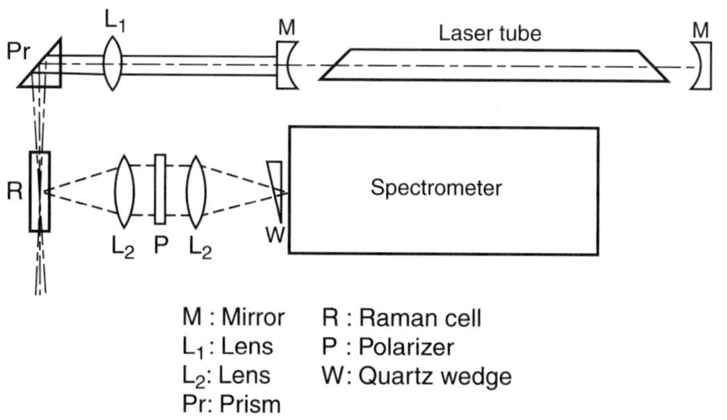

Fig. 1.12 Schematic Raman experiment.

3. But not quite all. A tiny fraction is scattered in all directions. Gather a cone of this with a pick-up lens and send it to a spectrometer.
   a. The main feature observed is light of exactly the same frequency as the original radiation. This is due to *Rayleigh scattering,* or to elastic collisions. It is not interesting for our purposes—in fact, it is a serious problem.
   b. Raman's great discovery was to find some extremely weak additional features in the scattered light at slightly different frequencies. They are displaced from the exciting line by exactly the vibrational energy spacings. These new features are called *Raman "lines,"* although they are really bands.
   c. Raman bands are extremely weak. Very approximately, the relative intensities are:

Incident light	:	Rayleigh scattering	:	Raman scattering
1	:	10^{-6}	:	10^{-10}

D. Collision model for scattering. See Figure 1.13.
1. Assume that a photon of the exciting radiation having frequency ν_0 and energy $h\nu_0$ collides with a molecule. Figure 1.13 shows that:
   a. An elastic collision gives Rayleigh scattering—a change in direction but no change in frequency.
   b. An inelastic collision leaves energy in the molecule—an amount $h\nu_{vib}$. Therefore the scattered photon has less energy than the incident one. Because it has the energy $h\nu_0 - h\nu_{vib}$, it will appear at the new position $\nu_0 - \nu_{vib}$. There is a change in both direction and frequency.
2. Thus for a Raman band, $\nu_{obs} = \nu_0 - \nu_{vib}$. This can be rearranged to $\nu_{vib} = \nu_0 - \nu_{obs}$. Since the latter two can be measured, one can obtain vibrational frequencies from Raman scattering.

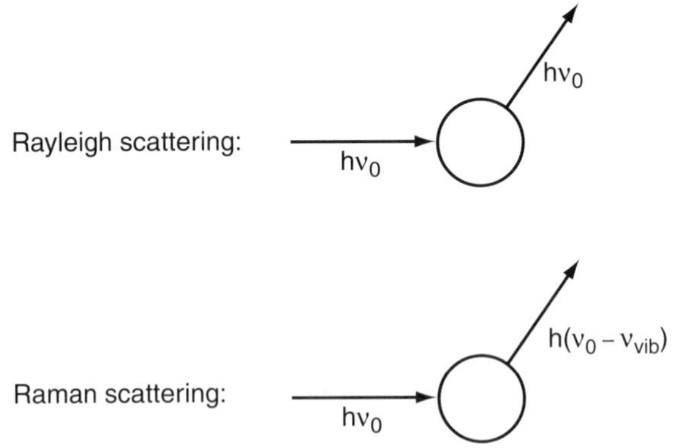

Fig. 1.13 Collision model for scattering.

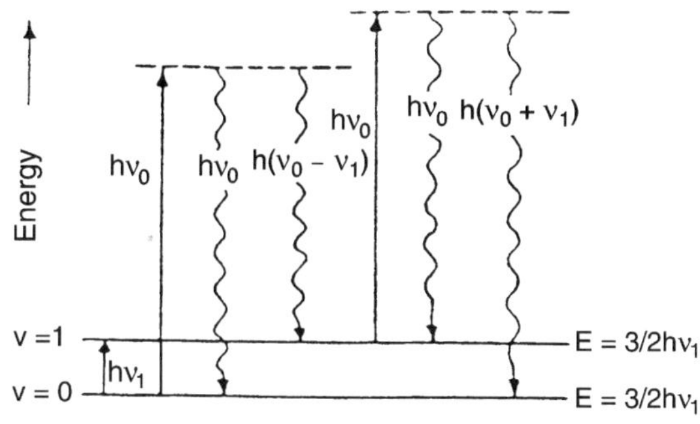

Fig. 1.14 Energy levels involved in scattering.

3. By convention, Raman frequencies are the *displacements* from the exciting line, $\nu_0 - \nu_{obs}$. They are always given in reciprocal centimeters, which is just ν (sec^{-1})/c (cm·sec^{-1}).

E. Energy levels involved in scattering. See Figure 1.14.
   1. Originally most molecules are in the lowest vibrational state, v = 0.
   2. *Infrared absorption* is the transition from energy level v = 0 to v = 1. V is the vibrational quantum number = 1, 2, 3.
   3. In Raman scattering a photon of energy $h\nu_0$ collides with a molecule, and at the moment of impact their combined energy is also $h\nu_0$, represented by the dashed line on the left half of the figure. This is not a real energy state (a "stationary state") and is therefore called a virtual level. Its energy depends on the energy of the exciting photon.
   4. Usually the collision is elastic. The scattered photon retains its original energy $h\nu_0$ after the collision has occurred, leaving the molecule in its original state v = 0. This is Rayleigh scattering—a change of direction but not energy.
   5. In a very small fraction of the collisions, part of the energy of the photon is used to increase the vibrational energy of the molecule from level v = 0 to v = 1. The scattered photon then has energy $h\nu_0 - h\nu_{vib}$ and appears at $\nu_0 - \nu_{vib}$. This is Raman scattering.
   6. Sometimes the scattered photon *acquires* energy from the molecule. (See the right half of the figure.)
      a. A fraction of the molecules are in an excited vibrational state with v = 1. During a collision there is a small chance of giving up this excess vibrational energy to the photon, which then has energy $h\nu_0 + h\nu_{vib}$. It is therefore scattered at the *higher frequency* $\nu_0 + \nu_{vib}$.

b. Raman shifts to higher frequencies are called *anti-Stokes* Raman bands; the ones that shift to lower frequency are called *Stokes* shifts.
c. According to the Boltzmann distribution law, the population of an excited energy level is less than that of the ground state, and furthermore it diminishes exponentially as the energy gap increases. Because the population of the initial state is smaller for anti-Stokes than for Stokes scattering, the shifts to frequencies higher than the exciting line give weaker Raman bands than the corresponding shifts to lower frequencies. At room temperature the population of levels greater than 600 cm^{-1} is so small that the anti-Stokes bands are seldom seen.
d. Because anti-Stokes shifts add no new information and are weaker, one nearly always starts at the exciting line and scans to lower frequencies (longer wavelengths).

F. *Example of a Raman spectrum*: CCl_4. See Figure 1.15.
1. Note the very intense exciting line in the center (marked 0 cm^{-1}). Both Stokes and anti-Stokes Raman bands are shown.
2. It is the *displacements* from the exciting line that are important. They are always given in reciprocal centimeters. The numerical values are shown in the figure.
3. Note that the same numerical displacements are obtained on each side of the exciting line but that the bands are weaker on the high-frequency side. They get rapidly weaker as the displacement gets larger, due to the exponentially decreasing population of the starting state. (Note especially 459 cm^{-1} and the 762–791 cm^{-1} pair.)
4. The same pattern of displacements occurs around *any* exciting line. As the exciting laser frequency is changed, the pattern of displacements moves with it.

G. Nature of Raman scattering
1. It is *not* an absorption process. It is a scattering effect and is closer to emission than to absorption.
2. It gives a difference spectrum.
3. The displacement in wavenumbers, $\Delta\tilde{\nu}$, corresponds to the IR range. Therefore Raman scattering offers a method of measuring IR-like frequencies by experiments done in the visible and UV regions.

H. Comparison of IR and Raman spectra
1. Einstein is said to have compared IR and Raman spectra in the following way:
   a. IR is analogous to putting a dime in a Coke machine and getting a Coke.
   b. Raman is analogous to putting \$1.00 in a Coke machine and getting a Coke plus 90¢ change. Clearly one can determine the cost of the Coke by taking the difference between the money put in and the change received.

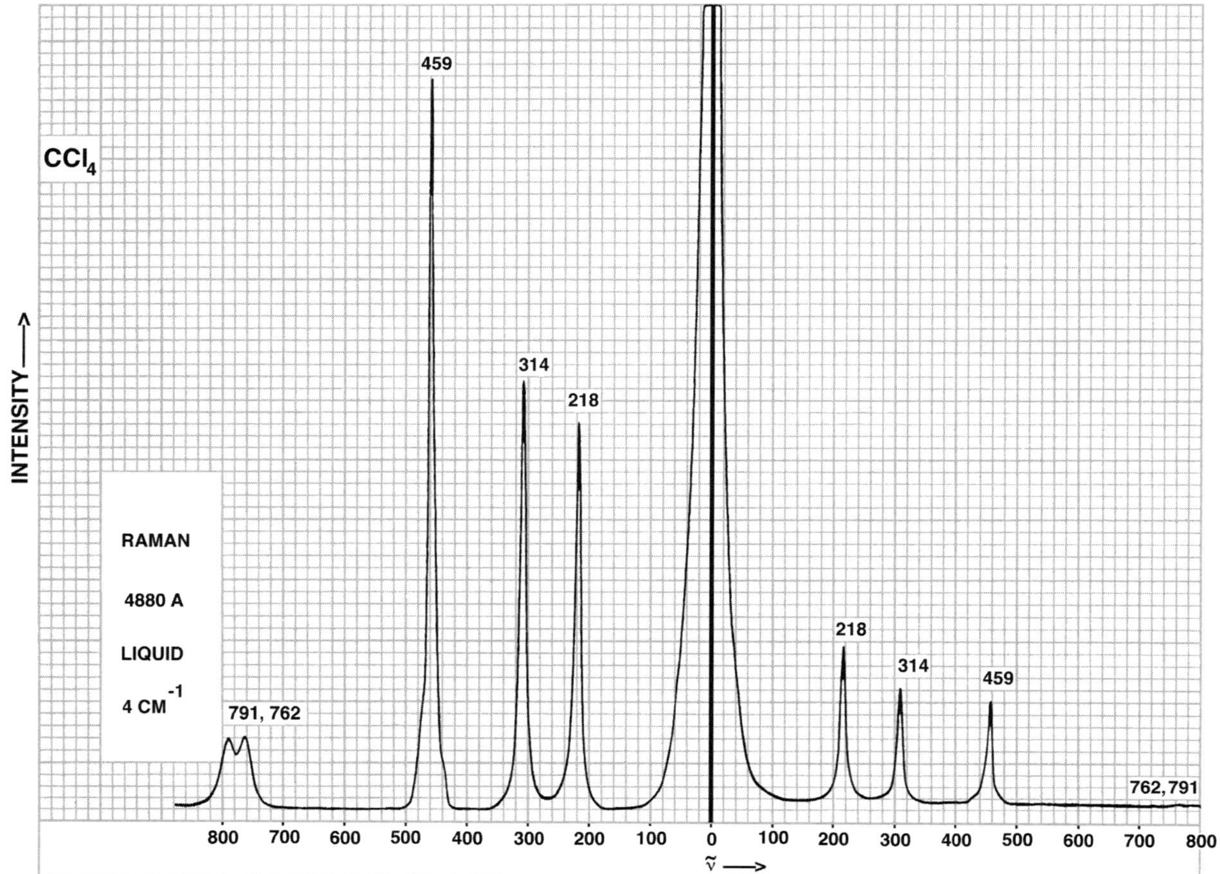

Fig. 1.15 Raman spectrum of CCl_4.

2. Both IR and Raman spectra are *vibrational* spectra.
3. They provide the same kinds of numbers—vibrational frequencies. For the same vibrational transition, the frequencies are exactly the same provided both spectra are measured in the same physical state.
4. However, their intensities may be quite different because they are due to very different physical mechanisms. Some bands that are easily seen in one type of spectrum may be very weak or missing in the other.
5. Therefore IR and Raman spectra are complementary.

I. *Example*: Cubane. See Figure 1.16.
1. In the Raman spectrum, note the intensity of the exciting line (at 0 cm^{-1} displacement). It goes far off scale.
2. Note how bands are intense in one spectrum and missing in the other.

XIII. Infrared and Raman intensities

A. Intensity in the IR

1. It is determined by the *change in the dipole moment during the vibration:*

$$I \propto \left(\frac{\partial \mu}{\partial Q}\right)^2 \qquad \text{More exactly,} \ I_k = \frac{\pi}{3c}\left(\frac{\partial \mu}{\partial Q_k}\right)_0^2$$

where μ = electrical dipole moment
Q = normal coordinate, a mathematical description of the vibration
k = particular vibration

Subscript 0 means that $\partial\mu/\partial Q$ is evaluated at the equilibrium position (zero displacement).

2. If the dipole moment μ does not change during the vibration, $I = 0$. The band is "forbidden" in the IR. The vibration still occurs, but it cannot be excited by absorption of IR radiation.
3. If $\partial\mu/\partial Q$ is large, the band is intense. (Note the squared dependence.)
4. One can often predict qualitatively when a band will be intense.
   a. *Stretching of polar bonds: O—H, C—O, C=O, C—F, C—Cl.* The atoms have very different electronegativities, and the dipole moment of the bond is therefore large. As the bond stretches, the electrons tend to follow the more electronegative atom, causing a large change in the dipole moment and intense absorption.
   b. *Out-of-plane wags of unsaturated C—H, as in olefins and phenyls.* When *all* the hydrogens move up out of the plane and all the carbons move down, there is a large change in the dipole moment.

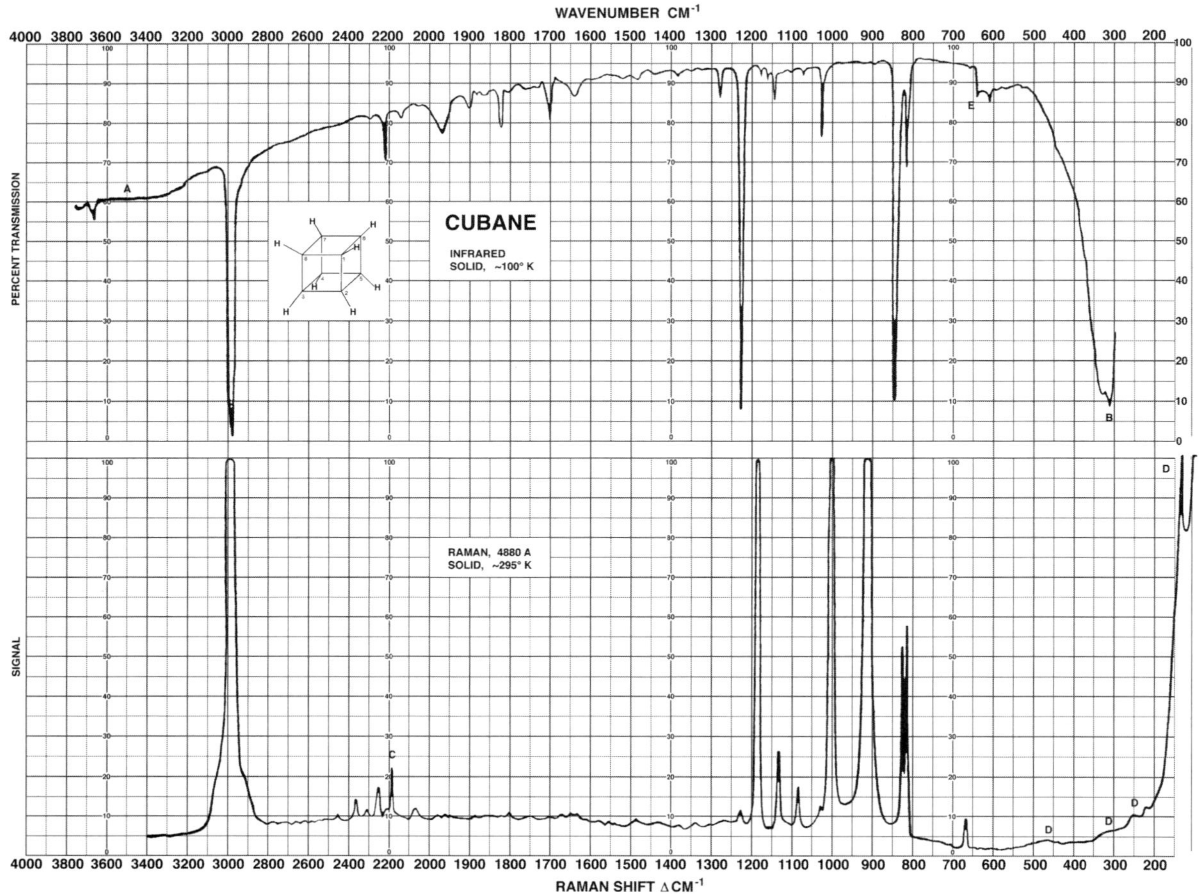

Fig. 1.16 Infrared (upper) and Raman (lower) spectra of cubane, C_8H_8.

5. Conversely, if the electronegativities of the two atoms forming the bond X–Y are nearly equal, stretching it usually gives a weak IR band. *Examples*:

   H–H, C–C

   C≡C stretch in H–C≡C–H, R–C≡C–H, or R–C≡C–R′

   C–H (The intensity per C–H is usually low.)

B. Intensity in Raman spectra

1. This is determined by the *change in polarizability during a vibration.*

$$I \propto \left(\frac{\partial \alpha}{\partial Q}\right)^2$$

   where α = polarizability
   Q = normal coordinate

2. The polarizability α measures the ease with which an external electrical field can *induce* a temporary dipole moment. Thus it measures the ease with which electrical charges in a molecule can be displaced.
3. The polarizability is large when there is a large concentration of loosely held electrons in a molecule. Empirically it is found that $\partial\alpha/\partial Q$ is also large then, giving relatively intense Raman bands. *Examples*:
   a. *Multiple bonds: C≡C, –C≡N, C=C, C=O.* There is a large concentration of electron density between the atoms, so stretching these bonds gives strong Raman bands.
   b. *Cl, Br, and I atoms.* They have outer shell electrons which are held relatively loosely and can be displaced by an applied electrical field. Stretching and bending vibrations involving these atoms are usually intense.
   c. *Other many-electron atoms: S, Hg, other heavy metals.* Again outer shell electrons are held relatively weakly. Bond stretches involving these atoms are relatively intense.
4. Conversely, in bonds that are strongly polar, Raman intensities tend to be weak because electrons are held tightly by the electronegative atom. Stretching of O–H and C–F are very weak in Raman spectra, whereas they are very intense in the IR.

C. Rule of mutual exclusion

1. *Statement*: For a molecule with a center of symmetry, vibrations which are IR active are forbidden in the Raman spectrum, and vice versa. (Given without proof.)
   a. Center of symmetry: A point around which everything in a molecule is mirrored.

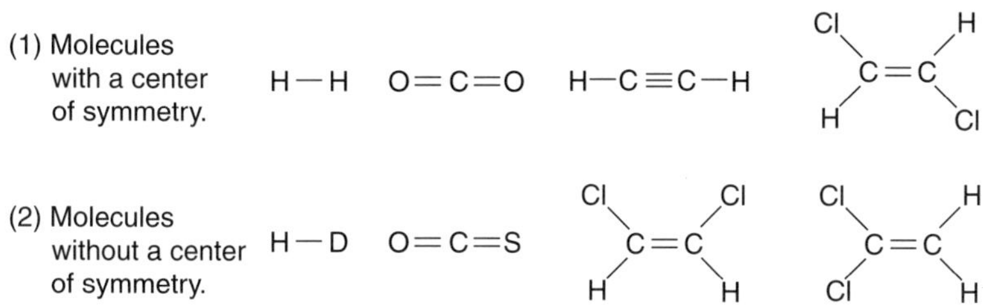

Fig. 1.17 Molecules with and without center of symmetry.

1) *Examples of molecules with and without a center of symmetry*. See Figure 1.17.
2) Do the molecules in Figure 1.18 have a center a symmetry? (Answers are at the end of the chapter.)
3) There may or may not be an atom at the center of symmetry. Compare CO_2 and O_2.
4) We shall use the concept of a center of symmetry on several occasions.

2. *Repeat*: For a molecule with a center of symmetry, vibrations which are IR active are Raman forbidden, and vice versa. There can also be vibrations forbidden in both but *none* allowed in both effects.
3. *Example*: Cubane. (See Fig. 1.16.) It has a center of symmetry, so we now know why the observed IR and Raman bands do not coincide.
4. For a molecule with a center of symmetry, it is especially useful to have both IR and Raman spectra.

XIV. Introduction to Raman group frequencies

A. Comparison of some IR vs. Raman group frequencies

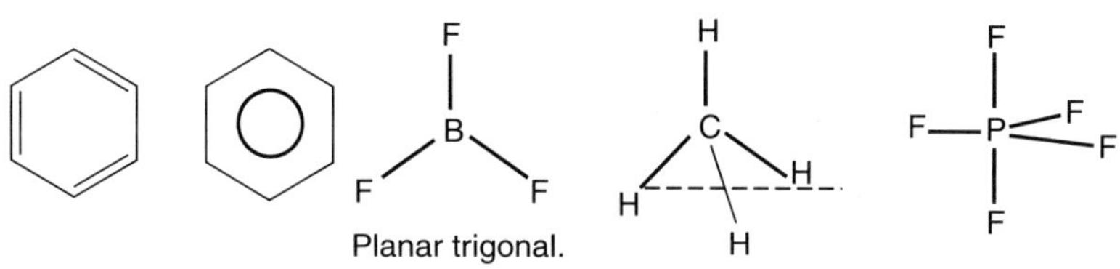

1 2 3 4 5

Fig. 1.18 Do these molecules have a center of symmetry?

1. Summary table:

IR VS. RAMAN GROUP FREQUENCIES

Weak in IR, Intense in Raman	Intense in IR, Weak in Raman	Intense in Both
C≡C stretch	C—F stretch	—C≡N stretch
C=C stretch	O—H stretch	>C=O stretch
N=N stretch	N—H stretch	C—Cl stretch
S—H stretch	C—H out-of-plane bend	—NO_2 stretch
C=S stretch		
C—S stretch		
S—S stretch		
CH_2 twist and wag		

2. Raman spectra tend to be more sensitive to the molecular framework. Infrared spectra are more sensitive to the functional groups attached to the framework.

B. Some Raman group frequencies that will be useful in the problem sections
   1. C=C stretch, 1680–1630 cm^{-1}
      a. In the Raman spectrum, always strong. Very reliable.
      b. In the IR, the intensity is highly variable. The band may be intense or it may not be seen.
   2. Phenyl compounds
      For mono-, 1,3-, and 1,3,5-substituted compounds only: $1000 \pm 10\ cm^{-1}$ (vs) in Raman.
   3. C—CH_3 symmetric deformation at $1378 \pm 5\ cm^{-1}$
      a. If the methyl is on a *saturated* carbon, $-\overset{|}{\underset{|}{C}}-CH_3$, the Raman band is extremely weak and is usually not seen.
      b. If the methyl is on an *unsaturated* carbon, e.g., aromatic —CH_3, =C—CH_3, or ≡C—CH_3, 1378 cm^{-1} is weak but detectable in the Raman.

XV. Coupling of vibrations

We will often be talking about the coupling of vibrations and about Fermi resonance, so these effects will now be described. They are both illustrated exceptionally well by CO_2.

A. First-order coupling of vibrations in CO_2
   1. CO_2 consists of two identical oscillators (C=O) sharing a common atom. If the C=O oscillators were separated, they would have exactly the same energy and frequency of vibration, as shown at the left of Figure 1.19.
   2. Let the oscillators be coupled by sharing the carbon atom. When one bond stretches, the motion of the carbon compresses the second

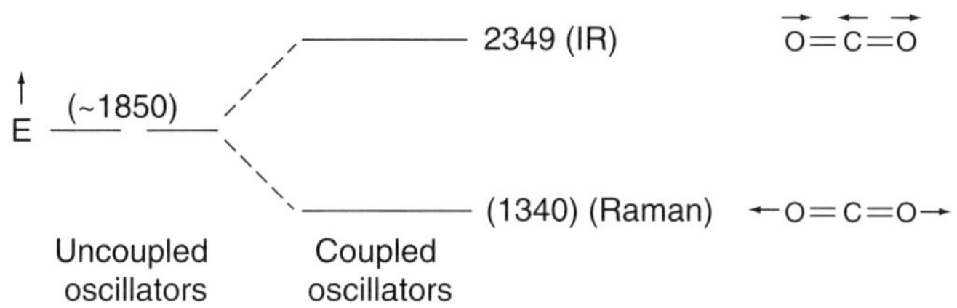

Fig. 1.19 First-order coupling in CO_2.

bond, and it too becomes involved in the vibration. The result of the interaction is that the two original energy states split to give two new ones, one higher and one lower than the original.

a. The splitting is symmetrical about the average of the two original states. (The original energies need not be identical, although in CO_2 they are.) From this we deduce that the isolated C=O oscillators would have had a frequency of about 1850 cm^{-1}. (They are *not* carbon monoxide.)
b. Figure 1.19 shows the two resulting states, their vibrational frequencies, their spectroscopic activity, and the forms of the two vibrations.

3. There are two requirements for such interaction:
   a. The energies of the original unperturbed states must be very nearly the same. The extent of the splitting is very sensitive to the original energy difference. In general, it must be less than about 50 cm^{-1} for vibrational splitting to be significant.
   b. There must be a physical mechanism for the splitting.
      1) In CO_2 there is a very effective one because the two oscillators are colinear and share the same central atom. As a result, the 1009-cm^{-1} splitting in CO_2 is exceptionally large.
      2) In SO_2, in contrast, the two S=O oscillators are at 119° to one another, so the stretching of one has much less effect on the other. Also the common atom, S, is heavier than the C in CO_2, which further reduces the interaction. As a result, the two bands of SO_2 are separated by only 211 cm^{-1} (1362 and 1151 cm^{-1}).
4. First-order coupling is an interaction between two (hypothetical) fundamental vibrations. It occurs even in the harmonic approximation.
5. First-order coupling is involved in several important group frequencies:
   a. The stretches of all cumulated double bonds, X=Y=Z
   b. The stretches in XY_2 groups, including $-CH_2-$
   c. The stretches in XY_3 groups, including $-CH_3$

d. The deformations of XY_3 groups
e. The N—H in-plane bend of secondary amides

6. The interaction of two states, with a resulting separation of their energies, occurs frequently in many branches of physics. Examples are coupled pendulums, electrical circuits, and the resonance energy of benzene.

B. Fermi resonance (a second-order coupling)

1. CO_2 also provides a good example of this. It is again an interaction of two states with a resulting energy separation, but with the difference that one of the states is a sum tone or an overtone.
   a. A sum tone is $\nu_a + \nu_b$, $2\nu_a + \nu_b$, $\nu_a + \nu_b + \nu_c$, etc.
   b. An overtone is $2\nu_a$, $3\nu_a$, $4\nu_a$, etc. It is therefore a special case of a sum tone where the frequencies are identical.
   c. Sum tones and overtones are usually very weak. As a crude rule of thumb, the intensity drops by more than 100-fold for each step in the sequence: fundamental, binary, ternary, etc.
   d. A sum tone or overtone is usually observed at somewhat less than its calculated value, typically about 1% less. Thus for the combination $1000 + 500 = 1500\ cm^{-1}$, one expects the band to actually occur about $1485\ cm^{-1}$.
   e. Sum tones and overtones are due to anharmonic terms in the potential function. Fermi resonance therefore also depends on anharmonic terms, which is why it is a second-order interaction.
2. In CO_2 the hypothetical Raman-active band at $1340\ cm^{-1}$ is really two bands of appreciable and nearly equal intensity at 1286 and $1388\ cm^{-1}$. Why? Enrico Fermi gave the explanation. The unperturbed symmetric stretch would be about $1340\ cm^{-1}$ (from hindsight). The bending mode of CO_2 is at $667\ cm^{-1}$, and its first overtone would be slightly less than $2 \times 667 = 1334\ cm^{-1}$. Because this is close to $1340\ cm^{-1}$, the two states mix and split apart, as shown in Figure 1.20. This splitting of $102\ cm^{-1}$ is unusually large for a Fermi resonance.
3. Because the wave functions of the two states mix, they share properties. Hence the intensity of the fundamental is shared with the overtone. (The overtone is said to rob the fundamental of intensity.) Therefore two bands of moderate intensity are observed rather than the expected strong fundamental and a weak overtone.

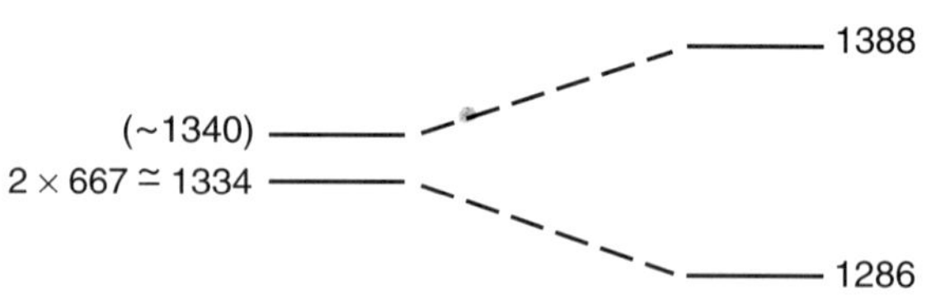

Fig. 1.20 Second-order interaction, or Fermi resonance, in CO_2.

4. Three requirements for Fermi resonance:
   a. The zero-order frequencies must be close together (typically within 30 cm^{-1}).
   b. The fundamental and the overtone or sum tone must have the same symmetry in the group theory sense, not elaborated here.
   c. There must be some mechanism for the interaction.
      1) Therefore the vibrations cannot be localized in different parts of the molecule.
      2) An interaction is especially easy to picture in CO_2. The bending motion of the molecule pushes the oxygen atoms apart a little every time the carbon goes through the equilibrium configuration, which is twice for each bending cycle.

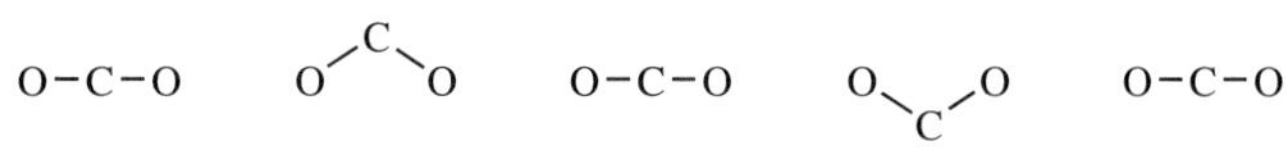

Twice the bending frequency happens to come very close to the symmetric stretching frequency. It therefore interacts with the symmetric stretch.
5. Results of Fermi resonance
   a. Bands are displaced from their normal position.
   b. Overtones are unexpectedly intense. There may be two strong bands where only one is expected. *Example*: In some solvents, cyclopentanone appears to have two carbonyl bands near 1750 and 1730 cm^{-1}, as shown in Figure 1.21, even though it has only one carbonyl group. This is due to Fermi resonance.
      1) The identity of the sum tone is known:

$$1280\,\text{cm}^{-1}(\text{a CH}_2\text{ wag}) + 470\,\text{cm}^{-1}(\text{a C=O bend}) = \sim 1750\,\text{cm}^{-1}$$

      2) If the solvent is changed, the interacting zero-order frequencies shift slightly but differently and the relative intensities of the two bands change a lot. In inert solvents, 1750 cm^{-1} is the more intense. In polar and hydrogen-bonding solvents, 1730 cm^{-1} is the more intense.[6]
6. Tests for Fermi resonance
   a. Fermi resonance is very sensitive to the separation of the zero-order frequencies (i.e., before mixing). If the separation is changed by only a small amount, there is a large effect on the mixing. Therefore anything that will change one of the zero-order frequencies more than the other will have a large effect on the Fermi resonance.
   b. The separation can be altered by:
      1) *Changing the physical state or the solvent*. The latter was used for cyclopentanone above.

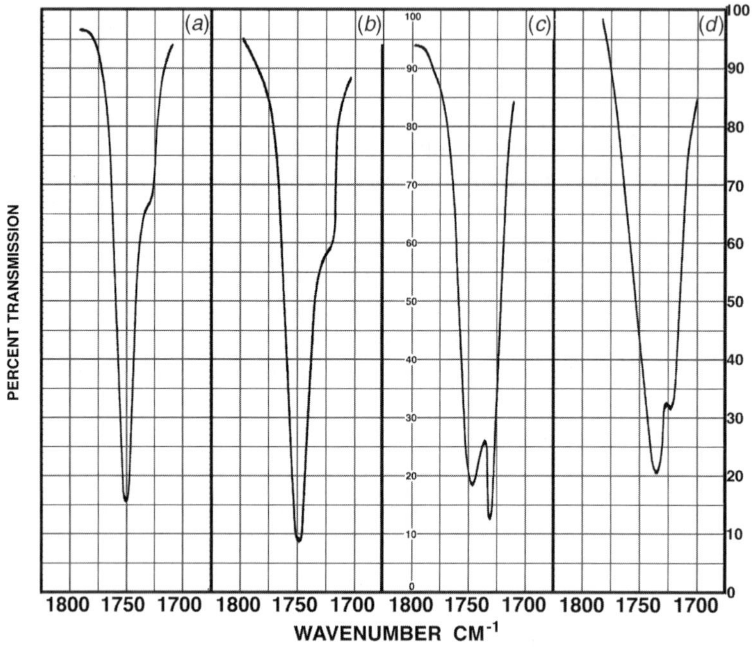

Fig. 1.21 Carbonyl region of cyclopentanone in various media: (*a*) carbon tetrachloride, 0.015 *M*; (*b*) carbon disulfide, 0.023 *M*; (*c*) chloroform, 0.025 *M*; (*d*) pure liquid (thin film). Spectral slit width 2 cm^{-1}.

2) *Deuteration.* In cyclopentanone, deuterating the α-carbon atoms completely removes the Fermi resonance because it shifts the sum tone away from the fundamental. See Figure 1.22.

7. Fermi resonance occurs quite often. *Examples*:
   a. *Ethylene carbonate.* Has only one carbonyl group but has two intense bands in the carbonyl region.
   b. *Benzoyl chloride, C_6H_5COCl, in CS_2 solution.* Has two bands in the carbonyl region.
   c. *CCl_4.* There is a sum tone 459 (Raman) + 314 (IR) = 773 cm^{-1}. This interacts with a fundamental near 775 cm^{-1} to give Raman bands at 762 and 790 cm^{-1}. See the Raman spectrum of CCl_4, Figure 1.15.
   d. *Benzene.* There is a sum tone 606 (Raman) + 992 (Raman) = 1598 cm^{-1}. This interacts with a fundamental near 1595 cm^{-1} to give bands at 1585 and 1606 cm^{-1} in the Raman spectrum (see Figure 5.6).

XVI. Names for regions of the IR

In the past there was considerable confusion about the meanings of "near infrared" and "far infrared." The far IR was usually the low-wavenumber

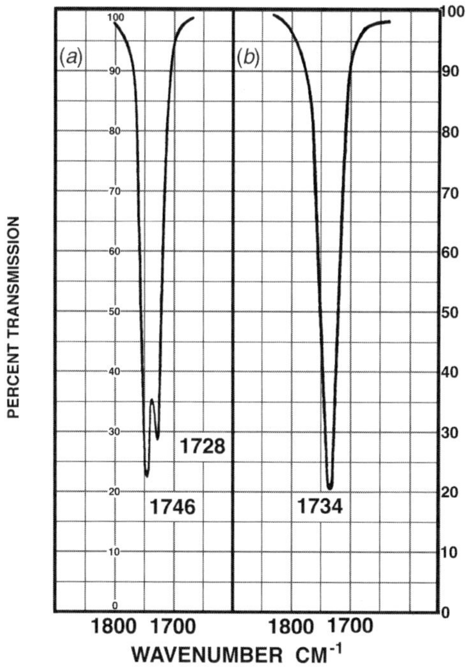

Fig. 1.22 Effect of deuteration on C=O band of cyclopentanone: (*a*) cyclopentanone; (*b*) cyclopentanone-α:α:α':α'-d_4. Chloroform solution, 2 cm^{-1} spectral slit.

range which was inaccessible to an instrument. Depending on optical components, it might be the region below 650 or 400 or 200 cm^{-1}. In 1962 the Triple Commission for Spectroscopy recommended the definitions given below[7]:

	cm^{-1}	μm
Near IR	12,500–4000	0.8–2.5
Mid IR	4000–200	2.5–50
Far IR	200–10	50–1000

XVII. Tabulations of characteristic group frequencies

A. There are many. See the Bibliography at the end of the book.

B. The best for infrared:

1. Bellamy, *Infrared Spectra of Complex Molecules*, Vol. 1.
2. Jones and Sandorfy, in A. Weissberger, (Ed.), *Technique of Organic Chemistry*, Vol. 9, Chapter 4.
3. Colthup, Daly, and Wiberly, *Introduction to Infrared and Raman Spectra*. Very good for physical interpretation of the vibrations.

C. The best for Raman—Dollish, Fateley, and Bentley, *Characteristic Raman Frequencies of Organic Compounds*.

D. Combined treatment of both infrared and Raman group frequencies: Lin-Vien, Colthup, Fateley, and Grasselli, *Infrared and Raman Characteristic Group Frequencies*, Academic, New York, 1991.

XVIII. *Answers to Section V, Desirable qualities of a good group frequency*

A. Reliable: Always present when the group is present.

B. Intense: It will not be overlooked or missed with a thin sample.

C. Unique (or isolated or free of interferences).

D. Narrow wavenumber range. (Ideally ± 5 cm^{-1}, usually ± 50 cm^{-1}.)

E. Ability to understand variations within this range (as one can for carbonyls).

F. Has some other identifying characteristic besides its wavenumber value, e.g., very strong, very sharp, very broad, or a doublet.

G. More than one characteristic frequency for the group. *Examples*:

1. $-NO_2$ 2 bands
2. —COOH 4 bands
3. —Phenyl 4–6 bands

XIX. *Answers to Section XIII. C. 1. a. 2, Do the molecules in Figure 1.18 have a center of symmetry?*

1. Kekule form of benzene. *No*. The atoms satisfy a center of symmetry but the bonding does not. Notice that a C=C double bond is opposite a C—C single bond.
2. Modern representation of benzene. *Yes*. All the C—C bonds are equivalent.
3. BF_3. *No*.
4. CH_4. *No*. It has very high symmetry (tetrahedral), but no center of symmetry.
5. PF_5. *No*.

REFERENCES

1. F. W. Herschel, *Philos. Trans. Roy. Soc. (Lond.)* **90**, 255–283, 284–292, 293–326, 437–538 (1800).
2. W. J. Herschel, *Nature* **23**, 76 (Nov. 25, 1880).
3. (a) Kyle and Shampo, *Medicine and Stamps*, Vol. 1, p. 162. (b) Encyclopedia Britannica, 1966, article on Fingerprints.
4. N. B. Colthup, L. H. Daly, and S. E. Wiberley, *Introduction to Infrared and Raman Spectroscopy*, 2nd ed. Academic, New York, 1975, p. 261.
5. R. L. Weber, *Droll Science*, Humana, Clifton, NJ, 1987, p. 160.
6. Bertran and Ballester, *Spectrochim. Acta* **39A**, 123 (1983).
7. *J. Opt. Soc. Am.* **52**, 476 (1962).

2 Characteristic Frequencies of Alkanes

DANA W. MAYO

I. General appearance of alkane spectra
  A. Vibrational modes present in *saturated* hydrocarbons.
    1. Because these systems have a simple elemental composition consisting of only hydrogen and carbon, they will possess a relatively simple distribution of normal modes which can be divided into:
       Carbon–hydrogen vibrations and carbon–carbon vibrations
       These two groups of vibrations can be further subdivided roughly into C—H and C—C stretching modes and C—H and C—C bending modes.
    2. Primary interest in the IR spectral data of *saturated* alkanes is focused on the C—H stretching and bending modes. The symmetry associated with C—C stretching and bending modes leads to very weak absorption associated with these vibrations. In addition, these latter weak modes often are mechanically highly coupled, and this coupling leads to a complex distribution of band peaks in an already inherently weakly absorbing system.
  B. A comparison of the spectra of *n*-hexane with mineral oil (Nujol) appears to indicate that Nujol, which is a complex mixture of high-molecular-weight alkanes, has a much simpler absorption spectrum than does the low-molecular-weight C_6 straight-chain hydrocarbon (Figures 2.1 and 2.2).
    1. This is an inconsistent conclusion, because the number of normal modes of vibration is given by the $3N - 6$ rule (where N is the number of atoms in the system). Thus *n*-hexane, C_6H_{14}, with 20 atoms, has 54 normal modes while *n*-pentadecane, $C_{15}H_{32}$, with 47 atoms, has 135 normal modes. In addition, while there are 5 structural isomers of

Course Notes on the Interpretation of Infrared and Raman Spectra, by Dana W. Mayo, Foil A. Miller, and Robert W. Hannah.
ISBN 0-471-24823-1

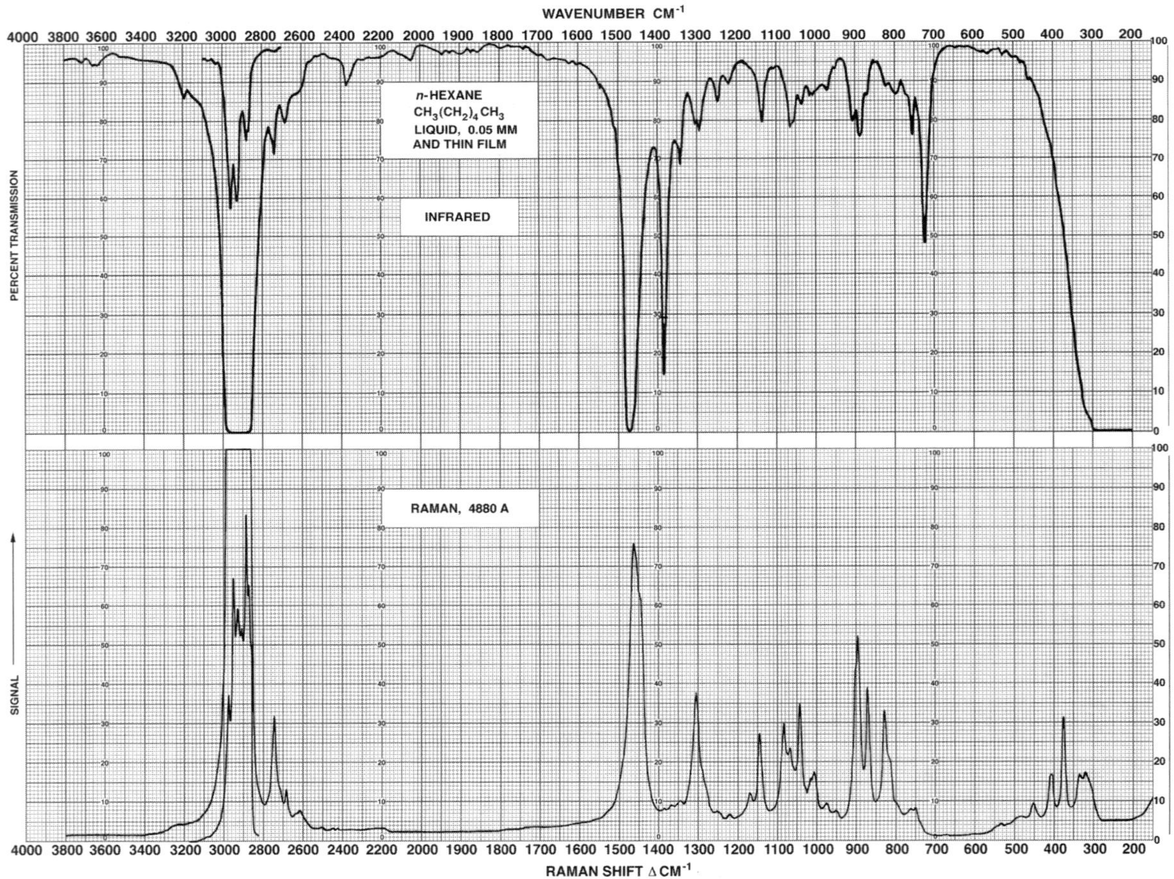

Fig. 2.1 Infrared and Raman spectra of *n*-hexane.

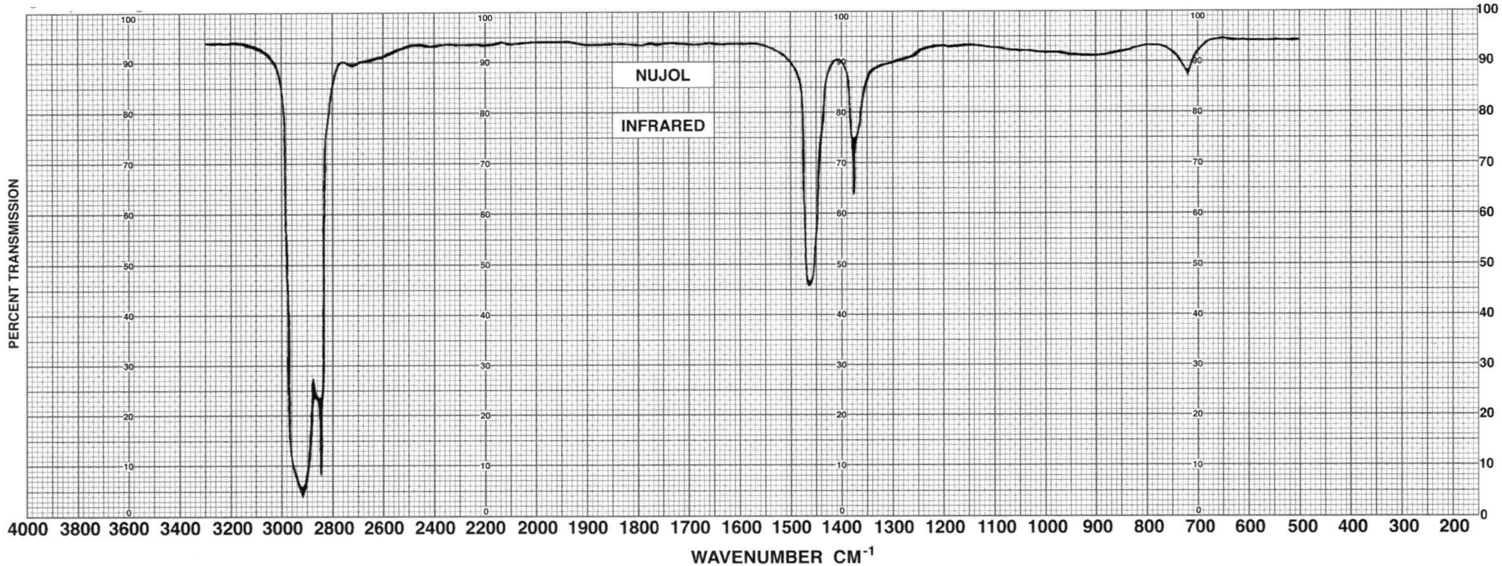

Fig. 2.2 Infrared spectrum of Nujol.

hexane (one of which is the normal isomer) that all possess different spectra, pentadecane has 4347 structural isomers, many of which are present in the mineral oil mixture and most of which have different spectra (the differences can be reasonably large or very small or in the case of optical isomers zero).

2. A further complication of both spectra is the presence of rotamers. Each rotational isomer would be expected to possess a slightly different spectrum from the other rotamers of a particular structural isomer (with the exception of optical isomers), and thus the spectrum of Nujol, in which these isomers abound, would be expected to be much more complex than expected on the basis of structural isomers alone.
3. Why is the Nujol IR spectrum so simple? In fact, it consists of only four narrow absorption regions. The four ranges are (1) a group of strong bands centered near 2900 cm^{-1} which with good resolution is shown to be a collection of four peaks, (2) a medium–strong band near 1465 cm^{-1}, (3) a relatively sharp weak band near 1380 cm^{-1}, and (4) a much weaker band near 720 cm^{-1}. The entire spectrum is generally limited to a total of seven peaks, whereas the *n*-hexane spectrum at normal sampling thickness has a minimum of several dozen peaks immediately observable.

   The great simplification of the Nujol spectrum over that of *n*-hexane can be satisfactorily explained by assuming that in the spectrum of a large mixture of compounds only those absorption bands which possess highly consistent wavenumber values give rise to bands which grow above the background. In effect, *the Nujol mixture establishes the group frequency band systems for the saturated hydrocarbons* (alkanes).

   Thus, hydrocarbon group frequencies are the only recognizable absorption bands in the Nujol spectrum. These group frequencies, of course, also are found in the *n*-hexane spectrum. The remaining bands in the hexane spectrum become merged into background absorption in the mixture of alkanes in the mineral oil and a very simple spectrum emerges.

II. Normal hydrocarbon spectra

A. Carbon–hydrogen stretching vibrations

1. The frequency range of these vibrations was originally established by rigorous analysis of very simple hydrocarbon molecules such as methane. Based on these investigations the C—H stretching region was located close to 3000 cm^{-1}. The assumption was then made that if these vibrations possessed group frequency characteristics, the force constants derived for the simpler systems could be transferred directly to the more complex molecules which were not amenable to theoretical analysis, such as normal coordinate calculations.

2. Based on an empirical survey of normal hydrocarbons in which the hydrogen atoms are all bound to sp^3-hybridized C—H systems, it was established that the C—H stretching range lies between 3000 and 2800 cm^{-1}. There are no exceptions to this range in *linear* hydrocarbons.
3. Infrared data have been collected over several decades. There have been three basic improvements to the routine resolution available in the 3000 cm^{-1} region of the spectrum. Initial measurements were made via dispersion spectra employing alkali halide prisms. In those early days NaCl prisms were employed primarily in the survey instrumentation. Those instruments had low resolution (usually only one or two bands were observed) in this region. Prism instruments were followed by instruments utilizing diffraction gratings which significantly improved the routine resolution accessible in this region. Usually three or four peaks were observed by the latter dispersion system (Figure 2.3). The most recent instrumental innovation has been the development of Fourier transform (FT) spectrometers. Standard survey FT systems give resolution in the 3000 cm^{-1} region equivalent to good diffraction grating instruments and therefore also generally observe four bands in the normal hydrocarbon series. [LiF prisms do give near grating resolution, but these prisms are limited to

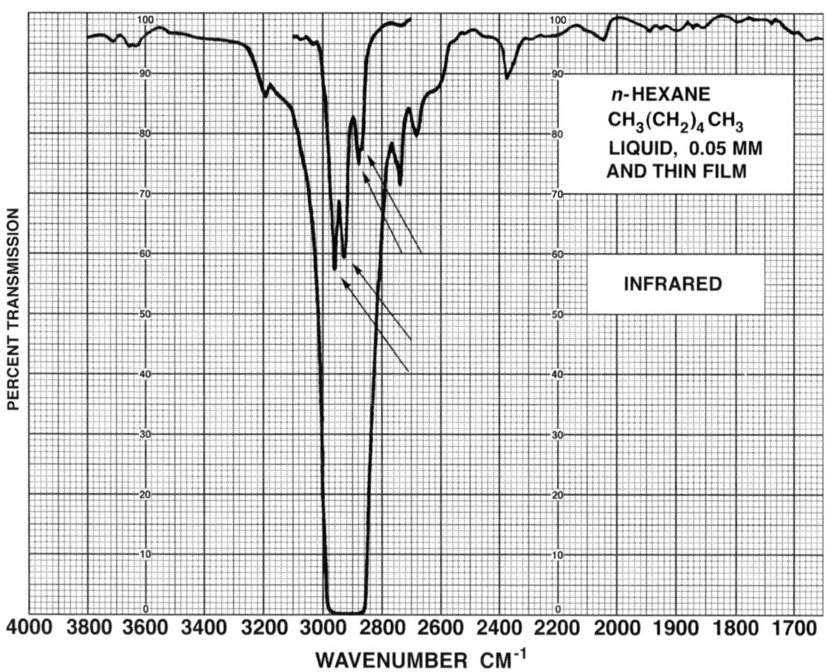

Fig. 2.3 Infrared spectrum of *n*-hexane in C—H stretching region obtained with grating (or FT) resolution.

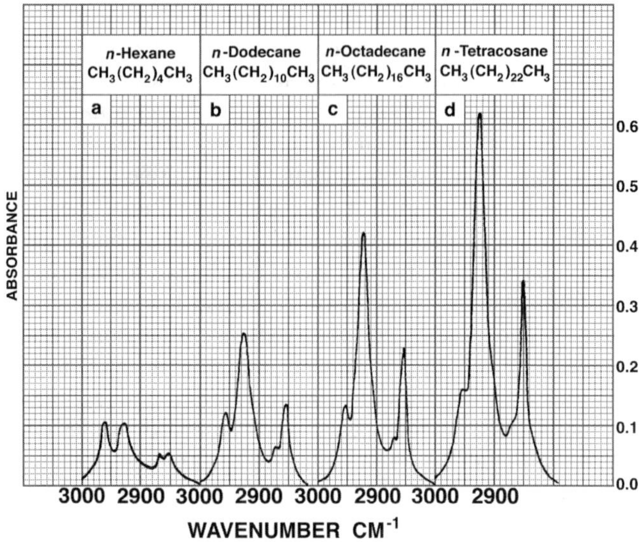

Fig. 2.4 Infrared spectrum of four normal hydrocarbons in C—H stretching region plotted as absorbance and obtained with LiF prisms.

measurements around the 3000 cm^{-1} region and were expensive (Figure 2.4).]

4. Because the C—H stretching modes exhibit the characteristics of good group frequencies, this collection of vibrations is found to occur over a relatively narrow wavenumber range of about 130 cm^{-1} (2970–2840 cm^{-1}). Indeed, in the normal series the individual peaks occur within ranges as narrow as ± 5 cm^{-1}. This is, of course, a rather limited number of substances.

B. Description of C—H stretching modes

1. First, note that the four absorption peaks in the 3000–2800 cm^{-1} region can be assigned as two pairs of doublets with one of the pairs easily shown to be associated with methylene groups and the other pair associated with methyl group vibrations. The assignments to methyl and methylene groups can be carried out by examining systems in which the relative number of methylene groups to methyl groups changes in a regular fashion (Figure 2.4).
2. Methine (tertiary hydrogen) systems do not usually give rise to observable absorption bands in this region because their relative population is very low and they have intrinsically weak absorbtivities.
3. The methyl C—H stretching doublet is found in normal hydrocarbons to occur near

$$2962 \pm 10\,\text{cm}^{-1} \quad \text{and} \quad 2872 \pm 10\,\text{cm}^{-1}$$

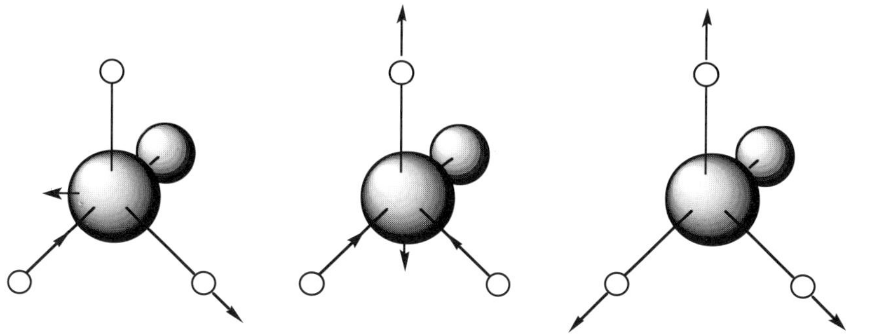

Fig. 2.5 Atomic displacements involved in normal methyl C—H stretching modes. Note the antisymmetric pair has required or accidental degeneracy.

The atomic displacements involved in these normal modes are pictured in Figure 2.5.

The antisymmetric methyl C—H stretching modes (Figure 2.5) are assigned to the higher of the two peaks based on mechanical analysis while the symmetric methyl stretch (Figure 2.5) is assigned to the lower wavenumber band.

In methyl groups there are, therefore, three C—H stretching oscillators, which in turn give rise to three absorption bands, two of which are degenerate (occur at the same frequency) or nearly degenerate. There is a symmetry requirement for true degeneracy of the antisymmetric modes which is seldom achieved in complex organic molecules. Therefore, although these modes may not be rigorously degenerate, they nevertheless occur in very similar wavenumber ranges and are only rarely resolved.

4. The methylene C—H stretching doublet is found in normal hydrocarbons to occur near

$$2926 \pm 10\,\mathrm{cm}^{-1} \quad \text{and} \quad 2853 \pm 10\,\mathrm{cm}^{-1}$$

The atomic displacements involved in these normal modes are described as shown in Figure 2.6.

There are two C—H oscillators and two methylene stretching vibrations. The antisymmetric methylene C—H stretching mode is assigned to the higher of the two peaks based on mechanical analysis, and the lower wavenumber band is assigned to the symmetric methylene stretch.

C. Factors which lead C—H oscillators to exhibit the characteristics of good group frequencies

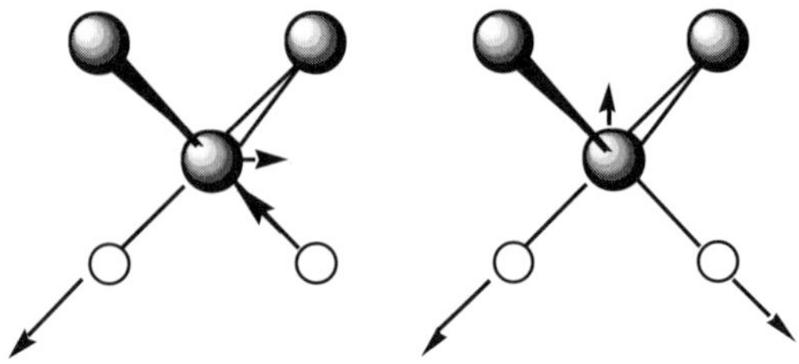

CH_2 antisymmetric stretch **CH_2 symmetric stretch**

Fig. 2.6 Atomic displacements involved in methylene C—H stretching modes.

1. During the atomic displacement of the C—H stretching vibration (an oscillator constructed with a light terminal atom), the hydrogen atom, which is attached only to the carbon atom, undergoes the very large majority of the motion. Thus only a single bond (the C—H bond) is directly involved in the vibrational distortion. The required displacement of the carbon atom to maintain a stable center of gravity is relatively small. Thus, the other bonds attached to the carbon atom (which normally consist of varying chemical characteristics) are perturbed only in minor fashion during the C—H oscillation.
2. Associated with the above factor is the fact that, as a result of the heavily skewed displacement, C—H vibrations can be described to a first approximation by a Hooke's law expression in which only the mass of the hydrogen atom is used in the expression ($\tilde{\nu} = 1/2\pi c\sqrt{k/m_{\mathrm{H}}}$), where $\tilde{\nu}$ is in wavenumbers and reduced mass μ is not required and

$$\mu = m_1 \cdot m_2/(m_1 + m_2)$$
$$c = \text{speed of light, cm/sec}$$
$$k = \text{force constant, dyn/cm}$$
$$m_{\mathrm{H}} = \text{mass of hydrogen atom, g}$$

3. The displacement of hydrogen therefore results in little or no significant vibrational coupling with the rest of the molecular system (no interaction with adjacent modes, i.e., that is displacement of atoms other than the C—H groups being directly considered is not required).
4. A light mass and strong force constants result in high wavenumber values for C—H stretching vibrations compared to most other vibrational modes. This property isolates the C—H stretching modes in the spectrum and makes identification relatively reliable and easy. It also is an additional contributing factor to reducing coupling to the rest of the molecular system as it leads to large frequency mismatch with most other modes.

5. The C—H stretching modes occur over a very narrow wavenumber range (±5–10 cm^{-1}), and although the molecular extinction coefficients are relatively low, the large number of these groups per molecule usually leads to strong bands in the C—H region.

D. Coupling in C—H stretching vibrations

1. There is some relatively weak coupling between C—H groups attached to the same carbon atom. Although the hydrogen undergoes the majority of the displacement in these vibrations, the direct linkage to other C—H oscillators which share the same carbon atom and that have identical frequencies leads to observable vibrational coupling. When three identical oscillators on tetrahedral carbon are coupled, the result is three mixed vibrational modes. Two of these will be higher in energy (the antisymmetric stretches) than the original oscillators and with sufficient symmetry present this pair will be degenerate (or when not degenerate they will be very close in frequency). The third mode is lower in energy (the symmetric stretch). Thus, a fundamental mode occurs for each oscillator in the coupled system (although in this case we usually observe only two bands, not three). The same analysis applies to the methylene case with two identical oscillators. Here the oscillators give rise to one mode occurring at a higher wavenumber value than the isolated oscillator (the antisymmetric stretch) and one mode is found at a lower wavenumber value (the symmetric stretch).
2. While coupling between two or more C—H oscillators attached to the same carbon atom is observed, coupling to other C—H oscillators one or more carbon atoms removed from the oscillators under consideration is essentially zero. (For example, consider the case of the C—H stretching modes on two adjacent methylene groups.) This is, in fact, one of the reasons why these modes are good group frequencies.
3. There are some exceptions to the above case (no coupling between C—H oscillators on different carbon atoms), but we are required to leave sp^3 C—H systems to find them. The case of acetylene, H—C≡C—H, is a ready example of coupled but not jointly bound C—H groups. In this case the two C—H oscillators are directly in line with each other, which maximizes the effect of the small carbon displacements. In addition, the two carbons are triply bonded, which, although it does not raise the C≡C stretching mode directly into the same region as the C—H stretching modes, does bring it much closer (1974 cm^{-1}) than in the single-bond systems (~900 cm^{-1}). The triple-bond stretch is, in fact, close enough that weak coupling between the C—H stretches can be observed (symmetric stretch 3374 cm^{-1} and antisymmetric stretch 3287 cm^{-1}; $\Delta\tilde{\nu} = 87$ cm^{-1}). In the case of diacetylene, H—C≡C—C≡C—H, the C—H coupling would be expected to collapse as the C—H oscillators are insulated by the low-frequency C—C single bond (because there is some delocalization of the π-system over the central bond, a small amount of residual

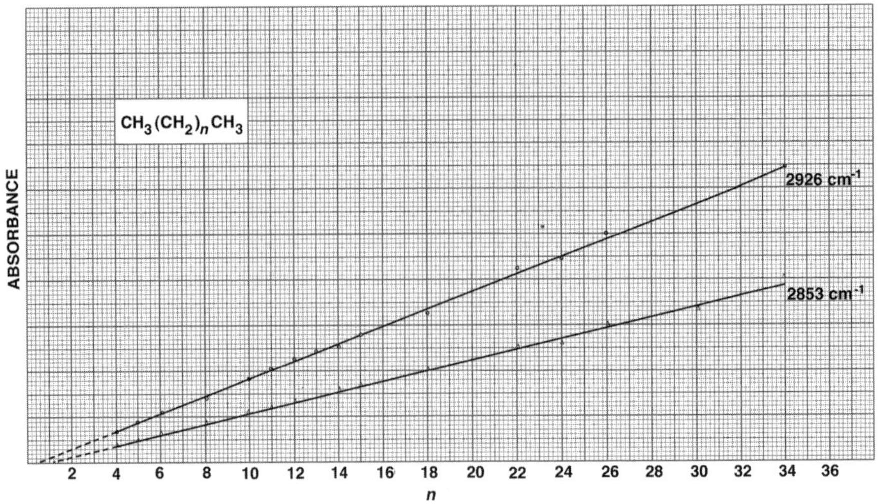

Fig. 2.7 Evidence for linear relationship of absorbance with increasing chain length in *n*-alkanes.

coupling might be expected to be found). Thus, the two C—H stretching modes of diacetylene would be expected to be nearly accidentally degenerate. Unfortunately, while the IR active mode has been observed at 3293 cm^{-1}, the gas-phase Raman mode does not appear to have been located.

E. Intensity of C—H stretching modes in the straight-chain hydrocarbons

As might have been expected, a linear intensity relationship is found for the methylene C—H stretching modes with increasing chain length (Figure 2.7).

F. Carbon–hydrogen bending vibrations

1. The wavenumber regions where the group frequencies associated with the bending vibrations can be identified by the same approach described above for the stretching of normal modes. This approach involves rigorous analysis of simple systems followed by the transfer of the derived force constants to complex molecules by empirical assignment. The success of this technique to the assigment of the bending modes occurs primarily for the same reasons discussed above for the stretching vibrations. There are just a few variations on the general scheme that need to be discussed throughout this section.
2. Description of the C—H deformation (bending) vibrations of methyl groups

   As in the case of the stretching modes, the three C—H oscillators possess three internal H—C—H bond angles that give rise to three locally coupled bending modes; see Figure 2.8.

   a. The degenerate (or near-degenerate) high-energy pair assigned to the antisymmetric deformation modes have been located near

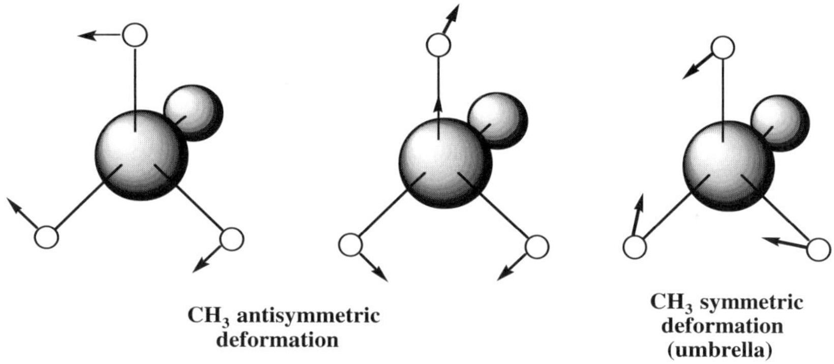

Fig. 2.8 Atomic displacements involved in methyl C—H bending modes.

$1460 \pm 10\ \text{cm}^{-1}$. As in the stretching case, little coupling to the rest of the molecular system is observed.

b. A single mode corresponding to the symmetric and lower energy deformation (or umbrella) vibration is found near $1375 \pm 10\ \text{cm}^{-1}$.

3. Description of the C—H deformation modes of methylene groups

The methylene group has a single H—C—H bond angle and therefore a single deformation mode (scissoring, Figure 2.9) directly analogous to that of the bending mode in the water molecule. This occurs near 1450 cm^{-1} in linear hydrocarbons (Figure 2.1) and is very close to the antisymmetric deformation modes of methyl groups ($\sim$1460 cm^{-1}).

4. Three additional bending modes also are available to the *methylene group* (Figure 2.9)

When the methylene group is fused into a molecule, three new vibrational degrees of freedom develop around this group, which have their origin in the rotational degrees of freedom in the isolated three-atom system. These modes can be extracted by considering the rotational motion of a water molecule about its three principal axes and then freezing the ability of the carbon atom to rotate. These three fundamentals (termed the *wag*, *twist*, and *rock*) are all subject to significant coupling to adjacent methylene groups. In addition, the majority of the transitions associated with these modes are also rather

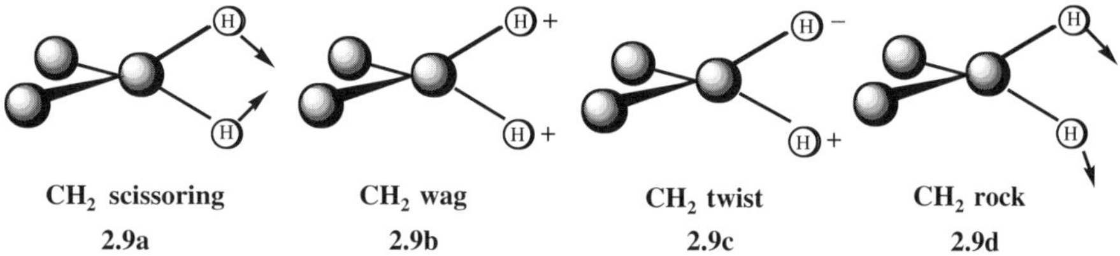

Fig. 2.9 Atomic displacements involved in methylene C—H bending modes.

weak in intensity. Thus, in general, the wag, twist, and rock are not useful group frequencies because they display varying wavenumber values and a multitude of weak bands. They are fingerprint vibrations.

a. The CH_2 wag is shown in Figure 2.9*b*. It essentially involves an oscillatory bending motion of the plane defined by the C—H bonds about an axis perpendicular to the carbon–carbon chain axis and passing through the carbon of the methylene group. These hydrogen displacements require that the two carbon atoms of the adjacent methylene groups be displaced in the opposite direction to keep the system from undergoing a rotation. The system is therefore sensitive to the nature of adjacent carbons and any change will affect the wagging mode. The wag is now coupled to the rest of the molecule.

   In the case of the wagging modes associated with structures in which several (>4 —CH_2— in a row) methylene groups form a chain, a weak band is found in the IR *near 1305 cm^{-1}*. This absorption band has been assigned to the all-in-phase methylene wag (see Figure 2.10). This mode is quite frequency stable in saturated systems, but it is also rather weak and falls on the edge of the fingerprint region. It is therefore of limited diagnostic value as a group frequency. In some cases where the coupled methylene system is adjacent to a very polar functional group (no longer in the hydrocarbon class of substances), the wagging mode appears to rob the functional group modes of some intensity. In this situation a series of weak to very weak wagging bands, one for each pair of coupled methylene groups in the system, often termed *progression* bands are observed (Figures 2.11 and 2.12). This series of bands has been used effectively to approximate chain lengths. The sample must be crystalline.

b. The rocking deformation is shown in Figure 2.9*d*. This mode involves a bending motion of the C—H bonds (both in the same direction) within a plane perpendicular to the carbon–carbon chain axis. As in the case of the wagging vibration, this displacement also requires a countermotion by the adjacent carbon atoms which just cancels out the rotation of the molecule about an axis parallel to the chain axis. The system is therefore sensitive to the nature of adjacent carbons, and any change in their character will affect the rocking mode.

   The all-in-phase rocking mode has a similar behavior to the wag, i.e., it can be assigned to a weak stable band found near 720 ± 3 cm^{-1} (>4 —CH_2— in a row). This mode gives rise to a modest dipole moment change even in hydrocarbons which often allows this vibration to be identified (see Figures 2.1 and 2.10).

   The band may be split when observed with solid-state samples (Figure 2.13). The splitting is related to a crystalline lattice

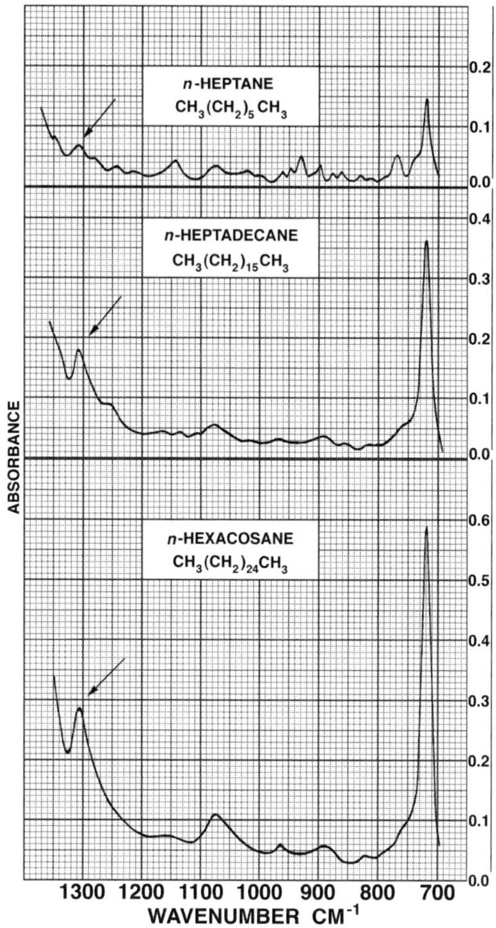

Fig. 2.10 Wagging (1305-cm^{-1}) and rocking (720-cm^{-1}) vibrations of *n*-alkanes in IR.

interaction between two adjacent chains because it disappears when the spectrum is obtained as a melt. If a frequency mismatch is developed by mixing a $(—CH_2—) >4$ compound with its deuterated analog, $(—CD_2—) >4$ (a perdeuteromethylene system), again the splitting collapses; see Figure 2.14. These bands were used in the past to establish the crystallinity of polyethylene samples (see Figure 2.15); better techniques have since been developed for making these measurements. See Chapter 10, p. 278.

c. The twisting mode (Figure 2.9*c*) involves the twisting of a plane defined by the two C–H bonds of the methylene group about an axis perpendicular to the chain axis and bisecting the methylene group. As in the case of the wagging and rocking vibrations, this

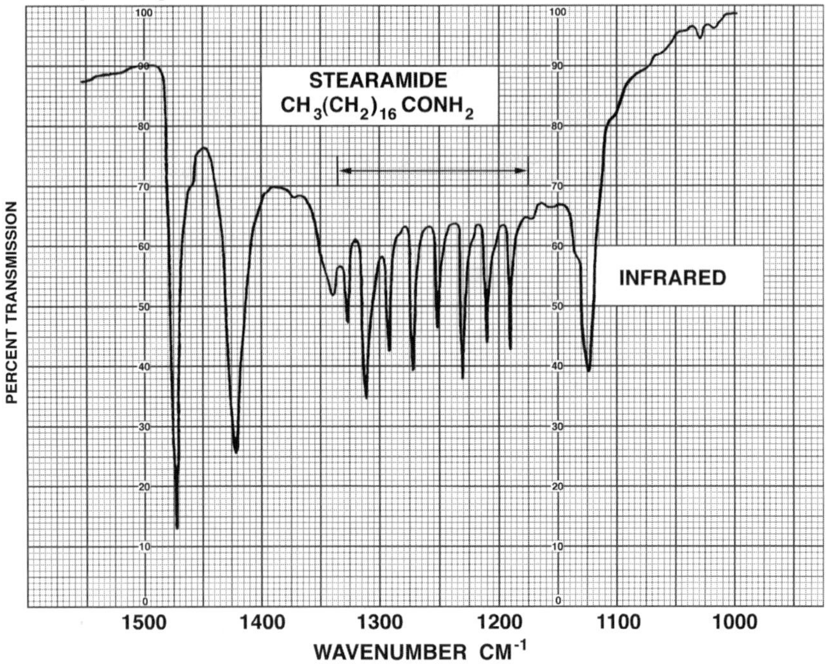

Fig. 2.11 Infrared spectrum of long-chain $[(\text{—}CH_2)_{16}\text{—}]$ crystalline stearamide which exhibits band progressions in wagging region.

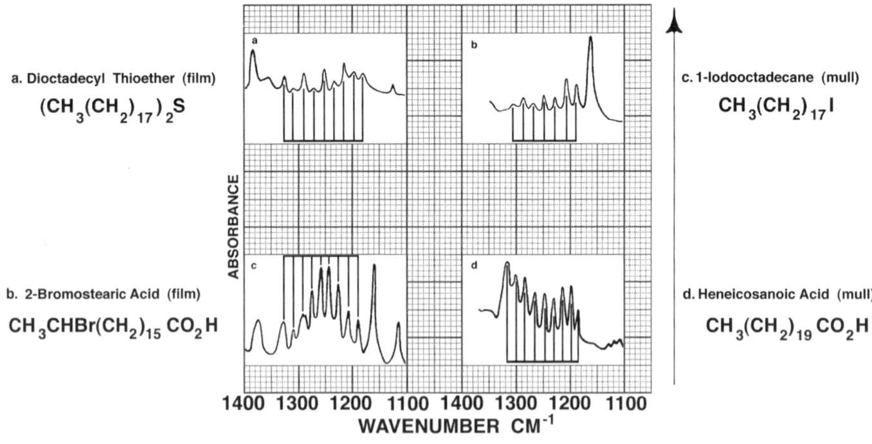

Fig. 2.12 Further examples of crystalline samples of long-chain methylene systems which demonstrate coupling of wagging modes.

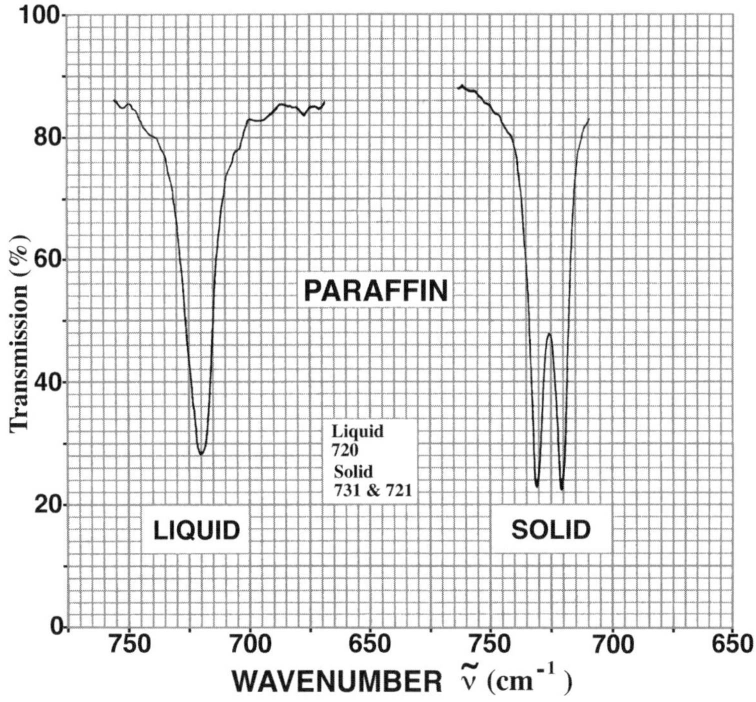

Fig. 2.13 Crystalline lattice interaction between two adjacent chains.

displacement also requires a countermotion by the adjacent carbon atoms which just cancels out the rotation of the molecule about the axis bisecting the carbon–carbon–carbon bond angle. The system is again sensitive to the nature of adjacent carbons, and any change in their character will affect the twisting mode. This mode is again coupled to the rest of the molecule.

The all-in-phase twisting mode for four or more $—CH_2—$ groups in a row is near 1300 cm^{-1}. It is very weak in the IR but has considerable intensity in the Raman effect (Figure 2.1). If Raman data are available, the all-in-phase twist band can be used as supporting evidence for assignment of the 720 cm^{-1} rocking mode in the IR.

5. The bending modes associated with the *methine hydrogen* are hard to identify in most hydrocarbons. (The in-plane bend of an isolated sp^2 C—H group can be particularly important, however, in certain heteroatom systems; see discussion of the aldehyde functional group.)

The C—H vibrational modes of the alkanes (or mixed compounds containing alkyl groups) that are characteristic and reliable group frequencies can be summarized as in Table 2.1.

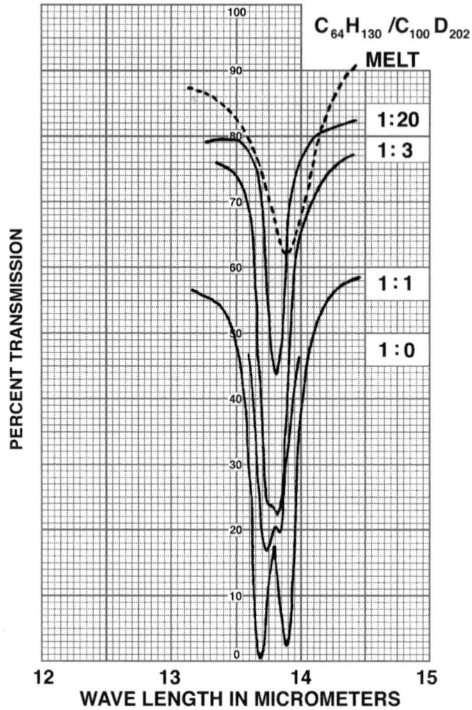

Fig. 2.14 A frequency mismatch is developed by mixing the methylene chain compound with the analogous deuterium substitute one.

TABLE 2.1 Alkane Normal Modes

C—H Vibrational Modes	$\tilde{\nu} \pm 10$ (cm^{-1})
Methyl groups	
Antisymmetric (degenerate) stretch[a]	2960
Symmetric stretch	2870
Antisymmetric (degenerate) deformation	1460
Symmetric (umbrella) deformation	1375
Methylene groups	
Antisymmetric stretch[a]	2925
Symmetric stretch	2850
Symmetric deformation (scissor)	1450
Wagging (all in phase)	1305
Twisting (all in phase, Raman active)	1300
Rocking mode (all in phase)	720

[a] The general rule is that sp^3 C—H stretching modes are all found below 3000 cm^{-1}. (The 3000 cm^{-1} division is a convenient reference value for evaluating C—H stretching modes.) Those modes that fall above 3000 cm^{-1} are bonded to *sp*- and sp^2-hybridized carbon, and those modes that fall below 3000 cm^{-1} are bonded to sp^3 carbon.)

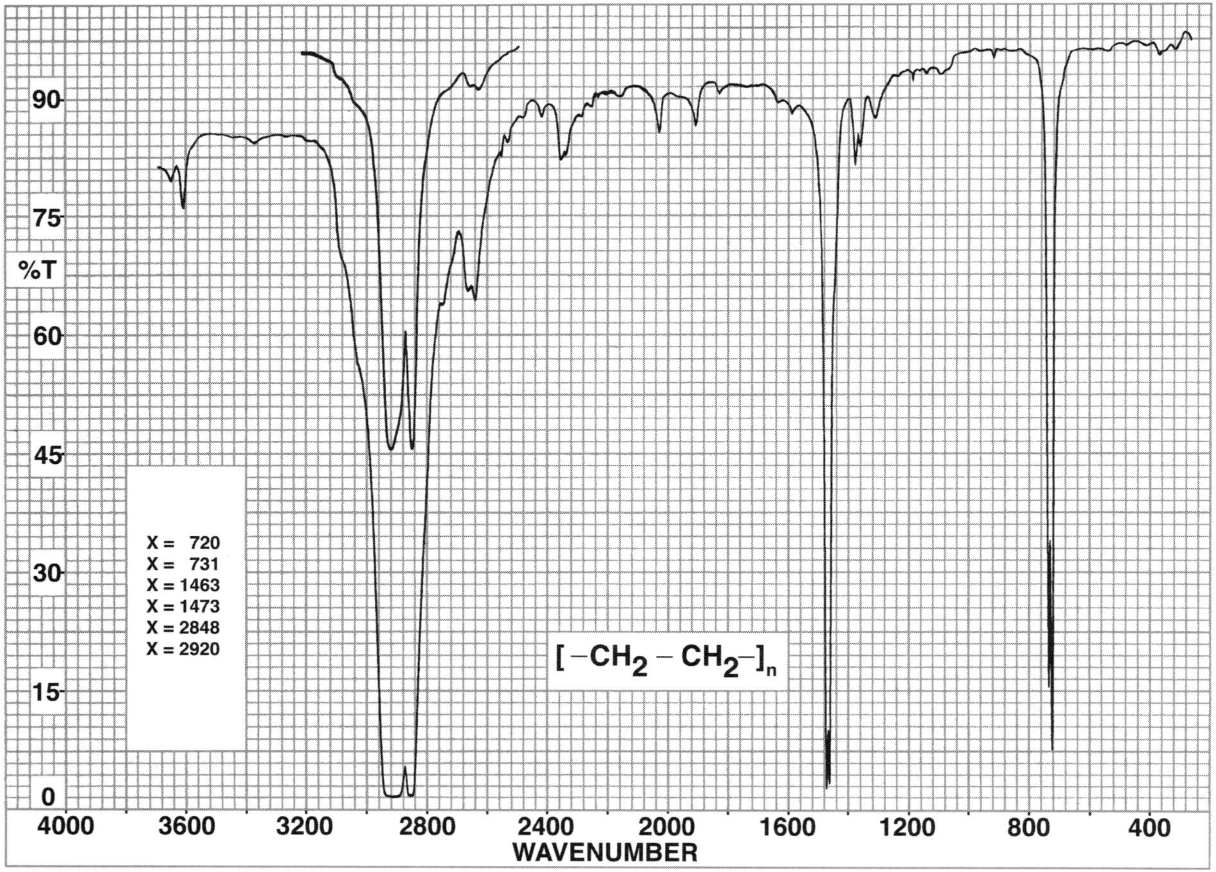

Fig. 2.15 Crystallinity in polyethylene.

G. Intensity of C—H bending modes in straight-chain hydrocarbons

1. As might have been expected, a linear intensity relationship is found for the scissoring, wagging, and rocking modes of the methylene, —CH_2—, group with increasing chain length (Figure 2.16).
2. The scissoring values in the Beer's law plot (Figure 2.16) do not pass through the zero point on extrapolation. It can be shown that this distortion is associated with the overlap of the scissoring mode found near 1450 cm^{-1} with the antisymmetric methyl deformation mode near 1460 cm^{-1}.

H. Carbon–carbon stretching and bending vibrations

Both of these vibrational categories lead to modes that are poor group frequencies for all the obvious reasons. Inherently, these modes possess low frequencies which place the vibrations in the fingerprint region. The connection of several nearly identical and very weak oscillators leads to strong coupling and many difficult-to-assign very weak bands. The C—C absorption pattern, in fact, becomes more characteristic of the substance than of a particular set of group frequencies. The Raman spectrum of complex multicarbocyclic systems which are efficient Raman scatterers has been explored as a route for identifying the skeletal ring system present with some modest success.

III. Branched-chain hydrocarbon spectra

A. Carbon–hydrogen stretching vibrations in branched alkanes

1. Most of the methyl and methylene C—H stretching modes that are good group frequencies in the linear alkanes transfer directly to branched alkanes with little change in wavenumber value. The relative intensity of the C—H stretch methyl doublet compared to the methylene doublet, however, often will be quite different from the linear chain isomer.
2. The concentration of methine C—H groups may be increased with branching, and these groups might be expected to be observed. However the methyl and methylene absorption bands still appear to overwhelm the methine groups, and thus there appear to be no viable group frequencies associated with the stretching modes of this type of C—H bond.

B. Carbon–hydrogen bending vibrations of branched alkanes

1. Methyl deformations

   In this class of vibrations the antisymmetric methyl deformation modes are found to be little perturbed by chain branching while the symmetric methyl modes can be modified in predictable ways that offer valuable group frequency information.
2. In the case of random methyl substitution the only change observed is that the intensities of the antisymmetric and symmetric deformations will reflect the change in concentration relative to the number of

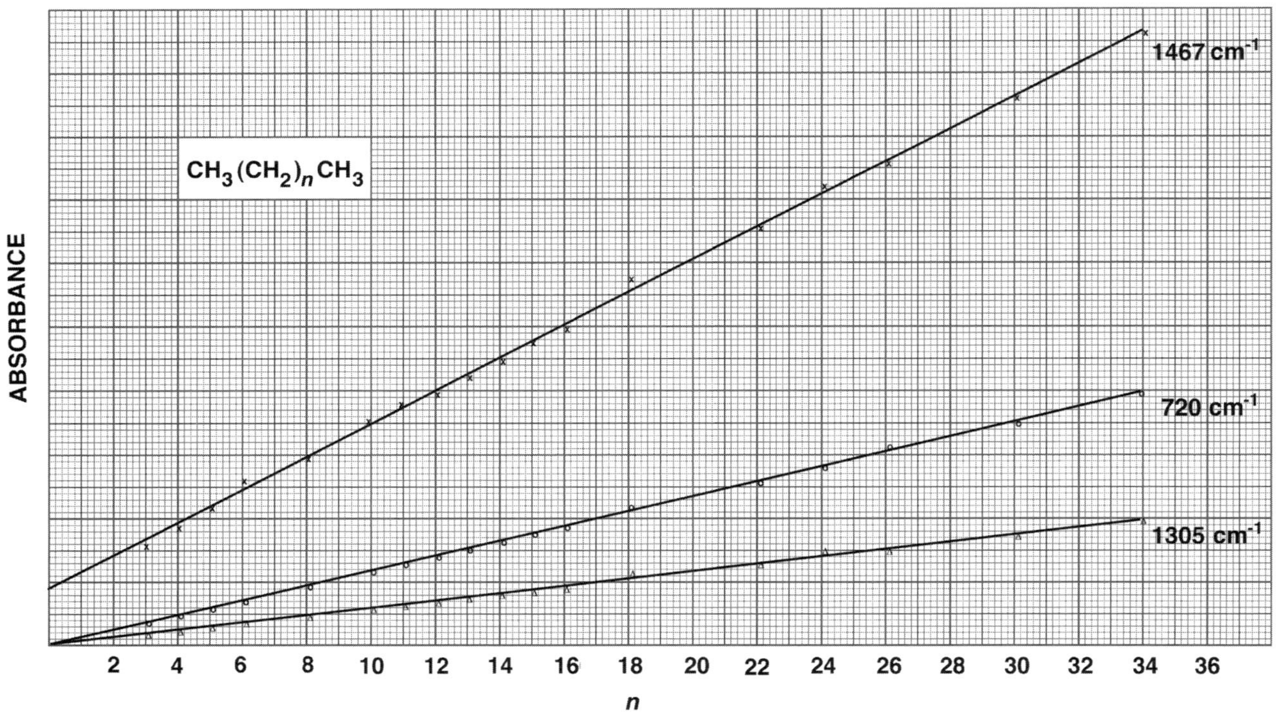

Fig. 2.16 Plot of linear relationship of intensity of methylene deformation modes with increasing chain length in *n*-alkanes.

methylene groups present (methine contributions are to a first approximation ignored). An example of this is the case of 3-methyl-heptane compared to *n*-octane (see Figure 2.17; the spectra were obtrained under identical conditions).

3. Significant spectral changes occur in the 1378 cm^{-1} symmetric methyl deformation mode (umbrella vibration) in structures which contain more than a single methyl group *attached to the same carbon* (or even in some cases where the methyl is attached to the adjacent or vicinal carbon atom). The most notable examples are in the case of isopropyl [—$CH(CH_3)_2$], *gem*-dimethyl [—$C(CH_3)_2$—], and tertiary butyl [—$C(CH_3)_3$] groups.
   a. In the case of isopropyl and *gem*-dimethyl systems the symmetric bend is split into a symmetrical doublet with a $\Delta\tilde{\nu} \sim 20$ cm^{-1} with the band peaks centered near 1385 and 1365 cm^{-1}. Good examples of this effect are shown in the case of 2-methylheptane and 3,3-dimethylhexane (see Figure 2.17). Why do we see three bands in the latter case?
   b. In the case of tertiary butyl systems the symmetric bending mode is split into a doublet with the higher leg (near 1395 cm^{-1}) often considerably weaker than the lower leg (near 1365 cm^{-1}). An example of this type of splitting is found in 2,2-dimethylhexane (see Figure 2.17). Again, why do we see three bands in this region in this case?
   c. It has been shown by studying *gem*-methyl ring systems (see Figure 2.18) that $\Delta\tilde{\nu}$ is inversely dependent on the CH_3—C—CH_3 bond angle and that it drops to very low values when the central carbon is sp^2 hybridized ($\sim$120°). Indeed, it is possible to roughly establish *gem*-methyl internal bond angles by measurement of the $\Delta\tilde{\nu}$ values (see Figure 2.19).

      The splitting does not involve a mechanical coupling of the two identical deformation modes through the valence bond system as originally expected. In that case, $\Delta\tilde{\nu}$ would have been expected to vary directly with the bond angle change. It has been shown to arise, however, via spatial interactions directly between the two methyl groups. As a result of this spatial coupling, the in-phase deformation mode occurs at higher wavenumber values than the out-of-phase modes (this also explains the increased intensity of the lower leg of the doublet in the case of tertiary butyl groups).

IV. Cyclic alkane C—H stretching and bending modes

A. Influence of change in hybridization on C—H stretching modes (see Table 2.2)

As the hybridization shifts from sp^3 toward sp^2, there is a gradual rise in both the antisymmetric and symmetric methylene stretching modes.

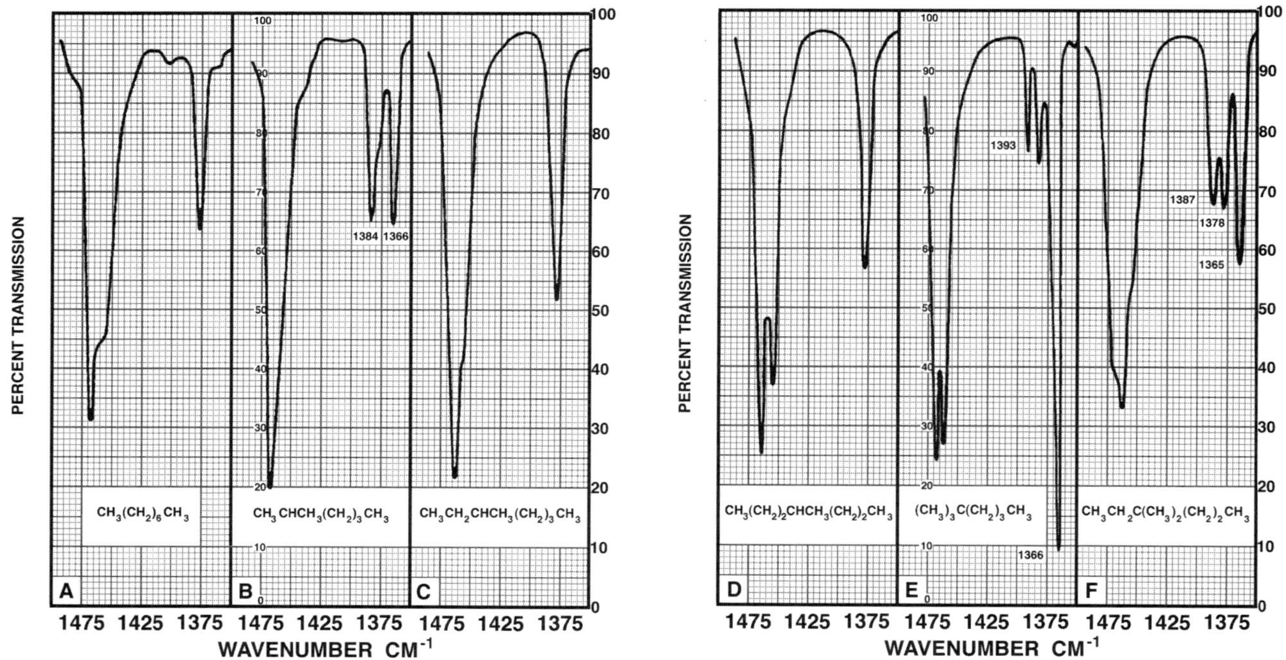

A. *n*-Octane
B. 2-Methylheptane
C. 3-Methylheptane

D. 4-Methylheptane
E. 2,2-Dimethylhexane
F. 3,3-Dimethylhexane

Fig. 2.17 Methyl bending region of six octanes (3-methyl-heptane compared to *n*-octane and 2-methylheptane compared to 3,3-dimethylhexane; also see the case of 2,2-dimethylhexane in this region).

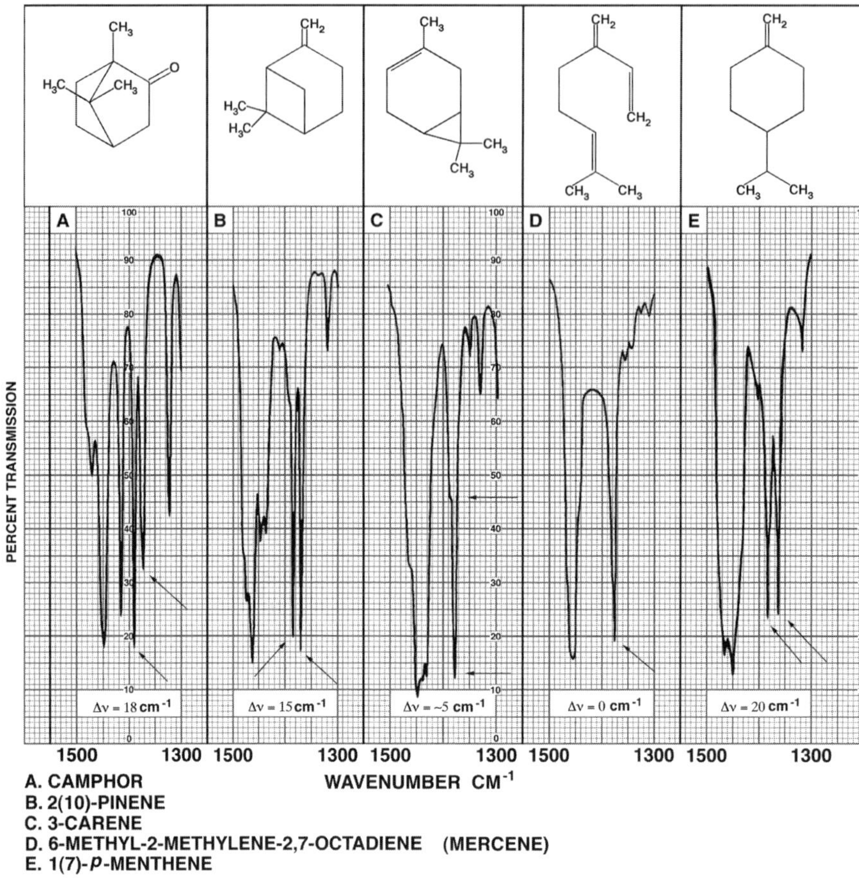

Fig. 2.18 The splitting of the *gem*-methyl symmetric bending modes is inversely dependent on the CH_3—C—CH_3 bond angle.

This shift is consistent with a rise in the C—H force constant values and with the frequencies observed for the alkene C—H stretching modes.

B. Hybridization effects on C—H bending modes

Hybridization appears to have less influence on the bending modes. Compare, for example (see Figure 2.20), the CH_2 scissoring of cyclopentane (1455 cm^{-1}) with cyclohexane (1452 cm^{-1}), methylcyclohexane (1452 cm^{-1}), and ethylcyclohexane (1452 cm^{-1}). (Note also that the antisymmetric methyl deformation mode is resolved and identified near 1466 cm^{-1} in methylcyclohexane and at 1464 cm^{-1} in the ethyl derivative. In more complex systems these modes become increasingly difficult to resolve and assign.)

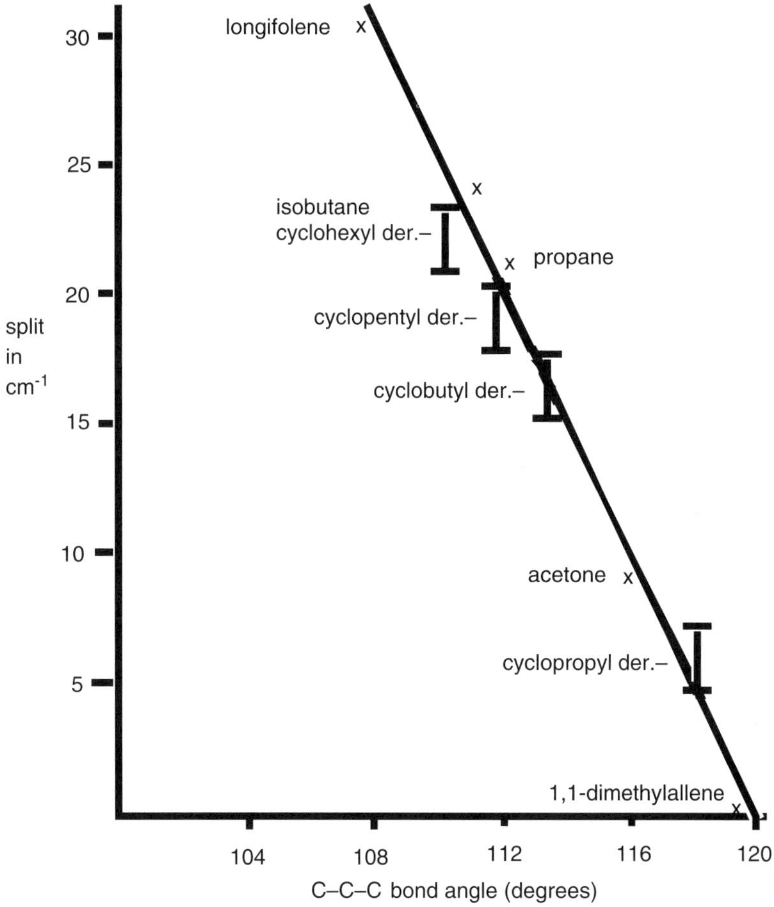

Fig. 2.19 Splitting of 1375 cm^{-1} symmetric methyl deformation mode as function of C—C—C bond angle. der = derivatives.

TABLE 2.2 Methylene Stretching Frequencies in Cycloalkanes

Ring Size	Antisymmetric Stretch (cm^{-1})	Symmetric Stretch (cm^{-1})
Three-membered[a]	3085	3028
Four-membered	2985	2907
Five-membered	2952	2866
Six-membered	2927	2854
n-Hexane	2926	2853

[a] The C—H stretching modes of cyclopropyl ring systems violate the 3000-cm^{-1} C—H stretching rule. The shift is not unexpected, however, as the hybridization of the ring carbons in these badly strained systems is between sp^2 and sp^3.

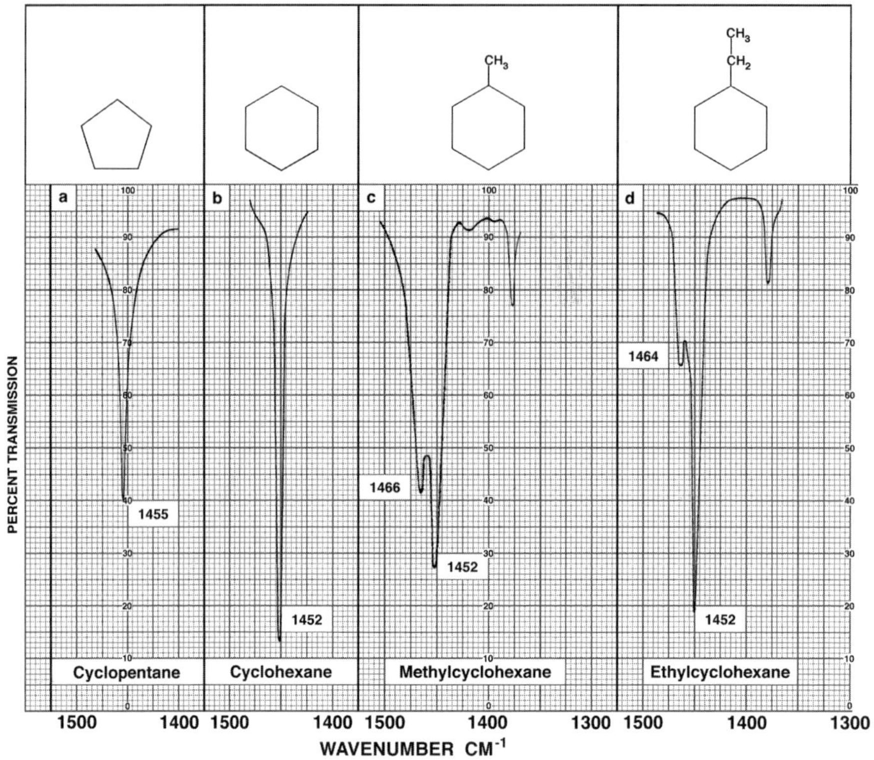

Fig. 2.20 Hybridization appears to have less influence on the C—H bending modes.

V. Effect of isotopic substitution on carbon–hydrogen group frequencies

Strongly associated with the good group frequency attributes of C—H bonds is, as has been noted, the fact that these vibrations can be described to a first approximation by a Hooke's law expression in which only the mass of the hydrogen atom is used in the expression $\tilde{\nu} = 1/2\pi c\sqrt{k/m_H}$, where $\tilde{\nu}$ is in wavenumbers and reduced mass μ is not required and

$$\mu = m_1 \cdot m_2/m_1 + m_2$$
$$c = \text{speed of light, cm/sec}$$
$$k = \text{force constant, dyn/cm}$$
$$m_H = \text{mass of hydrogen atom, g}$$

As a means of testing the above statement, the IR spectra of chloroform ($CHCl_3$) and deuterochloroform ($CDCi_3$) are considered (Figure 2.21):

A. First, the assumption is made (a good one) that the substitution of deuterium for hydrogen will not affect the bond strength and that the only change will be a doubling (or tripling in the case of tritium) of the

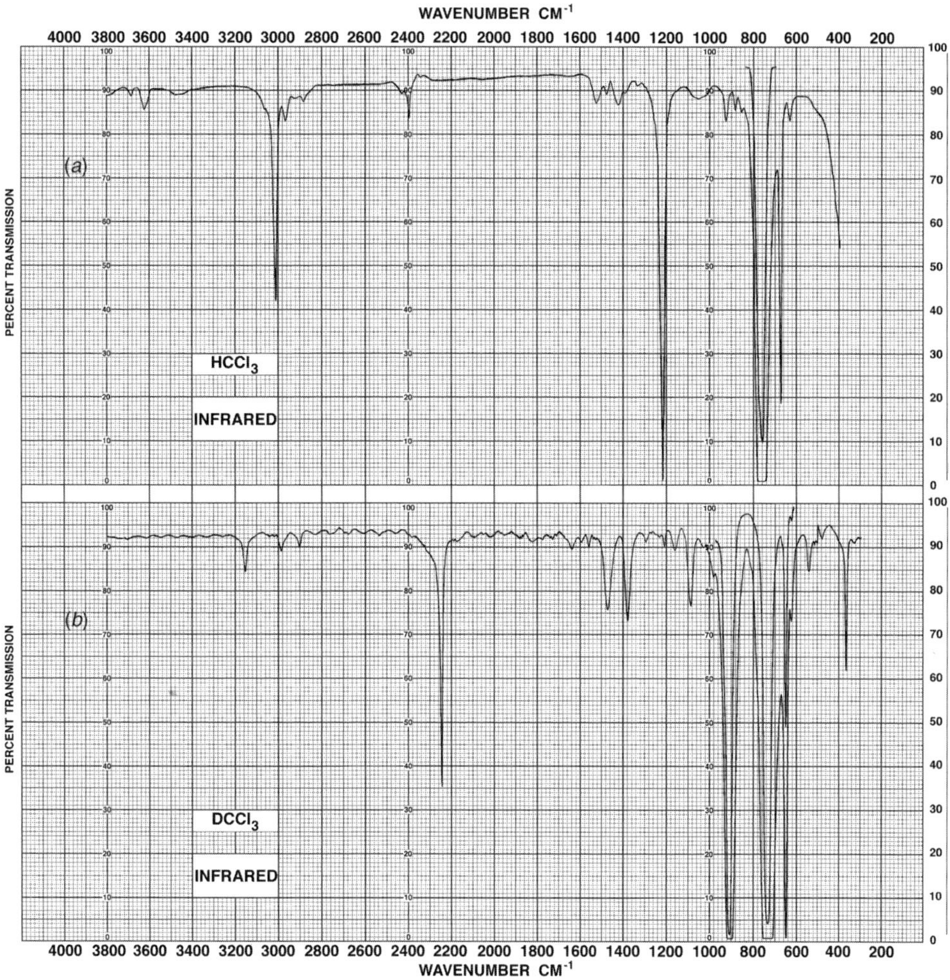

Fig. 2.21 Infrared spectrum of (*a*) chloroform ($CHCl_3$) and (*b*) deuterochloroform ($CDCl_3$).

mass. The introduction of the hydrogen isotopes results in the largest frequency shift of all isotopic labels. In the case of deuterium an inverse effect on the frequency will be a factor of slightly less than 1.41:

$$\tilde{\nu}_H/\tilde{\nu}_D = 1/2\pi c\sqrt{m_D/m_H} = \sqrt{2/1} = 1.4$$

B. The C—H stretching mode is easily identified near 3022 cm^{-1}.* It would be expected that the C—H stretching mode would drop by $\sqrt{2}$, or a factor

**Note*: Chloroform is another exception to the 3000 cm^{-1} rule for differentiating C—H stretching modes. In this case the substitution of three electronegative groups directly on to the carbon atom that is part of the C—H oscillator effectively shifts the hybridization of the carbon from sp^3 toward sp^2.

of 1.41, when the mass of the hydrogen atom is doubled and if the force constant (the bond strength) remains identical to the light system. This shift would move the C—H stretch into the 2143 cm^{-1} region. A new band is observed near 2256 cm^{-1} which is equivalent to a shift factor of 1.34. This band can be confidently assigned to the C—D stretching mode because the carbon atom has been assumed not to be displaced in the vibration. In fact, a value of 1.34–1.35 is a typical shift ratio for C—H systems in which little or no coupling is involved. The C—H group of chloroform will possess a bending mode which would be expected in the fingerprint region. A band near 1216 cm^{-1} in chloroform vanishes in the heavy compound and a new band appears near 908 cm^{-1} corresponding to a shift factor of 1.34. The 1216 cm^{-1} band can therefore be assigned with confidence to the C—H bending mode and the 908-cm^{-1} band to the C—D bending mode.

C. The case of ethylene and ethylene-d_4 (see Table 2.3)
   1. The antisymmetric out-of-phase IR-active C—H stretching mode is assigned to a band near 3106 cm^{-1}. A band near 2345 cm^{-1} in the heavy compound corresponds to a shift factor of 1.325 and indicates very little coupling to the rest of the system.
   2. The Raman-active C=C stretching mode is identified near 1623 cm^{-1}, and the corresponding mode in the heavy compound is located at 1515 cm^{-1}, which gives a shift factor of 1.072. Thus, the C=C stretch also has little coupling beyond the simple C=C oscillator.
   3. The Raman-active CH_2 symmetric in-phase deformation is found near 1342 cm^{-1}, and the associated mode in the heavy compound is assigned to a band at 981 cm^{-1} corresponding to a shift of 1.368 and indicates very little coupling to the other fundamental modes in the molecule.
   4. It is reasonable to conclude that all three of these normal modes of ethylene are reasonably described by the descriptive terms used.

TABLE 2.3 Frequency Shifts in Selected Deuterium Compounds

Compound	Nature of Group Vibration	H	D	Ratio
Allene	C—H stretching (IR)	3005	2230	1.348
	C=C=C stretching (IR)	1957	1921	1.019
	C=C=C stretching (Raman)	1076	874	1.231
	CH_2 deformation (Raman)	1440	1288	1.118
Acetylene	C—H stretching (Raman)	3374	2701	1.250
	C≡C stretching (Raman)	1974	1762	1.120
Ethylene	C—H stretching (IR)	3106	2345	1.325
	C=C stretching (Raman)	1623	1515	1.072
	CH_2 deformation (Raman)	1342	981	1.368

D. The case of allene and allene-d_4 (See Table 2.3)

1. The IR-active antisymmetric out-of-phase CH_2 stretching mode is assigned to a band near 3005 cm^{-1} and the corresponding mode in the d_4 compound is found at 2230 cm^{-1} with a frequency shift of 1.348. This C—H mode, therefore, is a relatively pure fundamental, as are most C—H stretching vibrations.
2. The IR-active antisymmetric C=C=C stretch is found at 1957 cm^{-1} in the light compound and at 1921 cm^{-1} in the heavy compound, giving a ratio of 1.019. Little mixing of this stretch seems to be taking place.
3. The symmetric C=C=C stretch mode (only Raman active) is located at 1076 cm^{-1} in the light compound and 874 cm^{-1} in the heavy one. These values lead to a ratio of 1.231, which indicates that one or both of these fundamental modes must involve coupling and cannot be described as purely C=C=C stretching.
4. In the deuterated molecule the C=C=C symmetric stretching mode is most likely interacting with the Raman-active in-phase symmetric —CH_2— deformation mode which is assigned to a band near 1440 cm^{-1} in the light compound and to a band at 1288 cm^{-1} in the deuterated material. This gives a ratio of 1.118, well below the expected 1.35 range.
5. Thus, the bands at 1076 and 1440 cm^{-1} cannot be described as simply C=C=C stretching and —CH_2— bending, respectively. In the heavy compound, particularly, vibrational energy derived from both fundamental modes is making contributions to both absorption bands.

E. The case of acetylene and acetylene-d_2 (see Table 2.3)

1. The Raman-active symmetric C—H stretching mode occurs near 3374 cm^{-1} in the light compound and has been identified at 2701 cm^{-1} in the heavy material to give a ratio of 1.250. This low shift value represents a relatively rare example of a mixed C—H stretching mode (note that they are on *sp*, not sp^3, hybridized carbons).
2. The —C≡C— stretch is located near 1974 cm^{-1} in the light compound and at 1762 cm^{-1} in the heavy one. This represents a shift ratio of 1.120, an unexpectedly large value which indicates that the symmetric C—H stretch is mixed weakly with the —C≡C— stretch in acetylene. This coupling was discussed earlier (II, D, 3) as an example of coupling between C—H oscillators not attached directly to the same carbon, but where they are bonded to vicinal carbon atoms. If the isotopic change of mass of the end atoms of acetylene could be adjusted over the mass range 0.1–100, the interaction of the two fundamentals can be estimated as shown in Figure 2.22. Thus, while the C—H and —C≡C— stretching modes are relatively pure fundamentals in acetylene, in the deuterium derivative (and also the tritium derivative) they are heavily mixed.

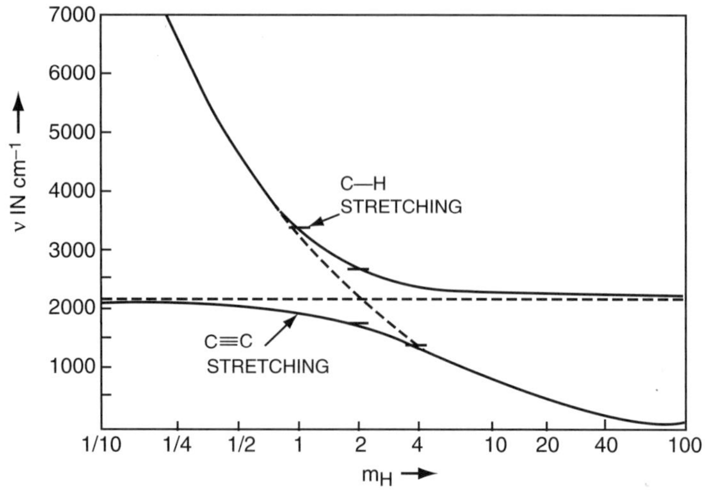

Fig. 2.22 Interaction in hypothetical acetylenes of the symmetric C—X stretch with —C≡C— stretch as function of mass of X group.

VI. Effect of hetero atom substitution on C—H group frequencies. Bohlmann bands.

Effect of lone-pair interactions with adjacent antibonding C—H molecular orbitals—the lowering of C—H stretching frequencies due to lone-pair interactions.

A. The effect results in unusually low wavenumber values for methyl symmetric C—H stretching modes in methoxyl and related groups possessing a lone pair of electrons in atomic orbitals of heteroatoms which can overlap with adjacent C—H bond molecular orbitals. Meakins at Manchester during the 1950s first reported this effect in the case of methoxycyclohexane (Figure 2.23). Note the 2815 cm^{-1} band.

B. Norman Sheppard then reported data which showed rather convincingly that lone-pair electrons were involved in the interaction with the symmetric stretching mode. It was first established by Hill that the effect was absent in aniline ($C_6H_5NH_2$) but that it was observed in *N*-methylaniline ($C_6H_5NHCH_3$) and in *N,N*-dimethylaniline [$C_6H_5N\ (CH_3)_2$] and that the band in question is the most intense band in the C—H stretching region of the latter material (Figure 2.24).

Sheppard then made the methiodide salt of the tertiary amine, *N,N,N*-trimethylanilinium iodide, in which the lone pair is bound to a methyl group and found that the low-frequency C—H stretch disappeared. A further example from this early work of Sheppard was a comparison between the spectra of trimethylamine and tetrametylammonium iodide

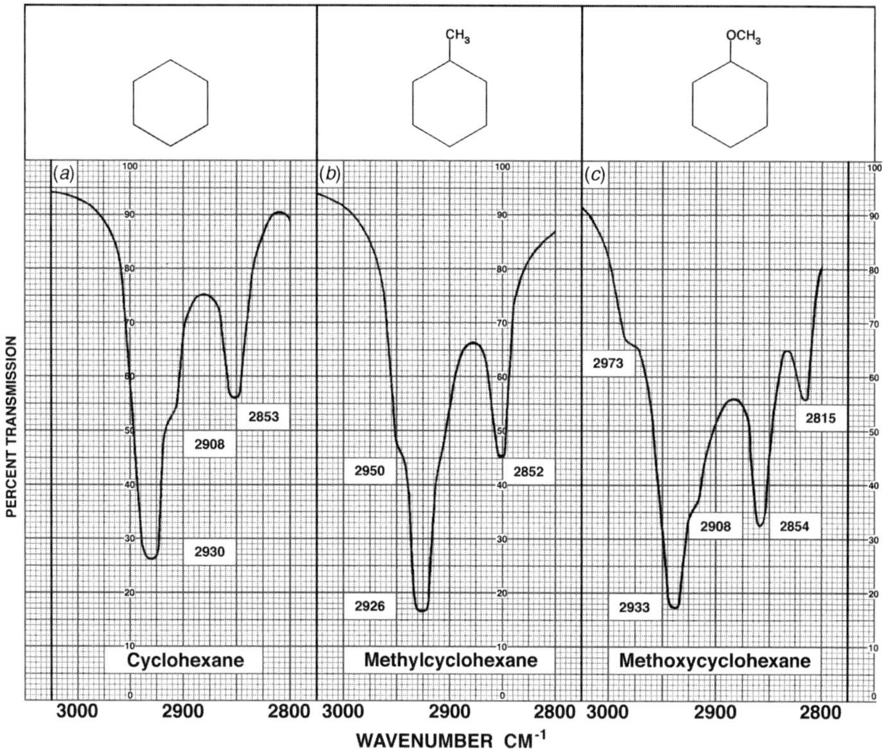

Fig. 2.23 The C—H stretching region of IR spectra of cyclohexane, methylcyclohexane, and methoxycyclohexane. Note the unusually low C—H stretch at 2815 cm^{-1} in methoxycyclohexane.

which showed exactly the same behavior (Figure 2.25). Cyclohexylamine and *N*-methylcyclohexylamine are other examples.

C. In the late 1950s Bohlmann published a series of papers which carefully solidified and extended this correlation.

Two particularly important facets of the lone-pair interaction were discovered by Bohlmann. First, the interaction is not limited to methyl C—H bonds, but most C—H bonds adjacent to lone pairs of electrons demonstrate the effect. Second, there is a distinct stereochemical requirement that all interactions have the lone-pair orbital and the involved C—H bond trans diaxial. Two good examples of this latter requirement are the quinolizidine (bicyclic) and yohimbine (pentacyclic) ring systems in which the stereochemistry of the heteroatom ring fusion can be established by the single C—H stretching mode which will be equitorial to the nitrogen lone pair in the cis fused systems and trans diaxial in the trans fused derivatives.

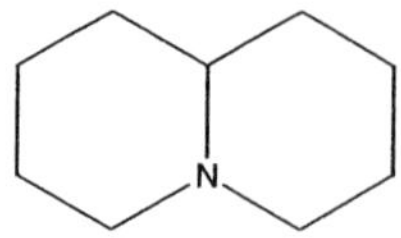

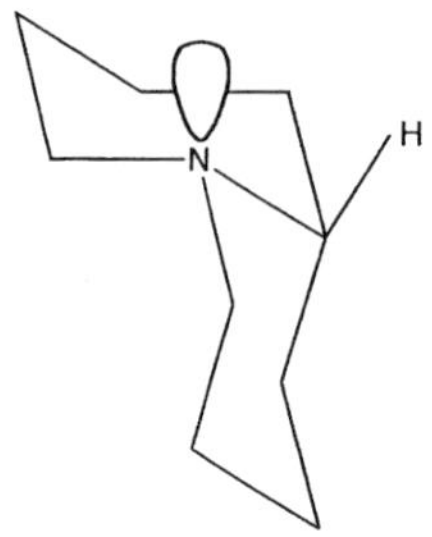

cis-equitorial-axial

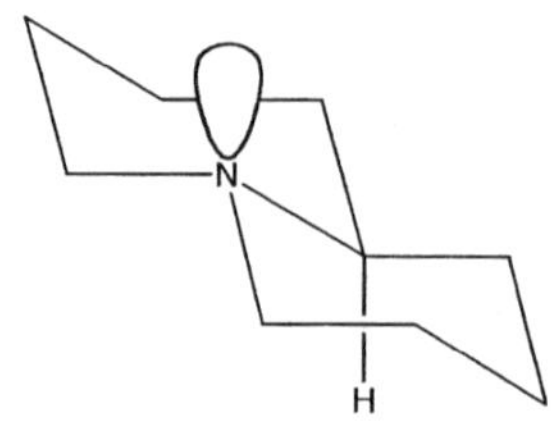

Trans-diaxial

Quinolizidines

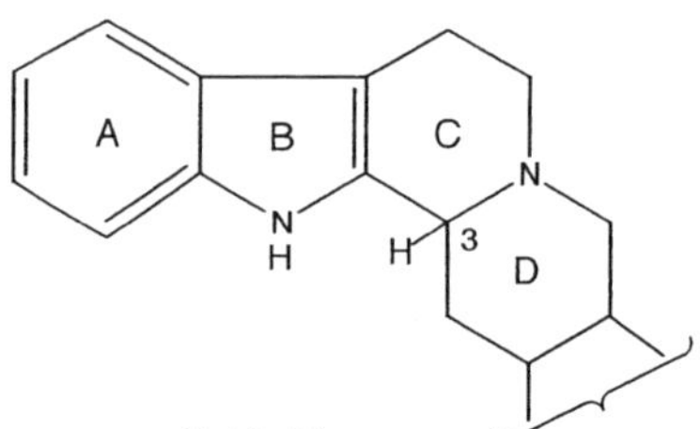

Yohimbines

The lone-pair orbital and the C—H bond involved must be trans diaxial.

D. The C—H stretch of the aldehyde group

1. The C—H stretch in the aldehyde group, —CHO, has been shown to occur as a relatively symmetric doublet with peaks centered near 2840 and 2740 cm^{-1}. The lower leg of this doublet falls below the normal C—H stretching region (3000–2850 cm^{-1}) and is a very good diagnostic band for the presence of the aldehyde functional group even in alkane systems where the higher leg is fully or nearly fully masked (see Figure 2.26*a*). In molecules devoid of sp^3 C—H groups, it is often possible to observe the entire doublet, e.g., in benzaldehyde (see Figure 2.26*b*).
2. The splitting of the single C—H stretching mode in aldehydes involves a second-order interaction, Fermi resonance, in which the first overtone of the in-plane C—H bending mode (located near 1395 cm^{-1}) would be expected to fall near 2790 cm^{-1} or a little lower depending on the amount of anharmonicity. If the uncoupled stretching mode were located in the 2790-cm^{-1} region, the requirements for strong Fermi resonance are satisfied and a symmetric doublet with $\Delta\tilde{\nu} = \sim 100\ cm^{-1}$ is not unreasonable and, in fact, is observed.

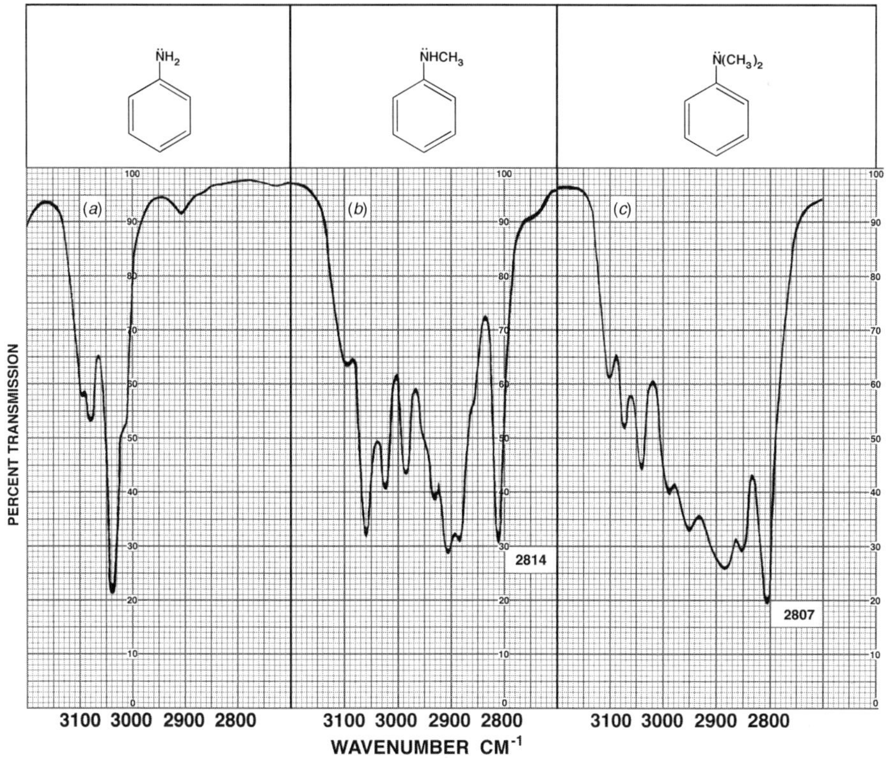

Fig. 2.24 The C—H stretching region of IR spectra of (*a*) aniline, (*b*) *N*-methylaniline, and (*c*) *N*, *N*-dimethylaniline.

3. The location of the uncoupled C—H stretching vibration just below 2800 cm^{-1}, however, has been a problem because the group is directly bonded to an sp^2 carbon. As we have seen, we would expect this stretching mode to occur at considerably higher values (3100–3000 cm^{-1}). The discovery of the lone-pair interaction resolves this problem. In the aldehyde group, the C—H bond is held firmly in the plane of the two lone pairs of electrons on the carbonyl oxygen atom, one pair of which is precisely trans to the C—H bond. Thus, the aldehyde group should exhibit strong lone-pair interaction, and a band shift of the C—H fundamental to below 2800 cm^{-1} is reasonable.

E. Lone-pair interactions involving double heteroatom substitution

In those cases where C—H bonds are adjacent to two trans-diaxial lone pairs of electrons, further perturbation of the symmetric stretching modes might be expected and it does seem to be observed.

1. Several good examples of the strongly perturbed and intensified C—H stretching of a single oscillator are found in the dioxolanes,

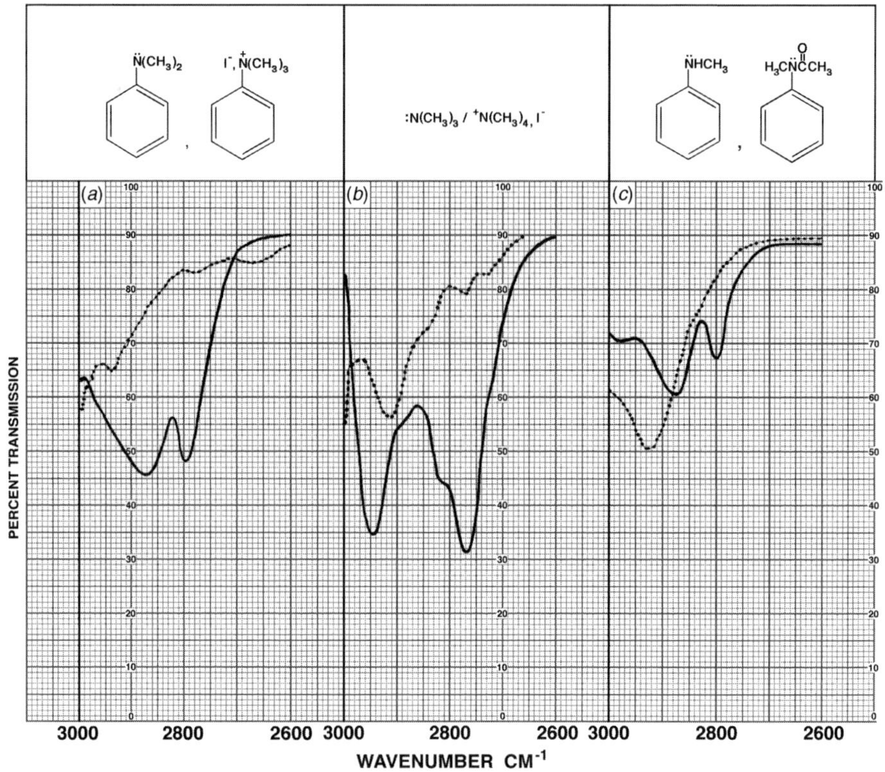

A. Dimethylaniline (_____) and its methiodide (....).*

B. Trimethylamine (_____) and Tetramethylammonium iodide (....).***

C. N-Methylaniline (_____) and N-Methylacetanilide (....).*

Substances in liquid film, or in hexachlorobutadiene mull (marked *), or in vapor phase (marked **).

Fig. 2.25 The C—H IR stretching region of (*a*) *N*,*N*-dimethylaniline and its methiodide (*b*) trimethylamine and tetramethylammonium iodide and (*c*) *N*-methylaniline and *N*-methylacetanilide.

dioxanes, and dioxepanes (see Figure 2.27), where a double lone-pair trans-diaxial interaction occurs with a single C—H stretching mode.

2. The double lone-pair interaction has been shown to perturb the C—H electron distribution to the point where chemical reactivity is controlled. For example, in the ozonolysis of glucopyranoses the β-anomer (double trans-diaxial system) reacts while the α-anomer (single trans-diaxial system) is unreactive. This effect clearly has an influence on the stereoelectronic control of reaction mechanisms, and IR spectroscopy can be a sensitive measure of that interaction.

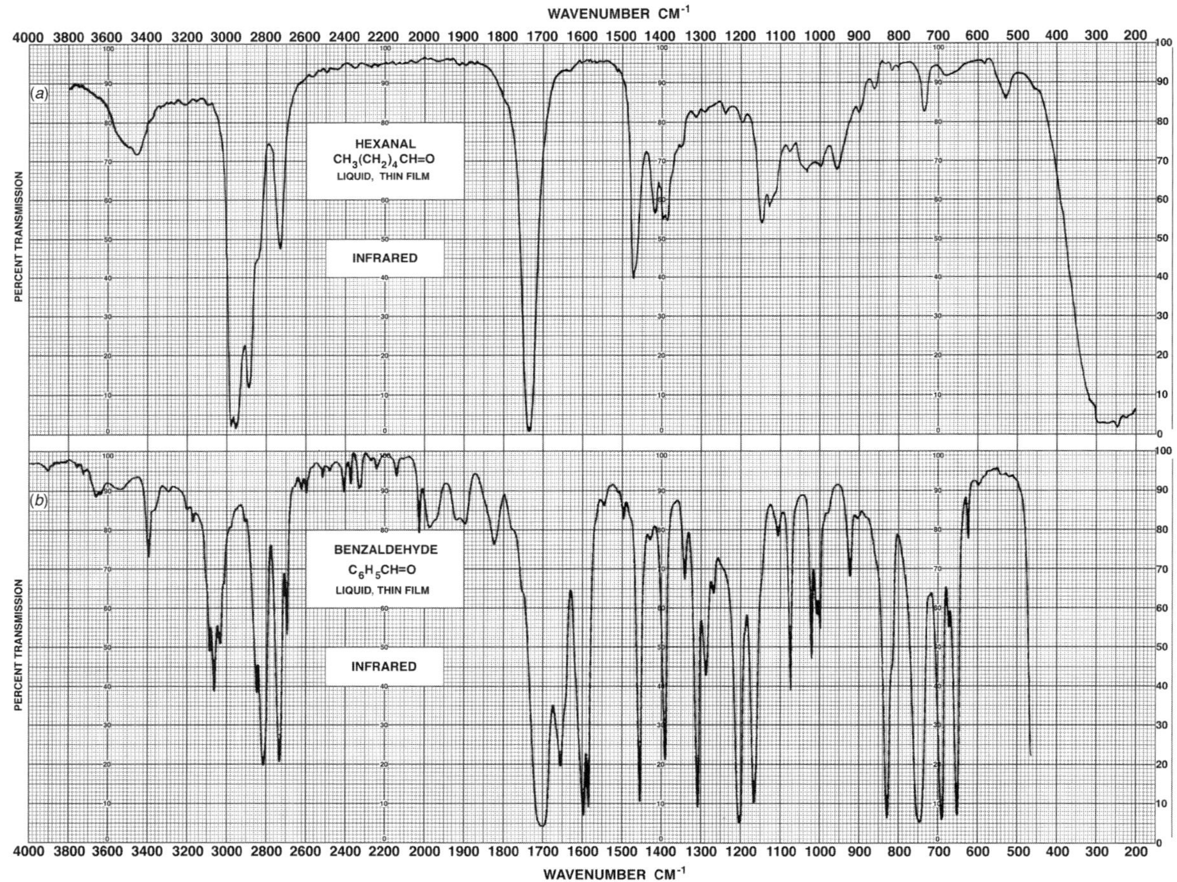

Fig. 2.26 Infrared spectra of (*a*) hexanal and (*b*) benzaldehyde.

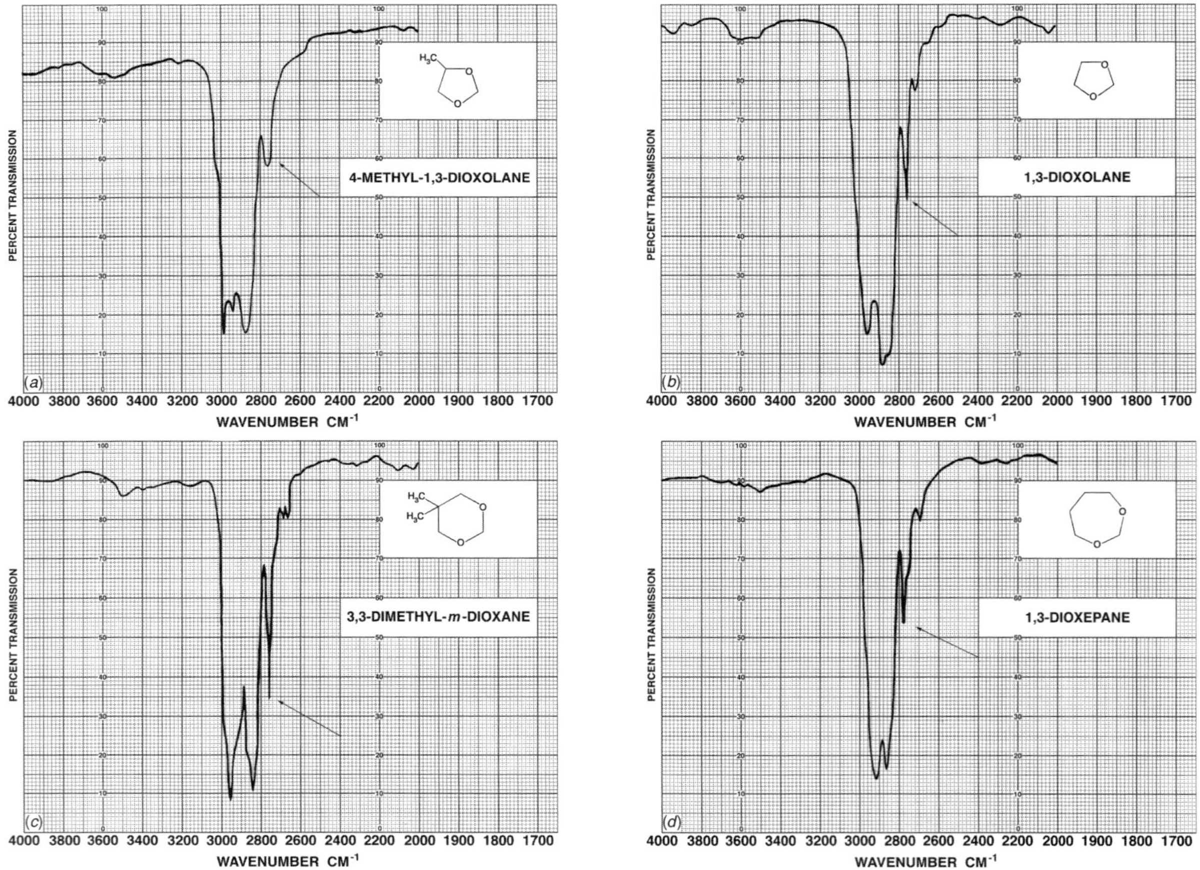

Fig. 2.27 Infrared spectra of 4-methyl-1,3-dioxolane, 1,3-dioxolane, 3,3-dimethyl-*m*-dioxane, and 1,3-dioxepane.

TABLE 2.4 Symmetric Methyl Bending Correlation with Electronegativity of Attached Atom

		cm^{-1} ⟶			
B–CH_3	C–CH_3[a]	N–CH_3	O–CH_3	F–CH_3	
1320 ±10 (gas) 1295 (liquid)	1378±5	1425	1455	1475	
	Si–CH_3	P–CH_3	S–CH_3	Cl–CH_3	
	1255±5	1300	1325	1355	
	Ge–CH_3	As–CH_3	Se–CH_3	Br–CH_3	↑
	1235±10	1250	1280	1305	cm^{-1}
Hg–CH_3	Sn–CH_3	Sb–CH_3		I–CH_3	
1200±10	1190±10	1200		1252	
	Pb–CH_3				
	1165				

Note: The layout is similar to the periodic table.
[a]If CH_3–CO–, drops from 1375 to 1355 cm^{-1}.

References

N. Sheppard, *Trans. Faraday Soc.* **51**, 1465 (1955). For gases.
Rochow, *Spectrochim. Acta* (1960). For Ge, Sn.

VII. The effect of heteroatom substitution on the methyl symmetric bending mode

A. The correlation with electronegativity is given in Table 2.4.

The symmetric deformation mode of the methyl group is sensitive to the electronegativity of the atom to which the methyl group is bonded. The carbon case, CH_3—C at 1378 cm^{-1}, has already been considered in detail. When methyl groups are attached to other elements, information about the heteroatom can be derived. (It has been shown that these systems are sensitive to the H—C—X bending force constant plus the C—X distance and require that nonbonding interactions be involved.) The symmetric deformation mode shifts to higher wavenumbers with increasing electronegativity of the substituting element. Thus, methyl fluoride has the highest value observed, coming near 1475 cm^{-1}.

B. In the case of *N*-methyl groups the umbrella mode is shifted close to 1425 cm^{-1}. In amines, therefore, a band at 1425 cm^{-1} can be used as supporting evidence for the assignment of a sharp band near 2800 cm^{-1} to the *N*-methyl symmetric stretch perturbed by the lone-pair interaction. It should be pointed out that in amides, while the 1425 cm^{-1} band is generally observed, the 2800 cm^{-1} band will be missing. The latter band vanishes because the lone pair of the amide nitrogen is heavily conjugated to the amide carbonyl group, which radically weakens the lone-pair effect.

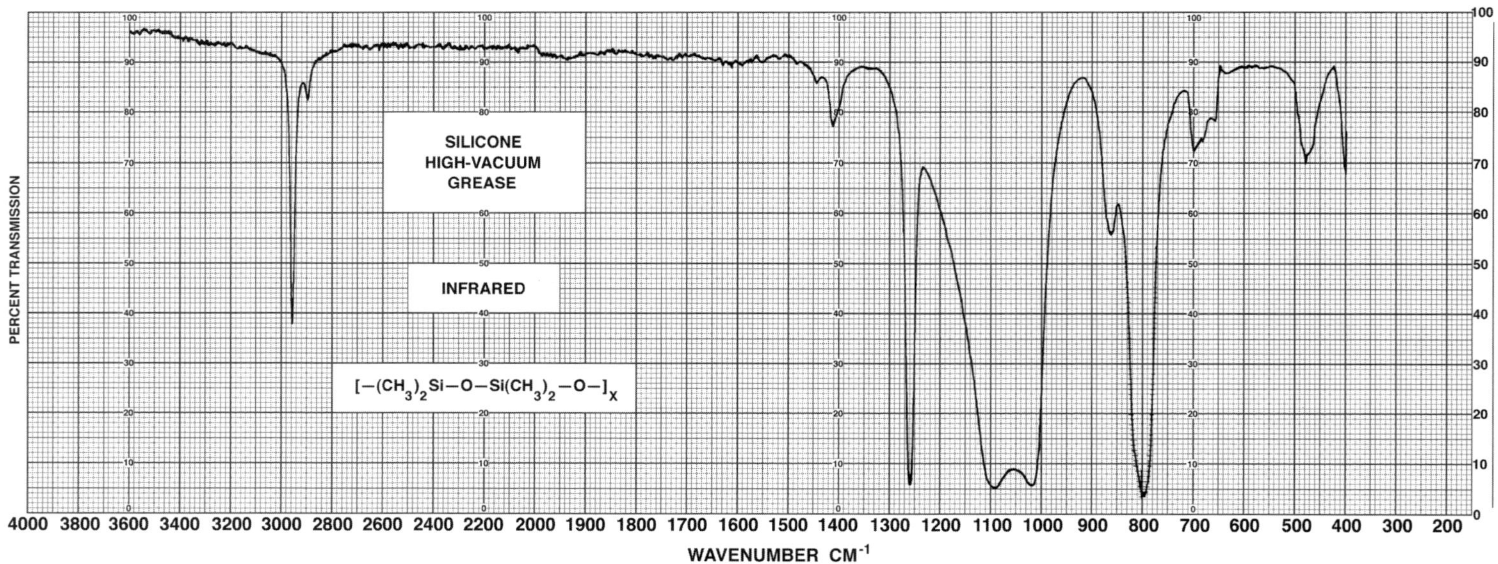

Fig. 2.28 Infrared spectrum of silicone high-vacuum grease.

C. In the case of *O*-methyl groups the band is shifted into the 1455 cm^{-1} region. Unfortunately this is just where the antisymmetric deformations of CH_3—C groups are found so that unambiguous assignments are rarely possible.

D. In the case of *Si*-methyl groups the band is found near 1255 cm^{-1}. It can be quite easily identified in silicon-containing systems such as stopcock grease (Figure 2.28).

VIII. The effect of hyperconjugation on C—H bending modes

A. Methyl and methylene groups alpha to carbonyl systems undergo weak hyperconjugation, which weakens the C—H bond and drops the frequencies. The effect is best observed on the symmetric bending modes.

1. In hexyl acetate the umbrella mode of the methyl group adjacent to the carbonyl drops to ~1355 cm^{-1} and is intensified (see Figure 2.29*a*). Note that the methyl group at the end of the hexyl chain remains in the normal wavenumber range near 1375 cm^{-1}.
2. In 3-heptanone (Figure 2.29*b*) the carbonyl group is bracketed by two methylene groups. The scissoring vibration mode of those groups is near 1420 cm^{-1} and is shifted down from the 1455 cm^{-1} region where the bending mode of the two remaining methylene groups is still found.

IX. The analysis of complex molecular structure

The structure of *cyclo*-artenol:

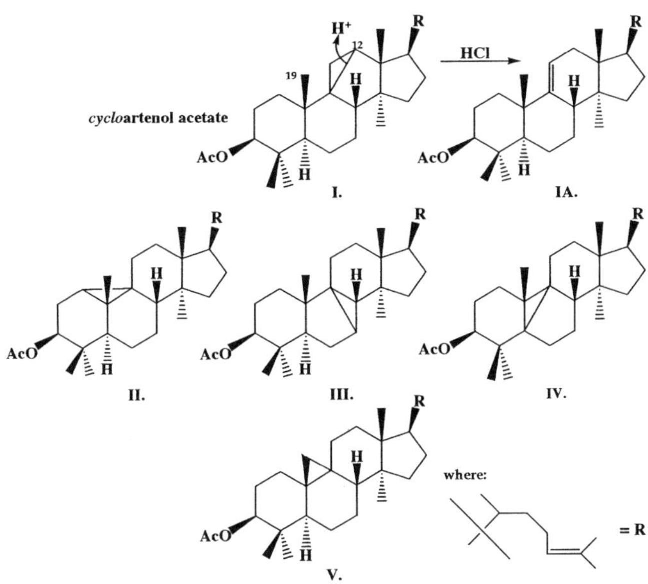

The case of *cyclo*artenol is an interesting example of the use of IR spectroscopy to establish molecular structure.

The acetate derivative of *cyclo*artenol on treatment with HCl gave an olefin (**IA**). The structure of the olefin was first established, and then the structure of the acetate was deduced from the reaction with acid. Because *cyclo*artenol is an isomer of the olefin and does not contain a double bond, the presence of a three-membered ring which cleaves under acidic conditions was proposed. Attack by acid as shown would be expected to give the tertiary carbocation, which undergoes elimination to form the alkene. The location of the ring must involve the tertiary position, but five possible isomers could lead to the known product. Initial studies of *cyclo*artenol considered only the isomers **I–IV** as possible structures. Based on the IR spectrum of *cyclo*artenol, Cole at Perth proposed structure **I** as the correct isomer. The argument was based on the observation that a weak but identifiable band was located in the 3085 cm^{-1} region of the *cyclo*artenol. Only structure **I** would be expected to possess a band in this region as it is the only isomer with at least one unsubstituted methylene group as part of the three-membered ring. It is only the antisymmetric stretch of the *cyclo*-propyl $—CH_2—$ group which rises above 3060 cm^{-1}.

Cole's group did not, however, consider structure **V** as a possible isomer, but D. H. R. Barton did. Barton recognized that the IR spectrum did not eliminate one more possible isomeric structure, structure **V**. Barton then used IR techniques to differentiate between structures **I** and **V** and establish the correct structure. To do this, Barton carried out the acid cleavage of the *cyclo*propyl group with DCl, which necessarily introduces a deuterium isotopic label at either carbon position 12 in structure **I** or carbon position 19 in structure **V**. If the label occurs at position 19, it is substituted on a methyl group. If substituted at position 12, it is part of a methylene group. There are seven methyl groups in both *cyclo*artenol (**I**) and the olefin derivative. Barton assumed that as the local structural environment of the methyl groups is similar in the two isomers and that the relative intensities of the symmetric methyl bending regions would, therefore, be (similar or) identical in the case where the label was introduced at position 12 but that the olefin would exhibit a drop of approximately one-seventh in the intensity of the symmetric methyl bending band in the case when the tag is found on the angular methyl group (C_{19}-*cyclo*artenol (**V**)). This drop in intensity is expected because the C—D bending frequency will not match the two C—H bending modes and the methyl bending vibration will be disrupted. The experiment was performed and the olefinic product was shown to possess a symmetric methyl bending band of intensity only six-sevenths that of the olefin formed on treatment of the starting acetate with HCl (the acetate methyl bending was shifted enough to be ignored). Thus, **V** is the correct structure for *cyclo*artenol.

*cyclo*Artenol was the first steroidal structure to be shown to possess a *cyclo*propyl group constructed from the C_{19} angular methyl group. Following this discovery a large family of molecules possessing these structures has been isolated.

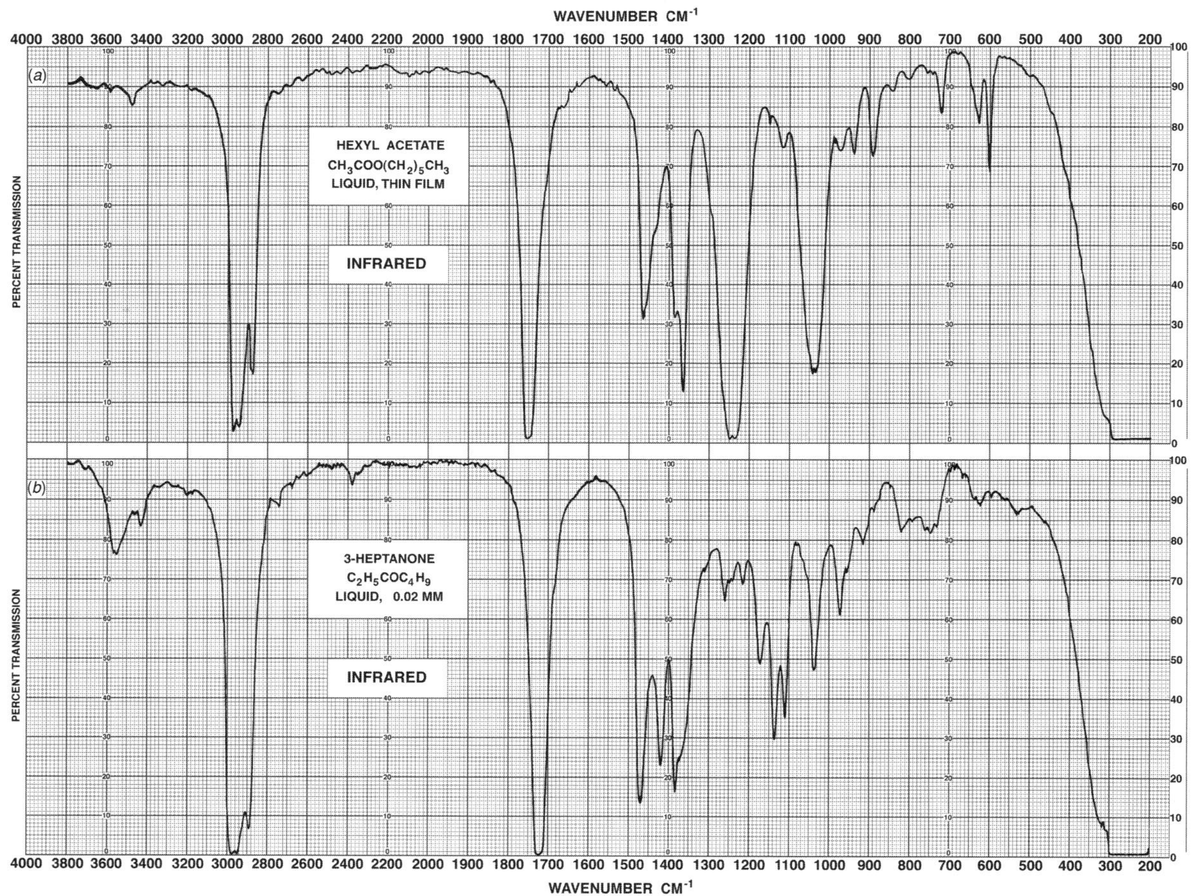

Fig. 2.29 Infrared spectra of (*a*) hexyl acetate and (*b*) 3-heptanone.

REFERENCES

1. Bellamy, Vol. 1, Chapter 2.
2. Bellamy, Vol. 2, Chapter 1.
3. R. N. Jones, In A. Weissberger (Ed.), *Technique of Organic Chemistry*, Vol. 9, Interscience, New York, 1956, Chapter 4.
4. Lord and Miller, *Appl. Spectrosc.* **10**, 115 (1956).
5. Bellamy and Mayo, *J. Phys. Chem.* **80**, 1217 (1976).
6. Bohlmann, *Chem. Ber.* **91**, 2157 (1958).
7. Mallinson and McKean, *Spectrochim. Acta* **39A**, 1133 (1974)
8. Barton, Page, and Warnhoff, *J. Chem. Soc.* 2715 (1954).

3 Characteristic Frequencies of Alkenes (Olefins)

FOIL A. MILLER

Alkenes (or olefins) have all the characteristic bands of saturated hydrocarbons, plus new ones for (a) the C=C stretch and (b) the olefinic C—H stretches and bends.

I. Olefinic C—H stretch, 3100–3000 cm^{-1}
  A. An important generalization
    C—H stretches for a hydrogen on an *unsaturated* carbon atom are above 3000 cm^{-1}; those for a hydrogen on a *saturated* carbon are below 3000 cm^{-1}.
    1. *Examples:*

sp carbon	≡C–H	Acetylenic C—H	~3300 cm^{-1}
sp^2 carbon	=C(–)–H	Olefinic C—H Aromatic C—H	3100–3000 cm^{-1} 3100–3000 cm^{-1}
sp^3 carbon	—C(–)(–)–H	Saturated C—H	2975–2800 cm^{-1} (<3000 cm^{-1})

    2. Therefore 3000 cm^{-1} is an important number to remember.
  B. Olefinic C—H stretches occur in two ranges:

Terminal $=CH_2$	$>C=CH_2$	~3080 cm^{-1}	As in $R'R''C=CH_2$
Nonterminal =CH–C	>C=CH–C	~3030 cm^{-1}	As in $R'HC=CHR''$

    1. The vinyl group, $-CH=CH_2$, has both bands.
    2. These bands are usually weak because there are not many olefinic hydrogens.
    3. The band near 3030 cm^{-1} may be overlapped by the saturated C—H stretch near 2975 cm^{-1} and appear only as a shoulder.

Course Notes on the Interpretation of Infrared and Raman Spectra, by Dana W. Mayo, Foil A. Miller, and Robert W. Hannah.
ISBN 0-471-24823-1

4. Why is the frequency so high in $=CH_2$? Because the two C—H stretches couple and split. The antisymmetric stretch moves up to about 3080 cm^{-1} and is usually evident. The symmetric one drops to about 3000 cm^{-1} and is usually hidden by strong saturated C—H stretches.
   Example: Ethylene, $H_2C{=}CH_2$ (R = Raman, IR = infrared):

Antisymmetric stretches	3103 (R) and 3106 cm^{-1} (IR)
Symmetric stretches	3026 (R) and 2989 cm^{-1} (IR)

II. C=C stretch

In the Raman spectrum this is always intense and reliable. In the IR it is useful, but the intensity varies enormously.

A. Frequency

1. For open-chain alkyl olefins, there are two ranges (see Figure 3.1):

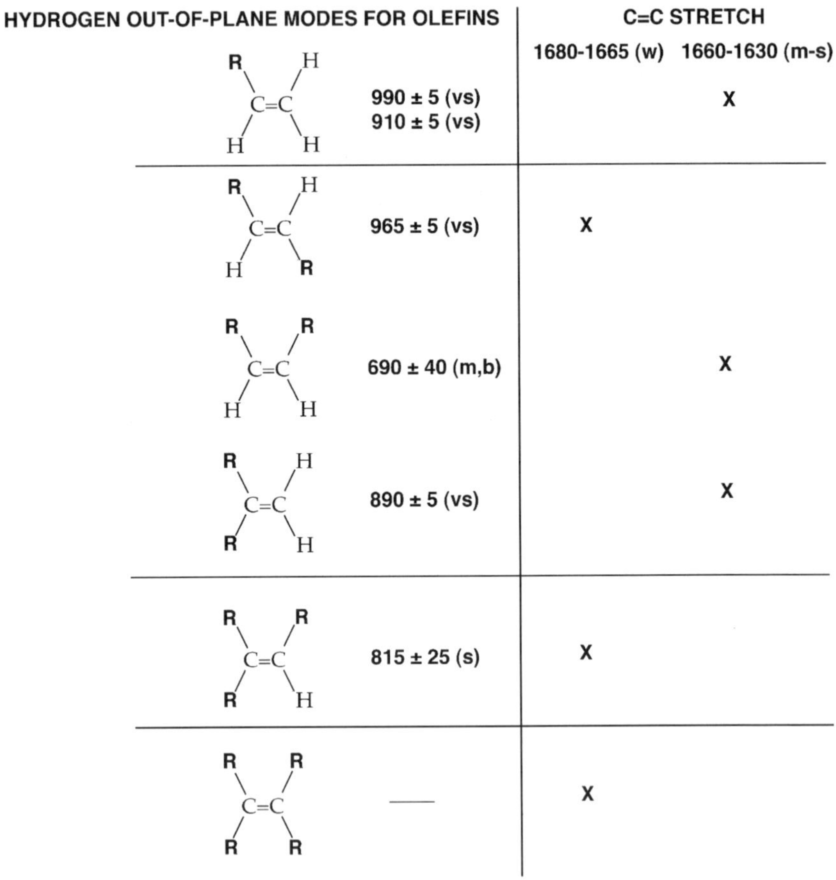

Fig. 3.1 C—H out-of-plane bends and C=C stretches of olefins.

a. 1680–1665 cm^{-1} (w) — In trans, trisubstituted, and tetrasubstituted olefins

b. 1660–1630 cm^{-1} (m to s) — In vinyl, cis, and vinylidene olefins

2. In the IR, the first group gives weak bands, the second group gives medium to strong ones.
3. The ranges apply to hydrocarbons with no conjugation. Both conjugation and polar substituents change the frequencies.
4. Rationalization for the frequencies.

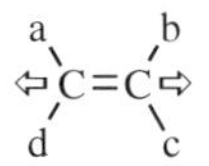

a. As the C=C bond is stretched, the adjoining single *bonds* are compressed a little. If a substituent is H, the compression of the single bonds is minimal because H is light and tends to ride with the olefinic carbon. If a substituent is heavier, there will be more resistance to moving it because of its larger inertia. Hence, as the C=C bond stretches, more work is done on the adjacent single bonds. This increases the effective force constant and raises $\tilde{\nu}_{(C=C)}$.

b. Note that compounds giving $\tilde{\nu}_{(C=C)}$ in the higher range have alkyl groups on *both* olefinic carbons. Of compounds in the lower range, vinyl and vinylidene have hydrogen atoms on only one of their olefinic carbons, so they are weighted less. Cis-disubstituted compounds, also found in this lower range, do not fit this simple model.

5. Other factors affecting the frequency

a. Conjugation

1) Lowers $\tilde{\nu}_{(C=C)}$ by 20–30 cm^{-1}. *Examples:*

CH_2=CH–CH	1651 cm^{-1} (Raman, gas)	Our reference value
CH_2=CH–CO–CH_3	1617 cm^{-1}	$\tilde{\nu}_{(C=O)}$ = 1684 cm^{-1}
CH_2=CH–CH=CH_2	1643 (Raman), 1599 cm^{-1} (IR)	Average = 1621 cm^{-1}

(See later discussion of butadiene.)

CH_2=CH–Phenyl	1630 cm^{-1}
CH_2=CH–C≡N	1609 cm^{-1}

2) This lowering is due to resonance forms which give some single-bond character to the two double bonds.

$$\text{C=C-C=C},\quad \overset{+}{\text{C}}\text{-C=C-}\overset{-}{\text{C}},\quad \text{and}\quad \overset{-}{\text{C}}\text{-C=C-}\overset{+}{\text{C}}$$

3) Conjugation to phenyl has about the same effect as conjugation to a double bond.

4) Conjugation on both sides of the double bond has little additional effect.

Examples:

$H_2C{=}CH{-}Ph$	$Ph{-}CH{=}CH{-}Ph$	(Ph = phenyl)
1630 cm^{-1}	1625 cm^{-1} (cis, IR)	
	1641 cm^{-1} (trans, Raman, solid)	
$H_2C{=}CH{-}C{\equiv}N$	$N{\equiv}C{-}CH{=}CH{-}C{\equiv}N$	
1609 cm^{-1}	1612 (trans, IR forbidden)	

5) In the C=C–C=O system, both $\tilde{\nu}_{(C=C)}$ and $\tilde{\nu}_{(C=O)}$ are lowered.

b. Bond angle effects

1) C=C external to a ring

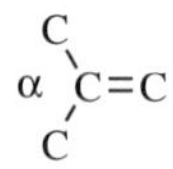

a) As α is decreased, $\tilde{\nu}_{(C=C)}$ is raised.
Reason: In a normal C=C stretch, there is some compression of the adjacent C–C single bonds. This adds additional resistance (increases the effective force constant). The resulting $\tilde{\nu}_{(C=C)}$ is about 1650 cm^{-1}. As α is made smaller, the C–C single bonds are more nearly in line with the C=C double bond and offer more opposition to its stretch. The $\tilde{\nu}_{(C=C)}$ therefore rises. (See Figure 3.2.)

b) Examples are given in Figure 3.2.

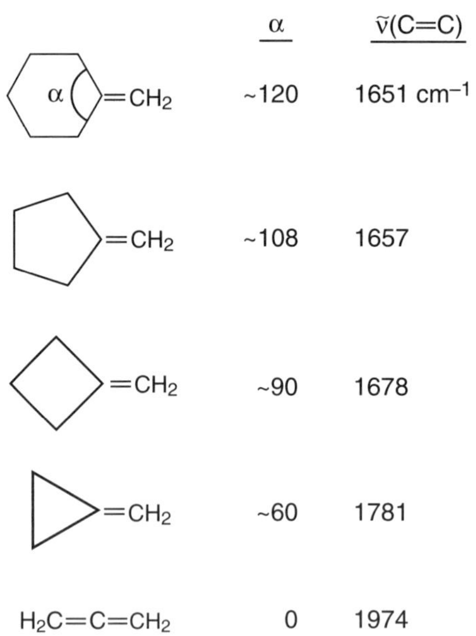

Fig. 3.2 C=C frequency when C=C is external to a ring.

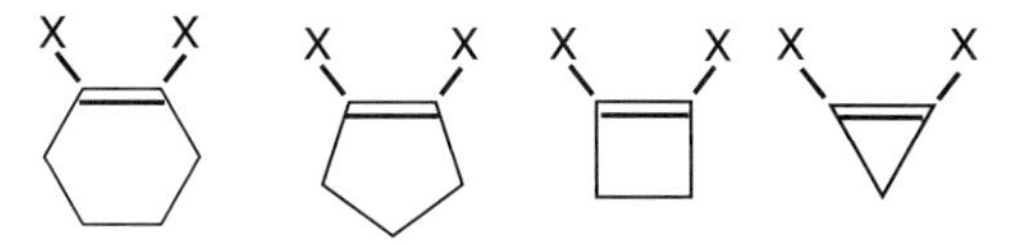

X				
–H	1653	← 1611	1565	1647 liq. 1656 gas
$-CH_3$	1683	1686	1695	1887
–F (perfluoro)	1746	1771 →	1799	1945

Fig. 3.3 C=C frequency when C=C is in a ring.

c) The same trend occurs for the stretch of a carbonyl external to a ring, as we shall see later when considering carbonyls.

2) C=C in a ring (cyclic olefins) (See Figure 3.3)

a) As the ring becomes smaller, the C=C stretching does *less* work on the adjacent C—C single bonds of the ring because they are more nearly perpendicular to the C=C. Therefore $\tilde{\nu}_{(C=C)}$ decreases. The lowering is greatest for the cyclobutene ring (where $\tilde{\nu}_{(C=C)}$ is all the way down to 1565 cm^{-1}) because the C—C bonds of the ring are nearly perpendicular to the C=C line.

b) In cyclopropene the C=C stretch again involves work on the C—C bonds (this time to stretch them), so $\tilde{\nu}_{(C=C)}$ goes back up to 1647 cm^{-1}, about where it was originally. This result shows that the lowering for the other compounds is not due to ring strain.

c) For the dimethyl derivatives, $\tilde{\nu}_{(C=C)}$ is nearly constant for the first three compounds. Why?

(1) In the first three compounds, as the ring gets smaller, the CH_3 groups are apparently flared outward so that they are more nearly in line with the C=C bond. This compensates for pulling the ring carbons more nearly perpendicular to the C=C line, and $\tilde{\nu}_{(C=C)}$ is scarcely changed because the two effects are approximately equivalent.

(2) In dimethyl cyclopropene all four C—C single bonds oppose the C=C stretch, and $\tilde{\nu}_{(C=C)}$ is exceptionally high (1887 cm^{-1}).

d) With fluorine on the olefinic ring carbons, the frequency goes *up* as the ring becomes smaller. Why? First, the mass of the F atom approximates that of a methyl group. Second, the C—F bond is about four-thirds times stiffer than the C—CH_3 bond. Therefore the flaring effect overcompensates compared to the methyl analog when pulling the ring carbons in, and the frequency goes *up* as the ring becomes smaller. It is amazingly high for perfluorocyclopropene (1945 cm^{-1}).

6. $\tilde{\nu}_{(C=C)}$ for other substituents
   a. For substituents other than H or C, the C=C stretch is altered. The effect is both mass and electrical. The latter has two components: electronegativity, which raises the C=C frequency, and resonance or mesomerism, which lowers it. The situation is therefore complicated, and it is difficult to make useful generalizations that are valid for more than a limited class of compounds.
   b. *Example*: When two or more fluorines are on a double bond, the frequency and intensity of the C=C stretch are so high that the band can be mistaken for a carbonyl. However, when chlorine is attached to a C=C, the stretch is *lower* than for corresponding alkyl compounds.
   c. Vinyl ethers, C–O–CH=CH_2
      1) $\tilde{\nu}_{(C=C)}$ is raised to about 1680 cm^{-1} and greatly intensified, so that it may be confused with a conjugated carbonyl. There are usually multiple peaks.
      2) There is a C–H stretch *above 3100* cm^{-1} which provides a warning.

B. Intensity of $\tilde{\nu}_{(C=C)}$
   1. In Raman spectra: intense and always present. A good, reliable band.
   2. In the IR the intensity is highly variable and may range from very strong to completely missing. The reason? Selection rules.
      a. Requirement for a vibration to be IR active: It must produce a change in the dipole moment.
      b. If a C=C bond is across a center of symmetry, the C=C stretch *cannot* produce a change in the dipole moment. The band is forbidden in the IR.
      Examples: ethylene, *trans*-1,2-dichloroethylene.
      Therefore we do not expect to see the C=C stretch in the IR for a symmetrically substituted ethylene.
      c. Butadiene

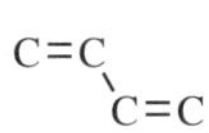

This has the trans conformation as shown. Because there are two C=C bonds, there are two C=C stretching vibrations:

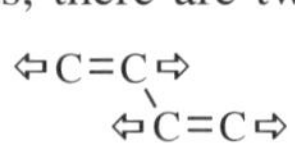

In-phase, or symmetric,
C=C stretch
1643 cm^{-1} (Raman, gas)

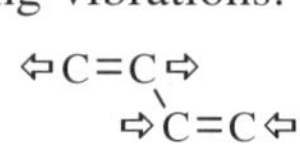

Out-of-phase, or antisymmetric,
C=C stretch
1599 cm^{-1} (IR, gas)

The in-phase stretch does not produce a change in the dipole moment, so it is forbidden in the IR. Therefore only one of the C=C stretching modes is observed in the IR. Similarly the out–of–phase stretch is forbidden in the Raman spectrum. Note that the rule of mutual exclusion applies.

d. Pseudo center of symmetry

The symmetry argument is rigorously valid only if there is a real center of symmetry, which requires that opposite substituents be identical. However, even if they are not identical but their electrical behavior is nearly the same, selection rules often apply rather well. For example, in *trans*-R′CH = CHR″, the substituents can be almost any alkyl groups and the C=C stretch will be weak or missing. It matters very little that one substituent might be ethyl and the other *n*-hexyl, for example. The molecule can be regarded as possessing a *pseudo center of symmetry* as far as the C=C stretch is concerned.

e. *A general rule:* The more symmetrically substituted an alkene is, the weaker is its stretching band in the IR.

f. *Conclusion:* If a sharp band is seen in the IR at 1680–1630 cm^{-1}, medium or stronger in intensity, one can rather safely conclude that a C=C is present. However, if there is no such band, one can *not* conclude that C=C is absent. The band may merely be weak or forbidden. If it is known by other means that there is a C=C group, the intensity provides some information about how it is substituted.

III. Hydrogen out-of-plane bending modes

Remember that the following system is planar.

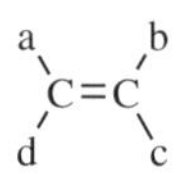

A. Two types of =C—H bending motions in alkenes:

1. In-plane bends: 1450–1200 cm^{-1}.

   Rather weak in the IR. Not good group frequencies.

2. Out-of-plane bends: 1000–650 cm^{-1}.

   a. Excellent group frequencies—not coupled with other vibrations and not affected by conjugation.

   b. There are as many out-of-plane *bends* as there are olefinic C—H *bonds*. The one where all the H atoms move upward out of the plane in unison is very intense in the IR and is the group frequency. These are among the strongest bands in the spectra of alkyl olefins.

B. Out-of-plane C–H bends for alkyl substituted olefins

1. Ethylene, the prototype molecule: $H_2C{=}CH_2$.

   a. We expect four C—H out-of-plane modes because there are four C—H groups. They are CH_2 wags and twists because CH_2 rockings and scissorings are in-plane. Since there are two CH_2 groups, there are two wags and two twists, for a total of four as expected.

b. Of these four out-of-plane C–H bends, only one is IR active because:
   1) One twist is an overall rotation of the molecule.
   2) One twist is the torsion, forbidden in both IR and Raman. (You are not expected to know that. It comes from a group theory argument.)
   3) One wag is only Raman active.
   4) The other wag is the only IR-active out-of-plane bend.

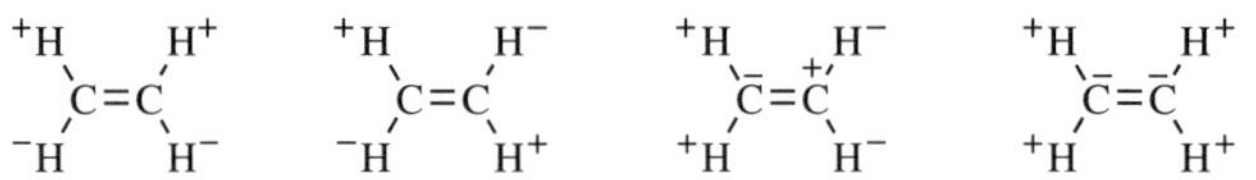

1) Twist rotation | 2) Twist torsion | 3) Wag Raman only | 4) Wag IR only 950 cm^{-1}, vs

2. The C–H out-of-plane bends of olefins. See Figure 3.4.
3. Vinylidene, $R_1R_2C{=}CH_2$
   a. 890 ± 5 cm^{-1} (vs)
   Due to:

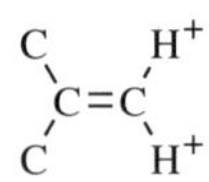

   Let us call this the "*=CH₂ wag*."
   b. The overtone of the wag (at about $2 \times 890 = 1780$ cm^{-1}) is often strong enough to be seen and is a useful confirmation. It usually shows negative anharmonicity; i.e., it is equal to or slightly *greater than* the calculated value. This is unusual and confirms the assignment of both fundamental and overtone.
   c. The second bend is not useful.

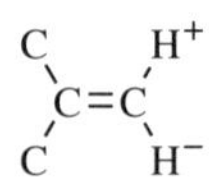

   Reason: The change of dipole moment is small because of the counter motions of the two olefinic hydrogens, so the band is weak.
4. Trans R–HC=CH–R′
   a. 965 ± 5 cm^{-1} (vs). The mode is mainly:

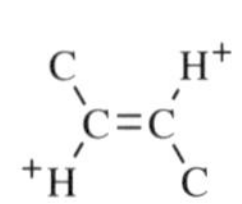

HYDROGEN OUT-OF-PLANE MODES FOR OLEFINS		C=C STRETCH	
		1680-1665 (w)	1660-1630 (m-s)
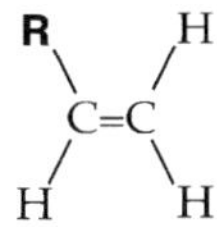	990 ± 5 (vs) 910 ± 5 (vs)		X
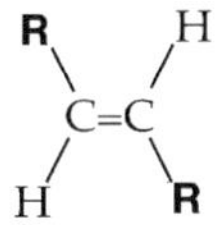	965 ± 5 (vs)	X	
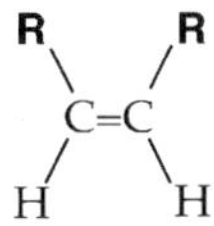	690 ± 40 (m,b)		X
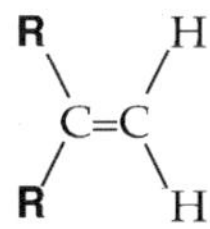	890 ± 5 (vs)		X
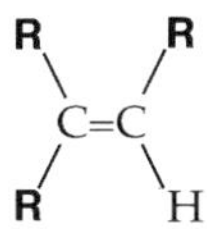	815 ± 25 (s)	X	
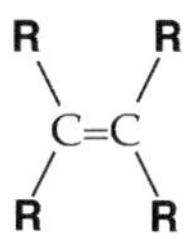	——	X	

Fig. 3.4 C—H out-of-plane bends and C=C stretches of olefins.

Let us call this the "*trans wag*" for convenience. Its overtone is usually not observed.

b. The second out-of-plane C—H bend,

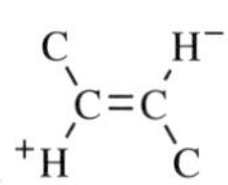

is very weak because of the countermotions of the two hydrogens and is not useful.

c. If trans —HC=CH— is part of a conjugated system, the 965 cm^{-1} band is raised. *Example:* All-trans $H_2C{=}CH{-}(HC{=}CH)_n{-}CH{=}CH_2$: 1000–990 cm^{-1}.

5. Vinyl, $-CH{=}CH_2$
   a. There are three C—H groups and therefore three out-of-plane bends. Only two of them are good group frequencies, and they are analogous to the two good ones above.
      1) 990 ± 5 cm^{-1} (vs) (trans wag of vinyl). Frequency is similar to the trans wag of R—CH=CH—R′ at 965 cm^{-1}. Therefore represented as shown below. The overtone is usually not seen.

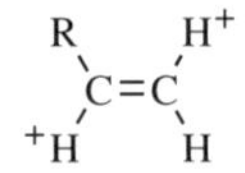

      2) 910 ± 5 (vs) ($=CH_2$ wag of vinyl). The frequency is similar to the vinylidene $=CH_2$ wag at 890 cm^{-1}. Therefore represented as shown below. The overtone $2 \times 910 = 1820$ cm^{-1} is often seen, usually at or slightly above the calculated value. Good confirmation.

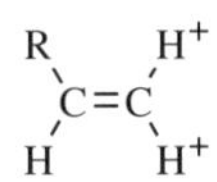

   b. In these representations, all three vinyl hydrogens are *not* moving in unison. The reason for showing them this way is the analogy to trans R—CH=CH—R′ and vinylidene.
   c. Note that in the three classes of olefins considered above the overtones of the $=CH_2$ wags are often seen (with negative anharmonicity) but the overtones of the trans wags seldom are observed.
6. Cis —HC=CH—

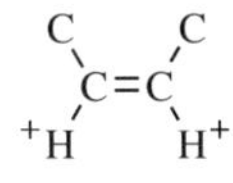

   a. 690 ± 40 cm^{-1} (m to w, often broad)

      Not a good group frequency. The countermotion (to avoid rotation) is not solely within the group but also involves the substituents. Therefore this vibration is more sensitive to the nature of the substituents than the C—H bends of other olefins. Note the broad range. The band is also weaker and broader than the preceding group frequencies.
   b. Sometimes the overtone of the cis mode is seen as a weak, sharp band at 1420–1400 cm^{-1}, usually slightly lower than calculated. If observed, it is modest confirmation of the assignment. A band at 1400 cm^{-1} is not often found for trans —CH=CH—.

c. The second out-of-plane bend of cis is weak because of the opposite motions of the two H atoms. It is not useful.

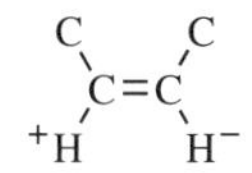

d. Cis-substitution is often deduced by eliminating other possibilities. If an olefinic C—H *stretching* band is present (showing the presence of a =C—H group) but no strong, reliable *bending* band due to one of the other types of olefinic substitution is found, it is concluded by elimination that the structure is cis.

7. Trisubstituted alkenes, R′R″C=CHR, $820 \pm 20\ cm^{-1}$ (m to s)

   Not as intense because there is only one =C—H. Sometimes difficult to pick out from other bands in this region.

8. Tetrasubstituted alkenes

   There is no analogous band because there is no olefinic C—H.

C. Effect of polar substituents

1. The above ranges—in most cases very narrow—are only for hydrocarbon substituents. Polar substituents upset the rules in complex but orderly ways.
2. Good references
   a. Potts and Nyquist, *Spectrochim. Acta* **15**, 679 (1959).
   b. Book by Colthup, Daly, and Wiberly. (See Bibliography.)
3. The trans wag ($965\ cm^{-1}$ of trans R—HC=CH—R′, $990\ cm^{-1}$ of vinyl)
   a. Essentially independent of resonance with a substituent X on C=C
      1) *Example:* If X = phenyl, C=C, >C=O, or $-NO_2$, there is little effect.
      2) A carbonyl can be ketone, ester, —COOH, or amide.
   b. But it is sensitive to the inductive power of a group. If substituent X has high electronegativity, the trans wag drops.
      1) For the trans wag of trans R—CH=CH—R′ (originally $965\ cm^{-1}$):
         a) If X = —O—, —Cl, or —Br, the wag drops $\sim 35\ cm^{-1}$ to $\sim 930\ cm^{-1}$.
         b) If there are two chlorines or two bromines, it drops $\sim 70\ cm^{-1}$ to $\sim 895\ cm^{-1}$.
      2) For the trans wag of vinyl (originally $990\ cm^{-1}$):
         a) If X = >N—, the wag drops 10–20 cm^{-1} to 980–970 cm^{-1}.
         b) If X = —O—, —Cl, —Br, —CN, or —SCN, it drops 30–50 cm^{-1} to 960–940 cm^{-1}.
4. The $=CH_2$ wag ($890\ cm^{-1}$ of vinylidene, $910\ cm^{-1}$ of vinyl)

   The effect of a polar substituent is quite different from that for the trans wag.
   a. Those groups which mesomerically *withdraw* electrons from the $=CH_2$ group raise the $=CH_2$ wag.

Example: If X = >C=O or –C≡N, the $=CH_2$ wag *rises* 40–60 cm^{-1}.

b. Those groups which mesomerically *donate* electrons to the $=CH_2$ group lower the wag.

$$H_2C{=}CH{-}O{-}C \longleftrightarrow H_2\bar{C}{-}CH{=}\overset{+}{O}{-}C$$

1) *Examples:*
   If X = –O–CO–R, the $=CH_2$ wag *drops* ~40 cm^{-1}.
   If X = –O–R, it drops ~60–100 cm^{-1}.
2) For an electronegative atom *except O or N* attached directly to the C=C group (e.g., Cl, Br), the $=CH_2$ wag drops 10–15 cm^{-1}. If there are two such substituents, the effect is approximately additive.
3) For vinylidenes, $RR'C{=}CH_2$, if there is more than one substituent, the effects are approximately additive.

$CH_2{=}CH{-}R$	~890 cm^{-1}	
$CH_2{=}CH{-}C{\equiv}N$	960	70 cm^{-1} higher
$CH_2{=}C(C{\equiv}N)_2$	985	95 cm^{-1} higher

4) –O– or halogen *on an* α-carbon raises the wag 5–25 cm^{-1}. Two halogens *on an* α-carbon raise it somewhat more.

5. *Examples:*

Compound		Trans Wag (cm^{-1})	$=CH_2$ Wag (cm^{-1})
$H_2C{=}CH{-}CO{-}O{-}CH_3$	Expect	~990	970–950
Methyl acrylate	Observe	988	964
$H_2C{=}CH{-}O{-}CO{-}CH_3$	Expect	960–940	~870
Vinyl acetate	Observe	977, 950	848 (poor)
$H_2C{=}CH{-}O{-}CH_3$	Expect	960–940	850–810
Methyl vinyl ether	Observe	960	813
$H_2C{=}C(CH_3){-}CO{-}O{-}CH_3$	Expect	—	950–930
Methylmethacrylate	Observe	—	942

There are many more examples in the Potts and Nyquist paper cited in C.2 above.

IV. Recommended procedure for alkenes

A. Ascertain that there is a C=C group (from Raman, from IR if observed, from the formula, or from other information).

B. Go to 3100–3000 cm^{-1} to see whether there is a hydrogen on the double bond. If so:

C. Go to the *strongest band(s)* in the range 1000–650 cm^{-1} to determine the type of substitution.

4 Characteristic Frequencies of Molecules with Triple Bonds and Cumulated Double Bonds

ROBERT W. HANNAH and FOIL A. MILLER

I. Acetylenes, —C≡C—

A. Summary for mono- and disubstituted acetylenes (condensed phases):

	≡C—H stretch (cm^{-1})	C≡C stretch (cm^{-1})
R—C≡C—H	~3330 ± 50 (m to s in IR, w in Raman, sharp)	2120 ± 20 (w in IR, vs in Raman)
R—C≡C—R′	— (none)	2225 ± 35 (vw in IR, vs in Raman)

In principle it is easy to distinguish between the two types, but the very low IR intensity of C≡C is a handicap. Raman is much better.

B. ≡C—H stretch

1. ~3330 ± 50 cm^{-1} (m in IR, w in Raman). Slightly higher in the vapor phase because there is some H bonding in condensed phases. In acetylene itself there are two C—H stretches. The antisymmetric stretch, which is IR active/Raman inactive, is found at 3287 cm^{-1} and the symmetric stretch near 3374 cm^{-1} (Raman active, IR inactive). The latter is weakly coupled to the C≡C stretch and therefore raised above the antisymmetric stretch.
2. Atypically intense in the IR even though there is usually only one such group per molecule. Much more intense than a single C—H on a saturated carbon.
3. Highest of all the C—H stretches.
4. *Much* sharper than any other bands here (which would be due to OH or NH).

Course Notes on the Interpretation of Infrared and Raman Spectra, by Dana W. Mayo, Foil A. Miller, and Robert W. Hannah.
ISBN 0-471-24823-1

C. ≡C–H bend

1. ~630 cm^{-1} (m to s in IR, m to w in Raman). Occurs at a slightly lower frequency in the vapor phase.
2. Not a good group frequency because it is in the fingerprint region. If one knows that the group is present, the band can often be identified, but not conversely.
3. If there is a threefold or higher axis of symmetry, this vibration is doubly degenerate. If not, the degeneracy is removed and there may be two different bands. In phenyl acetylene, where the phenyl ring defines a plane, there is only a twofold axis of symmetry. The in-plane ≡C–H bend is at 648 cm^{-1} and the out-of-plane ≡C–H bend is at 611 cm^{-1}. See Figure 4.1. For molecules like methylacetylene there is a threefold axis of symmetry present and only a single band appears. It is usually found near 630 cm^{-1}.
4. *Example:* Phenylacetylene (Figure 4.1).
5. The overtone of the ≡C–H bend may be seen in the IR near to or lower than $2 \times \sim 630 = \sim 1260$ cm^{-1}. It is often broad. See Figures 4.1–4.3. This overtone falls in the fingerprint region, but its breadth helps identify it.

D. C≡C stretch

1. Frequency

   a. About 100 cm^{-1} higher in disubstituted than in monosubstituted acetylenes:

H–C≡C–H	1974 cm^{-1}	Forbidden in the IR, s in Raman
R–C≡C–H	~2120 cm^{-1}	Weak in the IR, vs in Raman
R–C≡C–R′	~2225 cm^{-1}	Extremely weak in the IR, vs in Raman

   1) *Reason:* H is light and tends to ride with the C atom as the latter vibrates. R is heavier and has more inertia. Therefore when the C≡C bond is stretched, more work has to be done on an adjacent C–C single bond than on a C–H bond. The effective force constant is higher, making the C≡C stretch in C–C≡C–C higher than in C–C≡C–H. Because the system is linear, the effect is maximized.
   2) Note that $\tilde{\nu}_{(C\equiv C)}$ for acetylene is atypically low. The first member of an homologous series is often so far out of line that it is not included in making up tables of group frequencies. This is because of the relatively greater change in going from H to CH_3 than in going from CH_3 to any other alkyl group.

   b. For disubstituted acetylenes, there are often one or more sharp bands near 2225 cm^{-1} in both IR and Raman spectra. They are sum tones which have acquired intensity by Fermi resonance. These bands are usually more useful in the Raman because of the intense scattering by the C≡C group compared to the IR.

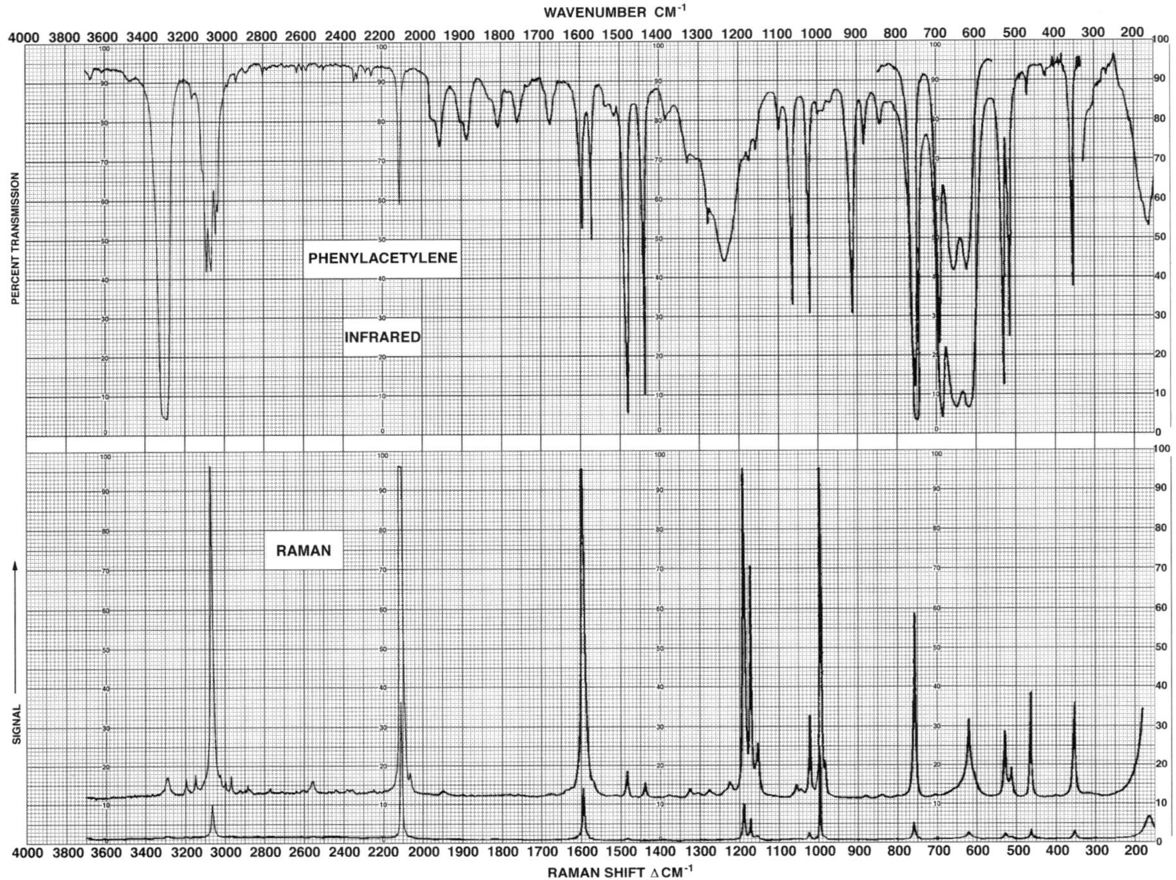

Fig. 4.1 Infrared and Raman spectra of phenyl acetylene, C_6H_5—C≡C—H.

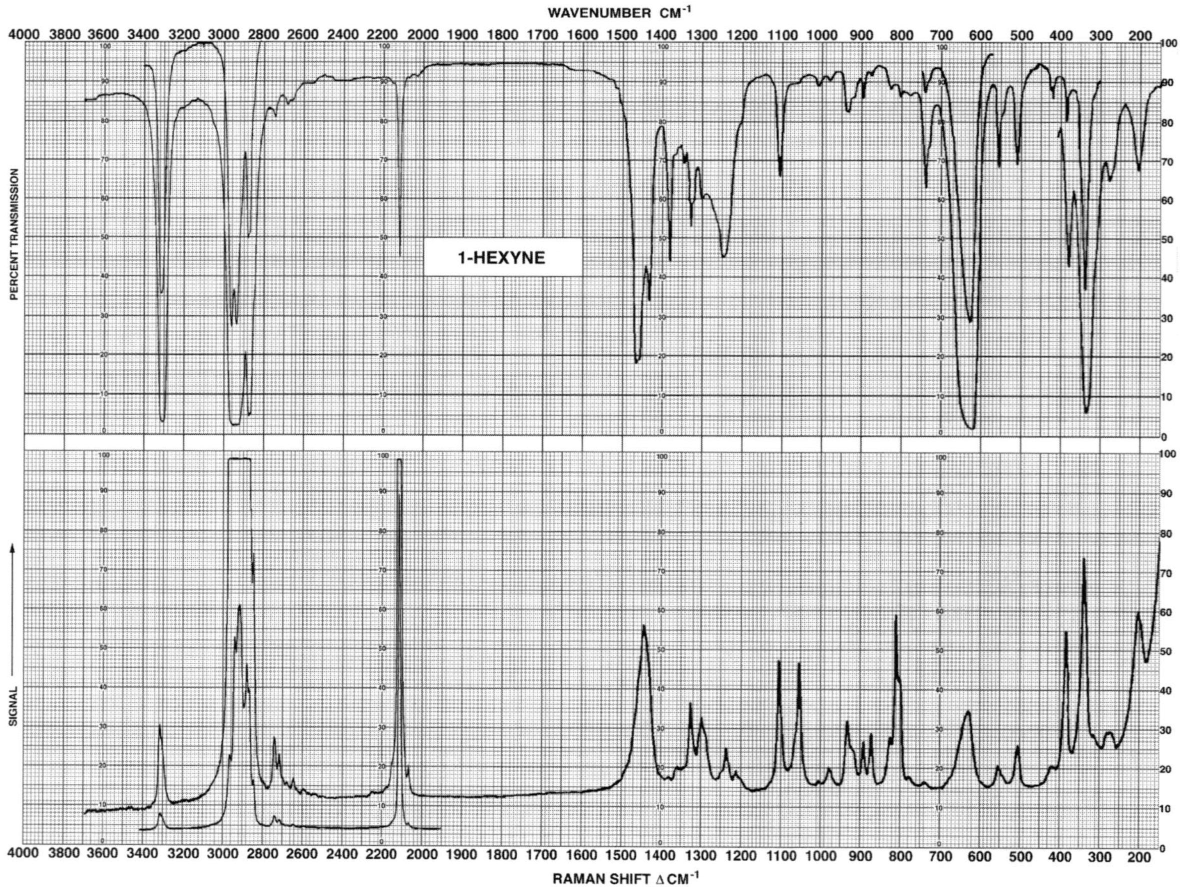

Fig. 4.2 Infrared and Raman spectra of 1-hexyne, $n\text{—}C_4H_9C\equiv CH$.

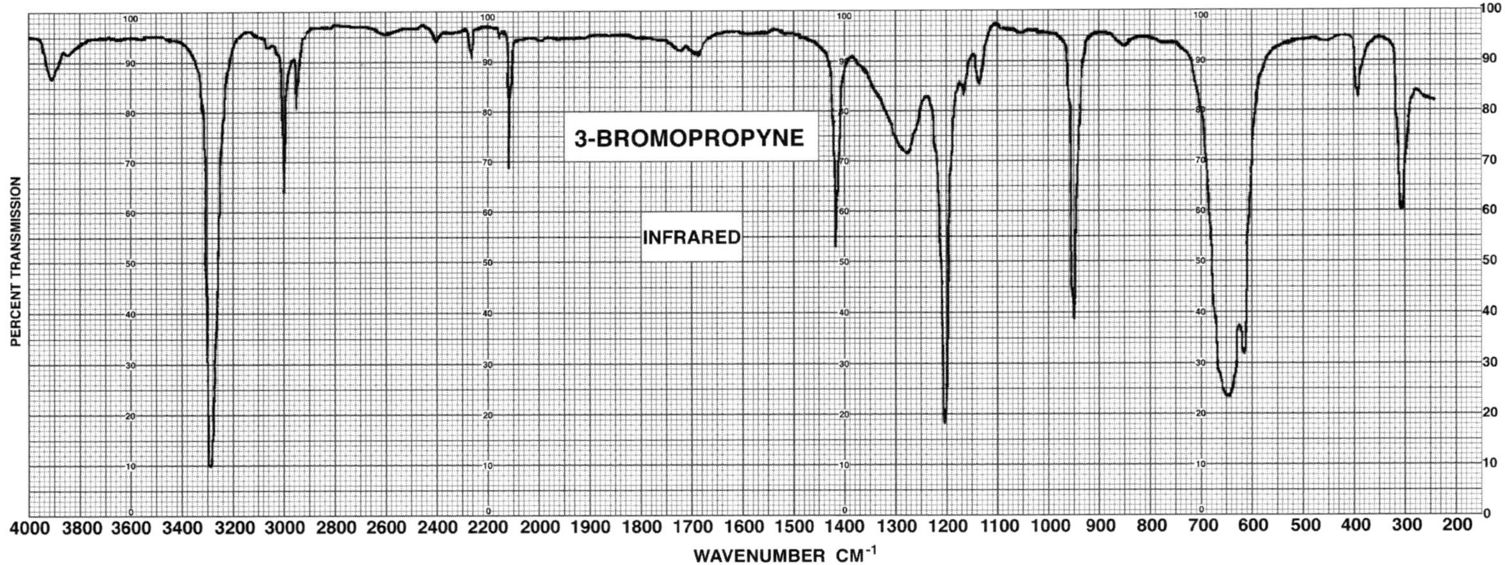

Fig. 4.3 Infrared spectrum of 3-bromopropyne, H—C≡C—CH_2Br.

c. $\tilde{\nu}_{(C\equiv C)}$ is only slightly affected by conjugation. It lowers other frequencies, but its own lowering is small and variable. Olefinic conjugation lowers the C≡C stretch while the effect of carbonyl conjugation is less predictable.

2. Intensity of C≡C stretch
   a. In the IR, the C≡C stretch is notoriously weak and undependable. This weakness is due to the presence of a real or a pseudo center of symmetry.
      1) If the C≡C is across a center of symmetry, $\tilde{\nu}$ (C≡C) is rigorously forbidden in the IR. *Examples:* H–C≡C–H, H_3C–C≡C–CH_3, Ph–C≡C–Ph.
      2) If the substituents are not identical, there may still be a pseudo center of symmetry and the C≡C stretch will be weak or missing in the IR. *Example:* Some heptynes, CH_3–$(CH_2)_i$C≡C –$(CH_2)_j$–X, with $i + j = 4$ and with X = H, OH, Cl, Br, –CN, or –$COOCH_3$. Unless the C≡C is adjacent or next adjacent to X, the C≡C stretch is so weak it is scarcely seen in the IR.
   b. In Raman spectra, $\tilde{\nu}$ (C≡C) is *always* very strong and reliable.
   c. *Conclusion:* The IR is a poor place to look for the C≡C stretch; often it will not be seen. The Raman spectrum is dependable.
3. C≡C–C bend, 355–335 cm^{-1}, m to s in both IR and Raman. See Figures 4.1 and 4.2. Not useful except for confirmation because it is so far down in the fingerprint region.

E. Other examples of acetylenic spectra
   1. 1-Hexyne. See Figure 4.2.
      a. 3313 cm^{-1} ≡C–H stretch
      b. 2119 cm^{-1} C≡C stretch. Compare the IR and Raman intensities. In the IR this is unusually strong because of the unsymmetrical substitution. However, the band still has only medium intensity.
      c. 627 cm^{-1} C≡C–H bend. Its overtone is observed near 1270 cm^{-1}, broad.
      d. 339 cm^{-1} C≡C–C bend. Strong in both IR and Raman.
   2. 3-Bromopropyne, H–C≡C–CH_2Br. See IR spectrum in Figure 4.3.
      a. Note the ≡C–H stretch at 3290 cm^{-1}.
      b. There are only two bands for saturated C–H stretches, as expected for –CH_2–. They are at 3000 and 2960 cm^{-1}.
      c. The C≡C stretch is at 2120 cm^{-1}, as expected. Even though the substituents –H and –CH_2Br are very different, the C≡C stretch still has only moderate intensity.
      d. The 1420-cm^{-1} band is the –CH_2– scissoring, although it is lower than the normal range of 1470–50 cm^{-1}. This is due to the adjacent C≡C and/or the adjacent Br.
      e. The 650- and 615-cm^{-1} bands are the ≡C–H bend. There are two because the bend is not degenerate in this molecule. The overtone of 650 cm^{-1} is observed near 1280 cm^{-1}.

F. *Problem:* A compound has one of two structures:

(1) $R_3C{-}C{\equiv}C{-}CR_3$ **I** or (2) $R_2C{=}\underset{R}{\overset{R}{C{-}C}}{=}CR_2$ **II** where $R = Si(CH_3)_3$

Spectral measurements show:

cm^{-1}	IR	Raman
2500–1500	Nothing	2135 cm^{-1} s

Which structure is correct and why? What vibrational bands are expected at 2500–1500 cm^{-1} for structure **I** and for structure **II**?

The answer is at the end of this chapter.

II. Nitriles, R—C≡N

A. Frequency of —C≡N stretch

1. Unconjugated — $2250 \pm 10\ cm^{-1}$. Strong in both IR and Raman.

   Conjugated (to phenyl, C=C, C≡C, etc.) — $2225 \pm 15\ cm^{-1}$. Strong in both IR and Raman.

   a. Thus the two types can be distinguished with confidence providing the wavenumber is determined accurately.

   b. If a $—CH_2—$ is adjacent to the C≡N, its scissors is lowered from ~1460 to 1435–1405 cm^{-1}.

   c. *Example:* 2-Methylene glutaronitrile. See Figure 4.4.

   $$N{\equiv}C{-}CH_2{-}CH_2{-}\underset{\underset{CH_2}{\parallel}}{C}{-}C{\equiv}N$$

   1) The spectrum shows both unconjugated and conjugated nitrile at 2262 and 2237 cm^{-1} respectively.

   2) The vinylidene C=C stretch should be between 1660 and 1640 cm^{-1}. Its location in the spectrum of this molecule is 1625 cm^{-1}, lowered due to conjugation with the C≡N. One also expects a vinylidene wag near 890 cm^{-1} (vs) due to the $=CH_2$ out-of-plane wag. It is actually at 950 cm^{-1}. This 60 cm^{-1} rise is the effect of the polar —C≡N. (See Chapter 3.)

   3) There are two $—CH_2—$ scissoring bands, a normal one near 6.95 μm or 1440 cm^{-1} and one due to the $—CH_2—$ adjacent to the —C≡N near 7.1 μm or 1410 cm^{-1}.

   4) Finally, the high antisymmetric vinylidene C—H stretch can be seen (weakly) near 3100 cm^{-1}.

2. There is no exception to these frequency ranges for —C≡N stretch except HCN:

HCN	2089 (gas), 2062 cm^{-1} (liquid)
CH_3CN	2249 cm^{-1}
RCN	$2250 \pm 10\ cm^{-1}$

   Thus again we see that the first member of a homologous series is anomalous.

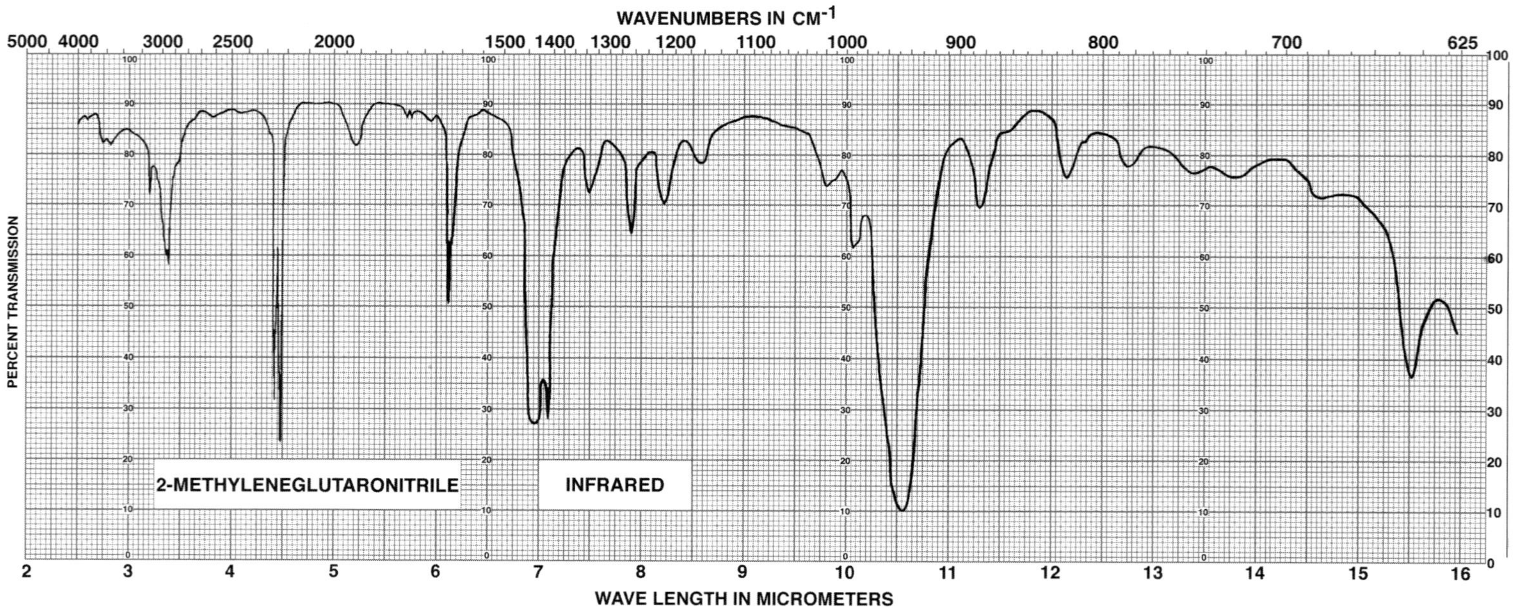

Fig. 4.4 Infrared spectrum of 2-methylene glutaronitrile, $N{\equiv}C{-}CH_2{-}CH_2{-}C({=}CH_2){-}C{\equiv}N$.

3. In the Raman spectrum —C≡N always gives a good, reliable band. In the IR it usually does, but see below for a few exceptions.
4. *Example: n*-Hexanenitrile. See Figure 4.5.
   a. The —C≡N stretch is at 2246 cm^{-1}, just where expected.
   b. The 1735 cm^{-1} band is due to a carbonyl impurity, possibly from partial hydrolysis to the acid.
   c. The 1434 cm^{-1} band is due to the scissoring of the —CH_2— which is adjacent to the nitrile.
   d. Note 2734 cm^{-1} in the Raman spectrum. It is the overtone of the C—CH_3 symmetric deformation at 1368 cm^{-1} which is observed in the IR but not the Raman.

B. Intensity of —C≡N stretch
1. In the IR:
   a. If the compound contains only C, H, and N, the band is strong.
   b. α-Substitution by an electronegative atom reduces the intensity markedly, and the C≡N band may not be seen.
      1) *Examples:*
         X—CH_2—C≡N, with X = O, F, or Cl
         F—CH_2—C≡N, 2257, vw in IR
         Cl—CH_2—C≡N, 2260, w in IR
         Br—CH_2—C≡N, 2254 m in IR
         I—CH_2—C≡N, 2244, s in IR
         (all are strong in the Raman)
      2) *Example*: CH_3—O—CH_2—C≡N (methoxyacetonitrile). See Figure 4.6.
         a) Note that the C≡N band is not seen in the IR but is very strong in the Raman spectrum.
         b) Note that there is a strong Bohlmann band at 2837 cm^{-1} in both spectra.
   c. But if X = N, the IR intensity of —C≡N is *increased* enormously!
   d. *Note:* if a C=O group is substituted on the α-carbon, the C≡N band rises in frequency and is very weak in the IR.
2. In the Raman spectrum, $\tilde{\nu}$ (C≡N) is always intense and evident, but in the IR the intensity is occasionally very low. Nitriles are generally difficult to recognize by nuclear magnetic resonance (NMR). Raman is the most dependable technique. (S. K. Freeman, *Applications of Raman Spectroscopy*, p. 73.)
3. Phenyl conjugation of nitrile groups: Substituents on the ring will affect the IR intensity and frequency of the C≡N stretch.
   a. Electron-withdrawing substituents, e.g., NO_2, —F, —OH. The band intensity is somewhat weaker and the frequency higher.
   b. Electron-releasing substituents, e.g., —NH_2. The band intensity increases and the frequency decreases.

C. Related groups
1. Isonitrile, $R—\overset{+}{N}\equiv C^-$, 2150 ± 50 cm^{-1}, about 100 cm^{-1} *lower* than C—C≡N

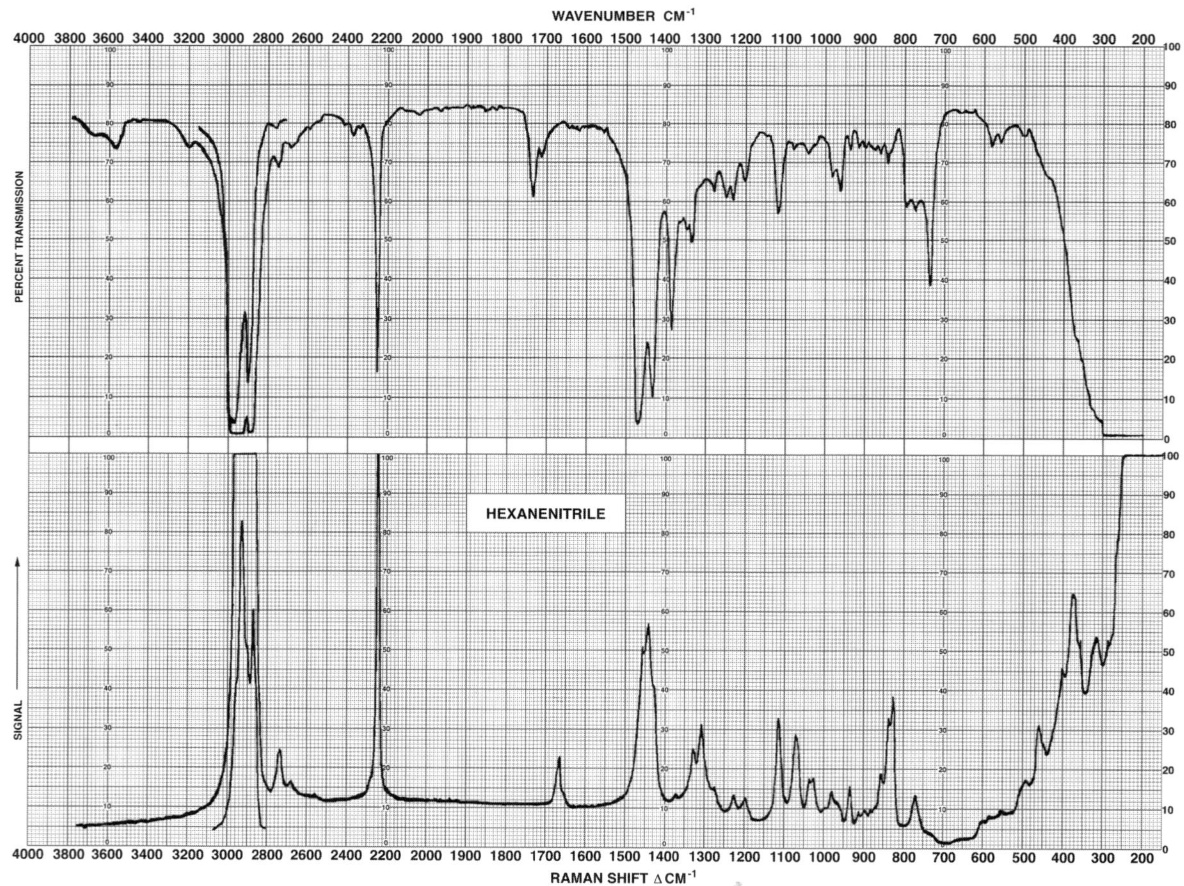

Fig. 4.5 Infrared and Raman spectra of *n*-hexanenitrile, *n*-C_5H_{11}—C≡N. The 1735 cm^{-1} band is due to a carbonyl impurity.

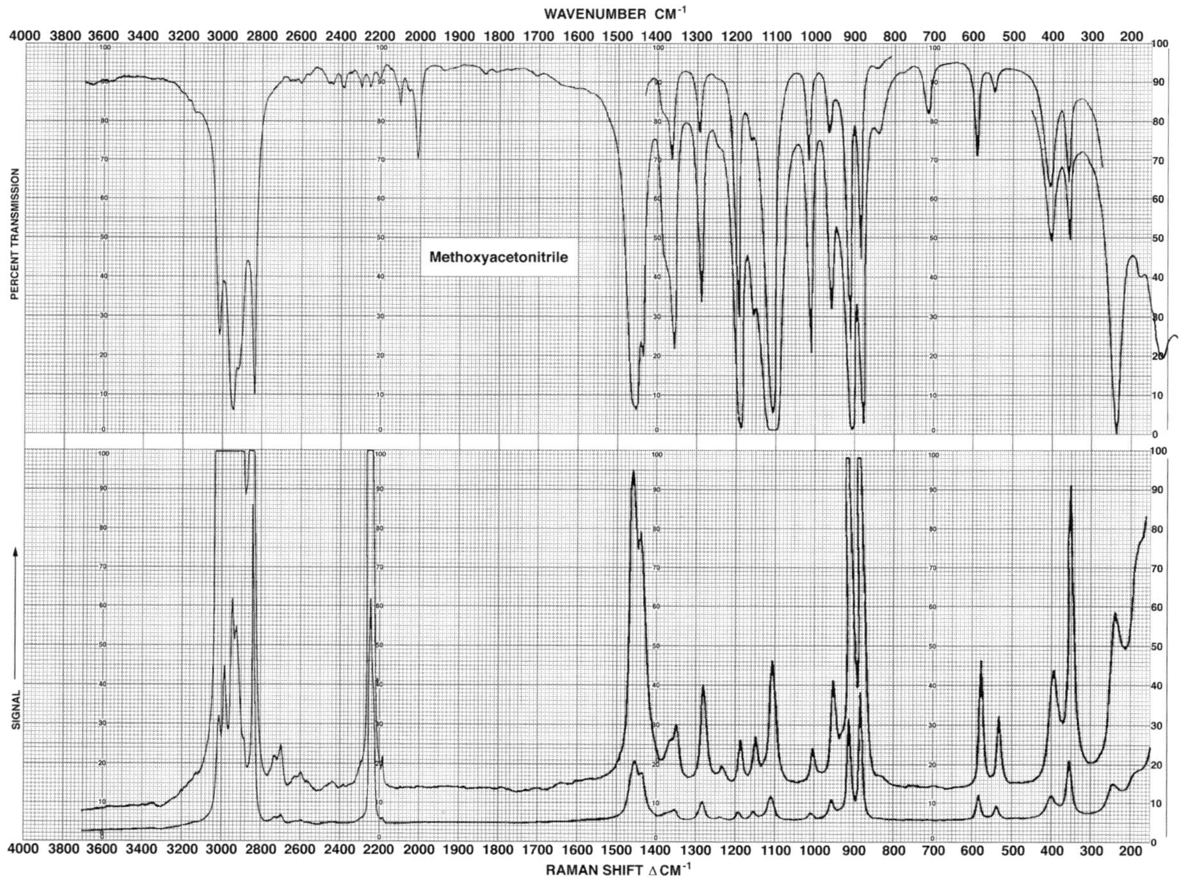

Fig. 4.6 Infrared and Raman spectra of methoxyacetonitrile, $CH_3{-}O{-}CH_2{-}C{\equiv}N$.

   a. When R = alkyl group, 2200–2130 cm^{-1}, s in IR, s in Raman.
   b. When R = aryl group, 2125–2110 cm^{-1}, s in IR, s in Raman.
2. Thiocyanate, R—S—C≡N, ~2150 cm^{-1} (vs), about 100 cm^{-1} *lower* than C–C≡N
   a. When R = alkyl group, 2160–2150 cm^{-1} s in IR, s in Raman.
   b. When R = aryl group, 2175–2160 cm^{-1} s in IR, s in Raman.
3. Nitrile oxide, —C≡N → O, ~2350 cm^{-1}, about 100 cm^{-1} *higher* than C–C≡N
   Note the near coincidence with the CO_2 antisymmetric stretch at 2349 cm^{-1}. There may be an interference, particularly for unpurged instruments.
4. Inorganic cyanide
   a. Cyanide ion, $[C{\equiv}N]^-$, 2080 ± 15 cm^{-1} for both aqueous solutions and crystals.
   b. Cyano complexes, $[M(C{\equiv}N)_x]^{y-}$, 2200–2000 cm^{-1}.

III. Cumulated double-bond systems, X = Y = Z
   A. Linear systems
      1. CO_2 is the prototype, with stretching frequencies at 2349 and 1340 cm^{-1}. All of the following groups are linear and therefore have very large splittings like CO_2.
      2. The higher band comes at 2300–1950 cm^{-1}, which is a relatively uncluttered region of the spectrum. It is very strong in the IR and is an excellent group frequency, but it is only weak to medium in the Raman spectrum.
      3. The lower band (the in-phase stretch) for all cumulated double-bond systems is near 1000 ± 200 cm^{-1}. It is not useful because it occurs in the fingerprint region and is often weak in the IR.
      4. >C=C=C Allenes 2000–1950 cm^{-1} (vs)
         In allene itself the frequencies are found near 1980 cm^{-1} (–, IR) and 1071 cm^{-1} (R, –). The latter is lowered by first-order interaction with a CH_2 scissoring.
      5. Substituted allenes, R_a R_b C=C=C R_c R_d
         a. Out-of-phase stretch, 2000–1900 cm^{-1}, s in IR, vw in Raman.
         b. In-phase-stretch, 1130–1060 cm^{-1}, vw in IR, vs in Raman.
         c. Both stretches are lowered by conjugation and halogen substitution on the α-carbon.
         d. If the allene system contains a terminal $=CH_2$ group, the out-of-plane wag of the latter is ~850 cm^{-1} (vs), and its overtone is often seen near 1700 cm^{-1}. Compare with the alkene vinylidene wag at 890 ± 5 cm^{-1} and its overtone. The CH_2 scissors of the terminal $=CH_2$ will be near 1440 cm^{-1} and the C—H stretch will be near 3070 cm^{-1}.
   B. Other linear systems
      1. >C=C=O Ketenes 2200–2000 cm^{-1} (vs)

a. Out-of-phase stretch, 2197–2085 cm^{-1}, s in IR, variable in Raman.
b. In-phase stretch, 1420–1120 cm^{-1}, m-s in IR, s in Raman.
c. *Note:* The out-of-phase stretch is near the C≡C stretch but the ketenes are much more intense in the IR.
d. *Note:* The in-phase stretch is in the fingerprint region and is less useful as a result.
e. Both bands are sensitive to substitution:
   1) Ketene — 2153 and 1130 cm^{-1}
   2) R′R″ C=C=O — 2134 and 1374 cm^{-1}
   3) $(Phenyl)_2$>C=C=O — 2105 and 1120 cm^{-1}
f. *Note:* The C=C=O bend is ~670 cm^{-1}, m to w in the IR. Not a useful group frequency.

2. –N=C=O Isocyanates
   a. Out-of-phase stretch, 2300–2250 cm^{-1}, vs, vb in IR, vw in Raman.
   b. In-phase stretch, 1450–1400 cm^{-1}, w in IR, s in Raman.
   c. The out-of-phase stretch generally has a complex band structure with a number of side bands. Conjugation and physical state have little or no effect. The band is near the nitrile position, but in the IR the isocyanate band is much broader and about 40 times more intense so there is little chance of confusion.
   d. *Example:* Hexyl isocyanate. See Figure 4.7. Note how strong and broad the –N=C=O out-of-phase stretch near 2280 cm^{-1} appears. Note also the weak side band near 2100 cm^{-1} and the unresolved shoulder near 2440 cm^{-1}.
      1) Note that the 2280 cm^{-1} band is absent in the Raman spectrum.
      2) In the Raman spectrum, the sharp feature at 2400 cm^{-1} is a noise spike due to a change in gain.
3. –N=C=S Isothiocyanates ~2200 cm^{-1} (vs, IR), ~2100 cm^{-1} (vs, IR)
   a. Out-of-phase stretch, two bands at 2300–2000 cm^{-1}, vs, vb, in IR, m to w in Raman.
   b. In-phase stretch, 1090–925 cm^{-1}, w in IR, s in Raman.
   c. Fermi resonance is probably responsible for the presence of two bands in the 2150 cm^{-1} region.
4. $-N=\overset{+}{N}=\overset{-}{N}$ Azides 2160–2120 cm^{-1} (s)
5. $>C=\overset{+}{N}=\overset{-}{N}$ Diazo compounds 2200–2000 cm^{-1} (s)
6. –N=C=N– Carbodiimides 2160–2100 cm^{-1} (vs)

   Example: With a cyclohexyl group at each end 2100 cm^{-1} (vvs)
7. >C=C=N– Ketene imines 2060–2000 cm^{-1} (s)

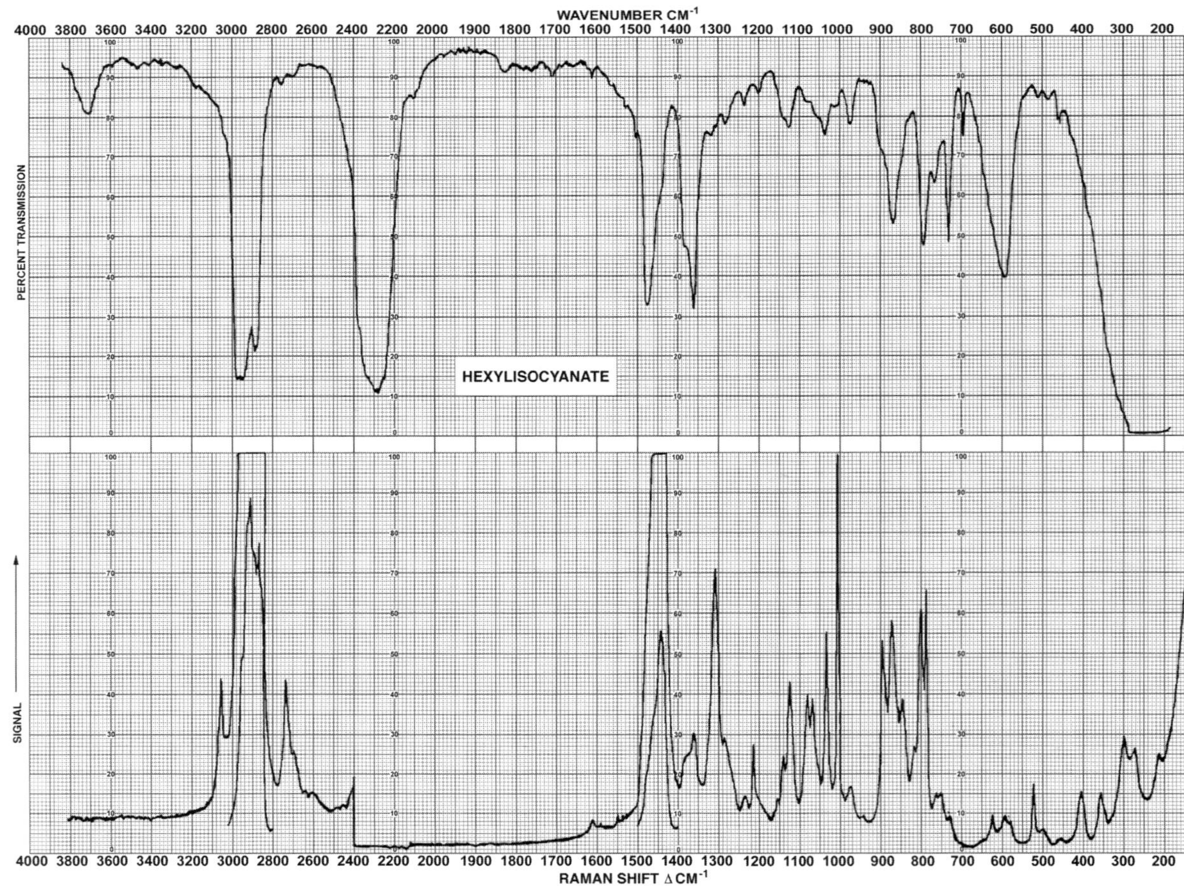

Fig. 4.7 Infrared and Raman spectra of *n*-hexyl isocyanate, $n\text{-}C_6H_{13}\text{—N=C=O}$.

C. Nonlinear systems

1. If the bonds are not colinear, the splitting is much smaller because the interaction is less effective. (It would be nearly zero for two bonds at right angles to one another.)
2. *Examples:*

   Sulfur dioxide, SO_2:

   Bond angle 119°. Stretches 1362 and 1151 cm^{-1}. Difference 211 cm^{-1}.

   Sulfones, R–SO_2–R:

Antisymmetric stretch	1330–1295 cm^{-1} vs in IR, m-w in Raman.	Midpoint ~1312 cm^{-1}
Symmetric stretch	1150–1125 cm^{-1} vs in both	Midpoint ~1137 cm^{-1}
		Difference ~175 cm^{-1}

3. R–NO_2 ~125° 1565 ± 35 and 1345 ± 35 cm^{-1} Difference ~220 cm^{-1}

   The difference between the antisymmetric and symmetric modes is larger for the –NO_2 group than for the –SO_2– group because:

   a. The bond angle is larger.

   b. The N atom is lighter than the S atom and will have a larger displacement during the vibration.
4. *N*–sulfinylamines, R–N=S=O

H–N=S=O	1261 and 1090 cm^{-1}	Difference 171 cm^{-1}
Ph–N=S=O	1306 and 1155 cm^{-1}	Difference 151 cm^{-1}

IV. Answer to the Problem on page 91.

A. The structure is certainly I because the strong Raman band at 2135 cm^{-1} is good evidence for a C≡C group. The band will not be seen in the IR because the C≡C straddles a center of symmetry. The value is about 100 cm^{-1} lower than usual for doubly-substituted triple bonds, but that is not surprising because the R group, –$Si(CH_3)_3$, is an unusual substituent.

B. If the structure were II, which is a butadiene derivative, one would expect to see both an IR and a Raman band in the C≡C stretching region, roughly 1670–1630 cm^{-1}, having somewhat different wavenumbers. See the discussion of butadiene on page 78.

5 Characteristic Frequencies of Aromatic Compounds (Group Frequencies of Arenes)

DANA W. MAYO

I. The general appearance of the spectra of aromatic compounds

A. Aromatic ring systems represent the final class of hydrocarbons to be considered. The discussion will center on the benzene ring, but many more complicated systems have been examined in detail, and it has been found that most of the arguments used in the case of the simple six-membered ring can be successfully extrapolated to those systems including the aromatic heterocyclic ring compounds.

The IR spectra of aromatic systems have been found to possess many needle-sharp bands. This characteristic sets these spectra apart from the spectra of aliphatic compounds, which generally possess bands of broader width. The sharpness of the aromatic (arene) absorption bands arises from the fact that aromatic systems are rigid molecules having little opportunity for rotational isomerism. With aliphatic compounds the observed IR spectra is, in reality, often the spectrum of a complex mixture of rotamers. These isomers all exhibit very similar but not identical spectra; thus band broadening results (see Figure 5.1).

B. Vibrational modes present in aromatic systems (such as the phenyl group, C_6H_5—)

The vibrational modes associated with phenyl systems can be classified as:

1. Carbon–hydrogen stretching and bending vibrations

   The bending modes are further classified as:

   a. In-plane bending
   b. Out-of-plane bending
   (where the plane of reference is the plane of the aromatic ring)

Course Notes on the Interpretation of Infrared and Raman Spectra, by Dana W. Mayo, Foil A. Miller, and Robert W. Hannah.
ISBN 0-471-24823-1

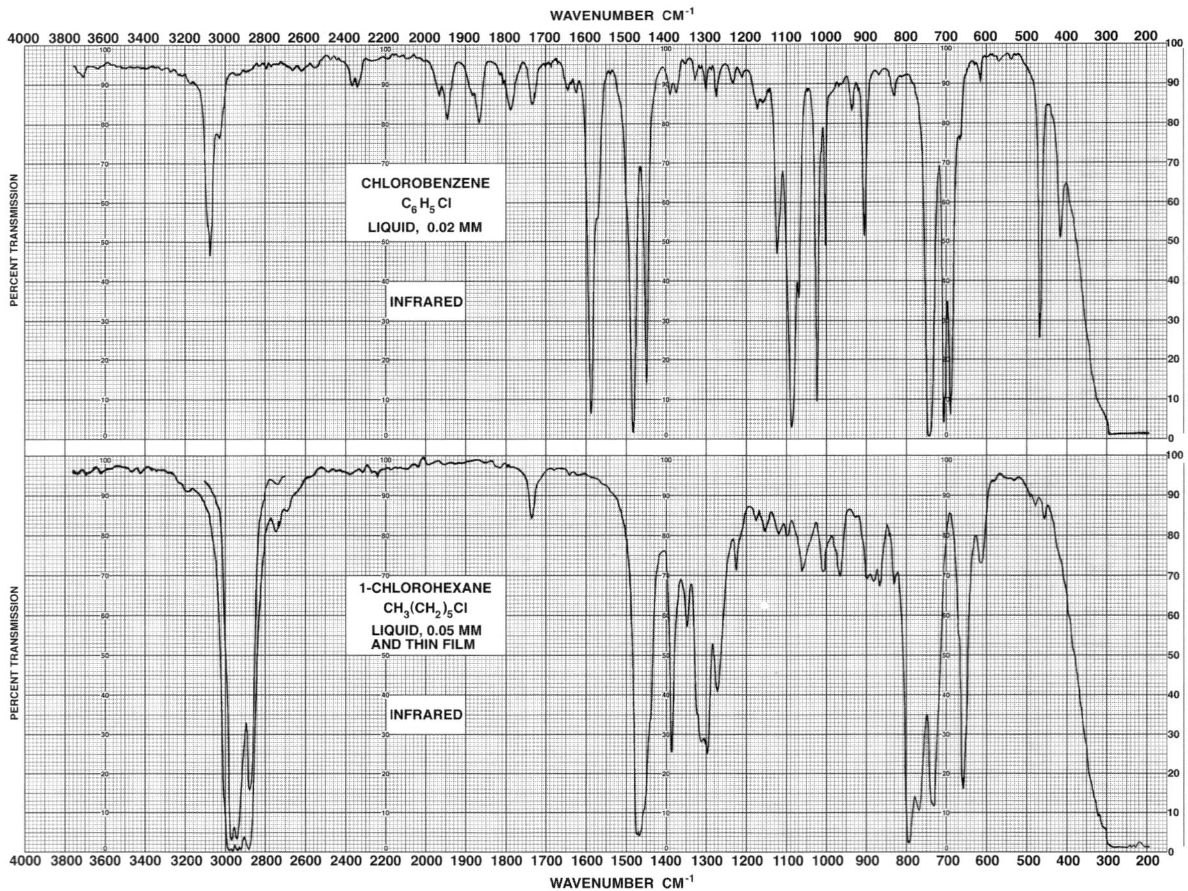

Fig. 5.1 Infrared spectra of chlorobenzene and 1-chlorohexane.

2. Carbon–carbon ring stretching and bending modes

   The bending modes are, as with the C—H modes, classified as in-plane and out-of-plane (ring puckering) deformation vibrations.
3. The ring substituent C_6H_5—X (where the hydrogen of benzene has been replaced by —X) will introduce the additional modes:
   a. C—X stretching
   b. C—X bending (in plane)
   c. C—X bending (out of the plane)
4. If the X— group is multiatomic, such as $-CH_3$, —CN, $-NO_2$, —OH, or —COOH, then the internal vibrations of —X (including its group frequencies) will be present in the spectrum of that particular substance.

C. There are two ways to establish which of the ring modes are potentially good group frequencies:
   1. It is possible to calculate the displacements of each mode:
      a. If the —X substituent is monoatomic, there will be 30 normal modes $[3(12) - 6 = 30]$ associated with the vibrational motion of the benzene ring (Figure 5.2).
      b. From the calculated displacements, the modes can be divided into two general classes:
         1) The X-sensitive vibrations
         2) The X-insensitive vibrations
      c. The X-insensitive modes have the potential to be good group frequencies because the substitution of the —X group will leave the rest of the ring modes unaffected (uncoupled).
   2. It is also possible to empirically establish the phenyl group frequencies by the same reasoning that was used in the analysis of the Nujol spectrum:
      a. In this approach a composite spectrum is generated from a large mixture of phenyl-containing substances (several thousand). The regions where peaks rise above background are assigned to group frequencies because these are the only regions where X-insensitive bands fall.
      b. It is not required that an actual mixture be used in the experiment (as in the case of Nujol). A spectrum can be generated by coadding spectra kept in IR library data files. This approach has, indeed, been carried out in the case of phenyl-containing substances, and the results are exactly equivalent to the results obtained by calculation.
      c. Group frequencies identified in the above fashion can be assigned with considerable confidence. On the other hand, although these calculated modes may be frequency stable, the resulting bands may vary unpredictably in intensity, and therefore some of them are of little empirical value. The experimental method is better.

D. Based on the above analysis, the group frequencies of the phenyl group can be classified as:

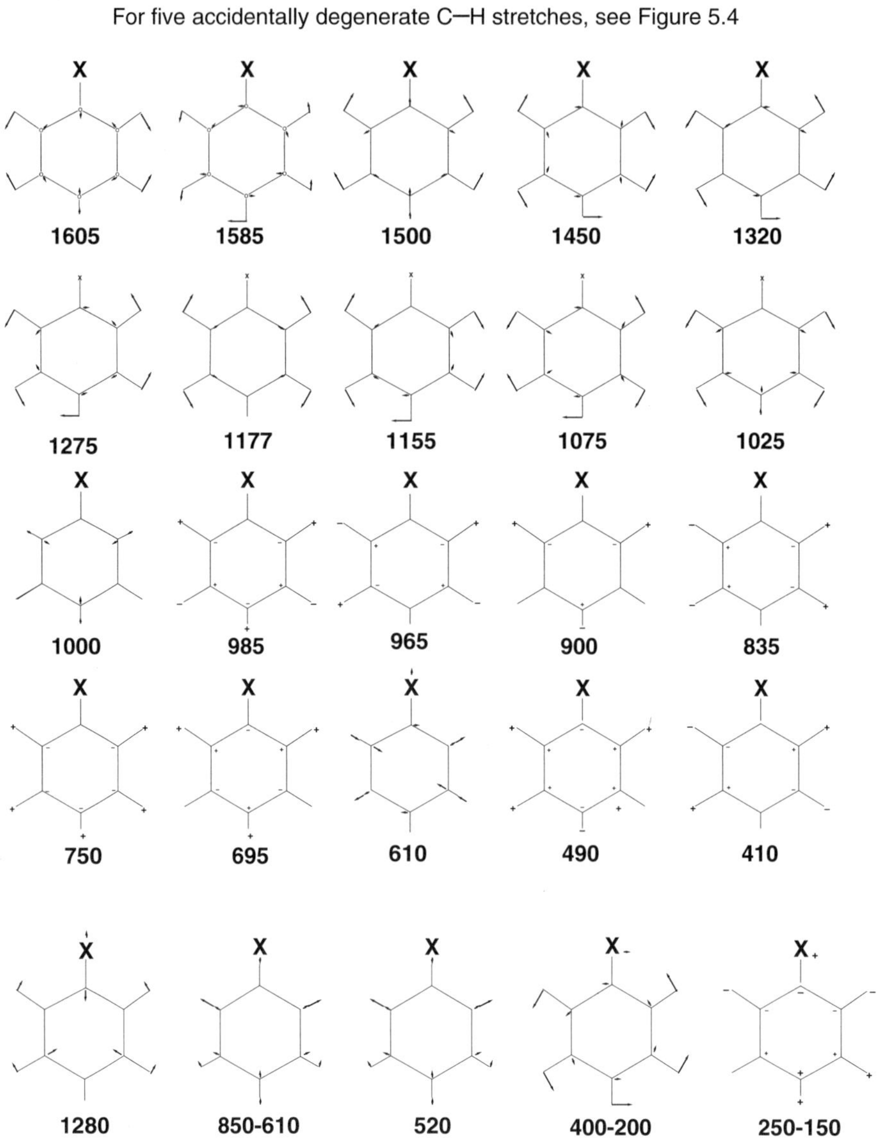

Fig. 5.2 Calculated displacements for normal modes of vibration for benzene substituted by a single atom.

1. Carbon–hydrogen vibrations consisting of stretching and out-of-plane bending modes
2. Carbon–carbon ring stretching and out-of-plane bending modes
   Most of the in-plane bending C—H and C—C modes are not good group frequencies.

II. Carbon–hydrogen stretching vibrations
   A. These modes occur in the region 3200–3000 cm^{-1}, but the very large majority of the bands occur in the lower half, 3100–3000 cm^{-1}.
      1. The location of these stretching modes is expected because the carbon atoms are sp^2 hybridized and essentially equivalent to alkene carbons which already have been shown to have C—H stretching modes in this region.
      2. The aromatic C—H stretch follows the 3000-cm^{-1} rule (only sp^3 C—H stretch below 3000 cm^{-1}).
      3. Potential for overlap with alkene absorption bands should be considered.
   B. These are generally weak and sharp bands.
      1. They are weak because the ratio of C—H/carbon has dropped considerably compared to aliphatic systems, and the molecular extinction coefficients are inherently low.
      2. They are sharp for two reasons:
         a. There is little or no rotational isomerism present in this class of substances, as noted earlier.
         b. The C—H oscillators are isolated and give little evidence of strong coupling between adjacent groups.
   C. Instrument resolution is important because the modes are in close proximity.
      1. When observed with NaCl prism resolution, this region is generally recorded as a single symmetric band (Figure 5.3*a*).

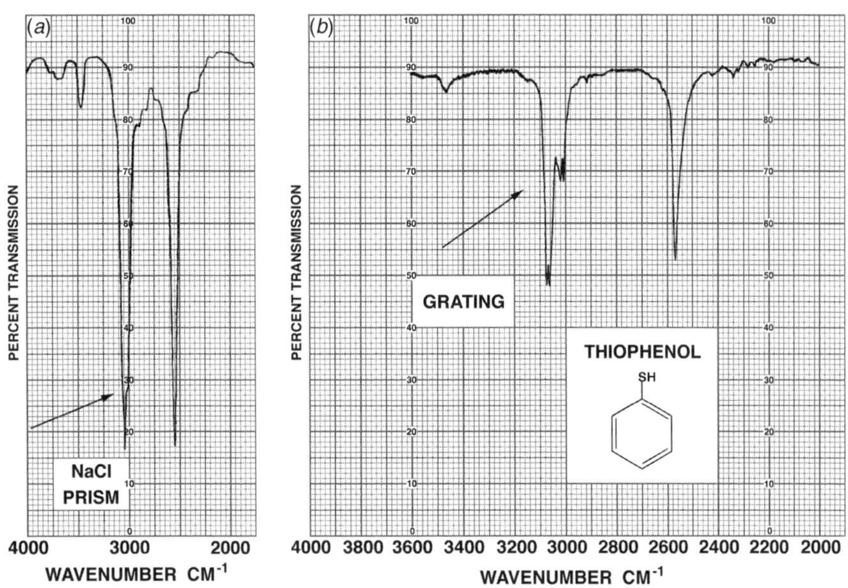

Fig. 5.3 The C—H stretching region of IR spectra of thiophenol with (*a*) NaCl prism resolution and (*b*) grating resolution.

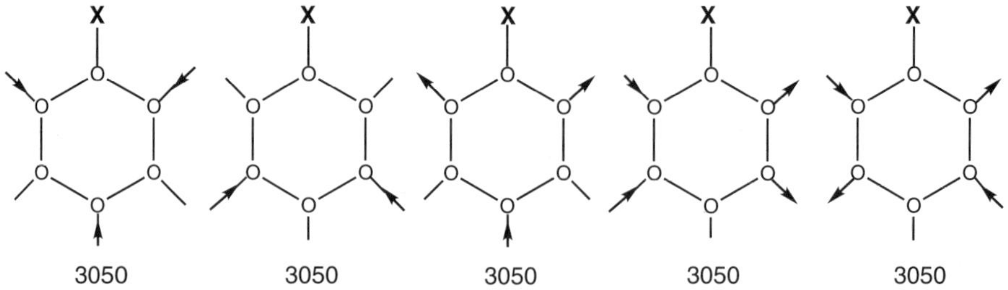

Fig. 5.4 Calculated displacements for normal modes of vibration for five accidentally degenerate C—H stretching vibrations of monosubstituted benzene.

2. When observed with grating resolution, this region generally will give evidence of more than a single band (splittings, shoulders, and unsymmetric band shapes; Figure 5.3*b*).

D. Coupling in aromatic C—H stretching modes
   1. Note that while there should be a band for each independent C—H oscillator, the weak coupling results in accidental degeneracy in most cases and a very simplified appearance to the region (see C—H stretching modes, Figure 5.4).
   2. In the highly symmetrical system of mesitylene,

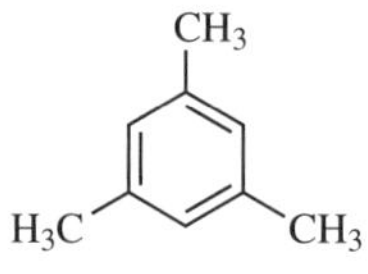

Mesitylene

where the three isolated C—H oscillators have identical frequencies, a high C—C stretch might have been expected to have enhanced coupling between the C—H groups. The vibrational isolation of the C—H oscillators, however, leads to only a very weak interaction. *Note*: In this compound with a threefold axis of symmetry the three identical oscillators will give rise to an IR-active degenerate pair plus a Raman-active symmetric C—H stretch. While the coupling present in three identical oscillators in a methyl group leads to a $\Delta\tilde{\nu}$ of $\sim$100 cm^{-1}, in mesitylene the shift between the IR and Raman bands is only a $\Delta\tilde{\nu}$ of $\sim$10 cm^{-1} (Figure 5.5).

E. Problems in the assignment of the modes

As the aromatic C—H stretching bands are generally weak, this absorption is often masked by strong aliphatic absorption bands which overlap the lower part (3030–2990 cm^{-1}) of this region in mixed aromatic–aliphatic compounds. This becomes more of a problem as the aromatic mode falls closer to 3000 cm^{-1}. Obviously, great care must be taken in the assignment

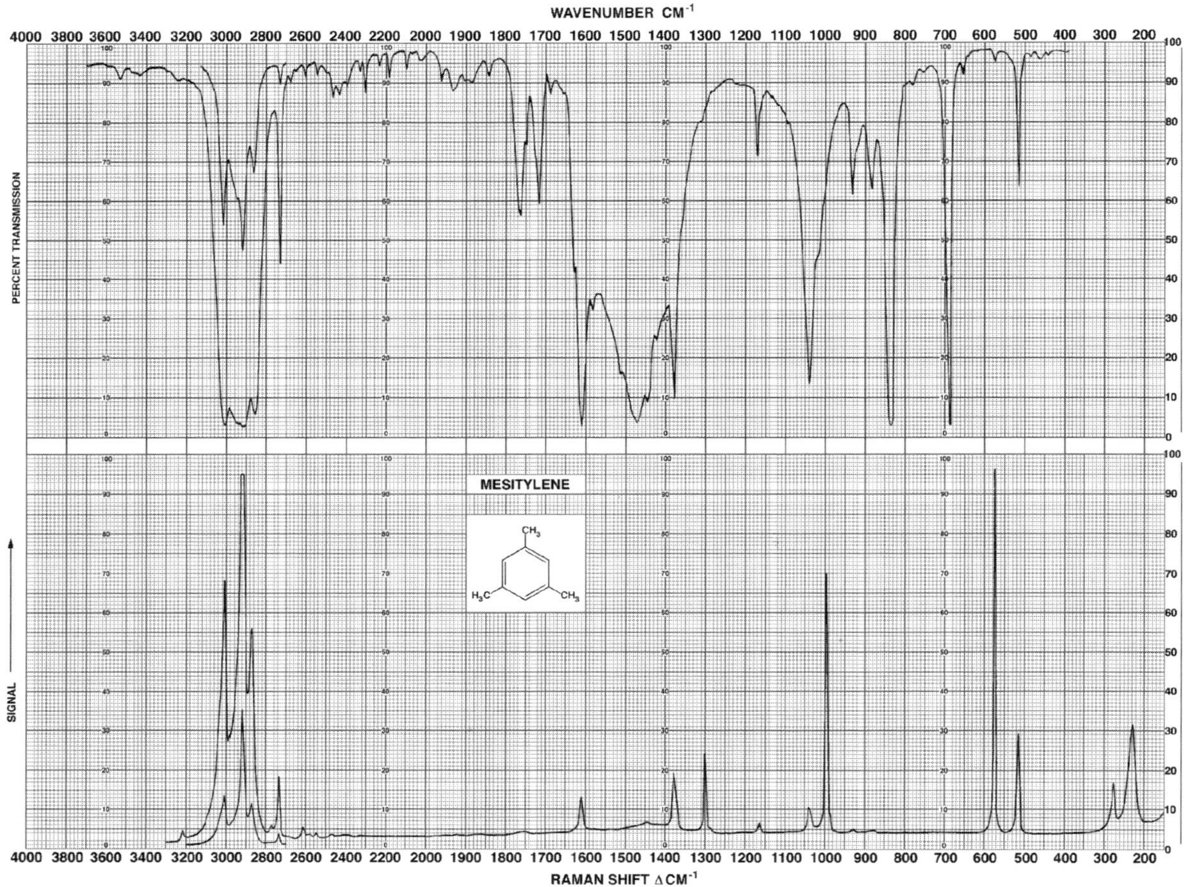

Fig. 5.5 Infrared and Raman spectra of mesitylene.

of these modes, and the region should be examined at the highest resolution available.

III. Carbon–carbon ring stretching modes

The aromatic ring system possesses C⩵C bond force constants which are intermediate between C=C and C–C. The result is that several of these vibrations move up out of the fingerprint region so that they become both isolated and much easier to identify.

In sorting out the phenyl group frequency modes it is convenient to refer to the fundamental modes of benzene which are roughly related to the phenyl system modes.

Note: In these notes we will use a numbering system for the benzene modes developed by Wilson. There are several others in the literature; this just happens to be the one we have used.

Note: Benzene has a center of symmetry. Thus IR-active bands will not occur in the Raman effect and Raman-active bands will not occur in the IR (rule of mutual exclusion). It is also found that occasionally some modes will not be active in either effect.

A. The 1600 cm^{-1} degenerate pair

1. Phenyl compounds possess two characteristic bands which appear in the 1600 cm^{-1} region. These bands have their origin in a doubly degenerate pair of fundamentals in benzene.
2. In benzene the 1600 cm^{-1} vibrations are assigned as ν_{8a} and ν_{8b} (R, –). That is, they are Raman active but infrared inactive.
3. In the Raman spectrum the modes should appear near 1594 cm^{-1}, but because of a second-order coupling (Fermi resonance), the band splits and the 1594 cm^{-1} band is observed as a doublet. Thus the band in benzene is a split peak even though it arises from a degenerate band system (Figure 5.6).
4. The description of the ν_{8a} and ν_{8b} degenerate modes is given in Figure 5.7. The atomic displacements involve C⩵C stretching. Mode ν_{8a} can be viewed as a combination of two waterlike molecules which are simultaneously undergoing symmetric stretching in phase with each other. The overall displacements in benzene will not result in a change in dipole moment and the mode will not be IR active. Mode ν_{8b} can also be viewed as a combined motion of two waterlike molecules, but in this case the displacements involve the out-of-phase combination of the two antisymmetric stretching motions of the triatomic systems. Note that this also leads to no change in dipole moment and no activity in the IR.
5. Substitution of the benzene ring by an –X group removes the degeneracy by destroying the high symmetry of the ring. The ring vibrations of the substituted system related to the ν_{8a} and ν_{8b} modes of benzene now become active in the IR.

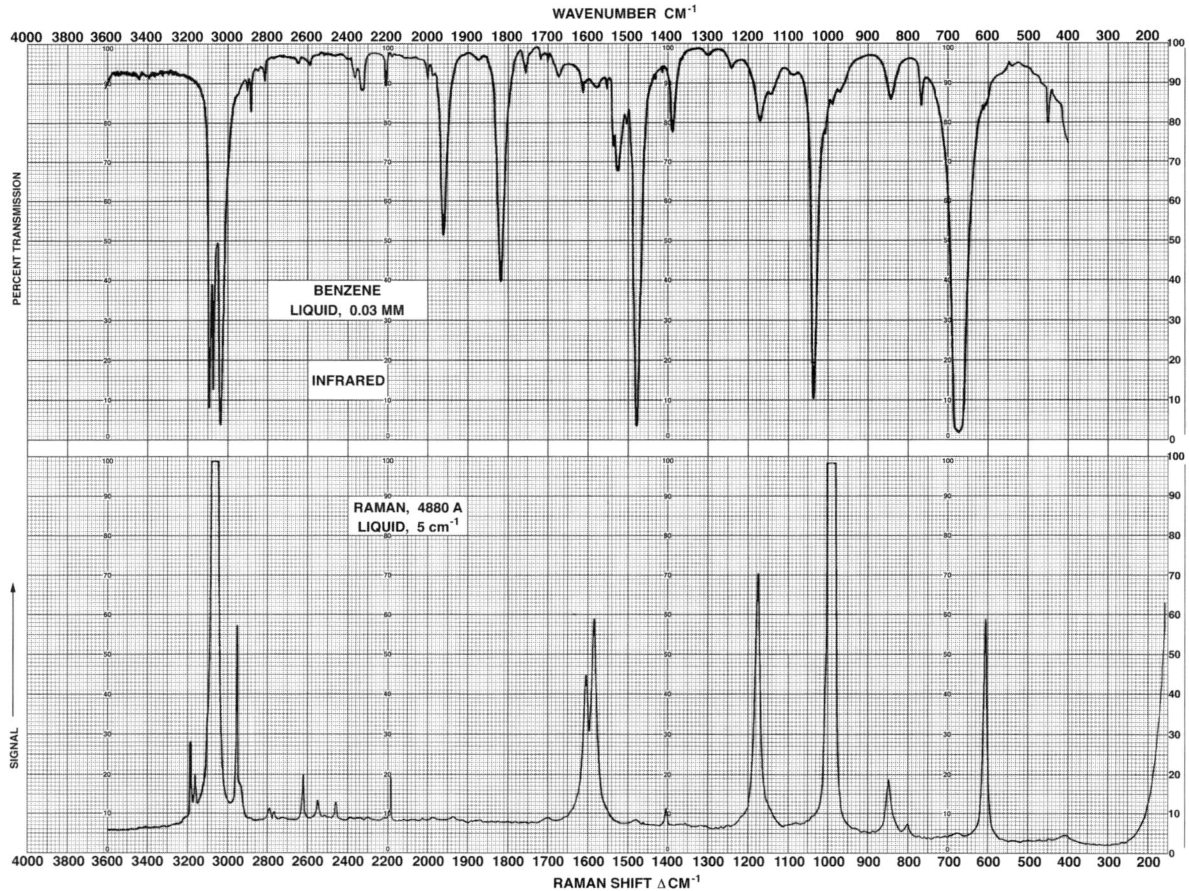

Fig. 5.6 Infrared and Raman spectra of benzene.

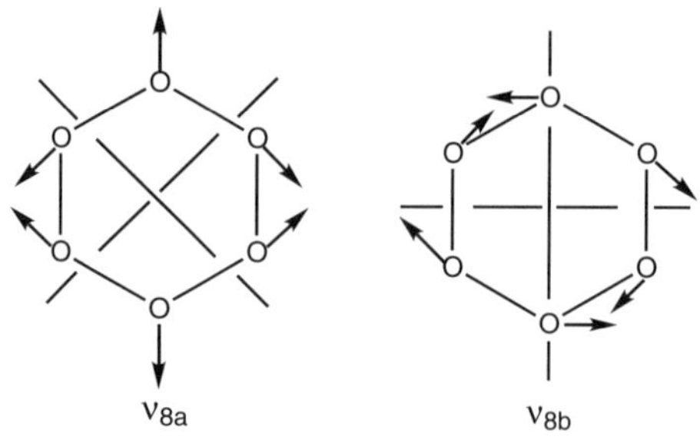

Fig. 5.7 In benzene both ν_{8a} and ν_{8b} will have identical frequencies and are degenerate vibrations.

6. Two bands near 1600 cm^{-1} have been shown to have their origins in the ν_{8a} and ν_{8b} ring modes (see Figure 5.8).
   a. ν_{8a} occurs very close to 1600 cm^{-1}.
   b. ν_{8b} usually occurs near 1580 cm^{-1}, lower and weaker than ν_{8a}.
   c. Both modes are observed in the IR and Raman and give rise to relatively weak bands. ν_{8a} and particularly ν_{8b} have quite variable intensities in the IR, and these variations can be useful in the diagnostic interpretation of complex molecular spectra.
7. If the —X substituent is conjugated to the phenyl ring, the intensity of ν_{8b}, which is weak to very weak when —X is unconjugated (see Figure 5.8), can be increased by a significant amount. The conjugation can involve either extended π-systems or heteroatom lone-pair interactions. A good example of this effect is in the spectrum of methyl salicylate, which has both carbonyl and heteroatom conjugating substituents (see Figure 5.9). (This example also demonstrates that the ν_{8a}–ν_{8b} pair is observed in this region in the IR in multiply substituted benzene ring systems.)

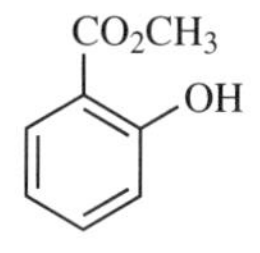

Methyl salicylate

8. The case of para-disubstitution of the benzene ring system:

   a. If —X and —Y are quite different in electrical character, ν_{8a} and ν_{8b} will behave normally; i.e., they will be active and observed (see,

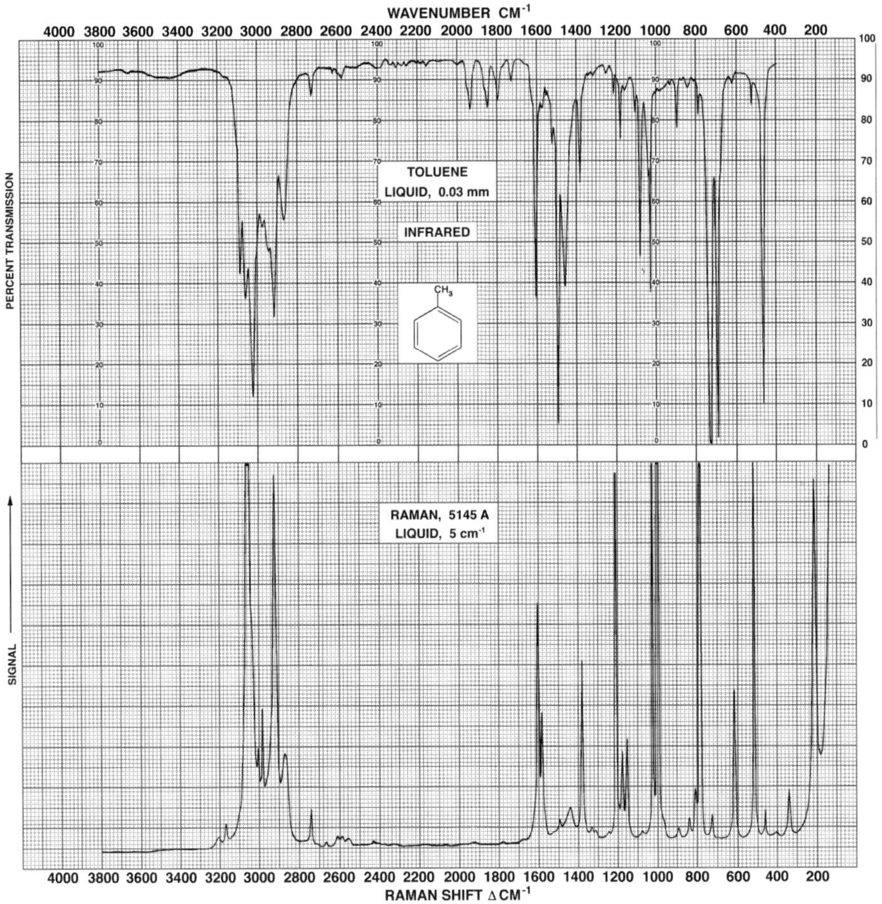

Fig. 5.8 Infrared and Raman spectra of toluene.

e.g., the IR and Raman spectra of *p*-nitrotoluene; Figure 5.10).

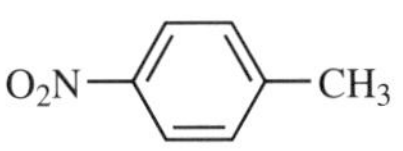

p-Nitrotoluene

b. If —X and —Y are identical, however ν_{8a} and ν_{8b} vanish from the IR because the para system now has a center of symmetry and there is no change in dipole moment during the vibration. The rule of mutual exclusion operates.

An example is *p*-xylene. No IR absorption is observed in the 1600 cm^{-1} region because ν_{8a} and ν_{8b} are IR inactive (no change in dipole moment; see Figure 5.11). (There is a weak absorption band found

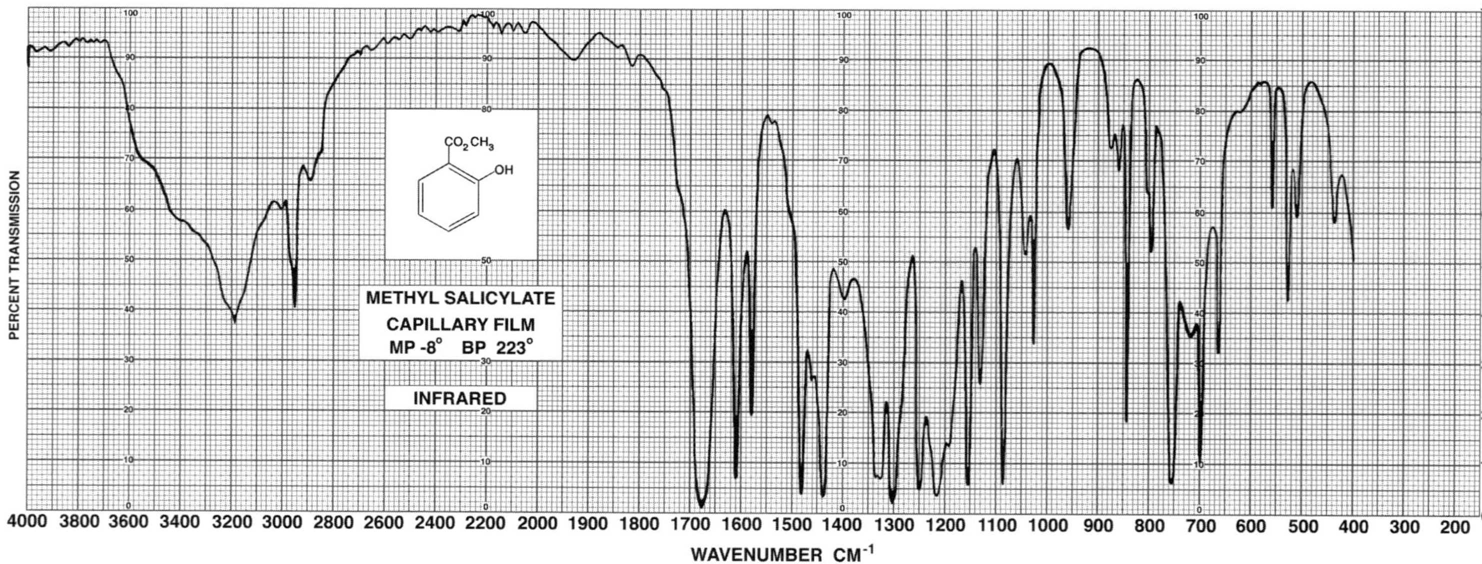

Fig. 5.9 Infrared spectrum of methyl salicylate.

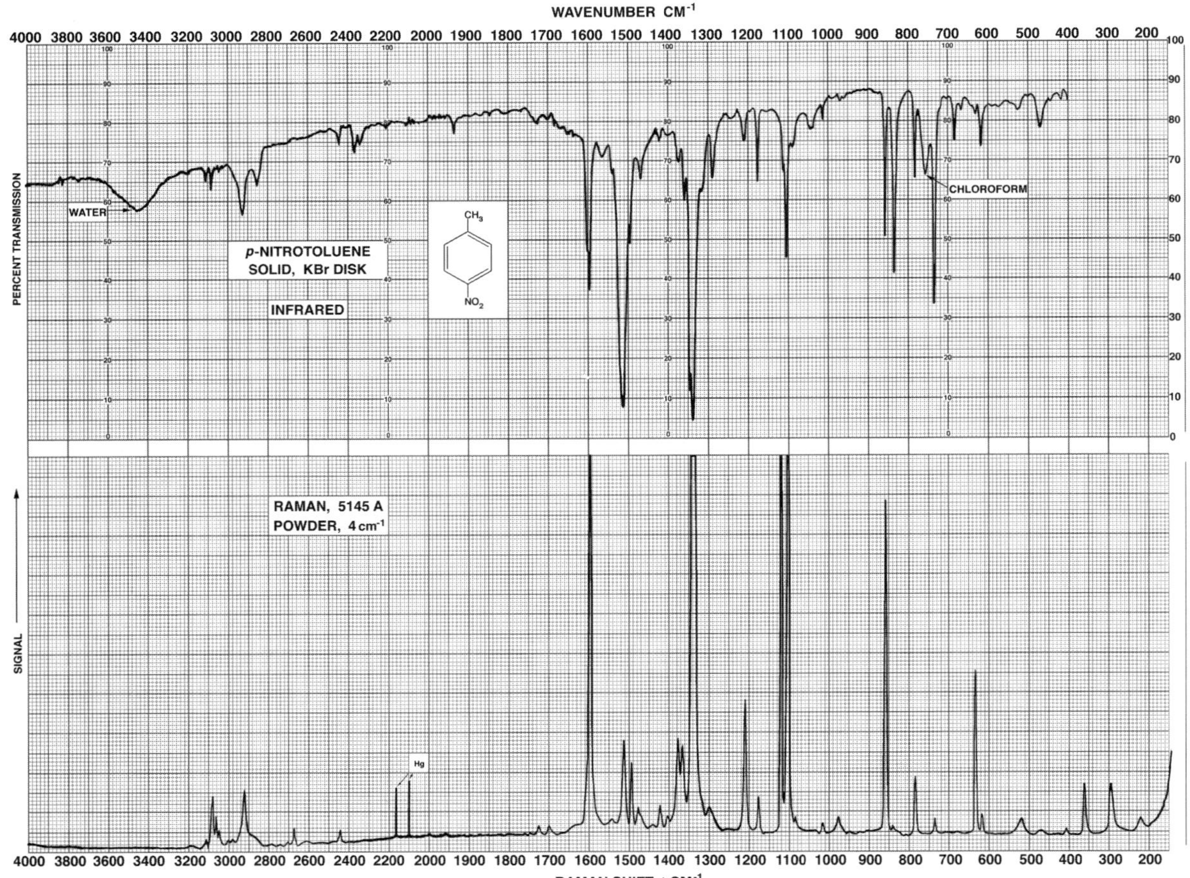

Fig. 5.10 Infrared and Raman spectra of *p*-nitrotoluene.

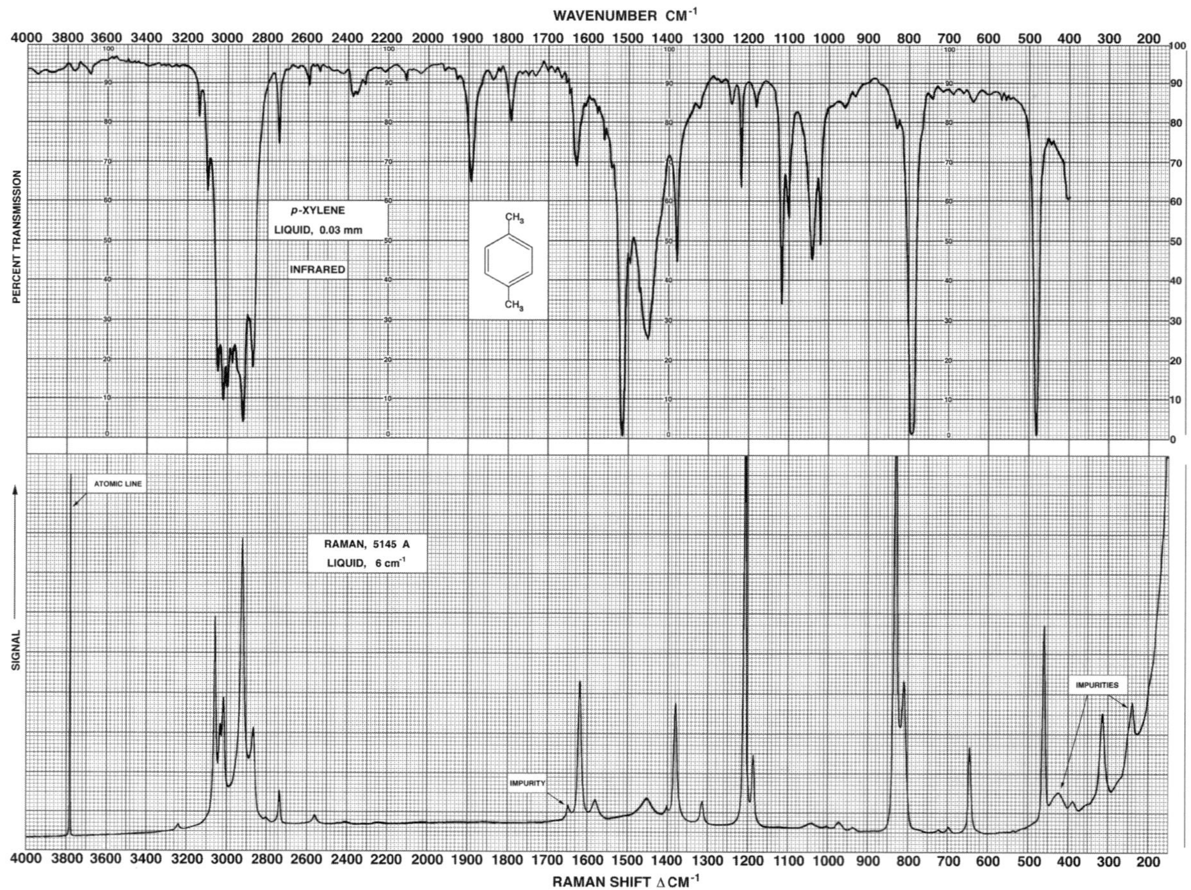

Fig. 5.11 Infrared and Raman spectra of *p*-xylene.

near 1630 cm^{-1} which is assigned to a sum tone of two low-lying fundamentals.)

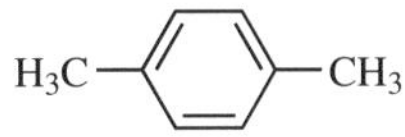

p-Xylene

c. If —X and —Y are nearly but not quite identical, e.g., the case of 4-ethyltoluene, the electrical nature of the two groups will be very similar and the dipole moment developed during the vibration greatly reduced (the presence of a pseudo center of symmetry). Under these conditions the 1600 cm^{-1} pair becomes very weak.

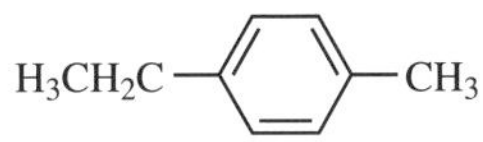

4-Ethyltoluene

d. It is therefore possible to predict some of the electrical properties of the para substituents based on the relative intensities of the two 1600 cm^{-1} bands. Indeed, careful examination of this region can suggest the potential presence of para-substitution itself.

B. The 1500 and 1450 cm^{-1} degenerate pair

1. Phenyl compounds have a second characteristic pair of bands in the 1500 cm^{-1} region. They have their origin in a degenerate pair of fundamentals in benzene.
2. In benzene these vibrations are assigned as ν_{19a} and ν_{19b} (–, IR). That is, they are IR active and Raman inactive.
3. They produce a very strong IR band near 1485 cm^{-1} in benzene.
4. The description of the ν_{19a} and ν_{19b} degenerate modes is given in Figure 5.12. The atomic displacements involve mainly carbon–carbon stretching. Mode ν_{19a} can be viewed as a combination of two waterlike molecules which are simultaneously undergoing symmetric stretching out of phase with each other. The overall displacements in benzene

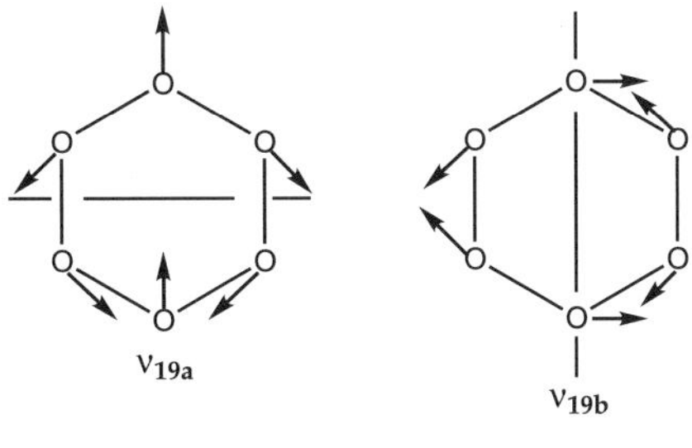

Fig. 5.12 Both ν_{19a} and ν_{19b} will have identical frequencies in benzene and are degenerate vibrations.

result in a change in dipole moment and the mode will be IR active. Mode ν_{19b} can also be viewed as a combination of two water molecules, but in this case the displacements involve the antisymmetric stretching mode of the two triatomic systems in phase with each other. Note that the overall displacements lead to a change in dipole moment and activity in the IR.

5. Substitution of the benzene ring by an —X group removes the degeneracy by destroying the high symmetry of the ring. The ring vibrations of the substituted system related to the ν_{19a} and ν_{19b} modes of benzene now also become active in the Raman effect and are split in both spectra.
6. Two bands near 1500 and 1450 cm^{-1} have been shown to have their origins in the ν_{19a} and ν_{19b} ring modes (see the IR and Raman spectrum of toluene; Figure 5.13).

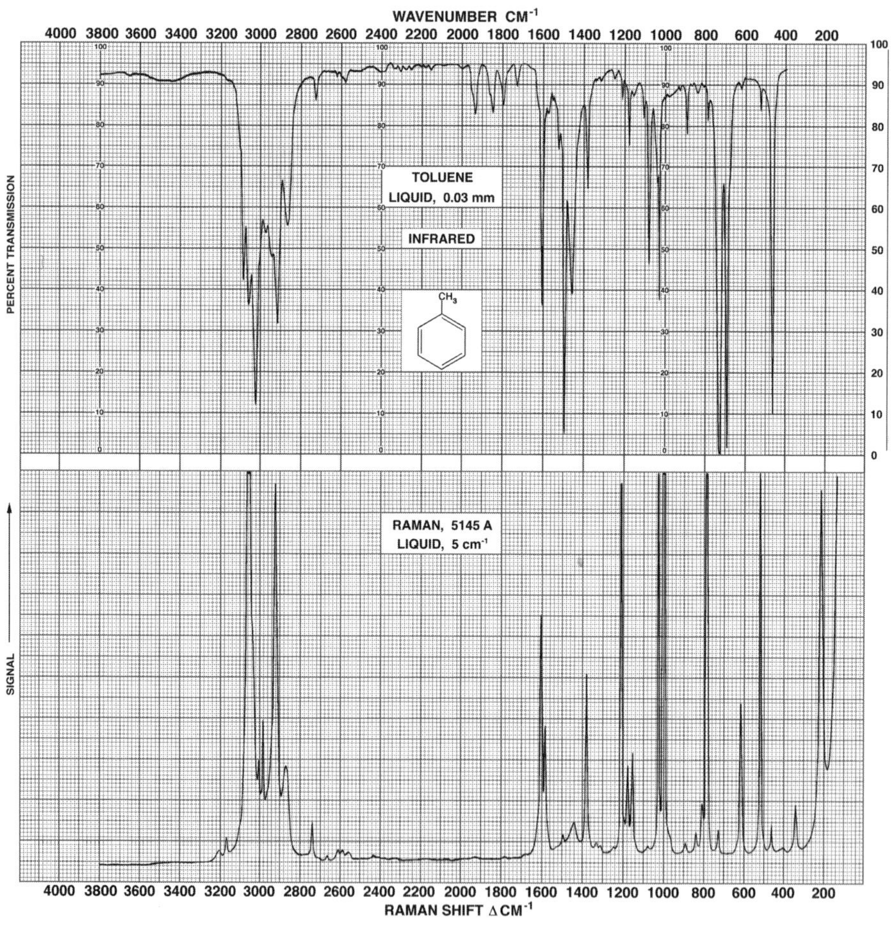

Fig. 5.13 Infrared and Raman spectra of toluene.

a. ν_{19a} occurs very close to 1500 cm^{-1}.
b. ν_{19b} usually occurs slightly lower than ν_{19a} near 1450 cm^{-1}.
c. Both modes are allowed in the IR and Raman. The higher wavenumber ν_{19a} is usually associated with a very strong band in the IR spectrum. The second mode, ν_{19b}, has a variable intensity in the IR because it is directly overlapped and masked by the methyl antisymmetric and methylene deformation vibrations. Thus, this frequency-stable group frequency has little diagnostic value.

7. The intensity of ν_{19a}, while generally quite intense, can be quite weak for exactly the same reasons that the 1600 cm^{-1} modes become weak. In this case the generation of an *electrical center of symmetry*, however, is not quite as obvious, but apparently it is possible. See, e.g., the case of 4-cyanobenzaldehyde discussed earlier (Chapter 1, pp. 4–7; Figure 5.14).

C. The 1010 cm^{-1} (ν_{12}) Raman band

1. In the case of benzene there is an in-plane ring deformation assigned as ν_{12} which involves displacing alternate carbon atoms around the ring, as shown in Figure 5.15.
2. In ν_{12} there is neither a change of dipole moment nor a change in polarizability during the vibration and the mode is not active in either the IR or Raman spectrum (–, –) and the frequency of the mode has to be obtained by other means.
3. On ring substitution, the vibration becomes active in the IR and Raman effects.
4. For mono-, meta-, and 1,3,5-trisubstituted benzenes, this is an excellent Raman group frequency. It always occurs at 1010 ± 10 cm^{-1} and is always one of the strongest bands. For other types of substitution it is not useful.
   a. In the Raman spectrum this mode is *very* strong and often gives rise to the strongest band. The absorption is located at 1010 ± 10 cm^{-1} and is highly stable. Because of the high intensity, this carbon–carbon ring mode makes an excellent group frequency even though it is found in the middle of the fingerprint region.
   b. The mode is weak and of little value in the IR.
5. The mode that becomes active on substitution is related to the benzene ν_{12} mode, but it only depends on a ring breathing motion of a single set of three alternate carbon atoms. Thus, it has been shown to be substituent independent as long as the substituents are introduced at the 1-, 1,3-, or 1,3,5-positions.
6. Any substitution that also involves the 2-, 4-, or 6-position requires the added substituents to be attached to both sets of three alternate carbons. Under these conditions, the band becomes X sensitive and will be moved out of the 1010-cm^{-1} region.
7. Thus, a strong Raman band near 1010 cm^{-1} is good evidence for 1-, 1,3-, or 1,3,5-substitution of the aromatic ring (see Figure 5.16), and conversely.

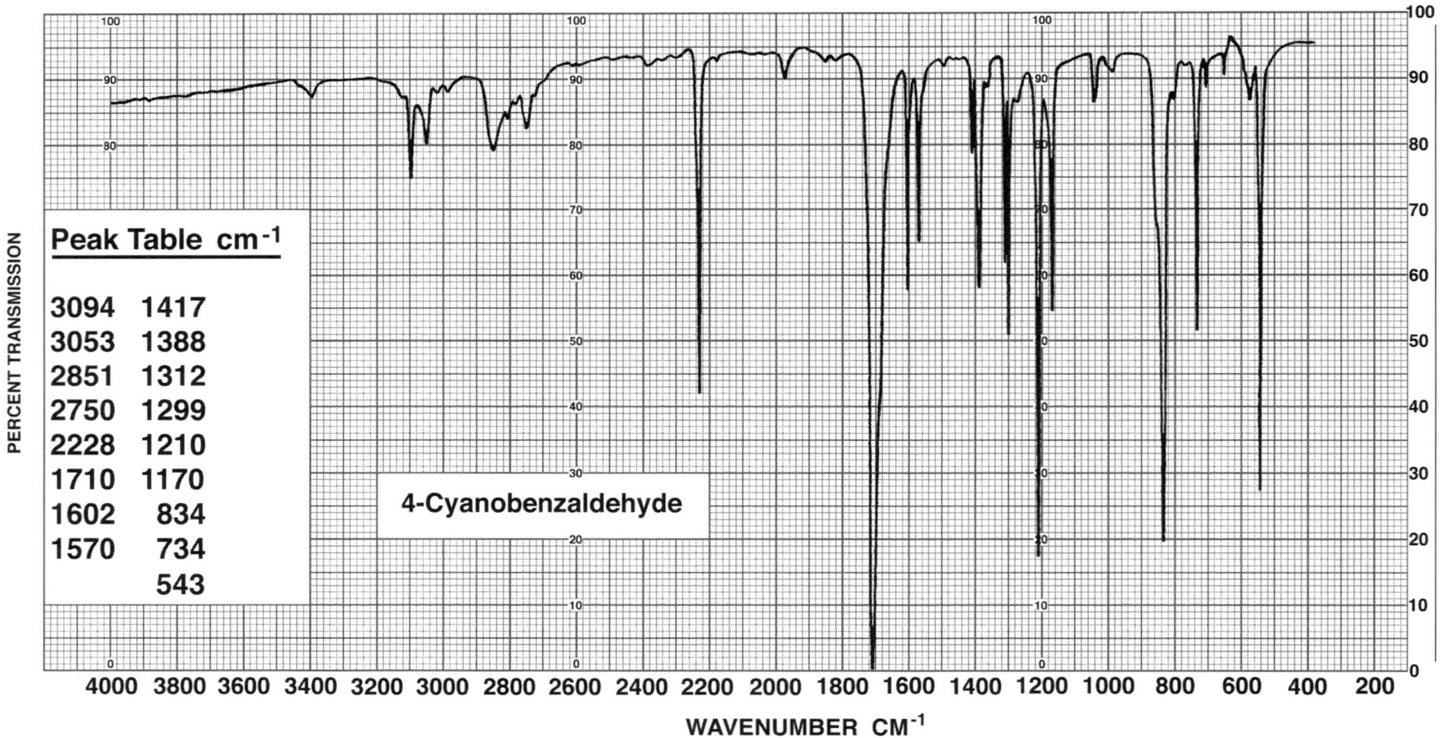

Fig. 5.14 Infrared and Raman spectra of 4-cyanobenzaldehyde.

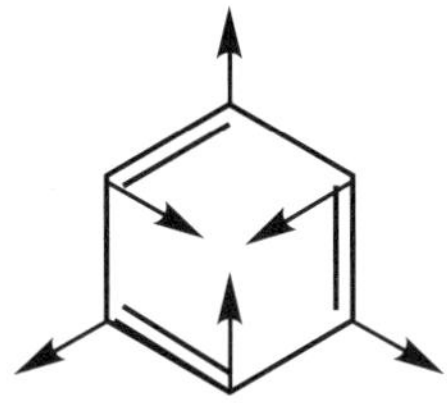

Fig. 5.15 In-plane ring deformation assigned to ν_{12}.

D. The symmetric ring breathing mode, ν_1
   1. In the case of benzene there is a totally symmetric ring stretching mode, ν_1, which involves all six carbons moving in and out in unison (see Figure 5.17).
      This mode will be Raman active and IR inactive (R, –) in benzene (no change in dipole moment).
   2. As all the bonds are being stretched in phase, the change in polarizability is very large, and this is reflected in the observation of a very intense band associated with this mode near 992 cm^{-1} in the Raman spectrum (see Figure 5.18).
   3. *Note*: The ν_1 ring breathing mode (992 cm^{-1}) lies within 20 cm^{-1} of the ν_{12} in-plane ring deformation mode (1010 cm^{-1}). This is not unexpected as the two modes have many similar features.
   4. The substitution required to make ν_1 IR active is also the undoing of its group frequency characteristics. The form of the vibration requires that the C—X displacement be involved and therefore the mode has the potential to be sensitive to the nature of X.
   5. When the atom that directly connects the X group to the ring is a C, N, O, or F atom, the stretching mode of the isolated oscillator is usually at 1000 ± 200 cm^{-1}. The mass and force constants of the C—X group, therefore, are similar to the ring C$\overset{...}{=}$C bonds, and strong coupling of the —X group with the ring modes occurs. In this environment the ring breathing mode will undergo wide and unpredictable coupling and frequency shifts. Thus, ν_1 is not a good group frequency even though it is the strongest band in the Raman spectrum of benzene.

E. The 606-cm^{-1} degenerate in-plane bend (ν_{6a} and ν_{6b})
   1. In benzene this mode is a doubly degenerate ring in-plane deformation. The atomic displacements are shown in Figure 5.19.
      Because no dipole moment will develop during the vibration in either mode the vibration is IR inactive. There is, however, a change in polarizability, and the degenerate modes are active in the Raman effect

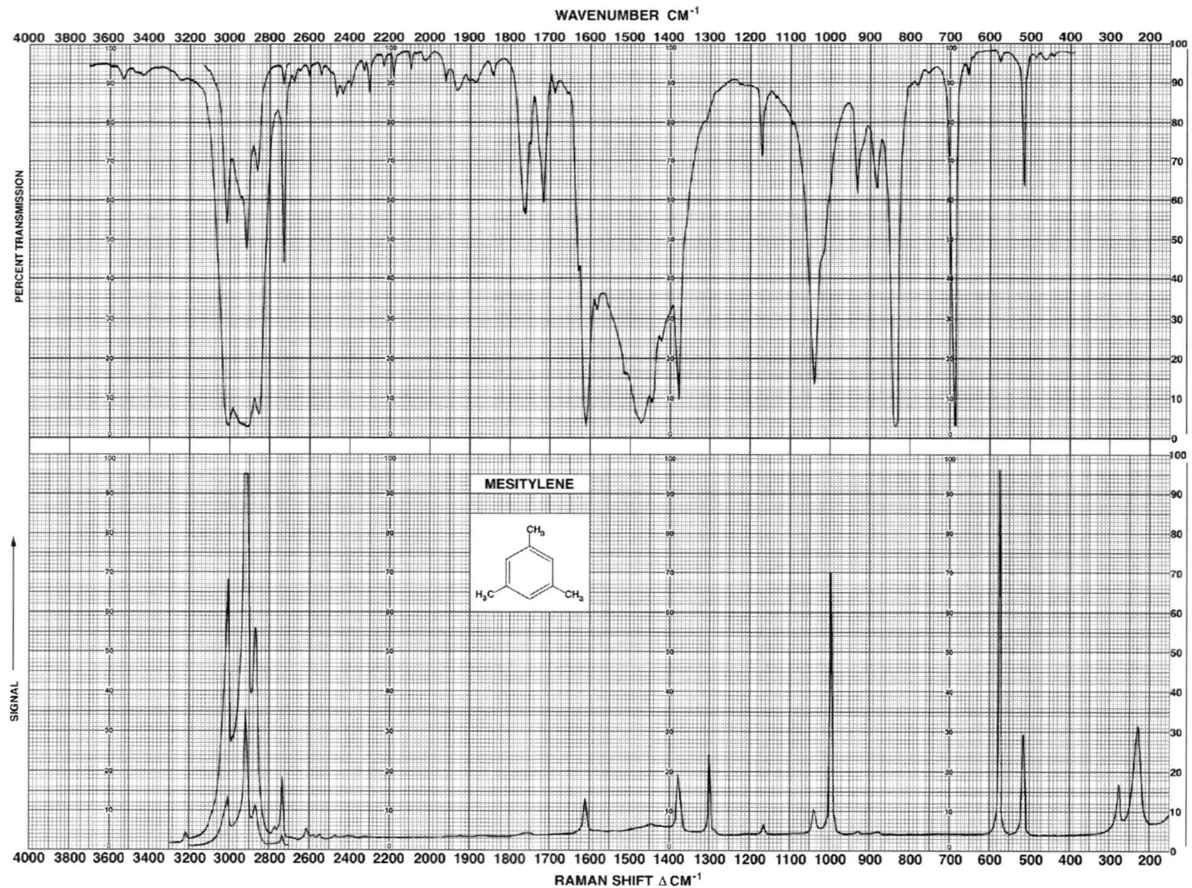

Fig. 5.16 Infrared and Raman spectra of mesitylene.

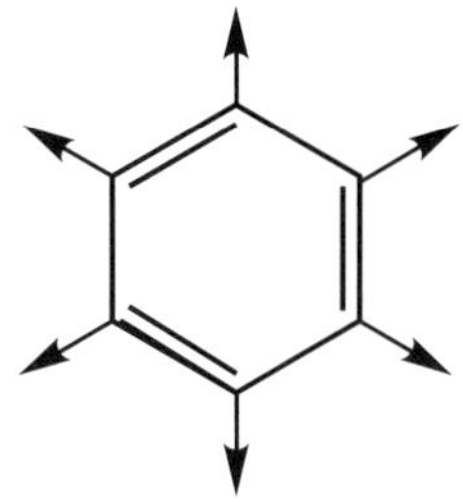

ν_1

Fig. 5.17 Totally symmetric ring stretching mode, ν_1.

(R, –). The vibration is assigned in benzene to a medium–strong band near 606 cm^{-1} (see Figure 5.18).

2. The two modes behave in quite different fashion on substitution.
   a. In the case of ν_{6a} the ring system is —X sensitive to substitution as all six ring atoms are displaced during the vibration.
   b. In the case of ν_{6b} carbons 1 and 4 undergo very little displacement. Thus, substitution with —X at carbon 1 (monosubstitution) or carbons 1 and 4 (para disubstitution) would be expected to leave the basic mode unperturbed. This appears to be the case, as a consistent weak-to-medium band is observed in the Raman effect near the original benzene position of 606 cm^{-1} in mono- and para-disubstituted aromatic compounds (see Figure 5.20, *p*-xylene, 648 cm^{-1}).
   c. Because of coupling, the location of ν_{6a} is not easily identified and the mode has not been correlated.
3. The degeneracy is broken by substitution on the ring. Both modes are, therefore, active in the Raman and the IR with the exception that substances that are para-substituted with identical substituents remain IR inactive or nearly so.
4. It is often possible to differentiate between mono- and para-substitution from the frequency of ν_{6b}. See Table 5.1 and the spectrum of toluene in Figure 5.13.
5. Other types of substitution such as 1,2, 1,3, 1,3,5, etc., give rise to coupled modes with both ν_{6a} and ν_{6b} and therefore do not generate good group frequencies.

IV. Carbon–hydrogen bending modes
   A. In-plane C—H bending modes
      1. The in-plane C—H bending modes give rise to rather weak IR bands, and although these modes are not highly coupled through the valence bond system, there does appear to be significant spatial coupling and a

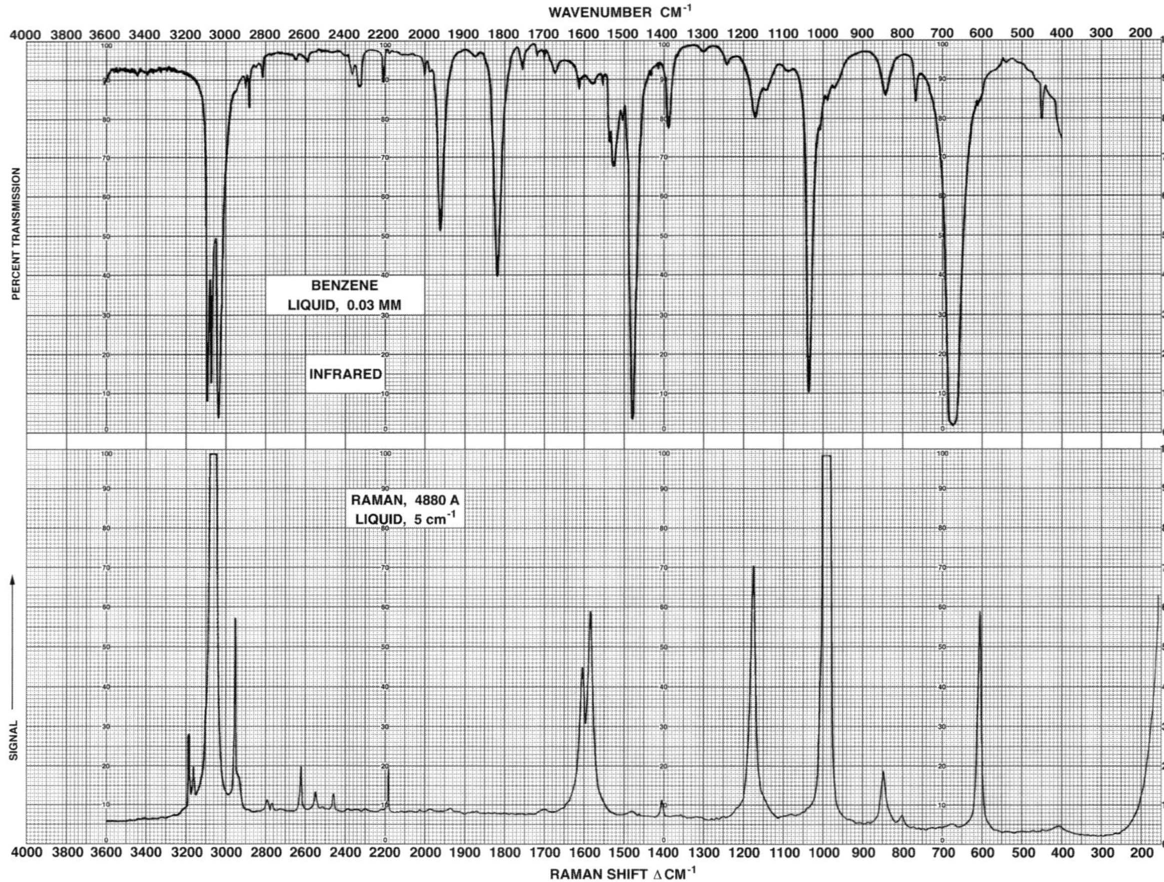

Fig. 5.18 Infrared and Raman spectra of benzene.

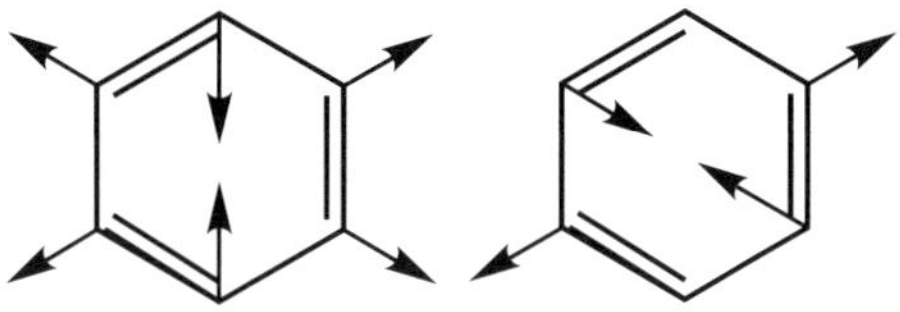

ν_{6a} ν_{6b}

Fig. 5.19 Benzene doubly degenerate ring in-plane deformation ν_{6a} and ν_{6b}.

related sensitivity to C—X substitution. These factors result in what is usually observed as a series of weak needlelike bands spread through the fingerprint region. These modes tend to be more intense in the Raman effect but are diagnostically no more powerful.

2. A particular pattern may be useful in identifying a particular compound (or closely related series), but this pattern will be relatively ineffective as a group frequency (or combination of group frequencies).

B. Out-of-plane C—H and C⩶C bending modes

1. There are as many C—H out-of-plane vibrations as there are ring C—H groups. The coupled modes in which all adjacent hydrogens are deformed together (the all-in-phase out-of-plane bend) develop strong dipole moment changes and give rise to very strong bands in the IR. The high intensity of these modes helps *greatly* in their assignment because the frequencies lie in the fingerprint region of the spectrum.
2. Only the *all-in-phase* C—H bending mode has strong intensity because in the other C—H bending modes, charge development during the vibration tends to be canceled. This results in a further simplification of the spectrum and improves the confidence of the assignment.
3. These C—H modes have little intensity in the Raman effect and have not been correlated.
4. The out-of-plane C—H bends are weakly coupled and largely C—X insensitive. Inasmuch as the all-in-phase mode is modestly sensitive only to the number of C—H oscillators mutually adjacent, it is often possible to establish the substitution pattern of a ring system simply by the assignment of this mode alone.
5. The sequence of all-in-phase C—H out-of-plane bending modes is shown in Table 5.2.
6. There is considerable overlap of the ranges because the coupling is low. The uncertainty of the substitution pattern can be greatly reduced, however, by the identification of an additional strong band in the 690-cm^{-1} region. This band is assigned to a C⩶C ring puckering vibration, ν_4, in which alternate carbons are displaced above and below the plane of the ring; see Figure 5.21. The substituted extension of this mode, as in the case of ν_{12}, is insensitive to C—X groups introduced at carbons

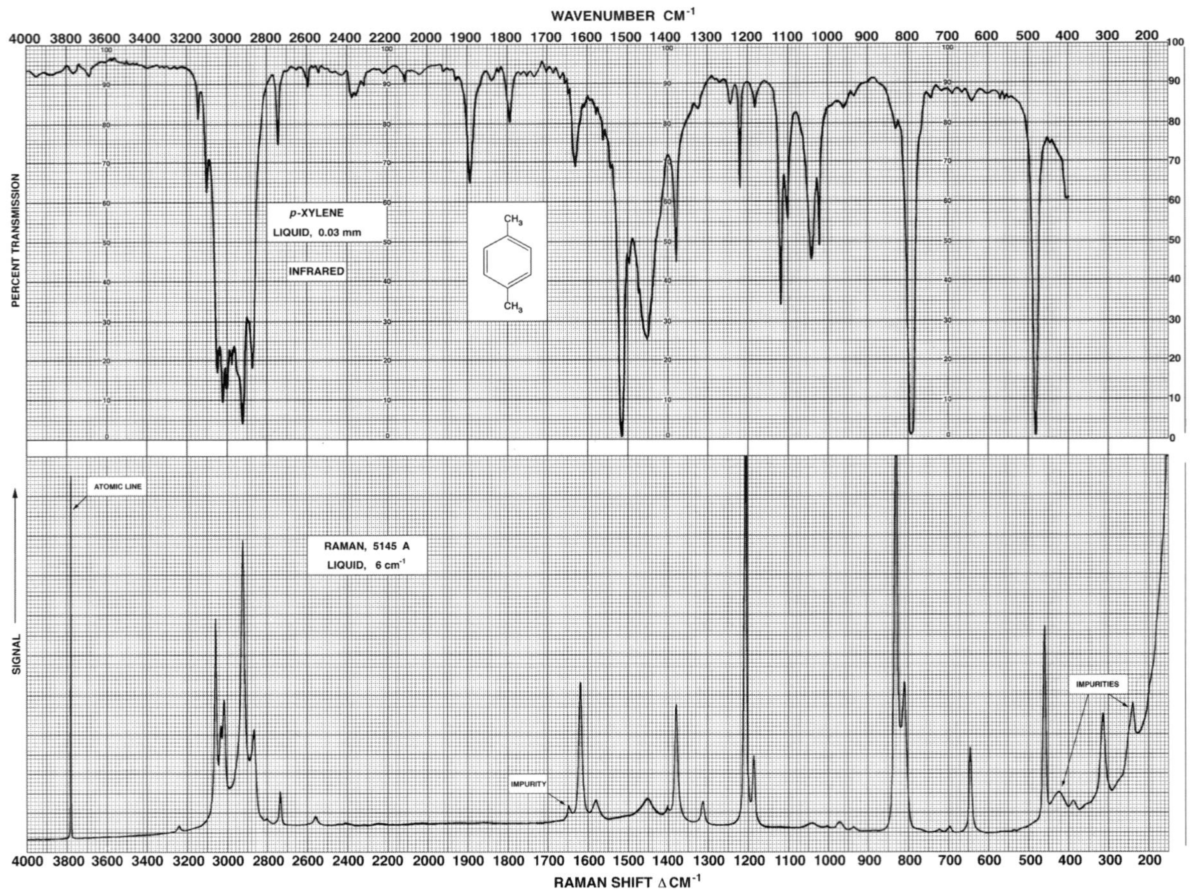

Fig. 5.20 Infrared and Raman spectra of *p*-xylene.

TABLE 5.1 ν_{6b} Ring In-Plane Deformation

ν_{6b}	1-Substituted	1,4-Substituted
Wavenumber values (cm^{-1})	630–605	650–620
Raman	Medium–weak	Medium–weak
Infrared	Weak	Weak–very weak

TABLE 5.2 All-in-Phase C—H Out-of-Plane Bending Modes

Adjacent C—H Groups	Range (cm^{-1})
5	770–730
4	770–735
3	810–750
2	860–800
1	900–860

1-, 1,3- (meta-disubstituted), or 1,3,5-trisubstituted. Thus, the presence of the 690 cm^{-1} band in addition to a strong band in the 750–730 cm^{-1} region would support monosubstitution over 1,2-disubstitution (ortho-substitution). If the latter band is above 770 cm^{-1}, then the substitution is more likely meta. The assignment remains ambiguous if the band falls in the 770–750 cm^{-1} range.

7. In benzene the ring puckering mode, ν_4, like ν_{12}, is not active in either the IR or Raman effects (–, –). On substitution, it becomes strongly active in the IR spectrum but remains a very weak band in the Raman effect.

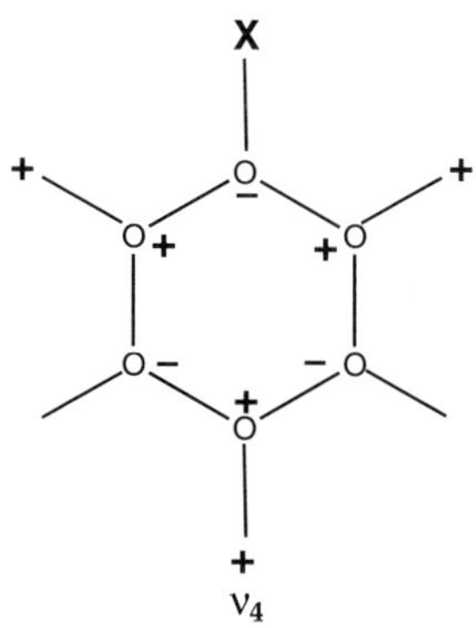

Fig. 5.21 The displacements are assigned to a C⋯C ring puckering vibration, ν_4, in which alternate carbons are displaced above and below the plane of the ring. The band is usually found near 690 cm^{-1}.

TABLE 5.3 C—H Out-of-Plane Bending and Ring Puckering Vibrations of Benzene Derivatives

Benzene		671	
Monosubstituted benzene		770–730	710–690
1,2-Disubstituted (ortho)		770–735	
1,3-Disubstituted (meta)		810–750	710–690
1,4-Disubstituted (para)		860–800	
1,2,3-Trisubstituted		810–750	(725–680)
1,2,4-Trisubstituted	900–860 (w)	860–800	
1,3,5-Trisubstituted		900–860	700–650
1,2,3,4-Tetrasubstituted		860–800	
1,2,3,5-Tetrasubstituted		900–860	
1,2,4,5-Tetrasubstituted		900–860	
1,2,3,4,5-Pentasubstituted		900–860 (w)	

8. A number of tables of these bending modes have been developed to aid in making the assignment of substitution patterns in a large number of aromatic ring systems. Table 5.3 is one such table for the benzene ring system.

C. Sum tones in the 2000–1650 cm^{-1} region

1. The C—H out-of-plane bending modes have been shown to contribute to the origin of a series of sum tones that are observed in the 2000–1650 cm^{-1} region of aromatic ring systems. These weak bands can be very useful as further confirmatory evidence to support assignments of substitution patterns made from fundamentals occurring in the fingerprint region.
2. Norman Wright of the Dow Chemical Company was the first to report that a series of weak absorption bands in the 2000–1650 cm^{-1} region could dependably be used to assign substitution patterns in benzene ring systems. The approach is as follows:
   a. If the IR spectrum of the sample is obtained with a normal path length, it may be necessary, at times, to employ a second sample of greater than normal path length. The thicker sample enhances the weak bands to be examined. Because these modes are related to the out-of-plane C—H bending modes, it is not surprising that these sum tones have very little intensity in the Raman effect.
   b. The *pattern* of absorption in the region is determined rather than the exact wavelengths of the weak peaks. A pattern match against a reference standard is required for the assignment of substitution. A set of patterns for the simple benzene ring system is given in Figure 5.22. These patterns are particularly useful as they were derived by averaging the region over a series of compounds all possessing the same substitution pattern. Thus, a perfect match would not be expected, but differentiation from other substitution patterns will be

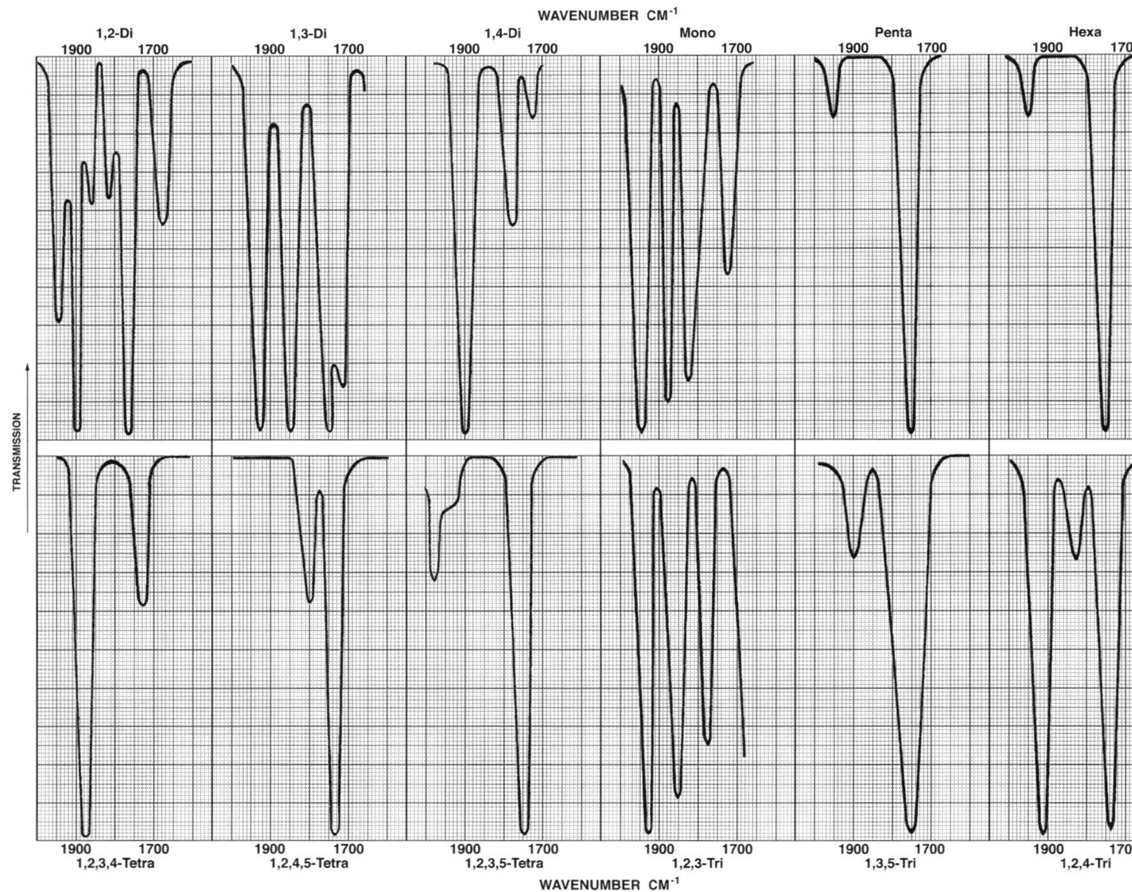

Fig. 5.22 Infrared sum tone and overtone patterns for substituted benzene derivatives in 2000–1600 cm^{-1} region.

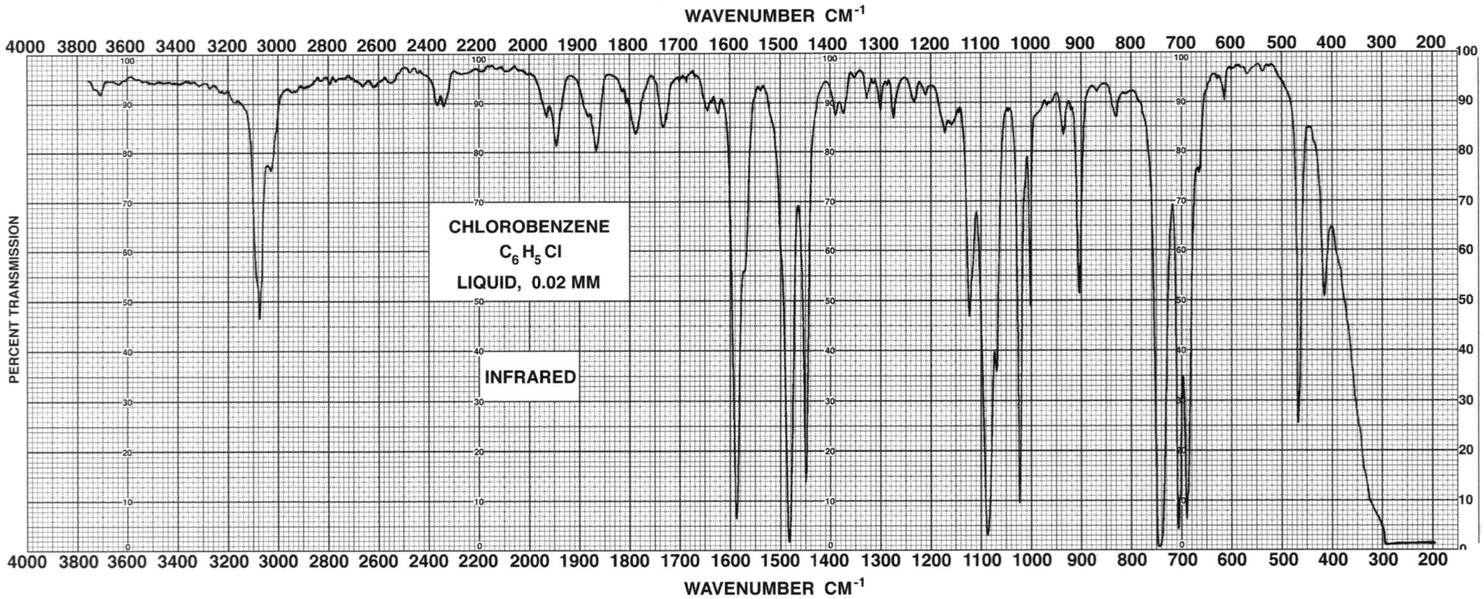

Fig. 5.23 Infrared spectrum of chlorobenzene.

enhanced (the averaged set was developed by Norman Jones at the Canadian Research Council).

c. The spectrum of chlorobenzene is a good example of a typical pattern match with the reference set (see Figure 5.23).

d. Compounds containing carbonyl groups are generally difficult to analyze by this approach because a good deal of the pattern region may be masked by absorption arising from the carbonyl stretching mode.

e. Sum tone patterns for a large variety of aromatic ring systems have been catalogued so that it is now possible to make substituent assignments to a large number of aromatized ring systems (Figure 5.24 is an example).

V. Aromatic heterocyclic spectra

A. Carbon–hydrogen stretching vibrations

1. Many of these modes are identical to those of the carbocyclic aromatic C—H stretching modes. They are weak and occur above 3000 cm^{-1} (substituted on sp^2-hybridized carbon) with most values lying in the 3100–3000 cm^{-1} region. An example of the appearance of this region is the spectrum of pyridine (see Figure 5.25).

2. In some cases the C—H stretching modes are perturbed well above 3100 cm^{-1}. A good example of a system with a high C—H stretching mode is furan, the oxygen-containing aromatic five-membered ring (see Figure 5.26).

3. The rise of the C—H mode in furan may be related to the same effect which perturbs vinyl ethers, CH_2—CH_2—O—CH=CH_2, which also have C—H stretching modes above 3100 cm^{-1}.

B. Ring C⋯C stretching vibrations

1. Six-membered heterocyclic systems will behave in rather similar fashion to the benzene analog. For example, pyridine will mimic a monosubstituted phenyl system:

a. The ν_{8a}–ν_{8b} pair (no longer degenerate) near 1595 and 1580 cm^{-1} is very similar to the benzene system. The strong intensity of ν_{8b} is consistent with the presence of a lone-pair heteroatom built into the ring system (Figure 5.25).

b. The ν_{19a}–ν_{19b} (no longer degenerate) pair is located a little lower but very close to the benzene values at 1482 and 1439 cm^{-1} with the lower mode again intensified (Figure 5.25).

2. In five-membered aromatic heterocyclic systems like furan or pyrrole, the system mechanics requires two pairs of ring stretching vibrations in this region of the spectrum. The modes, however, are somewhat perturbed relative to the six-membered series. For example, in furan bands at 1586 and 1490 cm^{-1} are related to the $\nu8$ pair of phenyl near 1600 cm^{-1}. Similarly furan bands at 1460 and 1384 cm^{-1} are related to

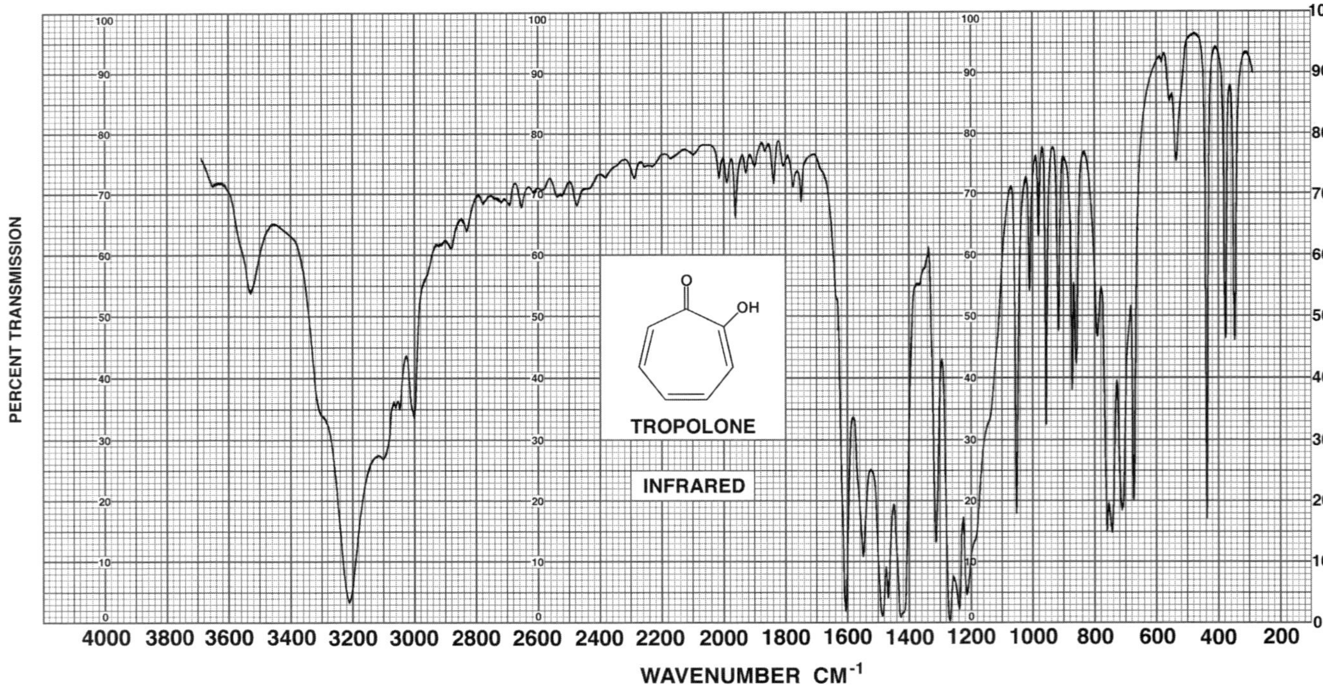

Fig. 5.24 Infrared spectrum of tropolone.

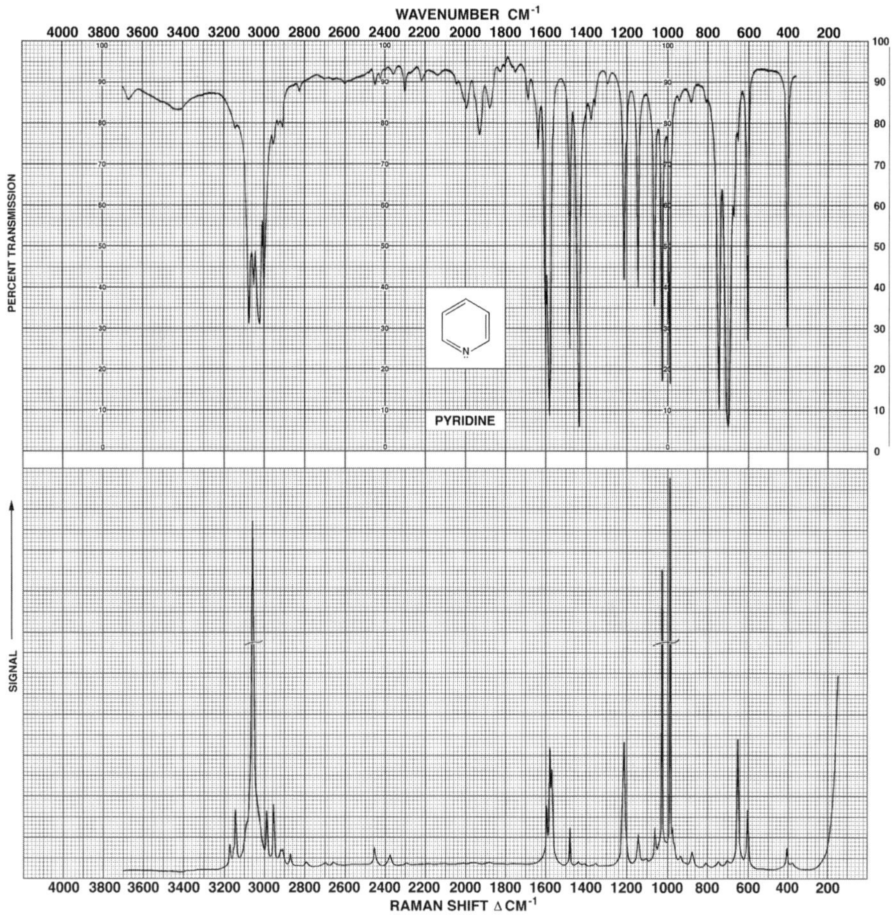

Fig. 5.25 Infrared and Raman spectra of pyridine.

the $\nu 19$ pair of phenyl near 1500 and 1450 cm^{-1}. The 1490 and 1460 cm^{-1} bands of furan are very strong and not easily resolved (see Figure 5.26).

C. Carbon–hydrogen out-of-plane bending vibrations

In this case similar patterns are found for both the heterocyclic five- and six-membered rings compared to the carbocyclic benzene system. The number of adjacent C—H oscillators effectively determines the position of the all-in-phase out-of-plane bend location in the spectral region. Thus we find that:

a. Pyridine acts like a monosubstituted benzene system, and furan appears similar to an ortho-disubstituted benzene system.

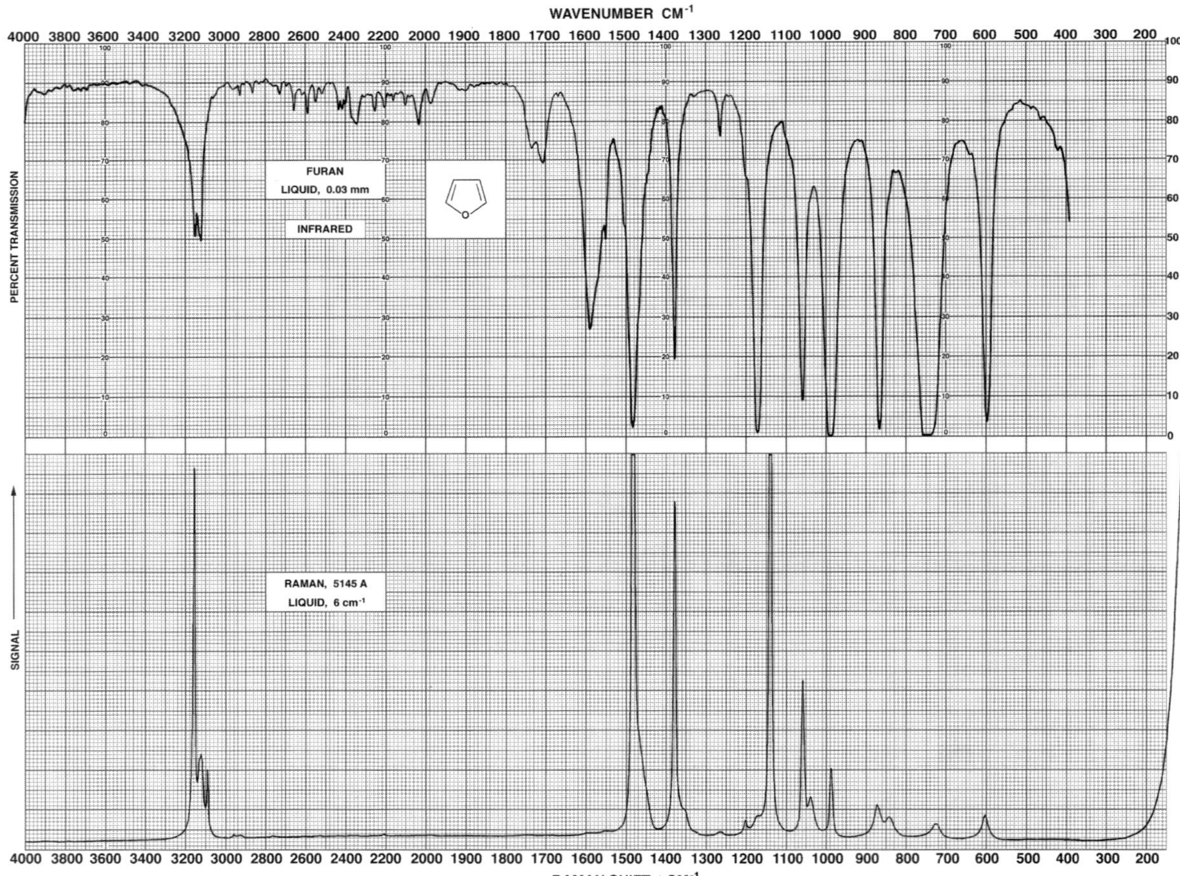

Fig. 5.26 Infrared and Raman spectra of furan.

b. The all-in-phase out-of-plane bending mode of the five C—H groups (ν_{11}) in pyridine is inverted relative to the ring puckering mode (ν_4), but they still are located very close to the values established for these modes in the benzene ring. Thus, ν_{11} is found near 700 cm^{-1} and the puckering mode is found at 749 cm^{-1}. The assignment of the C—H bending mode to the 700 cm^{-1} band is consistent as that band is the most intense band in the spectrum of pyridine (Figure 5.25).

c. Pyridine is a rare case where the system behaves like a monosubstituted six-membered ring but in which there is no group coupled to the nitrogen atom. Thus, ν_1, which is strongly Raman active, is not shifted at all from its position in benzene near 992 cm^{-1}. While ν_1 is now active in both spectra, it is the strongest band in the Raman spectrum of pyridine (Figure 5.25).

d. The same argument holds for ν_{12}, which is the origin of a very intense Raman band in pyridine only slightly perturbed at 1030 cm^{-1} (benzene, 1010 cm^{-1}; see Figures 5.18 and 5.25).

e. Finally, ν_{6a} and ν_{6b} are located in the Raman effect as very weak and weak bands, respectively, and very close to the benzene values. In this case ν_{6b} (652 cm^{-1}) occurs slightly higher than that usually identified in the benzene series for monosubstitution. Even though ν_{6a} has been assigned to a very weak Raman band in pyridine, the band has no value as a group frequency correlation (Figure 5.25).

D. Sum tone bands in the region 2100–1650 cm^{-1}

1. These weak bands are derived in exactly the same fashion as the weak bands that appear in the same region in the spectra of benzene derivatives. Characteristic patterns related to the substituent position can be developed in much the same way as for the benzene systems. The corresponding pyridine patterns are given here as an example:
   a. Pyridine sum tone bands, 2100–1650 cm^{-1} (Figure 5.27).
   b. As an example, compare the set of patterns given in Figure 5.27 with the spectrum of 4-picoline (the sample is slightly contaminated with water, Figure 5.28). The spectrum should have been obtained with a longer path length cell.
2. The use of sum tone patterns for substitution orientation also has been successfully extended to a number of polynuclear aromatic systems.

VI. Interpretation of aromatic spectra:

A. The polymerization of styrene to polystyrene (see Figure 5.29).

$$n\mathrm{CH(C_6H_5)}{=}\mathrm{CH_2} \rightarrow \left[\mathrm{CH(C_6H_5)}{-}\mathrm{CH_2}\right]_n$$

To get a sense of how powerful the use of infrared group frequencies is, an examination of the polymerization of styrene to polystyrene by infrared will be carried out. The starting monomer is the upper spectrum shown in Figure 5.29. It will be treated as if the spectrum were an unknown substance.

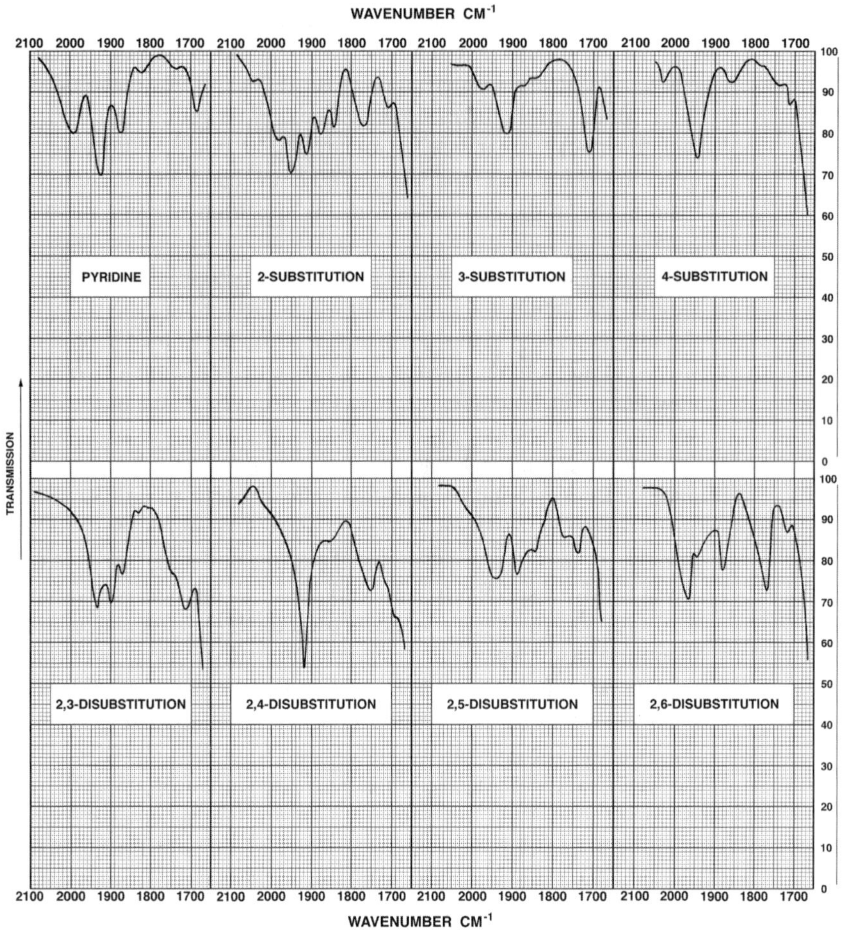

Fig. 5.27 Infrared sum tone and overtone patterns for substituted pyridine derivatives in 2100–1650 cm^{-1} region.

B. The IR spectrum of the monomer.

Start at the high wavenumber end of the spectrum and examine the 3000 cm^{-1} C—H stretching region. A complex series of bands is found that all fall above 3000 cm^{-1} except for a weak shoulder near 2990 cm^{-1}. Thus, it can be confidently assumed that the molecule contains only unsaturated C—H groups (either aromatic or olefinic) and that a reasonable number of different structural locations are present because of the complexity of the band system. Also it is quite certain that there are no saturated C—H groups present because of the lack of absorption below 3000 cm^{-1} in the region between 3000-2800 cm^{-1}. Note that the absence of a band can often allow a more unambiguous assignment than the presence of a band.

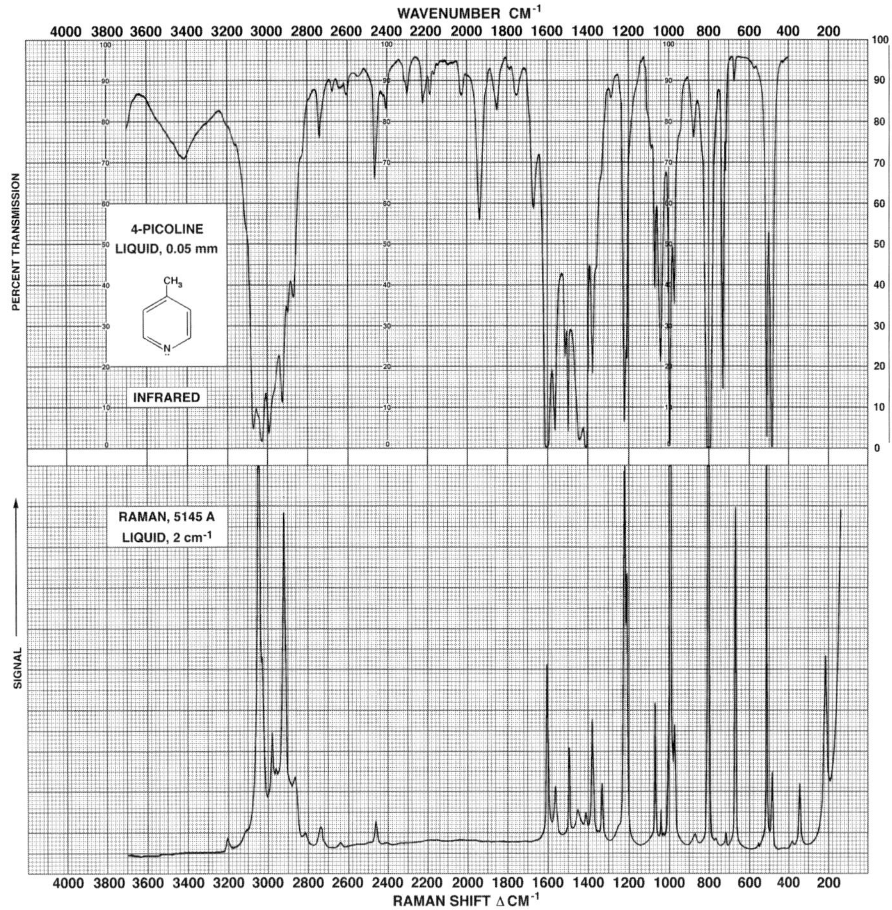

Fig. 5.28 Infrared and Raman spectra of 4-picoline.

Moving to lower wavenumbers, the next absorption occurs as a series of four weak bands in the region 2000-1700 cm^{-1}. These bands occur in the region where aromatic systems have weak sumtone and overtone patterns which have their origin in the out-of-plane ring C—H bends. Thus, the presence of this series of bands is more evidence that at least some of the C—H absorption above 3000 cm^{-1} may be from an aromatic system. For the moment this region will be left but it will be returned to later to see if any further structural information can be extracted from it.

A strong band is observed near 1630 cm^{-1}. This band is somewhat high to belong to an aromatic system and falls in the low end of the range for an olefinic double bond stretch. The high intensity and low wavenumber value suggest that the double bond is unsymmetrically substituted and likely

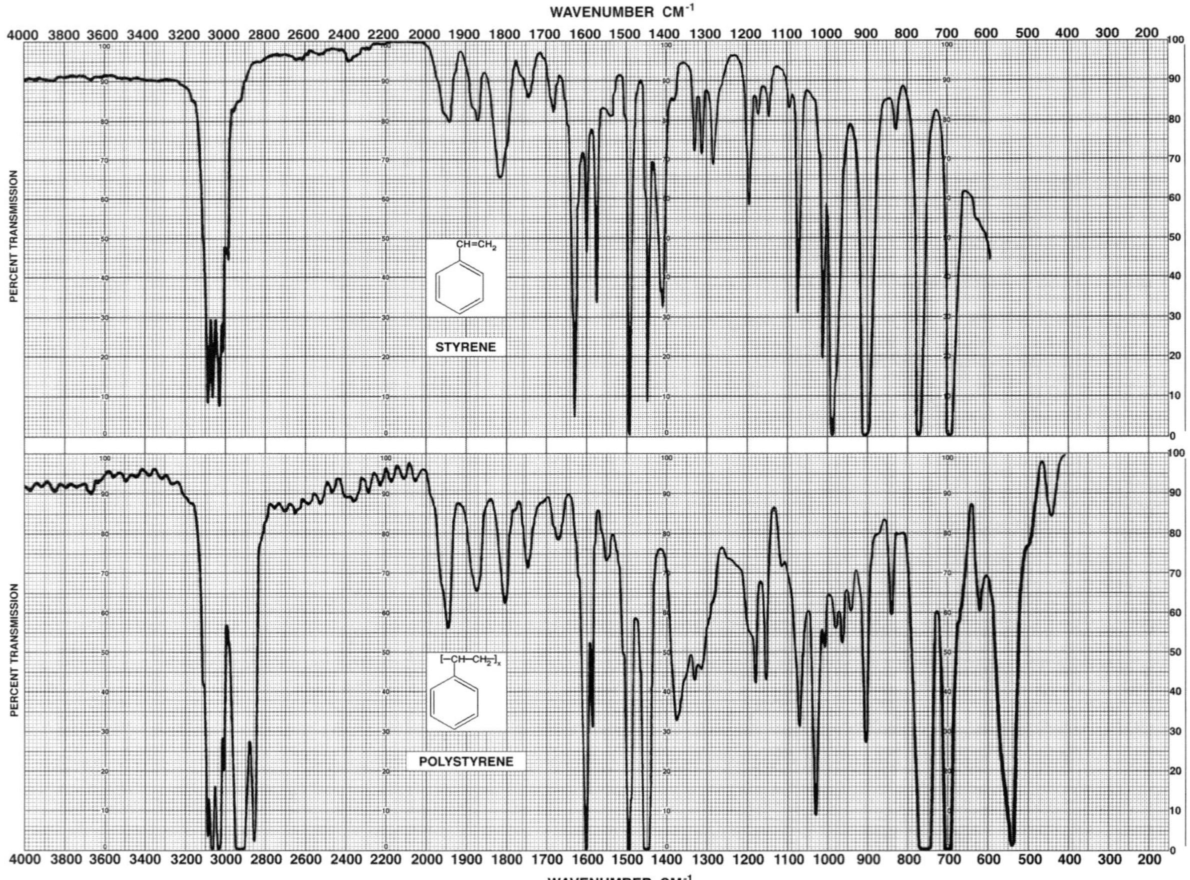

Fig. 5.29 Infrared spectra of styrene and polystyrene.

conjugated. The information up to this point is consistent with an aromatic ring substituted by a double bond.

Next a pair of bands near 1600 and 1575 cm^{-1} are observed. These bands are further evidence for the presence of an aromatic ring which now is likely to be a benzene type system and these bands can be assigned to the in-plane ring stretches (ν_{8a} and ν_{8b}). The 1575 cm^{-1} band is more intense than the 1600 cm^{-1} band. This intensity relationship is further evidence that conjugation is occurring and that it *must* involve the ring.

Just below ν_{8a} and ν_{8b} another pair of intense bands at 1493 and 1450 cm^{-1} are found. These bands are often observed in benzene ring derivatives and can be assigned to ν_{19a} and ν_{19b} in-plane ring stretches. The evidence now strongly suggests that a benzene ring is part of the structure.

Below 1450 cm^{-1} is a series of weak to medium bands running down to nearly 1000 cm^{-1}. Many of these bands have their origin in the in-plane C—H ring bending modes, but they are highly coupled and do not allow for confident assignments. They will be ignored. As the presence of an unsymmetrically substituted >C=C< group is being seriously considered, the region between 1000-800 cm^{-1} should be examined with some care (see Chapter 4). Two very strong bands fall just below 1000 cm^{-1} at 990 and 910 cm^{-1}. These bands can be assigned to the out-of-plane C—H bends of a vinyl group. Thus, it appears that tentatively it is possible to project a partial structure with a benzene ring substituted by a vinyl group.

As it is now suspected that a substituted benzene ring is likely to be present, attention is turned to the region 900–650 cm^{-1} where the out-of-plane C—H bends and the ring puckering vibration occur. There are two remaining intense bands which appear in the spectrum in this region near 770 and 695 cm^{-1}. These bands are characteristic of mono or meta substitution of the ring.

To decide between these two possibilities the weak series of bands falling between 2000–1700 cm^{-1} is again examined. The concern now is with the pattern of absorption, not the exact frequencies in this region. On comparison the pattern is much closer to that of a mono-substituted ring (four weak bands in a row with slightly decreasing intensity going from high to low wavenumber values) than to a meta-substituted ring, except for the region around 1800 cm^{-1} where the band appears more intense than would be expected. This unexpected intensification can be explained. If it is assumed that the argument for monosubstitution is correct then the structure must be vinylbenzene (styrene). If that is the case and a vinyl group is part of the structure, one of the out-of-plane C—H bends of this group that falls near 910 cm^{-1} has been identified. This mode is known to possess a weak overtone near 1820 cm^{-1} which could lead to further intensification of the ring sumtone band occurring in this region. Thus, a monosubstituted benzene structure with a vinyl group satisfies the spectrum completely and the starting monomer is identified as styrene.

C. The IR spectrum of the polymer.

The spectrum of polystyrene is given in the lower spectrum in Figure 5.29.

The analysis starts at the high wavenumber end of the spectrum. On examination of the 3000 cm^{-1} region it is found that the region looks rather different than that found in styrene. There is still considerable absorption above 3000 cm^{-1} and this would be expected because a benzene ring is in the repeat unit of polystyrene. Now, however, intense absorption is also observed below 3000 cm^{-1} in the region 2930–2850 cm^{-1}. This indicates that aliphatic C—H stretch is now, as expected, present in the system because in the polymerization process both methylene and methine repeat units were formed. Thus, the C—H stretching region is consistent with what would have been anticipated.

On moving to the 2000–1700 cm^{-1} region the sumtone–overtone pattern observed is much closer to that which would have been expected for a monosubstituted benzene ring than was observed in the case of styrene. This change can be explained by recognizing that the vinyl group which gave rise to the intensity anomaly has disappeared in the polymerization reaction. Again, what is observed in the spectrum is consistent with the changes in the monomer on polymerization.

The ring stretches ν_{8a} and ν_{8b} are again found (1605 and 1585 cm^{-1}) near where they were in styrene. However there has been a major intensity reversal because ν_{8a} is now significantly more intense than ν_{8b}. Again, this consistent with the loss of conjugation during reaction of the vinyl group in the polymerization process.

Next it is possible to identify ν_{19a} near 1495 and ν_{19b} at 1455 cm^{-1}, which is again consistent with the presence of a benzene ring system. Note that the 1455 cm^{-1} band is considerably broader than in styrene itself. This band broadening also can be explained as a result of the polymerization. When the vinyl group polymerizes, formation of methylene groups occur in the repeat unit. The scissoring mode of the methylene group occurs very close to 1455 cm^{-1} and it is this accidental degeneracy that is responsible for the increased band width.

Again the region between 1400 and 1000 cm^{-1} is ignored because there are no group frequency modes found here.

The most intense band in the spectrum is found near 765 cm^{-1} and a companion band near 698 cm^{-1}. These bands are again identical to those modes observed in the monomer and are consistent with a monosubstituted ring system.

The spectrum of the polymer is significantly different from that of the monomer and accurately reflects the chemistry that has occurred during the polymerization. Clearly, the use of infrared group frequencies is a powerful route to understanding the transformations that take place during chemical reactions.

A summary of the more important aromatic ring (phenyl) group frequencies is given in Table 5.4.

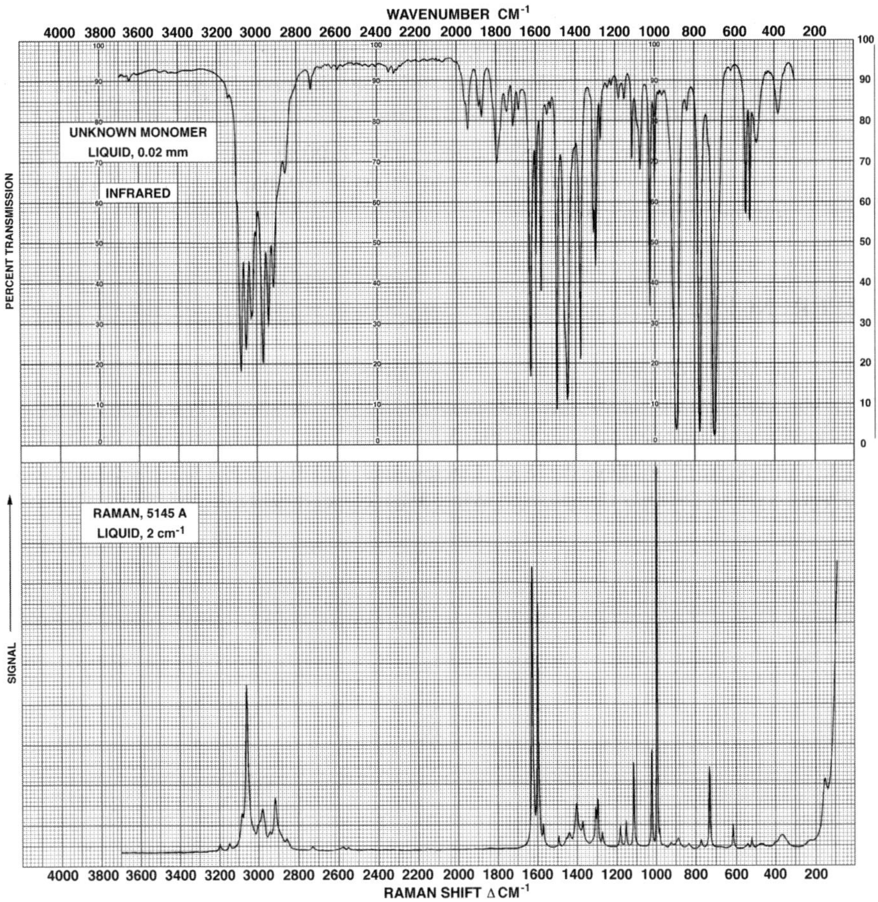

Fig. 5.30 Infrared and Raman spectra of an unknown simple monomer.

TABLE 5.4 Summary of the More Important Aromatic Ring Group Frequencies

Arene Fundamentals	$\tilde{\nu}$ cm^{-1}
C—H stretch	3200–3000
C=C ring stretch (ν_{8a})	1600 ± 10
C=C ring stretch (ν_{8b})	1580 ± 10
C=C ring stretch (ν_{19a})	1500 ± 10
C=C ring stretch (ν_{19b})	1450 ± 10
C—H out-of-plane bend (1H)	900–860
C—H out-of-plane bend (2H) 2 adjacent C—Hs	860–800
C—H out-of-plane bend (3H) 3 adjacent C—Hs	810–750
C—H out-of-plane bend (4H)	770–735
C—H out-of-plane bend (5H)	770–730
C—C ring out-of-plane bend (1-; 1,3-; 1,3,5-substituted)	690 ± 10
C—H out-of-plane bend sum tones	2000–1650

D. Unknown sample analysis (see Figure 5.30):

The reader is now asked to review the material in Chapters 1–5 and to undertake the interpretation of the unknown spectrum shown in Figure 5.30. Both the infrared and the Raman spectrum are given. After the assignments have been made and a proposed structure developed, readers may check their answers by referrng to Chapter 15 where a detailed interpretation is given.

REFERENCES

1. Bellamy, Vol. 1, Chapter 5.
2. G. Varsanyi, *Vibrational Spectra of Benzene Derivatives*, Academic Press (1969). Esp. pp. 141–397, and summary table on p. 394–5.
3. (Raman spectra) F.R. Dollish, W.G. Fateley, and F.F. Bentley, *Characteristic Raman Frequencies of Organic Compounds*, Wiley-Interscience (1974), p. 162–283.

Introduction to Exercises

ASSIGNMENT OF EXERCISES

In the exercise sections the reader will find a number of cases where spectra are distributed as pairs of IR (FT grating quality) and Raman (mostly argon ion excitation) spectra. In addition, there are a few examples of prism spectra, which are introduced to acquaint the student with the appearance of earlier spectra. These spectra are presented as problems in identification and structural analysis to illustrate by practical examples the procedures followed in the interpretation of vibrational spectra.

The exercises are placed at three stages in the development of the text and are based on the material which has been discussed in the precedings chapters. Section I covers material in Chapters 1–5, Section II material from Chapters 1–9, and the largest section, Section III, is based on all the group frequency material in Chapters 1–12.

In the large majority of cases the pairs of IR and Raman spectra are presented on one page on the same scale in reciprocal centimeters, with both spectra in a pair having the same identification number. Those IR spectra that are not accompanied by Raman spectra may be presented two to a page. Each pair of IR and Raman spectra and each unpaired IR spectrum constitutes a problem. The objectives of these problems are explained in the adjacent discussion.

Exercise Section I

In the first three exercises in this section the spectral data are presented as pairs of IR and Raman spectra set to the same scale. The fourth exercise utilizes only infrared data.

Exercise 1 (Curves 9-C, 8-B, 8-C, 9-B, and 3-C.) These problems are an introduction to the identification of pure hydrocarbons by the use of the group frequencies appearing in the IR and Raman spectra. The exercise contains the spectra of five

Course Notes on the Interpretation of Infrared and Raman Spectra, by Dana W. Mayo, Foil A. Miller, and Robert W. Hannah.
ISBN 0-471-24823-1

compounds that can be satisfactorily assigned to a variety of alkene structures based on their vibrational spectra.

Exercise 2 (Curves 6-A, 18-B, and 6-C.) This expands the hydrocarbon spectral data to include three problems involving aromatic systems.

Exercise 3 (Curves 16-A and 14-B.) This contains two additional hydrocarbon spectra. The spectra of sample 16-A are of a low-molecular-weight hydrocarbon, while the spectra of unknown sample 14-B are of a hydrocarbon polymer film.

Exercise 4 (Curves fractions A & B and 14-C.) This consists of two problems in which only the IR spectra are available and in which the unknown sample structures depart for the first time from being hydrocarbons. Both unknowns have nitrogen incorporated into the hydrocarbon structure. The first problem involves relatively low molecular weight molecules, while the second problem considers a high-molecular-weight polymer film.

On working through the problems, when sufficient information points to the compound's identity, the proposed structure should be checked against hard copy in one of the reference collections and then the reader should refer to the detailed discussions in the Answers to the Exercises at the end of the book.

Exercise Section II

This section consists of six exercises. In the first five exercises the spectral data are presented as pairs of IR and Raman spectra set to the same scale. Exercise 10 utilizes very early prism IR data.

Exercise 5 (Curve 2-B.) This problem involves the identification of a compound of simple molecular structure. The major group frequency assignments for this system were discussed in Chapter 5.

Exercise 6 (Curve 12-C.) This problem is another example of the application of group frequencies to the elucidation of the structure of a high-molecular-weight polymer.

Exercises 7 and 8 (Curves 1-B & 19-B and 3-B.) These curves contain three unknown samples that possess carbonyl groups located in different molecular environments.

Exercise 9 (Curve 2-A.) This unknown is another example of a pure organic liquid. In this case the reader may find that the absence of a band is more important than the presence of a band in determining the molecular structure.

Exercise 10 (Curves M-12 and M-40.) These spectra contain examples of two practical problems from the organic research laboratory.

As in Exercise Section I, when sufficient information points to the compound's identity, the proposed structure should be checked against hard copy in one of the reference collections and then the reader should refer to the detailed discussions in the Answers to the Exercises.

Exercise Section III

This set of nine exercises consists of spectra representative of all areas of group frequencies discussed in the text. There are also a few exercises (Exercises 15–17) that involve spectra of materials not discussed directly in the text material. They will allow the reader to tackle new spectral data using the approach developed in the text.

Exercise 11 (Curves 3-A and 5-C.) These spectra contain two sets of IR and Raman spectra that have path lengths (see experimental data) that are only consistent with gas-phase spectra in the IR. As vapor pressure values are also given with the Raman data for these samples, it would appear that the IR and Raman spectra of both materials were obtained in the gaseous state.

Exercise 12 (Curves 1-A and 20-C.) Here are two sets of IR and Raman spectra of pure substances that possess polar functional groups other than carbonyl groups.

Exercise 13 (Curves "benzene?" and "di-*N*-butyl amine?") This exercise consists of two unrelated "real-life" problems which involve assessing the accuracy and/or purity of the samples. Only the IR spectra are available for the determination.

Exercise 14 (Curves 6-B and 19-A.) These unknowns are pure samples that have spectra representative of the more difficult, but still interpretable, spectra of complex molecules.

Exercise 15 (Curve 18-A.) This exercise contains a set of IR and Raman spectra of a relatively polar organic solvent.

Exercise 16 (Curves 21-A and 21-B.) Here we have early examples of IR prism reference standards. These samples are examples of organic esters of inorganic acids (common mineral acids). There are no Raman spectra.

Exercise 17 (Curves 7-A and 7-B.) This problem contains another set of two early prism IR spectra. The samples are both inorganic materials. There are no Raman data available.

Exercise 18 (Curves 15-C & 17-C, 16-C & 16-D, 10-C, and 17-A.) This exercise contains a set of four different polymer samples plus a polymer prototype molecule (17-A).

Exercise 19 (Curves 34E and 35E.) This final exercise involves another set of polymers, but with both the Raman and IR data available to aid in the interpretation.

As in Exercise Sections I and II, when sufficient information points to the compound's identity, the proposed structure should be checked against hard copy in one of the reference collections and at that point the reader should refer to the detailed discussions in the Answers to the Exercises.

Exercise Section I

EXERCISE 1

Spectra Index Nos. 9-C, 8-B, 8-C, 9-B, **and** 3-C

Description of Samples: All are samples of olefinic (alkene) hydrocarbons with boiling points in the C_6 range whose spectra were obtained as pure liquids.

Objective: Identify the kinds of olefinic unsaturation and, if possible, the specific compounds whose spectra are shown. Note especially the difference between the intensities of the characteristic group frequencies in the IR and Raman spectra. The basic structural features of each compound can be determined from its group frequencies, but final identification often may not be made in this way alone. Using the structural units suggested by the group frequencies (such as cis double-bond, terminal methylene, vinyl, isopropyl, etc.), propose compounds in the indicated molecular weight range which fit the suggested structures.

Once the spectrum has been established by comparison to catalog data such as the *Coblentz Society Desk Book* by C. D. Craver (Part 1 in the Bibliography) or the *Aldrich Handbook*, refer to the Answers to the Exercises for a detailed discussion.

Proposed Structures

9-C?
8-B?
8-C?
9-B?
3-C?

Comments

Course Notes on the Interpretation of Infrared and Raman Spectra, by Dana W. Mayo, Foil A. Miller, and Robert W. Hannah.
ISBN 0-471-24823-1

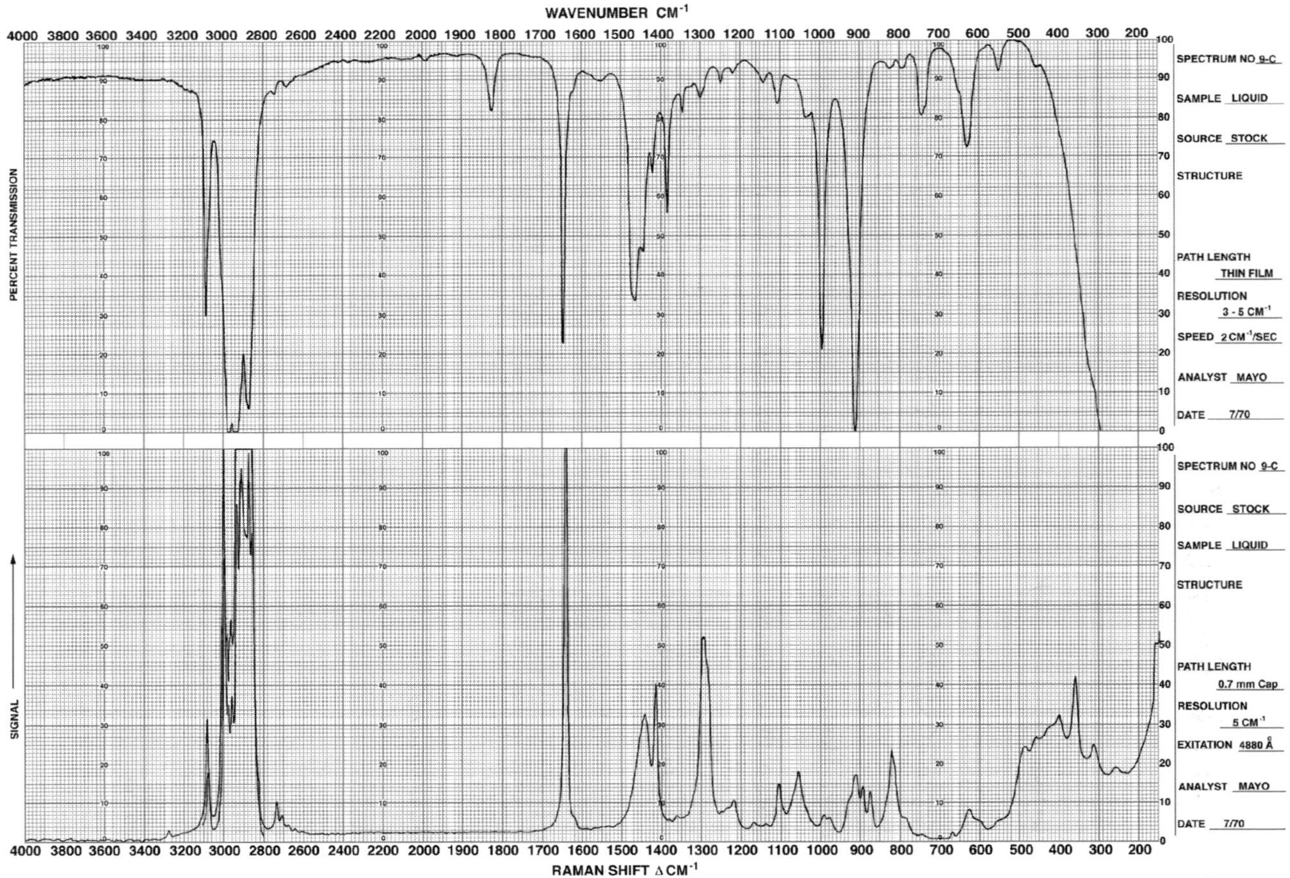

Fig. 1 Unknown 9-C.

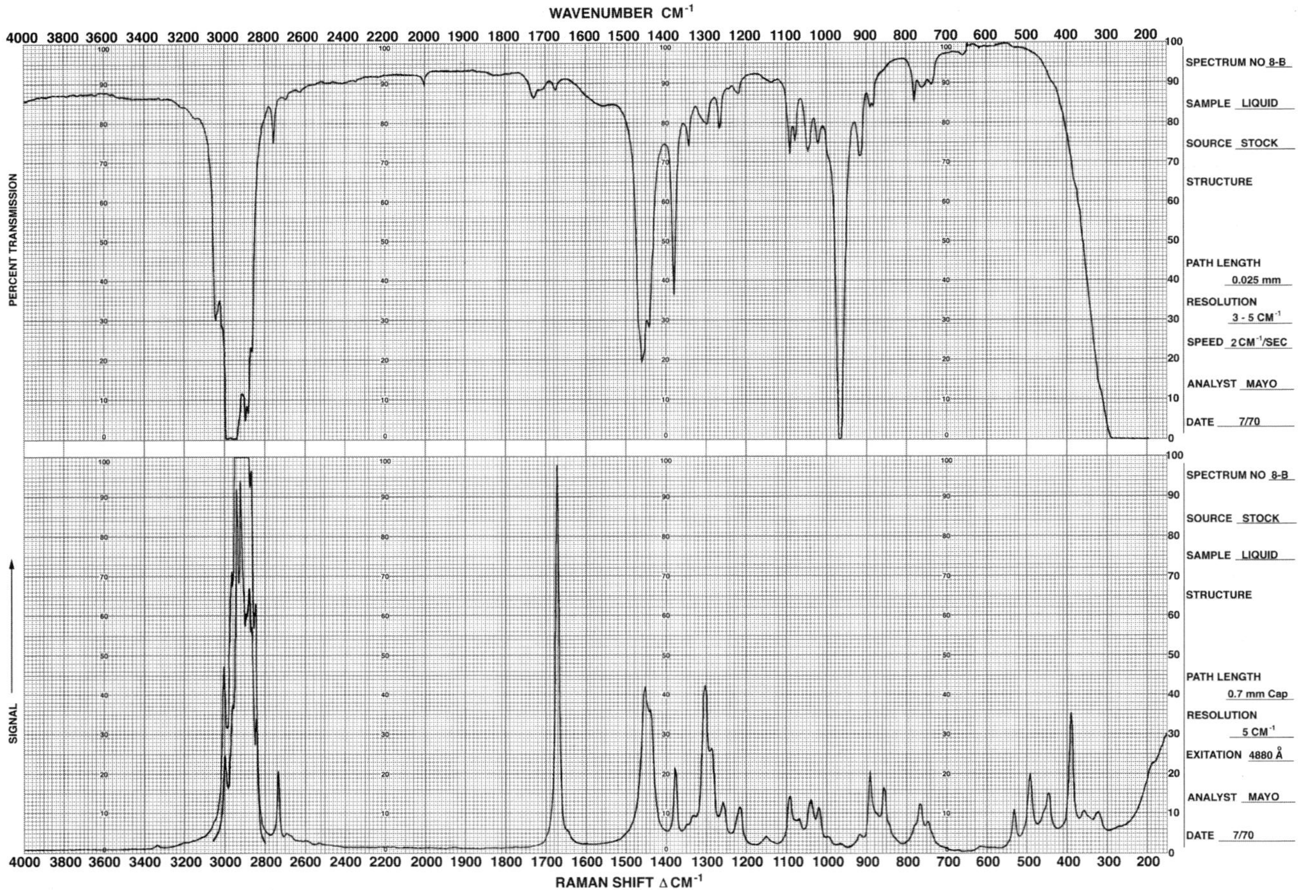

Fig. 2 Unknown 8-B.

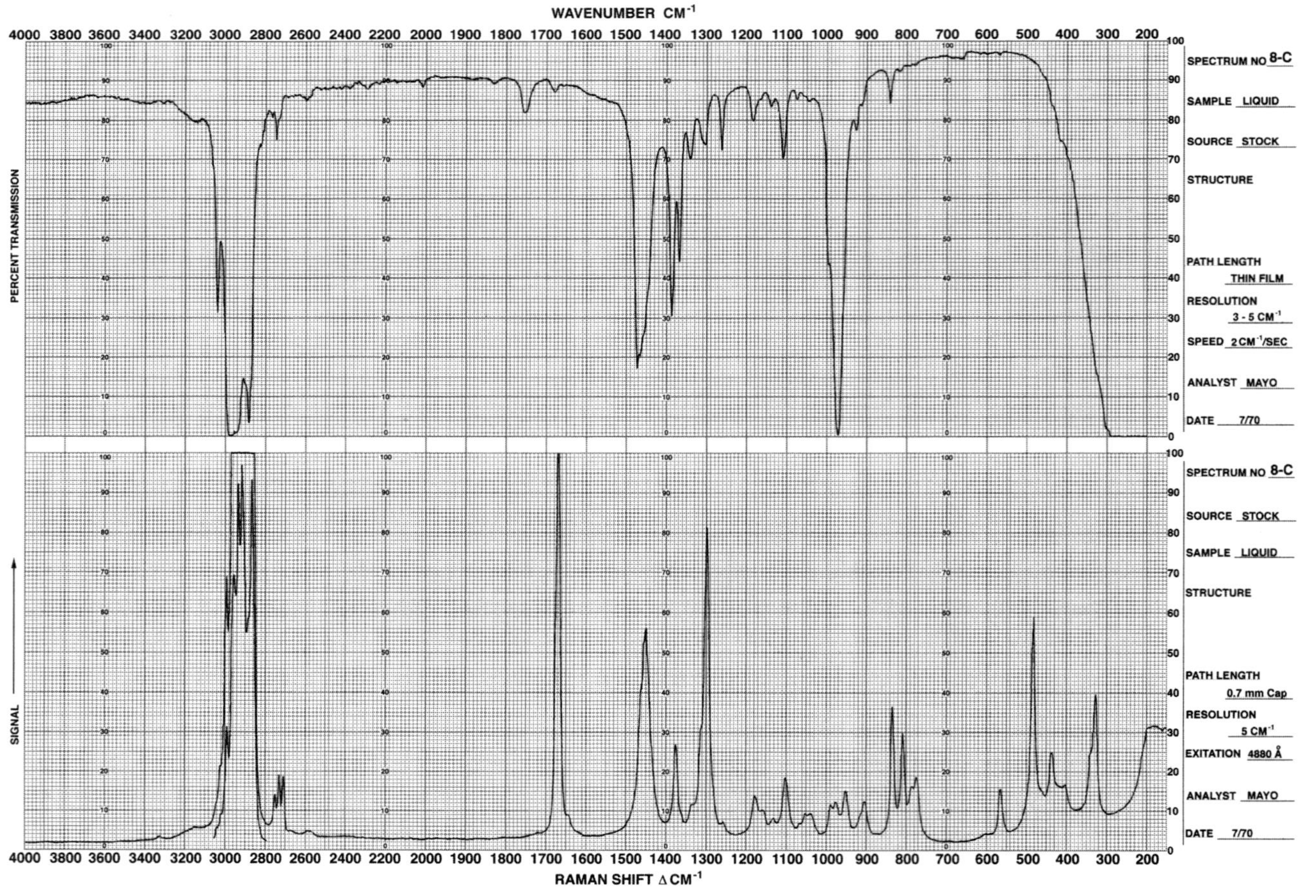

Fig. 3 Unknown 8-C.

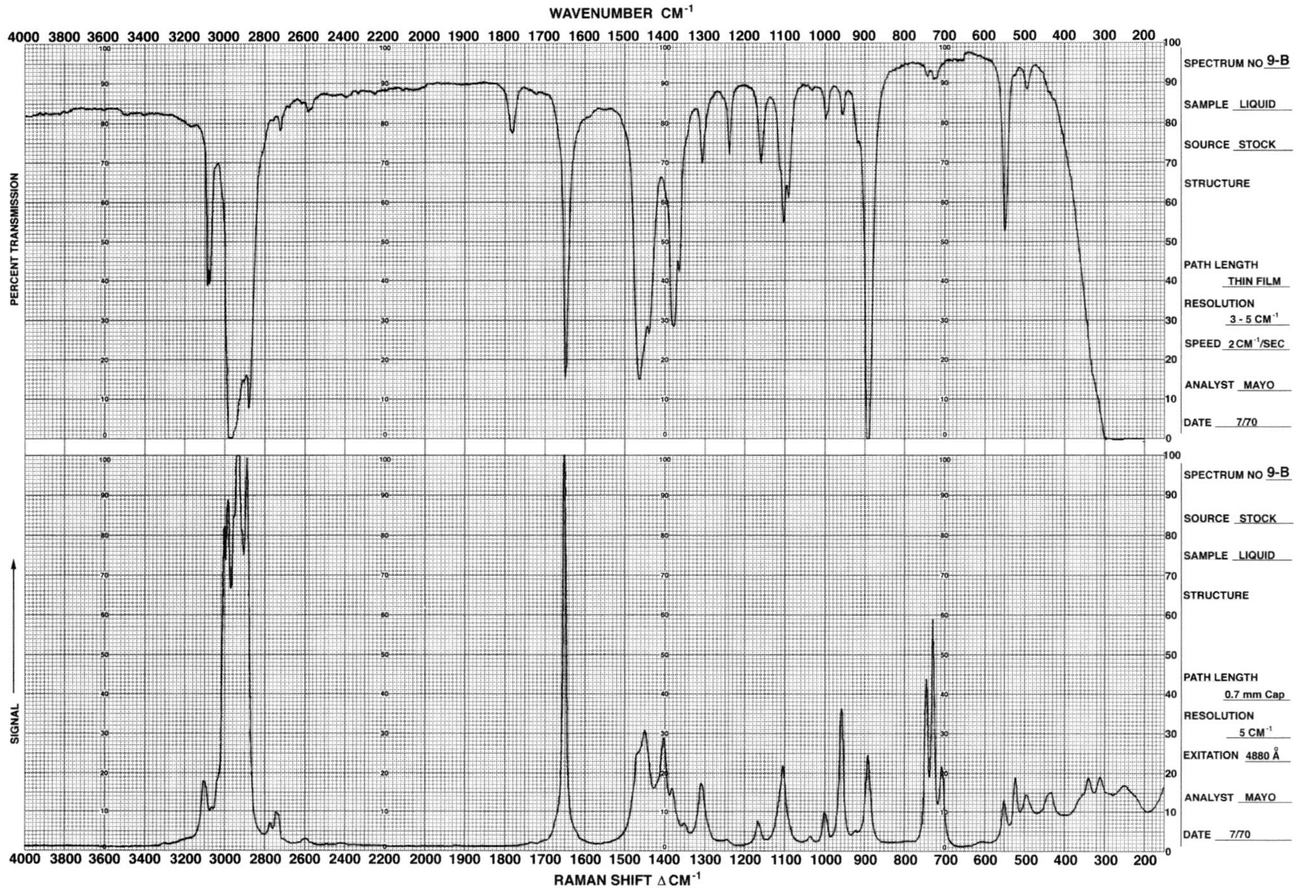

Fig. 4 Unknown 9-B.

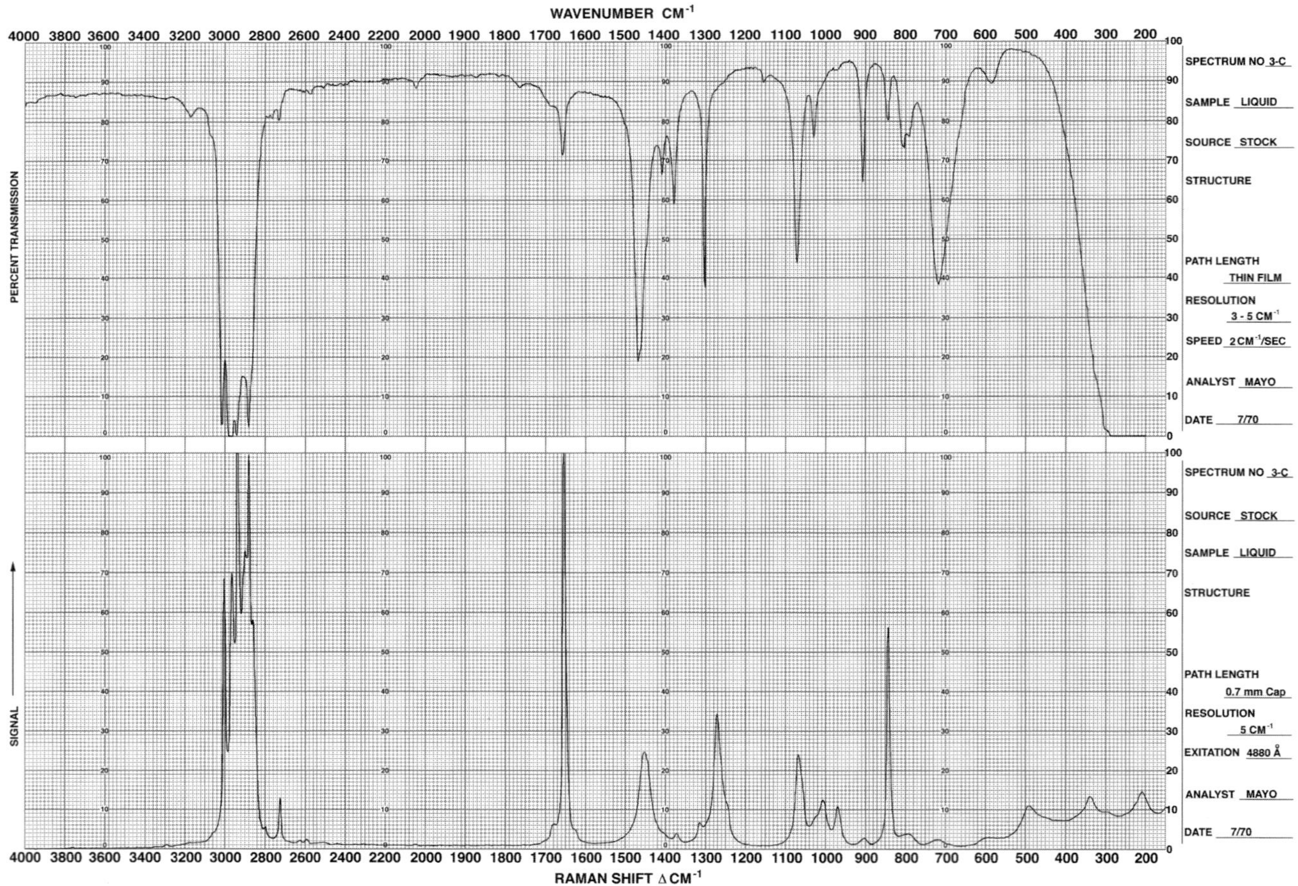

Fig. 5 Unknown 3-C.

EXERCISE 2

Spectra Index Nos. 6-A, 18-B, **and** 6-C

Description of Samples: These sets of IR and Raman spectra are those of pure organic liquids. All three compounds were shown to be hydrocarbons by elemental analysis.

Objective: Identify the compounds so far as possible with the help of their characteristic frequencies as discussed in Chapters 1–5 and correlation charts.

Once the sample has been identified by comparison with catalog data such as the *Coblentz Society Desk Book* or the *Aldrich Handbook*, refer to the Answers to the Exercises for a detailed discussion.

Proposed Structures

6-A?
18-B?
6-C?

Comments

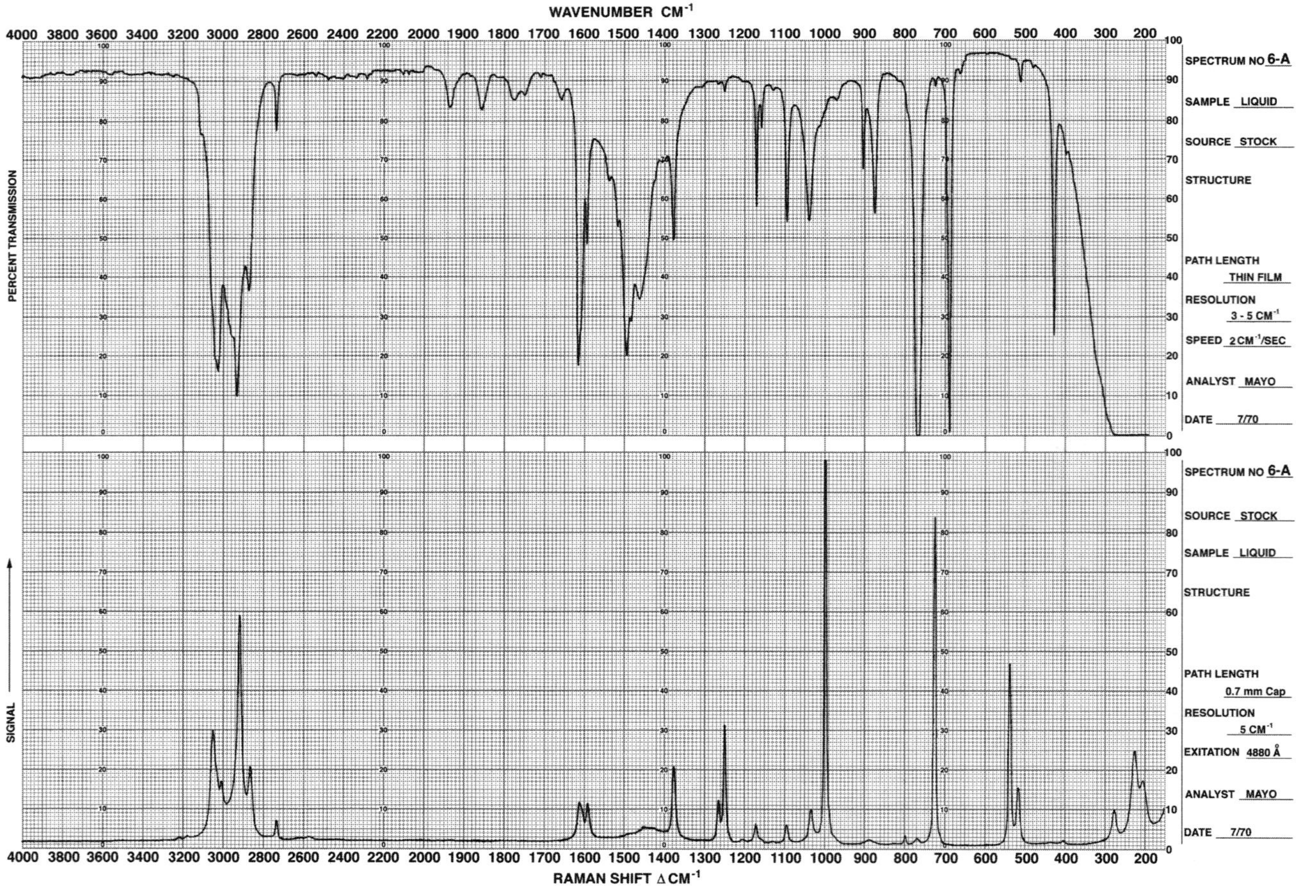

Fig. 6 Unknown 6-A.

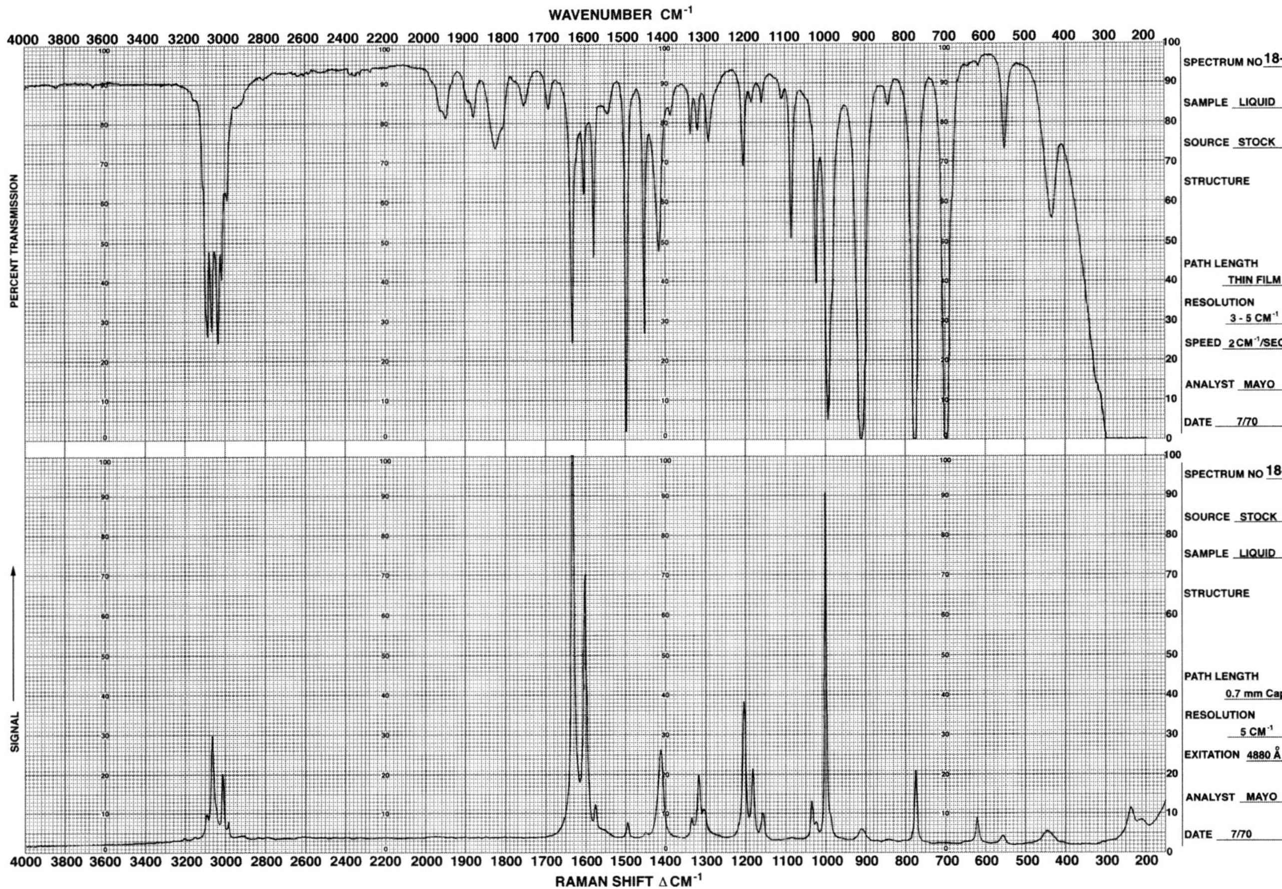

Fig. 7 Unknown 18-B.

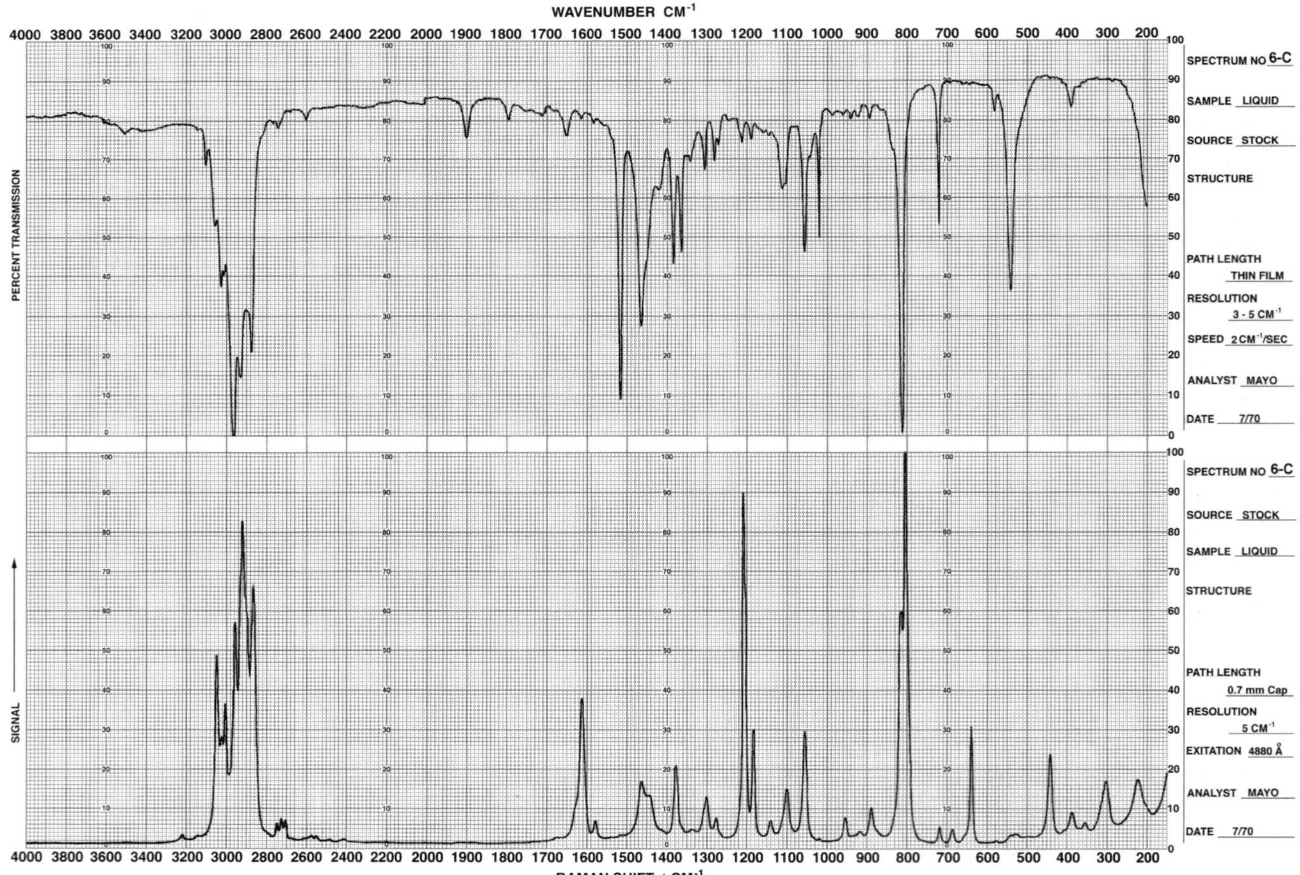

Fig. 8 Unknown 6-C.

EXERCISE 3

I. Spectrum Index No. 16-A

Description of Sample: The sample of spectrum 16-A was obtained a low-boiling liquid by-product, **IV** (11% yield), of the pyrolysis of methyl *n*-propyl isoamyl amine oxide (**I**). Other products from the pyrolysis were propylene (**II**) and isoamylene (**III**). *Molecular weight data obtained on the unknown gave a value of 70, which among other things indicated that the sample was unlikely to contain nitrogen.*

Once the spectrum has been established by comparison to catalog data such as the *Coblentz Society Desk Book* or the *Aldrich Handbook*, refer to the Answers to the Exercises for a detailed discussion.

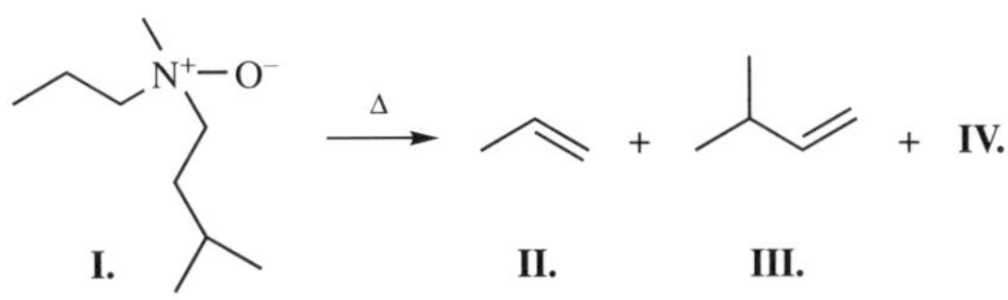

I. **II.** **III.**

Objective: Identify compound **IV** with the help of its characteristic frequencies.

Proposed Structure for Reaction Product

16-A?

Comments

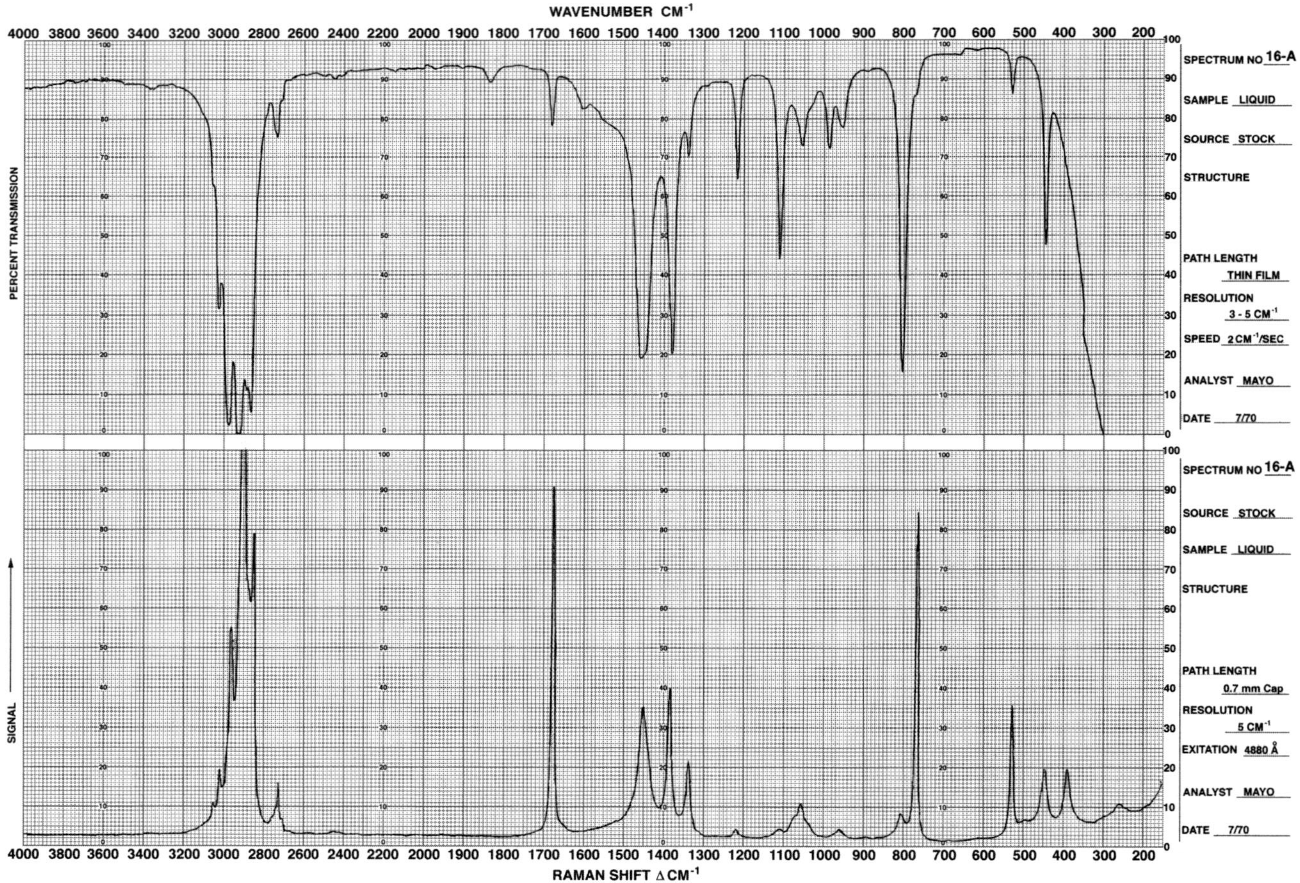

Fig. 9 Unknown 16-A.

II. Spectrum Index No. 14-B

Description of Sample: Curve 14-B is the spectrum of a standard well-known high-molecular-weight polymer. It was obtained with a grating instrument.

Objective: Identify the functional groups in the polymer film from its spectrum and infer the monomeric units from which the polymer was formed.

After a monomer has been proposed, check the tentative structure by referring, for example, to Hummel's *Atlas of Polymer and Plastics Analysis*, Vol. I, if it is available to you. Once the spectrum has been established by comparison, refer to the Answers to the Exercises for a detailed discussion.

Proposed Structure (Both Monomer and Polymer)

14-B?

Comments

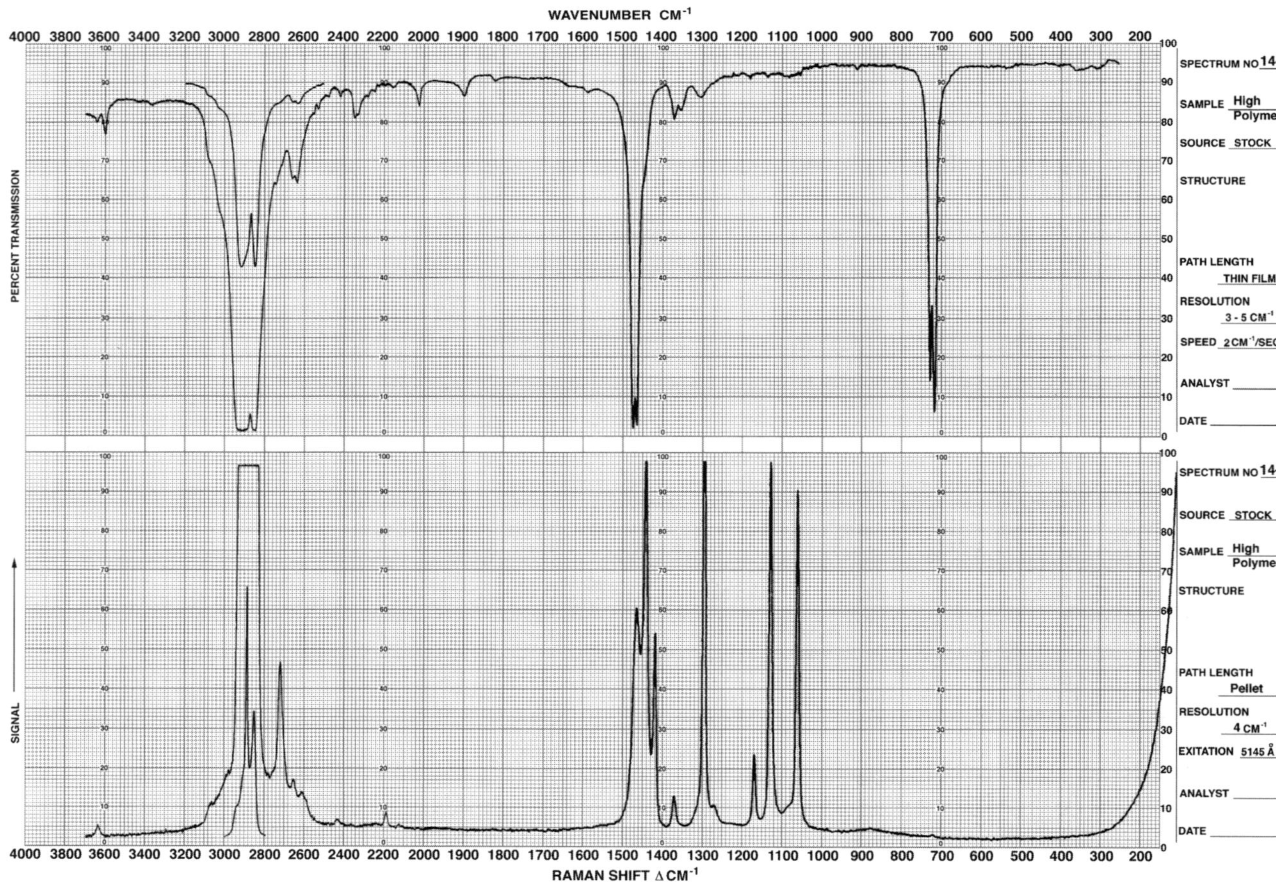

Fig. 10 Unknown 14-B.

EXERCISE 4

I. Spectrum Index Nos.: Fractions A and B

Description of Sample: Liquid mixture of reaction products which possess structures **I** and **II**. Upon distillation fractions A and B were obtained.

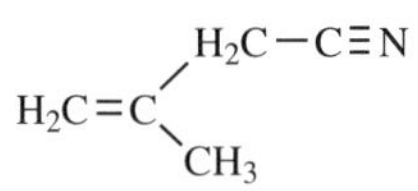

Structure **I**

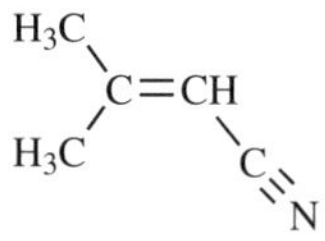

Structure **II**

Objective: Which fraction has structure **I** and which fraction has structure **II**, and why?

Answer

	Fraction A	Fraction B
Structures		

Comments

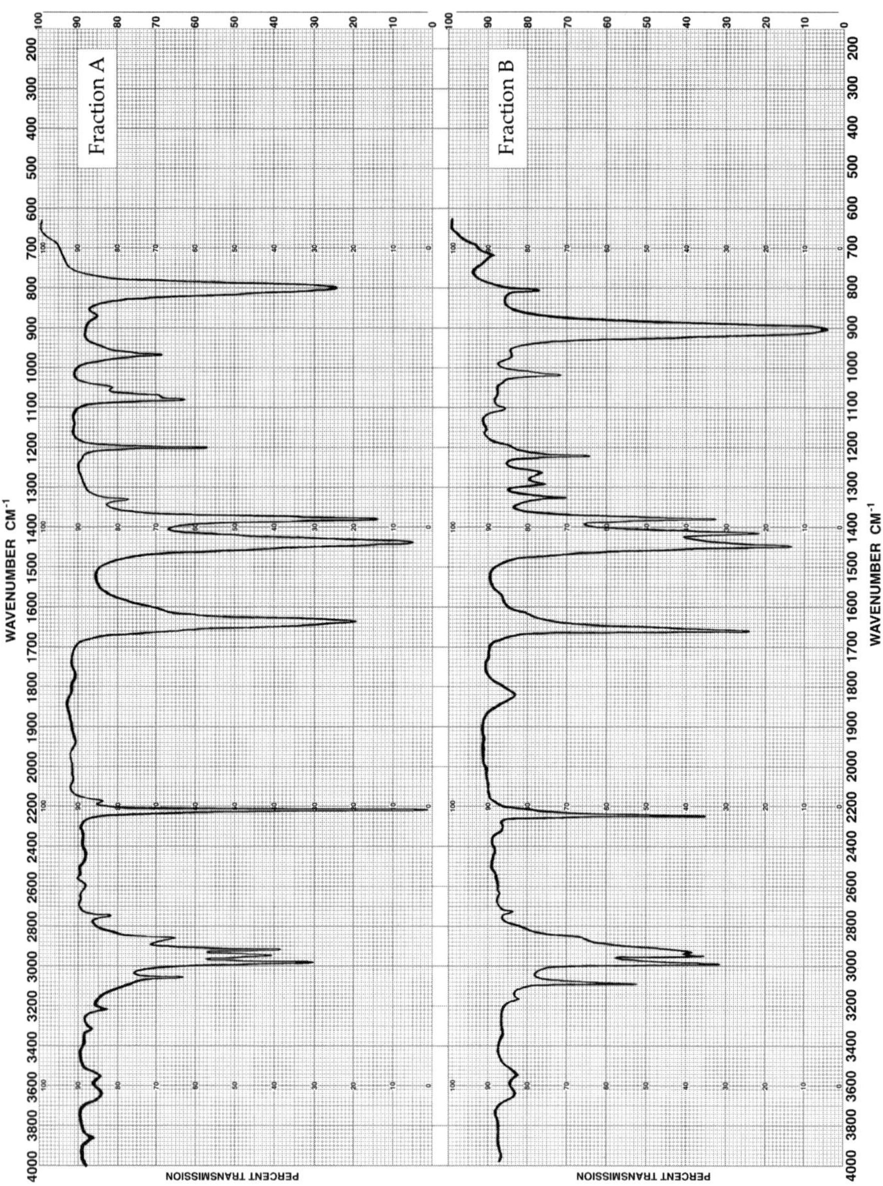

Fig. 11 Unknown fractions A and B.

II. Spectrum Index No. 14-C

Description of Sample: High polymer.

Objective: Identify the functional groups in the polymer film from its spectrum and infer the monomeric units from which the polymer was formed. The polymer spectra in the exercises have been arranged in order of increasing structural complexity.

After a monomer has been proposed, it is possible to check the tentative structure by referring, for example, to Hummel's *Atlas of Polymer and Plastics Analysis*, Vol. I, if it is available to you. Once the spectrum has been established by comparison, refer to the Answers to the Exercises for a detailed discussion.

Proposed Structure (Both Monomer and Polymer)

14-C?

Comments

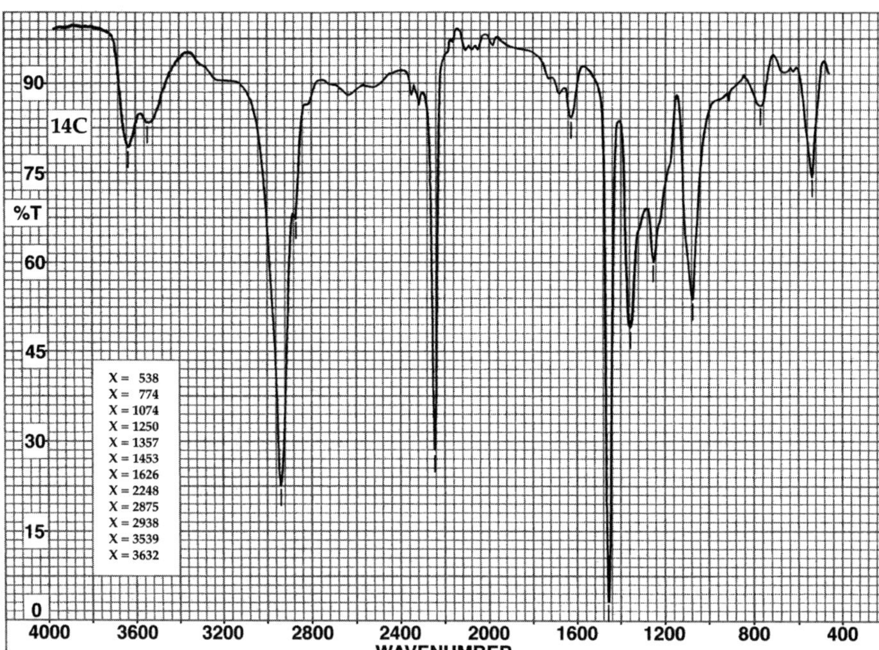

Fig. 12 Unknown 14-C.

6 Spectra of X–H Systems (With Emphasis on O–H and N–H Groups)

FOIL A. MILLER

I. General remarks on X–H stretches
  A. These occur in the high-frequency region, nearly all above 2000 cm^{-1}. The most important exceptions are stretches of B–H bridge bonds, as in diborane, which are at 2100–1600 cm^{-1}.
  B. Because X–H stretches are so high, they normally do not interact with any vibrations except other X–H stretches (e.g., two C–H groups on the same carbon atom).
  C. They are highly diagnostic and reliable. The stretching frequency of a *non-hydrogen-bonded* group X–H is different from that of any other group Y–H, so X can be identified reliably.
  D. Some X–H groups form hydrogen bonds in condensed states. This lowers the stretching frequency, sometimes dramatically. Fortunately hydrogen bonding can be recognized easily because it also produces great broadening and intensification of the bands.
  E. An X–H stretch is an end-atom vibration—i.e., the hydrogen is at the periphery of the molecule and has most of the amplitude because it is so much lighter than X.
    1. An X–H stretch is approximately a two-body harmonic oscillator, so

$$\tilde{\nu}(\text{X–H}) \approx \frac{1}{2\pi c}\sqrt{\frac{k}{\mu}}$$

where c = speed of light.

Course Notes on the Interpretation of Infrared and Raman Spectra, by Dana W. Mayo, Foil A. Miller, and Robert W. Hannah.
ISBN 0-471-24823-1

H-H
(4160)

→ Increasing frequency

Li-H 1360	Be-H 1960	B-H Terminal 2500 Bridge 1600-2100	C-H 2950 w s in Ram.	N-H 3400* s m in Ram.	O-H 3600* vs w in Ram.	F-H 3950*
Na-H 1133	Mg-H 1433	Al-H 1700-1900	Si-H 2180 s	P-H 2350 s s in Ram.	S-H** 2575 w s in Ram.	Cl-H 2886*
K-H 956	Ca-H 1260	Ga-H 1830	Ge-H 2120 s	As-H 2150	Se-H** 2300	Br-H 2560
Rb-H 908	Sr-H 1172	In-H 1425	Sn-H 1580	Sb-H 1890	Te-H 2070	I-H 2230

* - Approximate monomeric frequency. Greatly decreased by hydrogen bonding.
** - Smell suffices.

Fig. 6.1 Approximate values for X–H stretching frequencies, cm^{-1}.

μ = reduced mass = $(m_H \times m_X)/(m_H + m_X)$. Since m_H is usually much less than m_X, $m_H + m_X \approx m_X$. Therefore $\mu \approx m_H$ and is about the same regardless of the nature of X.

k = force constant. This is the variable parameter that changes significantly as atom X is changed.

2. In the periodic table, k increases and hence $\tilde{\nu}$ (X–H) increases as one:
   a. Goes to the right in any row
   b. Goes upward in any column

   Therefore atoms in the upper right corner have the highest X–H stretches.

F. Summary table. See Figure 6.1.

G. H_2 has the highest of all X–H frequencies (4160 cm^{-1}), but it is transparent in the IR. The H–F monomer is the next lower (3950 cm^{-1}), but it is seldom met in practice. This brings us to O–H and N–H. Their frequencies are strongly affected by hydrogen bonding, so we will make a short digression to describe its effects.

II. Hydrogen bonding

A. This is a weak bond which occurs in the system X—H······Y, where X and Y are strongly electronegative atoms. In practice they are only F, O, N, and Cl.
   1. The energy of a hydrogen bond is only 4–8 kcal/mol; that of a single bond is 40–110 kcal/mol. (17–34 kJ/mol vs. 170–460 kJ/mol).
   2. But a hydrogen bond produces striking changes in vibrational spectra.

B. Four vibrations of a hydrogen-bonded system, e.g. —O—H······O<
   1. O—H stretch 3400–2500 cm^{-1}, vs, vb
   2. R—O—H in-plane bend, 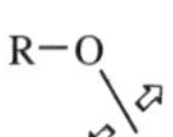 1400–1300 cm^{-1}, vs, b

3. R—O—H out-of-plane bend, or torsion 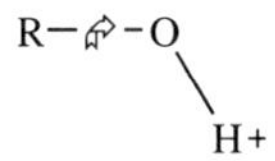 ~650 cm^{-1}
4. —O—H······O< stretch (stretch of the dotted bond) ~175 cm^{-}

(Items 3 and 4: Not useful group frequencies)

C. Analogous bands will occur for any X–H······Y system when X and Y are F, O, N, or Cl. Thus the important hydrogen-bonding systems are

O–H······O O–H······N O–H······Cl O–H······F
N–H······O N–H······N N–H······Cl N–H······F

D. Effects of H bonding on vibrational spectra
The stronger the H bond, the greater is each of the following effects.
1. Hydrogen bonding *lowers* the X–H stretch.

	Free X–H (usually gas, cm^{-1})	H-bonded X–H (liquid or solid, cm^{-1})
HF	3950	1450 $[F–H–F]^{-}$
Alcohols	3625 ± 25	~3350
–COOH acids	~3550	~3000
Amines	3400 ± 80	~3300

H–F is the most extreme case. In $[F–H–F]^{-}$ it produces the strongest hydrogen bond known, with a bond strength of 27 kcal (113 kJ)/mol.

2. Hydrogen bonding *raises* the X–H bends.
Example: CH_3OH 1345 cm^{-1} (vapor) vs. 1418 cm^{-1} (liquid)
Increased 73 cm^{-1}
3. Hydrogen bonding broadens all X–H bands, both stretches and bends.
   a. For O–H stretches, the half widths may become 300–500 cm^{-1}!
   b. *Compare the spectra of hexane and hexanol*; see Figures 6.2*a*,*b*. Note the huge change made by replacing one C–H by an O–H. All the O–H modes in hexanol have been broadened and intensified by hydrogen bonding. Note especially the enormously broad O–H stretch at ca. 3370 cm^{-1}.
4. Hydrogen bonding greatly intensifies all X–H bands, both stretching and bending. One sees this most clearly by noting the band area—the integrated intensity.

E. It is risky to do quantitative analysis using a band that is affected by hydrogen bonding because both its position and intensity are affected by dilution. See Figures 6.3*a*,*b*.

We now leave hydrogen bonding and return to O–H group frequencies.

III. O–H group frequencies
A. O–H stretch
1. Free O–H—i.e., *not* hydrogen bonded
   a. This requires either the gas phase or an extremely dilute solution in an inert solvent (<0.001 M).

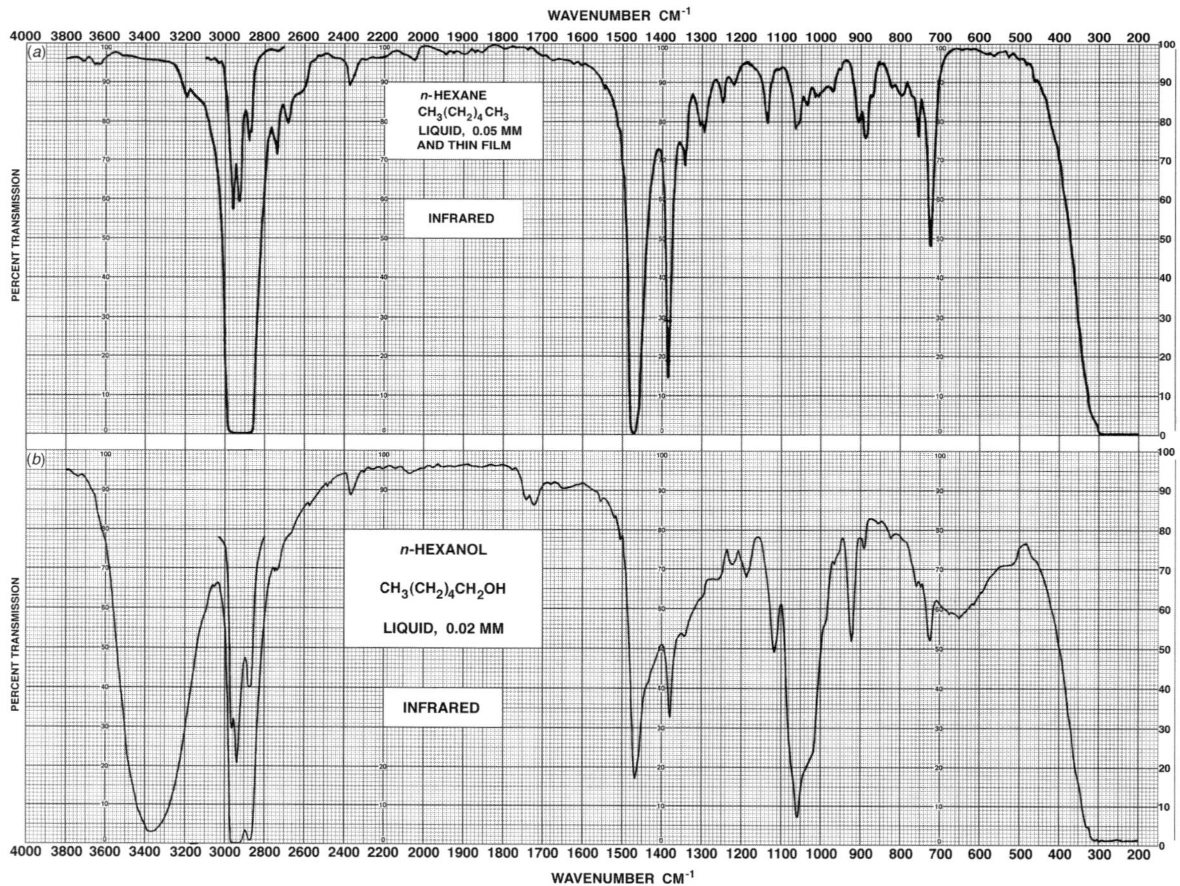

Fig. 6.2 Infrared spectra of (*a*) *n*-hexane and (*b*) *n*-hexanol.

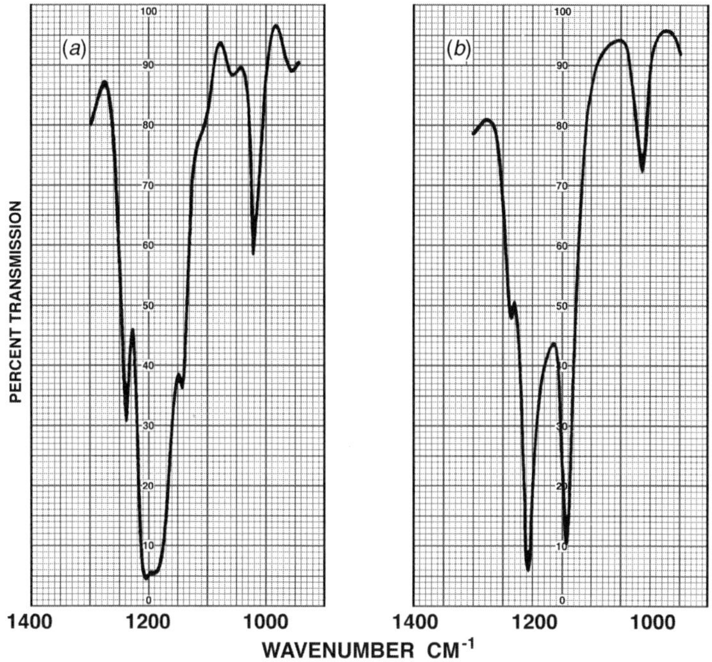

Fig. 6.3 Infrared spectrum of *tert*-butanol at two different dilutions to illustrate effect of variable hydrogen bonding: (*a*) 10% CS_2 solution in a 0.1-mm cell; (*b*) 1% CS_2 solution in a 1-mm cell.

b. A free O–H stretch is the highest fundamental known other than for H_2 or the HF monomer. Any *fundamental* IR band above 3500 cm^{-1} in an unknown can be confidently assigned to the —OH group.

c. The band is sharp and varies within narrow limits according to the nature of the OH:

Primary alcohols	∼3640 cm^{-1}	
Secondary alcohols	∼3630 cm^{-1}	
Tertiary alcohols	∼3615 cm^{-1}	
Phenols	3600 ± 15 cm^{-1}	The higher the acidity,
–COOH acids (monomers)	3520 ± 20 cm^{-1}	the lower the frequency.

d. But one seldom meets a sample with a free OH. An example is an OH group which is sterically blocked so that it cannot hydrogen bond. Then its stretch is high and sharp. *Examples:*
2,6-Di-*t*-butylphenol: 3645 cm^{-1}.
2,6-Di-*t*-butylcresol: 3640 cm^{-1}. See Figure 6.4.

2. Hydrogen-bonded O–H stretch
   a. Can come anywhere between 3450 and 2500 cm^{-1}

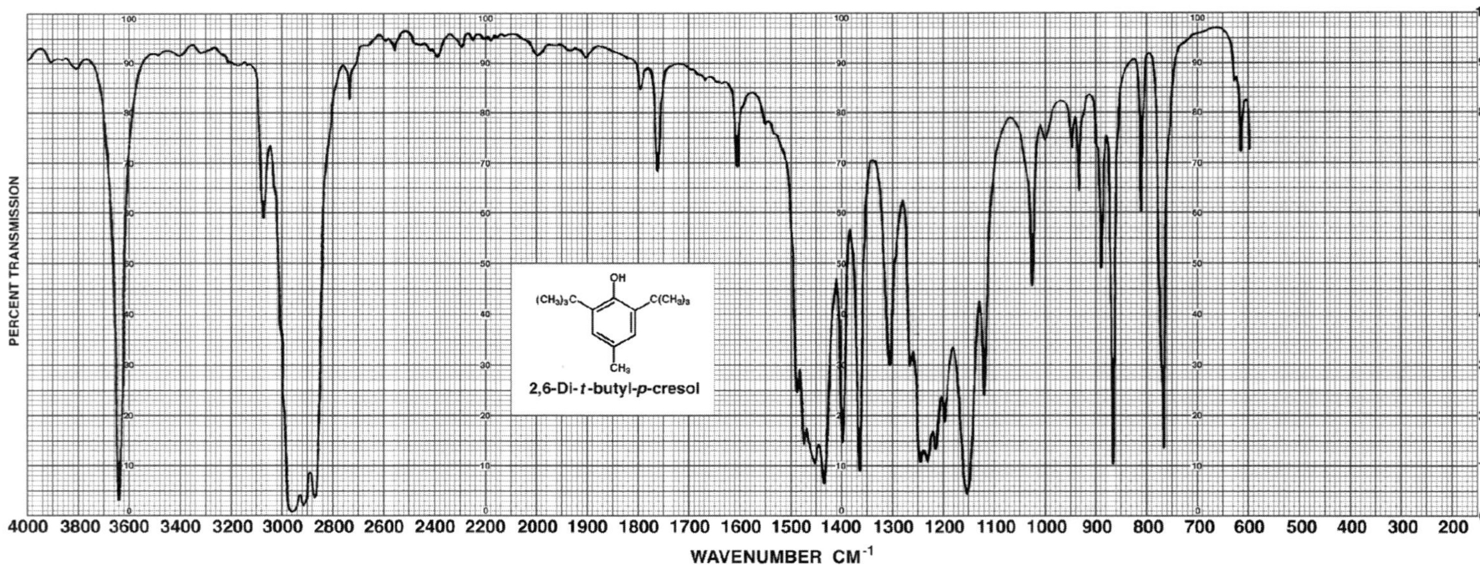

Fig. 6.4 Infrared spectrum of 2,6-di-*t*-butylcresol. Note the very high and sharp free O–H stretch at 3640 cm^{-1}.

b. Specific types of OH have much narrower ranges. *Examples*:
Alcohols: $\sim$3350 cm^{-1}.
–COOH acids: $\sim$3000 cm^{-1}.
$[HCO_3]^-$ (really $[HO\text{–}CO_2]^-$) $\sim$2500 cm^{-1}. Half width about 1500 cm^{-1}.
c. Inter- vs. intramolecular hydrogen bonds cannot be distinguished by the position or the intensity of the O–H stretch. Dilution studies are required.

B. O–H in-plane bend
1. This is *raised* by hydrogen bonding.
2. It is in the range 1350 ± 50 cm^{-1}.

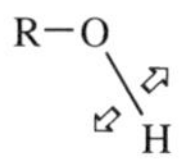

C. C–O–H out-of-plane bend, or torsion
1. This also is raised by hydrogen bonding.
2. All liquid and solid alcohols have a broad band at 700 ± 50 cm^{-1} (m, b) due to this. It is not a useful group frequency. If there is no hydrogen bonding, as in the vapor, the band is much lower, sharper, and weaker.

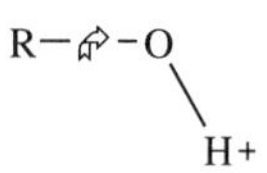

3. In –COOH dimers, the –O–H out-of-plane bend is $\sim$950 cm^{-1}, m, *broad*. Useful for confirming an acid.

IV. Two important classes of HO-containing compounds: alcohols and carboxylic acids

A. Alcohols, 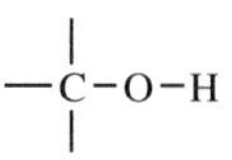

1. One can look at the spectrum of an alcohol and pick out all the bands due to O–H from their breadth. See Figure 6.5.
a. O–H stretch, $\sim$3350 cm^{-1}, vs, vb. Note that the huge intensity in *n*-hexanol is due to just one O–H group.
b. C–O–H in-plane bend
Alkyl alcohols: 1410 ± 10 cm^{-1} s, b. In hexanol, picture a very broad band underlying the 1470 and 1380 cm^{-1} bands.
Phenols: 1360 ± 30 cm^{-1} s, b.
c. The broad feature in hexanol near 1050 cm^{-1} is not an O–H mode. It is an overlap of two bands, one of which is a C–O stretch discussed in Chapter 8 on C—O single bonds.
d. C–O–H out-of-plane bend, or torsion, 700 ± 50 cm^{-1}, m, b
2. Note that in the Raman spectrum all the O–H bands are very weak.
3. Distinguishing between primary, secondary, and tertiary alcohols. At least two methods have been proposed for this, but neither is dependable over a large variety of alcohols.
a. Using the O–H stretch in a non-hydrogen-bonding situation
b. Using the 1160–1040 cm^{-1} band (due to C–O stretch) in a hydrogen-bonding system

B. Carboxylic acids, –COOH. See Figure 6.6.

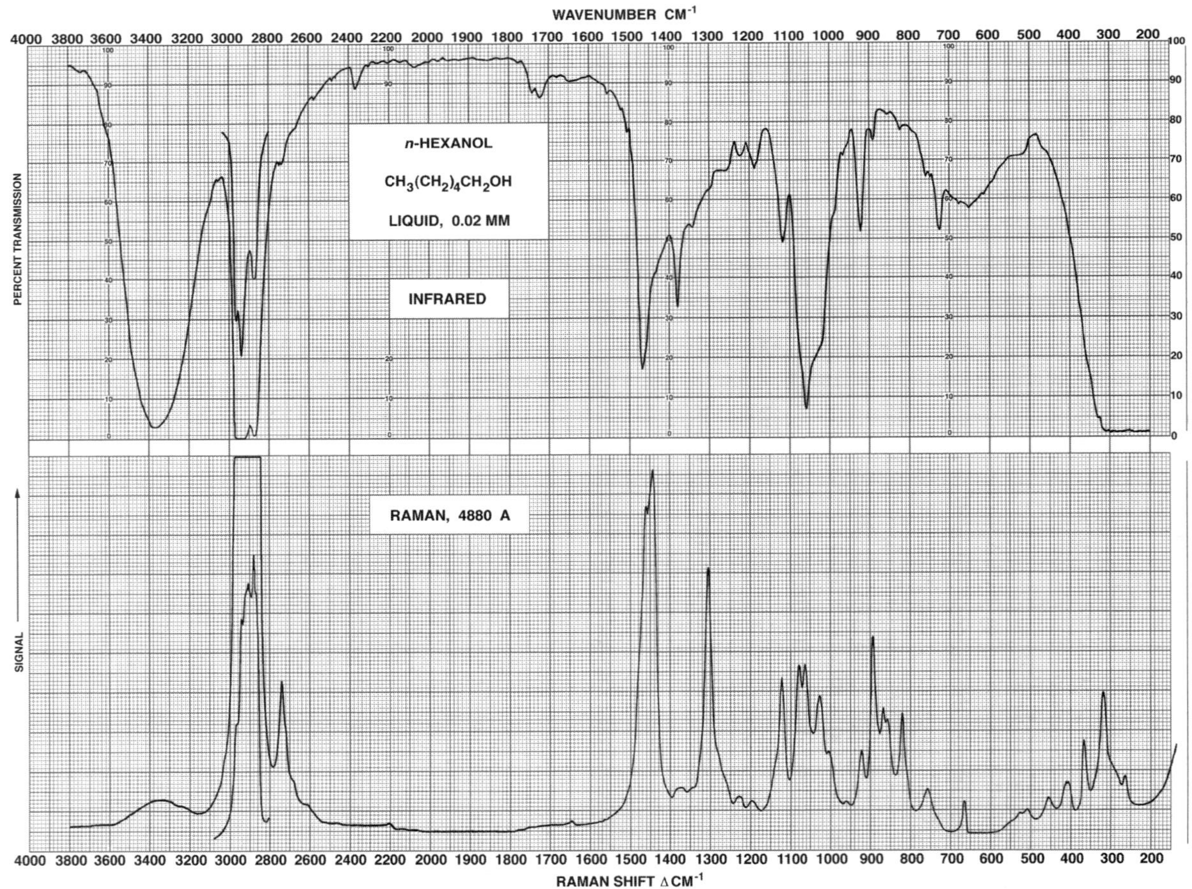

Fig. 6.5 Infrared and Raman spectra of *n*-hexanol.

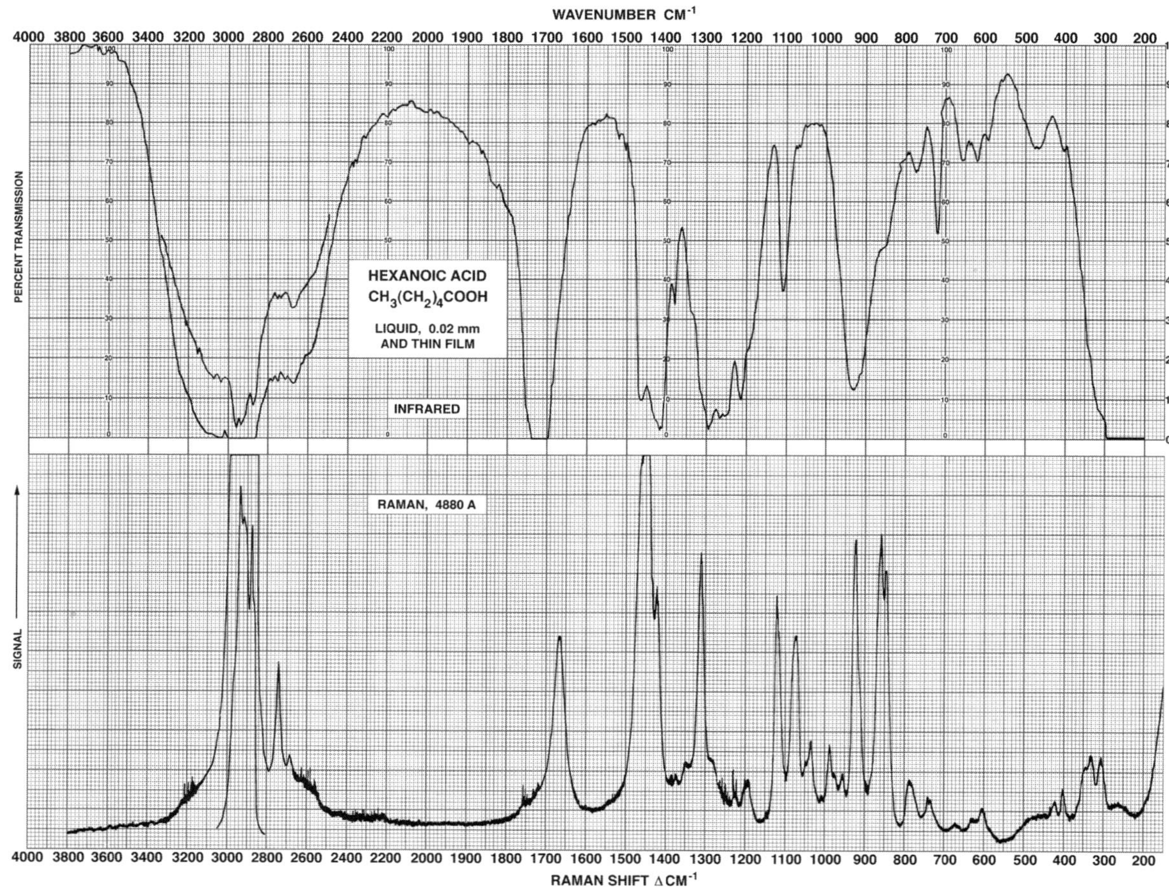

Fig. 6.6 Infrared and Raman spectra of *n*-hexanoic acid, *n*-$C_5H_{11}COOH$.

1. Carboxylic acids form dimers in condensed states and even to a considerable extent in the vapor.

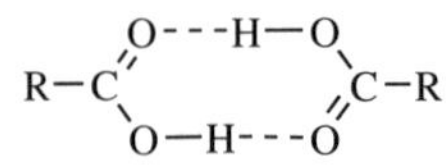

2. The dimers have four characteristic frequencies.
   a. O–H stretch, ~3000 cm^{-1}, vb, vs. Note that this is distinctly lower than for alcohols (ca. 3350 cm^{-1}), indicating a stronger hydrogen bond.
   b. "2600-cm^{-1}" bands. There are one to three weak bands about 2600 cm^{-1}. These are sum tones. They occur in an open portion of the spectrum, and they are sharper and more easily seen for solids than for liquids. Although weak, they are very characteristic and provide useful confirmation of a carboxylic acid.
   c. Carbonyl stretch, ~1720 cm^{-1}, vs
   d. The "dimer band," ~950 cm^{-1}, w, *broad*. This is the out-of-plane motion of the two hydrogens in the dimer ring in which both hydrogens move up out of the plane in unison. The band occurs *only* for the dimer.

V. N–H group frequencies
   A. General remarks
      1. These also are subject to hydrogen bonding, but the bonding is weaker than for O–H so the effects are less extreme.
      2. In the IR, N–H bands are less intense than O–H bands, but in Raman spectra they are more intense.
   B. N–H stretches
      1. Dominant factor: The more basic the N–H, the weaker is the N–H stretch in the IR.
      2. Typical frequencies for unassociated (free) N–H groups:

	$-NH_2$, cm^{-1}	>NH, cm^{-1}
Amide	3500 and 3410	3450
Aryl	3480 and 3395	3430
Alkyl	3390 and 3330	3355

Again, samples seldom have free N–H. Therefore, *H-bonded values will be given in the following discussion.*

      3. It is a bit difficult to tell an N–H stretch from an O–H stretch in condensed phases.
         a. The free O–H stretch is higher than free N–H, but hydrogen bonding is stronger and the band often shifts to the same region as H-bonded N–H.
         b. In the IR the band for a hydrogen-bonded N–H band is weaker and less broad than an O–H band at the same frequency. With experience one can usually distinguish between them.

c. The Raman spectrum is a great help because the N–H stretch is distinctly stronger in it than is O–H stretch.

C. Primary amines, R–NH_2. *Example: n*-Hexylamine. See Figure 6.7.

1. N–H stretches

The two N–H groups couple to give two N–H stretching bands.

	Out of Phase, or Antisymmetric Stretch	In Phase, or Symmetric Stretch
Aromatic amines	3450 ± 60 cm^{-1}, m in IR	3375 ± 45 cm^{-1}, m in IR
Aliphatic amines	3375 ± 25 cm^{-1}, m in IR	3300 ± 30 cm^{-1}, m in IR

2. –NH_2 scissoring 1620 ± 30 cm^{-1} s to m, b in IR
3. –NH_2 wagging ~800 cm^{-1} s, b in IR
4. The twist (or torsion) and the rocking are not useful group frequencies.

D. Secondary amines, R_1R_2NH

1. >N–H stretch. There is only one, ~3300 cm^{-1}, b.
2. >N–H bend. ~1500 cm^{-1}, m to w in IR.

E. Distinguishing between primary, secondary, and tertiary amines. There are three bands to use:

	Primary	Secondary	Tertiary
1. Number of N–H stretches	2	1	0
2. N–H bend	1620 ± 30 cm^{-1}	~1500 cm^{-1}	—
3. C–H stretch of CH_2–N<	Normal C–H	2850–2720 cm^{-1}	2850–2720 cm^{-1}

a. The low frequencies in the last two classes are apparently a case of the Bohlmann effect. To have a low C–H stretch, a hydrogen on a carbon atom must be *trans diaxial* to the lone-pair electrons on nitrogen.

b. Reference: Song Guo-nan, Peoples Republic of China, personal correspondence with FAM.

F. Amine salts, R–NH_3^+, $R'R''NH_2^+$, etc.

1. The N–H stretch is much lower than for amines, possibly because the positive charge on the N greatly increases the strength of the hydrogen bonding.
2. There are often several bands in the region 2800–2200 cm^{-1}, sometimes weak and sometimes strong. They may be superimposed on a very broad background absorption.
3. *Example:* 1-Hexanamine hydrochloride, CH_3–$(CH_2)_5$–$NH_3^+ \cdot Cl^-$. See Figure 6.8.
4. *Example:* Diethylamine hydrochloride, $(C_2H_5)_2$–$NH_2^+ \cdot Cl^-$. See Figure 6.9.
5. *Example:* Triethylamine hydrochloride, $(C_2H_5)_3$–$NH^+ \cdot Cl^-$. See Figure 6.10.

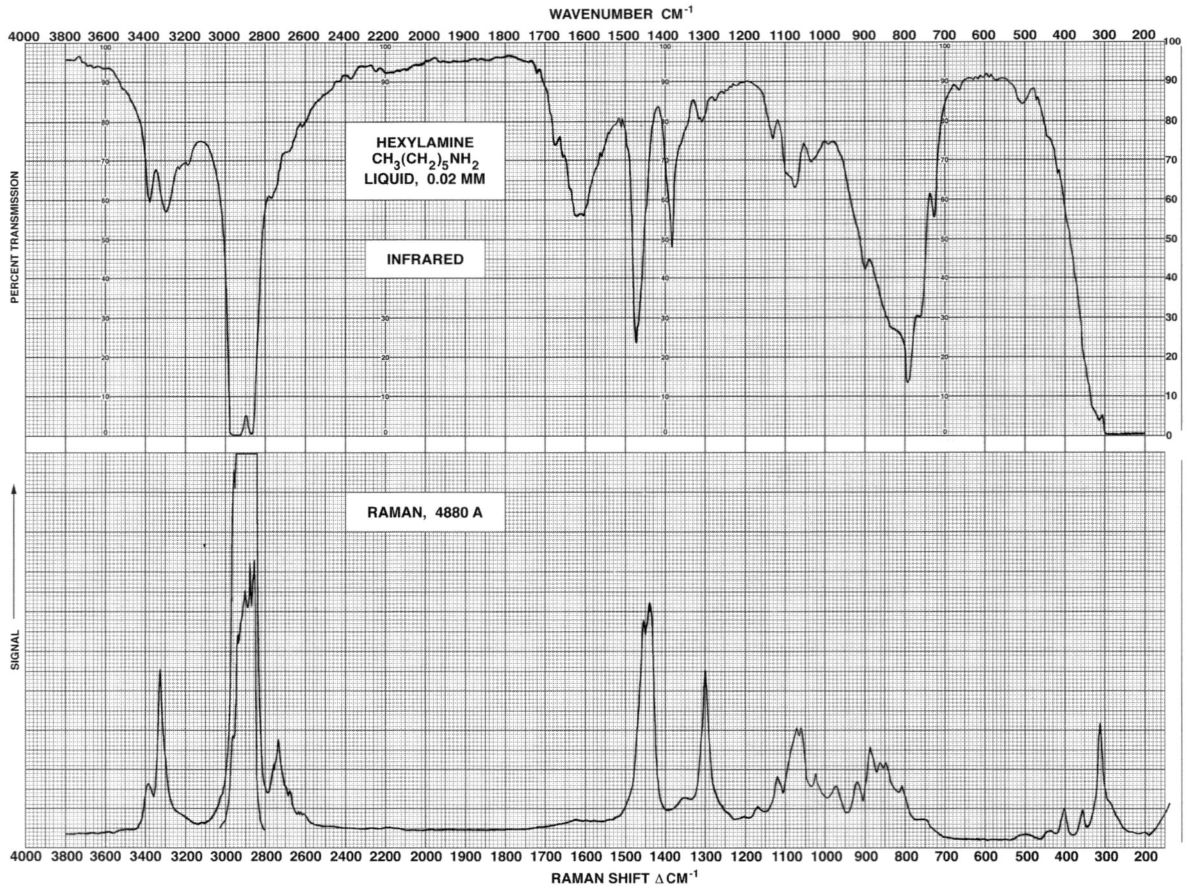

Fig. 6.7 Infrared and Raman spectra of *n*-hexylamine, $C_6H_{13}NH_2$.

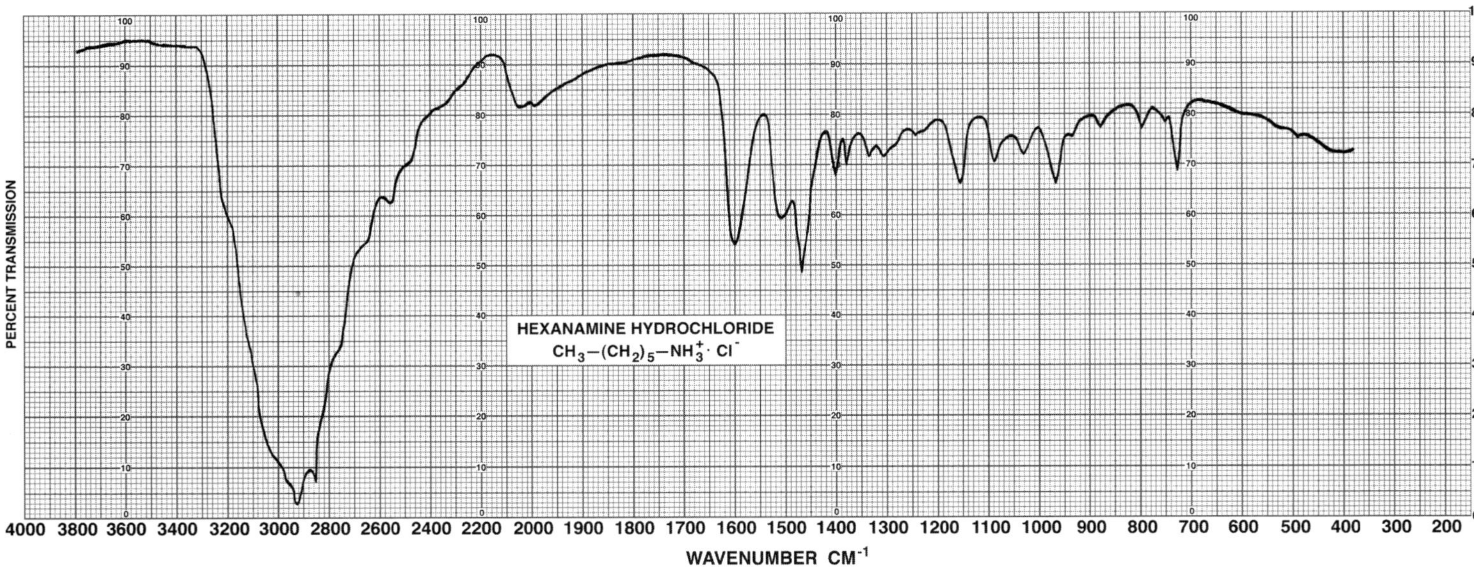

Fig. 6.8 Infrared spectrum of 1-hexanamine hydrochloride, $CH_3-(CH_2)_5-NH_3^+ \cdot Cl^-$.

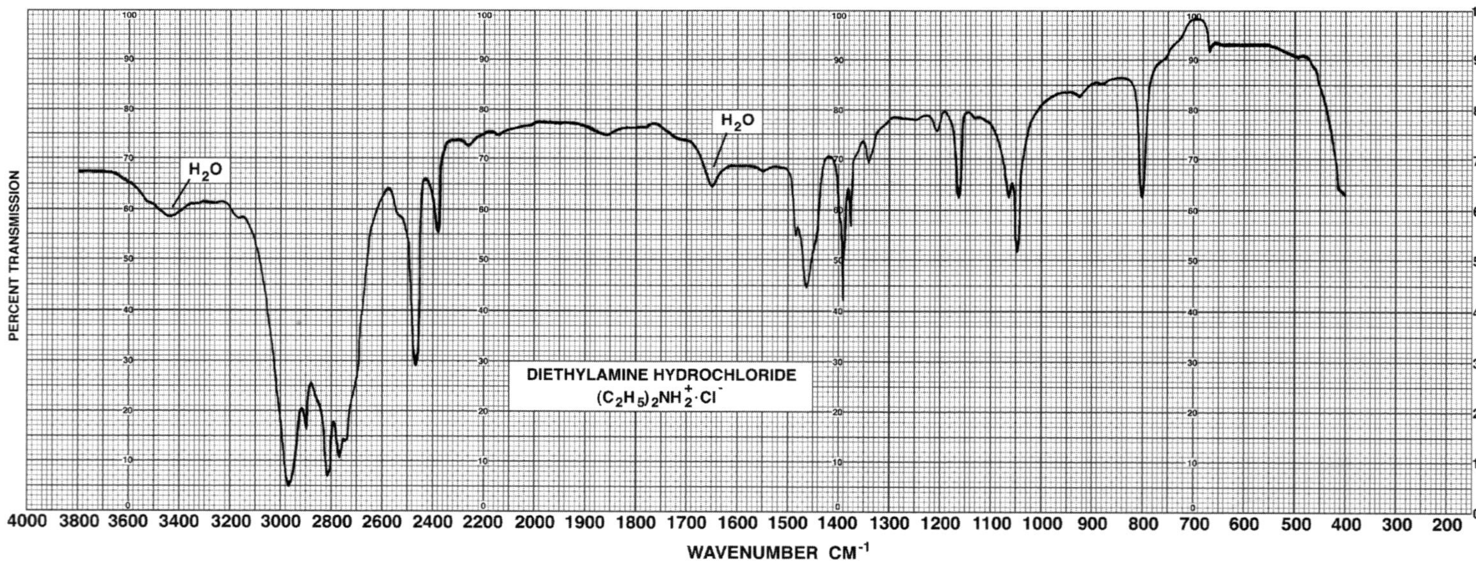

Fig. 6.9 Infrared spectrum of diethylamine hydrochloride, $(C_2H_5)_2{-}NH_2^+ \cdot Cl^-$.

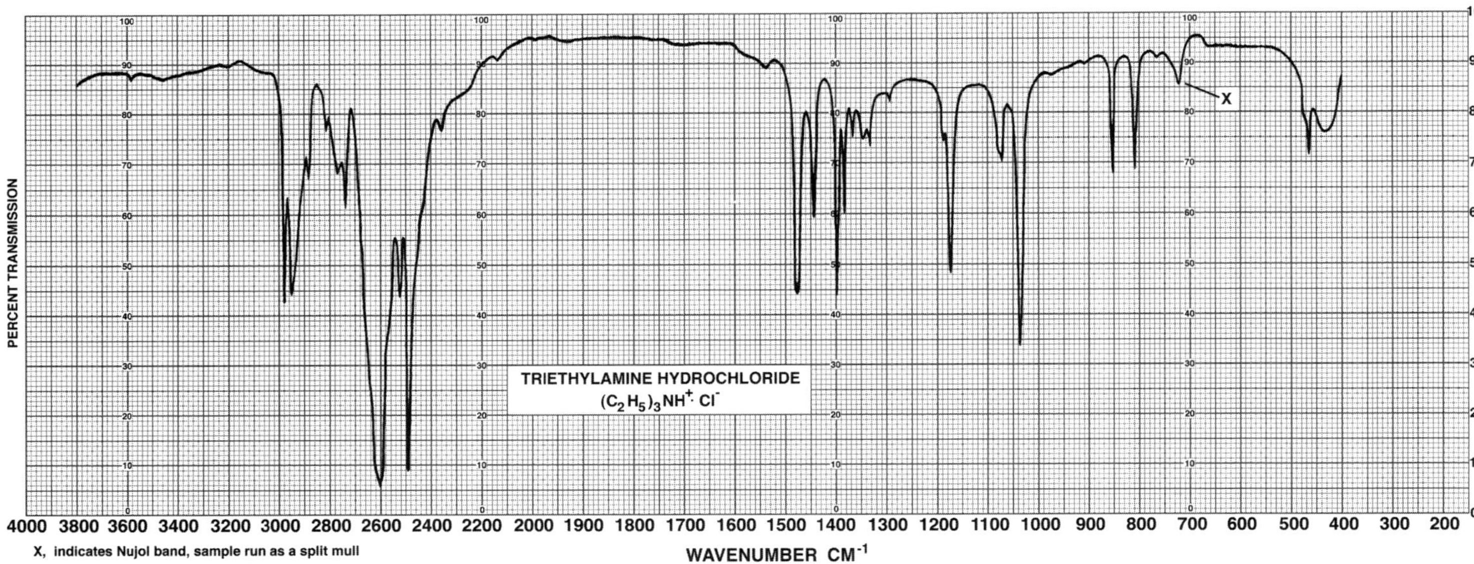

Fig. 6.10 Infrared spectrum of triethylamine hydrochloride, $(C_2H_5)_3–NH^+ \cdot Cl^-$.

6. It is often possible to distinguish between primary, secondary, and tertiary amine salts. See references.

VI. Other X–H stretches

A. C–H stretches have been discussed in Chapter 2.

B. Si–H, P–H, and S–H stretches will be discussed later in Chapter 9 on Si, P, and S compounds.

C. For other X–H groups, see the summary table in Figure 6.1.

7 Spectra of Carbonyl Compounds of All Kinds (Factors Affecting Carbonyl Group Frequencies)

DANA W. MAYO

I. Introduction

The carbonyl group is perhaps the single most important functional group in organic chemistry. It is certainly the most commonly occurring functionality. Infrared spectroscopy can play a powerful role in the characterization of the carbonyl, and therefore in organic chemistry, because this group possesses all of the properties that give rise to an excellent group frequency.

A. The carbonyl group has a large dipole moment derivative, which gives rise to very intense absorption bands.

B. As a result of the large force constant, the C=O group has a stretching frequency that occurs at high values and above the fingerprint region. In addition, nature has added the further good fortune in leaving the carbonyl region of the spectrum nearly devoid of other fundamental modes.

C. The carbonyl stretching fundamental occurs in a range that is reasonably narrow, $1750 \pm 150\ cm^{-1}$, but sensitive enough to the local environment to allow for considerable interpretation of the surrounding structure. (This range would be unacceptably broad if it fell in the middle of the fingerprint region)

D. Where the stretching vibration of a particular C=O group falls in the $300\ cm^{-1}$ range of frequencies is determined by a number of factors that are now well understood in terms of the effects outlined below:

1. Mass effects
2. Geometric effects

Course Notes on the Interpretation of Infrared and Raman Spectra, by Dana W. Mayo, Foil A. Miller, and Robert W. Hannah.
ISBN 0-471-24823-1

3. Resonance and inductive effects
4. Interaction effects

E. The carbonyl is usually only weakly coupled to the rest of the molecule:
   1. The mode of interest involves a stretching motion along the internuclear axis involving a displacement of both the carbon and oxygen atoms of about the same magnitude.
   2. Both atoms are significantly displaced during the vibration because the masses are similar. Thus the frequency of the system is dependent on the reduced mass of the oscillator.
   3. The double bond plays a key role in establishing the properties of an excellent group frequency for the C=O group. The increased strength over the C—C single bond systems raises the natural frequencies of the carbonyls out of the fingerprint region and uncouples this mode from the rest of the molecular motion.
   4. The carbonyl is uncoupled because the vast majority of the carbonyls are bonded to the remainder of the system by low-frequency C—C, C—O, or C—N oscillators. These single bonds create a frequency mismatch between the C=O stretching mode and the rest of the molecule and thereby greatly reduce mechanical coupling to the system.
   5. Note that even though there is considerable displacement of the carbon atom which is directly connected to the rest of the molecule, the frequency mismatch controls the magnitude of the coupling, which is generally quite low.

F. In this discussion the standard reference frequency for the carbonyl group will be the wavenumber values for simple open-chain aliphatic ketones, such as acetone (1715 cm^{-1} neat, 1719 cm^{-1} CCl_4, 1712 cm^{-1} $CHCl_3$) or 3-heptanone (1720 cm^{-1} CCl_4; see Figures 7.1 and 7.2).

II. Factors affecting carbonyl group frequencies

A. Mass effects

The mode of principal interest is the stretching vibration. In this oscillator the C and O atoms undergo comparable displacements; thus, we must replace the simplified mass expression in the Hooke's law approximation that was applied to the C—H stretching fundamentals ($\tilde{\nu} = 1/2\pi c\sqrt{k/m_H}$) by the reduced mass μ (where $\mu = m_C m_O/(m_C + m_O)$).

1. On substitution of the isotopes ^{13}C and ^{18}O, the predicted small frequency shifts (~30–40 cm^{-1}) are observed. See, e.g., the case of the labeled di-isopropyl ketones: $(CH_3)_2CH^{12}C={}^{16}OCH(CH_3)_2$ with $\tilde{\nu}_{C=O} = 1712$ cm^{-1} and $(CH_3)_2CH^{13}C={}^{16}OCH(CH_3)_2$ with $\tilde{\nu}_{C=O} = 1674$ cm^{-1} to give $\Delta\tilde{\nu}_{C=O,\,obs} = 38$ cm^{-1}, and the calculated value using reduced mass is $\Delta\tilde{\nu}_{C=O,\,calc} = 38$ cm^{-1}.

These results are consistent with a relatively low degree of mechanical coupling to the rest of the system. This lack of coupling,

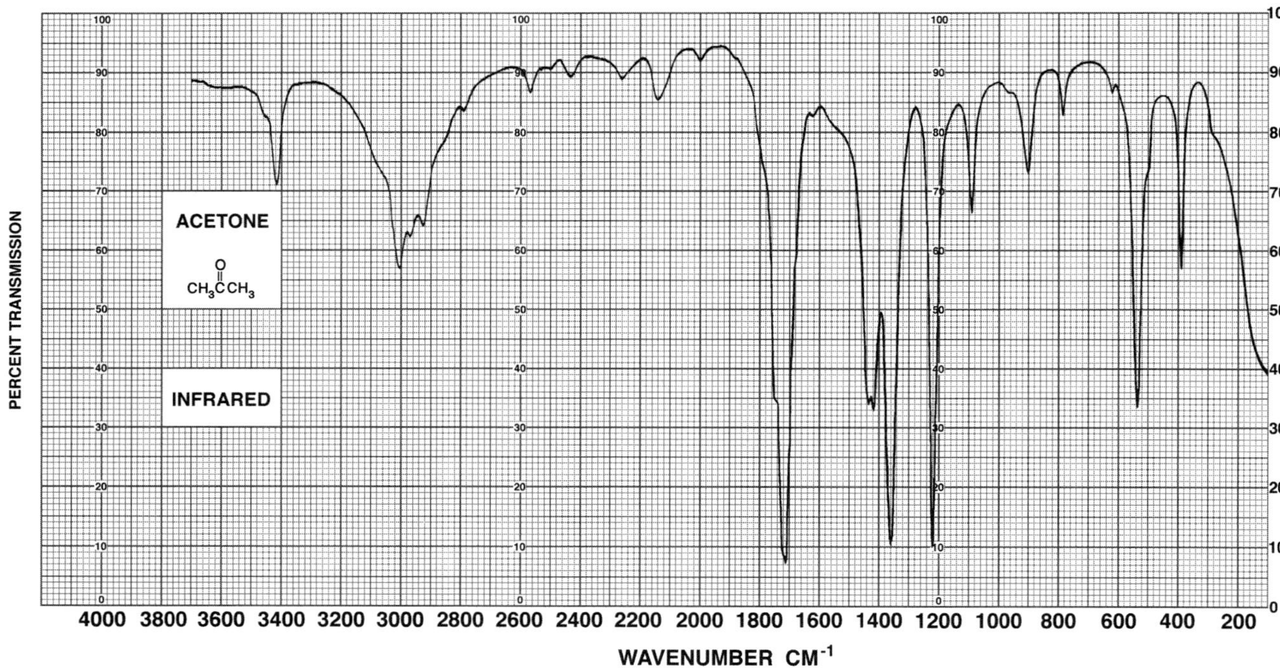

Fig 7.1 Infrared spectrum of acetone.

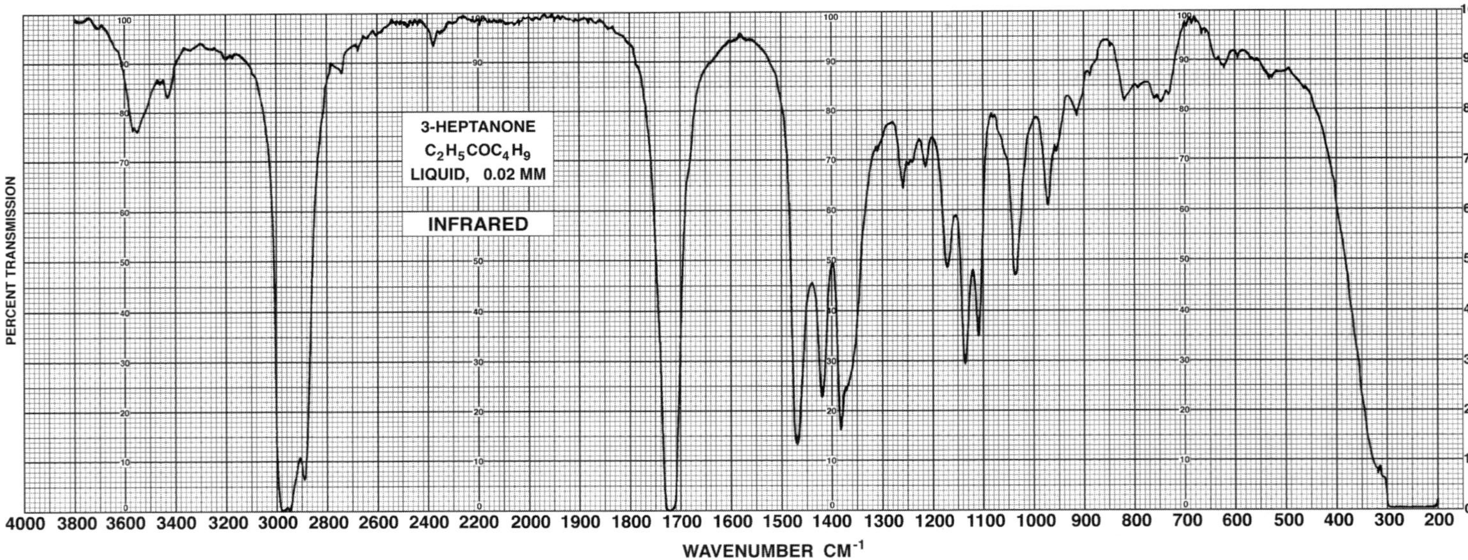

Fig 7.2 Infrared spectrum of 3-heptanone.

as pointed out above, is expected because the force constant of the multiple bond is significantly different from the values of the force constants of the bonds that connect the carbonyl to the rest of the molecule.

2. Gas-phase spectra of formaldehyde, $H_2C{=}O$, $\tilde{\nu}_{C=O} = 1746\ cm^{-1}$, and deuteroformaldehyde, $D_2C{=}O$, $\tilde{\nu}_{C=O} = 1700\ cm^{-1}$, ratio 1.027, indicate very little coupling of the system when the mass and frequency matches are poor. However as the mass of X in the $X_2C{=}O$ system begins to approach that of carbon, coupling can become large. The coupling is also dependent on the X–C–X bond angle, being a minimum when the angle is 180° and a maximum when it is 0°. See Figure 7.3.

B. Geometric effects on the carbonyl frequency

1. Geometric effects can play a major role in determining the location of the carbonyl frequency within the 1750 cm^{-1} region. Although the displacement of the oxygen atom involves simply stretching or compressing the C=O bond, the displacement of the carbon atom is more complex. This latter movement also contains a compression component in the two connecting single bonds as the carbonyl carbon is being stretched. In the opposite phase of the vibration, a stretching component in the two connecting bonds is required because the carbonyl carbon is being compressed against the oxygen atom.
2. The magnitude of these additional force constant components is angle dependent. As the angle between the single bonds (C–CO–C angle) decreases, the contribution of the single-bond components to the effective C=O stretching force constant will be raised. Since the frequency of the vibration is directly proportional to the square root of the force constant, a decrease in the internal carbonyl bond angle

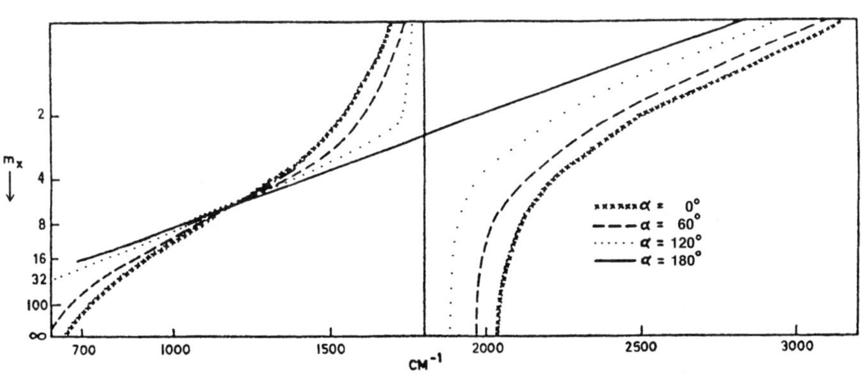

Fig 7.3 Variation in coupling interaction between symmetric X–C–X and the C=O stretching modes with changes in mass of X and angle α in 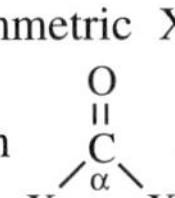 .

TABLE 7.1 Variation of Carbonyl Frequency (cm^{-1}) versus Bond Angle

	By Ring Size					
	7	6	5	4	3	(2)
Lactones	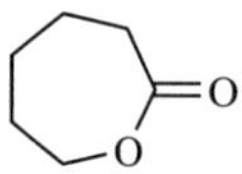 1727	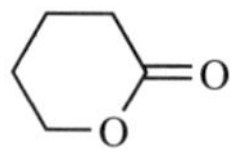 1740	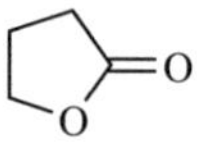 1775	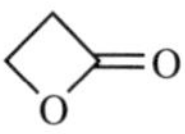 1832		
Ketones	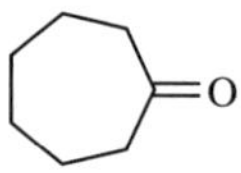 1699	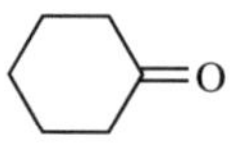 1710	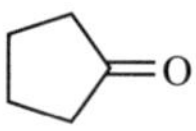 1744	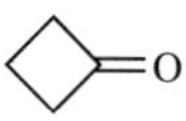 1782	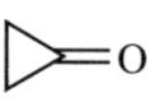 1906	$H_2C{=}C{=}O$ 2049
Lactams		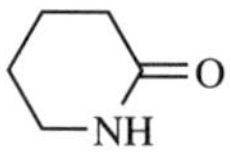 1670	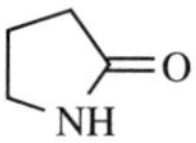 1695	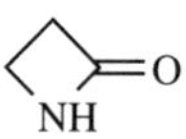 1750		

will raise the frequency (see Table 7.1). Alternatively, an increase in the internal carbonyl bond angle will lower the carbonyl frequency.

3. The internal carbonyl bond angle is steadily constrained to smaller and smaller angles as the ring size of cyclic carbonyls is diminished. As predicted, the carbonyl frequencies rise in the smaller ring systems (see Table 7.1). The local environment of the carbonyl needs to be held constant in order to observe this frequency shift. For example, comparisons can be made within series of cyclic ketones, lactones, or lactams but are not easily made across these three series.
4. A shift in internal bond angle is dependent not only on ring size but can also be controlled by other factors such as steric hindrance. In open-chain ketones the standard value is close to 1720 cm^{-1}, but in di-*t*-butyl ketone, $(CH_3)_3CC{=}OC(CH_3)_3$, the carbonyl stretch is found at 1690 cm^{-1}, and in *t*-butyltriphenylmethyl ketone, $(CH_3)_3CC{=}OC(C_6H_5)_3$, it is found near 1675 cm^{-1}. In the first case the shift is down ~30 cm^{-1} and in the second ~45 cm^{-1}. This drop in frequency results from the mutual repulsion of the two large bulky aliphatic groups attached to the central carbonyl group. The steric factors act to force the internal carbonyl bond angle to open up. This increase in angle results in a decrease in the compression component of the carbonyl stretching mode, which leads to a drop in carbonyl frequency.

C. Electronic effects (resonance and inductive)

Resonance and inductive effects can profoundly influence the vibrational frequency of the carbonyl group. In the following discussion, as mentioned above, the carbonyl stretching mode for acetone ($\tilde{\nu}_{C=O} = 1715$ cm^{-1}, liquid) will be used as a reference frequency representative of simple alkyl substitution on the carbonyl group. Effects that perturb this reference fundamental to either higher or lower values will be examined.

1. Effects which raise the carbonyl frequency (relative to acetone)
   a. When an alkyl substituent is replaced by a more electronegative system, the balance of contributing resonance forms to the electronic structure of the carbonyl is slightly shifted. The shift is away from dipolar forms to forms with increased double-bond character and results from the strong inductive effects of the substituent. The increased double-bond character results in a larger effective C=O force constant and higher C=O frequencies.

   For example, in hexanoyl chloride, $CH_3(CH_2)_4COCl$, $\tilde{\nu}_{C=O} = 1805\ cm^{-1}$ (Figure 7.4).
   b. When the carbonyl is doubly substituted with electronegative groups, the frequency may continue to rise (see the case of —F substitution), but if the bond angle opens up in response to increased steric effects, the rise in frequency may be offset by geometric lowering effects (see, e.g., Cl—CO—Cl compared with CH_3COCl, see Table 7.2):
2. Effects which lower the carbonyl frequency (relative to acetone)
   a. Direct conjugation of the carbonyl via α,β-unsaturation will introduce new dipolar carbonyl resonance forms that lower the effective force constant values (increase the single-bond character of the carbonyl) and thus decrease the carbonyl stretching frequency. For example, consider cyclohexene methyl ketone (**I**), $\tilde{\nu}_{C=O} = 1685\ cm^{-1}$, and acetophenone (**II**), $\tilde{\nu}_{C=O} = 1687\ cm^{-1}$ and examples listed in Table 7.3.

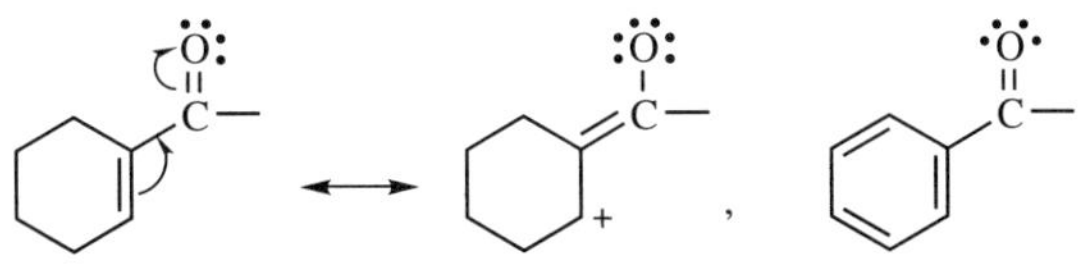

I **II**

3. Ester carbonyl frequencies (competing inductive and resonance effects)
   a. The substitution of the more electronegative oxygen for carbon in going from ketones to esters will raise the carbonyl frequency in esters as a result of inductive influences (see II.C.1 above, electronic effects which raise the carbonyl frequency). On the

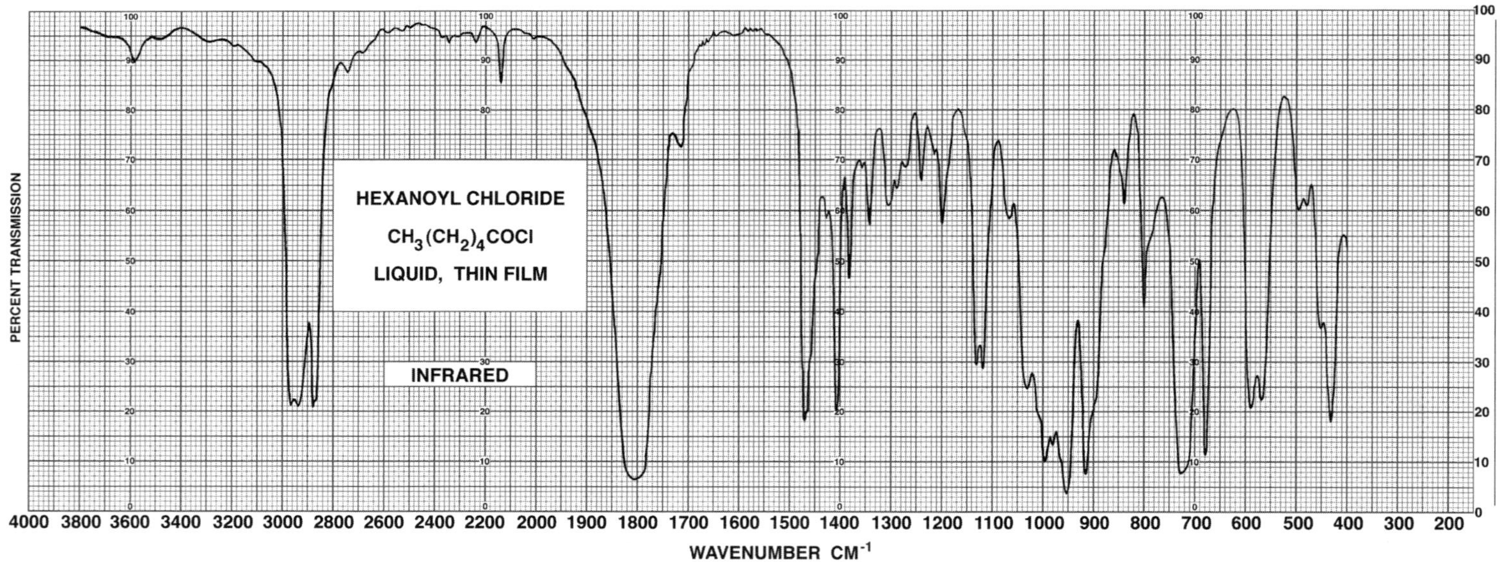

Fig 7.4 Infrared spectrum of hexanoyl chloride.

TABLE 7.2 Effect of Electronegativity

	cm^{-1}
CH_3–CO–CH_3	1715
CH_3–CO–OCH_3	1746
CH_3–CO–OH	1785 (gas)
CH_3–CO–CH_2Cl	1726, 1742
CH_3–CO–$CHCl_2$	1724, 1744
CH_3–CO–Cl	1807
Cl–CO–Cl	1807
CH_3–CO–F	1840
F–CO–F	1928 (gas)

other hand, the lone-pair electrons present on the ether oxygen of the ester will be in direct conjugation with the carbonyl. This latter interaction will generate dipolar resonance forms that will tend to drop the C=O frequency. The balance between these two competing effects in esters, which might be difficult to anticipate, favors the inductive effects. Thus ester carbonyls commonly are located 20–40 cm^{-1} higher than the simple aliphatic ketones and lie in the range 1755–1735 cm^{-1} (see the spectrum of hexyl acetate; Figure 7.5).

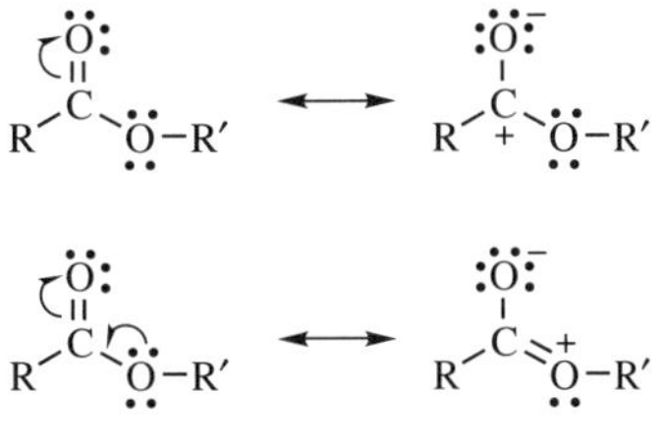

b. Examples of the IR spectrum of an aliphatic ester are given in Figure 7.5.
c. If the ester carbonyl is directly involved with α,β-unsaturation, the normal ester frequency is lowered by 20–30 cm^{-1}. Thus,

TABLE 7.3 Effects of Conjugation

	CCl_4 soln (cm^{-1})
CH_3–CO–CH_3	1719
C_6H_5–CO–CH_3	1692
CH_2=CH–CO–CH_3	1687
C_6H_5–CO–C_6H_5	1666

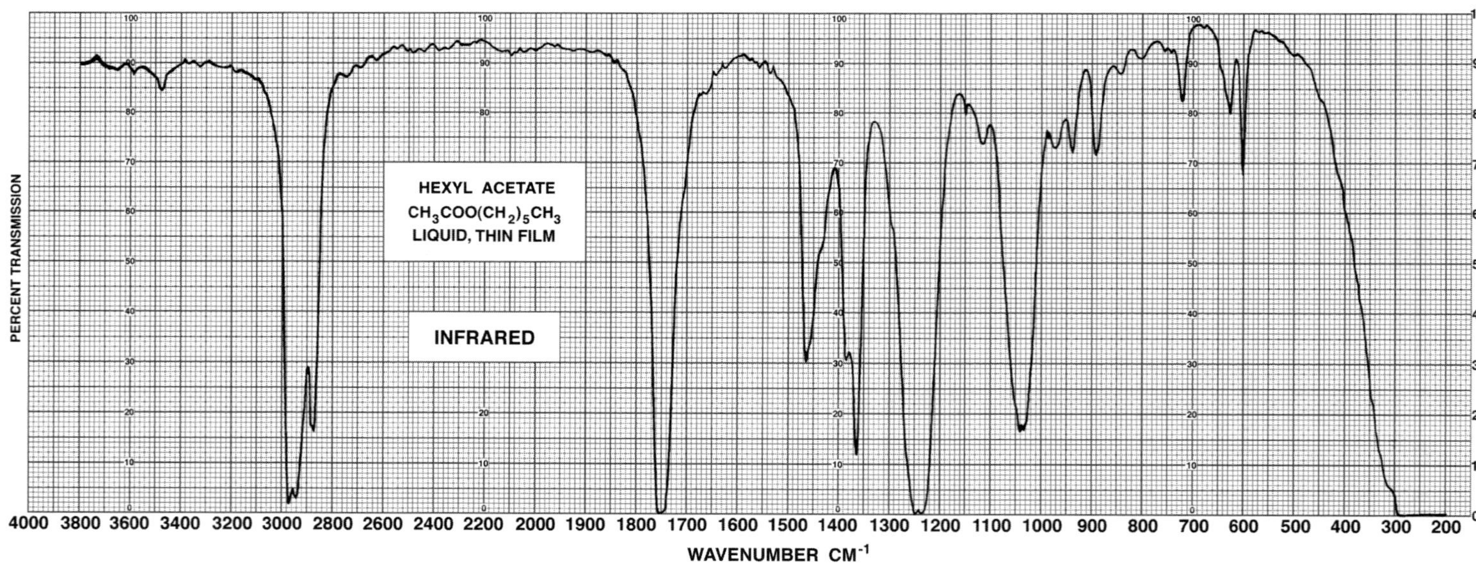

Fig 7.5 Infrared spectrum of hexyl acetate.

unsaturated ester carbonyl frequencies occur very nearly in the same region as simple aliphatic ketone frequencies. For example, in ethyl benzoate, $\tilde{\nu}_{C=O} = 1720\ cm^{-1}$.

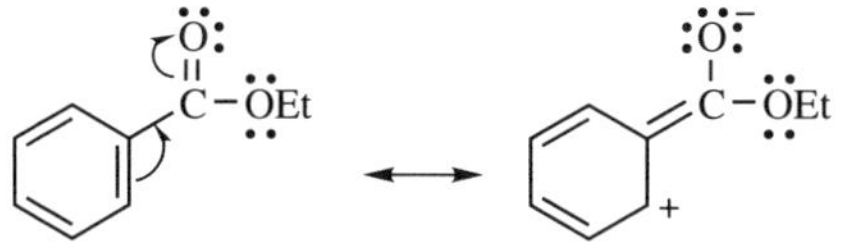

d. If the ester is conjugated but the conjugation is located adjacent to the ether oxygen rather than alpha to the carbonyl group, then the carbonyl frequency is raised. The higher $\tilde{\nu}_{C=O}$ results from resonance competition for the lone-pair electrons of the ether oxygen by the carbonyl and the new conjugating group, e.g., phenyl acetate, $\tilde{\nu}_{C=O} = 1769\ cm^{-1}$. These frequency shifts support the arguments concerning competition between inductive and resonance effects within the ester ether group. (For an example of a spectrum of this type of conjugation, see the spectrum of vinyl acetate; Figure 7.6)

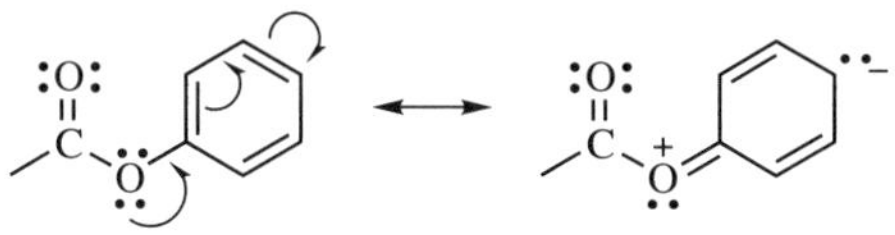

e. If the ester group is directly conjugated on both sides, then the resonance effects should cancel. We would expect this type of system to exhibit near-normal carbonyl frequencies, e.g., phenyl benzoate, $\tilde{\nu}_{C=O} = 1743\ cm^{-1}$, as compared with ethyl acetate, $\tilde{\nu}_{C=O} = 1742\ cm^{-1}$.

D. Interaction effects on the carbonyl frequency

Interaction effects vary from those that have a dramatic impact on the spectra to those that are barely detectable. They can be divided into intramolecular and intermolecular types of interactions.

1. Intramolecular carbonyl interactions

a. *First-order coupling effects* are rarely observed, for as has been noted, this oscillator is rather effectively decoupled from the rest of the molecule by differences in frequency. One of the few spectacular examples of heavy carbonyl coupling that has been observed is the case of carbon dioxide. In O=C=O the two coupled oscillators are aligned for maximum interaction (180°), possess identical frequencies, and share the single carbon atom. The splitting between the antisymmetric and symmetric levels in this case is very large, approximately $1000\ cm^{-1}$.

A second, and much less dramatic, example is the case of anhydrides. In this instance the two oscillators are joined through a central oxygen. Delocalization can occur across the connecting

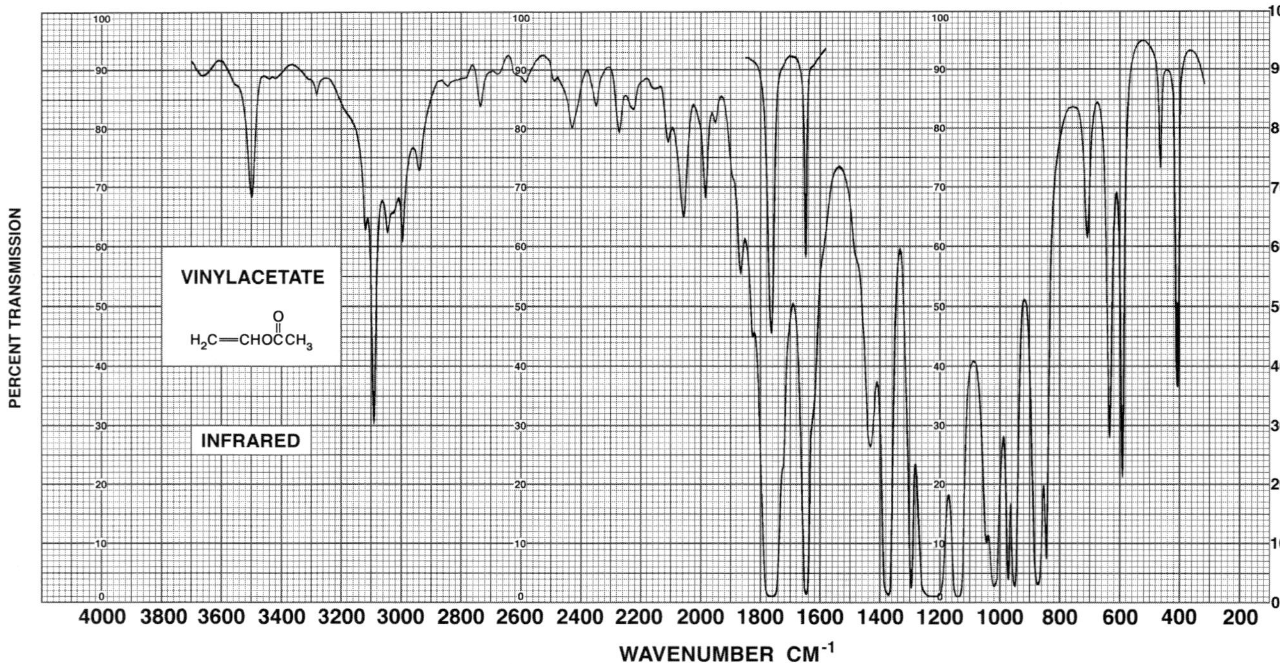

Fig 7.6 Infrared spectrum of vinyl acetate.

atom and operates to maintain planarity of the system and thereby to support the coupling. Even so, since the carbonyls are no longer held at the optimum angle and are vibrationally insulated by an intervening atom, the first-order coupling drops to $\Delta\tilde{\nu}_{C=O} \sim 70\ \text{cm}^{-1}$, less than 10% of the CO_2 value, e.g. hexanoic anhydride, 1817, 1750 cm^{-1}; see Figure 7.7.

b. The uncoupled anhydride C=O vibrations would be expected to occur near 1770 cm^{-1}. This latter value is consistent with an oxygen-substituted carbonyl in which conjugation of the lone-pair electrons on the ether oxygen has been essentially canceled. (The full inductive effect of the ether oxygen atom on the carbonyl stretching vibration can be inferred from these data.) Equalized resonance competition by both carbonyl systems of the anhydride for the ether lone-pair electrons might be expected to bring about just such an effect.
c. In open-chain anhydrides the high-frequency carbonyl band is more intense than the low-frequency component of the doublet. The observed intensities correlate with the known structure of this group in open-chain systems where the carbonyls lie nearly parallel to each other.

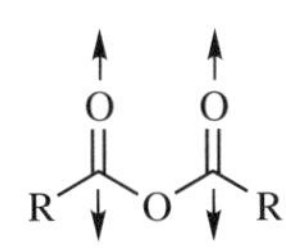

In this case the higher frequency (more intense) mode corresponds to the in-phase vibration and the out-of-phase vibration to the lower (less intense) mode. The reason for the reversal of the usual mechanics for these coupled oscillators has been suggested by Colthup. It is proposed that an additional interaction constant is required because of the heavy contributions of the dipolar resonance forms **II** and **III**. These forms enhance the single-bond character in the out-of-phase mode and thereby drop that mode below the in-phase stretch, which depends more heavily on contributions from form **I** and has more double-bond character.

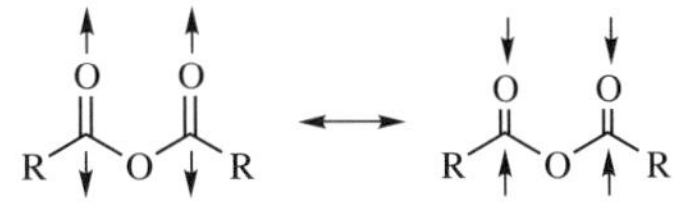

I

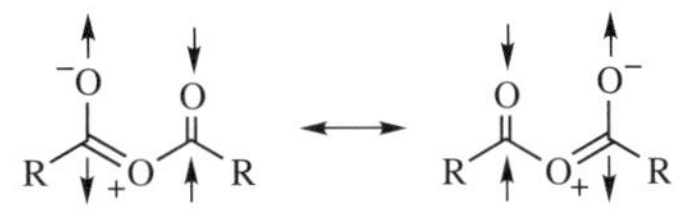

II **III**

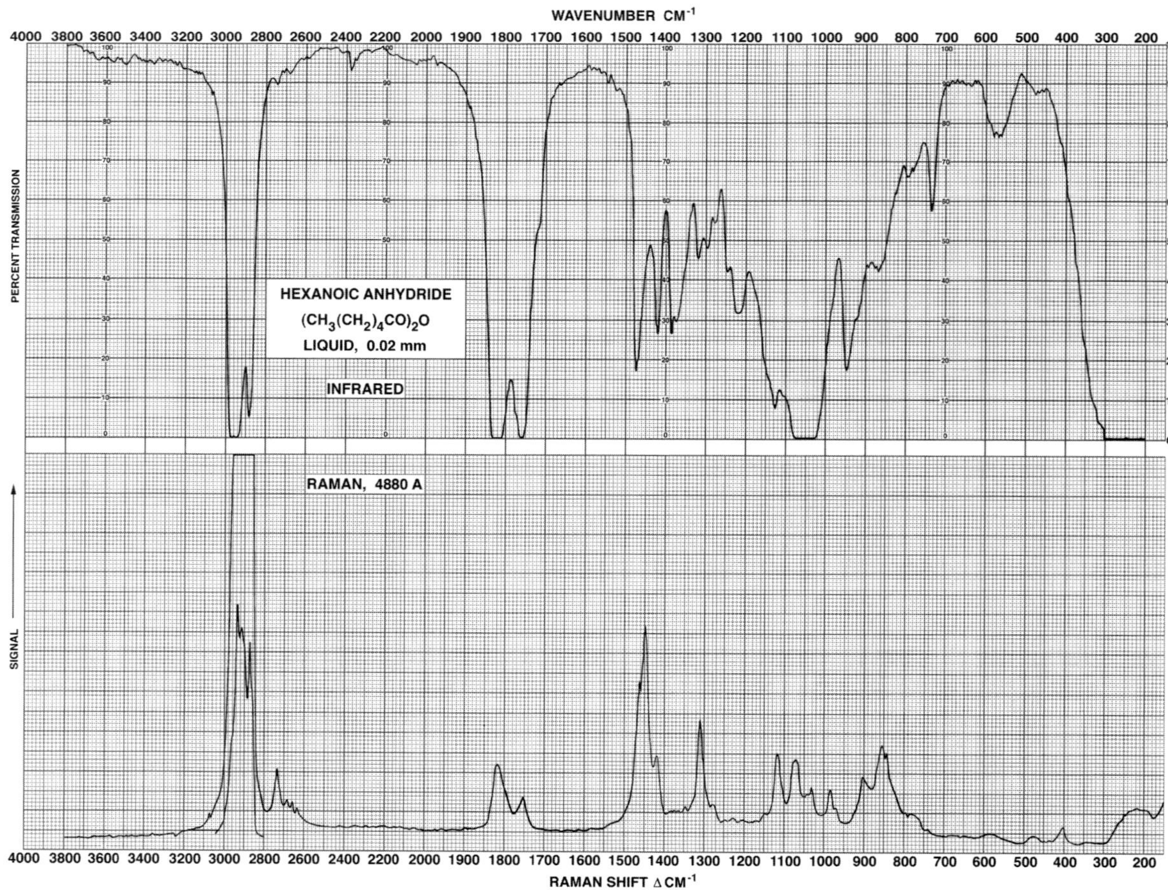

Fig 7.7 Infrared spectrum of hexanoic anhydride.

d. In cyclic anhydrides the intensity of the carbonyl doublet can be reversed. This intensity effect is explained as follows: In cyclic anhydrides, as the ring size decreases, the relative orientation of the two carbonyls becomes reversed compared to the open-chain conformation. In the common five-membered ring anhydrides the two oscillators are now nearly 180° back to back. Under these conditions the high-frequency in-phase mode drops significantly in IR intensity as the dipole moment collapses. See, e.g., Figure 7.8, the IR and Raman spectra of the five-membered ring anhydride maleic anhydride.

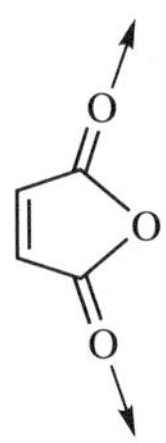

In the rare four-member ring anhydrides the in-phase vibration can completely vanish from the IR spectrum.

e. Intramolecular H bonding

Intramolecular H bonding of carbonyls can be enhanced by resonance interactions, and this significantly perturbs the C=O stretching frequency. For example, let us consider the anthraquinone series shown below:

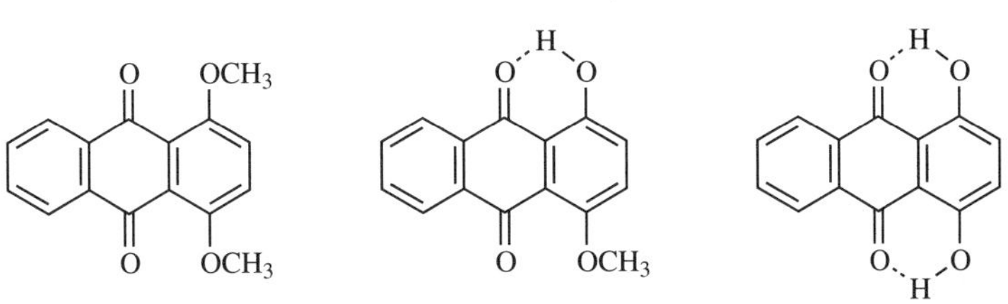

1675 cm^{-1} **I** | 1675, 1636 cm^{-1} **II** | 1627 cm^{-1} **III**

1) In structure **I**, H bonding is not present and a single stretching frequency for the quinone is observed at 1675 cm^{-1}. This value reflects direct conjugation with the aromatic ring and with the methoxyl groups (really vinylogous esters in which lone-pair conjugation can occur but where the inductive effect is not observed because the C=O and $-OCH_3$ groups are remote from each other) conjugated in equivalent fashion with both carbonyls. The *single* band at 1675 cm^{-1} indicates little coupling between the carbonyls through the ring system. It was expected and is observed.

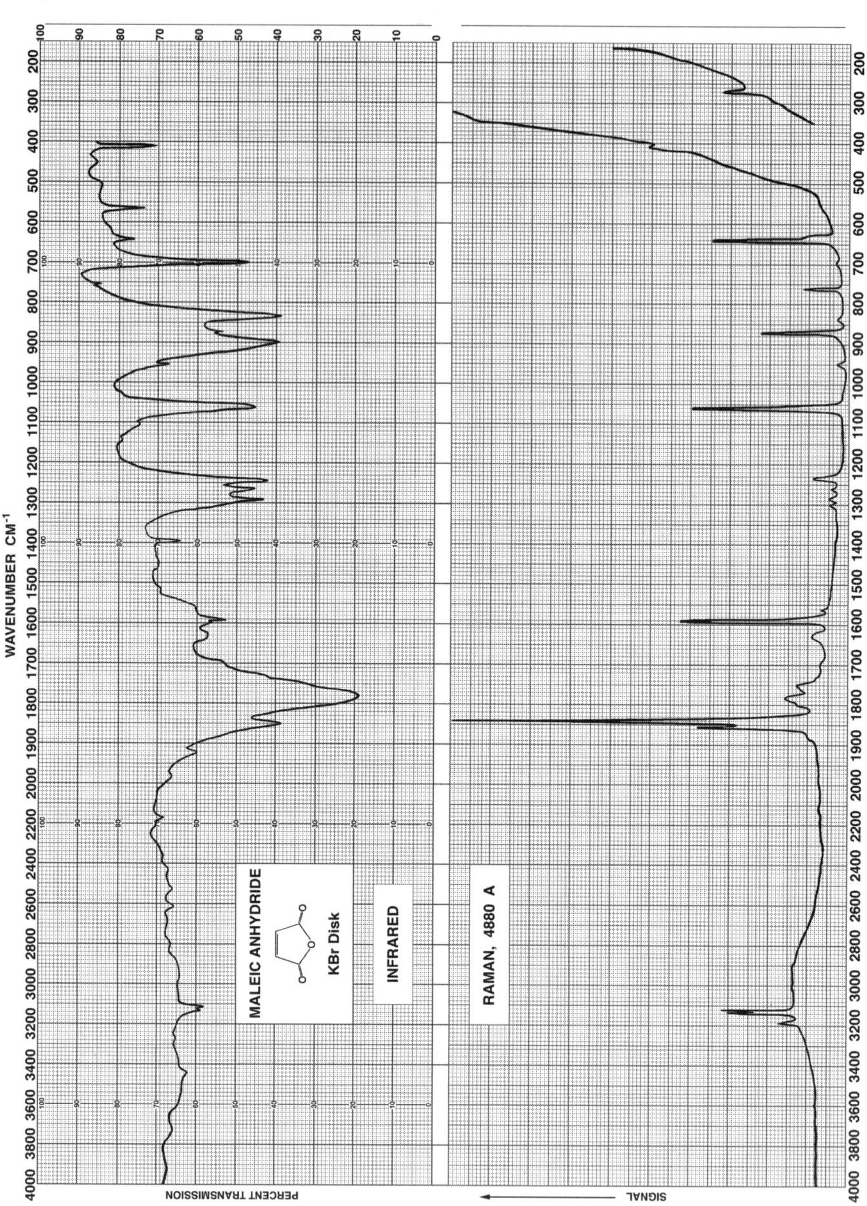

Fig 7.8 Infrared and Raman spectra of maleic anhydride.

In structure **II** one of the methoxyl groups is cleaved to yield a phenol group. Resonance forms operating through a tightly hydrogen-bonded six-membered ring act to reduce the effective force constant of the carbonyl, with the result that the observed frequency, $\tilde{\nu}_{C=O} = 1636\ cm^{-1}$, is lowered nearly 40 cm^{-1}. The remaining carbonyl spatially adjacent to a methoxyl group is unshifted in wavenumber value from the dimethoxy derivative. This unshifted carbonyl stretch indicates that *all* of the H bonding which occurs is *intramolecular* in nature and *all* the phenolic groups are tied to the adjacent carbonyl system.

In structure **III** both methoxyl groups have been replaced by phenolic groups. Now a single peak is observed, but it falls near 1627 cm^{-1}. This change indicates that both carbonyls are undergoing strong intramolecular H bonding. This interaction involves resonance forms similar to those found in structure **II**. Thus, a single band is observed at $\tilde{\nu}_{C=O} = 1627\ cm^{-1}$. This downward shift approaching 50 cm^{-1} can be ascribed primarily to strong internal H bonds present in the anthraquinone system.

2) Several other examples of very strong intramolecular H bonding are known. For example, in tropolone the carbonyl band is essentially missing from the spectrum. The absorption that occurs near 1605 cm^{-1} may involve some C⋯O displacement (Figure 7.9).

f. Second-order coupling (Fermi resonance)

This type of interaction involves the carbonyl stretching mode and an overtone or combination level of a second fundamental mode. Fermi demonstrated that this type of interaction could occur and that it explained a number of unresolved problems in the interpretation of IR and Raman spectra. Hence the effect is often referred to as Fermi resonance. Fermi interactions are symmetry restricted and require a very close frequency match between the fundamental and the overtone for strong coupling to occur. Weak interactions with strong modes can often be observed, however, if they fall in open regions of the spectrum.

Thus, second-order coupling often occurs with carbonyl vibrations in complex organic molecules of low symmetry. In the large majority of cases the frequency of the fundamental is not noticeably affected. The main evidence for the interaction will be weak shoulders associated with the main carbonyl peak. An example of nontrivial coupling is the case of cyclopentanone, where the carbonyl is often doubled, $\tilde{\nu}_{C=O} = 1746, 1728\ cm^{-1}$. The Fermi interaction involves the carbonyl stretching interaction with a combination tone (CH_2 wag, 1280 cm^{-1}, and >C=O

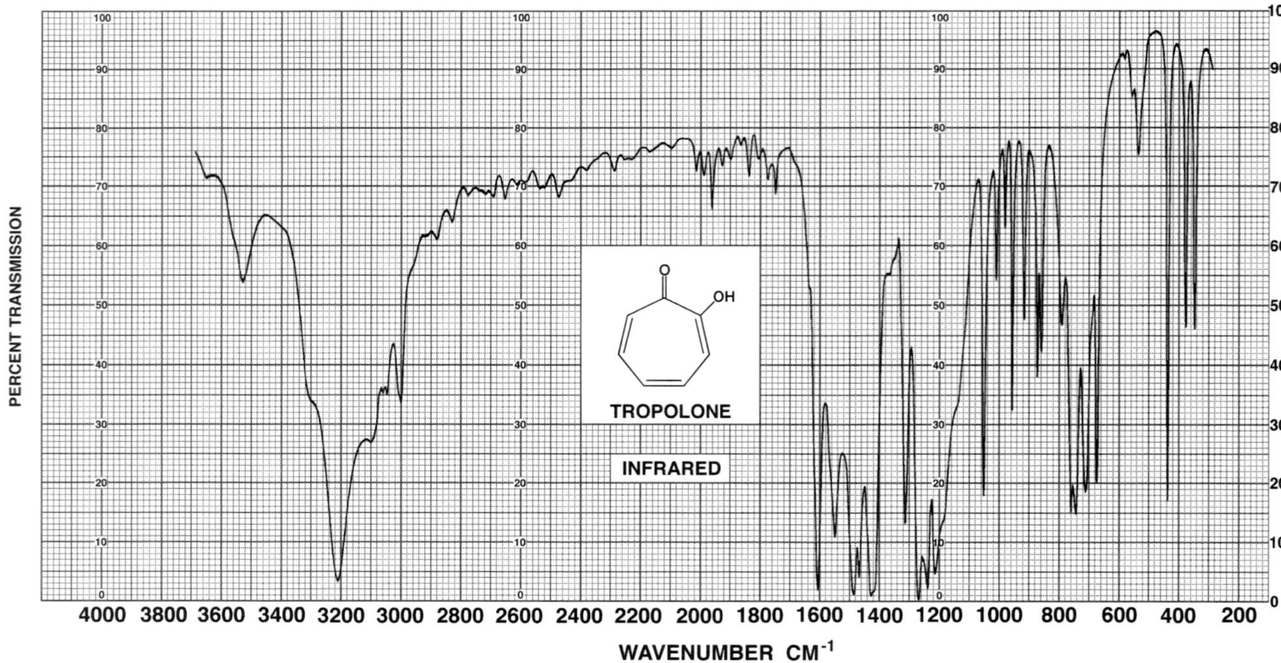

Fig 7.9 Infrared spectrum of tropolone.

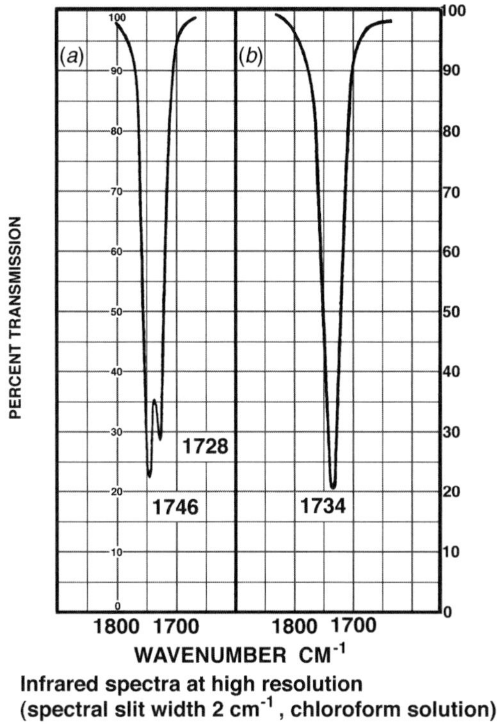

Fig 7.10 Infrared spectra of (*a*) *cyclo*-pentanone and (*b*) *cyclo*-pentanone-α,α,α′,α′–d_4.

bend, 470 cm^{-1}). This was demonstrated by showing that the splitting collapses to a singlet in 2,2,5,5-tetra-d_4-cyclopentanone (see Figure 7.10).

g. *Field effects* will also perturb the carbonyl frequency. The classic case of a field effect is that of the chloroacetones. In those rotamers in which the chlorine atom is in the eclipsed position with respect to the oxygen, repulsive lone-pair interactions occur. Chloro groups can generate *field* effects that result in suppression of the contribution of dipolar carbonyl resonance forms, and therefore the system undergoes a rise (∼30 cm^{-1}) in the stretching frequency.

These assignments are supported by the observation that two carbonyl frequencies are present in the mono- and dichloroacetones and single frequencies are present in acetone and trichloroacetone (see Table 7.4). *The small frequency rise observed in the low-frequency component (1715 → 1726 cm^{-1}) is attributed to inductive effects*, which therefore can have only a minimal influence on the high-frequency component.

TABLE 7.4 Field Effects in Chloroacetones

Compound	Structure	$\tilde{\nu}$ Range (cm^{-1})
Acetone	CH_3–CO–CH_3	1715
Chloroacetone	CH_3–CO–CH_2Cl	1742, 1726
1,1-Dichloroacetone	CH_3–CO–$CHCl_2$	1743, 1724
1,1,1-Trichloroacetone	CH_3–CO–CCl_3	1765

h. *Transannular interactions* occur when cyclic carbonyl groups are sterically positioned so that the carbon atom of the carbonyl is oriented toward an electron-rich center lying across the ring. The interaction can greatly enhance the dipolar resonance form of the carbonyl and result in a significant drop in the stretching frequency. For example, the effect has a major impact on the carbonyl frequency of the alkaloid protopine ($\tilde{\nu}_{C=O} = 1660$ cm^{-1}).

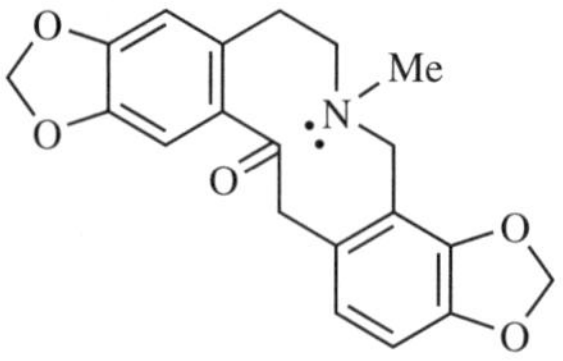

Most interesting, however, are the results obtained from a number of model compounds synthesized by Nelson Leonard at Illinois. In the case of cyclooctaaminoketone (**I**), $\tilde{\nu}_{C=O} =$ 1666 cm^{-1}, its perchlorate salt (**II**) exhibits no carbonyl absorption band at all!

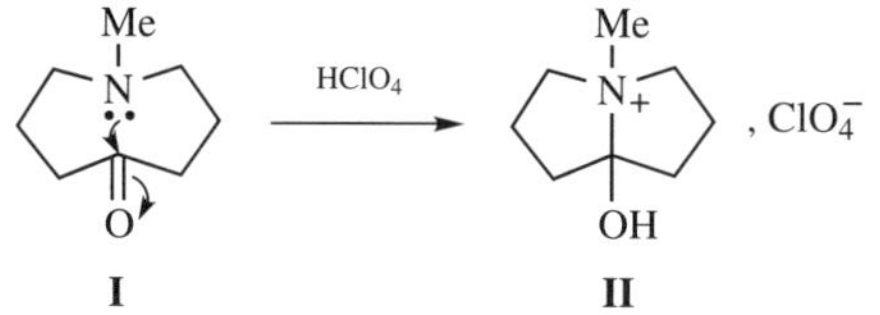

2. Intermolecular carbonyl interactions
   a. Strong intermolecular H bonding can significantly perturb carbonyl frequencies. It is known in the case of aliphatic carboxylic acids that these substances form strongly H-bonded dimers when neat or in highly concentrated solutions. Association through the carbonyl groups leads to the formation of a symmetric eight-membered ring containing two linear H bonds.

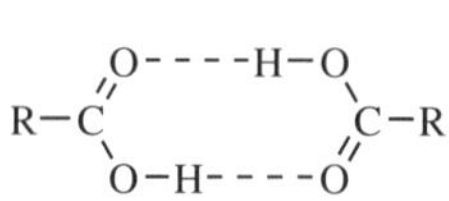

Coupling through the tightly bonded ring results in a splitting of the carbonyl levels of approximately the same magnitude as found in anhydrides ($\tilde{\nu}_{C=O} = \sim 70\ cm^{-1}$). Because the dimer possesses a center of symmetry, the rule of mutual exclusion will hold and the in-phase mode will be Raman active but not IR active ($\tilde{\nu}_{C=O} = \sim 1650\ cm^{-1}$). The out-of-phase stretch of the carbonyls, however, will be IR active. Thus, the antisymmetric C=O stretch gives rise to a strong band in the IR ($\tilde{\nu}_{C=O} = \sim 1720\ cm^{-1}$).

In very dilute solution, it is sometimes possible to observe these systems in the monomeric state. Under these conditions the carbonyl frequencies return to expected values ($\tilde{\nu}_{C=O} = \sim 1770\ cm^{-1}$).

As the discussion of group frequencies continues, several other examples of the effect of intermolecular H bonding on the carbonyl group frequencies will be noted. (For the neat spectrum of hexanoic acid see Figure 7.11.)

b. The interaction of weak H bonds is relatively hard to detect in the IR, as the shifts are measured in terms of a few wavenumbers. One of the better examples is the effect on the carbonyl stretch of acetone as measured in cyclohexane solution ($\tilde{\nu}_{C=O} = 1722\ cm^{-1}$). When the hydrocarbon solvent is replaced by chloroform, $HCCl_3$, weak H bonds (=O····H–C) develop, and the carbonyl mode drops 12 cm^{-1} to 1710 cm^{-1}.
c. Weak dipolar intermolecular interactions directly between carbonyls can also be observed in the IR. The frequency shifts caused by these interactions closely parallel the development of polarization in the carbonyl group, as can be judged by the data given in Table 7.5.
d. This discussion of interactions effects completes the survey of factors governing carbonyl stretching frequencies.

III. An example of the interpretation of other carbonyl-containing compounds utilizing the above parameters

Amides: The case of amides is very interesting because this functional group is a particularly complex system to analyze and because all of the

TABLE 7.5 Carbonyl Dipolar Interactions[a]

Compound	$\Delta\tilde{\nu}$ Range (cm^{-1})
Acetyl chloride	15
Phosgene	13
Acetone	21
Acetaldehyde	23
Dimethylformamide	50

[a]Shift measured between dilute nonpolar solution and neat sample. Similar effects occur on a change of state: gas to liquid.

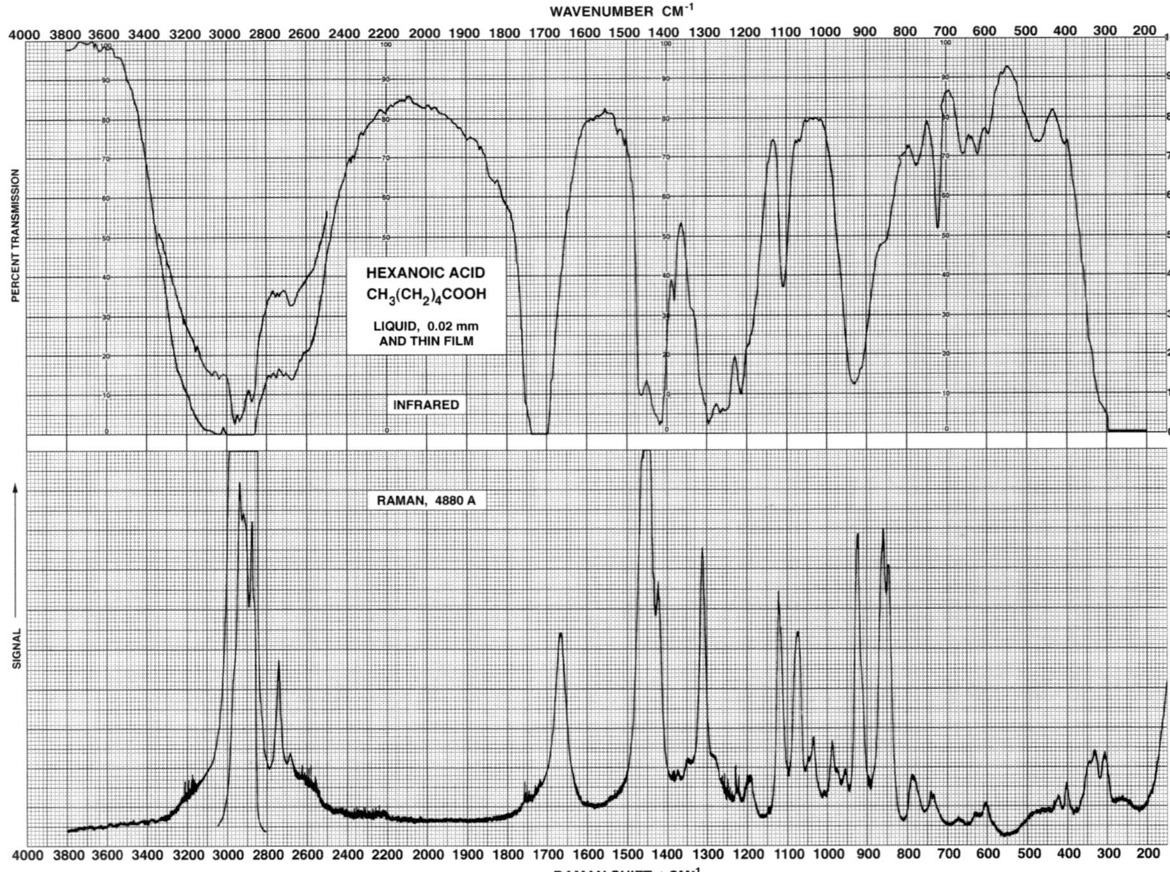

Fig 7.11 Infrared and Raman spectra of hexanoic acid.

factors considered above that perturb the carbonyl are present. Geometric effects will raise or lower the amide carbonyl as seen in lactams; electronic effects can both raise the carbonyl mode (electronegative nitrogen substitution) and lower it (nitrogen lone-pair conjugation to carbonyl), and the presence of both intra- and intermolecular H bonding can drastically lower the C=O stretch in amides because very strong N—H····O bonds can form.

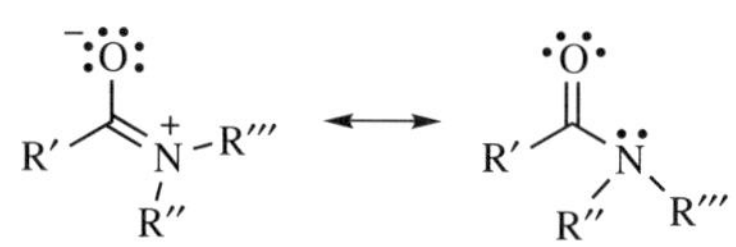

The spectrum of hexanamide is given in Figure 7.12 and that of *N*-methylhexamide in Figure 7.13.

Thus the prediction of the position of an amide carbonyl is one of the more difficult calls in the interpretation of an unknown spectrum. As these systems are seldom observed in solution (usually very insoluble in most nonpolar solvents used in IR sampling), the following comments may be of value. Studies of amide carbonyl frequencies in dilute nonpolar solution indicate that H-bonding effects are largely responsible for the low frequencies observed with primary and secondary amides but obviously play no role in tertiary amides.

The data indicate that when H bonding is removed in primary amides, the inductive effect of the nitrogen dominates over the influence of conjugation, but by not as much as in the case of esters. This is consistent with the relative electronegativities involved in esters and amides. In secondary amides with an electron-releasing *N*-alkyl group replacing a hydrogen, conjugation involving the nitrogen lone pair with the carbonyl begins to overcome the inductive effect. In tertiary amides with two *N*-alkyl substituents present, conjugation now dominates the inductive effect. Under these conditions polarized resonance forms make large contributions to the character of the carbonyl, and the tertiary amide C=O frequencies are observed to decrease to values in the same range as the frequency shifts generated via H bonding (see Table 7.6).

TABLE 7.6 Amide Carbonyl: Solution and Solid-Phase Data

Amide	Dilute Solution (cm^{-1})	Solid (cm^{-1})
R-CO-NH_2	~1730	~1690–1650
R-CO-NH**R**	~1700	~1670–1630
R-CO-N**R**$_2$	~1650	~1650

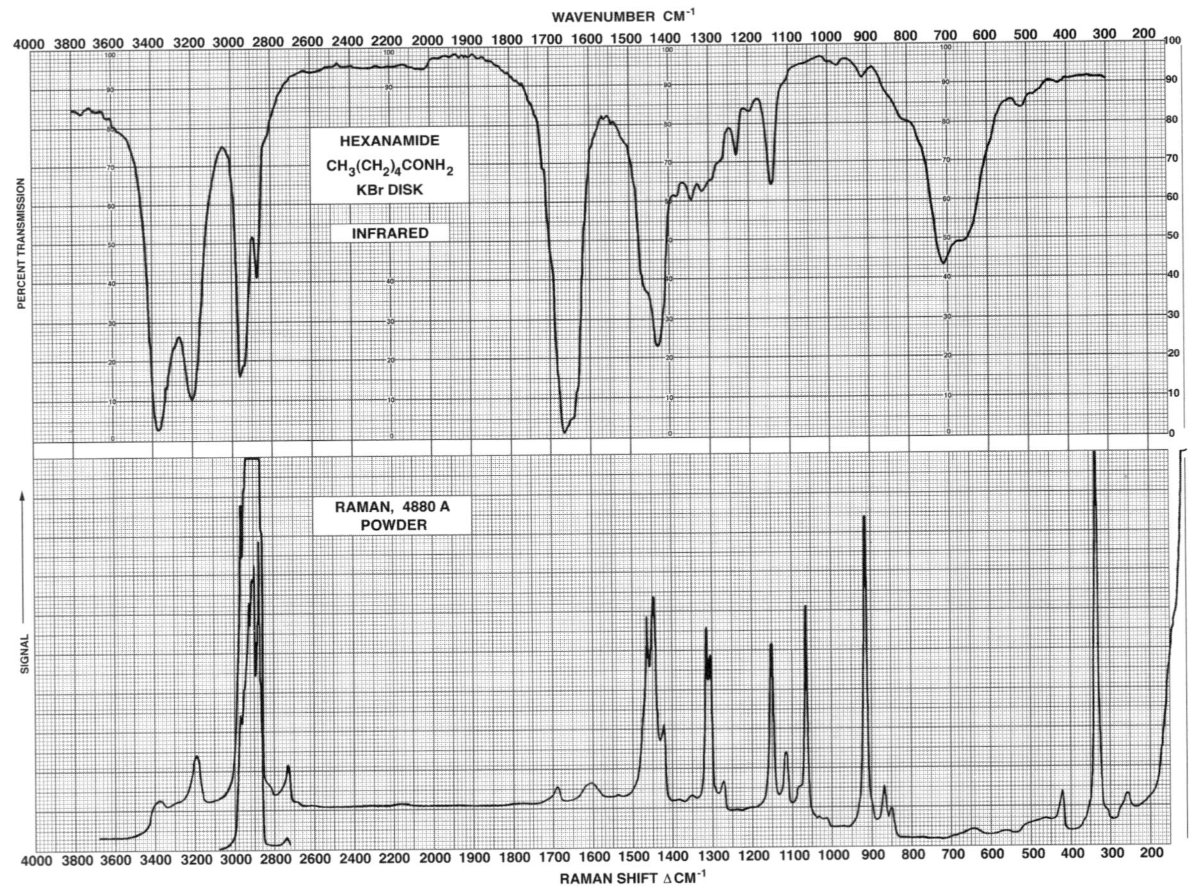

Fig 7.12 Infrared and Raman spectra of hexanamide.

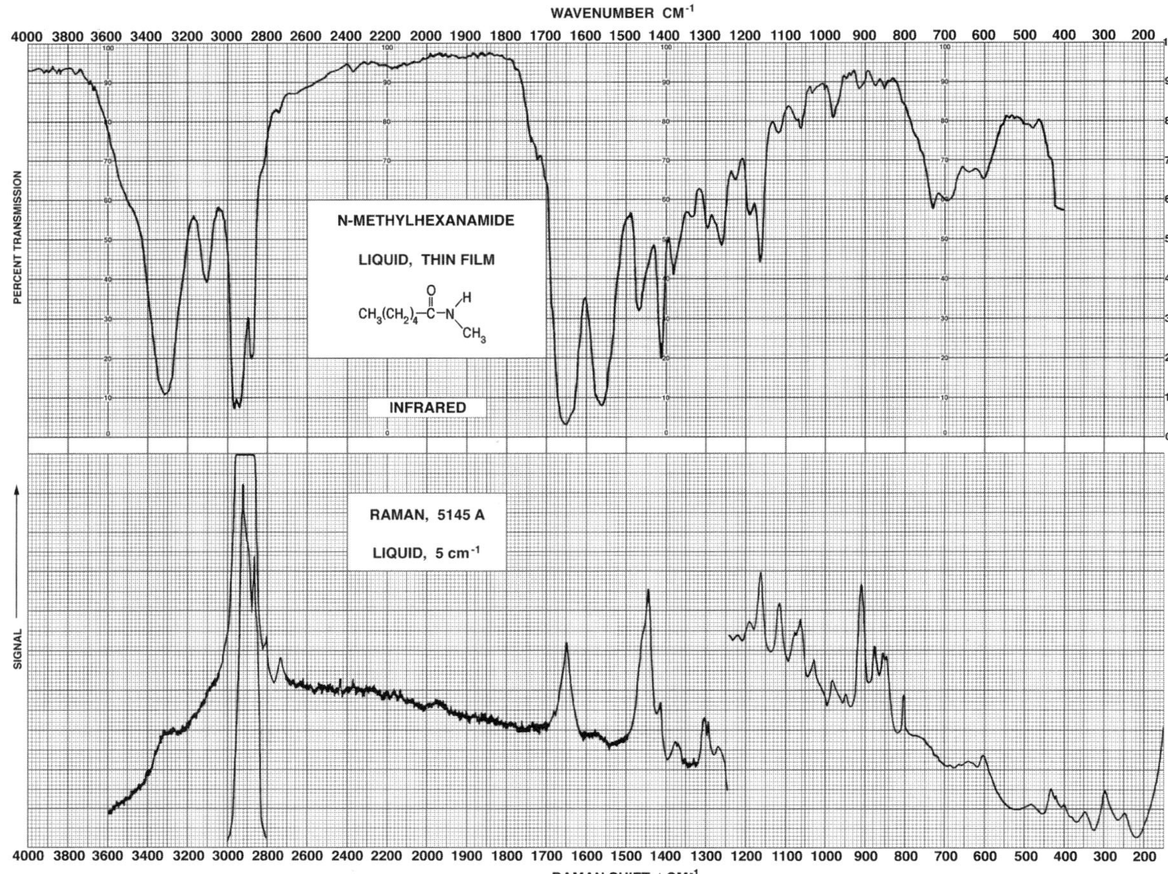

Fig 7.13 Infrared and Raman spectra of *N*-methylhexanamide.

IV. The major factors perturbing carbonyl frequencies can be summarized as follows:

Factors that raise the C=O frequency:

i. Electronegative substitution
ii. Decrease in C—CO—C internal bond angle

Factors that lower the C=O frequency:

i. Conjugation
ii. Hydrogen bonding

Because several of these factors may be operating simultaneously, careful judgment as to the contribution of each individual effect must be exercised in predicting carbonyl frequencies. This judgment develops rapidly with practice at interpretation.

A summary of carbonyl group frequencies is given in Table 7.7.

TABLE 7.7 Summary of Carbonyl Group Frequencies

Compound	$\tilde{\nu}$ Range (cm^{-1})
Ketones, aliphatic, open chain ($\mathbf{R}_2$C0)	1725–1700
Ketones, conjugated	1700–1675
Ketones, ring	See Table 7.1
Acid halides	>1800
Esters, aliphatic	1755–1735
Esters, conjugated	1735–1720
Esters (conjugated to oxygen)	1780–1760
Lactones	See Table 7.1
Anhydrides. aliphatic, open chain	1840–1810, 1770–1740
Acids, aliphatic	1725–1710
Amides	See text
Lactams	See Table 7.1
Aldehydes	1735–1720

REFERENCES

1. Bellamy, *Advances in Infrared Group Frequencies*, Chapter 5. Literature citations to 275 papers on carbonyl frequencies.
2. Lord and Miller, *Appl. Spectrosc.* **10**, 115 (1956).
3. J. O. Halford, *J. Chem. Phys.* **24**, 830 (1956).
4. Overend and Scherer, *Spectrochim. Acta* **16**, 773 (1960).
5. Overend and Scherer, *J. Chem. Phys.* **32**, 1296 (1960).
6. Miyazawa, Shimanouchi, and Mizushima, *J. Chem. Phys.* **24**, 408 (1956).
7. T. Shimanouchi, *Tables of Molecular Vibrational Frequencies: Consolidated Volume I*, Report NSRDS-NBS 39, U.S. National Bureau of Standards, Washington, DC, 1972.
8. Pimentel and McClellan, *The Hydrogen Bond*, 1980.
9. Kagarise and Whetsel, *Spectrochim. Acta* **18**, 315 (1962).

8 Amides, Carboxylate Ion, and C—O Single Bonds

FOIL A. MILLER

I. Amides, R–CO–N<

A. Summary of characteristic frequencies

Type of Amide	Number of N–H Stretch Bands	Amide I (C=O Stretch) (cm^{-1})	Amide II (cm^{-1})
Unsubstituted, or primary, R–CO–NH_2	2	1690–1650	1630–1620 (NH_2 scissors)
Monosubstituted, or secondary, R–CO–NH-R′	1	1650–1640	1570–1530
Disubstituted, or tertiary, R–CO–NR′R″	0	~1630	—

1. These bands are all vs or s in the IR.
2. They provide a good way to distinguish between the three types.
3. The frequencies are for liquids or solids, where hydrogen bonding operates.

B. N–H stretch

1. R–CO–NH_2. If hydrogen bonded, two bands between 3350 and 3180 cm^{-1}
2. R–CO–NH–R′
   a. With H trans to the C=O (the usual case) $3300 \pm 20\ cm^{-1}$
   b. With H cis to the C=O (as in lactams) $3160 \pm 20\ cm^{-1}$

C. "Amide I" band

1. This is the C=O stretch.
2. The standard C=O frequency of our reference ketone (1715 cm^{-1}) has been:
   a. Raised by the electronegative N atom

Course Notes on the Interpretation of Infrared and Raman Spectra, by Dana W. Mayo, Foil A. Miller, and Robert W. Hannah.
ISBN 0-471-24823-1

b. Lowered by resonance with the nonbonding electrons on the N atom
c. Lowered by hydrogen bonding

3. The net result is that the C=O stretch comes around 1650 cm^{-1}, about 65 cm^{-1} lower than our reference value of 1715 cm^{-1}.

D. "Amide II" band

1. In primary amides, R–CO–NH_2:
   a. Amide II is the NH_2 scissoring.
   b. It may be hidden under amide I or may be only a shoulder on it.
2. In secondary amides, R–CO–NH–R′
   a. Amide II is a mixture of the C–N stretch and the N–H in-plane bend:

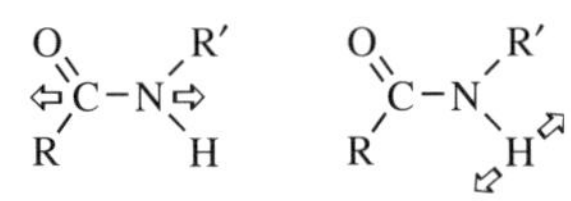

1) They interact to give two bands: ~1550 cm^{-1} (amide II) and 1300–1220 cm^{-1} (amide III). The first is very useful in the IR. The latter is not a useful group frequency. It is usually somewhat better in the Raman than in the IR.
2) This is another example of first-order interaction. Each of the two final modes has some stretching and some bending component.

   b. The 1550 cm^{-1} band is very distinctive.

1) Only two other groups have bands near here: –COO^- and –NO_2. Because each of the three groups has a second strong characteristic band, there is no chance of confusing the secondary amide with either carboxylate or nitro groups.

	cm^{-1}	cm^{-1}	cm^{-1}
Secondary amide	~1650	1550 ± 20	—
Carboxylate ion	—	1580 ± 30	1360 ± 60
Nitro group	—	1530 ± 30	1350 ± 20

2) Typical contours. See Figure 8.1.

The contour for a secondary amide is very diagnostic and should be impressed on one's memory.

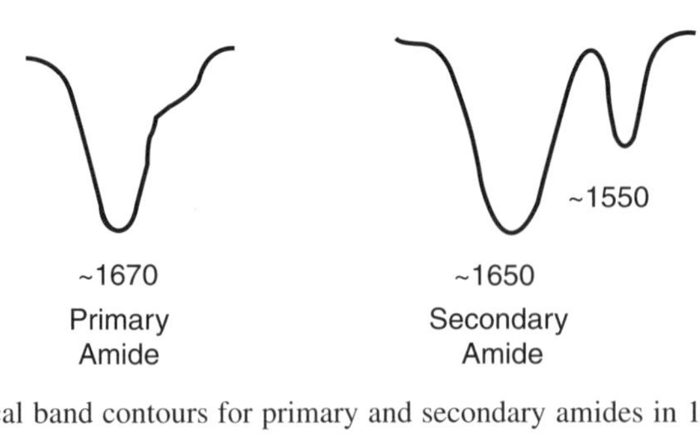

Fig. 8.1 Typical band contours for primary and secondary amides in 1600 cm^{-1} region.

c. 1550 cm^{-1} is characteristic of trans secondary amides only. Lactams, which must be cis, do not have this band.

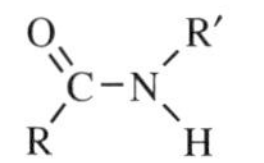

A trans amide

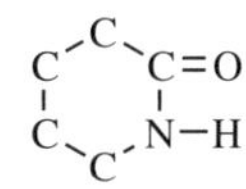

A lactam

Open-chain amides are nearly always trans and exhibit the 1550 cm^{-1} band.

d. Secondary amides are important and widespread. The peptide link

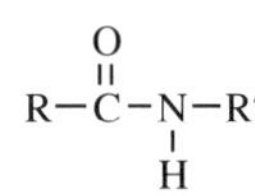

is a secondary amide: It occurs in proteins, polypeptides, and nylons.

3. Tertiary amides do not have amide II.

E. Examples of amide spectra
   1. 1-Hexanamide. See Figure 8.2.
      a. Note the two N–H stretches at 3370 and 3200 cm^{-1}.
      b. Amide I, the carbonyl stretch, is near 1670 cm^{-1}.
      c. Amide II, the NH_2 scissoring, is not resolved, but asymmetry of the 1670 cm^{-1} band hints that it is underneath the 1670 cm^{-1} (amide I) band.
      d. The broad band near 700 cm^{-1} is the $–NH_2$ wag. Another band at 710 cm^{-1} is superimposed on it. This may be the CH_2 rock of the chain of four methylenes seemingly displaced from its usual position of 720 cm^{-1} by the band overlap.
   2. *N*-methylhexanamide. See Figure 8.3.
      a. There is only one N–H stretch, near 3300 cm^{-1}.
      b. The weaker band near 3100 cm^{-1} is the overtone of the 1570 cm^{-1} amide II band. ($2 \times 1570 = 3140$ cm^{-1}.) The overtone probably picks up intensity from the N–H stretch by Fermi resonance. This band is often seen in IR spectra of secondary amides.
      c. Note the amide I and amide II bands near 1650 and 1570 cm^{-1}. This contour is very characteristic of secondary amides and is unique to them. Commit it to memory!
      d. Amide III is the unimpressive band near 1270 cm^{-1}. It is useless as a group frequency.
      e. The Raman spectrum is of very little help with these group frequencies. The N–H stretch is very weak, the 1570 cm^{-1} band is missing, and so is amide III in this case.

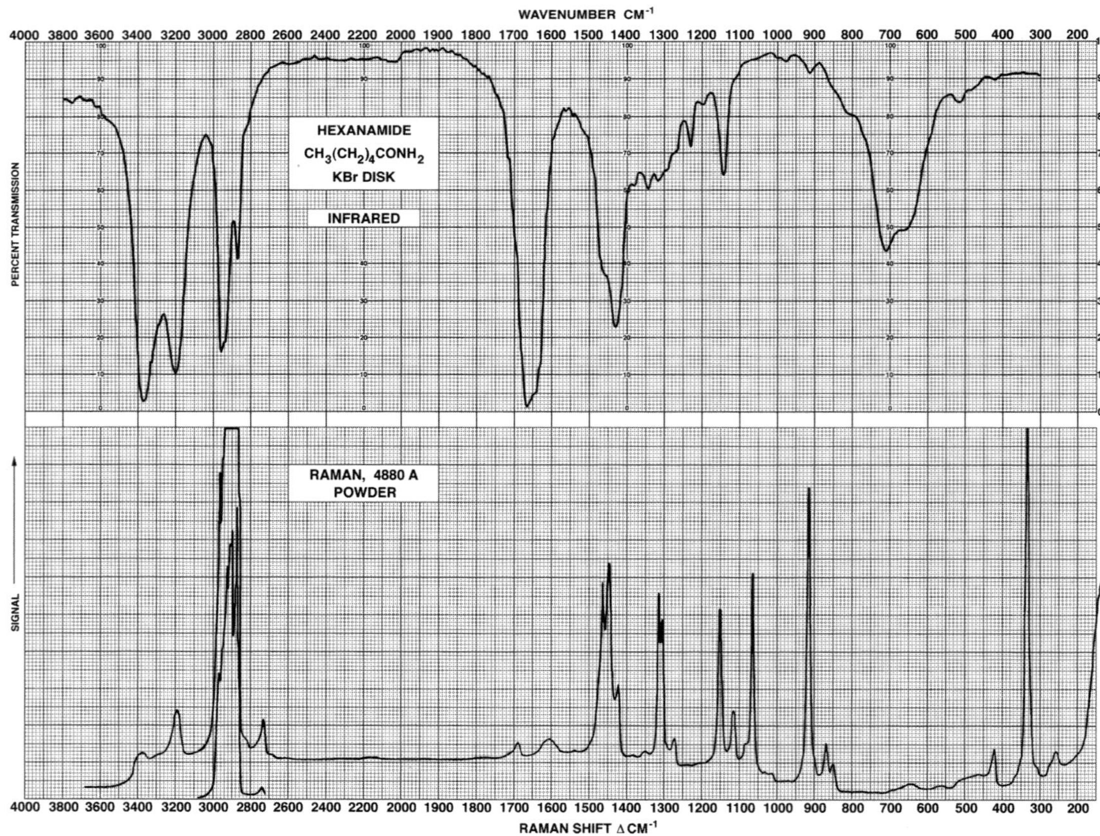

Fig. 8.2 Infrared and Raman spectra of 1-hexanamide, $CH_3(CH_2)_4CONH_2$.

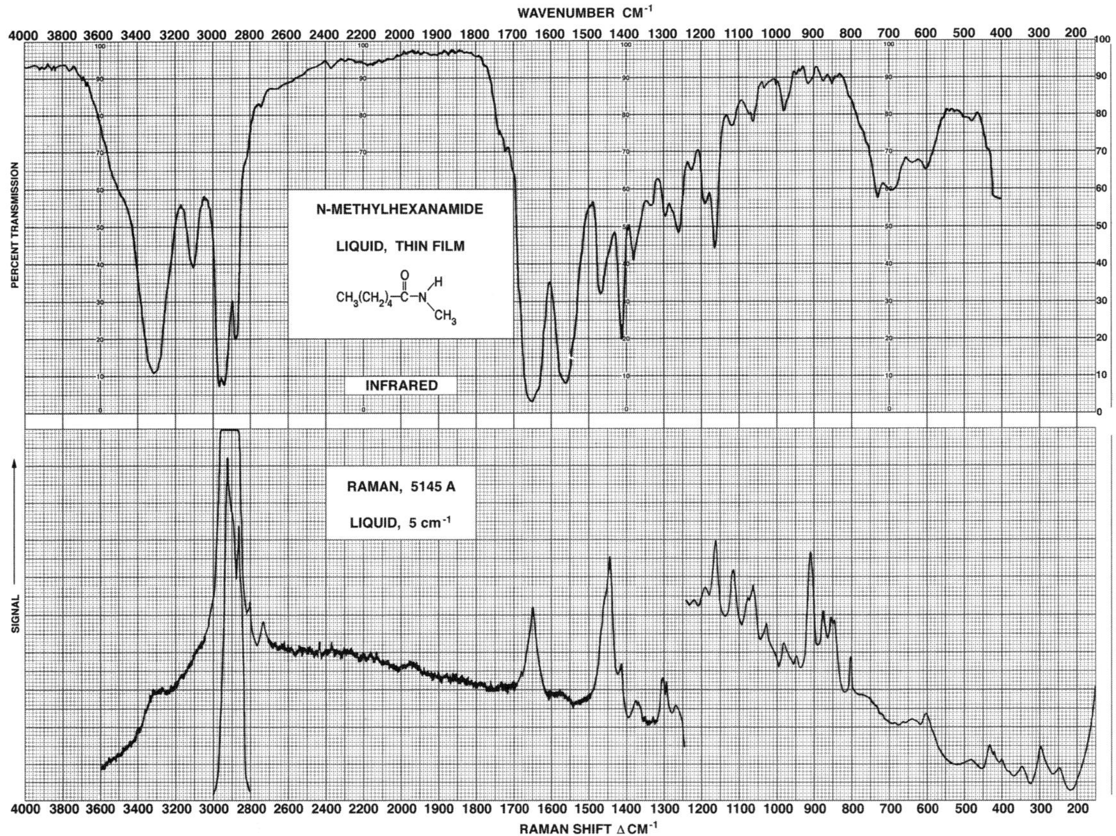

Fig. 8.3 Infrared and Raman spectra of *N*-methylhexanamide, $CH_3(CH_2)_4CO{-}NH{-}CH_3$.

II. Carboxylate ion, $-COO^-$

A. This has the structure

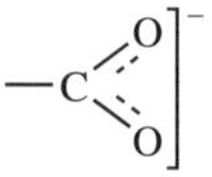

1. The two bonds are identical. The conditions for mechanical interaction are fulfilled, so two bands are expected.
2. The bonds are intermediate between single and double, so we anticipate two frequencies lower than for a carbonyl.

B. Observed: 1580 ± 30 cm^{-1} vs
1360 ± 60 cm^{-1} s to m. Widely variable in position and intensity. Often not helpful.

Both bands are rather broad.

C. *Example:* Sodium acetate. See Figure 8.4.

1. The two bands are about 1590 and 1420 cm^{-1}. The lower band is abnormally intense here.
2. Note that the 1420 cm^{-1} band completely overlaps the two $-CH_3$ deformations, the degenerate one near 1450 cm^{-1} and the symmetric one near 1378 cm^{-1}.
3. Changing the positive metal ion has only a small effect.

III. C–O single bonds

A. These are present in ethers, alcohols, esters, anhydrides, and carboxylic acids. Their bands occur in the fingerprint region (usually between 1250 and 1000 cm^{-1}) and are subject to erratic shifts from quite small structural changes. Fortunately they are very intense in the IR, which usually enables them to be identified. They are useful for confirming an ester, deciding on the type of ether or alcohol, or suggesting a ketone (by their absence).

B. Ethers, C–O–C

1. The antisymmetric stretch is a good group frequency; the symmetric one is not.
2. Frequency:
   a. Saturated ether —C–O–C— 1100 ± 50 cm^{-1} vs
   b. Doubly unsaturated ether (including aryl) =C–O–C= 1250 ± 20 cm^{-1} s. Often slightly broad.
   c. Mixed ether e.g., —C–O–C= Has *both* 1100 and 1250-cm^{-1} bands!
3. Why does unsaturation *raise* the C–O stretch? Resonance gives the C–O single bond a little more double-bond character:

$$\mathrm{C{-}O{-}CH{=}C{<}} \qquad \text{and} \qquad \mathrm{C{-}\overset{+}{O}{=}CH{-}\overset{-}{C}{<}}$$

4. Example of a saturated ether: *n*-Hexyl ether. See Figure 8.5.

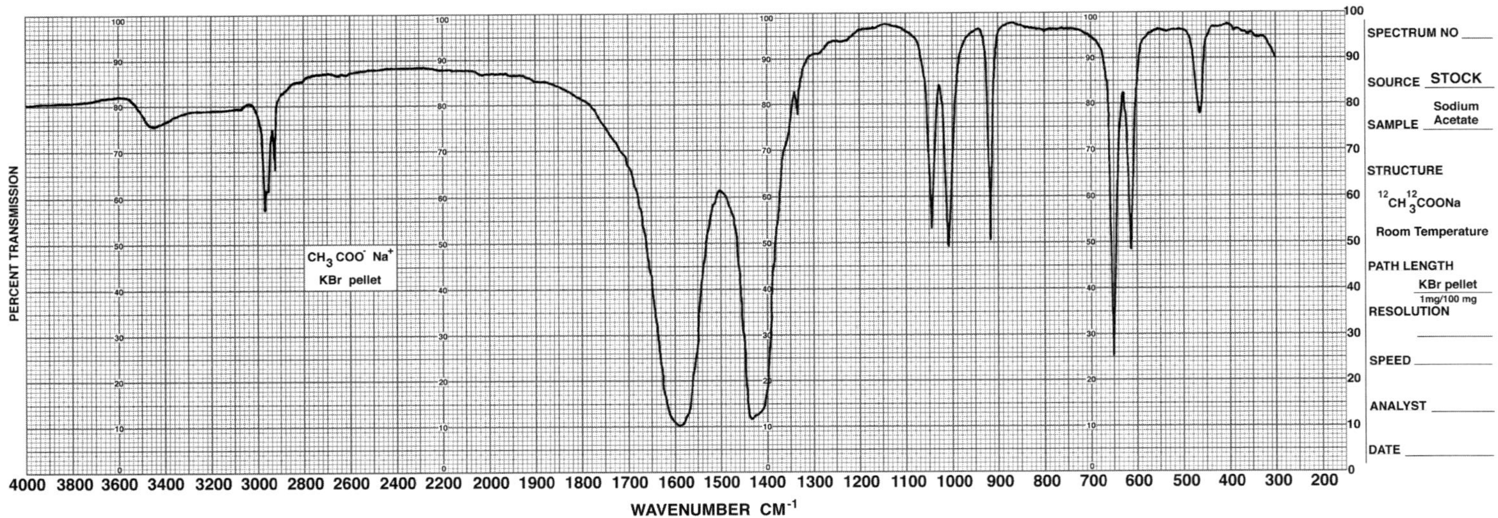

Fig. 8.4 Infrared spectrum of sodium acetate, $CH_3COO^- \cdot Na^+$.

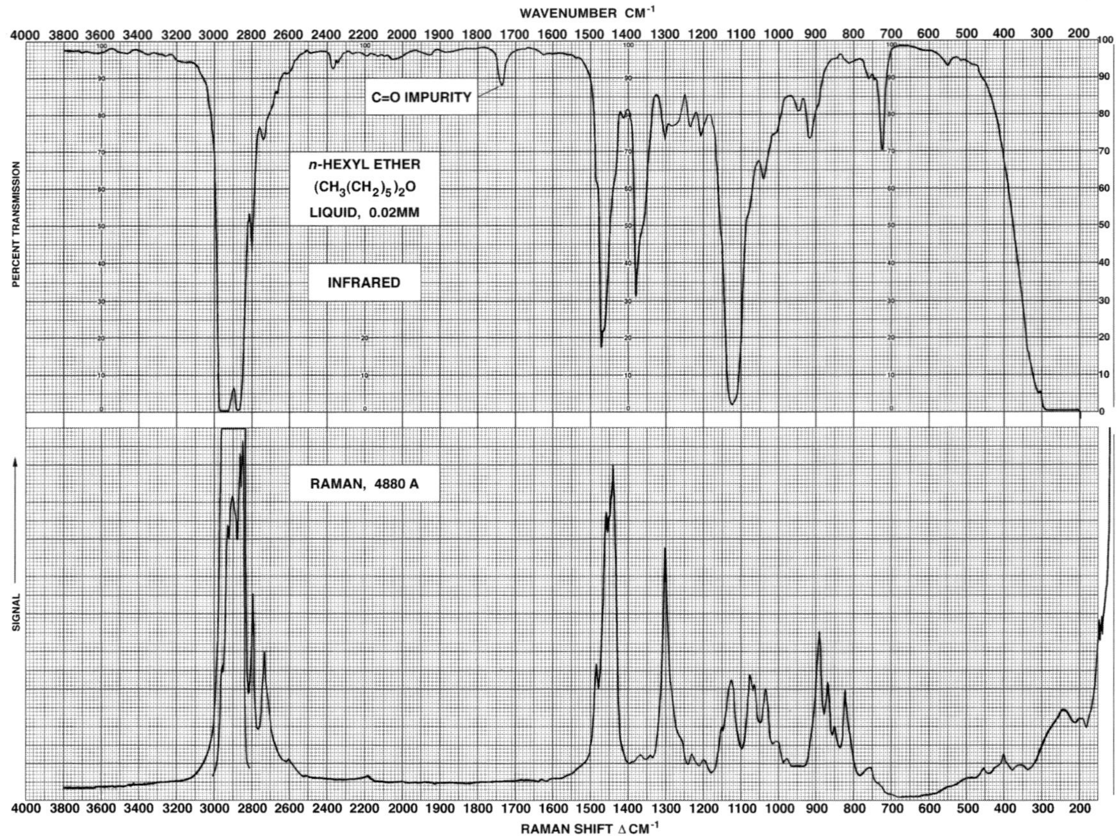

Fig. 8.5 Infrared and Raman spectra of *n*-hexyl ether, $(n\text{-}C_6H_{13})_2O$.

a. Note the very strong C–O–C antisymmetric stretch at 1125 cm^{-1}. It is the only band characteristic of ethers. The symmetric stretch is not obvious.
b. The C–O–C stretch is unimpressive in the Raman spectrum.

5. Epoxides, 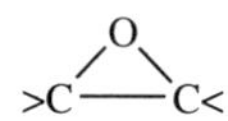

a. ~1250 cm^{-1}, s — Symmetric stretch
   ~860 cm^{-1}, s — Antisymmetric stretch (the better band)
b. This is not reliable unless one is asking whether or not the sample is an epoxide; then it is suggestive.
c. Note that the descriptions are reversed from those for conventional ethers.
d. Cis epoxide — ~885 cm^{-1}, vs
   Trans epoxide — ~845 cm^{-1}, vs

C. Esters, R–CO–OR′

1. This group is analogous to a mixed ether: $R{-}CO\overset{a}{-}O\overset{b}{-}R'$.
   Therefore we expect bond *a* to give a band around 1250 cm^{-1} and bond *b* to give one around 1100 cm^{-1}.
2. Observed:
   a. 1200 ± 50 cm^{-1}, vs. (About 50 cm^{-1} lower than for unsaturated ethers.) In acetates this band is near 1230 cm^{-1} and is useful.
   b. A second, weaker band is often found in the 1200–1000 cm^{-1} range. It is more variable in position and intensity and is not useful.
3. Ketones do not have these bands. Therefore these bands help distinguish an ester from a ketone
4. *Example:* Hexyl acetate. See Figure 8.6.
   a. In hexyl acetate the bands are near 1242 cm^{-1} (vvs) and 1044 cm^{-1} (s). Note the great intensity of the 1242 cm^{-1} band—comparable to that of the carbonyl.
   b. Neither band is useful in the Raman spectrum.

D. Anhydrides, R–CO—O—CO–R

1. The C–O stretch is similar to that of an ether.

Cyclic anhydride — ~1250 cm^{-1}, vs (cf. a doubly unsaturated ether)
Open-chain anhydride — ~1050 cm^{-1}, vs (cf. a saturated ether)

2. *Example:* Hexanoic anhydride. See Figure 8.7.
   a. Note the intense IR band at 1043 cm^{-1}, missing in the Raman spectrum.
   b. Anhydrides can always be identified with confidence from the two high carbonyl stretches.

E. Peroxides, C–O–O–C

1. It is very difficult to identify these from IR or Raman spectra.
2. ~1100 — Ether type C–O stretch.
   890–820 cm^{-1}, w — Said to be the O–O bond stretch. Of little use.

F. Alcohols and carboxylic acids have already been discussed in Chapter 6 on X–H groups.

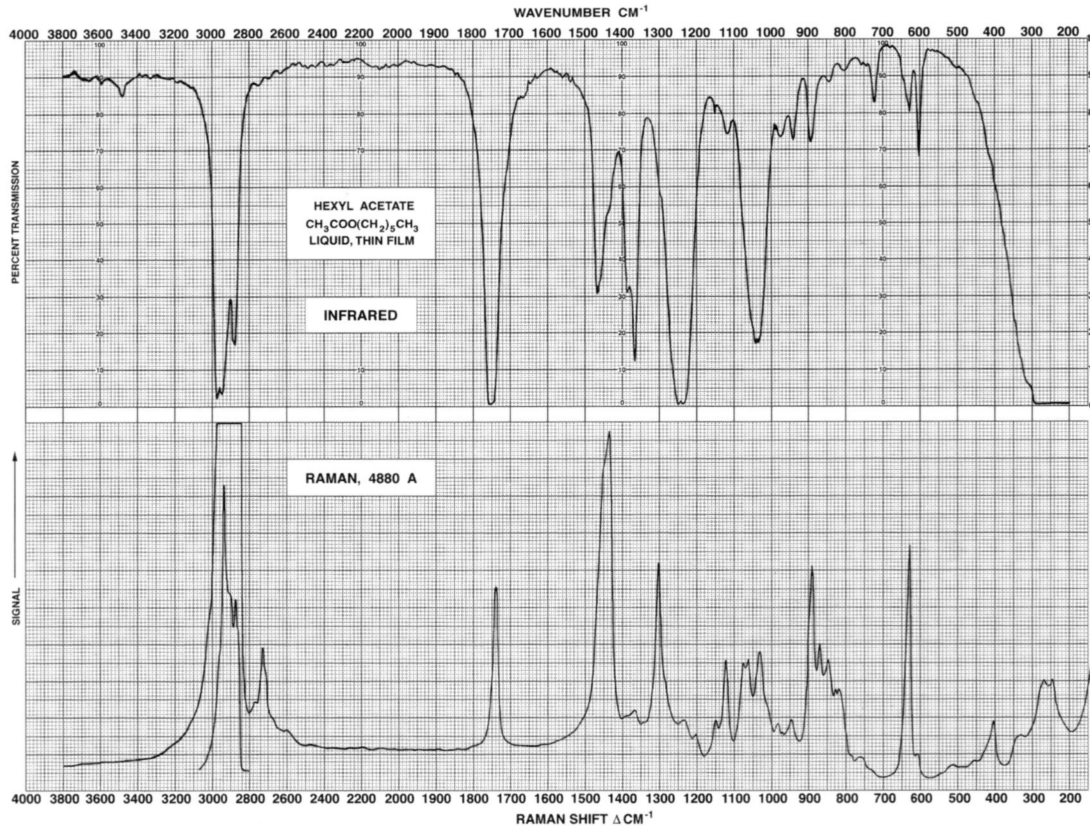

Fig. 8.6 Infrared and Raman spectra of *n*-hexyl acetate, $CH_3COO\text{-}n\text{-}C_6H_{13}$.

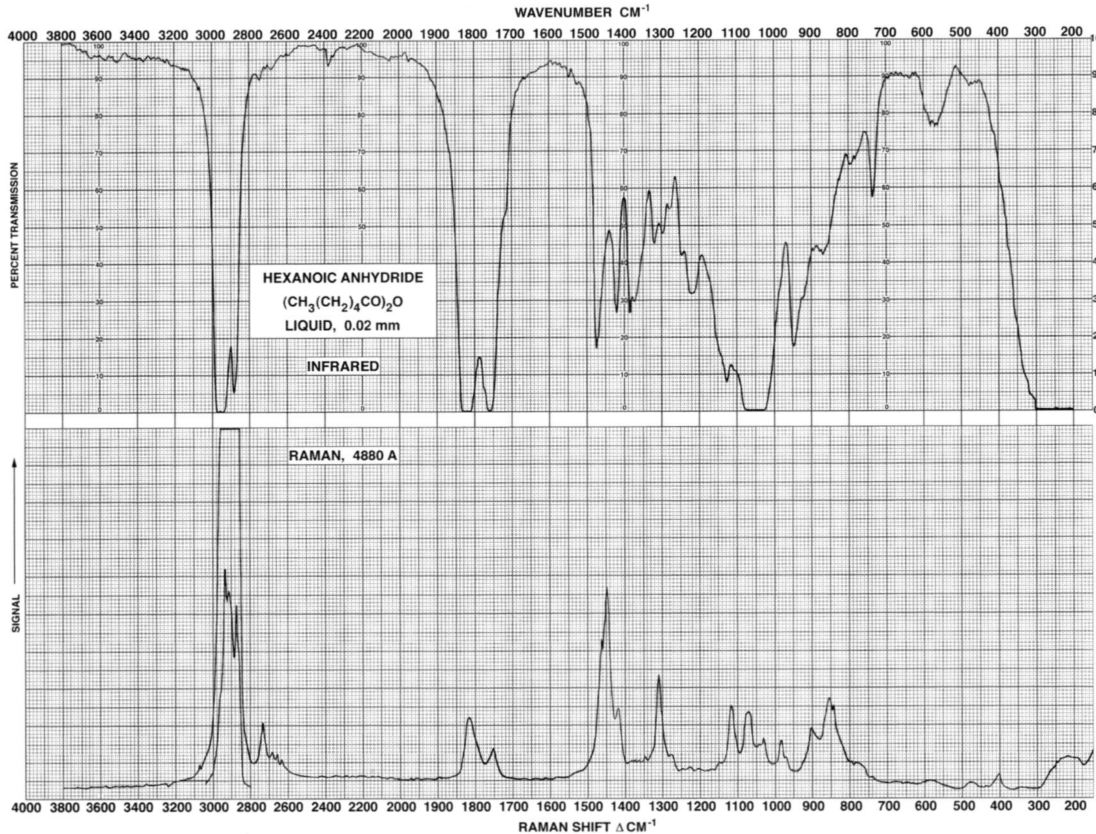

Fig. 8.7 Infrared and Raman spectra of *n*-hexanoic anhydride, n-C_5H_{11}–CO–O–CO-n-C_5H_{11}.

9 Groups Containing N=O Bonds, or Si, P, S, or Halogen Atoms

ROBERT W. HANNAH

GENERAL COMMENTS

I. All the group frequencies for these elements occur below 1500 cm^{-1} except for the X–H stretches and some N=O stretches. They are therefore among the fingerprint vibrations. They all give strong bands, but finding a band at the expected frequency is not adequate evidence for the presence of the group. If it is known by other means that one of these elements is present, then the bands are useful. Consequently, an elemental analysis is very helpful, and every IR laboratory should be prepared to do the Beilstein copper wire test for Cl, Br, and I and to do the sodium fusion test for N, S, and halogens

II. Bonding
  A. The bonds have considerable ionic character (high polarity), which leads to strong IR absorptions.
  B. The oxygen atom is quite basic and therefore subject to strong hydrogen bonding and perturbation by Lewis acid groups. Consequently, group frequencies may be quite sensitive to physical state; i.e., frequencies may be quite different for solid, liquid, and solution.
  C. Configuration is also important.
    1. R–N=O is a bent-planar system. Therefore, N=O groups lie in the plane of the substituent and can enter into resonance interaction.
    2. Bonding around P and S atoms is pyramidal. Therefore, P=O and S=O groups are *not in* the plane of the substituents and are *not subject* to resonance effects and strained ring effects.
  D. All the X=O groups are subject to inductive effects. This is an electronegative effect transmitted through σ-bonds.

Course Notes on the Interpretation of Infrared and Raman Spectra, by Dana W. Mayo, Foil A. Miller, and Robert W. Hannah.
ISBN 0-471-24823-1

E. In addition to the intense stretching bands of these groups, they also have other absorptions at lower frequencies. These latter bands often provide confirmation for the group. Only rarely can one rely on a single band as proof of the presence of a group.

NITROGEN–OXYGEN COMPOUNDS

I. The N=O bond behaves similarly to the >C=O bond but comes at a lower frequency.

II. N doubly bonded to one oxygen: —N=O

A. The normal frequency is about 1580 cm^{-1} (aliphatic) or 1500 cm^{-1} (aromatic).

B. Nitroso: C—N=O.

1. These dimerize readily, and then the frequency falls by 100–400 cm^{-1}.
2. The N=O stretching frequency is subject to resonance effects since the N=O group can lie in the plane of the substituent group. This is a necessary requirement for resonance to occur. The frequency is also subject to inductive effects related to the electronegativity of the R group.
3. For the monomer, the frequencies are:
   a. Aliphatic —N=O stretch, 1620–1540 cm^{-1}
   b. Conjugated —N=O stretch, 1525–1490 cm^{-1}
4. Primary and secondary alkyl nitroso compounds are unstable and isomerize (enolize) to oximes, $-CH_2-N=O \rightarrow -CH=N-OH$. This has led to substantial confusion in the assignments in the earlier literature. In solution, primary and secondary nitroso compounds exist as dimers and the principal absorption is near 1290 cm^{-1}.
5. Tertiary alkyl and aromatic nitroso compounds cannot isomerize (enolize) to oximes and in solution exist as dimers. In dilute solution, monomers exist.
   a. For the dimers, cis and trans configurations occur and a partial N=N bond is formed:

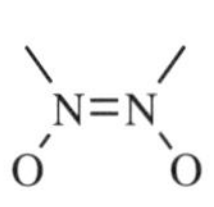

cis

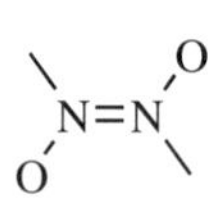

trans

   b. The trans form has a center of symmetry and, therefore, the N=N stretch will not appear in the IR but will appear in the Raman. Similarly, the in-phase N—O stretch is Raman active while the out-of-phase stretch is IR active. The similarity in behavior to cis and trans alkenes should be noted.

	Trans	Cis
Alkyl N=N	~1490 cm^{-1}, Raman	1420–1330 cm^{-1},
N—O in-phase stretch	1290–1175 cm^{-1}, Raman	IR and Raman
N—O stretch		1344–1323 cm^{-1}, IR and Raman
Aryl N=N	~1440 cm^{-1}, Raman	~1407 cm^{-1},
N—O in-phase stretch	1300–1253 cm^{-1}, Raman	IR and Raman
N—O		1397–1389 cm^{-1}, IR and Raman

c. C—N stretch

There are two bands due to coupling in the dimer.

Alkyl C—N stretch	~1100 cm^{-1}, IR, vs	860–750 cm^{-1}, IR, m
Aromatic	1340–1320 cm^{-1}, IR, vs	

The aromatic C—N stretching frequencies are higher than expected due to interactions with ring vibrations.

C. Nitrites, R—O—N=O

1. The N=O frequency is raised (in comparison with the nitroso N=O stretch) by the ether oxygen because it is an electronegative substituent (inductive effect). (The carbonyl group behaves similarly.)
2. The N=O stretch gives rise to strong bands in both IR and Raman in the 1680–1610-cm^{-1} region. These groups exist in trans and cis isomers:

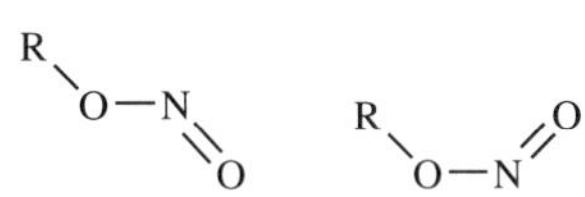

trans cis

These geometric isomers can give rise to two strong bands in this region in the IR:

N=O stretch 1680–1653 cm^{-1} (trans) 1625–1613 cm^{-1} (cis)

3. Inceased branching at the α-carbon which increases the acidity of the carbon of the C—O bond reduces the frequency of the N=O stretch, but still within the ranges quoted. However, the intensity ratio which is sensitive to isomer distribution changes markedly. For the gas phase the trans intensities compared to the cis intensities for the N=O stretch are stronger for the trans form by the following factors:

Primary	1–3
Secondary	5–10
Tertiary	35–50

For condensed phases, the literature is incomplete, but similar ratios should be expected. The interpretation of these ratios is that in going from primary to secondary to tertiary substitution of the α-carbon, there is an increase in the concentration of the trans configuration.

4. Spatial interaction with highly electronegative substituents on the R group raises the frequency. A similar "field effect" is observed with carbonyls.

D. Nitrosoamines, R_2>N—N=O

1. The frequency of the N=O stretch drops in comparison to the nitroso N=O stretch. This results from the substitution of the less electronegative singly bonded nitrogen atom for the singly bonded oxygen atom.
2. Alkyl monomer (vapor) N=O stretch ~1490 cm^{-1}
   Solution monomer ~1450 cm^{-1}
   Dimer ~1310 cm^{-1}
3. Conjugation of aromatic substituents with the singly bonded nitrogen results in increased double-bond character in the N=O group and causes slightly higher N=O stretching frequencies to occur.
4. N—N stretch:

Alkyl	1150–1030 cm^{-1}
Aromatic	1025–925 cm^{-1}

   The frequency drops with aromatic substituents for the same reason the N=O frequencies rise with conjugation, i.e., decreases the force constant of the N—N bond.
5. C_2>N—: The symmetric stretch of the C—N bonds (in disubstituted aliphatic nitrosoamines) gives rise to a strong Raman band near 850 cm^{-1}.

III. N bonded to two equivalent oxygen atoms, —NO_2 and —O—NO_2

A. Nitro and nitrate groups have symmetric and antisymmetric N=O stretches.

1. Many of the bands are very strong in both the IR and the Raman and are good group frequencies. The symmetric stretch may couple with other vibrations and has a wider frequency range. In the vapor-phase IR the antisymmetric band intensity will be stronger than the symmetric band by a factor of 2–16, and in condensed phases the antisymmetric band also will generally be the stronger band of the two.
2. The lower frequency band (symmetric stretch) may have several components; the reasons why are unknown.

B. Nitro, R—NO_2. (All band centers are given in reciprocal centimeters.)

	Antisymmetric Stretch	IR	Raman	Symmetric Stretch	IR	Raman
Aliphatic	1601–1531	vs	m-w	1381–1310	s	vs
Aromatic	1555–1487	vs	m-w	1357–1318	s	vs

1. Conjugation lowers both stretching bands ca. 30 cm^{-1}, as shown in the table, and increases the intensities. For aryl nitro compounds both —N=O stretches are mixed with some aromatic ring vibrations.
2. The wavenumber locations of —N=O stretches are dependent on the electronegativities of the substituents. With increasing electronegativity, the antisymmetric frequency is raised and the symmetric frequency is lowered. The antisymmetric $—NO_2$ frequency can be calculated from the sum of Taft's σ^* values for the substituents,

$$\tilde{\nu}_a = 1538 + 25.92 \sum \sigma^* \text{ values}$$

   The values are good to 10–12 cm^{-1}.
3. Another way of calculating the antisymmetric and symmetric stretches is to use the values for CH_3NO_2 reference and the shift factors tabulated below:

	$CH_3—NO_2$: $\tilde{\nu}_a(NO_2) = 1558\,cm^{-1}$ (antisymmetric), $\Delta\tilde{\nu}_a(NO_2)$	$CH_3—NO_2$: $\tilde{\nu}_a(NO_2) = 1375\,cm^{-1}$ (symmetric), $\Delta\tilde{\nu}_s$ (NO_2)
α-Substituents		
—H	0	0
$—CH_3$	−5	−8
$—C_2H_5$	−5	−8
$—C_6H_5$	—	−8
—C=O	+10	−8
—F	+17	−23
—Cl	+17	−23
—Br	+17	−23
$—NO_2$	+29	−29
β-Substituents		
—H	0	0
—OH	−4	−6
—Cl	+2	−4
—Br	+2	−4
$—NO_2$	+2	−15

Example: $CH_3—CCl_2—NO_2$. All the substituents are on the α-carbon.

$\tilde{\nu}_a(NO_2) = 1558 + 1 \times (-5)$ [for methyl] $+ 2 \times (+17)$ [for 2 Cl]

$= 1587\,cm^{-1}$. (Observed $1581\,cm^{-1}$.)

$\tilde{\nu}_s(NO_2) = 1375 + 1 \times (-8)$ [for methyl] $+ 2 \times (-23)$ [for 2 Cl]

$= 1321\,cm^{-1}$. (Observed $1325\,cm^{-1}$.)

4. Other vibrations
   a. Primary and secondary alkyl nitro compounds exist in trans and gauche conformations leading to two C—N stretches:

      1° Trans 915–900 cm^{-1}, w in IR, s in Raman
      1° Gauche 895–873 cm^{-1}, w in IR, s in Raman
      2° Trans 910–870 cm^{-1}, w in IR, s in Raman
      2° Gauche 863–847 cm^{-1}, w in IR, s in Raman

      Since these are in the fingerprint region, they are less useful as group frequencies.
   b. Aromatic C—N, 1177–865 cm^{-1}

      Often coupled with ring vibrations. Not useful.
   c. NO_2 deformations

      Because this is a planar three-atom terminal group, there are three deformation modes available to the system. An in-plane bend (scissors), an out-of-plane wag, and an in-plane rock. The twist mode of the methylene system is a rotation mode in the case of the nitro group.

	Aliphatic	Aryl
In-plane bend	660–610 cm^{-1}	857–830 cm^{-1}
Out-of-plane wag	560–477 cm^{-1}	
In-plane rock	495–470 cm^{-1}	

      All are medium to very weak in both the IR and Raman and are not useful group frequencies.
5. *Example*: *p*-Nitrotoluene, *p*-CH_3—C_6H_4—NO_2. See Figure 9.1.
   a. 3100 cm^{-1}, weak, sharp, two bands, sp^2 (aromatic) C—H stretch.
   b. 2940–2870 cm^{-1}, sp^3 aliphatic C—H stretches
   c. 1600 cm^{-1}, aromatic ν_8 ring vibration (Note the moderate intensity of this band in the IR for the unsymmetrically para-substituted ring.)
   d. 1515 cm^{-1}, antisymmetric NO_2 stretch
   e. 1340 cm^{-1}, symmetric NO_2 stretch
   f. 840 cm^{-1}, out-of-plane C—H bend, two adjacent hydrogens, para-substitution

C. Nitrate, R—O—NO_2
1. Organic nitrates are explosive. BEWARE!
2. NO_2 stretch (Band centers are given in reciprocal centimeters)
   a. Note that the antisymmetric stretch is 70 cm^{-1} higher than for the nitro group (~1565 cm^{-1}).
   b. Note the symmetric stretch is 70 cm^{-1} lower than in the nitro group (~1345 cm^{-1}).

Antisymmetric	IR	Raman	Symmetric	IR	Raman
1640–1620	vs		1285–1270	vs	vs

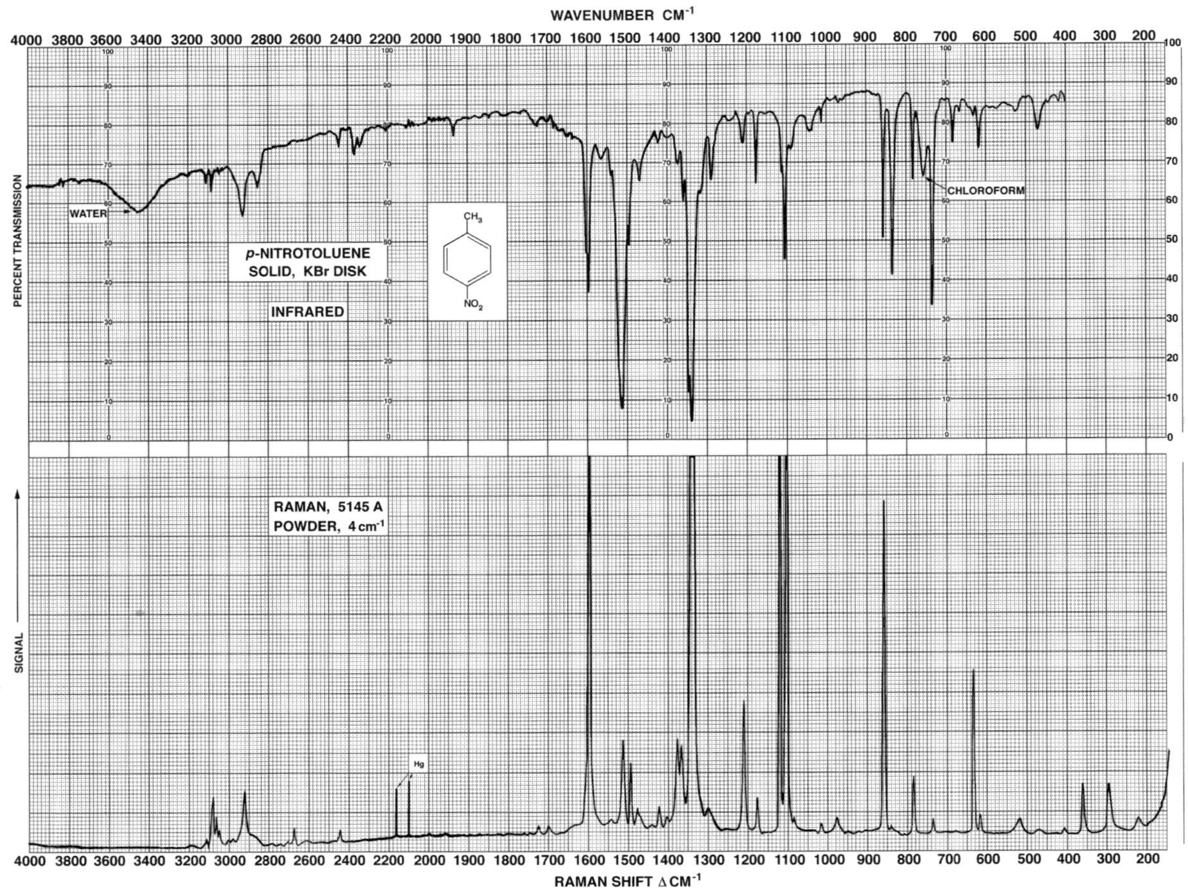

Fig. 9.1 Infrared and Raman spectra of *p*-nitrotoluene.

3. N—O stretch, 870–855 cm^{-1}, IR, s
4. NO_2 in-plane bend (scissors) 760–755 cm^{-1}, IR, s
   NO_2 out-of-plane wag 710–695 cm^{-1}, IR, s
5. The N—O stretch and NO_2 in-plane and out-of-plane bends for nitrates are in the fingerprint region and, while strong, are less useful as group frequencies. If the nitrate group is thought to be present based on the N=O stretches, then these vibrations are useful as evidence for the group.

IV. N—O single bonds do not have a useful group frequency.

SILICON COMPOUNDS

I. General remarks

A. The word "silicone" is a misnomer. It was originally used because silicon was postulated to form double bonds with oxygen analogous to ketones. However, silicon does not normally form double bonds, although silicon–oxygen bonds do show some properties of bonds with order > 1.

B. The electronegativity of Si is only 1.8 (carbon 2.5), and that of oxygen is 3.5. Therefore, Si—O bonds are quite ionic and absorb *very* strongly in the IR—about 3.5 times more than most IR bands. Consequently a small amount of silicone grease can dominate the spectrum in the 1300–600-cm^{-1} range. (Its spectrum is given later.)

II. Si—H bands

A. The Si—H stretch occurs in the range 2280–2050 cm^{-1}, and it can be predicted accurately from:
1. Modified electronegativities of other substituents. See A. L. Smith and N. C. Angelotti, *Spectrochim. Acta* **15**, 412 (1959).
2. The Taft σ^* values for the substituents; $\tilde{\nu}(\text{Si—H}) = 2106 + 17.5 \sum \sigma^*$. See H. W. Thompson, *Spectrochim. Acta* **16**, 238 (1960).
3. For example:

	Stretch	IR	Raman	Bend	IR
Alkyl	2110–2094	s	m to s	842–800	s
Aryl	2115–2103	s	m to s	842–800	s
F_3SiH	2282				
Br_3SiH	2236				
$(MeO)_3SiH$	2203				

B. Stretches in $—SiH_3$ and $—SiH_2—$
1. Si is so heavy that coupling between the Si—H stretches is small. Therefore the two Si—H stretches for each group are often not resolved. This is an example of accidental degeneracy.

2. $R–SiH_3$

	Stretch	Deformation	Rock
Alkyl	2153–2142 cm^{-1}	947–930 cm^{-1} (antisymmetric) 930–910 cm^{-1} (symmetric)	720–680 cm^{-1}
Aryl	2157–2152 cm^{-1}	Same as alkyl	720–680 cm^{-1}
$–C{\equiv}C–SiH_3$	2190–2170 cm^{-1}		

All are strong in the IR.

3. R_2SiH_2 (Band centers are given in reciprocal centimeters.)

	Stretch	Scissors	Wag	Twist	Rock
Alkyl	2138–2117	950–930	800–840	690–560	540–480
Aryl	2147–2130	940–928	870–843		

All are strong in the IR and weak in the Raman.

4. *Example*: Figure 9.2, diphenyl silane, $(C_6H_5)_2SiH_2$. (Band centers are given in reciprocal centimeters.)
   a. 2140, Si–H stretch
   b. 1950–1750, aromatic ring sum tones, monosubstituted.
   c. 1585, ν_8 aromatic ring vibration
   d. 1480, 1430, ν_{19a} and ν_{19b} ring vibrations
   e. 1120, Si-phenyl
   f. 935, SiH_2 scissors
   g. 840, SiH_2 wag
   h. 735, out-of-plane aromatic CH bend, monosubstituted
   i 695, out-of-plane aromatic ring bend, monosubstituted
   j. *Note* how weak the aromatic bands at 735 and 695 cm^{-1} appear compared to the SiH_2 bands.

III. $Si–CH_3$ deformations

A. Most organosilicon compounds contain Si–Me groups. They have a strong and very characteristic absorption at 1280–1255 cm^{-1} due to the symmetric deformation of CH_3 on Si. This is strong and sharp in the IR and weak in the Raman.

B. With two or more Me on Si, this band is always accompanied by an equally strong but broader (complex) band near 800 cm^{-1}. This results from the overlap of the Si–C stretch and the antisymmetric CH_3 rocking modes.

1. $>SiMe_2$, 1280–1255 cm^{-1}, s in IR; 800 $cm^{-1} \pm 10$ cm^{-1}, s in IR, broader than 1270 cm^{-1}.
   Example: $Me_3Si–O–(Me_2Si–O)_x–SiMe_3$; see Figure 9.3.
2. $SiMe_3$, 1280–1255 cm^{-1}, s in IR; ~840 and ~760 cm^{-1}, s in IR, broader than 1270 cm^{-1}. *Example*: $Me_3Si–O–SiMe_3$; see Figure 9.4.

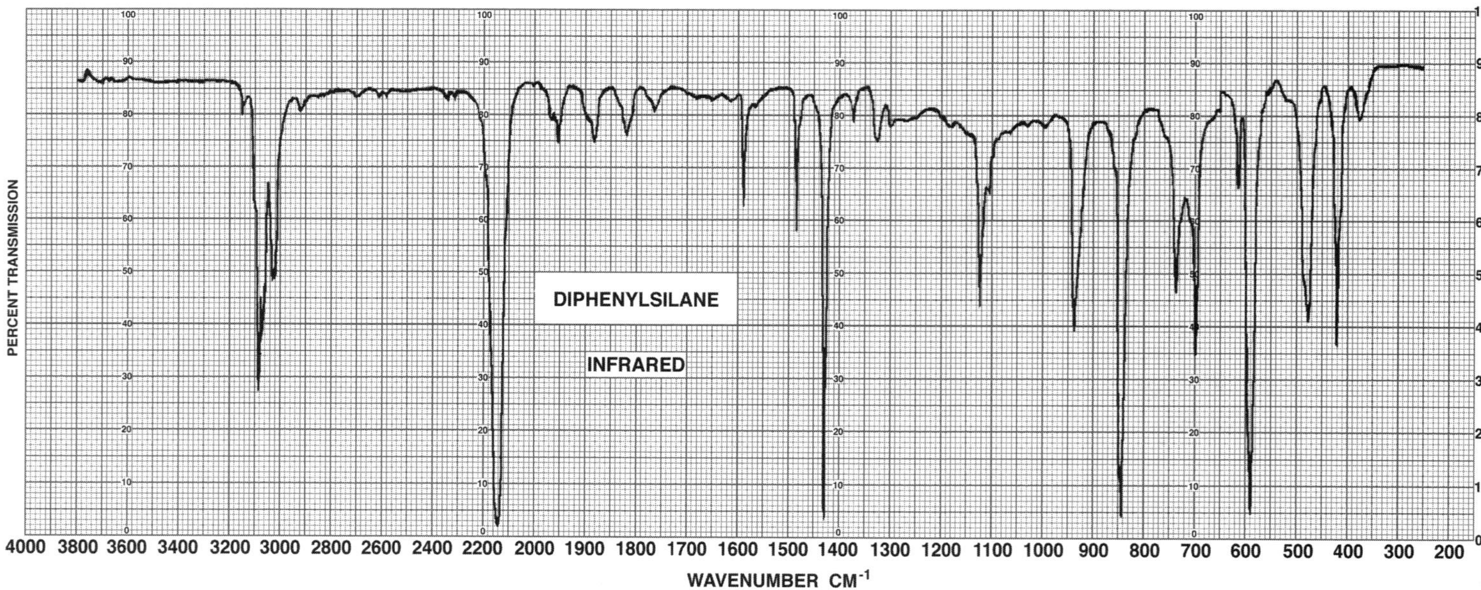

Fig. 9.2 Infrared spectrum of diphenylsilane.

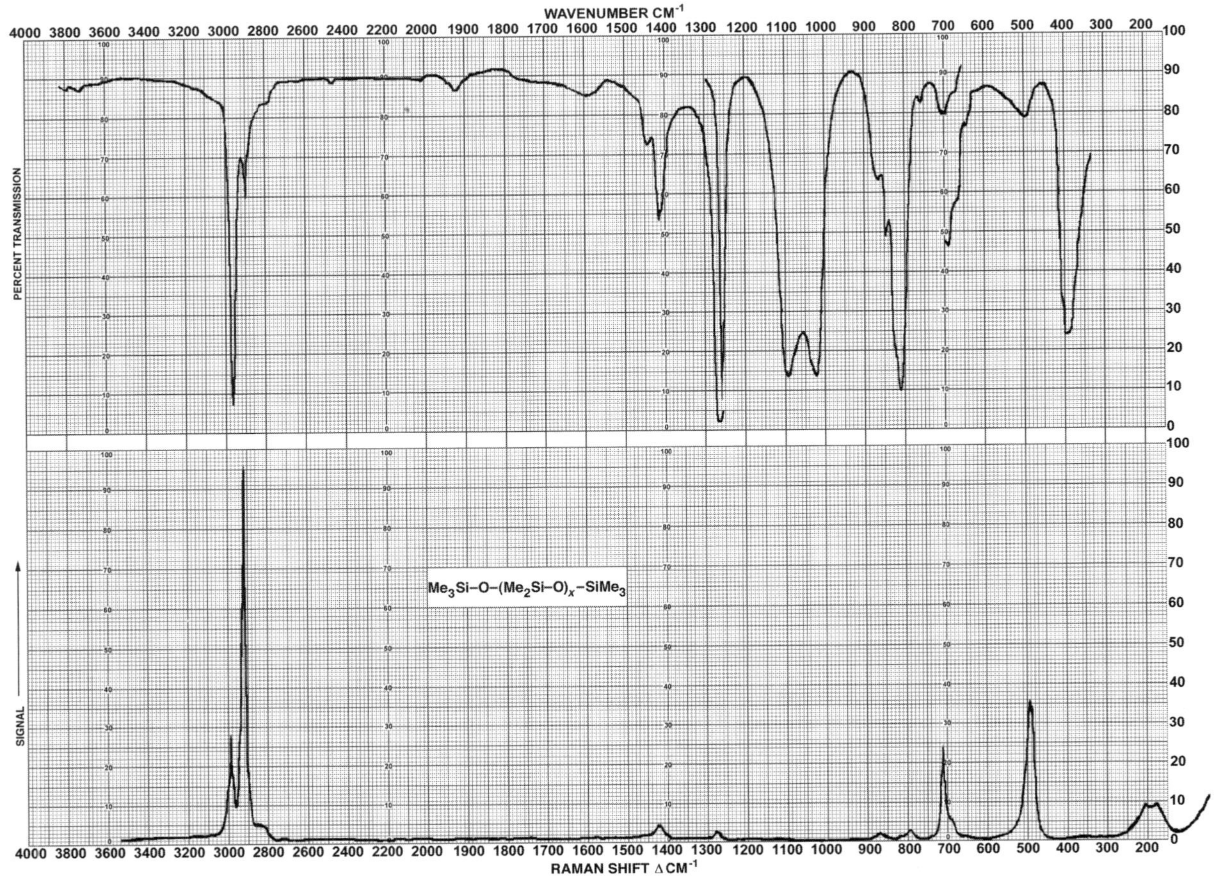

Fig. 9.3 Infrared and Raman spectra of $Me_3Si–O–(Me_2Si–O)_x–SiMe_3$. Note the useful IR bands near 1260, 1100 (doublet), and 810 cm^{-1}.

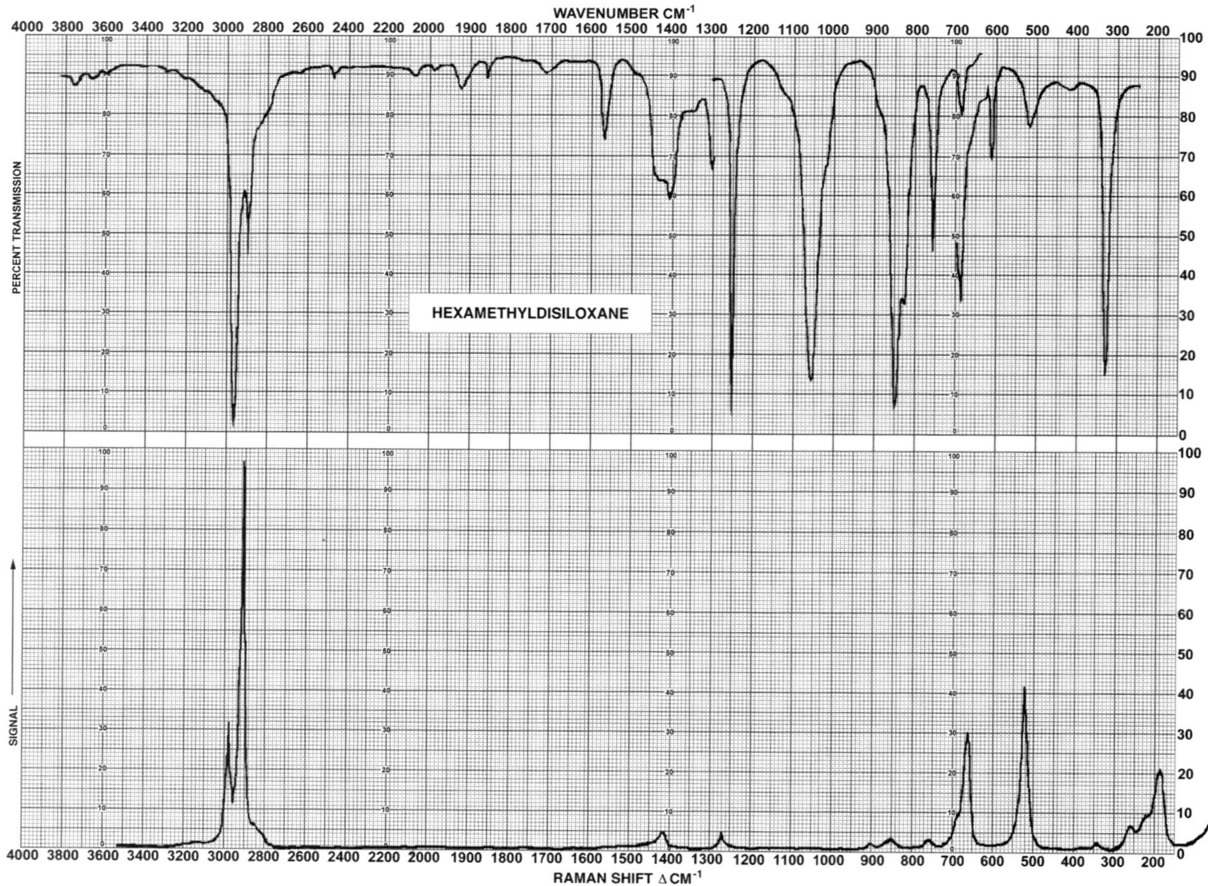

Fig. 9.4 Infrared and Raman spectra of Me_3Si–O–$SiMe_3$. Note the characteristic IR bands near 1250, 1050, 845, and 760 cm^{-1}.

C. Other Si–R groups
1. Si–CH=CH_2, 1615–1590 cm^{-1} (C=C stretch), 1430–1390 cm^{-1} (=CH_2 scissors), 1020–1000 cm^{-1} (trans CH=CH_2 wag), 980–950 cm^{-1} (terminal = CH_2 wag)
2. Si–phenyl, 1125–1100 cm^{-1} and aromatic bands near 1600, 1450, 730, and 700 cm^{-1}

IV. Siloxanes, Si–O–Si

A. 1100–1000 cm^{-1} (vvs, vb in IR, w in the Raman). Due to the antisymmetric Si–O–Si stretch.
1. *Example*: Spectrum of Me_3–Si–O–Si–Me_3. See Figure 9.4.
2. The 1100–1000-cm^{-1} band is a doublet for polymers containing more than eight Si atoms; frequencies are close to 1090 and 1020 cm^{-1} (vvs and vvb in IR and w in Raman).
   a. *Examples*: Spectrum of Me_3Si–O–(Me_2Si–O)$_x$–$SiMe_3$; see Figure 9.3.
   b. *Examples*: Silicone grease; see Figure 9.5. Beware of silicone grease! The 1100–1000-cm^{-1} band is vvs.
      1) It is a polymer with a repeat unit (–Si$(CH_3)_2$–O–Si$(CH_3)_2$–O–)$_n$ plus a little finely divided SiO_2 powder to thicken it that will also have broad absorption in the 1100-cm^{-1} region.
      2) The 1260-cm^{-1} band is the symmetric deformation of CH_3 on Si. The band at 800 cm^{-1} is due to the Si–C stretch of two CH_3 groups on Si, and the 1050-cm^{-1} complex band is assigned to the Si–O–Si antisymmetric stretch.
      3) Silicone grease is used widely as a lubricant. It is slightly soluble in $CHCl_3$ and many other organic solvents.
3. Si–OH

3. Si–OH	Si–O stretch	920–830 cm^{-1}
	O–H stretch	3700–3200 cm^{-1}

V. Halosilanes, Si–X

a. Si–Cl stretch	625–420 cm^{-1}	
$SiCl_3$	620–570 cm^{-1}, s in IR	530–450 cm^{-1}, m in IR
$SiCl_2$	600–535 cm^{-1}, s in IR	540–460 cm^{-1}, m in IR
Si–Cl	550–470 cm^{-1}, s in IR	
b. Si–F	1000–800 cm^{-1}, s in IR	

c. The Si–X compounds react with moisture to form siloxanes.

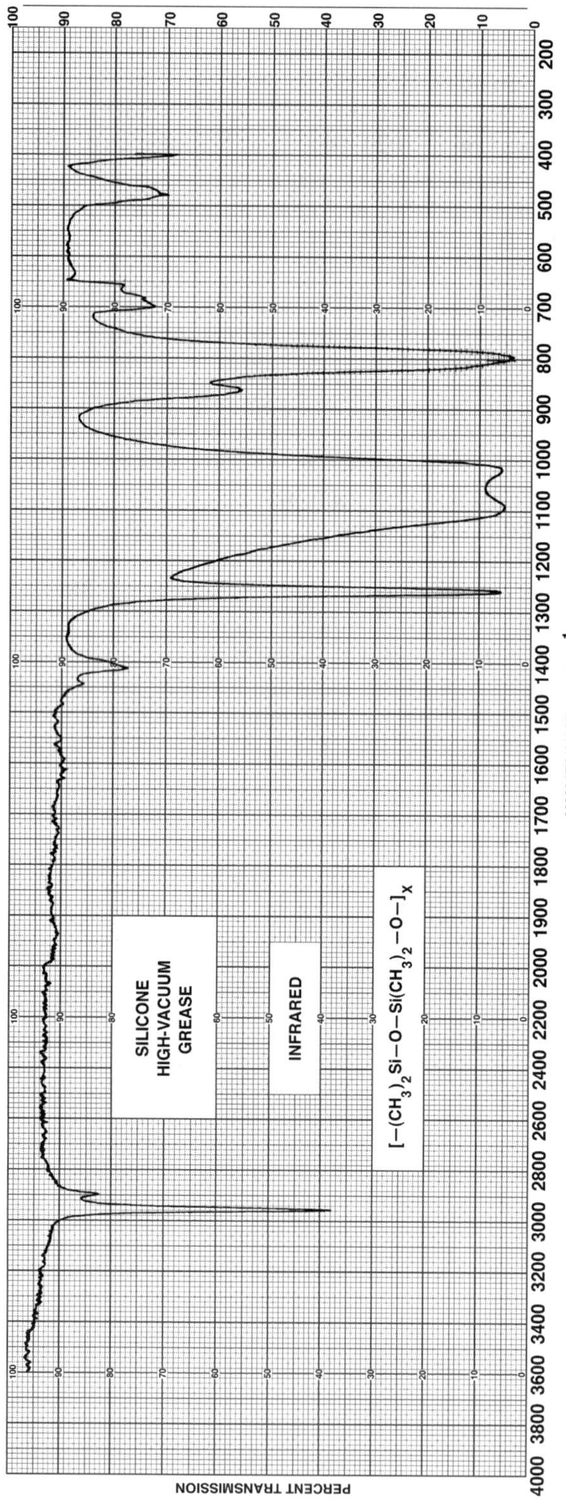

Fig. 9.5 Infrared spectrum of silicone grease.

VI. References

A. A. L.Smith and J. A. McHard, *Anal. Chem.* **31**, 1174–1179 (1959).
B. A. L. Smith, *Spectrochim. Acta* **16**, 87 (1960). Structure–spectra correlations.
C. D. R. Anderson, in A. L. Smith (Ed.), *Analysis of Silicones*, Wiley, New Yorks, 1974, pp. 247–286. (QD412 S6 S55).
D. E. D. Lipp and A. L. Smith, Silicones: Infrared, Raman, Near infrared, and Ultraviolet Spectroscopy. in A. L. Smith (Ed.), The Analytical Chemistry of Silicones, Wiley-Interscience, New York 1991, Chapter 11, pp. 305–345. (QD412 S6 A74).
E. A. L. Smith, *Coblentz Society Deskbook*, pp. 422–423.
F. There are many more references for Si compounds.

PHOSPHOROUS COMPOUNDS

I. General remarks

A. There is an enormous amount of information in the literature on the IR spectra of these compounds. The situation is quite complex because of the great variety and number of substances. For example, Thomas and Chittenden have reviewed the results on over 900 compounds containing a P=O group [*Spectrochim. Acta* **20**, 467 (1964)]. If interested in details, see also the chapter in Bellamy and references cited there.

B. One usually cannot identify the presence of P from the IR spectrum. If it is known by other means that P is present, IR bands are often useful.

II. P—H stretch. 2505–2222 cm^{-1}, strong and sharp in IR and Raman

Example of use: At one time it was thought that compounds $P(OH)(OR)_2$ existed, until it was found from the IR spectrum that they have a P—H bond. They are really $H{-}P(=O)(OR)_2$.

A. For PH_2, the bond angle is close to 90° (about 94°), and with the presence of the heavy central P atom coupling between the antisymmetric and symmetric P—H stretching bands is greatly reduced and the bands are not resolved.

B. P—H stretch

1. C—P(H)—C: The P—H stretch appears at the lower end of the quoted range, i.e., nearer to 2220 cm^{-1}.
2. O—P(=O)(H)—O (phosphonic derivatives) and O—P(=O)(H)—C (phosphinic derivatives): The P—H stretch appears at the upper end of the range, i.e., nearer to 2500 cm^{-1}.

III. PH and PH_2 bends and P—CH_3

A. P—H bend, 990–885 cm^{-1}, alkyl, w in IR; aryl, ms in IR

B. P<H_2 bends

1. H—P—H in-plane bend (scissors), 1090–1075 cm^{-1}, m to s in IR; m to w in Raman

2. H–P–H out-of-plane wag, 840–810 cm^{-1}, m to s in IR; m to w in Raman

C. P–CH_3

1. CH_3 symmetric deformation, 1305–1295 cm^{-1}, m in IR; w in Raman
2. CH_3 antisymmetric deformation, 1450–1395 cm^{-1}, w in IR; w in Raman
3. CH_3 rock, 977–842 cm^{-1}, m in IR; mw in Raman

IV. P=O stretch.

A. Four factors affect P=O frequencies: inductive effects, H bonding, rotational isomers, and to a smaller extent valence angle.

1. Substitution by elements more electrogenative than carbon or hydrogen should raise the P=O frequency (for exactly the same reason that the C=O frquency is raised)
2. H bonding will lower P=O stretching as in C=O systems.
3. Conformational isomers formed via rotational isomerism will lead to band broadening or complex band shape of P=O stretching modes.
4. Contraction of the X_2>P valence angle should result in a rise in the P=O frequency while an increase in the X_2>P angle should lead to a drop in the P=O frequency. But these shifts would be expected to be smaller than those exhibited by carbonyl systems because the displacement of the massive P atom is considerably smaller.

B. Free P=O stretch (as in dilute solution of P acids), 1350–1250 cm^{-1} (s).

C. Because these bands occur in the fingerprint region, they cannot be used without some other indication that P is present, because C–O and O–H groups can also absorb strongly in these ranges.

D. These bands are not, however, affected by conjugation and resonance because the bonds around P are not planar. Thus, P=O groups behave in a fashion similar to S=O but unlike C=O when the potential for conjugation exists. For example, $\nu_{(P=O)}$ in $Ph_3P=O$ is not lower than that in $Me_3P=O$; it is actually a little higher (this also suggests that geometric effects resulting from steric repulsion of the R groups attached to the massive P atom are minimal). This rise in frequency probably occurs because the sp^2-hybridized carbons of the conjugating group are slightly more electronegative than the sp^3 carbons of nonconjugating carbon systems.

E. As in carbonyls, however, attaching an electronegative atom to P raises the P=O stretch. The effect of each additional electronegative oxygen is roughly additive.

F. Aromatic phosphates, $(ArO)_3P=O$.

The P=O stretch is often a doublet, The reason for this doubling has not been established, but a likely possibility is rotational isomerism.

G. If one of the substituents is an OH group, then in condensed phases or concentrated solution the OH stretch is found in the strong H-bonding range 2725–1600 cm^{-1}, very broad and strong in IR.

H. The P=O stretch is very sensitive to H bonding—more so than in carbonyls because the oxygen atom in this case is considerably more basic than the C=O oxygen.

I. *Note also* that both ether and alcohol bands occur in this region and can complicate assignments.

V. P—O—C stretches

A. P—O—C (aromatic), 1240–1190 cm^{-1}, vs in IR
P—O—C (aliphatic), 1050–990 cm^{-1}, vs in IR

1. These bands mainly arise from stretching of the C—O section, with observed values similar to those found in ethers.
2. *Note* that the spectrum of an organic phosphorous ester or acid will exhibit bands arising from C—O stretching in addition to the P=O stretch.
3. In many of the compounds containing P—O—C there is also a band near 950 cm^{-1} (vs in IR). It may involve a component of P—O single bond stretch.
4. Bands in the 1080–850-cm^{-1} region of these compounds can be very intense and useful. There is much detailed knowledge known about them; see references to detailed data.
5. There is also an additional band found near ~800 cm^{-1} which has been assigned to mostly P—O stretch.

VI. P—O—P stretch, 1025–900 cm^{-1}, out-of-phase stretch, s in IR

VII. *Example*: Spectrum of tri-*p*-cresylphosphate, (p-CH_3—C_6H_4—$O)_3P$=O; see Figure 9.6.
Note that the P=O stretch is high because there are three electronegative oxygen atoms on P (in addition to the doubly bonded one). Note also the P=O band at 1300 cm^{-1} is a doublet.

VIII. References on IR spectra of P compounds

A. Thomas and Chittenden, *Spectrochim. Acta* **20**, 467 (1964).
B. Nyquist and Potts, in M. Halman (Ed.) *Analytical Chemistry of Phosphorous Compounds*, Wiley, New York, 1972, Chapter 5. Good reference.
C. R. A. Nyquist and C. D. Craver, *Coblentz Society Deskbook*, pp. 399–403. Includes a Colthup-type chart for phosphorous compounds.
D. D. Lin-Vien, N. B. Colthup, W. G. Fateley, and J. G. Grasselli, *Infrared and Raman Characteristic Frequencies of Organic Molecules*, Academic, New York, pp. 263–276, 1991.

SULFUR COMPOUNDS

I. S—H stretch, 2600–2550 cm^{-1}

A. Weak and sharp in the IR and may not be seen. Strong in the Raman.

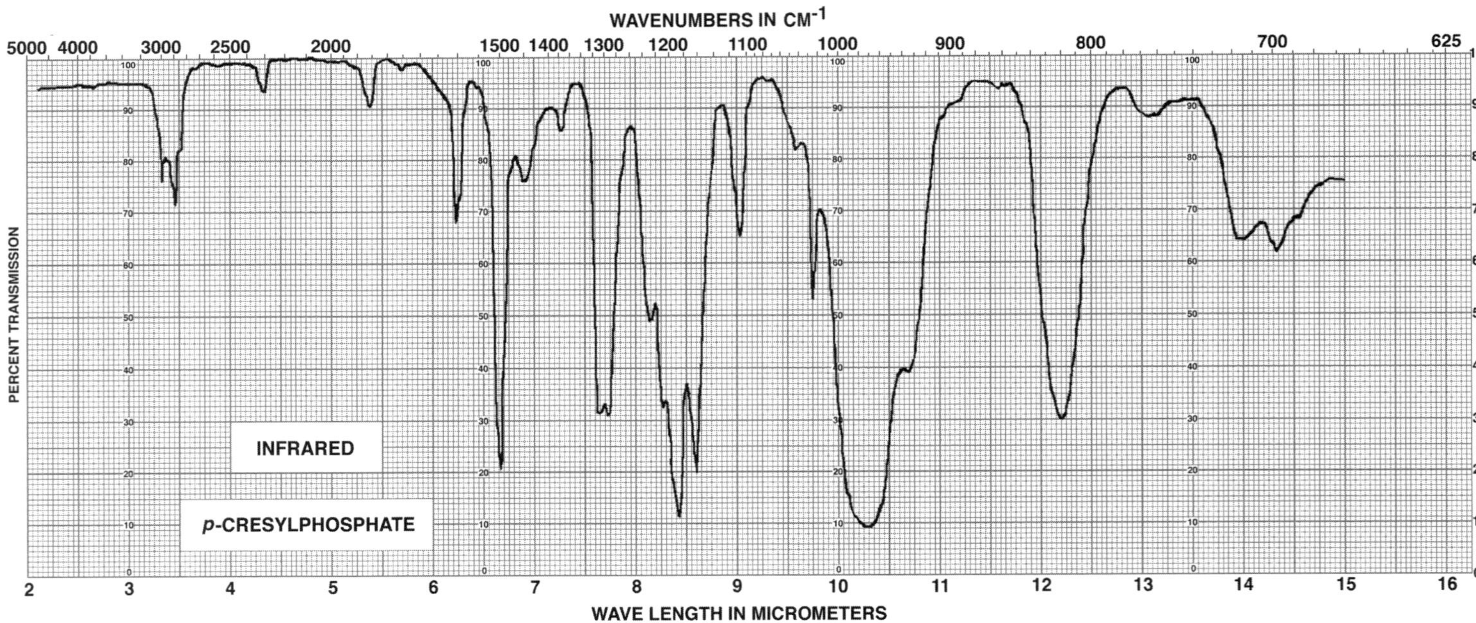

Fig. 9.6 Infrared spectrum of tri-*p*-cresylphosphate. Note that the spectrum is linear in micrometers. The label on the spectrum says *p*-cresylphosphate. A more accurate name would be "*p*-tricresylphosphate."

B. One does not need a spectrum to know that S—H is present; the odor suffices.

C. *Example*: Spectrum of *n*-hexanethiol; see Figure 9.7.

II. C—S and S—S stretches

A. C—S stretch, 735–590 cm^{-1}, m to s in Raman

B. S—S stretch, ~530 cm^{-1}, m to s in Raman; m to w in IR

C. Neither is particularly useful in the IR, but they are of considerable value in the Raman, where these modes have been extensively studied in proteins and other sulfur-containing natural products.

III. S=O stretch

A. The $C_1C_2S{=}O$ group is not planar; it is pyramidal. Consequently, the S=O stretch is insensitive to conjugation and resonance, as in the case of P=O systems, and is not very sensitive to ring size, as in P systems. These disubstituted S=O groups are also *less prone* to exhibit conformational isomerism.

B. The frequency is, therefore, mainly affected by the electronegativity of the substituents. Their effect is approximately additive. *Examples*:

1. Replacing C by a Cl raises the S=O stretch by ~100 cm^{-1} or more. Replacing the second C by another Cl raises it ~100 cm^{-1} more.
2. For the inductive effect of oxygen, see Section C.

C. Compounds containing one S=O bond

1.

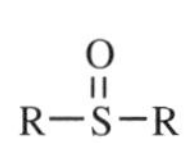

Sulfoxide

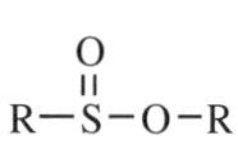

Sulfinic ester

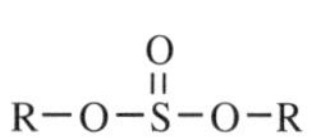

Sulfite

Sulfoxide	Sulfinic ester	Sulfite
1050 ± 20 cm^{-1}, s in IR, m in Raman	1130 ± 10 cm^{-1}, s in IR, vs in Raman	1210 ± 30 cm^{-1}, vs in IR, m-w in Raman

a. Other bands, sulfites
   1) Antisymmetric O—S—O stretch, ~690 cm^{-1}, vs in IR, m in Raman
   2) Symmetric O—S—O stretch, ~733 cm^{-1}, vs in IR, s in Raman

b. Other bands, sulfoxides
   1) Antisymmetric C—S—C stretch, ~689 cm^{-1}, m in IR, s in Raman
   2) Symmetric C—S—C stretch, ~672 cm^{-1}, m in IR, s in Raman
   3) C—S—C stretch is about 10 times stronger in Raman than the S=O stretch.

2. Note how the S=O stretch rises as additional single-bonded oxygen atoms are attached to S.

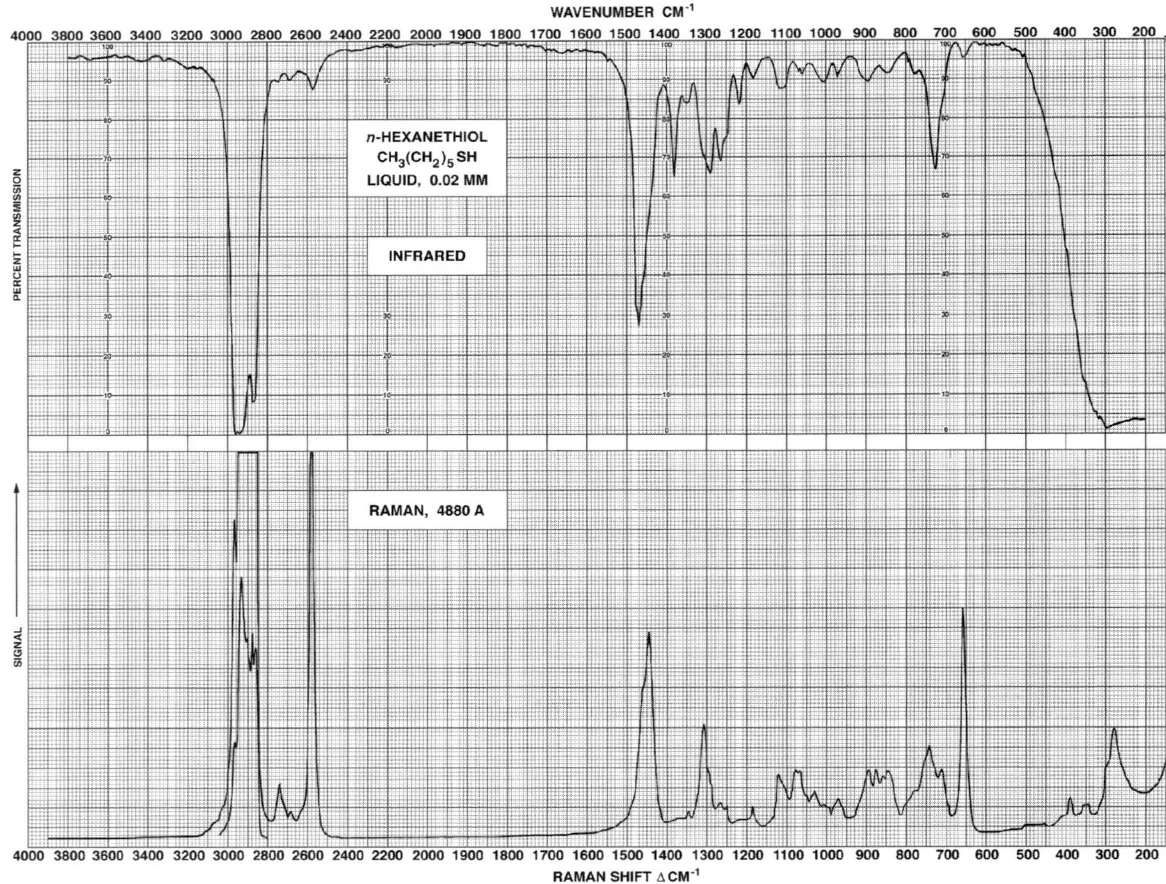

Fig. 9.7 Infrared and Raman spectra of *n*-hexanethiol.

3. *Example:* Dimethyl sulfoxide; see Figure 9.8.

D. Compounds containing an >SO_2 group (two S=O bonds).

1. There are two S=O stretching bands: symmetric and antisymmetric.
2. Both are vs in the IR and of variable intensity in the Raman.
3. The higher frequency band always has several maxima, giving a jagged appearance.
4. The separation (coupling) of the two bands can be correlated with the angle between the two S=O bonds.
5. The bonding is approximately tetrahedral around the sulfur, so the frequencies are not altered by conjugation.
6. Again, replacing a C on the sulfur by a more electronegative atom raises the two S=O frequencies. Some examples:

	Antisymmetric Stretch	Symmetric Stretch
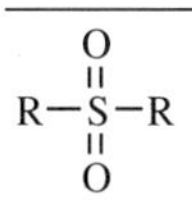 Sulfone	1340–1310 cm^{-1} vs in IR m to w in Raman	1165–1135 cm^{-1} vs in IR variable in Raman
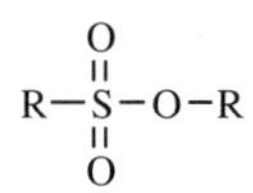 Sulfonate	1430–1330 cm^{-1} s in IR variable in Raman	1200–1140 cm^{-1} s in IR vs in Raman
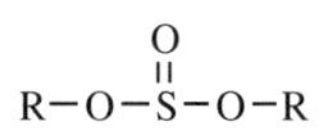 Sulfate	1450–1350 cm^{-1} s in IR s in Raman	1230–1150 cm^{-1} s in IR vs in Raman

7. *Example*: Dimethyl sulfone, Me—SO_2—Me; see Figure 9.9.
8. *Example:* Benzenesulfonyl chloride, C_6H_5—SO_2—Cl; see Figure 9.10.
   The Cl raises the frequencies from values typical of a sulfone to those of a sulfonate: ~1370 and ~1180 cm^{-1}.
9. Sulfonamides, R—SO_2—NH_2

Antisymmetric Stretch	Symmetric Stretch
~1340 cm^{-1}, s in IR, m in Raman	~1150 cm^{-1}, vs in IR, s to vs in Raman

IV. C=S stretch, 1400–1050 cm^{-1}

Difficult to use because of the wide range in the fingerprint region and because dimers form easily.

V. C—S single-bond stretch

CH_3—S—	705–685 cm^{-1}
R—CH_2—S—	660–630 cm^{-1}
R_1R_2—CH—S—	630–600 cm^{-1}
$R_1R_2R_3$—C—S—	600–570 cm^{-1}

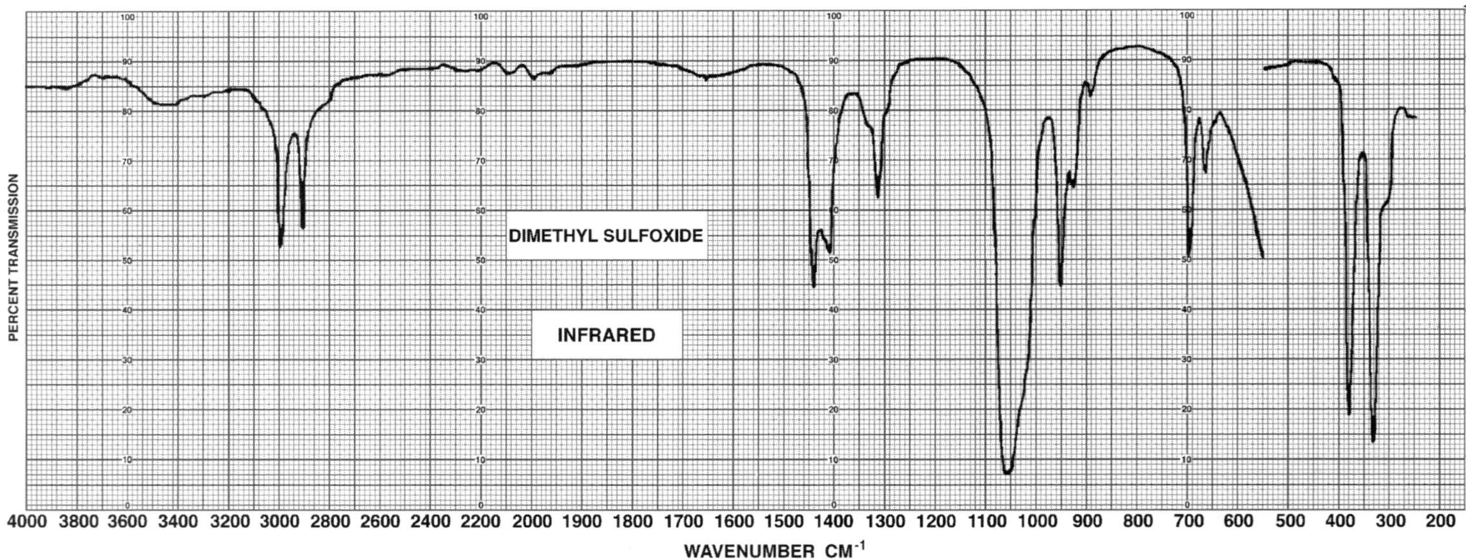

Fig. 9.8 Infrared spectrum of dimethyl sulfoxide. Note the very strong (>S=O stretch or $\nu_{(S=O)}$) at 1050 cm^{-1}.

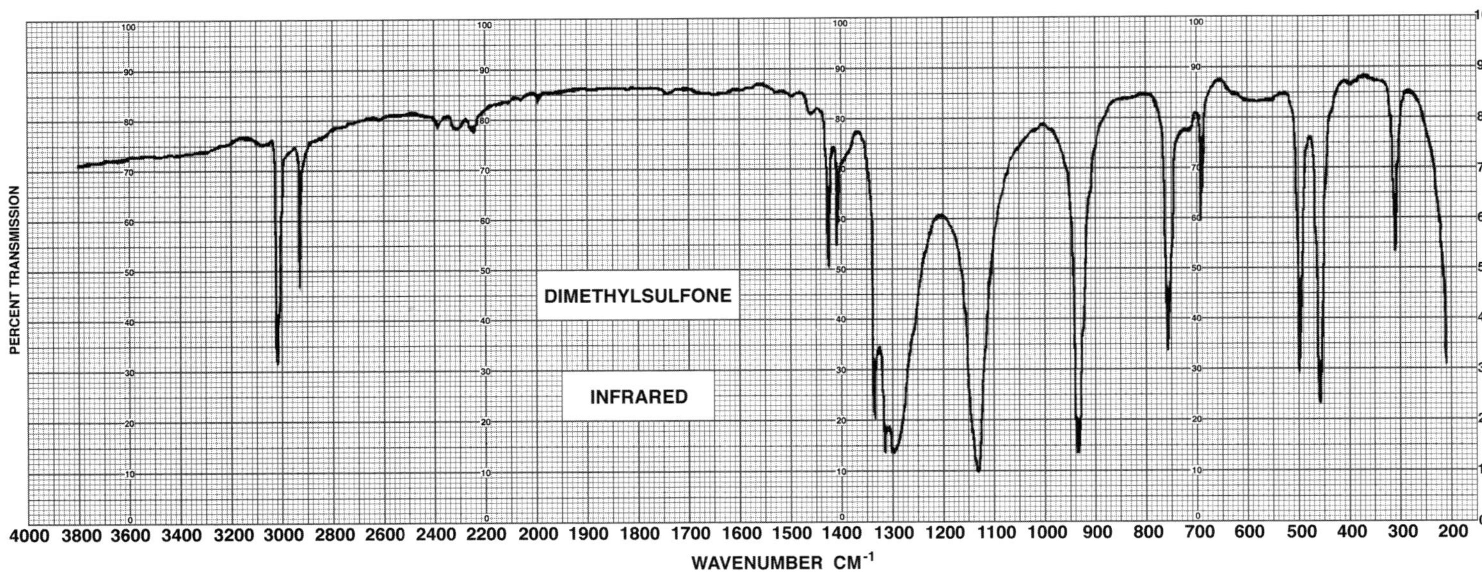

Fig. 9.9 Infrared spectrum of dimethyl sulfone. Note the very strong O=S=O stretches at 1300 (mutiple peaks) and 1130 cm^{-1}. Both are a little lower than the ranges cited for this group.

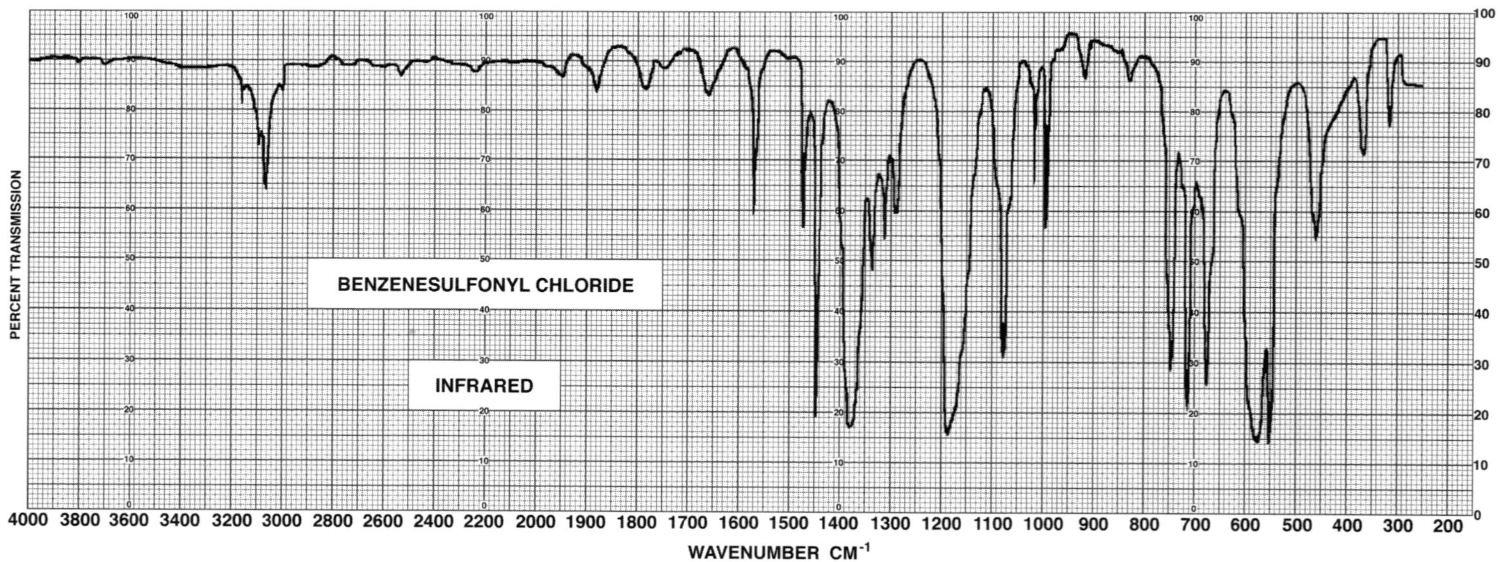

Fig. 9.10 Infrared spectrum of benzenesulfonyl chloride.

All are w in the IR and vs in the Raman. Values apply to —SH, sulfides, disulfides, and C—S—C in a ring.
Reference: N Sheppard, *Trans. Faraday Soc.* **46**, 429 (1950).

VI. —S—S—, 500–400 cm^{-1}, vw and not useful in IR, vs in Raman

VII. Raman spectra are useful for S compounds—e.g., for S—H, S—S, C=S, etc.

VIII. *Reference on S compounds*: L. J. Bellamy, The Infrared Spectra of Organo Sulfur Compounds, in N. Kharasch (Ed.), *Organic Sulfur Compounds*, Vol. 1, Pergamon, New York, 1961, Chapter 6, pp. 47–56. (QD412 S1 068).

HALOGEN GROUP FREQUENCIES

I. Except for fluorine, the halogens do not have good group frequencies, and the uses of the latter are limited. It is seldom possible to tell from the IR or Raman spectrum that a halogen is present. The Beilstein test for halogens is very useful in assisting assignments of the C—X vibrations.

II. C—F groups
   A. C—F stretch, 1400–1000 cm^{-1} (vvs)
      1. If there is only one F on a carbon, 1100–1000 cm^{-1}.
      2. As more fluorines are added, the frequency rises.
      3. —CF_2— gives two bands.
      4. IR, vvs. This is a useful confirmation. In Raman, w to m.
   B. C—F bend, <650 cm^{-1}, not useful
   C. The introduction of F perturbs other bands in both frequency and intensity. This is due, at least in part, to its large electronegativity. It has the highest value of any element.
   D. *Examples:*
      1. Perfluorocyclohexane, *cyclo*-C_6F_{12}; see Figure 9.11.
      2. Fluorolube, repeat unit (—CF_2—CFCl—$)_n$; see Figure 9.12.
   Note that Fluorolube has both C—F and C—Cl groups.

III. C—Cl, C—Br, and C—I

A.	C—Cl	800–600 cm^{-1}.	Usually s in the IR, vs in the Raman. Note the broad range.
	C—Br	600–500 cm^{-1}	Usually s in the IR, vs in the Raman. Note the broad range.
	C—I	$\sim$ 500 cm^{-1}	Usually s in the IR, vs in the Raman

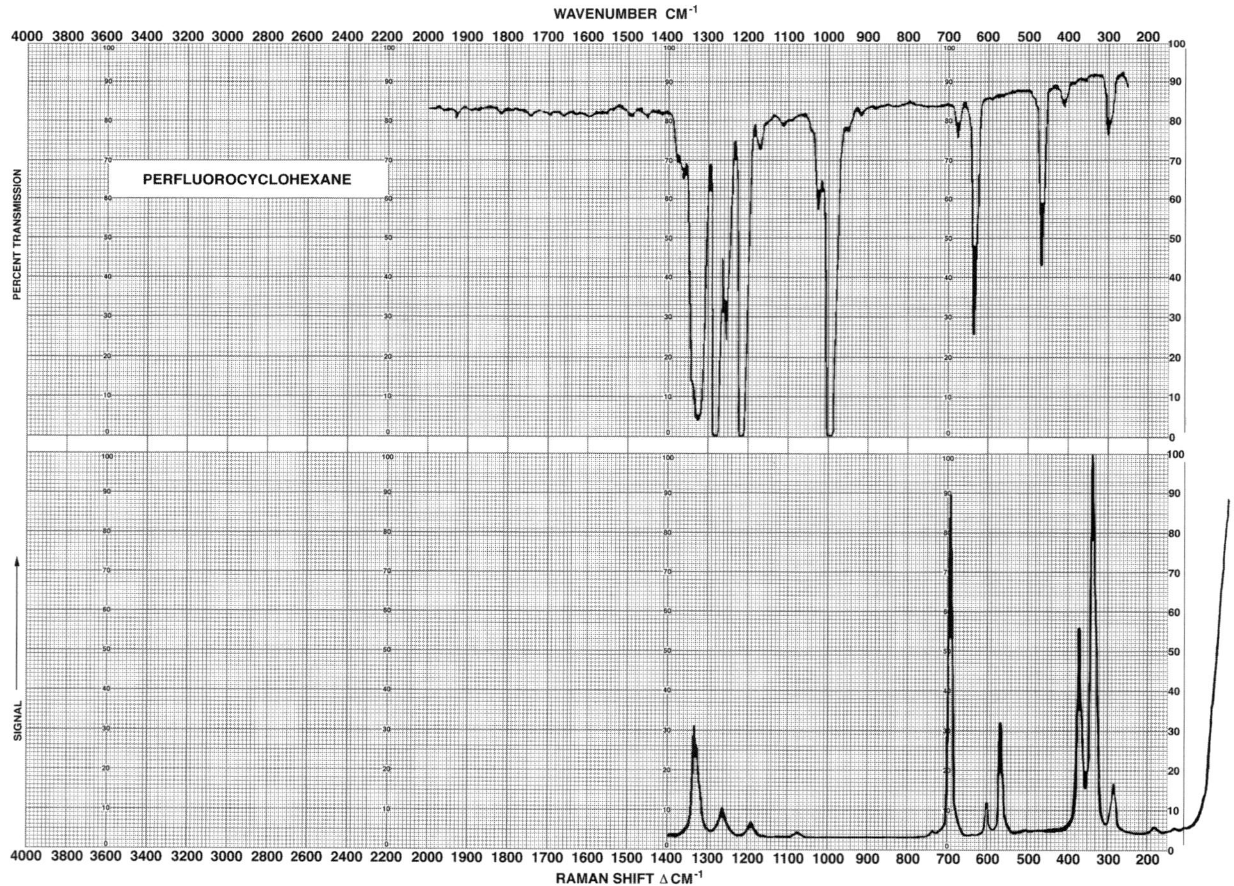

Fig. 9.11 Infrared and Raman spectra of perfluorocyclohexaane, *cyclo*-C_6F_{12}. Note the intense IR bands near 1325–1200 cm^{-1}.

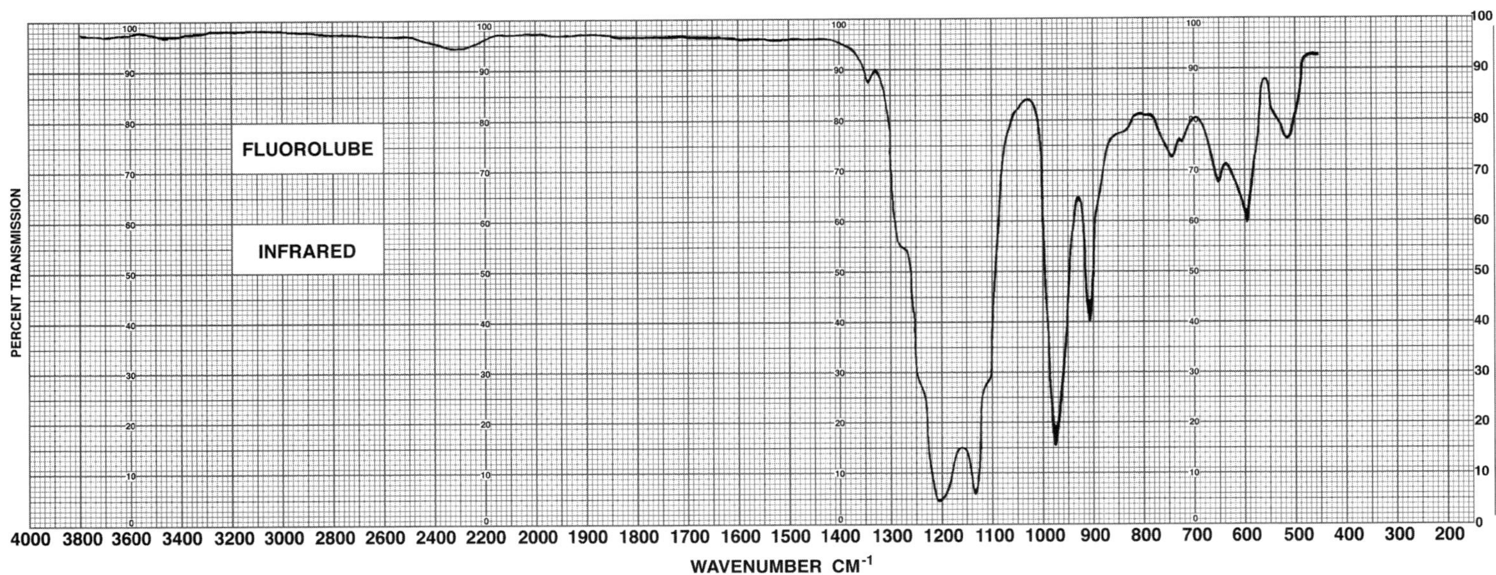

Fig. 9.12 Infrared Spectrum of Fluorolube, $(\text{—}CF_2\text{—}CFCl\text{—})_n$.

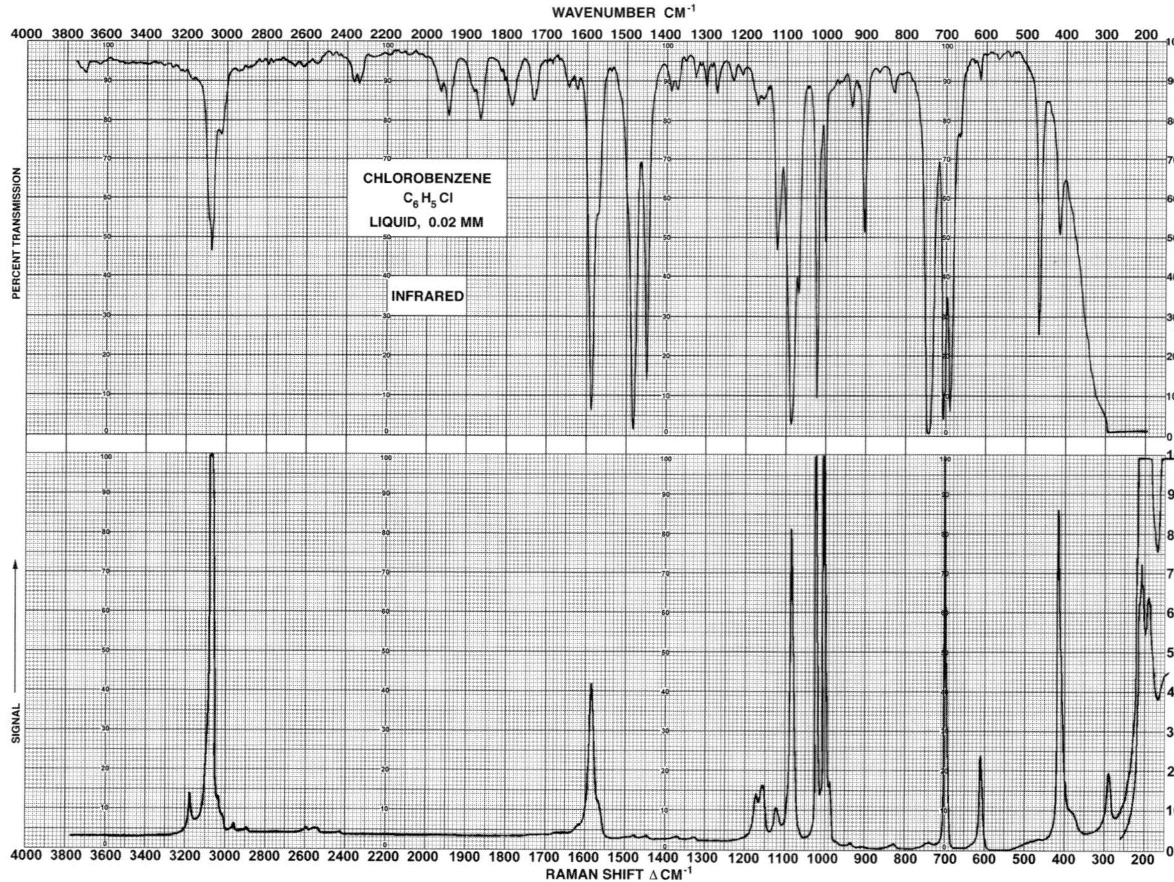

Fig. 9.13 Infrared and Raman spectra of chlorobenzene.

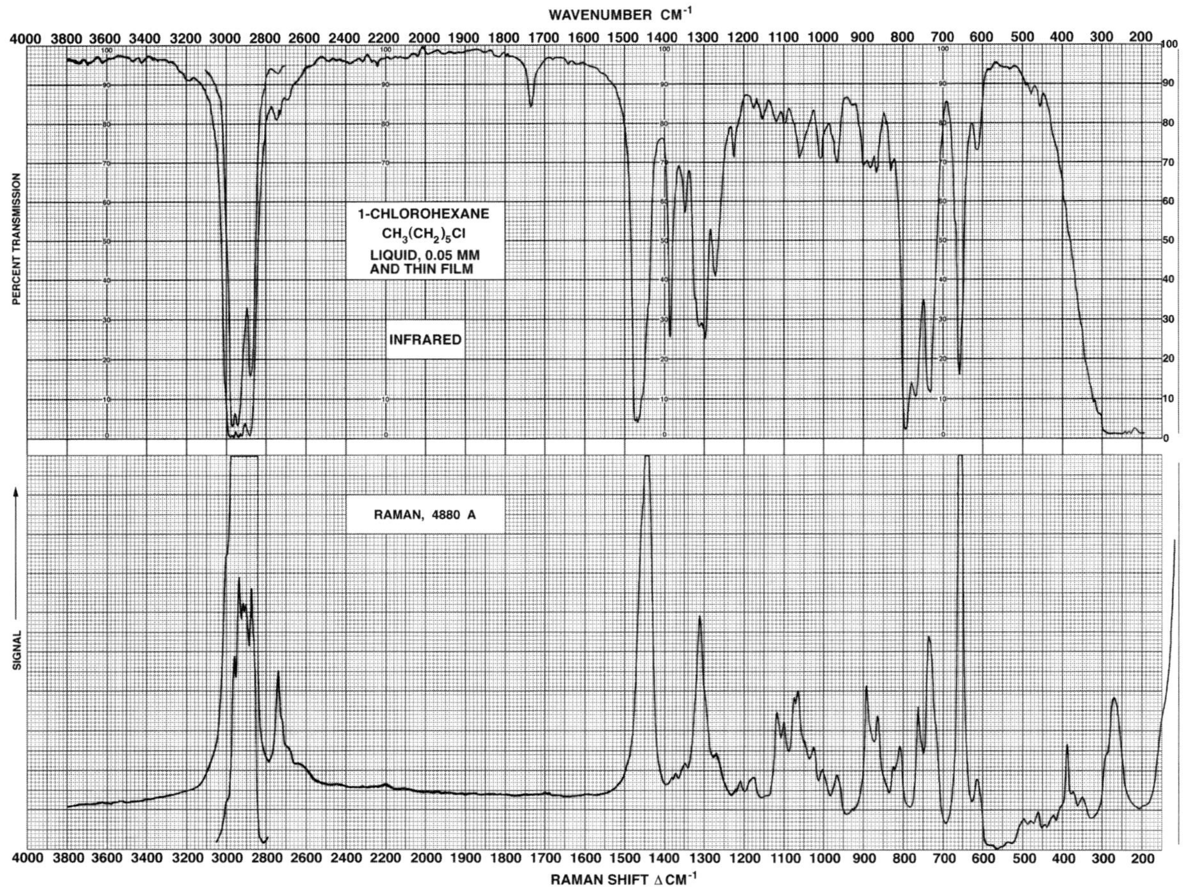

Fig. 9.14 Infrared and Raman spectra of 1-chlorohexane.

B. These are not good group frequencies because:
   1. They are far down in the fingerprint region and can be confused with other bands. For example, the C—Cl stretch in chlorophenyl compounds may come near the intense out-of-plane hydrogen and phenyl ring modes around 750–690 cm^{-1}.
   2. Much of the vibrational amplitude is in the C atom, so coupling to the remainder of the molecule is important.
   3. They may interact with other low frequencies and move around.

C. *Examples:* Fluorolube (spectrum above). The C—Cl stretch may be 600 cm^{-1}, but it is not intense compared to the C—F bands, nor is it sharp.

D. *Examples:* Chlorobenzene. see Figure 9.13.
   1. The C—Cl stretch is 700 cm^{-1} (see Raman spectrum in Figure 9.13).
   2. In the IR, the band at 690 cm^{-1} is the ring puckering mode and the band near 750 cm^{-1} is the out-of-plane hydrogen bending mode.

E. Rotational isomerism and the C—Cl stretch
   1. In 1-chlorohexane there are two strong bands at 730 and 660 cm^{-1} in both the IR and Raman spectra. (See Figure 9.14.)
   2. These two bands (730 and 660 cm^{-1}) are the C—Cl stretches of the trans and gauche rotamers, respectively.
   3. Description of the rotamers. Use the Newman projection below to look along the first C—C bond in $ClH_2C—CH_2—R$. Rotate the end group relative to the rest of the molecule.

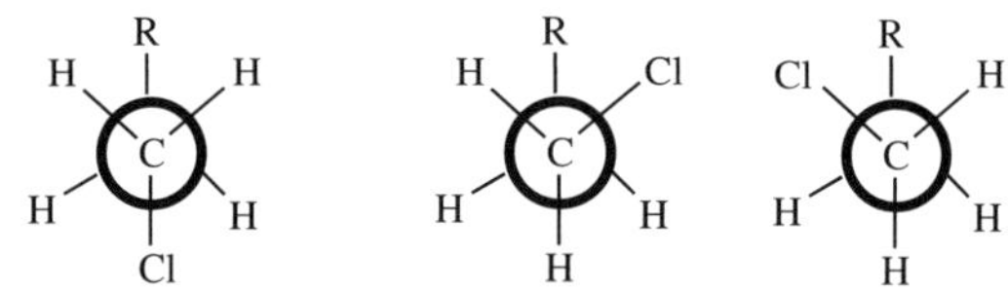

	Trans rotamer 730 cm^{-1}	Two equivalent gauche rotamers 660 cm^{-1}
Infrared	s	s
Raman	m	vs

IV. Aliphatic halocompounds

A. C—F 1400–1000 cm^{-1}

B. C—Cl 760–550 cm^{-1}

C. C—Br 700–510 cm^{-1}

D. C—I 600–480 cm^{-1}

E. The C—X frequencies are sensitive to rotational conformation as noted above.

Exercise Section II

EXERCISE 5

Spectrum Index No. 2-B

Description of Sample: This spectrum is that of a pure organic compound of simple structure. Unknown 2-B on sodium fusion yielded a positive Prussian blue test for the presence of nitrogen.

Objective: Identify the compound with the help of its characteristic vibrational frequencies as discussed in Chapters 1–9 and the correlation charts.

Once the spectrum has been established by comparison to catalog data such as the *Coblentz Society Desk Book* by C. D. Craver (Part 1 in the Bibliography) or the *Aldrich Handbook*, refer to the Answers to the Exercises for a detailed discussion.

Proposed Structures

2-B?

Comments

Course Notes on the Interpretation of Infrared and Raman Spectra, by Dana W. Mayo, Foil A. Miller, and Robert W. Hannah.
ISBN 0-471-24823-1

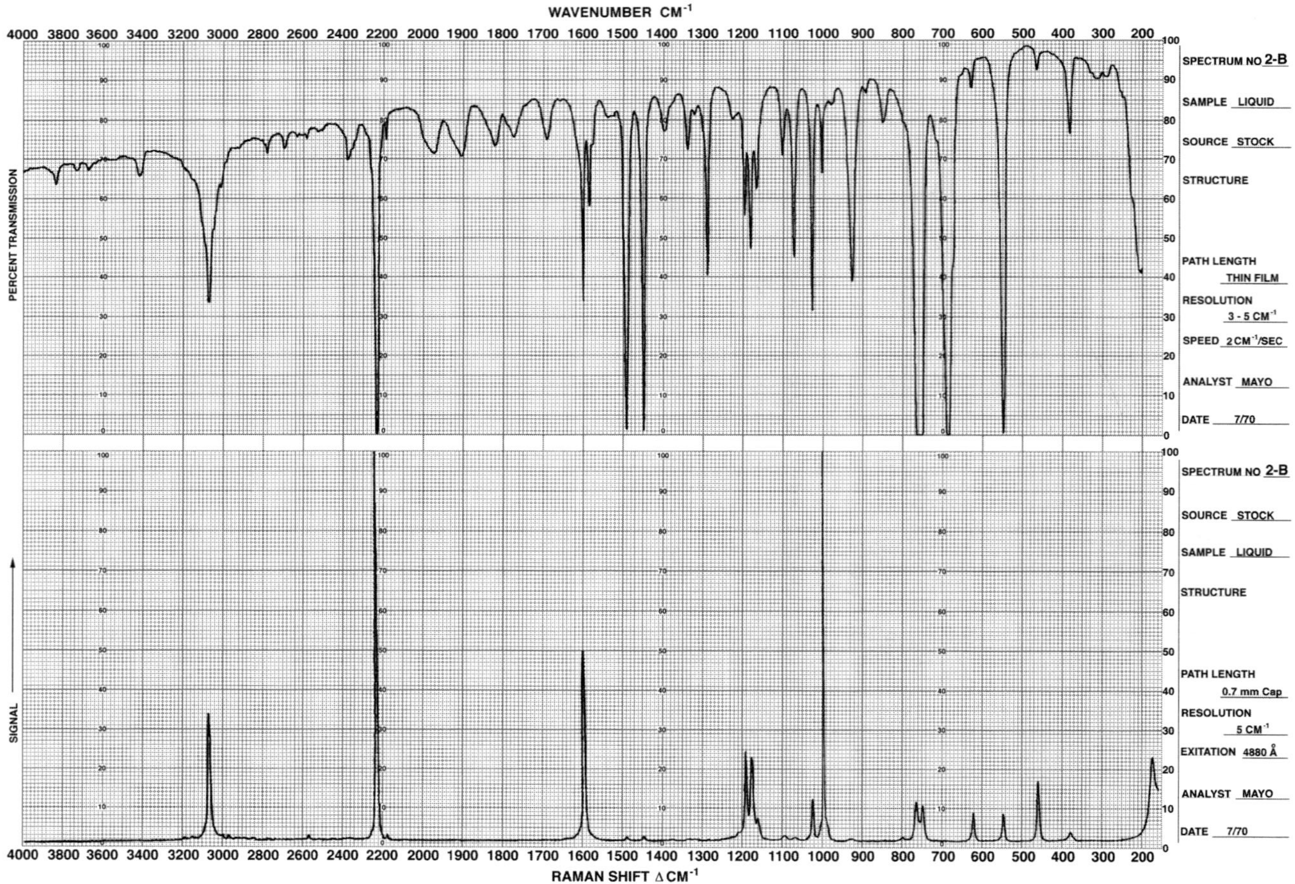

Fig. 1 Unknown 2-B.

EXERCISE 6

Spectrum Index No. 12-C

Description of Samples: Curve 12-C is a spectrum of a well-known high-molecular-weight polymer.

Objective: Identify the structural groups in the polymer film from its spectrum and infer the monomeric units from which the polymer was formed.

After a monomer has been proposed, it is possible to check the tentative structure by referring, for example, to Hummel's *Atlas of Polymer and Plastics Analysis*, Volume I, if it is available to you. Once the spectrum has been established by comparison, refer to the Answers to the Exercises for a detailed discussion.

Proposed Structure (Both Monomer and Polymer)

12-C?

Comments

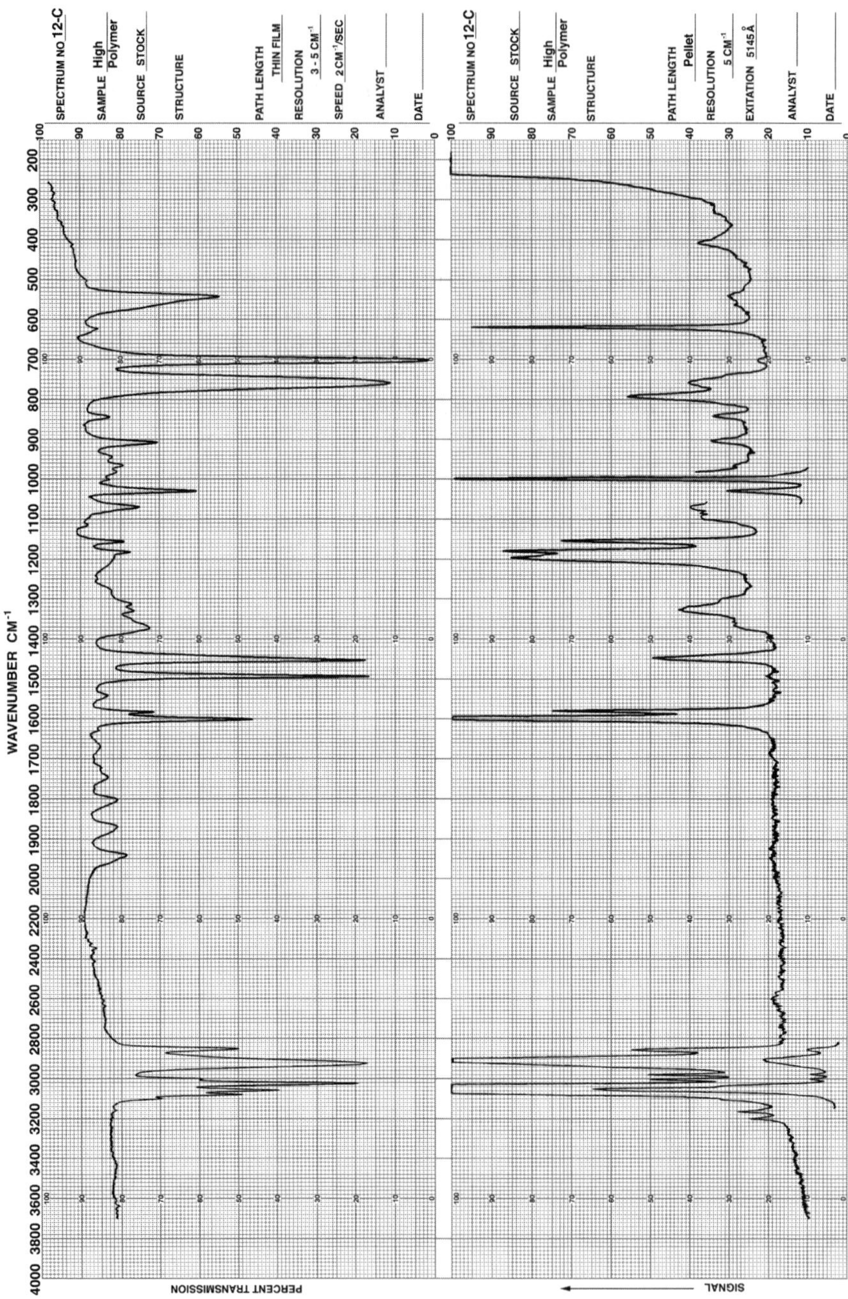

Fig. 2 Unknown 12-C.

EXERCISE 7

Spectra Index Nos. 1-B and 19-B

Description of Samples: Both sets of spectra are those of pure organic liquids of simple structure.

Objective: Identify the compounds with the help of their characteristic frequencies and the correlation charts. Spectrum 1-B belongs to a compound which gave a positive Prussian blue precipitate (color test for the presence of nitrogen). Spectrum 19-B belongs to a compound which gave a positive Beilstein test (flame test for chlorine—green). It is an extremely reactive material when in contact with moist air.

Once the spectrum has been established by comparison to catalog data such as the *Coblentz Society Desk Book* or the *Aldrich Handbook*, refer to the Answers to the Exercises for a detailed discussion.

Proposed Structures

1-B?
19-B?

Comments

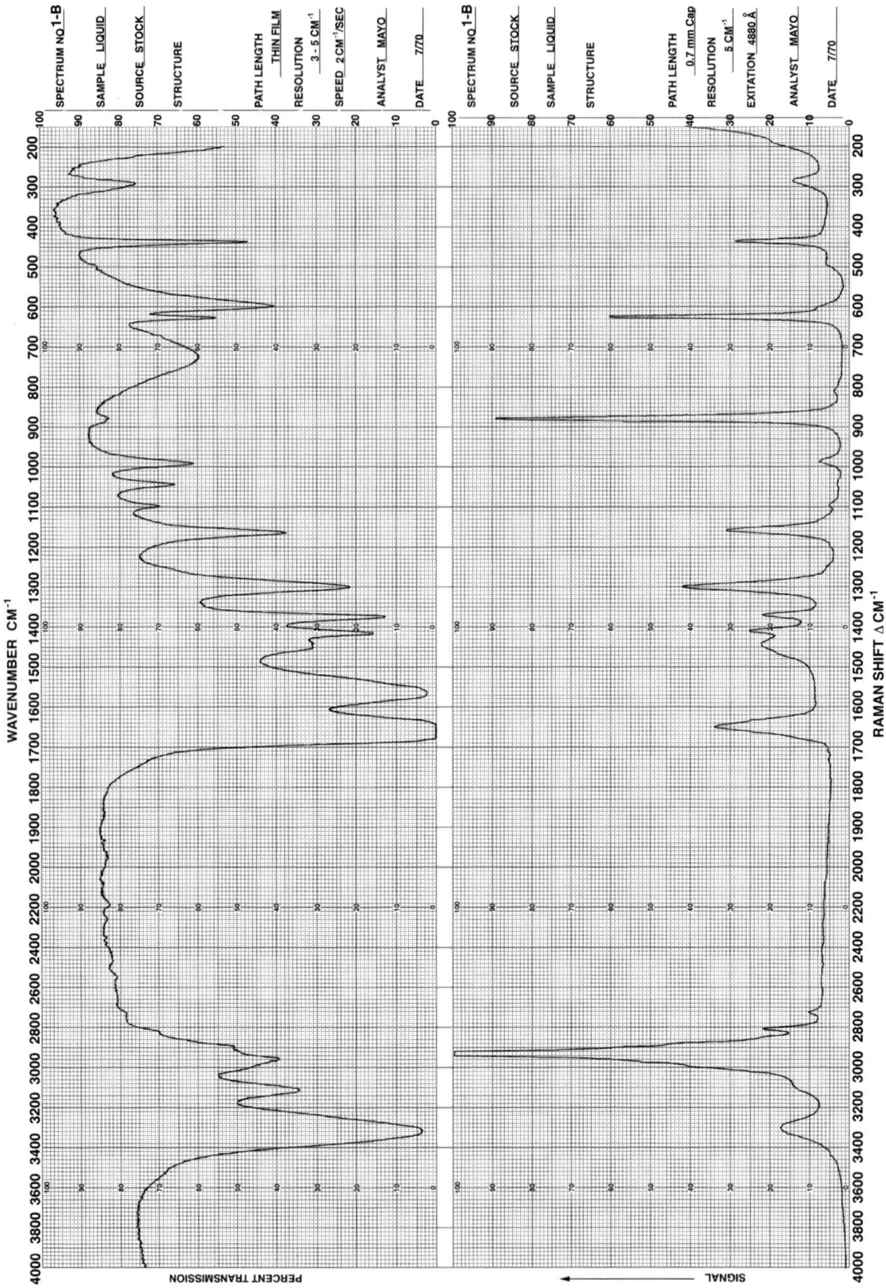

Fig. 3 Unknown 1-B.

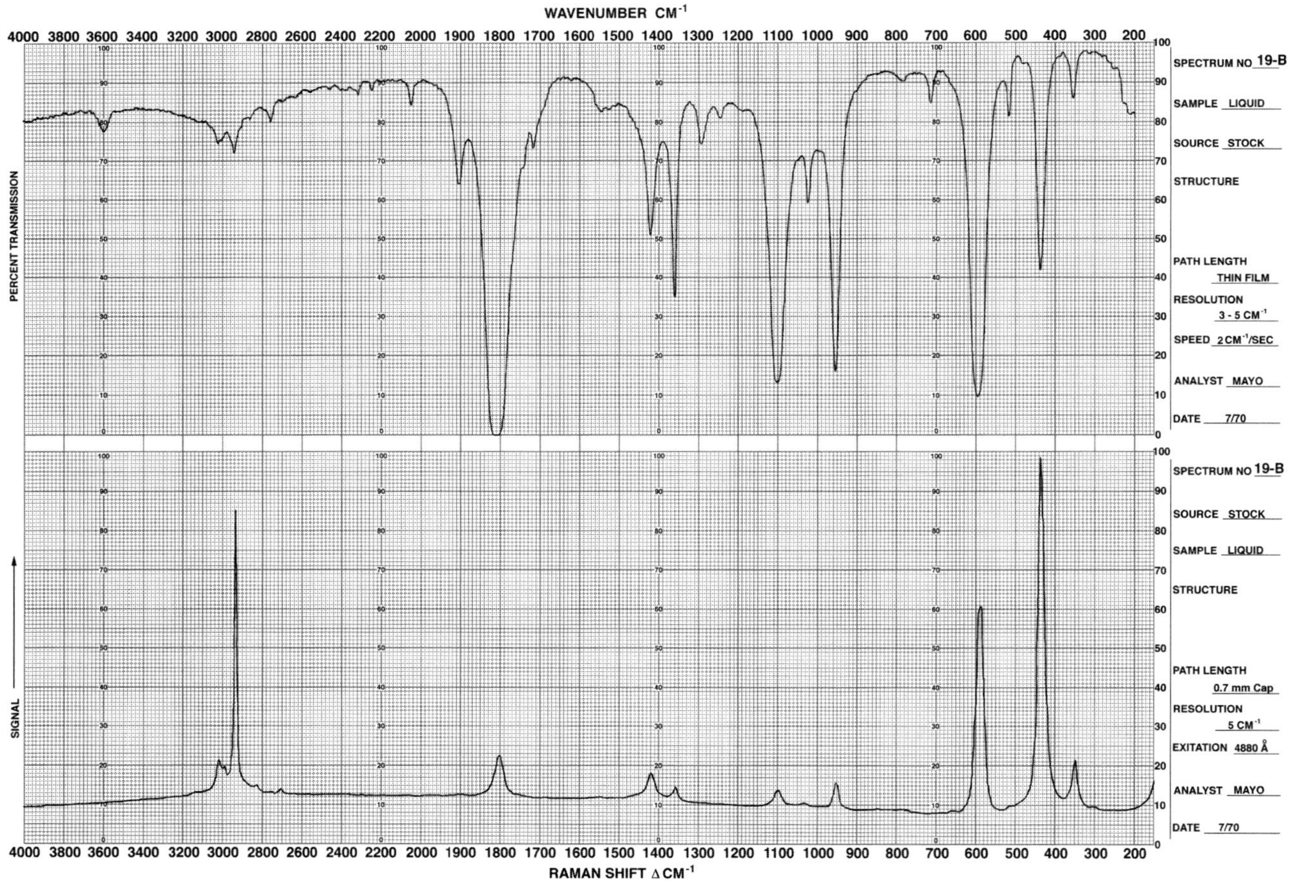

Fig. 4 Unknown 19-B.

EXERCISE 8

Spectrum Index No. 3-B

Description of Sample: The IR and Raman spectra are those of a pure organic liquid of simple structure. These belong to a compound which gave a negative Beilstein test (no halogen). The material, however, is a highly reactive substance when in contact with protic solvents.

Objective: Identify the compound with the help of its characteristic frequencies and the correlation charts.

The IR spectrum of compound 3-B, as presented in a prism IR data bank (a rather ancient collection), is also given on a following page. Note any differences between the standard spectrum (prism) and the unknown (grating) spectrum and explain. Once the spectrum has been established by comparison to catalog data such as the *Coblentz Society Desk Book* or the *Aldrich Handbook*, refer to the Answers to the Exercises for a detailed discussion.

Proposed Structure

3-B?
3-B reference

Comments

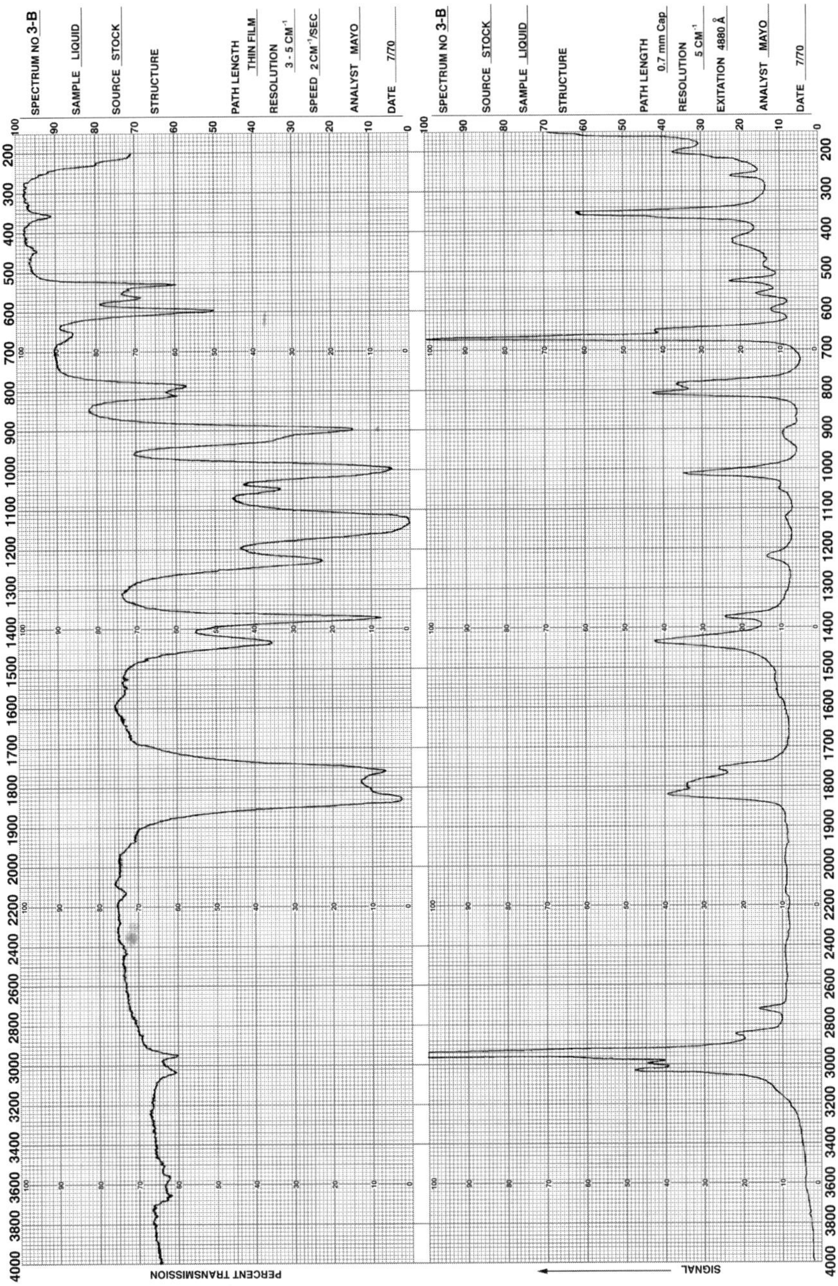

Fig. 5 Unknown 3-B.

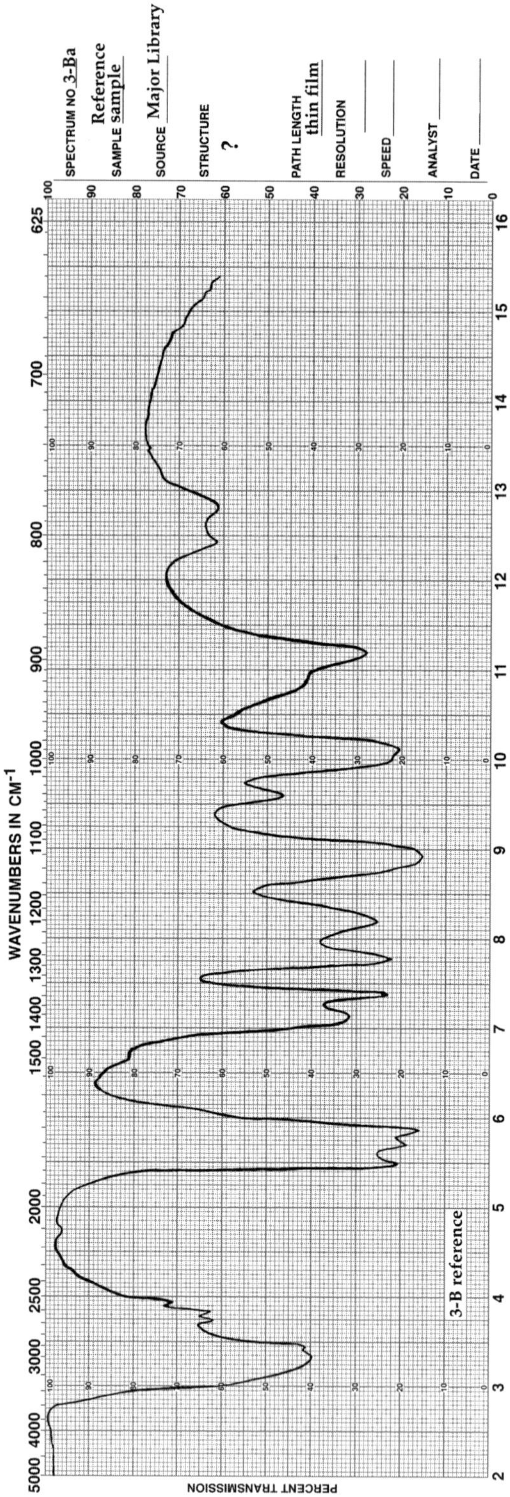

Fig. 6 Unknown 3-B reference.

EXERCISE 9

Spectrum Index No. 2-A

Description of Sample: This set of IR and Raman spectra belong to a pure organic compound of simple structure.

Objective: Identify the compound with the help of its characteristic frequencies and the correlation charts.

Once the spectrum has been established by comparison to catalog data such as the *Coblentz Society Desk Book* or the *Aldrich Handbook*, refer to the Answers to the Exercises for a detailed discussion.

Proposed Structure

2-A?

Comments

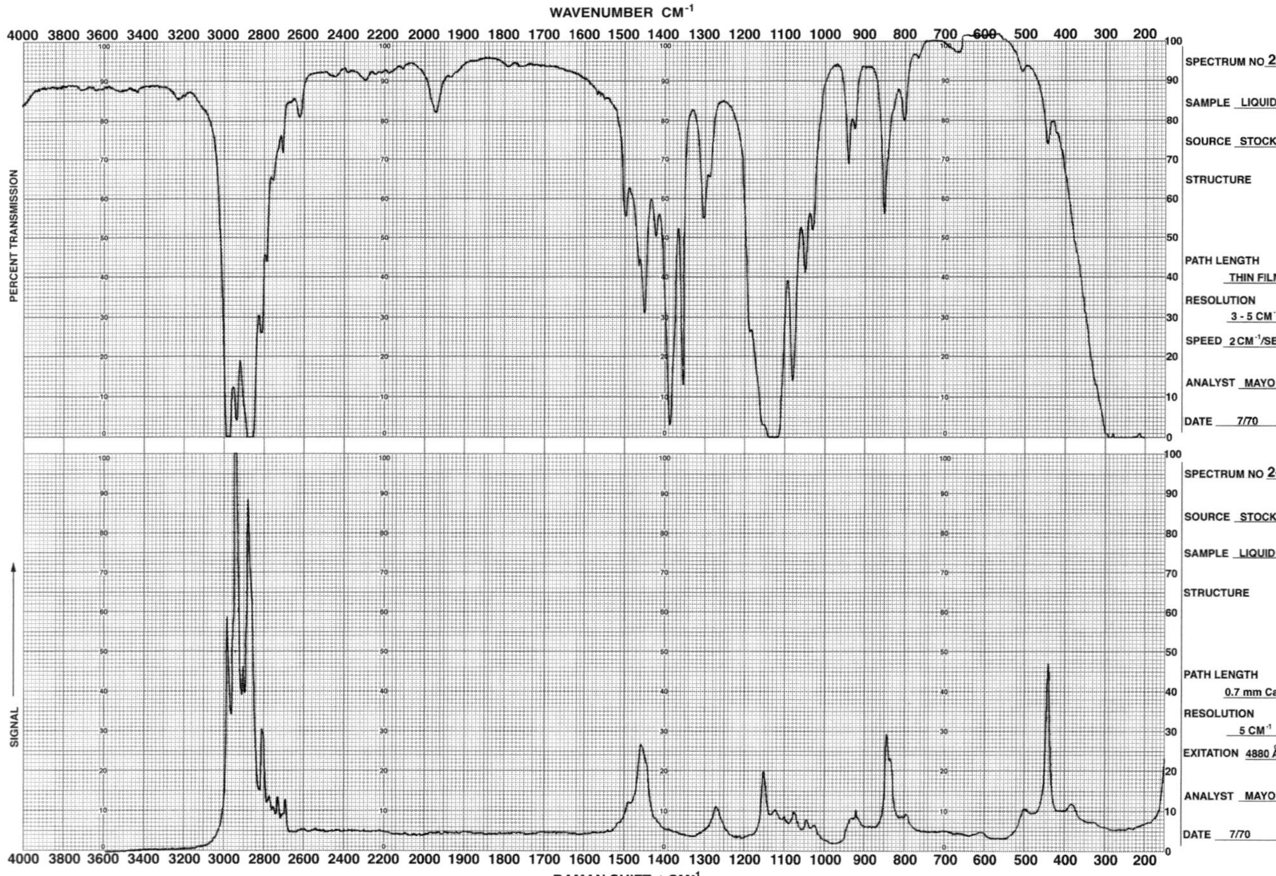

Fig. 7 Unknown 2-A.

EXERCISE 10
(PRISM SPECTRA FROM AN ANCIENT COLLECTION)

Spectra Index Nos. M-12 **and** M-40

These problems are from actual practice and from a period when many prism instruments recorded data linear in micrometers. Both spectra of these solid samples were obtained as Nujol mulls. No Raman data were collected.

I. M-12 was thought to be one of the following isomers:

1. $H_2C{=}C{=}CH{-}COOH$
2. $H_3C{-}C{\equiv}C{-}COOH$
3. $HC{\equiv}C{-}CH_2{-}COOH$

What evidence does the spectrum provide that will allow us to assign the correct structure to the unknown?

Once the potential structure has been established, refer to the Answers to the Exercises for a detailed discussion.

II. M-40 is the spectrum of a sample prepared via a reaction which should have yielded one of the following structures:

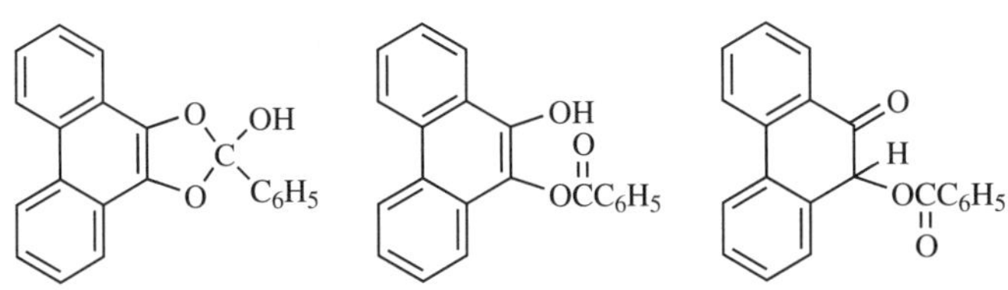

A. B. C.

Which structure does the spectrum support, if any?

Once the potential structure has been established, refer to the Answers to the Exercises for a detailed discussion.

Proposed Structures

M-12?
M-40?

Comments

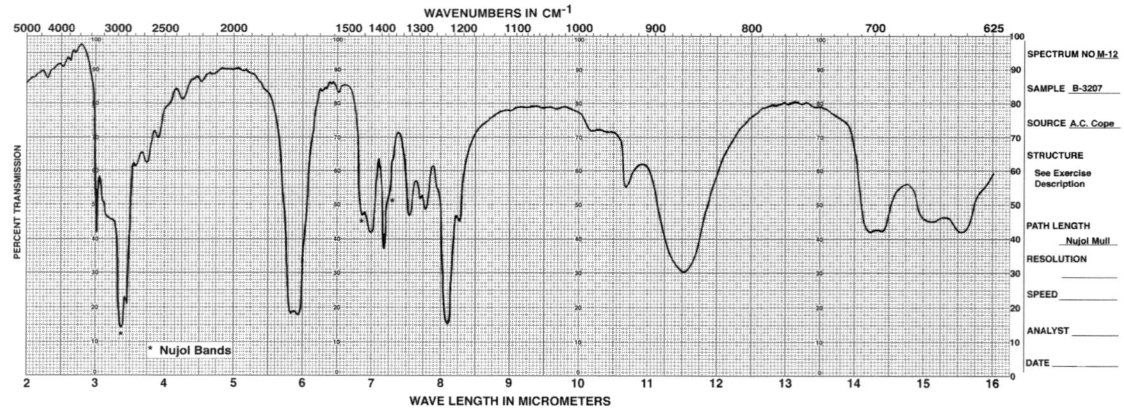

Fig. 8 Unknown M-12.

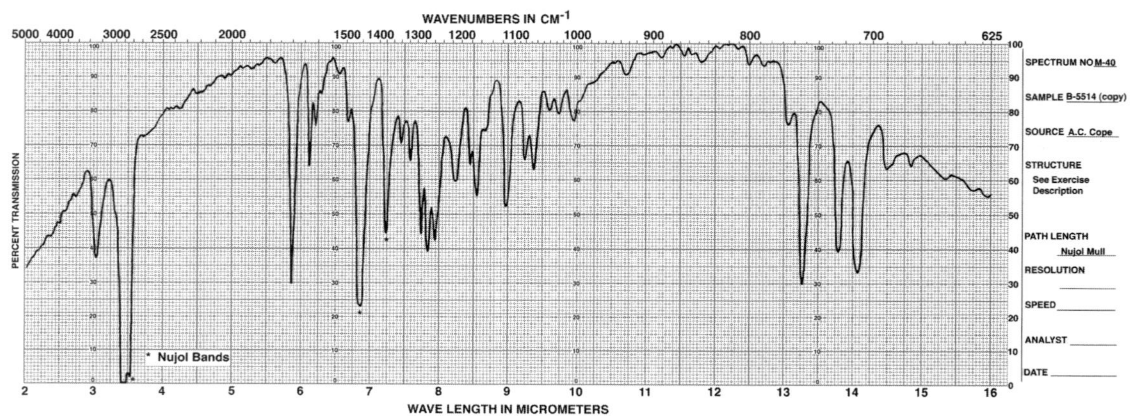

Fig. 9 Unknown M-40.

10 Infrared Spectra of Polymers: Introduction

ROBERT W. HANNAH and DANA W. MAYO

This chapter is concerned with the fundamentals of the IR spectra of polymers. Polymers are large, complex molecules, yet their spectra may be relatively simple. First, a few spectra of simple polymers will be reviewed, and an interpretation of these spectra will be carried out which clearly demonstrates that the techniques learned for interpreting the spectra of simple molecules can be extended directly to polymers. Second, how polymers are formed will be examined (a bit superficially), and this discussion will also illustrate that polymer spectra contain more information than can be extracted in simple fashion. Third, there are additional problems associated with the interpretation of some polymer spectra and a number of these will be considered. Fourth, a short introductory discussion about polymer sampling will be covered. More detailed consideration of sample preparation and the handling of additives and mixtures will be given later in the chapters on sampling and mixtures. Many standard sample preparation procedures can be applied directly to polymer systems, just as these techniques are used successfully for handling mixtures of relatively simple low-molecular-weight organic molecules. Fifth, a flow chart approach to the interpretation of polymer spectra will be explored.

I. Shown in Figures 10.1–10.3 are the IR spectra of the hydrocarbon polymers polyisobutene (Figure 10.1), polybutadiene (Figure 10.2), and polystyrene (Figure 10.3).

Butyl rubber (polyisobutene; Figure 10.1) has the relatively simple structure shown below:

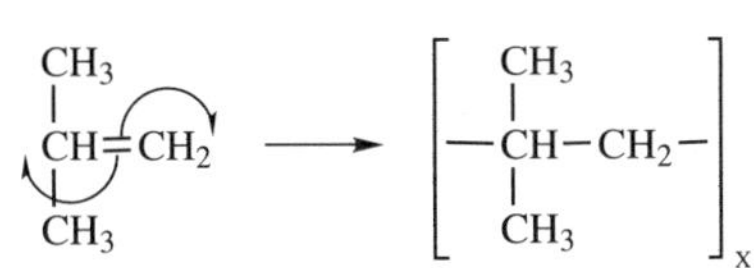

Isobutene Polyisobutene

Course Notes on the Interpretation of Infrared and Raman Spectra, by Dana W. Mayo, Foil A. Miller, and Robert W. Hannah.
ISBN 0-471-24823-1

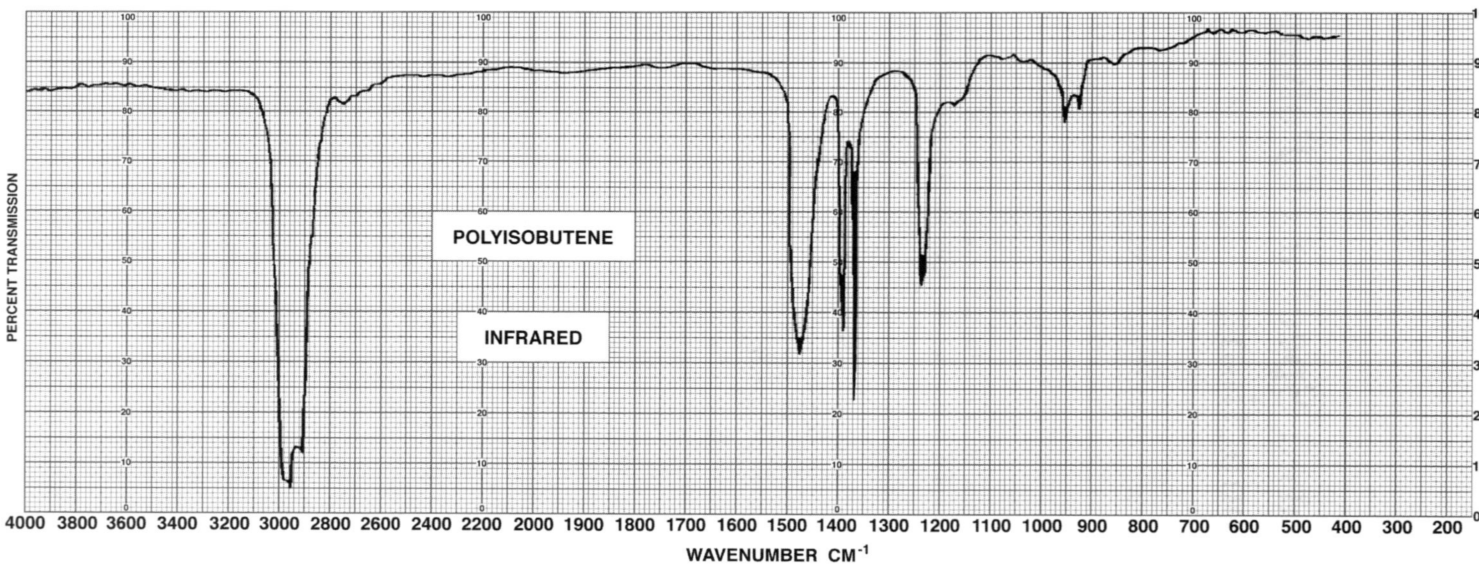

Fig. 10.1 Infrared spectrum of polyisobutene.

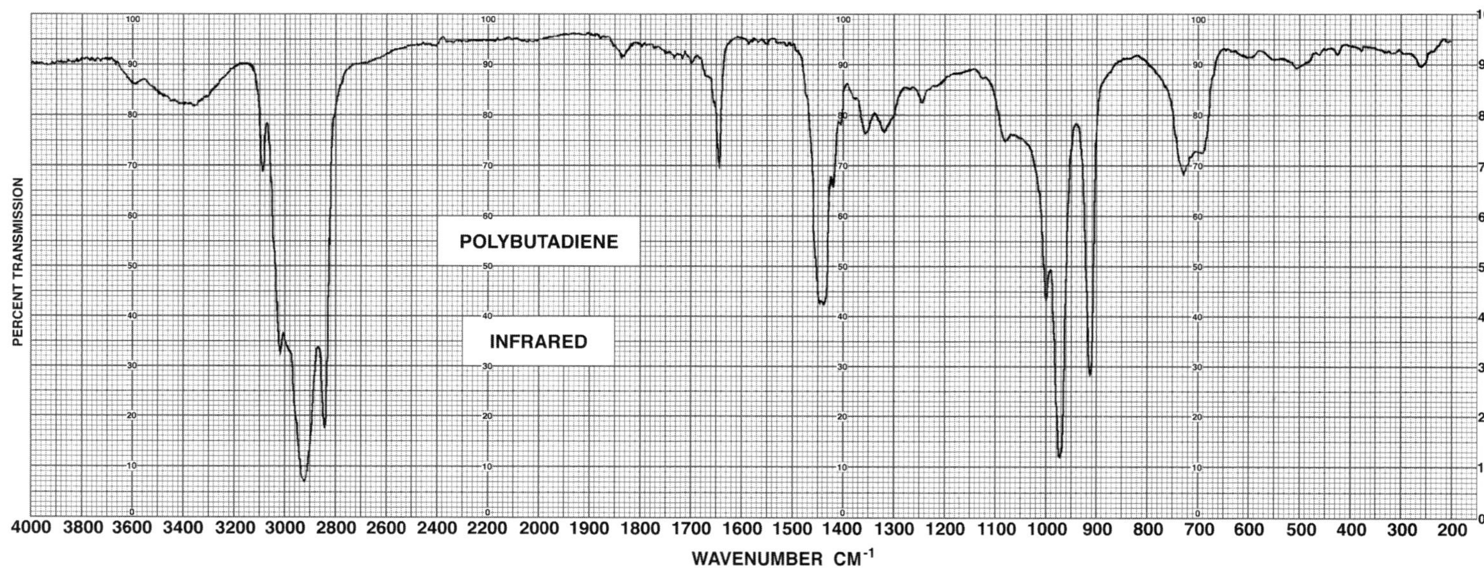

Fig. 10.2 Infrared spectrum of polybutadiene.

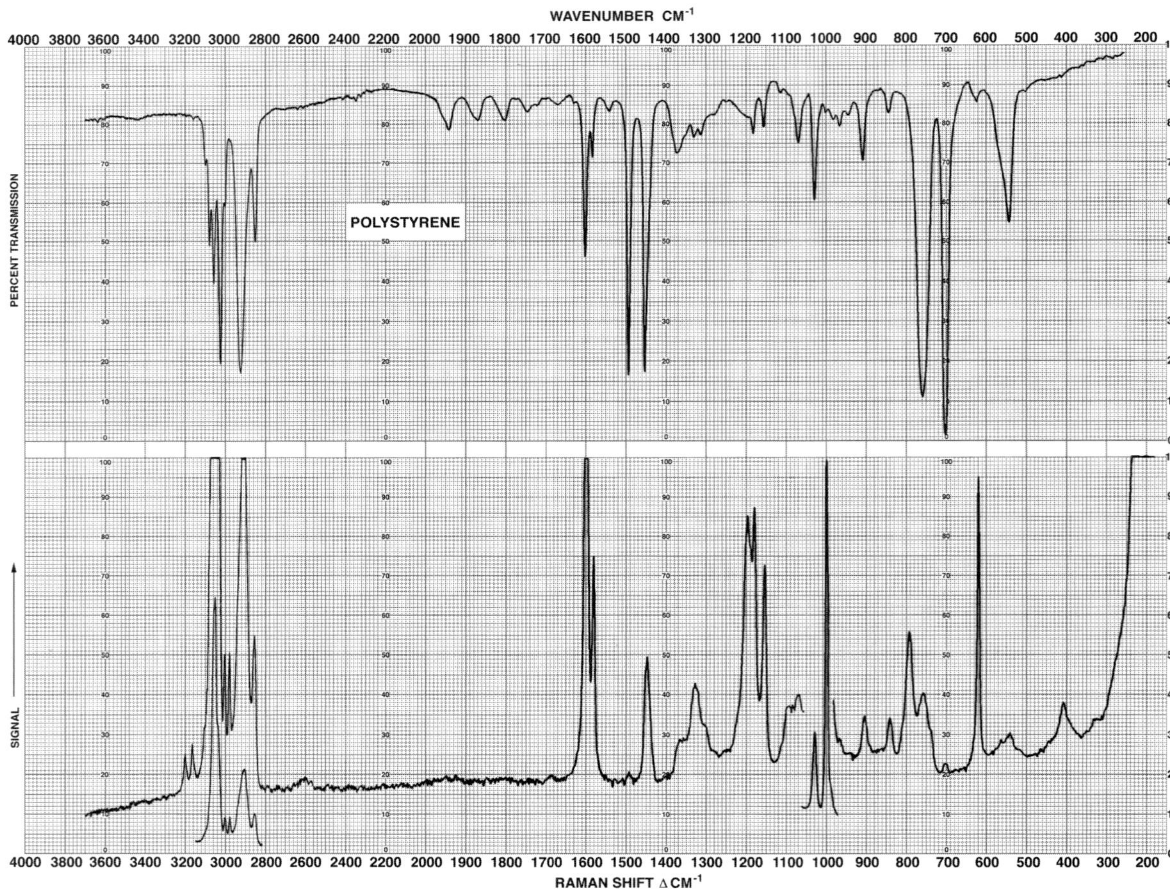

Fig. 10.3 Infrared and Raman spectra of polystyrene.

The polymerization takes place as shown, with the double bond opening and monomer units chaining together to form the polymer. What distinctive features would be expected to be observed in the IR spectrum of this polymer? First, the polymer has a *gem*-dimethyl group, and therefore a doublet ought to be found in the 1380 cm^{-1} region with the low-wavenumber component somewhat stronger than the high-wavenumber component. See Chapter 2, III B3. This band pair is clearly observed. Second, only group frequencies associated with CH_2 and CH_3 stretching and bending modes should be detected, and we see the bands assigned to these fundamental modes just below 3000 and near 1460 cm^{-1}. The rest of the absorption spectrum is associated with skeletal modes in the molecule and with the CH_2 rocking and twisting modes. These are the fingerprint bands of this particular molecule.

Polybutadiene is another hydrocarbon polymer similar to polyisobutene (Figure 10.2). In this case the precursor monomer butadiene may polymerize in three different ways, as shown here:

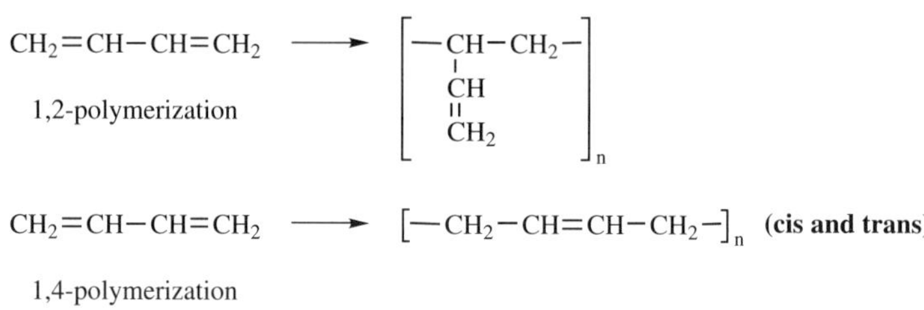

In the first case, this polymerization route yields a terminal vinyl group, so a pair of strong bands characteristic of the vinyl group would be expected to be observed near 990 and 905 cm^{-1} and they are clearly present. The second reaction path should yield either or both cis and trans polymers, and for these structures bands should be found near 700 cm^{-1} (± 40 cm^{-1}, w and b, cis) and 965 cm^{-1} (± 5 cm^{-1}, vs, trans), respectively. The rest of the polybutadiene spectrum would be expected to exhibit the C=C stretching modes near 1650 cm^{-1} (the band is split by the cis and vinyl modes, the trans mode is very weak and hard to identify) and C—H stretching vibrations for both sp^3- and sp^2-hybridized carbon atoms respectively just below and just above 3000 cm^{-1}. The sp^2 band is not well resolved, and it is usually observed as a high-wavenumber shoulder on the sp^3 band (the vinyl CH_2 shows near 3100 cm^{-1} with the rest of the unsaturated C—H stretch near 3010 cm^{-1}).

The spectrum of atactic polystyrene is shown in Figure 10.3. The term *atactic* will be defined later. This polymer is often used in checking calibration and other IR instrument performance. It also has a spectrum which is easily interpreted following the normal group frequency rules. See Chapter 5, VI C. The aromatic C—H stretching modes are located as weak bands above 3000 cm^{-1}. The out-of-plane C—H bending and ring puckering vibrations associated with monosubstituted aromatic systems are easily found in the 750–690 cm^{-1} region. The sum tones which correlate with monosubstitution in the 2000–1650 cm^{-1} region are present. (These weak bands are often distorted by underlying interference fringes.) The doublets near 1600 and 1500 cm^{-1} associated with aromatic in-plane ring stretching are also easily assigned.

In the above examples the rules of interpretation developed for lower molecular weight organic molecules can be applied directly to polymers. Obviously, there may be occasional differences between larger molecules and the smaller well-known examples, and many of these differences appear to arise as a result of the regular order in which the polymeric chains evolve as they are formed.

II. Commercial polymers for the most part do not occur as pure materials. Many are modified by additives which impart certain desired physical properties. An example is the case of polyvinylchloride, which in its pure state is a rigid insoluble material. Plasticizer additives can be used, e.g., di-2-ethylhexyl-phthalate (commonly referred to as dioctylphthalate), to soften this chlorinated hydrocarbon and make it workable.

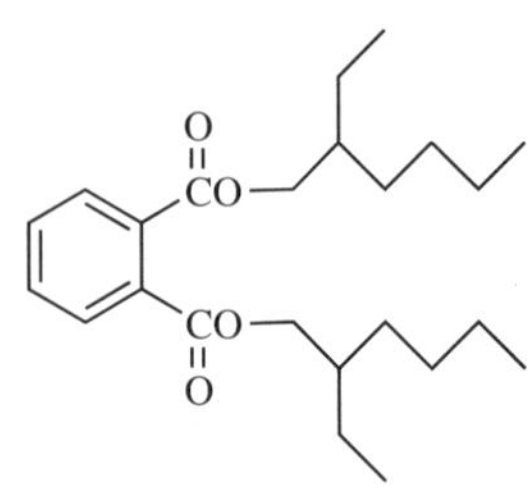

2-ethylhexylphthalate (dioctylphthalate)

Antioxidants and UV inhibitors are often added in low concentrations. Weak bands attributed to these low-molecular-weight molecules (which are essentially dissolved in the polymer) may be identified in the spectra. Under certain conditions these additives may be extractable and their unmasked spectra obtained. There also may be inorganic compounds added (in this case the additives are insoluble and are merely suspended in the polymer) to make the sample opaque or translucent, e.g., titanium dioxide, barium sulfate, and iron oxide. Other common additives include dyes and pigments. Polymer samples as you receive them, therefore, may be fairly complex mixtures, and full characterization of the sample may require extensive separation and application of other instrumental techniques.

Let us now consider the case of plasticized polyvinyl chloride (PVC). The polymer is formed from vinyl chloride monomer as shown below, and the spectrum is given in Figure 10.4.

Polyvinyl chloride is quite a simple polymer. What should we expect to see in the spectrum? There should be CH_2 bands near 3000 and 1450 cm^{-1} and C–Cl-related vibrations near 700 cm^{-1}. In addition, fingerprint bands would

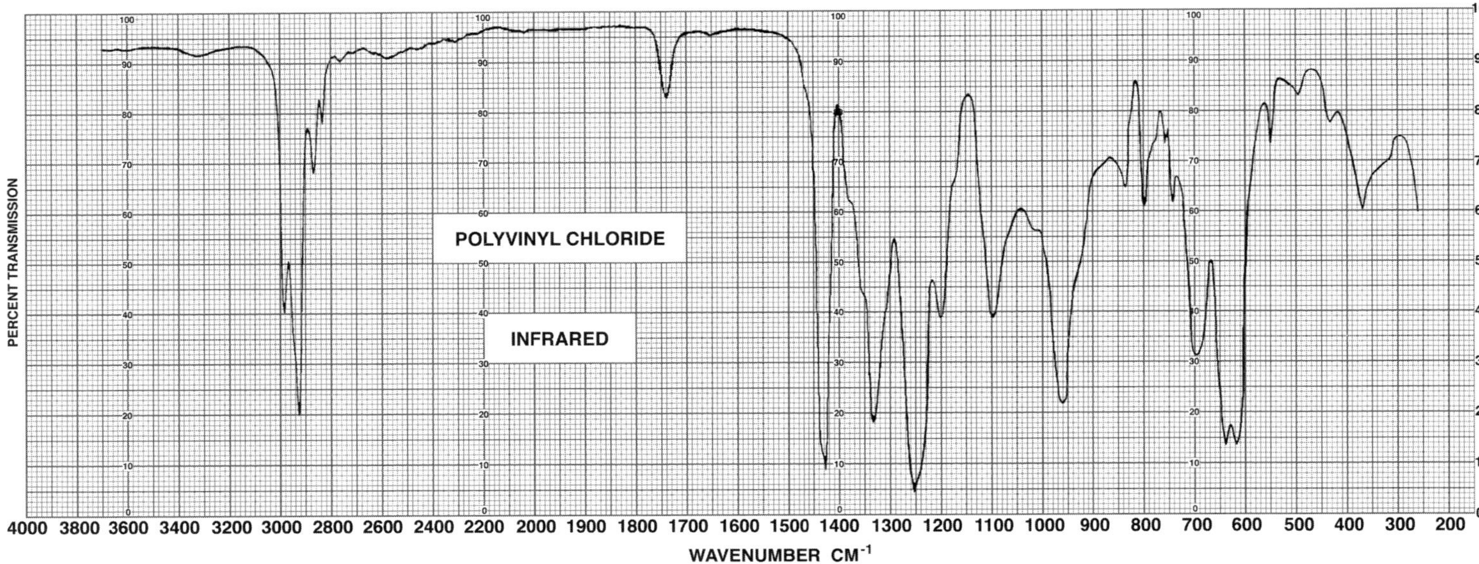

Fig. 10.4 Infrared spectrum of PVC.

be expected. The band assigned to the C—Cl stretch is much more complex than that expected for a simple C—Cl stretch. (Even in simple systems the C—Cl stretching mode is sensitive to its environment. In 1-chlorohexane, e.g., two bands are observed which both have been shown to have their origin in the C—Cl stretch present in two rotational conformational isomers; see Figure 10.5 and discussion at the end of Chapter 9.) Hummel and others have resolved this band mathematically into a number of individual components, and these components have been assigned to structures determined by the spatial configuration of the chlorine substitution. Spatial configuration, therefore, determines how the C—Cl groups are arranged with respect to each other along the polymer backbone, a property that imparts a certain tacticity to the polymer. A detailed definition of tacticity is again deferred until a little later in the discussion.

$$CH_2{=}CH{-}Cl \longrightarrow \left[-CH_2-\overset{\displaystyle Cl}{\overset{|}{C}H}- \right]_n$$

The spectrum of di-2-ethylhexylphthalate is shown in Figure 10.6. This is an aromatic diester in which the aryl conjugated carbonyl and aromatic ring bands are easily assigned, as are the remainder of the bands related to this structure. If a small amount of di-2-ethylhexylphthalate is used as a plasticizer for PVC, the spectrum in Figure 10.4 results. If the polymer contains modest amounts of the plasticizer, the strong absorption bands of the plasticizer actually mask many of the weak absorption bands of the PVC. The highly polar plasticizer spectrum is appreciably stronger than that of PVC, and this really complicates the interpretation of the overall spectrum. This type of problem can sometimes be handled by physically separating the additive from the polymer and obtaining its spectrum as a separate component. Computer subtraction may then be used to obtain the pure polymer spectrum.

III. Preparation of polymer samples often presents some unusual problems: The polymer may be insoluble, it may not grind, etc. Many polymers are prepared as free-cast films. New techniques such as diffuse reflectance, photoacoustic spectroscopy, and IR microscopy may be useful sampling techniques for polymers. While details of these techniques can be found in Chapter 13 on sampling, a few of the more important methods will be discussed at this point. The art of hot-pressing films is one of those techniques. Sampling of nylons is a good example of this method. These polymers are particularly difficult to prepare for sampling because they are fairly hard insoluble materials. One successful way of handling nylons (polyamides) is to make a hot-pressed film. The sample is placed between two steel blocks and the blocks are heated to the softening point of the polymer. Pressure is applied

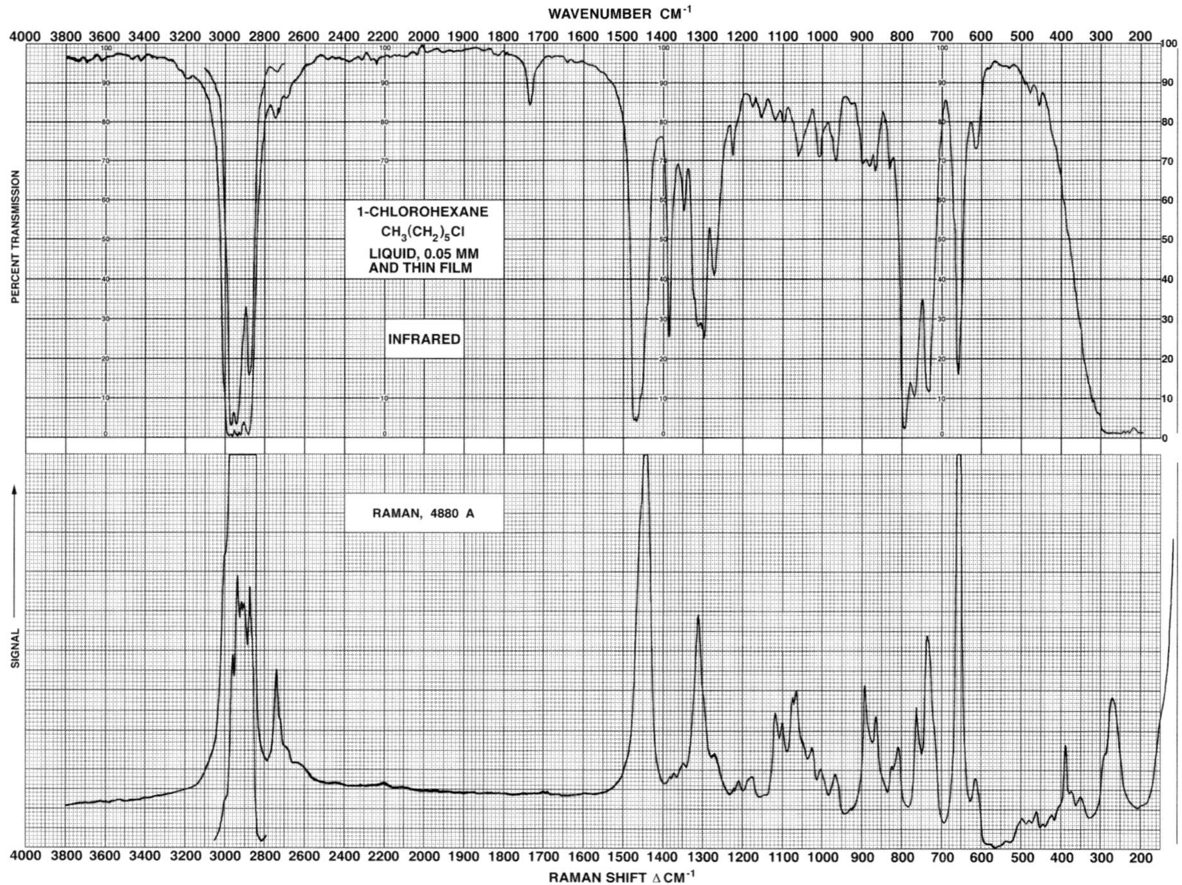

Fig. 10.5 Infrared spectrum of 1-chlorohexane.

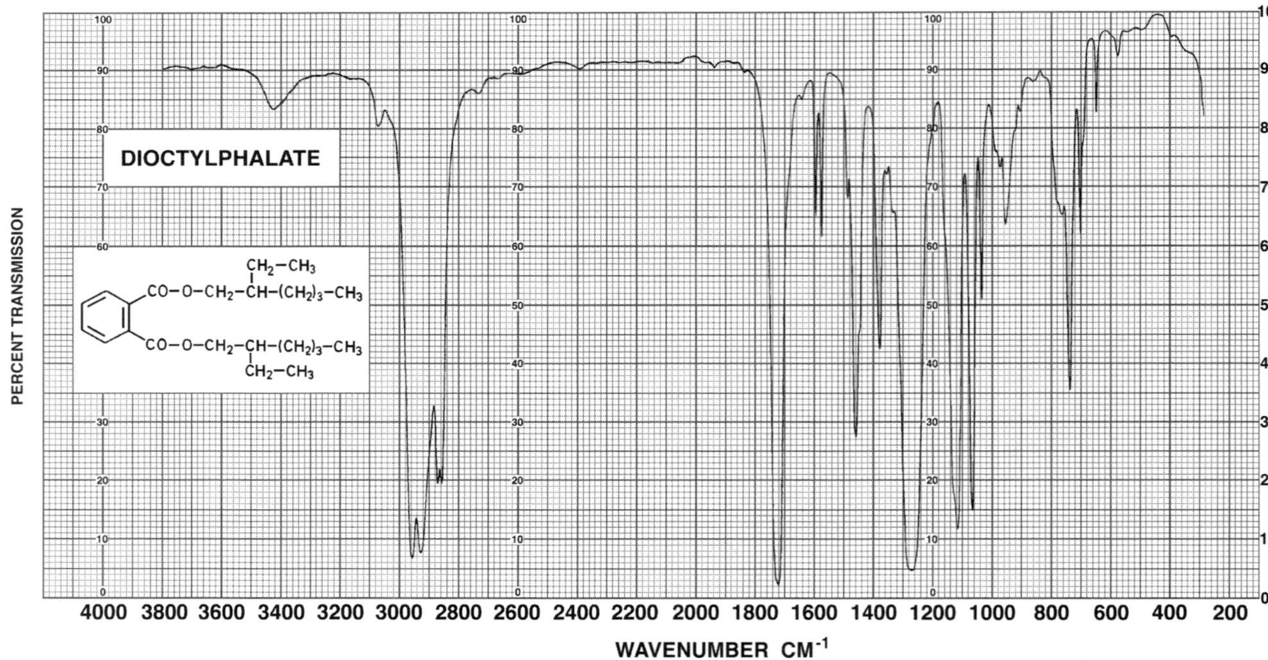

Fig. 10.6 Infrared spectrum of di-2-ethylhexylphthalate (known as dioctylphthalate).

using a hydraulic press. On cooling, the blocks are often difficult to get apart, which is an undesirable result since the sample is between the blocks. Some simple modifications have been developed to avoid this problem. Films of cellophane or Teflon are sometimes used between the sample and the steel blocks. Dr. Harry Willis has recommended that the steel blocks be modified by sand blasting and then coated with a thin film of Teflon (nonstick fry pan surface) on the steel surface. Hot blocks treated in this fashion can be easily separated from the sample, but of course, every sample that is pressed will have varying traces of Teflon on the outer surface. The spectrum of this very thin trace of Teflon impurity, however, can very often be subtracted from the sample spectrum, leaving a corrected spectrum of the polymer. Why sand blast the steel surface? Sand blasting creates a rough surface on the pressed film, and as a result, interference fringes are eliminated or greatly reduced in the spectra of samples prepared by this technique.

Internal reflection is another sampling technique which often can be applied to polymers with considerable success. Many laboratories already have ATR [or multiple internal reflection (MIR) or frustrated MIR (FMIR)] accessories available. In this case the polymer surface is generally pressed against two opposite sides of the crystal (Figure 10.7) and the IR surface spectrum obtained. There are some limitations and advantages to the ATR (MIR) technique which should be mentioned. At each refection, an evanescent wave is generated which penetrates very slightly into the sample. This depth of penetration is wavelength dependent and shallow, generally less than one wavelength. Thus, there is a finite distance over which there is sufficient amplitude in the evanescent electromagnetic wave to interact with the oscillating dipoles of the molecular vibrations of the sample, and hence an absorption band is produced. If the reflection crystal has a rough surface, then contact with this face may be inefficient and a weak spectrum is observed. If the sample has multiple layers, then the spectrum may be only that of the surface layer and not of the bulk polymer sample. It must be clearly understood that the spectrum measured is predominately the spectrum of the surface of the polymer, and this may not be the same as the bulk sample. For example, the polymer surface may have oxidized or a plasticizer may have been forced to the surface by pressure. Under these conditions the spectrum observed may not be the spectrum expected. Pigments which very often float to the polymer surface during an evaporation process (low

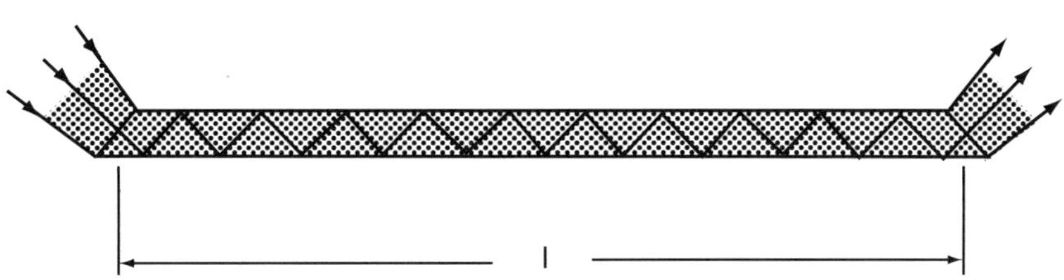

Fig. 10.7 Frustrated multiple internal reflection crystal.

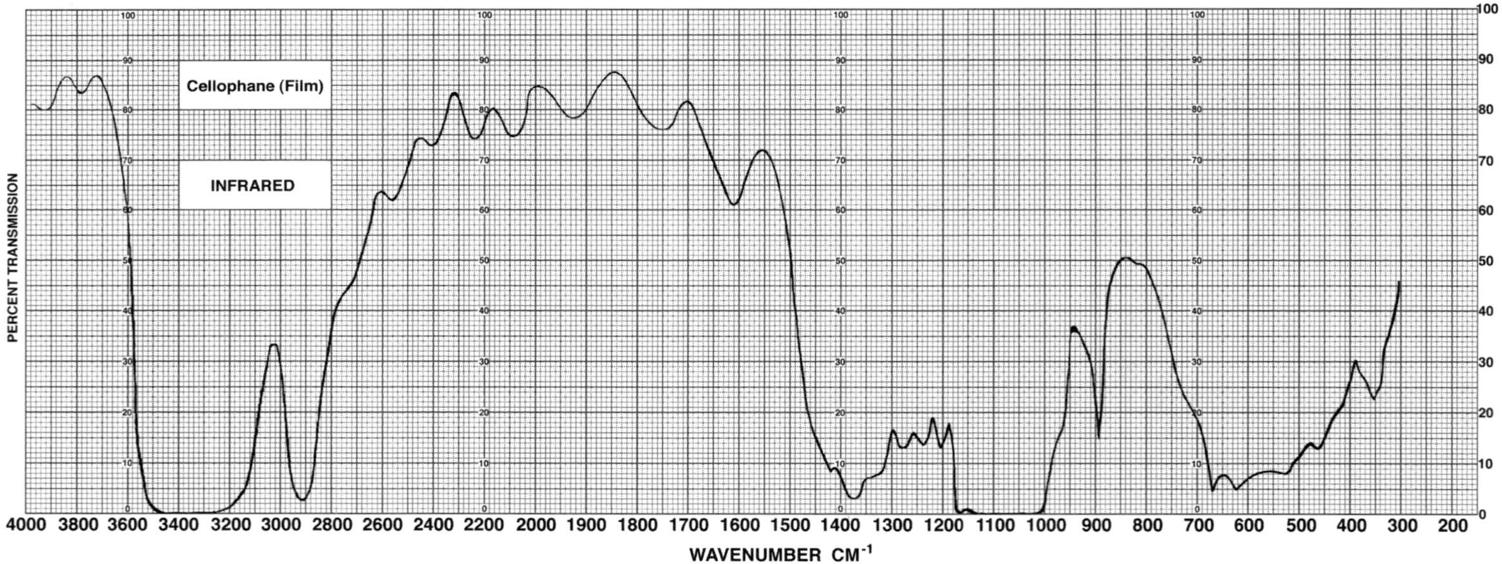

Fig. 10.8 Infrared spectrum of cellophane (film).

pressure) are another potential problem, and in this case the spectrum observed is primarily that of the pigment. Sometimes, however, the shallow depth of radiation penetration can be useful, as when studying surface oxidation or weathering of paint films where surface spectra are important to understand the chemistry taking place in this layer.

A good illustrative example of the value of ATR sampling is a cellophane overwrap from a package of chewing gum. That the primary component is cellophane (a cellulose derivative with a glycerol plasticizer) can be easily established from the transmission spectrum (see Figure 10.8).

If the spectrum of the outside surface is obtained by ATR/MIR, however, we obtain the spectrum shown in Figure 10.9, and that is not a spectrum of cellophane but rather of nitrocellulose. There are no bands in this spectrum corresponding to cellophane at all. The spectrum, not shown, from the inside of the same overwrap was also obtained by ATR/MIR, and this latter spectrum very closely matched that of a reference secondary amide, a commonly used vehicle for printing inks. It also appears that dioctylphthalate (or perhaps camphor) was used as a plasticizer on both the inside and outside surfaces.

IV. To understand polymer spectra, it is worthwhile to consider the chemistry of formation of these materials at least briefly. Polymers may be formed from *condensation* or *addition* reactions:

A. *A typical condensation reaction*:

Polymer Formation
Condensation Reaction

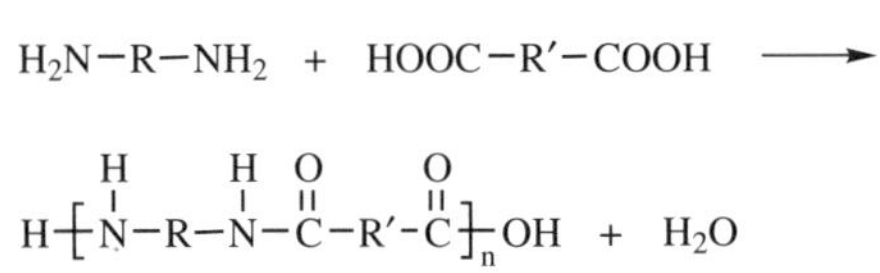

Spectrum is that of a secondary amide

Here, two bifunctional molecules (monomers) react (or *copolymerize*) to form the polymer. A diamine and a diacid react by splitting out water (a small stable molecule) to produce a repeating series of secondary amide groups. The spectrum of this polymer then should be that of an "emphasized" secondary amide.

The chemistry of formation of nylon 6/6, the classic example of a poly-secondary-amide, is given below and the spectrum of the polymer is shown in Figure 10.10.

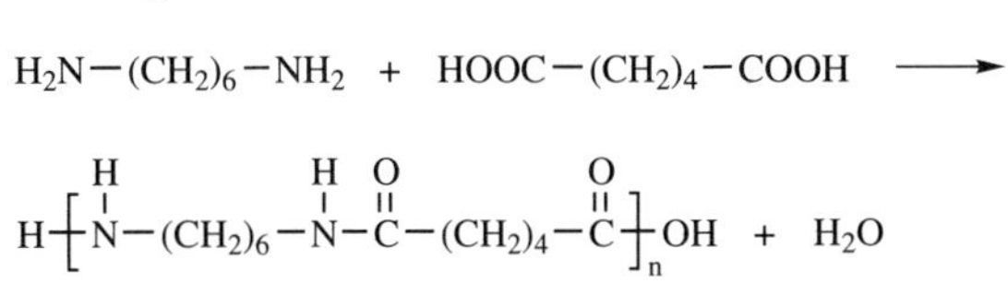

Nylon 6/6

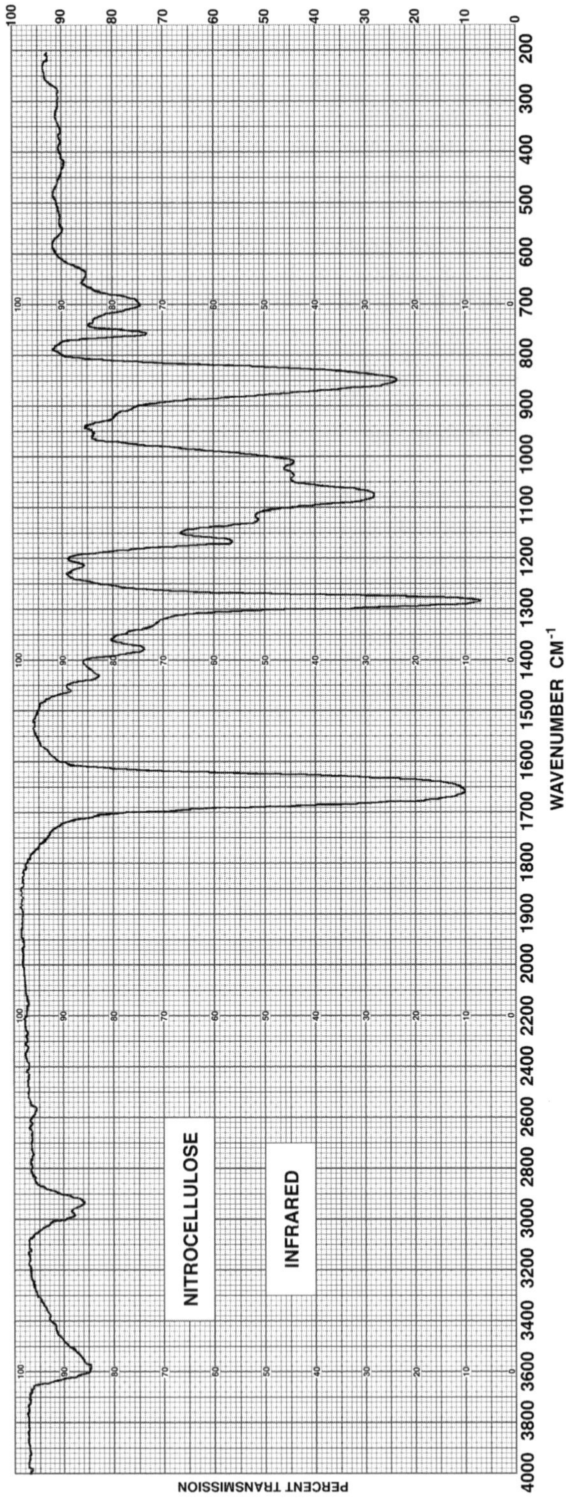

Fig. 10.9 Infrared spectrum of nitrocellulose.

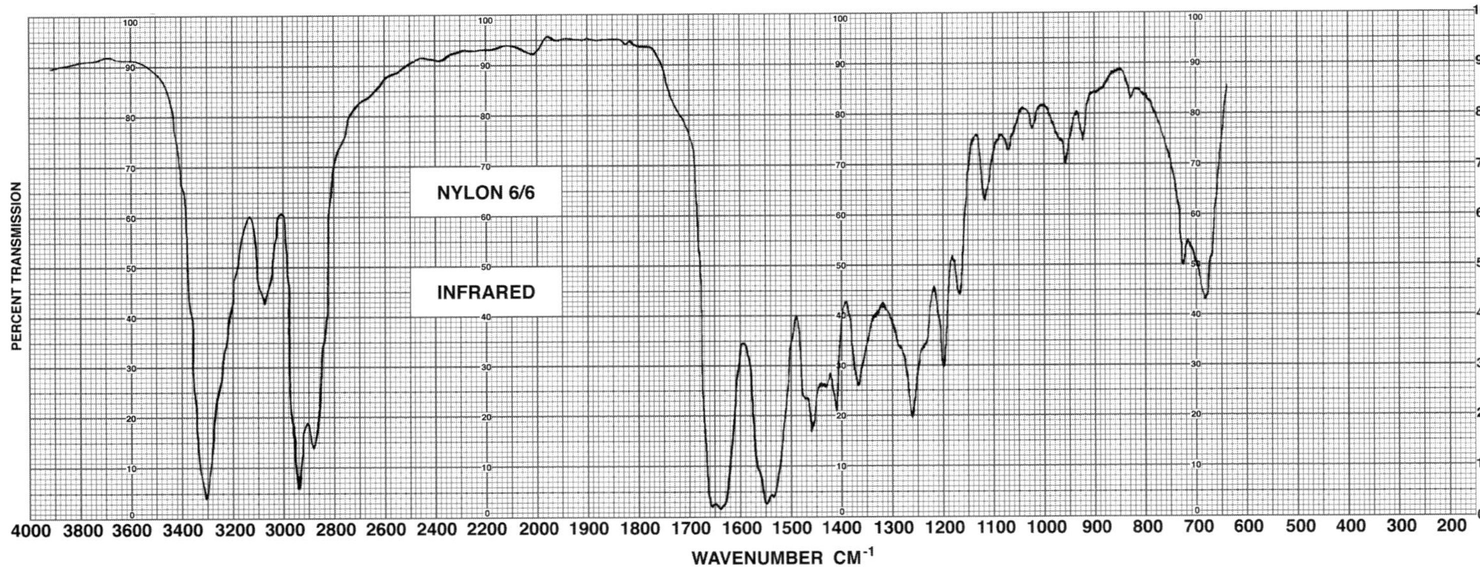

Fig. 10.10 Infrared spectrum of nylon 6/6.

The spectrum is exactly as predicted from the conventional low-molecular-weight secondary amide group frequencies. The typical absorption bands found in a secondary amide are observed in the polymer spectrum as follows: N—H stretching modes near 3300 cm^{-1}, Fermi resonance enhanced sum tone near 3070 cm^{-1}, sp^3 C—H stretching mode near 2950 cm^{-1}; amide carbonyl near 1630 cm^{-1} (the so-called amide I band), and the amide II band (a mixed mode of C$\cdots$N stretch and N—H in-plane bend) located near 1540 cm^{-1}. Other polyamide bands have been identified in the fingerprint region. For example, amide III (the low-frequency component of amide II) often falls near 1300 cm^{-1}, but it is variable in range and intensity, as are amide IV and amide V, which are located even deeper in the fingerprint region. If the spectra of the several kinds of polyamides are carefully examined, it is possible to detect small differences. It is not generally possible, however, to determine exactly what the monomers are from the spectrum of the polymer because many of the features that occur in the polyamide spectra are, in fact, due to the crystalline properties of that particular sample. The most successful route to determining the original monomers is to hydrolyze the sample, separate the acidic and basic components, and obtain their spectra as individual compounds.

B. Polymers derived from *addition reactions*

A second reaction mechanism leading to polymerization is the addition reaction. In this case the double bond opens and long polymer chains are produced. If the R group is hydrogen, then the resulting polymer is polyethylene (Ziegler received the Nobel Prize for his discovery that aluminum alkyls will catalyze this reaction in a highly ordered fashion to yield high-density polyethylenes); if the R group is a methyl group, then the polymer is polypropylene; if it is chlorine, then the polymer is polyvinylchloride (PVC); if the R group is an acetate group, then the polymer is polyvinylacetate; etc.

Polymer Formation
Addition Reaction

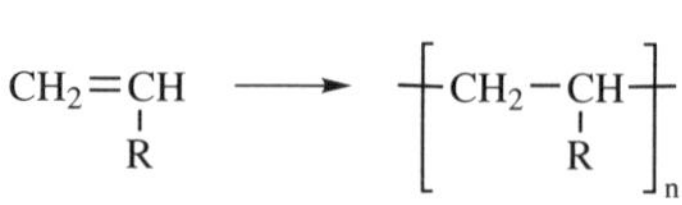

If R = H, then polyethylene
If R = CH_3, then polypropylene
If R = Cl, then polyvinylchloride
If R = OH, then polyvinylalcohol
If R = ϕ, then polystyrene

If R = —C(=O)—OCH_3, then Polymethylacrylate

If R = O—C(=O)—CH_3, then polyvinylacetate

The simplest of these addition polymers is polyethylene. The spectrum of the monomer ethylene is shown in Figure 10.11.

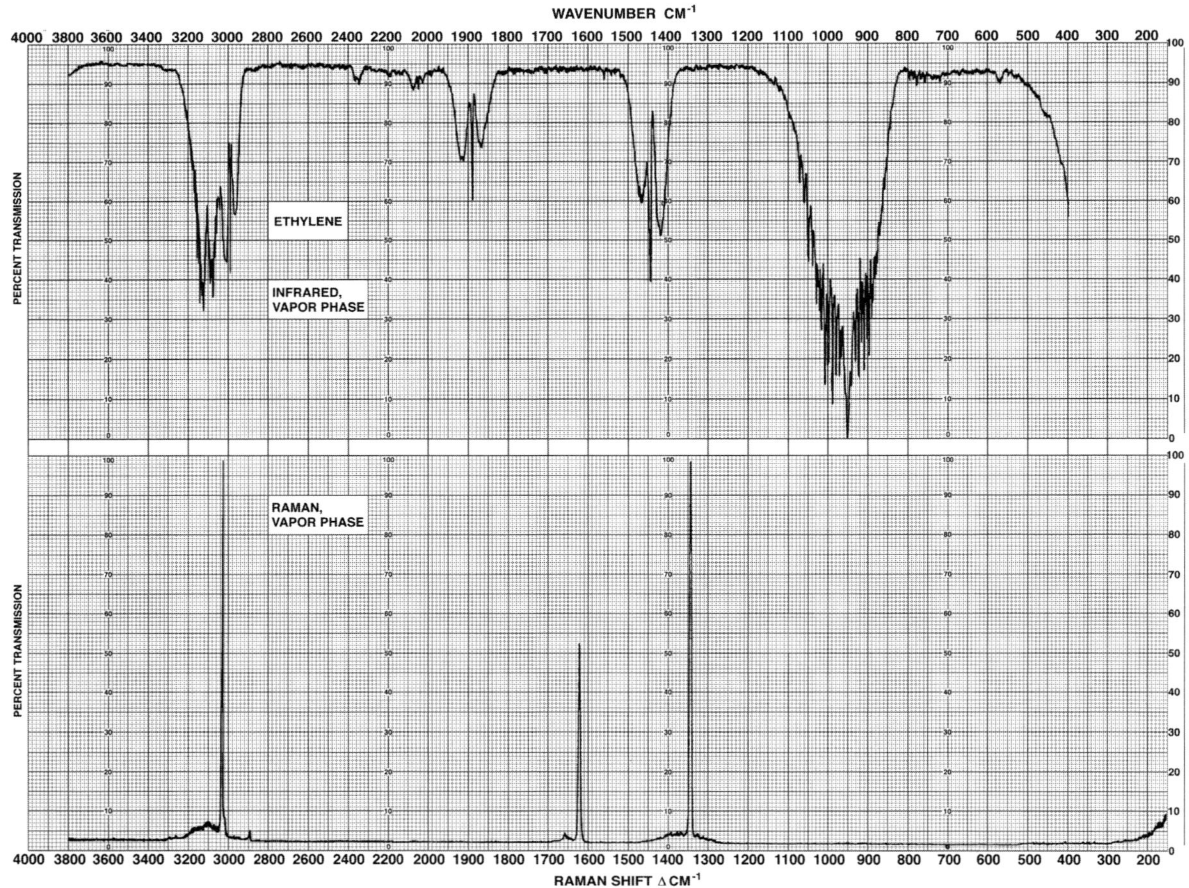

Fig. 10.11 Infrared spectrum of ethylene.

While ethylene has very few vibrational modes active in the IR spectrum, each vibrational band does possess substantial rotational fine structure. This fine structure is resolved only for very light molecules in the gas phase. When ethylene is polymerized to form polyethylene, there are significant changes in the spectrum and a substantially simpler spectrum results (see Figure 10.12).

In the polymer spectrum the two methylene C—H stretching vibrations are located near 2950 cm^{-1}, the C—H scissoring vibration is near 1460 cm^{-1}, and a CH_2 rocking mode is at 720 cm^{-1} . The splitting of the 720 cm^{-1} rocking mode is directly related to the spatial configuration of the polymer chains. In the case of polyethylene two chains pass through the unit cell, leading to interchain coupling and in-phase and out-of-phase absorption bands. To accurately measure the level of sample crystallinity, a band near 1894 cm^{-1} has been found to be more sensitive to chain order than the 730 cm^{-1} band. On the other hand, the intensity of a band near 1303 cm^{-1} appears to be directly associated with the amorphous character of the sample. In the present case the weakness of the 1894 cm^{-1} band and the strength of the 1303 cm^{-1} band indicate that this sample is not very crystalline (the product of a high-pressure process). The fact that the band assigned to CH_3 deformation near

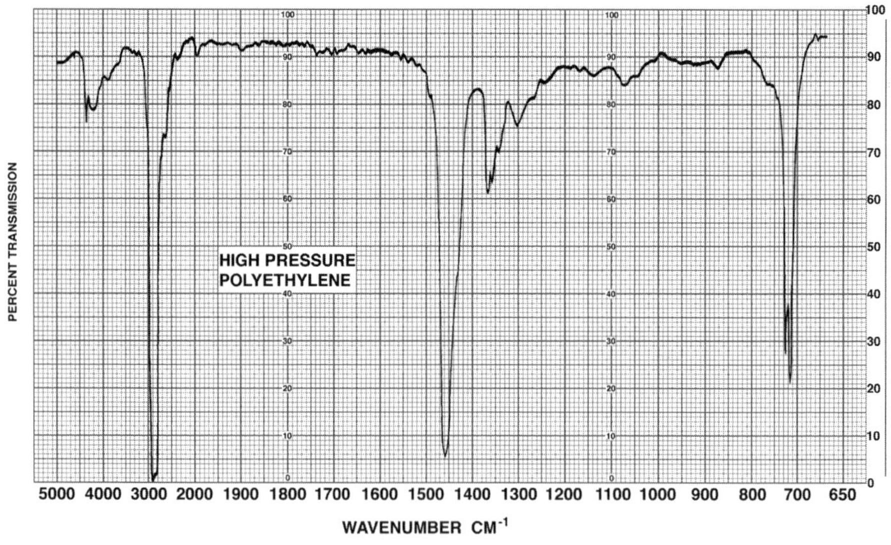

Fig. 10.12 Infrared spectrum of polyethylene.

1375 cm^{-1} is quite strong is a further indication of a high concentration of methyl groups and hence lots of side chains. This chain branching reduces the possibility for any regular order in the chains and therefore reduces crystalline content. In another example (the product of a low-pressure process) of polyethylene formation (see Figure 10.13), increased crystallinity appears to be present based on the relative intensity of the 1894 and 1303 cm^{-1} bands.

It is also possible to form polyamides through addition reaction mechanisms. For example, a major industrial process employs caprolactam as the starting material. This interesting substance can be induced to self-polymerize via addition as shown below:

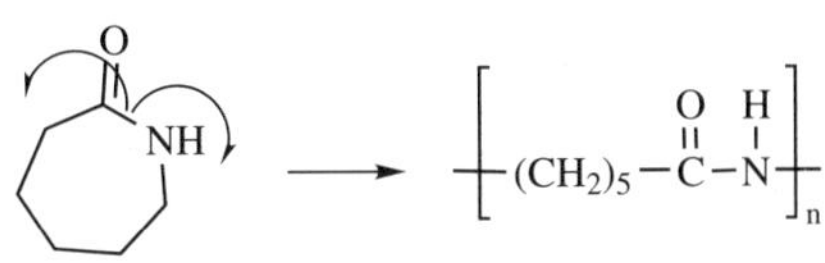

caprolactam polycaprolactam

These polymers possess very similar but not identical spectra to the polyamides formed from the condensation reactions illustrated earlier (see Figure 10.14).

C. As we have seen in the case of polyethylene, changes in reaction conditions can lead to changes in polymer properties, and these changes can often be detected in the IR spectrum of the material. A summary list of some of these important properties follows: First, as in the case of polyethylene, if the sample is very linear, the possibility of high order is present and, therefore, crystallinity may be increased. Second, this increased crystallinity can lead to a polymer for which the density is higher, the melting point is higher, the material is more translucent than transparent (i.e., the polymer scatters a little bit more light), and the material is harder. It has been found that one branch in 100 carbon atoms drops the melting point of polyethylene by 10°C, and in this case the IR spectral changes also indicate a higher concentration of methyl groups in side chains.

V. Copolymers

A further variation in the polymerization process is the formation of *copolymers* and even higher combinations of monomers. Two examples are illustrated below:

Polmer Formation:

a) Random Copolymer

-A-B-A-A-B-B-A-B-A-B

b) Block Copolymer

-A-A-A-A-A-B-B-B-B-A-A-A-A-B-B-B-B-B-

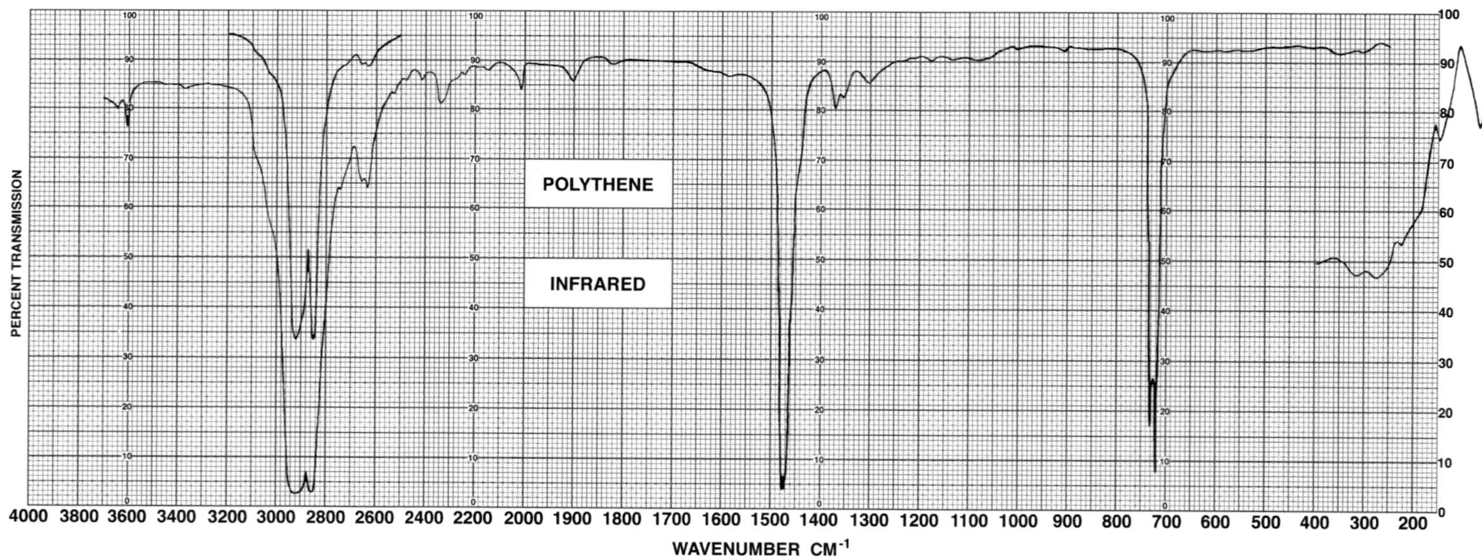

Fig. 10.13 Infrared spectrum of polythene, $[-CH_2-]_n$.

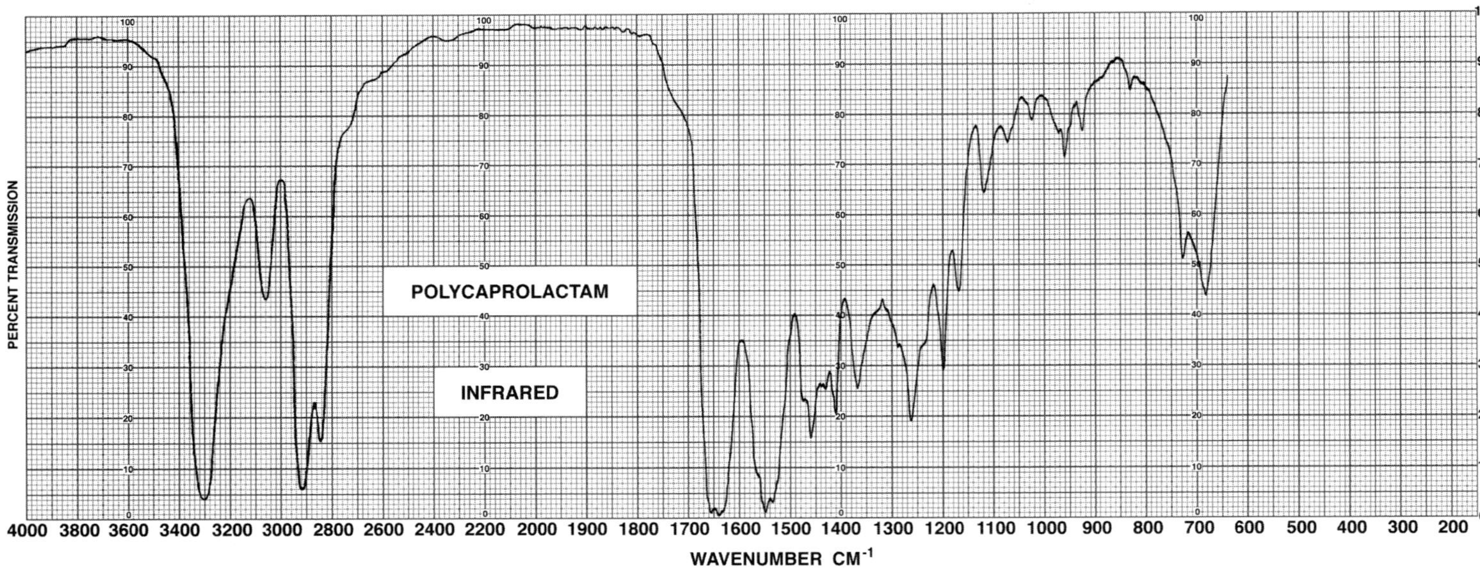

Fig. 10.14 Infrared spectrum of polycaprolactam.

As shown in example a, a random polymer can be made from two monomers.

On the other hand, it is possible to polymerize these same two monomers in such a way that relatively long chains of homopolymers are first formed and then subsequently these homomolecular chains are further polymerized together to form *block copolymers*, as shown in example b. The IR spectra of the materials formed by these two different types of polymerization have been shown to be substantially different in some cases.

Consider the case of a methylmethacrylate–styrene copolymer. First, the polymerization of the monomer, methylmethacrylate, is shown below. As in the above addition reactions, the double bond opens and polymerization occurs to produce chains containing pendant carbomethoxy groups. The major difference in the reaction, in this case, is that the polymerization is controlled to limit the length of the chains.

Polymer Formation
Polymethylmethacrylate

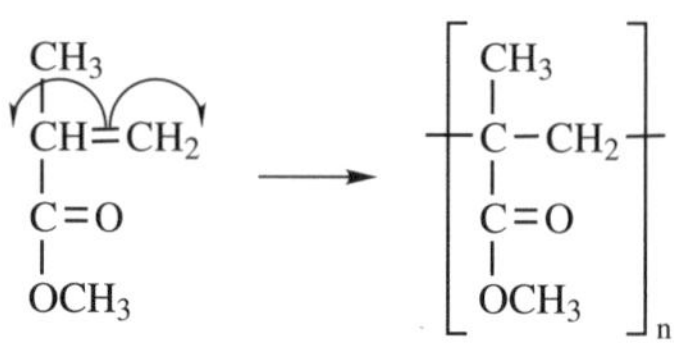

If long sequences,
then bands at 1270, 1240, 1190, 1150 cm^{-1}
due to C–O–C vibrations.
If short sequences,
then bands near 1210 cm^{-1} and 1130 cm^{-1}.

The IR spectra of the long-chain methacrylate polymers possess two pairs of highly characteristic absorption bands in the 1200 cm^{-1} region. These bands occur as partially resolved doublets with the first pair near 1270 and 1240 cm^{-1} and the second pair near 1190 and 1150 cm^{-1}. This set of absorption bands has been assigned to the C—O—C stretching vibrations of the ester part of the molecule. If the polymer is composed of short sequences, as you might expect in a random copolymer, these bands will change substantially in appearance. In this latter case the two pairs now appear as two single and rather broad bands near 1210 and 1130 cm^{-1}. If the spectrum of the homopolymer, polymethylmethacrylate, is examined, it is possible to observe the effect of the length of the long-chain sequence of pendant carbomethoxy groups on the IR spectrum (see Figure 10.15). The pair of doublets is clearly evident in the 1250 and 1170 cm^{-1} regions.

The effect on the polymethylmethacrylate spectrum when the polymer is incorporated into a copolymer is nicely demonstrated in the following

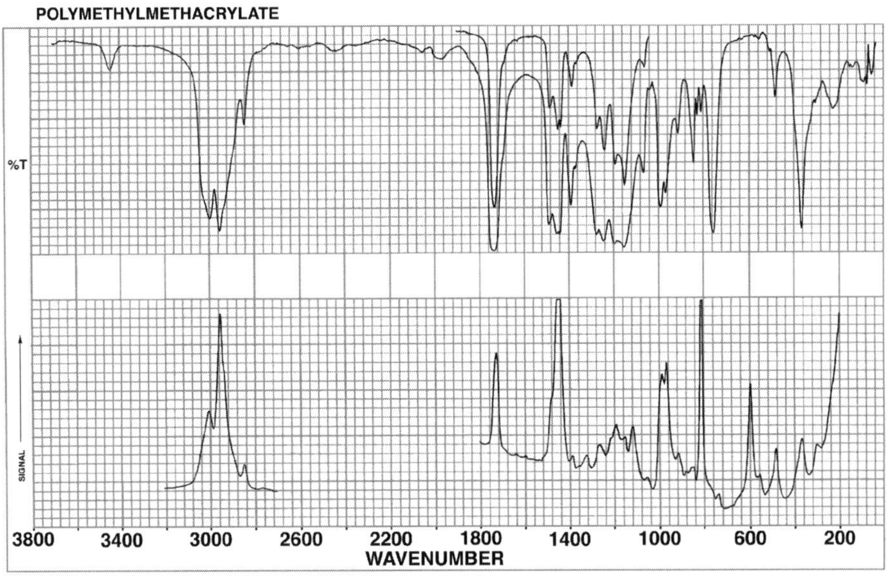

Fig. 10.15 Infrared and Raman spectra of polymethylmethacrylate.

example where styrene is used as the other monomer. It will be illustrative to consider both cases, the random copolymer and the block copolymer. In Figures 10.16*a*,*b* the spectra of both homopolymers (polymethylmethacrylate, *top*, and atactic polystyrene, *bottom*) are shown.

If these spectra are mentally overlaid and compared to the spectrum of the block copolymer shown in 10.17*a*, it is possible to see that, indeed, Figure 10.17*a* is effectively the summation of the spectra of the two homopolymers. The spectrum of the sample of the block copolymer (Figure 10.17*a*) can also be compared to the spectrum of the random copolymer, shown in Figure 10.17*b*. The pair of doublets discussed above are easily identified in the spectrum of the block copolymer. They are centered around 1250 and 1170 cm^{-1}. On the other hand, the random copolymer, which does not have a long sequence of polymethylmethacrylate chains, shows only two broad absorption bands centered around 1210 and 1140 cm^{-1}. Thus the spectrum of the random copolymer is quite different from that of the block copolymer formed from the same two monomers.

The polystyrene absorption bands come through clearly in both types of polymers and there is no trouble in detecting the presence of this aromatic material in the copolymer. One of the things that can happen in the formation of block copolymers is that a substantial amount of local order may occur within the structure of the polymer itself. This increased order can result in local crystallinity effects which can lead to a sharpening of the absorption bands, and this band narrowing, in fact, may be the only way to detect the

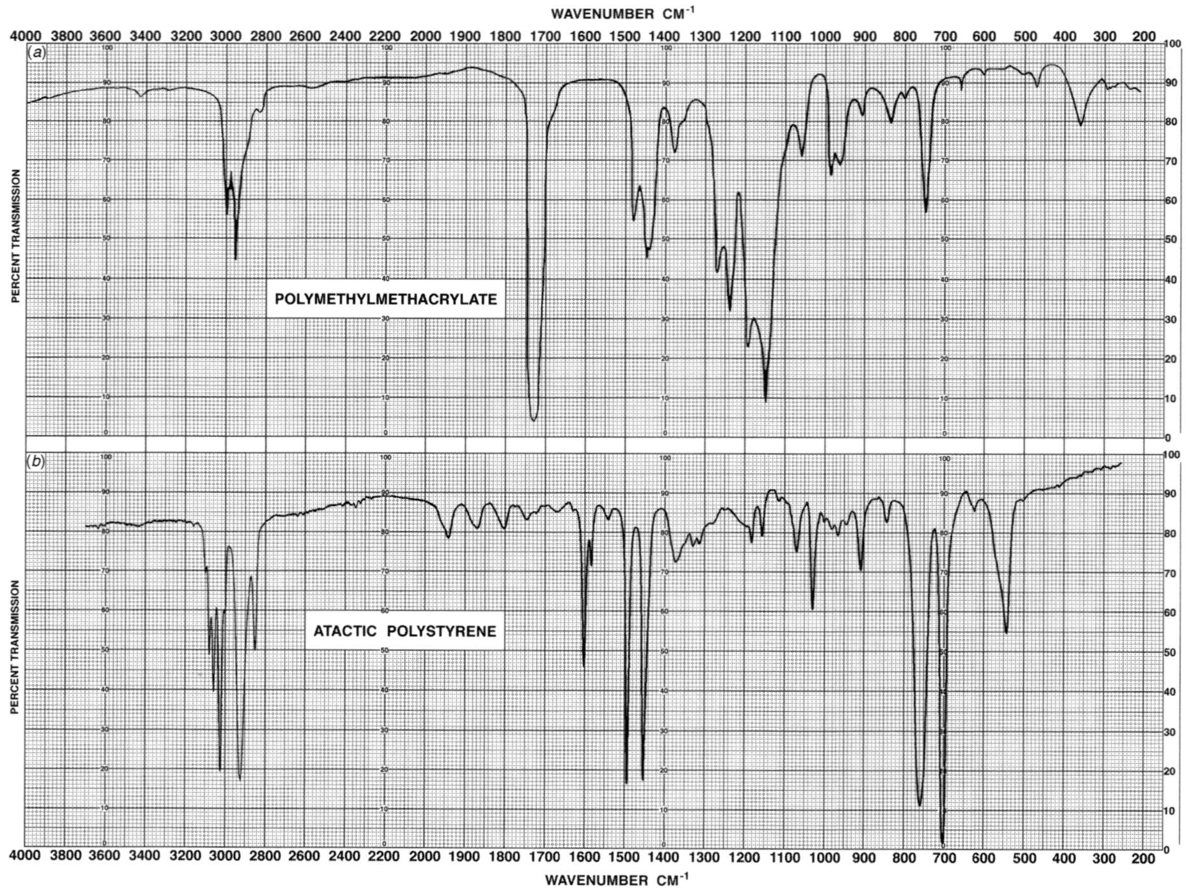

Fig. 10.16 Infrared spectra of (*a*) polymethylmethacrylate and (*b*) atactic polystyrene.

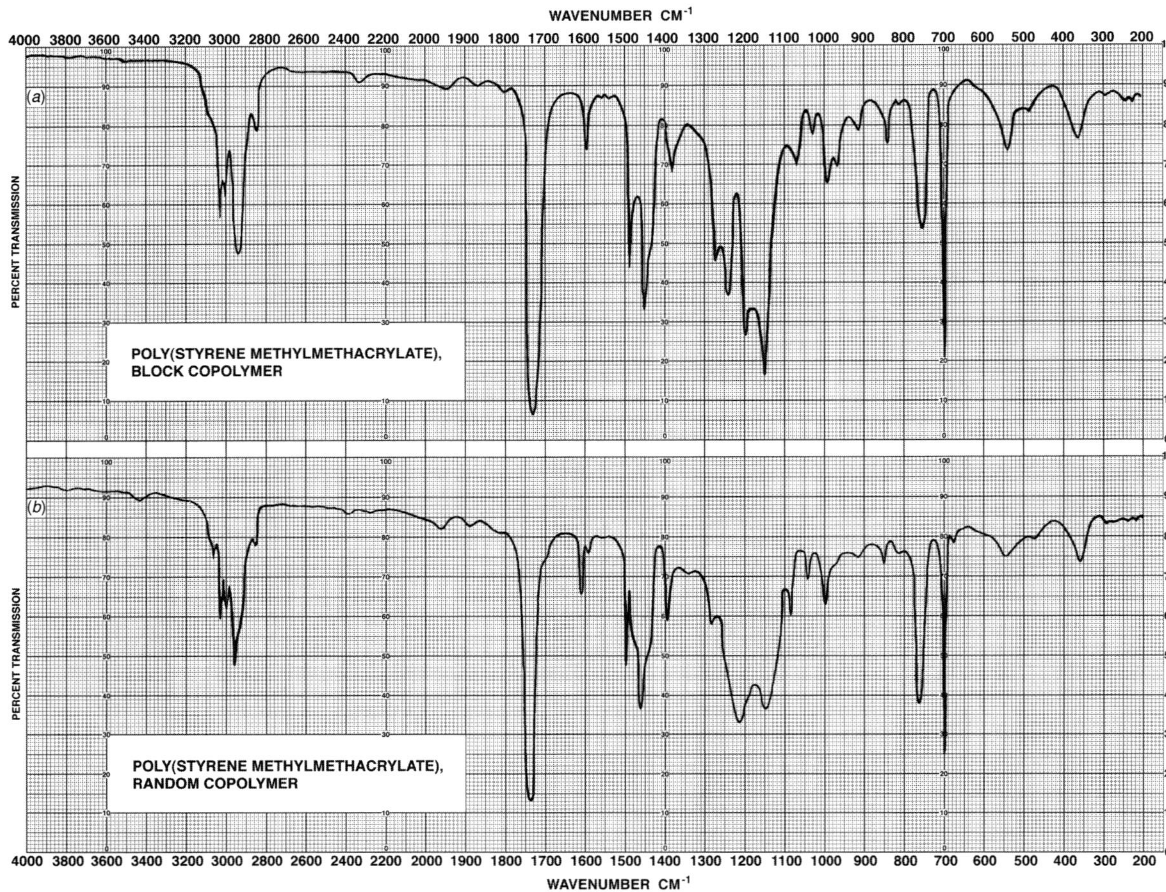

Fig. 10.17 Infrared spectra of (*a*) poly(styrene methylmethacrylate), block copolymer, and (*b*) poly(styrene methylmethacrylate), random copolymer.

development of this relatively subtle structural condition. Thus, it becomes quite clear that an understanding of band shapes is important to the correct interpretation of polymer spectra. This additional information available from the IR spectrum should be taken into account if at all possible.

VI. Crosslinked polymer formation

It is also possible to form polymers which possess very high molecular weights via a process known as crosslinking. The chemistry of one such process which involves the formation of a phenolic polymer is described below. The reaction outlined takes place between phenol and formaldehyde to form an alcohol derivative which then reacts further with additional formaldehyde to build an extremely high molecular weight, insoluble polymer. The crosslinking routes and major mechanisms are shown in the second diagram:

Crosslinked polymer formation:

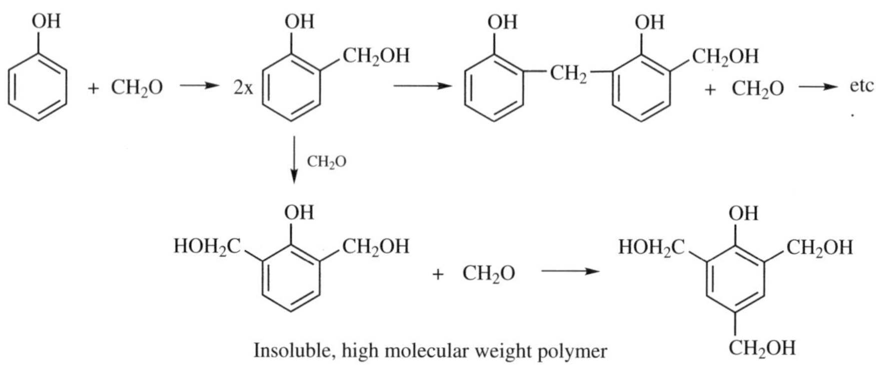

The mechanism of phenolic crosslinking polymerization

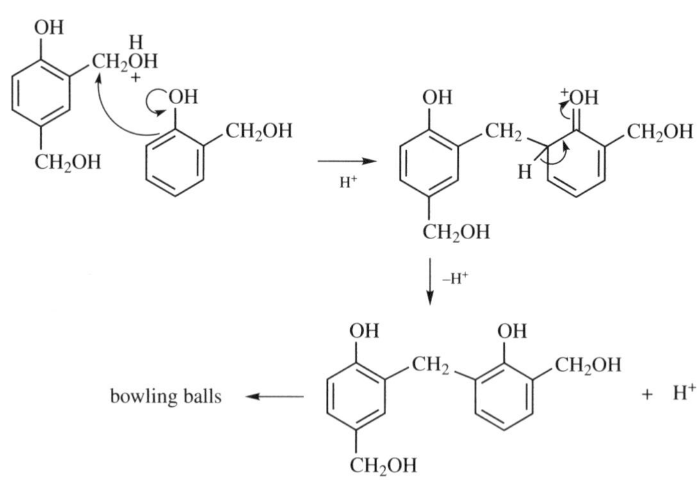

The phenol/formaldehyde condensation is a good example of a polymerization which takes place in three dimensions to yield materials useful in the manufacture of such products as bowling balls and telephones. These substances can be viewed as one large crosslinked molecular polymer with an extremely high molecular weight. As expected, any attempt at dissolving these materials leads to frustration. The molecular weight is simply too high. These polymers are very insoluble materials, indeed. Attempts to measure the melting points of these polymers are also quite unsuccessful and generally result in decomposition.

VII. Tacticity

To understand the subtle nature of tacticity, it is best at this point to examine the polypropylene system, the formation of which is shown below.

Polymer Formation
Addition Reaction

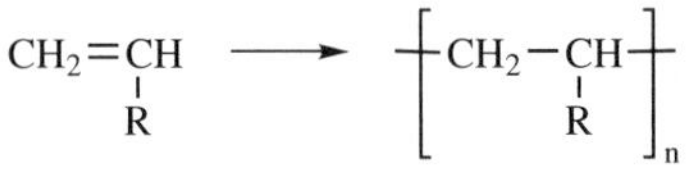

If R = CH_3, then polypropylene

Polypropylene is normally formed by the above addition reaction of propylene, and since the reaction normally occurs in a head-to-tail mode the structure can be quite uniform. However, it is possible to obtain three different configurational isomers during the polymerization of propylene. They are shown below.

Polymer Formation—Polypropylene

a) Isotactic

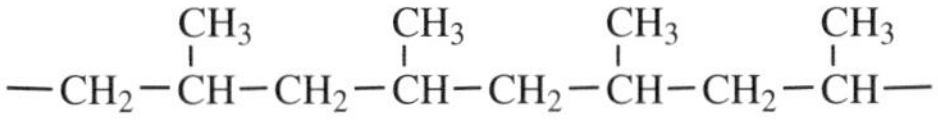

same

b) Syndiotactic

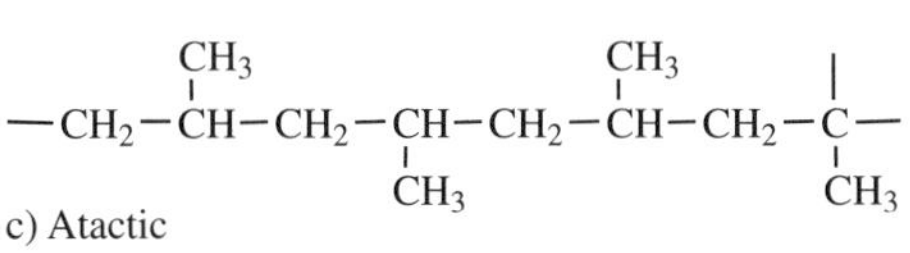

alternate

c) Atactic

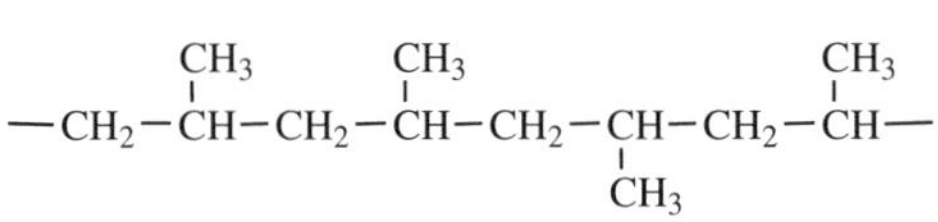

random

These are stereoisomers and are termed *configurational stereoisomers*. While the same atoms are bonded together in each isomer, it requires a bond to be broken to interchange one isomer with another. The three different polymeric stereoisomers are termed *isotactic*, *syndiotactic*, and *atactic*. In the isotactic case, all of the methyl groups are positioned on the same side of the chain and a very regular uniform structure results. On the other hand, the syndiotactic form also may have a relatively uniform structure, but in this isomer the methyl groups alternate from side to side in a regular fashion down the carbon–carbon backbone. The atactic form has the methyl groups randomly oriented with respect to each other along the polymer carbon chain. The first two stereoisomers, as noted above, possess a more uniform configuration than the atactic form, although the syndiotactic form does not appear to pack in the solid state as well as the isotactic form. The more uniform, higher order forms allow secondary forces such as hydrogen bonding or van der Waals forces to play a larger role and help to overcome thermal disordering effects. Thus the isotactic, and possibly syndiotactic, solid polymers become very ordered and these stereoisomers can actually be crystalline. The isotactic and syndiotactic polypropylenes tend to be harder and possess higher melting points than the atactic form, which is largely amorphous in character. Remember that covalent bonds are 50–200 kcal/M, whereas a dipole–dipole interaction or hydrogen bonding involves 1.5–7 kcal/M and van der Waals forces some 0.5–2 kcal/M. These latter types of intramolecular interaction have low bonding energies between adjacent molecules. Linear or branched-chain polymers of the polypropylene type will melt on heating because the dipole–dipole, van der Waals, or hydrogen-bonding forces which are holding them together in the solid phase are thermally sensitive. As noted before, however, a crosslinked polymer will degrade rather than melt. It is also true that solvents will dissolve the linear or simple branched polymers, as is the case for polypropylene, if the bonds to the solvent are stronger than the secondary forces holding the polymer together. Dissolution of many long-chain polymers may require some warming of the solvent. The first evidence that the system is beginning to dissolve may be only a swelling of the polymer, but eventually many polymers of this type will completely dissolve.

The spectra of the three forms of polypropylene possessing different tacticity are shown in Figure 10.18. The atactic, syndiotactic, and isotactic polymers, respectively, are shown from top to bottom in the figure. It is clear that the isotactic polymer has a substantially different spectrum from the other two, with many sharp needlelike bands and some of the band systems apparently being split. The atactic and syndiotactic forms look remarkably similar to each other with only differences in relative intensities of some of the weaker bands. Clearly, in the case of polypropylene it appears that the isotactic form packs considerably better (with higher order) than either the syndiotactic or atactic forms.

VIII. Conformational isomerism

A final twist to the interpretation of the IR spectra of polymers is that apparently *conformational* isomerism is also present in these materials. Thus,

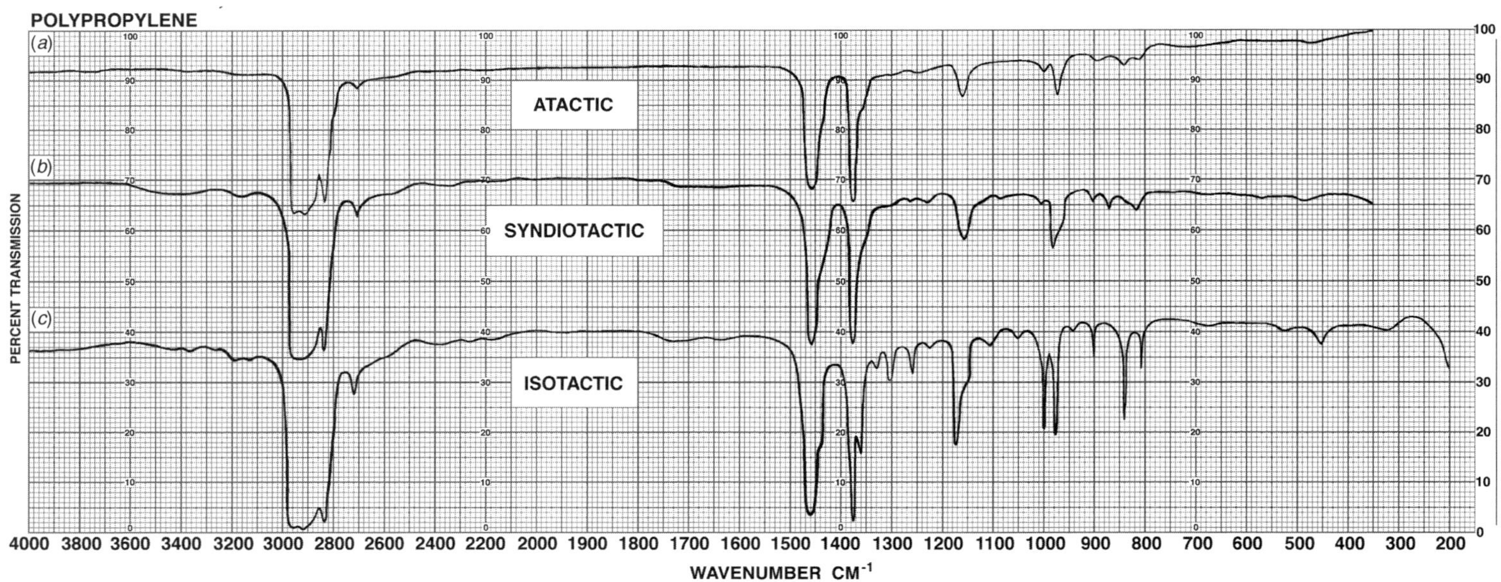

Fig. 10.18 Infrared spectra of (*a*) atactic, (*b*) syndiotactic, and (*c*) isotactic polypropylene.

the folding of the polymer chain in the formation of the solid phase can also play a significant role in determining both band shape, intensity, and position in the IR. For example, in the case of polypropylene, which has just been examined in the discussion, the conformational structure of isotactic polypropylene in space has been established to be that of a threefold helix, with three monomer units per turn of the helix. The IR spectrum, in fact, ultimately depends both on the tacticity, i.e., the orientation of the methyl groups down the carbon–carbon backbone, and on the helical *conformation* in space. When isotactic polypropylene is melted, the bands at 1165, 998, 899, and 840 cm^{-1} disappear. These latter bands have been shown to be related to the presence of the helix conformation in which this particular molecule (isotactic polypropylene) exists. If the melt is allowed to resolidify, these bands reappear. Thus, the spectrum depends not only on the tacticity of the polymer but also on the fact that a property of this tacticity is that it leads to a threefold helical conformation of the polymer chain in the solid state. The combination of the very regular chain conformation and isotacticity may actually allow for the presence of crystallinity in the sample. It turns out that there is often a great deal of confusion about whether crystallinity or local conformational structure is reflected in the IR spectrum that is being observed. The fact is that the IR spectrum may or may not reveal anything about the tacticity or crystallinity for a particular molecule. The crystal of isotactic polypropylene is monoclinic with four chains in the unit cell. Thus, there are 12 monomer units or 108 atoms per unit cell (which translates, $3n - 6$, to 318 vibrational modes). The spectrum, however, appears to be much simpler than the predicted number of normal modes. This simplification can be explained on the basis of a single helix rather than on the basis of tacticity alone or the fact that the sample simply might be crystalline.

IX. Summary

Crystallinity is a property that is not accurately measured by the IR spectrum without some problems associated with it. A regular chain structure, as for a stereoregular polymer, promotes crystallinity. The presence of side chains hinders crystallinity and also lowers melting points and decreases the density. The secondary forces holding the chains in the crystal must be stronger than the thermal energy disordering effects if we are going to have crystalline materials formed. Thus, hydrogen bonding or dipole–dipole interactions are normally necessary for crystallinity. Amorphous samples lack crystallinity by definition. As noted above, however, it is possible to have highly ordered local situations which may show evidence of substantial amounts of local microscopic crystallinity. In these latter cases it can be very hard to distinguish this subtle crystallinity from the spectrum. Another point to consider is that drawing or machining, i.e., any kind of processing that we do to the polymer, may increase the sample crystallinity.

The effect of increased crystallinity on the mechanical properties of a sample is, e.g., reflected in a higher density. High-pressure (low-density) polyethylene has more branches than low-pressure (high-density) polyethylene. There are approximately two to six carbon atoms per branch in high-pressure polyethylene. These side chains result in a lower density polymer. Branching interferes with crystallization, and this interference contributes to the fact that we have poorer packing of the polymer chains in the high-pressure product. Highly regular polymers will normally be of higher strength and higher rigidity (low-pressure polyethylene is a good example). These high-density polymers will also be more opaque and light scattering. Transparent polymers are generally amorphous whereas translucent polymers are normally very regular and may be crystalline.

X. Flow charts

Finally, another way of characterizing and identifying polymers will be considered. This approach is based on the fact that polymers do, in fact, follow the general rules for group frequencies and that often a very simplified codification of the group frequencies may be applied to the characterization of these materials. This approach utilizes the so-called *flow chart* or flow diagram in the identification process. In Figure 10.19 a representative flow chart is shown. Very simply, the key rule for proceeding down the chart to the ultimate identification is to identify the presence or absence of a band in the unknown sample spectrum at each stage in the diagram.

To get a feeling for how this approach operates, consider a sequence of questions that might be posed. The first question which is asked in the flow diagram of all samples is whether or not there is a band in the 1750 cm^{-1} region (a further requirement is that this band must be strong in intensity, as noted just to the right of the upper block). A band present in this region of the spectrum shows the presence of a carbonyl group, and in fact the question is really asking whether this group is present or absent. (There are, of course, a few other groups which give rise to strong bands in this spectral region, such as immonium ions and vinyl ethers, and the user of such a group frequency correlation system must be aware of the potential conflicts that could be lurking in the background.)

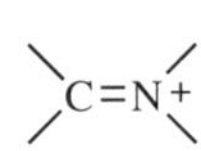

or

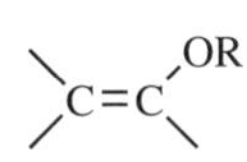

Moving to the left on the diagram, one enters an area of the diagram in which the carbonyl is present. In this case (carbonyl present) the next question to be considered is whether or not there are bands near 1600, 1580, and 1500 cm^{-1}. These are, of course, the very sharp bands which are characteristic of the aromatic group. If the three bands are located, the flow is again to the left, and the spectrum is examined for a band in the

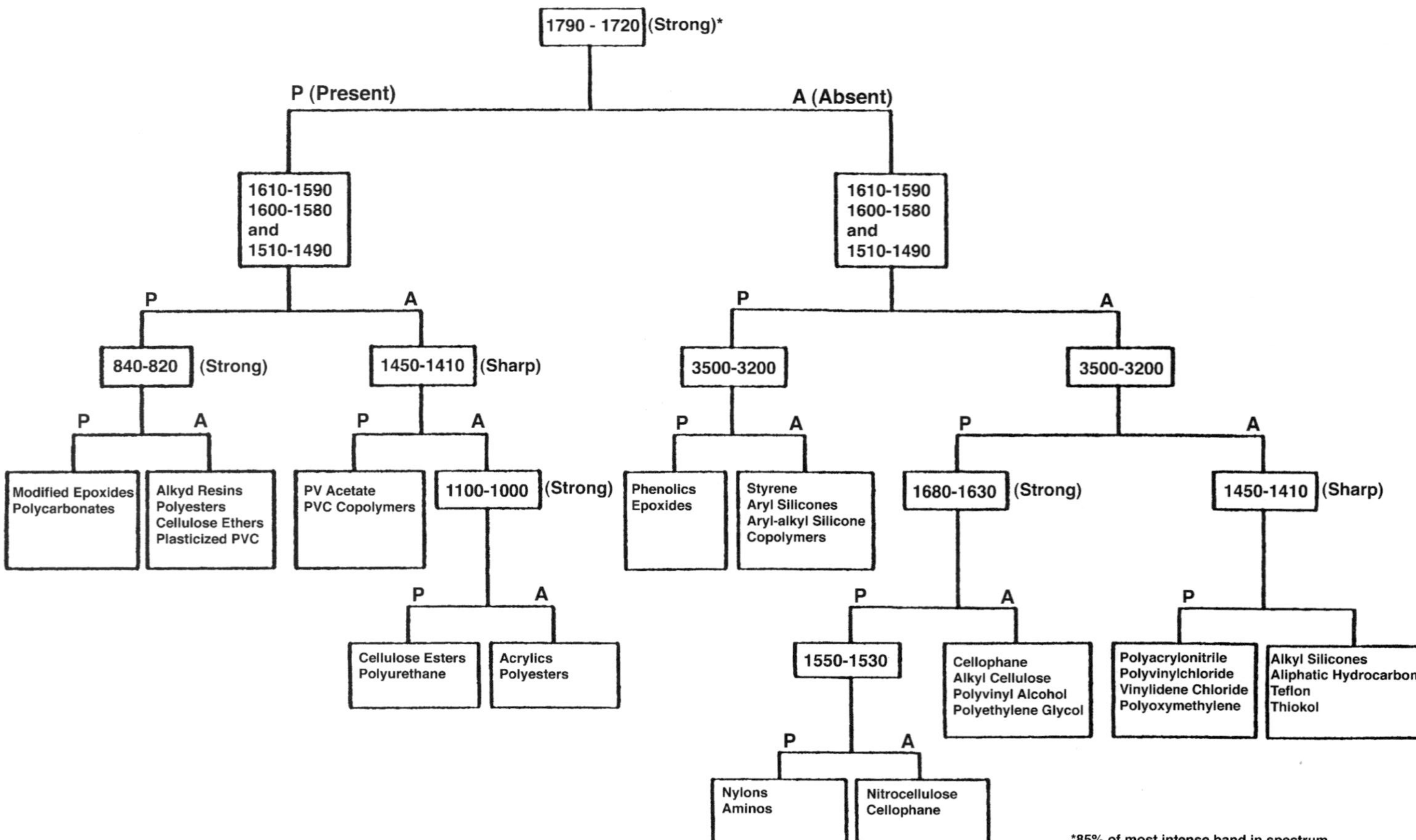

Fig. 10.19 Flow diagram for polymer identification using IR spectra.

840–820 cm^{-1} region (again there is an intensity restriction, i.e., the band must be very strong to strong in intensity). (The presence of this band gives a strong indication of simple para-substitution of the ring in the repeating unit.) If on examining the spectrum a band is found, the flow is again to the left. At this point the search requires comparison of the unknown with known polymers fitting the chart description, and a comparison of modified epoxides or polycarbonates is suggested. These flow diagrams should be taken with a grain of salt. The limits on the positions of the bands are reasonably good, but if a band is found near but outside the border of the frequency region suggested, it is probably prudent to consider the band as being present, at least for the moment. Sometimes it is possible to go through the diagram without success. In this case return to the point where a difficult judgment call had to be made and see if that is where the difficulty lies. Was the band really strong or broad or sharp or just outside or inside the frequency region? Then go down the other branch of the diagram and see if it is possible to find a better match for the unknown. Normally all a flow diagram of this type is meant to do is to place the search within the class of compounds belonging to the unknown. At this point, a comparison of the unknown spectra with a library set will be required for the final identification.

Two examples of the application of the flow diagram follow. They will help to make its usage clearer.

Figure 10.20 gives the spectrum of unknown polymer I.

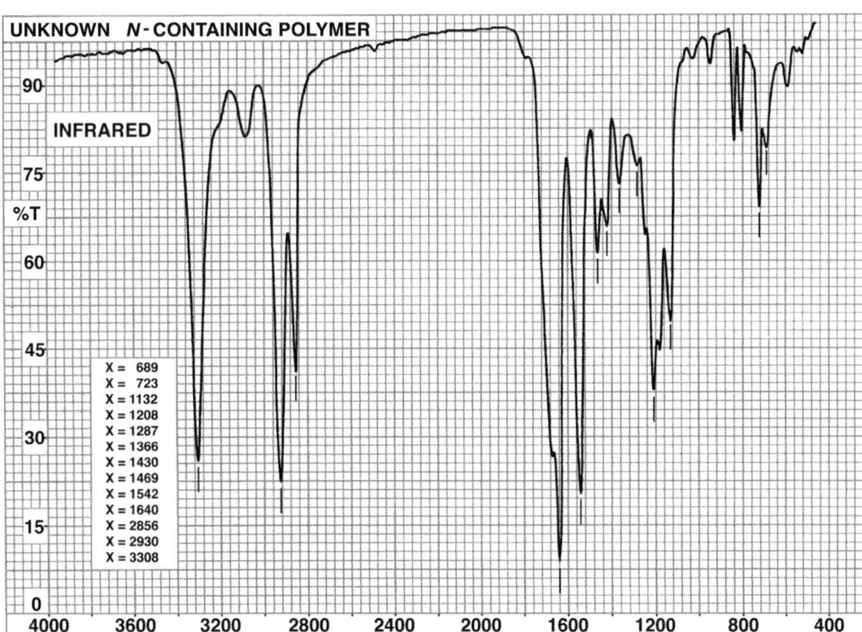

Fig. 10.20 Unknown polymer I.

Because these are relatively common polymers, it may be possible to identify the material directly. Nevertheless it will be illustrative to proceed through the application of the diagram. The first question that the diagram asks is whether or not there is a band in the 1790–1720 cm^{-1} region which is strong. When this region of the unknown spectrum is examined, it is found to be vacant. Thus, the path through the flow diagram is to the right side, i.e., the absent, or A, side. The next question requires the presence of three bands near 1600, 1580, and 1500 cm^{-1} and that these bands be relatively sharp. This question is clearly searching for the presence of an aromatic ring system. Again the region examined (1600, 1580, and 1500 cm^{-1}) in the unknown is empty of sharp bands and again the flow is to the right side. (There are some broad bands in the region of interest, but there are no sharp bands and hence the absorption which is present is not characteristic of what would have been expected with the presence of an aromatic system.) Thus, the flow continues down the absent branch of the diagram. The next question is whether the spectrum has a band between 3500 and 3200 cm^{-1}, and the answer to that question is yes, it does. Now the flow moves to the left for the first time and further down the branch. The next stage asks for a band between 1680 and 1630 cm^{-1} which is strong. An examination of the unknown spectrum reveals that, indeed, there is a broad strong band within that frequency region, and so again the move is to the left and down the current leg to the next question. This last question asks about a band between 1550 and 1530 cm^{-1}. A final examination of the spectrum reveals that the band in question is indeed present. Thus the move is for a third time to the left and down the tree. This last move leads to a block of suggested structures, such as nylons and amino compounds. On referring to the reference spectra of nylons, it is established that the unknown does, indeed, fall into this class of compounds. If a polyamide was the initial guess before starting down the diagram, it would have been correct!

Consider one more example: unknown polymer II as shown in Figure 10.21.

Once again the first question is concerned with the presence or absence of a band between 1790 and 1720 cm^{-1}. Again the region is vacant of strong bands, as in the previous example. The flow is therefore to the right and down the absent branch. The next question, as before, asks if there are three sharp bands between 1610 and 1590 cm^{-1}, 1600 and 1580 cm^{-1}, and 1510 and 1490 cm^{-1}. Again the region is vacant of the required set of three sharp bands (aromatic ring systems) and the flow continues to the right and down the absent branch. The next question concerns a band between 3500 and 3200 cm^{-1}. The answer is again no, which leads to a further move to the right and down the absent branch. At this stage on the flow diagram the question asked is whether there is a sharp band between 1450 and 1410 cm^{-1}. There is indeed a band in this region of the unknown spectrum and it is reasonably sharp. This obviously is a question of judgment. As a result of the presence of the sharp band between 1450 and 1410 cm^{-1}, the flow now

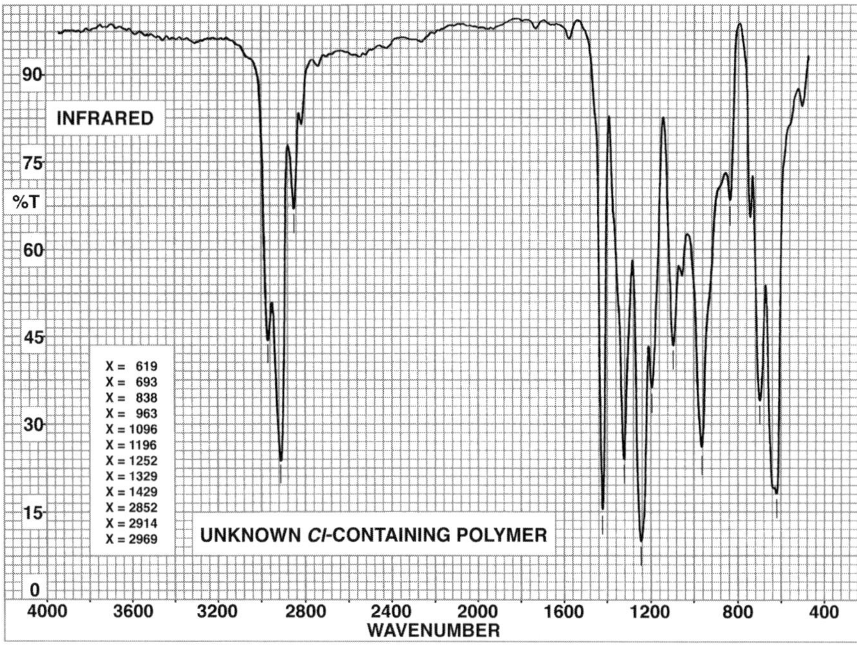

Fig. 10.21 Unknown polymer II.

moves to the left and down the present branch. The last move leads to a block which suggests that the unknown belongs to four possible structural classes: polyacrylonitriles, PVC, polyvinylidenechloride, and polyoxymethylene. On examining the spectrum of the polyacrylonitrile spectrum, there appears to be no match. This result could have been foretold from the spectrum of the unknown because a polyacrylonitrile would have a sharp band in the 2250 cm^{-1} region due to the presence of the nitrile group. When the comparison is made to a reference spectrum of PVC, an almost precise match results. Hence a positive identification of the class of polymers to which the unknown polymer belongs has been made.

XI. Conclusion

This concludes an introduction to the IR spectra of polymers. Clearly, the interpretation of polymer spectra can be reasonably straightforward based on the group frequency logic that has been developed to deal with relatively small organic molecules. There are, however, complexities which result from the fact that these solid systems may be associated with configurational isomerism, stereoisomerism, and conformational isomerism. These added complexities affect the spectra and must be remembered and taken into account in the interpretive process. Finally, there are aids such as the flow diagram which make it somewhat easier and more straightforward to carry out the characterization and identification of a particular polymer without requiring reference to computer searches or big spectral library searches. If only a few polymer spectra are to be identified from time to time, then the flow diagram can be an extremely useful device. As has been pointed out, however, the flow chart should be used with a certain amount of caution and the knowledge gained from the other chapters on group frequencies. The flow diagram described here obviously does not work with copolymers. Use it wisely, but keep it in mind that there can be problems in the identification which are brought about as a result of the peculiar properties associated with polymers themselves.

11 Infrared Spectra of Inorganic Materials

FOIL A. MILLER

I. Introduction
   A. *This chapter will deal almost entirely with IR spectra.* Raman spectroscopy is an excellent way to obtain vibrational spectra of inorganic materials, and there is much information in the literature, but very little of it is included here.
   B. Covalent bonds produce vibrational spectra.
      1. If two or more atoms are held together by covalent bonds, the system will have normal vibrations and a vibrational spectrum.
         a. It does not matter whether the system is organic or inorganic. Hence inorganic molecules have vibrational spectra: H_2O, NH_3, $SnCl_4$, PF_5, CrO_2Cl_2, etc.
         b. The system may be charged—may be a polyatomic ion.

Organic ions	CH_3COO^-, RNH_3^+, . . .
Inorganic ions	NO_3^-, CO_3^{2-}, SO_4^{2-}, NH_4^+, . . .

      2. Vibrations of molecules or polyatomic ions are called *internal vibrations* because they occur *within* the molecule or ion.
      3. *In crystals* there are also *external vibrations or lattice vibrations* (called *phonons* by physicists).
         a. They are due to motions of the molecules or ions relative to each other.
         b. They occur at low frequencies, usually less than 300 cm^{-1}.
         c. They are found only in crystals.
         d. They will be described at the end of the chapter.

II. General comments on the spectra to be shown
   A. Most of the IR spectra were obtained as a *split mull.* Fluorolube was used for 3800–1333 cm^{-1} and Nujol for 1333–400 cm^{-1}.

Course Notes on the Interpretation of Infrared and Raman Spectra, by Dana W. Mayo, Foil A. Miller, and Robert W. Hannah.
ISBN 0-471-24823-1

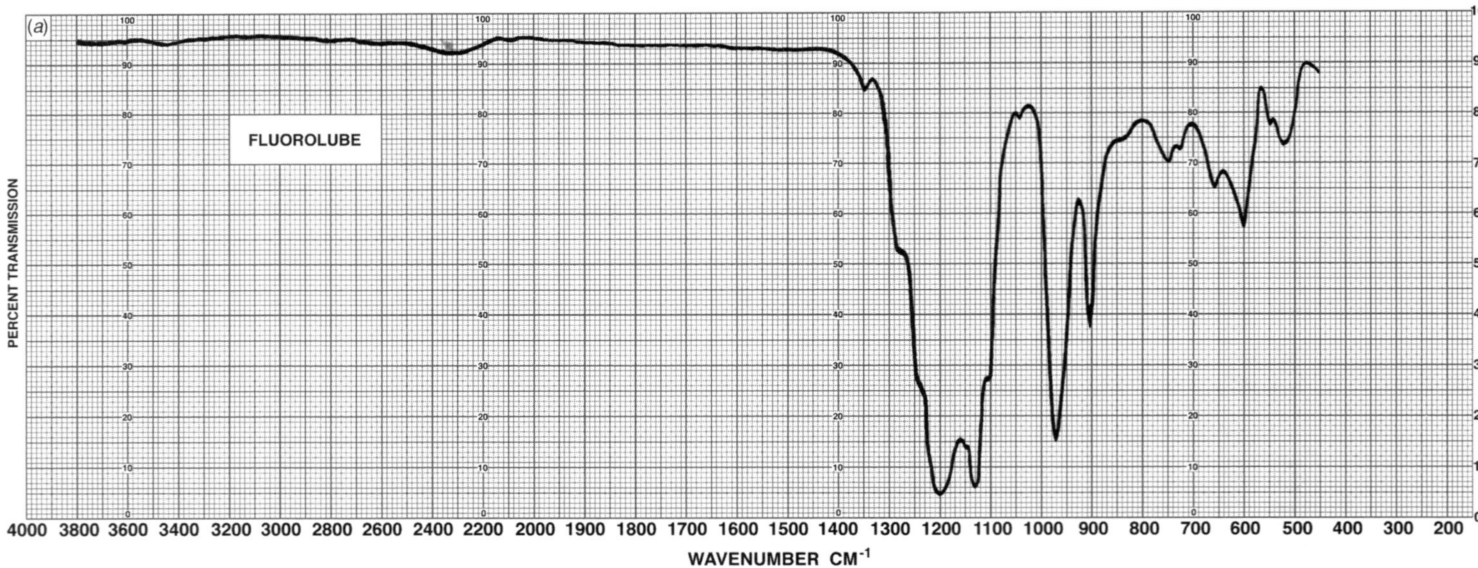

Fig. 11.1 Infrared spectra of (*a*) Fluorolube and (*b*) Nujol.

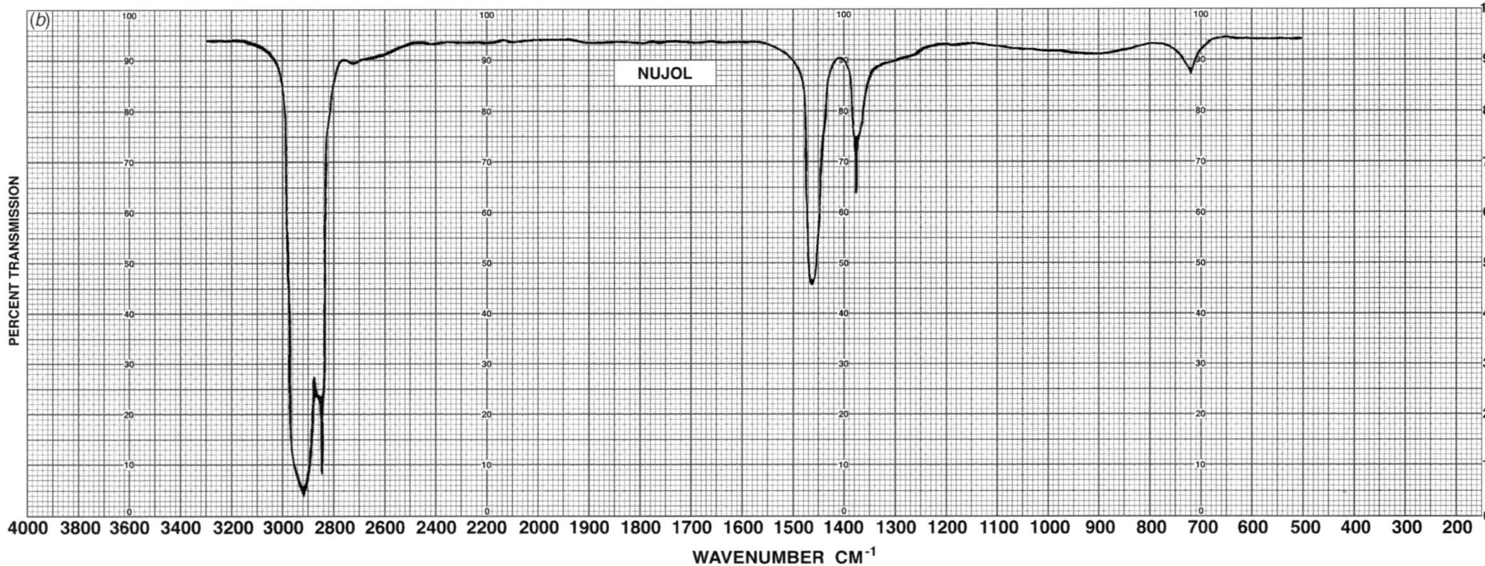

Fig. 11.1 (*Continued*)

See Figures 11.1*a,b*. Consequently there are no bands from the mulling agent.

B. BaF_2 windows were used in the higher wavenumber region and AgCl windows in the lower one to reduce the possibility of ion exchange.

III. Water occurs so often in inorganic materials that we must know its spectrum.

A. Liquid water

1. Infrared spectrum. See Figure 11.2.
2. Liquid water has IR bands at:

a. *3450–3200 cm^{-1} (vs, vb)*. This is due to superposition of the symmetric and antisymmetric O—H stretches.

b. *~1640 cm^{-1} (m, b)*. The scissoring frequency.

c. *~700–300 cm^{-1} (m, vvb)*. This is due to the hindered rotation (or *libration*) of water.

B. Water in inorganic samples absorbs in these same regions.

C. *Example*: $BaCl_2 \cdot 2H_2O$. See Figure 11.3.

1. The Ba^{2+} and Cl^- ions have no internal vibrations. Therefore all the bands seen above 400 cm^{-1} are due to H_2O.
2. Note that:

a. The scissoring band near 1600 cm^{-1} is a doublet. This is probably the result of the crystal packing. There may be two nonequivalent lattice sites for the water molecules which result in two different frequencies. Alternatively, the sites may be equivalent but the scissoring vibrations of two different molecules may interact to give the splitting.

b. There are three bands in the O—H stretching region, probably for similar reasons.

c. Librational bands are seen at 700 cm^{-1} and about 550 cm^{-1}.

d. A break in the curve at 1330 cm^{-1} shows where the mulling agent and windows were changed. There is no band near 1300 cm^{-1}.

D. In inorganic compounds, H_2O may be classified as:

1. *Lattice water.* The molecules occupy voids in the crystal lattice (in a regular arrangement in the ideal case). They are held rather weakly by hydrogen bonds and by dipole–dipole interactions and often are relatively easily removed. *Example* $BaCl_2 \cdot 2H_2O$.
2. *Coordinated water.* The water molecule is held to a central atom by covalent (directed) bonds. Often called *aquo complexes*. *Example*: $[Cr(H_2O)_6]^{3+} \cdot 3Cl^-$.
3. *Adsorbed water.* Held at the surface by either physical or chemical forces. No definite chemical composition is observed as there is in the preceding two cases.

E. Unfortunately it is usually impossible to distinguish between these three types from the vibrational spectrum.

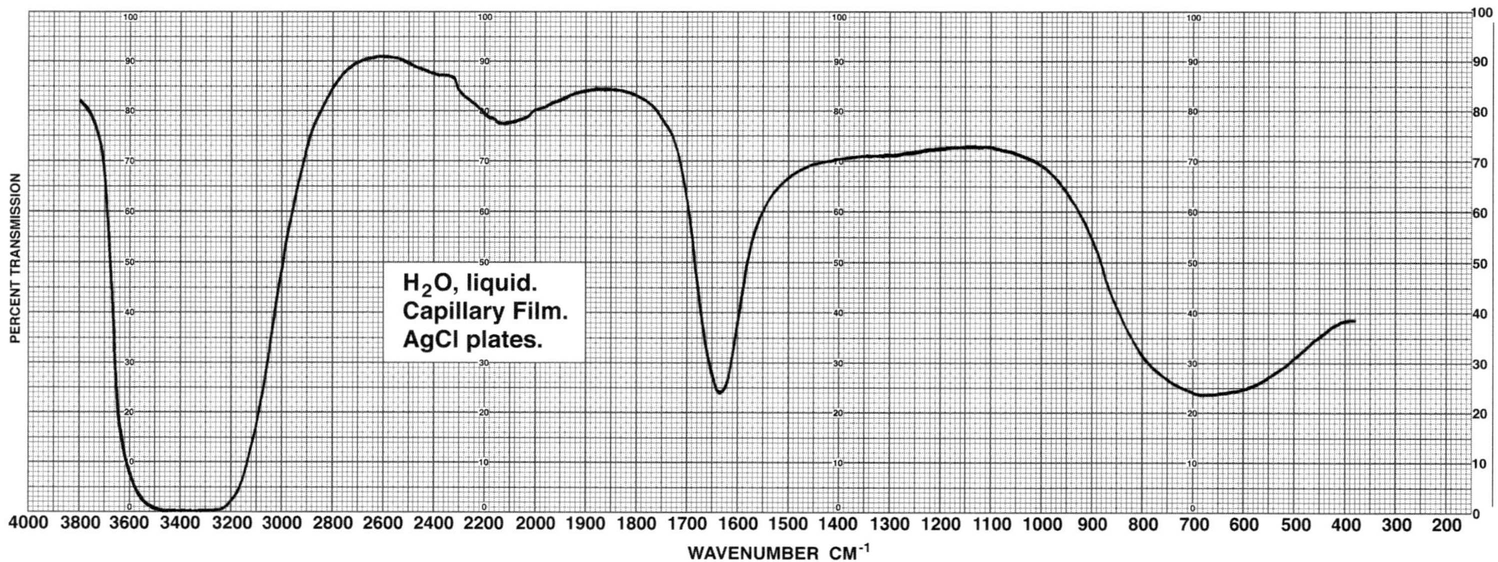

Fig. 11.2 Infrared spectrum of a capillary film of liquid water.

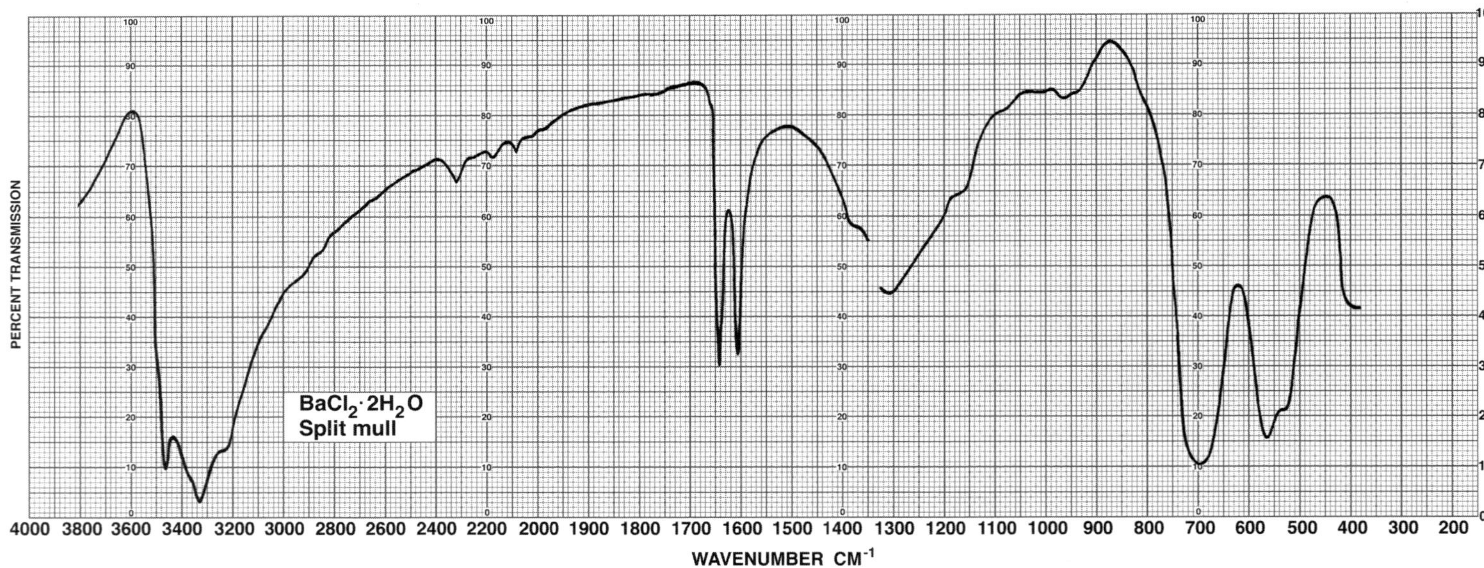

Fig. 11.3 Infrared spectrum of $BaCl_2 \cdot 2H_2O$.

IV. Infrared spectra of inorganic *salts*
  A. Inorganic salts give useful IR spectra. *Example*: Sodium sulfate. See Figure 11.4.
  B. Consider the spectra of a series of sulfates.
    1. Each salt will have bands characteristic of the SO_4^{2-} ion. These will be nearly the same for every sample because in each case they arise from the same vibrating entity, the SO_4^{2-} ion.
    2. However, there will be rather minor shifts of the bands between one salt and another because of the slightly different environment in which the SO_4^{2-} vibrates. These changes are due to different geometries of packing and different charges on the counterions, leading to different external potential fields. One can therefore sometimes narrow an unknown to two or three probable sulfate salts.
    3. Summary for 10 sulfates. See Figure 11.5.
      a. Note that they all have their strongest band near 1100 cm^{-1} and a weaker one between 600 and 700 cm^{-1}.
      b. For those salts containing water, the water scissoring band comes near 1650 cm^{-1}.
  C. A Colthup-type chart of group frequencies for inorganic ions can be prepared. See Figure 11.6.
    1. This is useful for qualitative analysis.
    2. Most polyatomic inorganic ions are anions: SO_4^{2-}, CO_3^{2-}, NO_3^-, etc. This is the very group for which there are not other good chemical or instrumental methods for identification. Thus IR spectroscopy fills a useful role.
    3. For cations (Na^+, Ca^{2+}, etc.) there are other excellent methods for identification, whereas IR has little to contribute unless they are polyatomic (e.g., NH_4^+).
  D. At first one might think that the vibrations of inorganic ions are low in frequency because heavy atoms are involved.
    1. Often they are not low, however. Some reasons:
      a. In CO_3^{2-}, NO_3^-, NH_4^+, and HCO_3^-, none of the atoms are heavy.
      b. In CrO_4^{2-}, MnO_4^-, SO_4^{2-}, PO_4^{3-}, etc., the central atom is heavy but the oxygen atoms are relatively light.
      c. The bonds are often multiple, which increases the force constants.
    2. The result is that most polyatomic ions have bands above 600 cm^{-1}, where they are readily observed.
  E. Some examples of IR spectra of simple polyatomic ions
    1. Overview for tetrahedral ions
      a. All *tetrahedral* XY_4 molecules—CCl_4, NH_4^+, SO_4^{2-}, PO_4^{3-}, CrO_4^{2-}, etc.—have two IR-active fundamentals:
        1) A triply degenerate X—Y stretch
        2) A triply degenerate deformation

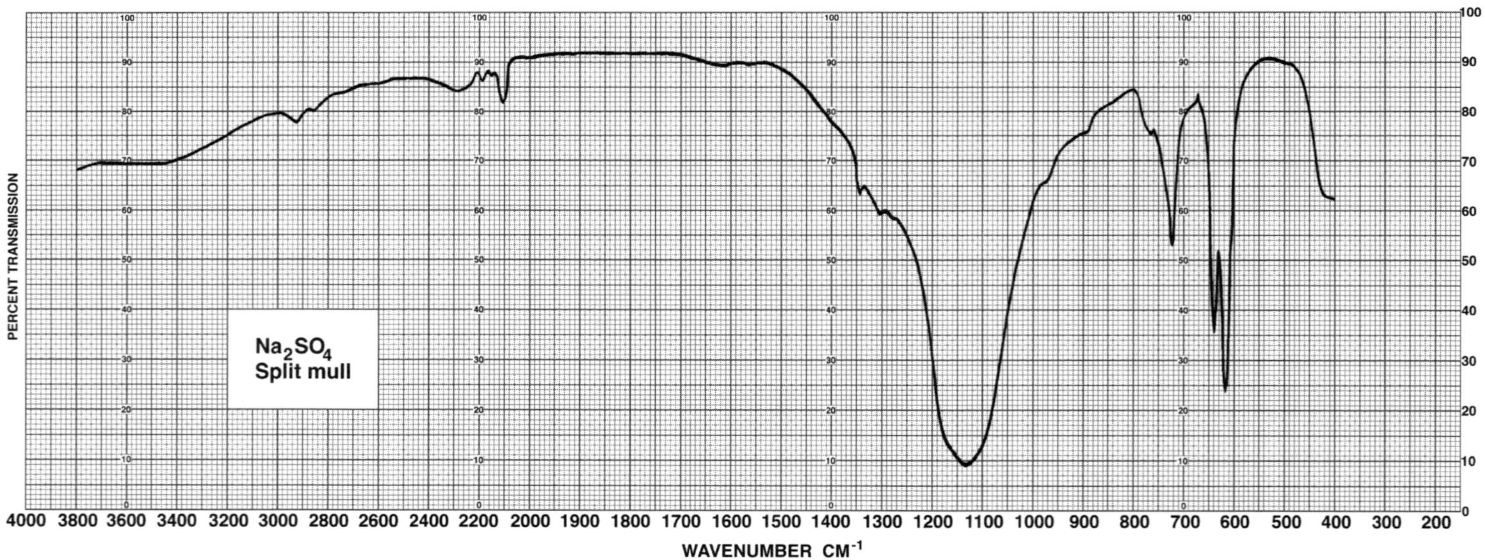

Fig. 11.4 Infrared spectrum of sodium sulfate, Na_2SO_4.

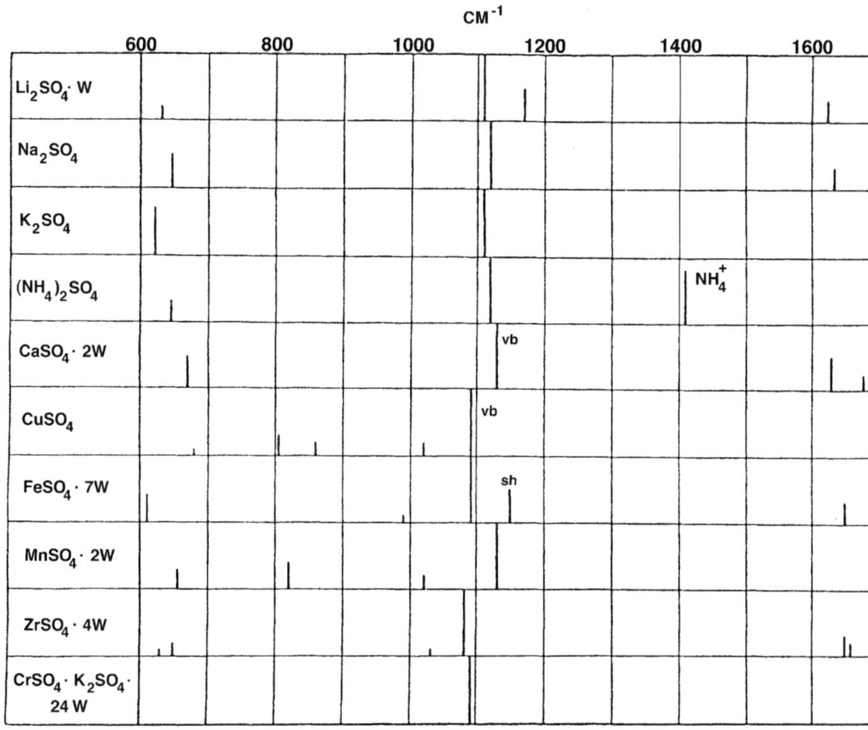

Fig. 11.5 Summary of IR bands for 10 inorganic sulfates. Note that the frequency scale increases to the right. W = water, e.g., $CaSO_4 \cdot 2W = CaSO_4 \cdot 2H_2O$.

b. Summary for some tetrahedral ions to be discussed:

Tetrahedral XY_4	NH_4^+	SO_4^{2-}	PO_4^{3-}	CrO_4^{2-}
X—Y triply degenerate stretch (cm^{-1})	~3100	~1100	~1020	~900
XY_4 triply degenerate deformation (cm^{-1})	~1400	~600	~500	~400

2. NH_4^+, ammonium ion
   a. *Example*: Ammonium chloride. See Figure 11.7. All the internal modes come from NH_4^+.
   b. The triply degenerate N—H stretch of NH_4^+ is ~3100 cm^{-1} (vs, vb).
      The triply degenerate deformation of NH_4^+ is ~1400 cm^{-1} (vs, vb).
3. SO_4^{2-}, sulfate ion
   a. Also tetrahedral.
      Triply degenerate S—O stretch is ~1100 cm^{-1} (vs, vb).
      Triply degenerate deformation is 670–580 cm^{-1} (m).

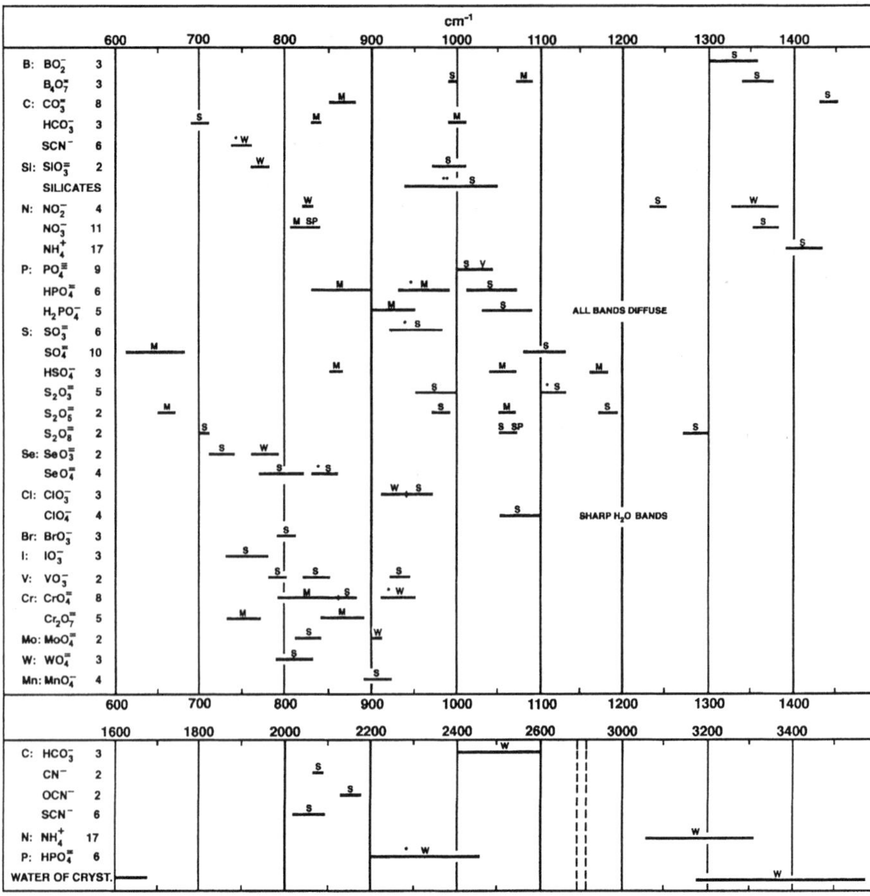

Fig. 11.6 Chart of characteristic IR frequencies for inorganic ions.

   b. *Examples*: Potassium sulfate and calcium sulfate. See Figures 11.8*a*,*b*. Note the added water bands in the latter.

4. HSO_4^-, bisulfate
   a. There is no H—S bond because there is no H—S stretch near 2575 cm^{-1}. This is really $[H—O—SO_3]^-$, with a bent H—O—S. The symmetry has been drastically lowered, which breaks the degeneracy (giving more frequencies) and relaxes the selection rules (allowing more vibrations to be IR active). In addition, there will be four new bands due to the H—O—S group: an H—O and an O—S stretch plus a bend and a torsion of the bent H—O—S system. Finally, hydrogen bonding will broaden and intensify some of the bands dramatically.
   b. The result is a spectrum with many bands, some very broad.

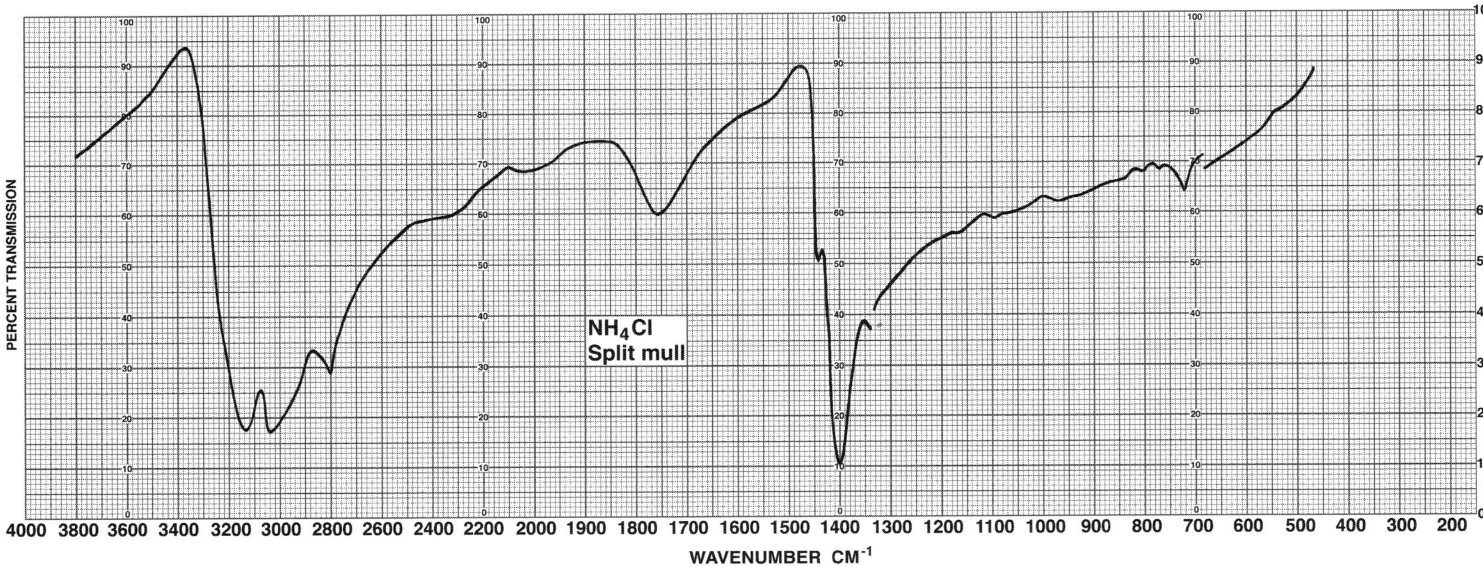

Fig. 11.7 Infrared spectrum of ammonium chloride, NH_4Cl.

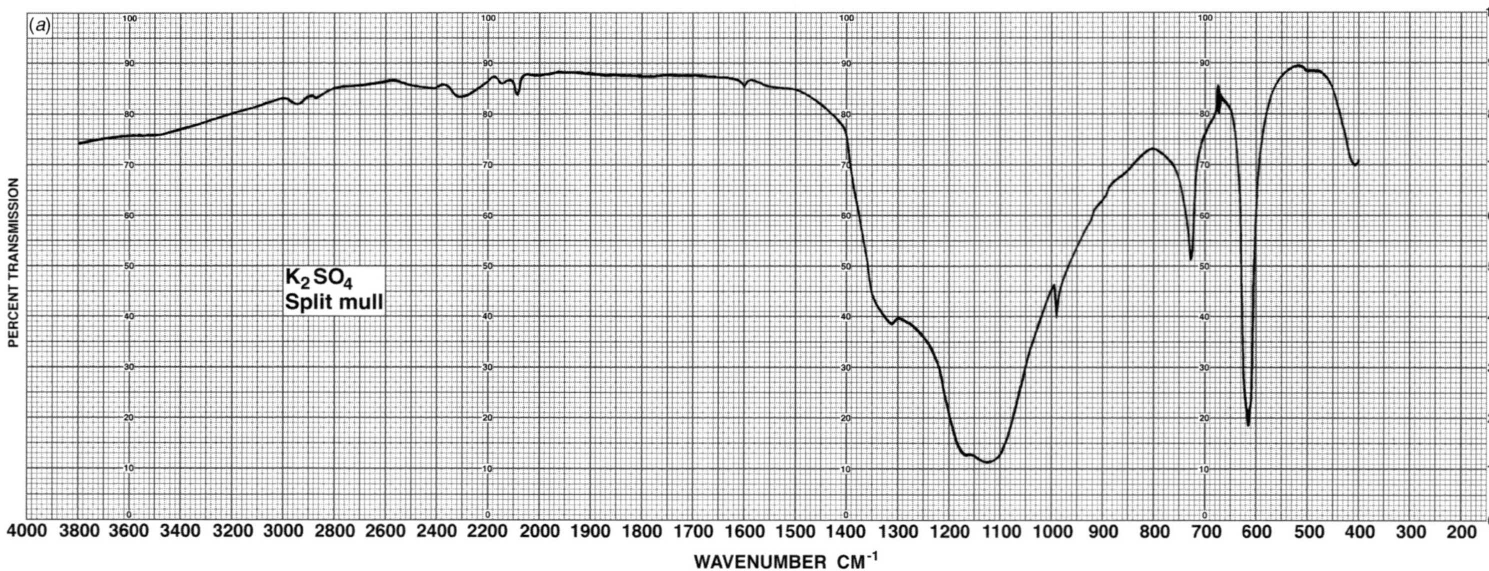

Fig. 11.8a Infrared spectrum of K_2SO_4.

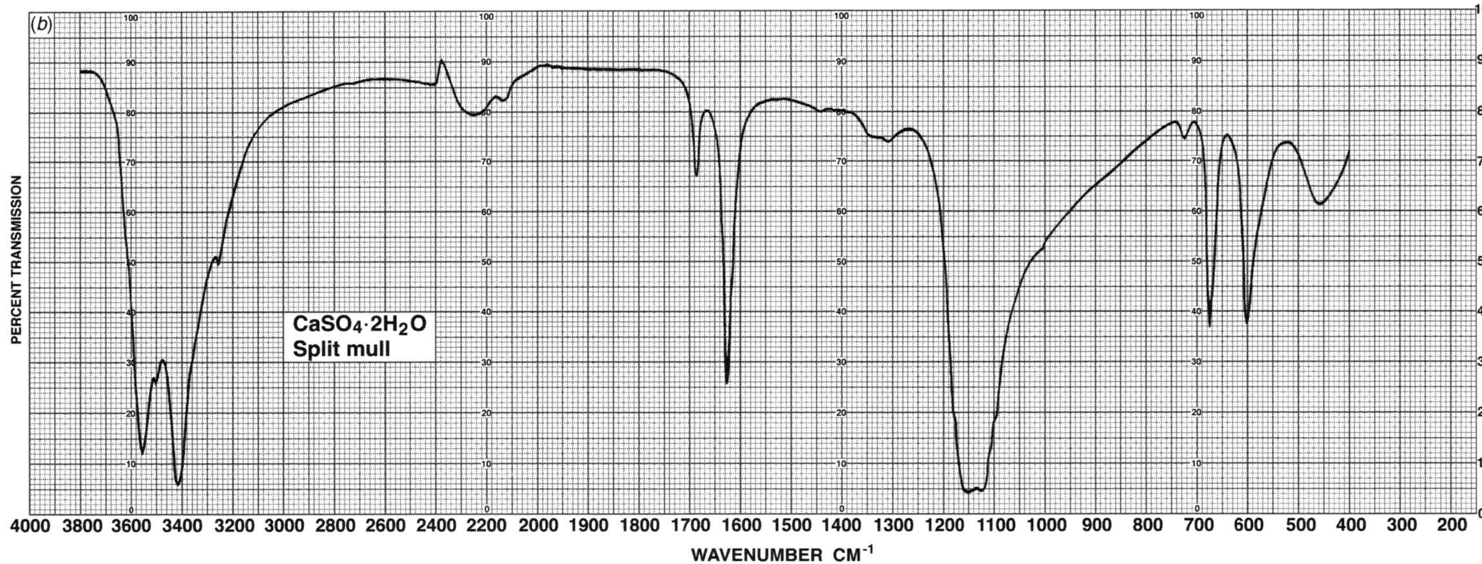

Fig. 11.8b Infrared spectrum of $CaSO_4 \cdot 2\ H_2O$.

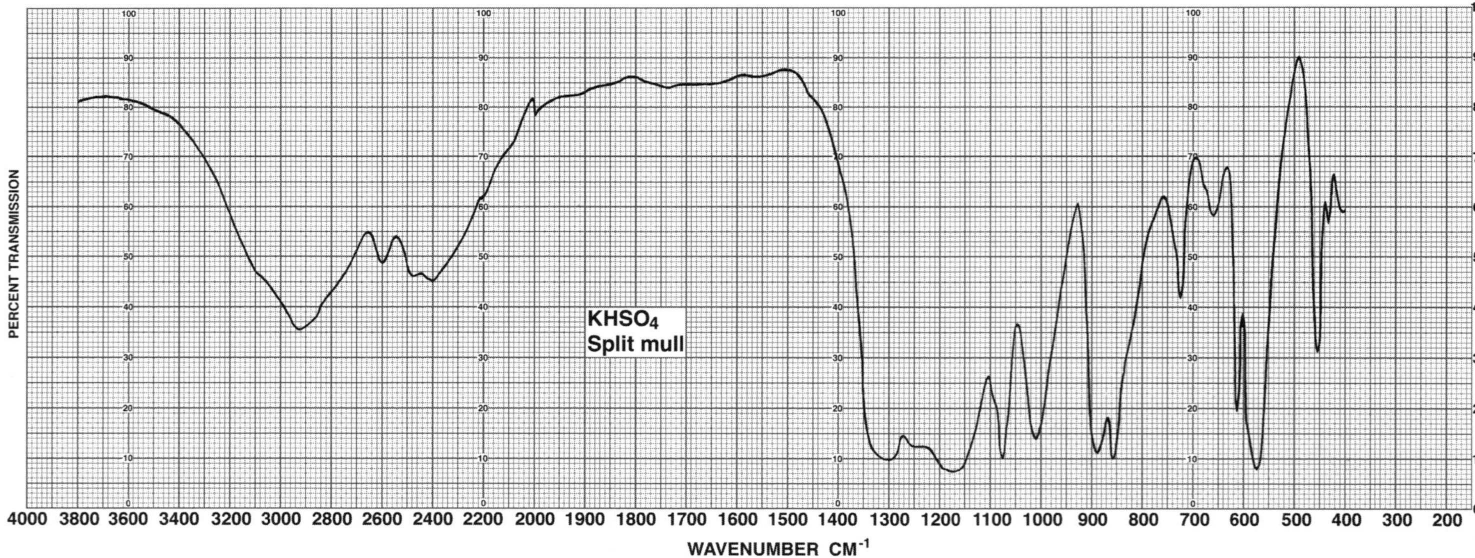

Fig. 11.9 Infrared spectrum of $KHSO_4$.

c. *Example*: Potassium bisulfate. See Figure 11.9.

d. Similar changes occur for $H_2PO_4^-$, HPO_4^{2-}, and HCO_3^-.

5. PO_4^{3-}, orthophosphate

a. It is tetrahedral.

P–O stretch	$\sim$1020 $\pm$ 20 cm^{-1} (vs, vb)
Deformation	$\sim$500 cm^{-1} (m)

b. *Example*: Lead orthophosphate, $Pb_3(PO_4)_2$. See Figure 11.10.

6. HPO_4^{2-}, dibasic orthophosphate

a. This is really $[H—O—PO_3]^{2-}$, with a bent H—O—P group. [Note that there is no H—P stretching band near 2350 cm^{-1} (s).] Therefore the symmetry has been lowered from that of PO_4^{3-}, which breaks the degeneracy and relaxes the selection rules. In addition there will be four new bands from the H—O—P group: H—O and O—P stretches and a bend and a torsion of the bent H—O—P group. Finally, hydrogen bonding will broaden and intensify some of the bands dramatically.

b. The result is a spectrum with many broad bands.

c. *Example*: Dibasic barium orthophosphate, $BaHPO_4$. See Figure 11.11.

7. CrO_4^{2-}, chromate

a. This too is tetrahedral.

b. The two IR-allowed bands are near 900 and 400 cm^{-1}.

c. Lead chromate is a common yellow pigment.

d. *Example*: Barium chromate, $BaCrO_4$. See Figure 11.12.

8. Planar trigonal ions (CO_3^{2-}, NO_3^-): Overview.

a. All planar trigonal XY_3 molecules have three IR-active fundamentals:

Planar Trigonal XY_3	CO_3^{2-}	NO_3^-
X—Y doubly degenerate stretch (vs, vb), cm^{-1}	$\sim$1450	$\sim$1370
Out-of-plane deformation (w, shp), cm^{-1}	$\sim$860	$\sim$850
In-plane doubly degenerate deformation (w), cm^{-1}	$\sim$740	$\sim$710

b. Out-of-plane deformations are usually *lower* than in-plane ones. Note that the last two are inverted from this order.

9. CO_3^{2-}, carbonate

a. *Examples*: Lithium carbonate and sodium carbonate. See Figures 11.13*a*,*b*.

b. Polymorphism

1) This is the existence of two or more crystalline forms for the same substance, called *polymorphs*. These will generally give slightly different spectra because the molecules or ions are in different environments in the crystal. Typically a band that is a

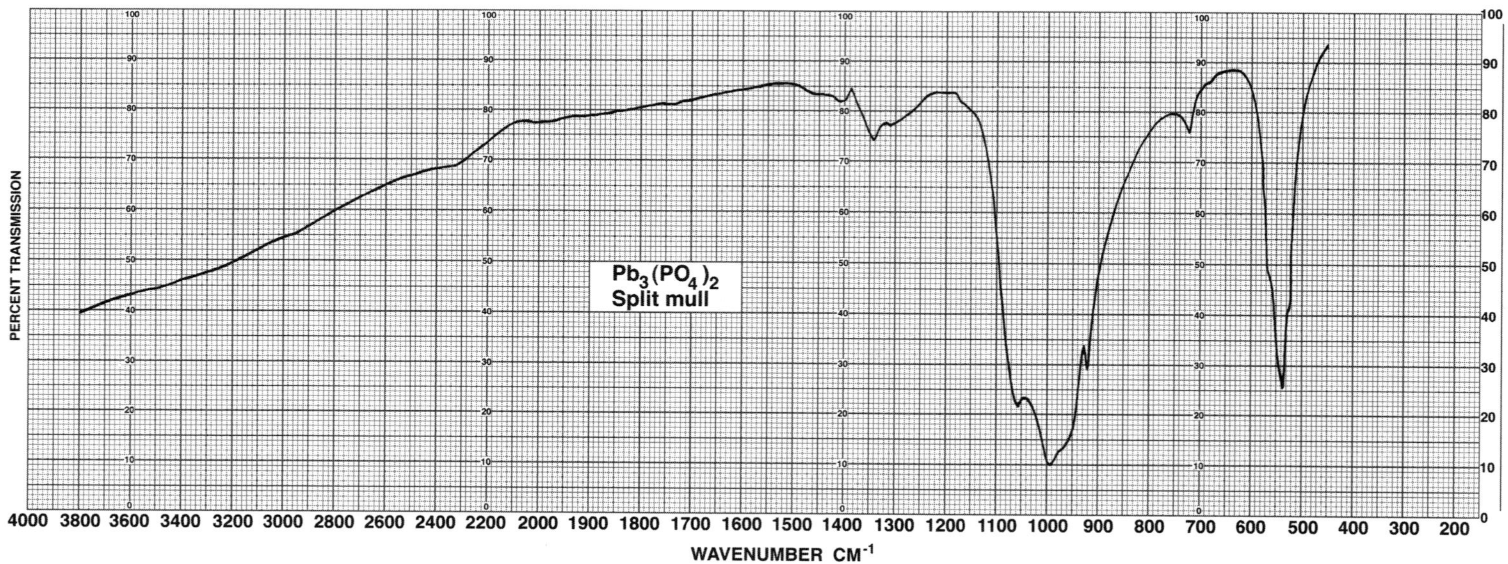

Fig. 11.10 Infrared spectrum of lead orthophosphate, $Pb_3(PO_4)_2$.

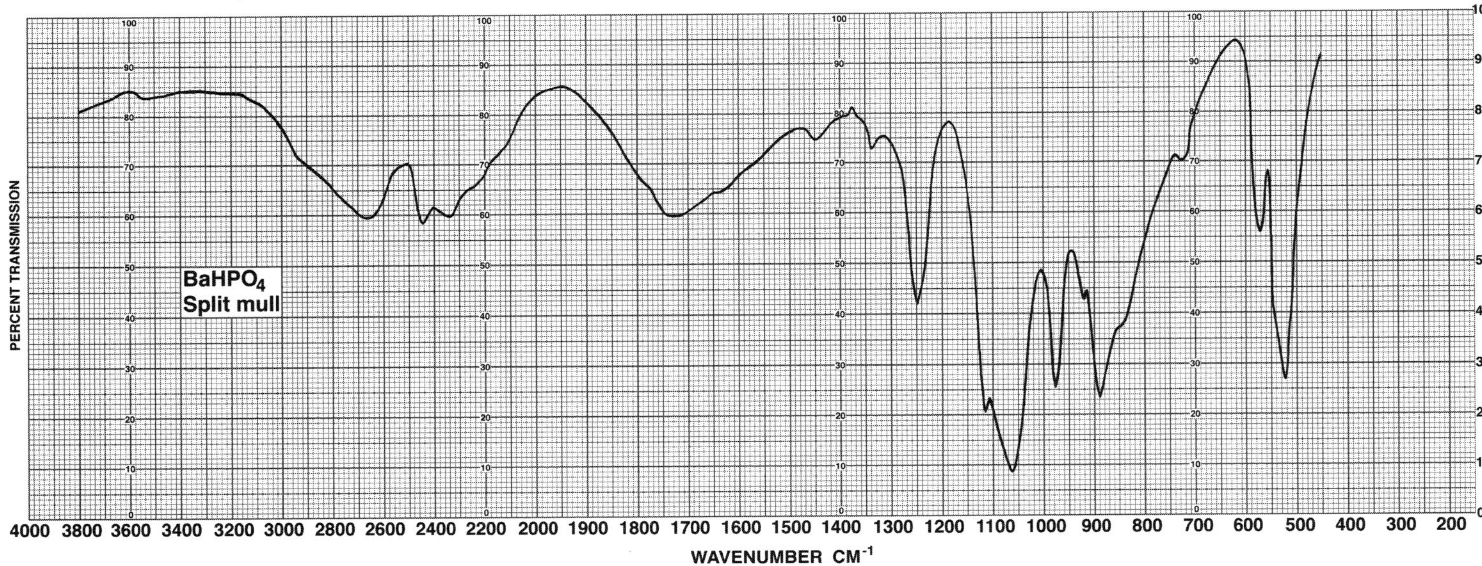

Fig. 11.11 Infrared spectrum of dibasic barium orthophosphate, $BaHPO_4$.

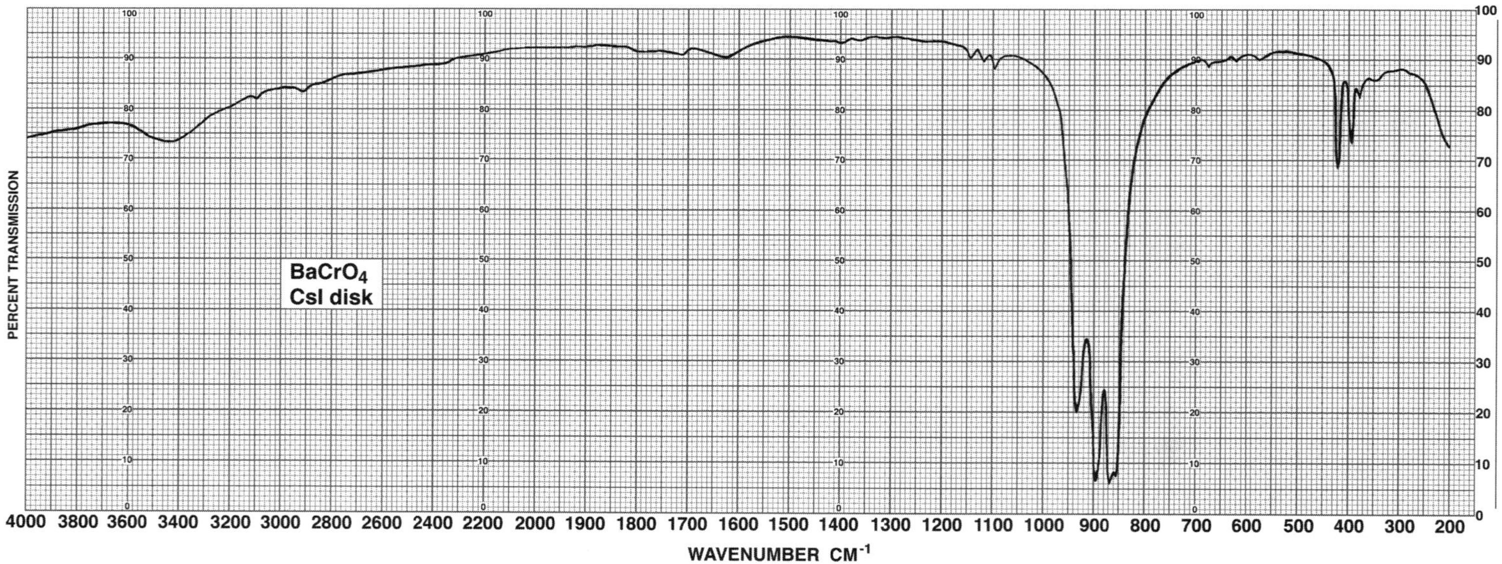

Fig. 11.12 Infrared spectrum of barium chromate, $BaCrO_4$.

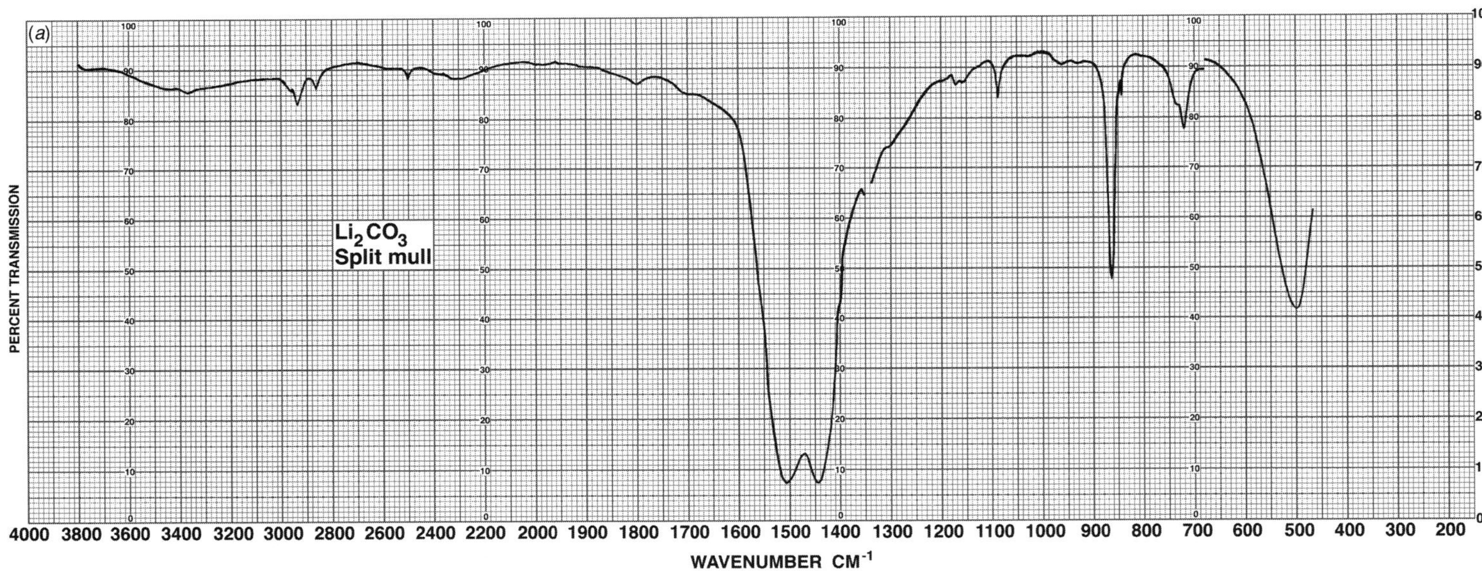

Fig. 11.13a Infrared spectrum of lithium carbonate (Li_2CO_3).

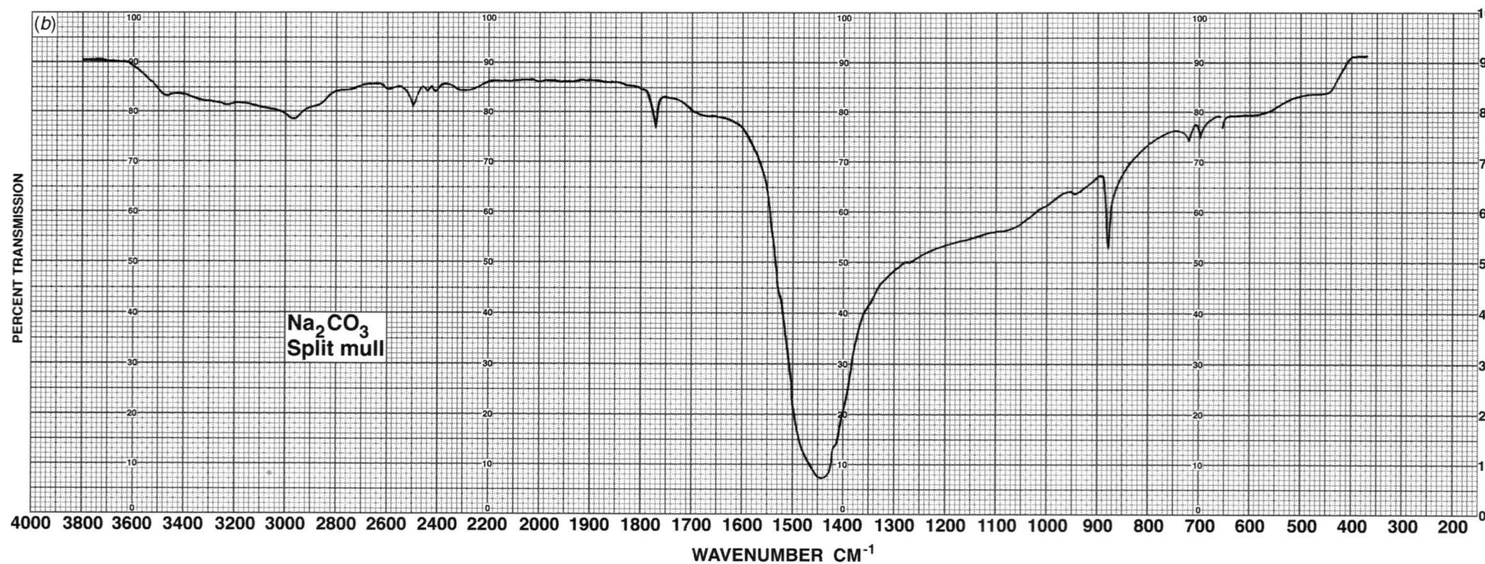

Fig. 11.13b Infrared spectrum of sodium carbonate (Na_2CO_3).

singlet in one may be a doublet in another or may be completely missing. There are usually small frequency differences of 5–30 cm^{-1}.

2) *Example*: Calcite and aragonite, two crystalline forms of $CaCO_3$.
   a) Spectra. See Figures 11.14*a*,*b*.
   b) In calcite the CO_3^{2-} ions retain their planar three-fold symmetry in the crystal, so the totally symmetric stretch near 1080 cm^{-1} is forbidden in the IR. (It cannot produce a change in the dipole moment.) In aragonite the environment around the CO_3^{2-} ions is of lower symmetry, and this vibration is now allowed (but is weak).
   c) The in-plane bend is doubly degenerate in calcite (at 706 cm^{-1}), but the degeneracy is split in aragonite to give 706 and 711 cm^{-1}.

		$\tilde{\nu}_1$ (Symmetric Stretch)			$\tilde{\nu}_4$ (In-Plane Bend)		
	Symmetry of CO_3^{2-}	IR	Selection Rule	Observed	IR	Selection Rule	Observed
Calcite	D_3	—	(Forbidden)	—	+	(Doubly degenenerate)	706 cm^{-1}
Aragonite	C_s	+	(Allowed)	1083 w	+	(Degeneracy removed)	706, 711 cm^{-1}

c. Composition of marine shells—an interesting application.
   1) Calcite: crabs, sand dollars, sea urchins.
   2) Aragonite, or a mixture of aragonite and calcite: Oysters, clams, conches, whelks.
      Reference: W. F. Kinard, *J. Chem. Ed.* **57**, 783 (1980).
   3) Pearls are aragonite in nearly all cases. Occasionally they may contain calcite. (See Encyclopedia Britannica.)

10. HCO_3^-, bicarbonate
    a. This is really $[H—O—CO_2]^-$. Therefore the symmetry is much lower than for CO_3^{2-}. We expect to find H—O and C—O stretches and a bend and a torsion of the bent C—O—H system. Hydrogen bonding will be strong.
    b. The O—H stretch is abnormally low: ~2600 cm^{-1} (vs, vvb).
    c. *Example*: Potassium bicarbonate, $KHCO_3$. See Figure 11.15.
11. NO_3^-, nitrate
    a. This is isoelectronic with CO_3^{2-} and likewise is planar trigonal. Therefore the spectra are similar.
    b. *Examples*: Sodium nitrate and barium nitrate. See Figures 11.16*a*,*b*.

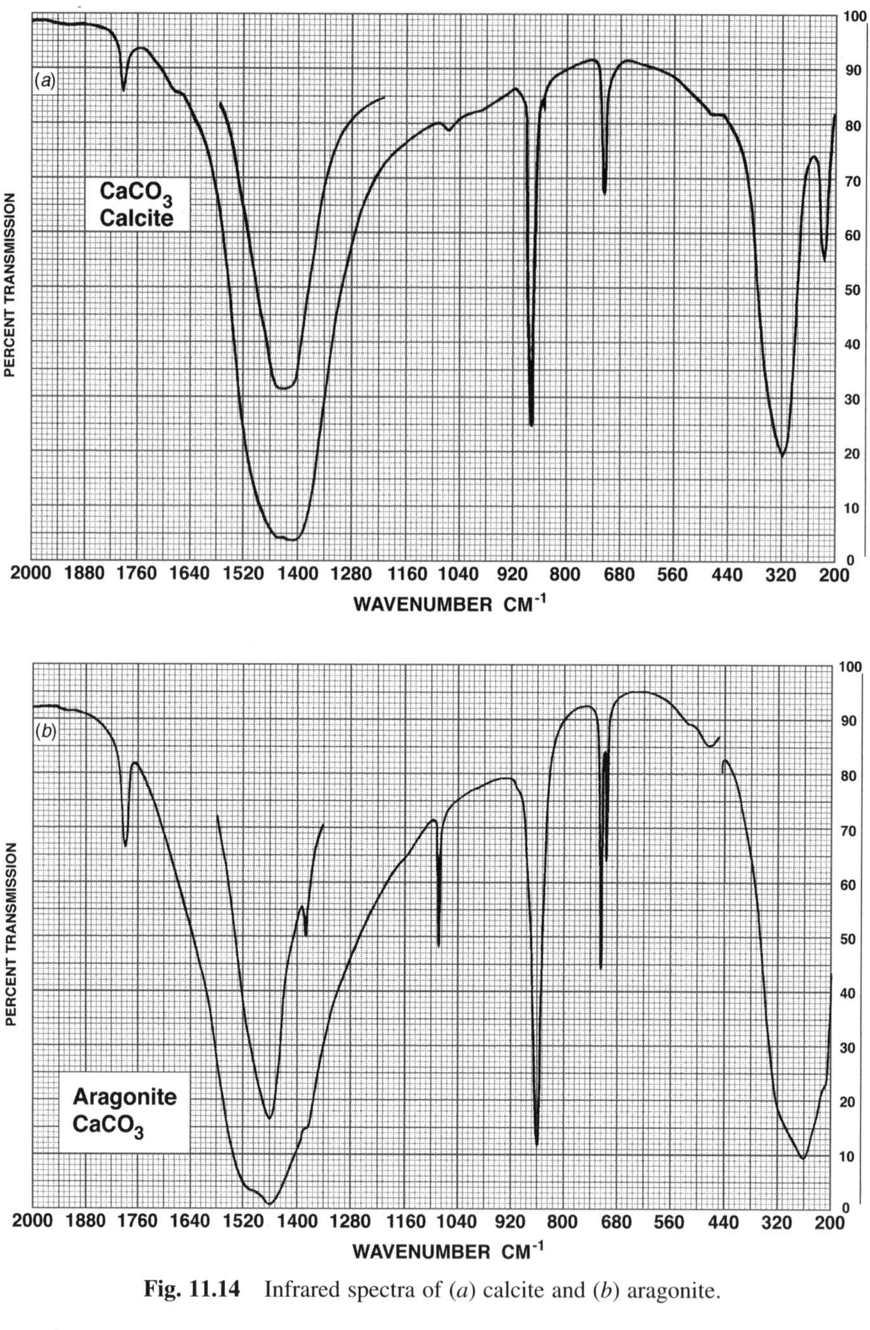

Fig. 11.14 Infrared spectra of (*a*) calcite and (*b*) aragonite.

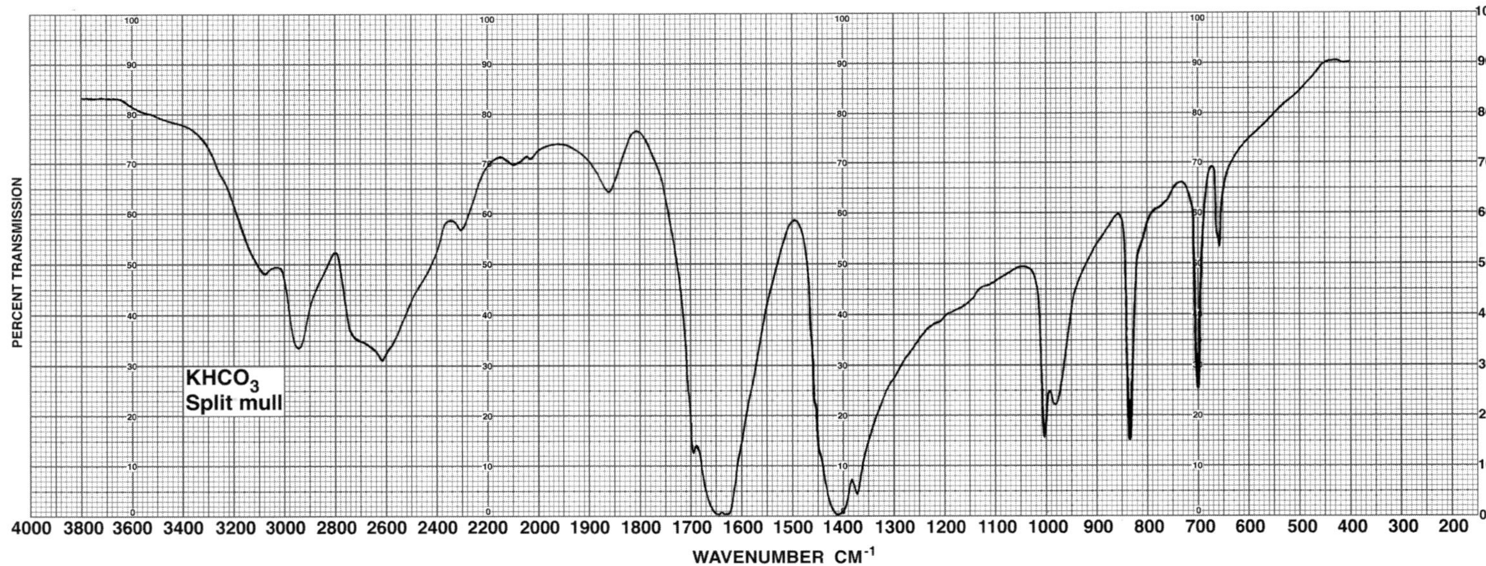

Fig. 11.15 Infrared spectrum of potassium bicarbonate, $KHCO_3$.

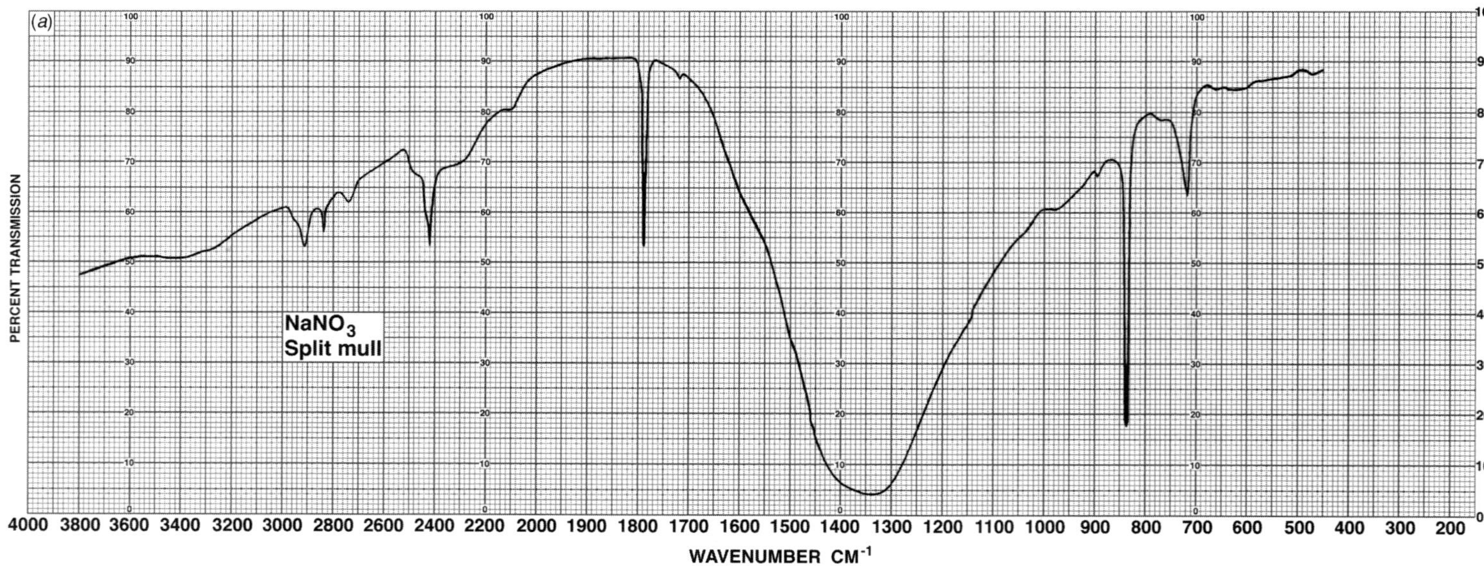

Fig. 11.16a Infrared spectrum of sodium nitrate, $NaNO_3$.

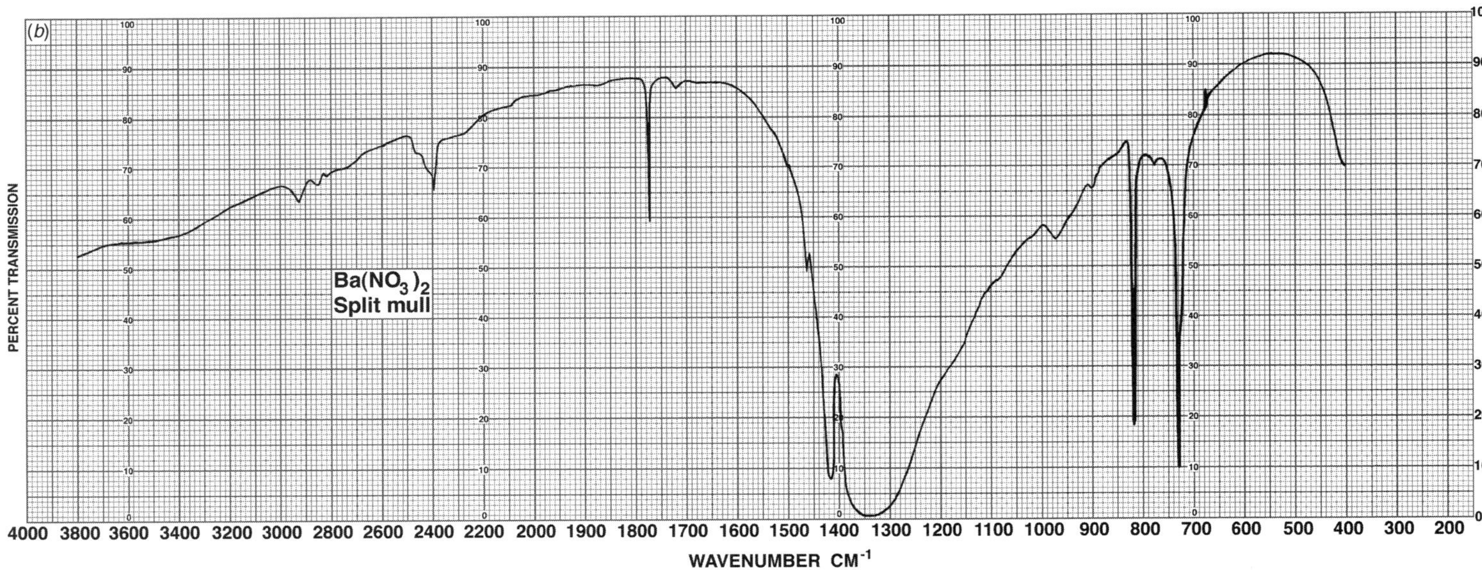

Fig. 11.16b Infrared spectrum of barium nitrate, $Ba(NO_3)_2$.

c. $\sim$1340 cm^{-1} (vs, b) N—O doubly degenerate stretch

$\sim$850 cm^{-1} (m)	Out-of-plane deformation
$\sim$710 cm^{-1} (w)	Doubly degenerate in-plane deformation

Again the last two are inverted from the usual order.

d. The sharp band near 1780 cm^{-1} is an IR-allowed sum tone of the totally symmetric stretch (not seen in the IR) and the in-plane deformation.

12. NO_2^-, nitrite ion

a. *Example*: Sodium nitrite. See Figure 11.17.

b. The nitrite ion has a symmetric bent structure 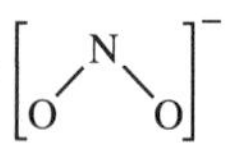 analogous to H_2O and therefore has three fundamentals:

$\sim$1320 cm^{-1}	(vw)	Symmetric N—O stretch
$\sim$1250 cm^{-1}	(vs)	Antisymmetric N—O stretch
$\sim$830 cm^{-1}	(w)	Scissors

1) Note that the first two invert the usual order of the antisymmetric being higher.

2) The spectrum is surprisingly similar to that of nitrate ion. That is accidental.

13. CN^-, cyanide

a. Because this is diatomic $[C{\equiv}N]^-$, only one band is expected, and it will be in the triple-bond region. It occurs at 2080 $\pm$ 15 cm^{-1} (s) in both crystals and aqueous solutions. If in a complex ion, the range is 2200–2000 cm^{-1}.

b. *Example*: NaCN. See Figure 11.18.

1) The spectrum is linear in micrometers.

2) The sample has a great deal of Na_2CO_3 and H_2O impurity. Nonetheless, one can see that the $[C{\equiv}N]^-$ band is near 2080 cm^{-1}.

c. *Example*: Zinc cyanide. See Figure 11.19.

1) Note that the strong band at 2230 cm^{-1} is too high to be a cyanide ion. Note also the strong band at 470 cm^{-1}, which is not compatible with $[C{\equiv}N]^-$.

2) The structure is such that each Zn is tetrahedrally coordinated with C≡N bridges, $Zn[-C{\equiv}N{-}Zn]_4$.

d. *Example*: Cuprous cyanide, CuCN. See Figure 11.20.

1) This spectrum is shown because the CN^- band provides an extreme example of the Christiansen effect: a rise in transmission when approaching the band from the high frequency side, a precipitous drop in transmission when going through the band, and then a slow increase in transmission when continuing to lower frequencies. The cause is well understood but will not be explained here. Changing the mulling agent can greatly alter the appearance of the band.

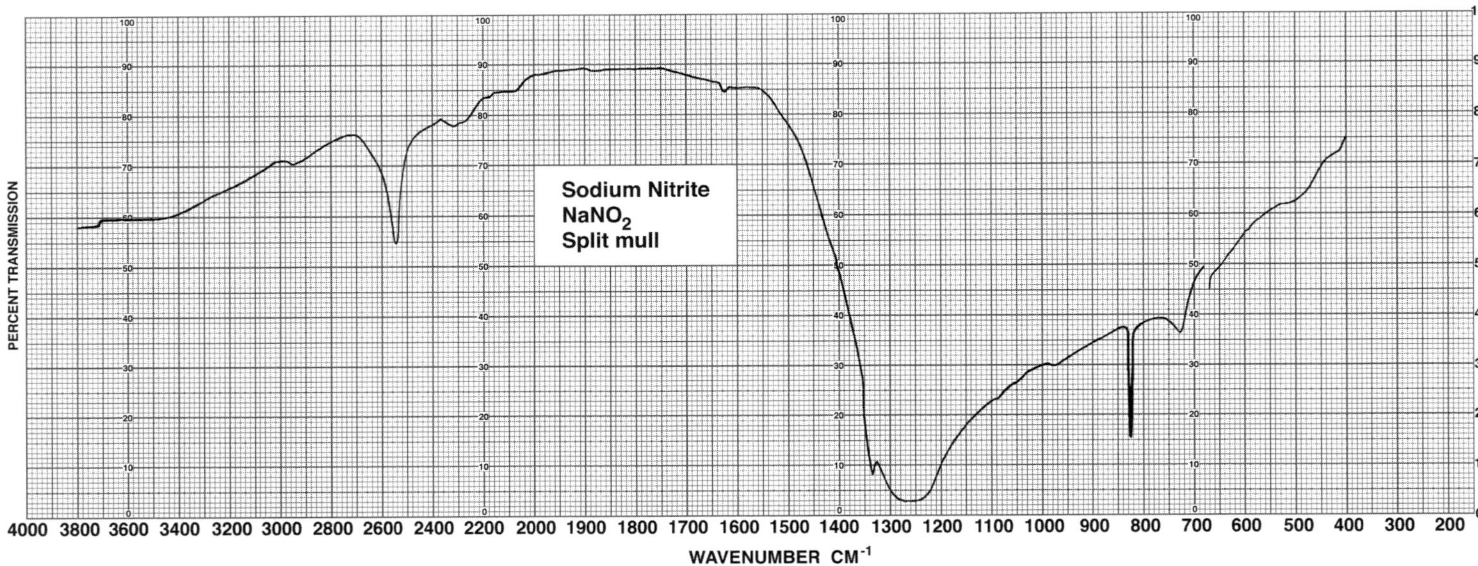

Fig. 11.17 Infrared spectrum of sodium nitrite, $NaNO_2$.

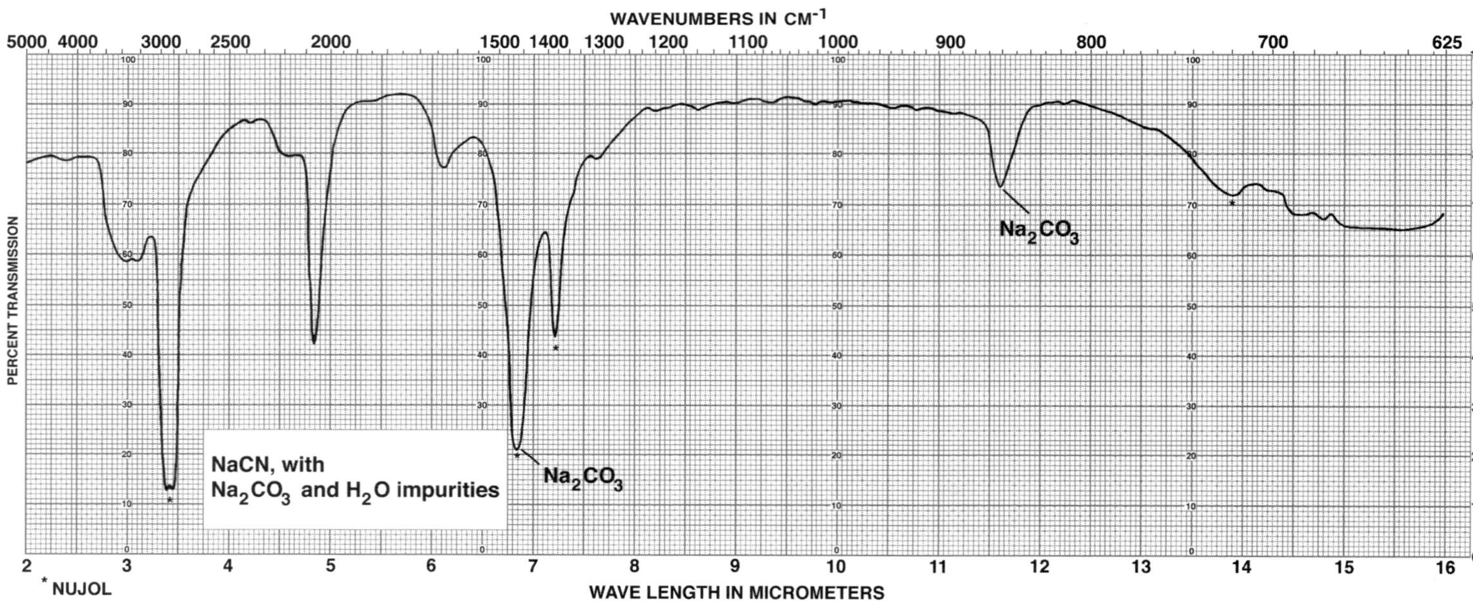

Fig. 11.18 Infrared spectrum of sodium cyanide, NaCN. Note that the spectrum is linear in micrometers. There is a great deal of Na_2CO_3 and H_2O impurity.

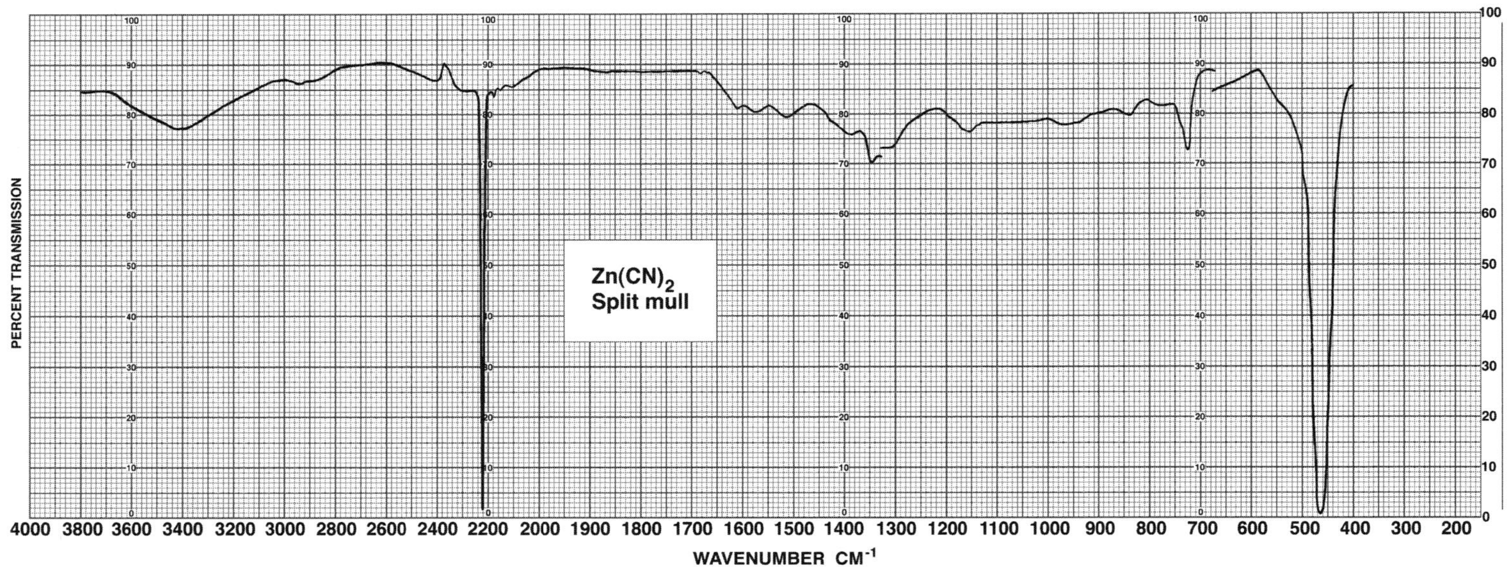

Fig. 11.19 Infrared spectrum of zinc cyanide, $Zn(C{\equiv}N)_2$.

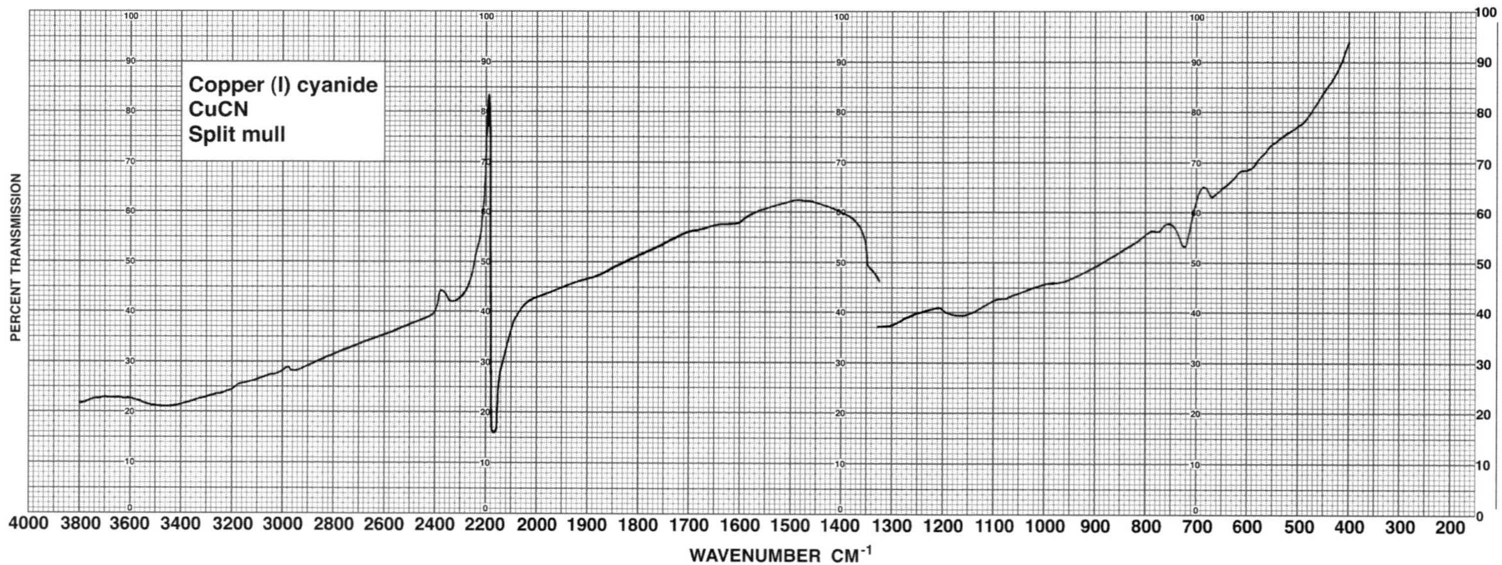

Fig. 11.20 Infrared spectrum of cuprous cyanide, CuCN.

2) Again the strong band (about 2180 cm^{-1}) is too high for a cyanide ion. This substance has a complex structure which contains the units –Cu–C≡N–Cu–.

14. $[S–C≡N]^-$, thiocyanate
   a. This ion has a linear structure. The C≡N stretch is about 2050 cm^{-1}, the C–S stretch about 740 cm^{-1}, and the S–C≡N bend about 470 cm^{-1}. The last vibration often gives a doublet if its degeneracy is removed in the crystal.
   b. *Example*: Mercuric thiocyanate. See Figure 11.21.
   c. This C≡N stretching band also shows Christiansen distortion.
   d. If $[SCN]^-$ is coordinated to a metal, the attachment may be through either the S or the N atom. The spectra are slightly different for the two cases. An example will be shown later when discussing the hazards of the KBr pressed disk technique.

V. Lower frequency IR bands
   A. Nearly all the bands discussed so far are above 700 cm^{-1}. Vibrational frequencies are often lower than this, of course, especially when atoms are heavy and/or bonds are weak.
   B. *Examples*:
      1. Sodium chlorate, $NaClO_3$. See Figure 11.22*a*. Note the bands at 627 and 484 cm^{-1}.
      2. Sodium bromate, $NaBrO_3$. See Figure 11.22*b*. Note the bands at 447 and 370 cm^{-1}.
      3. I_2Cl_6
         All the fundamentals are below 350 cm^{-1}.
         The lowest IR-active one is at 56 cm^{-1}.

VI. Ugly spectra
   A. Sometimes inorganic samples give ugly spectra. Some possible reasons:
      1. They may contain water or other O–H groups. The strong hydrogen bonding gives very broad bands.
      2. The sample is amorphous and the structure is therefore chaotic, giving broader bands. A well-crystallized sample usually gives a reasonable spectrum.
      3. The sample is impure.
   B. *Examples*:
      1. Dibasic calcium orthophosphate, $CaHPO_4 \cdot H_2O$. See Figure 11.23.
      2. Manganese(II) oxide, MnO. See Figure 11.24.

VII. When a solid sample gives a simple spectrum, one should suspect an inorganic salt, especially if one or more of the bands are very strong and broad.
   A. SO_4^{2-}, CO_3^{2-}, and NH_4^+ are particularly common. A sample may have been neutralized with dilute H_2SO_4 or aqueous NH_3 and the resulting inorganic salt later isolated by mistake.

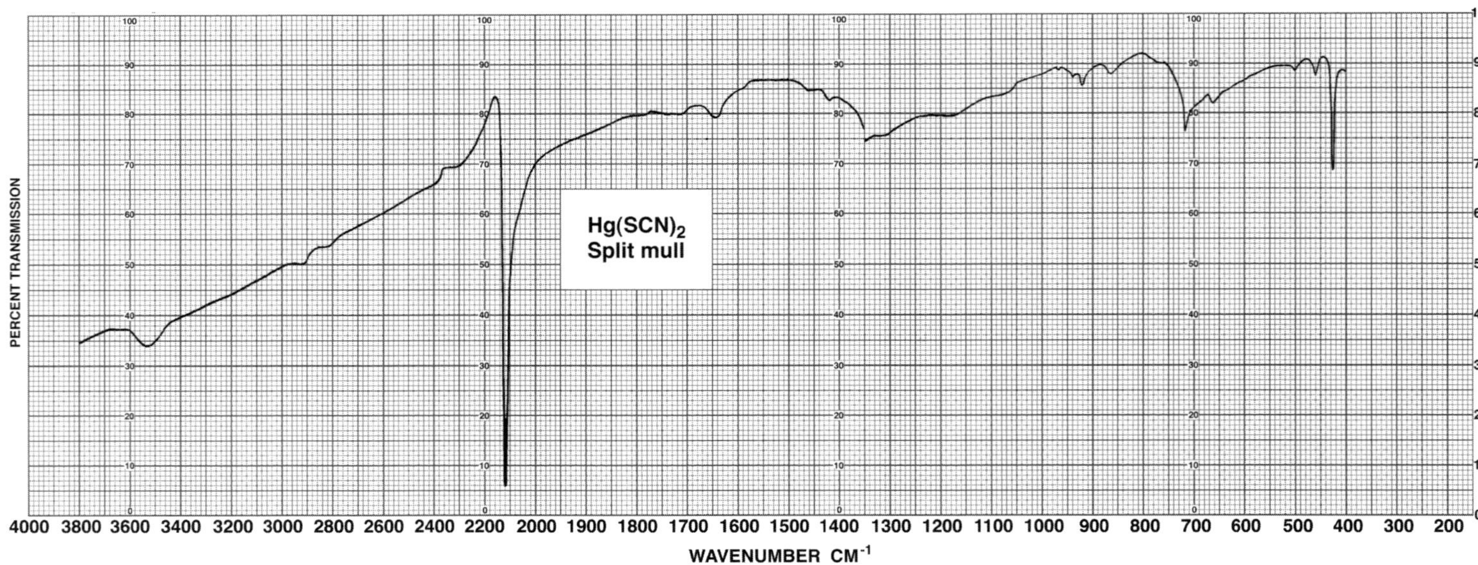

Fig. 11.21 Infrared spectrum of mercuric thiocyanate, $Hg(SCN)_2$.

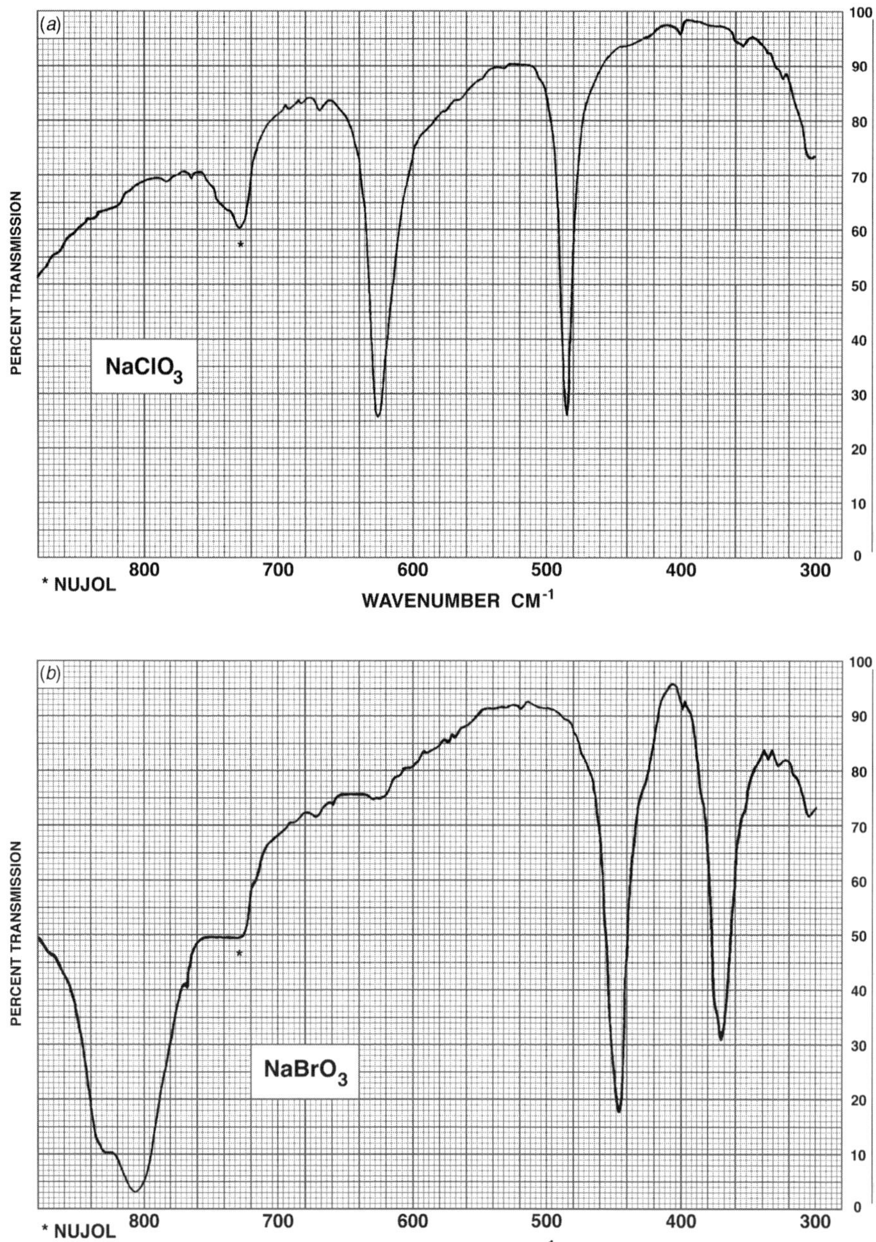

Fig. 11.22 Low-frequency IR spectra of (*a*) $NaClO_3$ and (*b*) $NaBrO_3$.

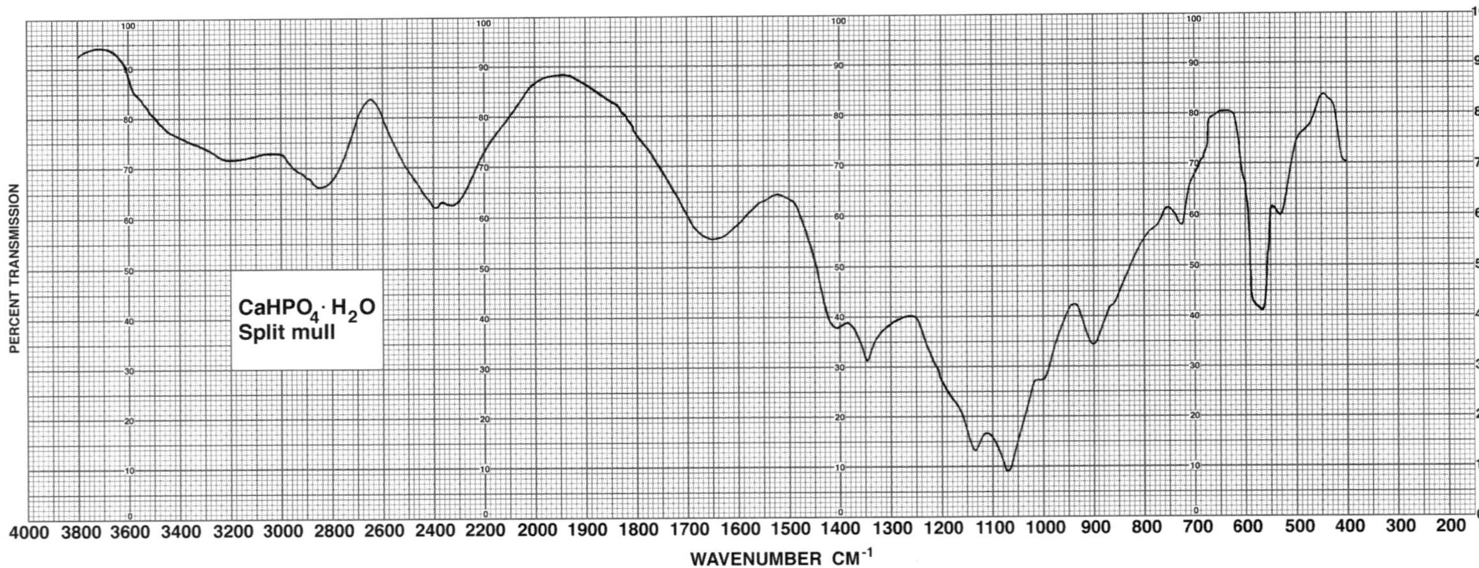

Fig. 11.23 Infrared spectrum of dibasic calcium orthophosphate, $CaHPO_4 \cdot H_2O$.

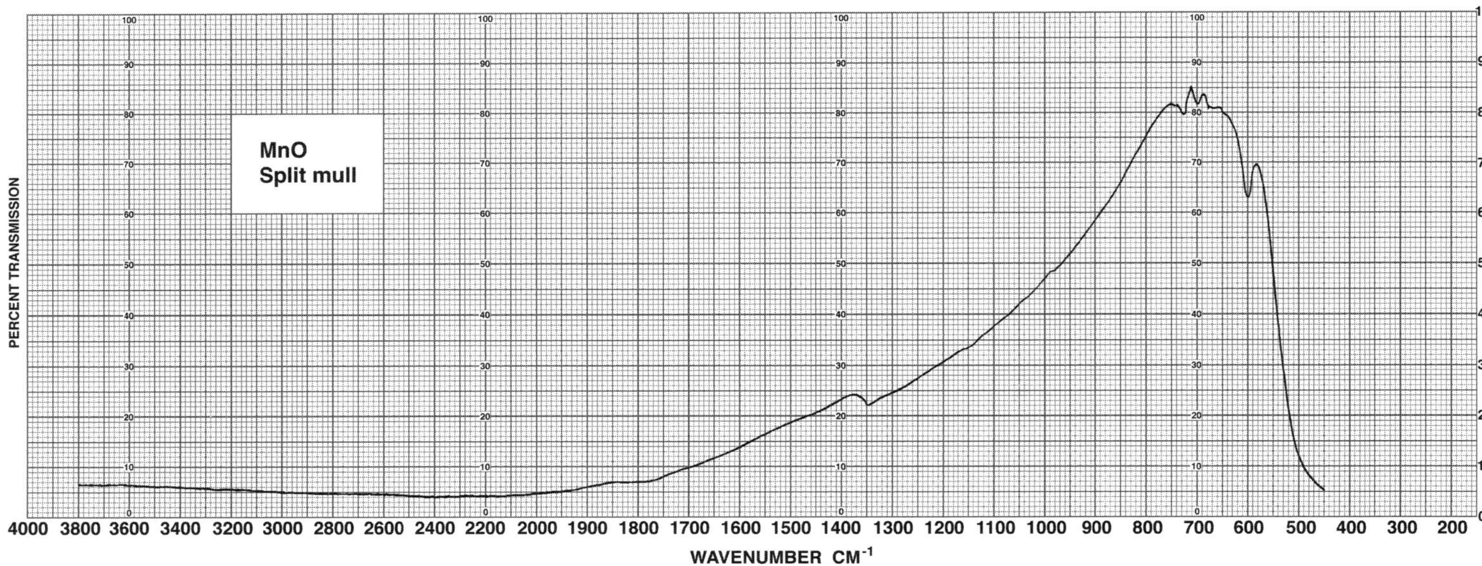

Fig. 11.24 Infrared spectrum of Mn(II) oxide, MnO.

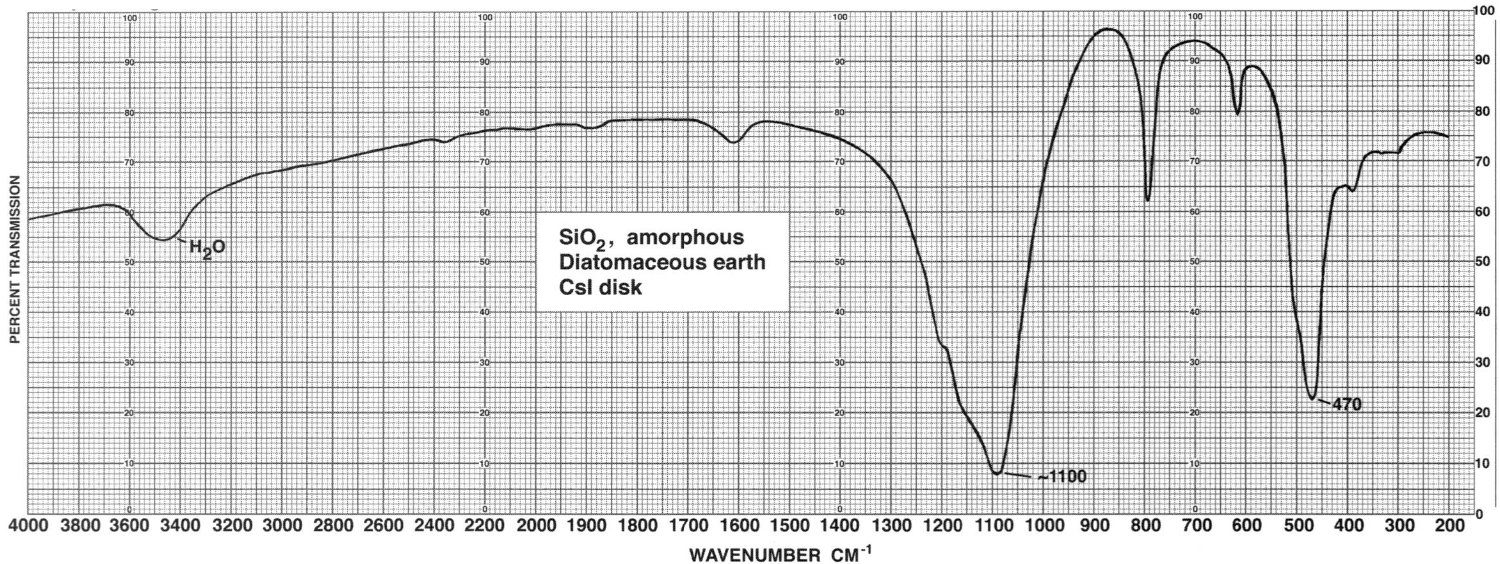

Fig. 11.25 Infrared spectrum of amorphous silica powder.

B. If an inorganic salt is suspected, it is often helpful to put a little of the sample on a spatula and heat gently in a flame. If it burns completely, it is organic, but if a residue remains, there is at least some inorganic component.

VIII. For the quantitative analysis of an inorganic powder by IR, one *must* use a weak band.

A. If a very strong band is used, there will be a problem with saturation. If a particle absorbs virtually all the radiation at a given reciprocal centimeter (e.g., silica at 1100 cm^{-1}), further changes in absorption by later particles will not be measured properly.

B. Even if the sample is ground very finely, this problem will still occur because of the great intensity of the band. The solution is to use a band of weaker intrinsic intensity.

C. *Example*: Determining silica in polyethylene. See Figure 11.25.

1. Do not use the 1100-cm^{-1} band of silica.
2. The 800 or 470 cm^{-1} band is satisfactory. The lower frequency also reduces scattering.

IX. Comment on the KBr pressed disk technique

A. It is particularly risky to use this with inorganic salts for two reasons.

1. A double decomposition reaction may occur. *Examples*:

a. $AgNO_3 + KBr$ ⇨ $AgBr + KNO_3$

b. $CuSO_4 + 2\,KBr$ ⇨ $CuBr_2 + K_2SO_4$

This reaction is fast enough at room temperature that one can see a brown color develop. ($CuBr_2$ is a black crystal.)

2. Ions of the sample may enter the KBr lattice to give the spectrum of their solution in KBr, which differs from that of the original salt.

B. Both effects are well documented. The result is an incorrect spectrum. Effects altering the spectrum may also occur with nonionic samples, but they are more likely with ionic ones. It is therefore safer with ionic samples to use mulls or diffuse reflectance rather than pressed KBr disks.

C. *Example*:

1. The following two compounds can be made. They differ in which end of $[SCN]^-$ is attached to Pd.

a. $Pd(SCN)_2\,(AsPh_3)_2$. Has Pd—S bonds. Unstable; prepared at 0° C.

b. $Pd(NCS)_2\,(AsPh_3)_2$. Has Pd—N bonds. The stable isomer.

2. Spectra of each can be obtained in a Nujol mull. See Figures 11.26*a–c*.
3. In a KBr disk, if one starts with compound *a*, it converts partially to *b* because the SCN bands of both appear. A reaction occurs in the KBr.

D. Even with mulls there is sometimes a reaction with the alkali halide plates. *Example*: $NaBO_2 \cdot H_2O$ + NaCl plate $\longrightarrow$?

X. Oxides

A. These are often used as fillers, extenders, or pigments. Silica and alumina are widely used for catalyst supports.

B. Some IR spectra

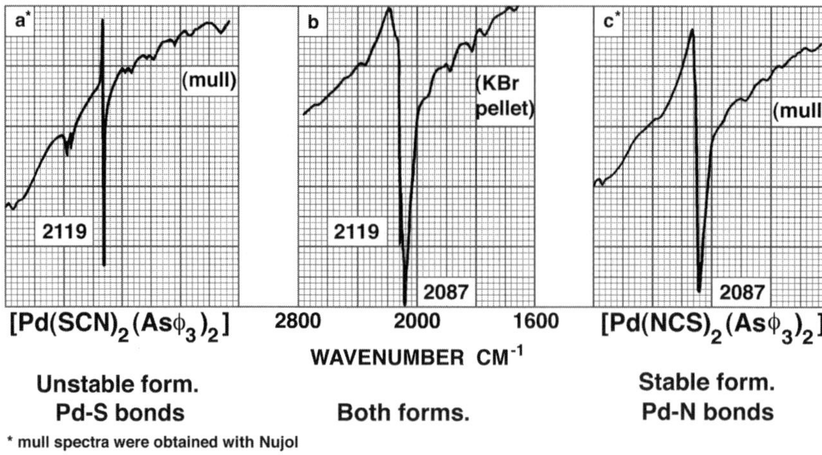

Fig. 11.26 Infrared spectra of $Pd(SCN)_2$ $(AsPh_3)_2$ and $Pd(NCS)_2$ $(AsPh_3)_2$ as Nujol mulls and their mixture formed in a KBr pressed disk.

1. Silica, SiO_2
   a. The spectrum depends on whether the sample is crystalline or amorphous. In every case there is a very, very strong and very broad band near 1100 cm^{-1} due to Si—O stretch.
   b. *Examples*: Crystalline quartz, amorphous silica, and silica gel. See Figures 11.27*a–c*.
2. Alumina, Al_2O_3
   a. It has good transmission above 2000 cm^{-1} and in the form of sapphire is a useful window material there for high temperature and pressure.
   b. The region around 600 cm^{-1} has strong bands.
   c. Comparison of α-, γ-, and ι-alumina. See Figures 11.28*a–c*.
3. TiO_2
   a. There are two common forms: rutile and anatase. See Figures 11.29*a,b*.
   b. Neither has significant absorption above 800 cm^{-1}. They differ at 400 cm^{-1}.
   c. Rutile is a widely used white pigment, whereas anatase is an undesirable contaminant in it. A mixture of the two can be analyzed quantitatively by either X-ray or Raman spectroscopy. For repetitive analyses, Raman is better.
4. Fe_2O_3

   Spectra of the red and yellow forms are strikingly different. See Figures 11.30*a,b*.
5. ZnO—an uninteresting spectrum, not shown
6. MnO_2 and CuO—no bands above 600 cm^{-1}

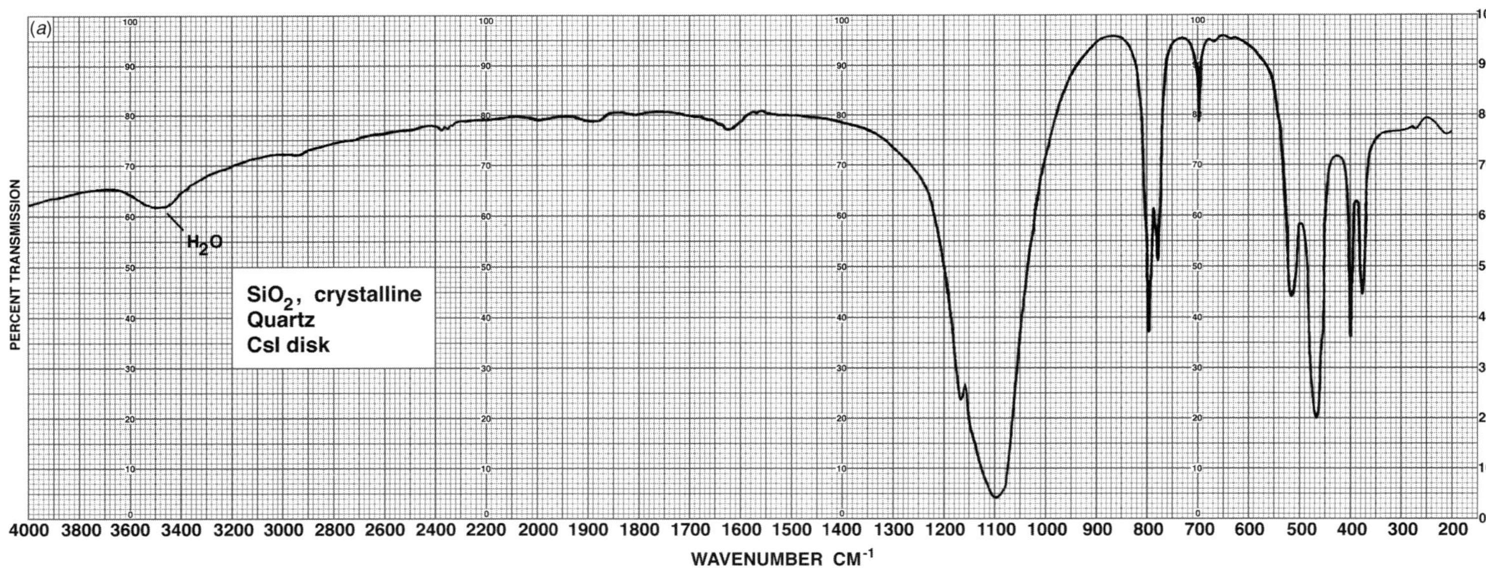

Fig. 11.27 Infrared spectra of three forms of silica: (*a*) crystalline quartz, (*b*) amorphous silica, and (*c*) silica gel.

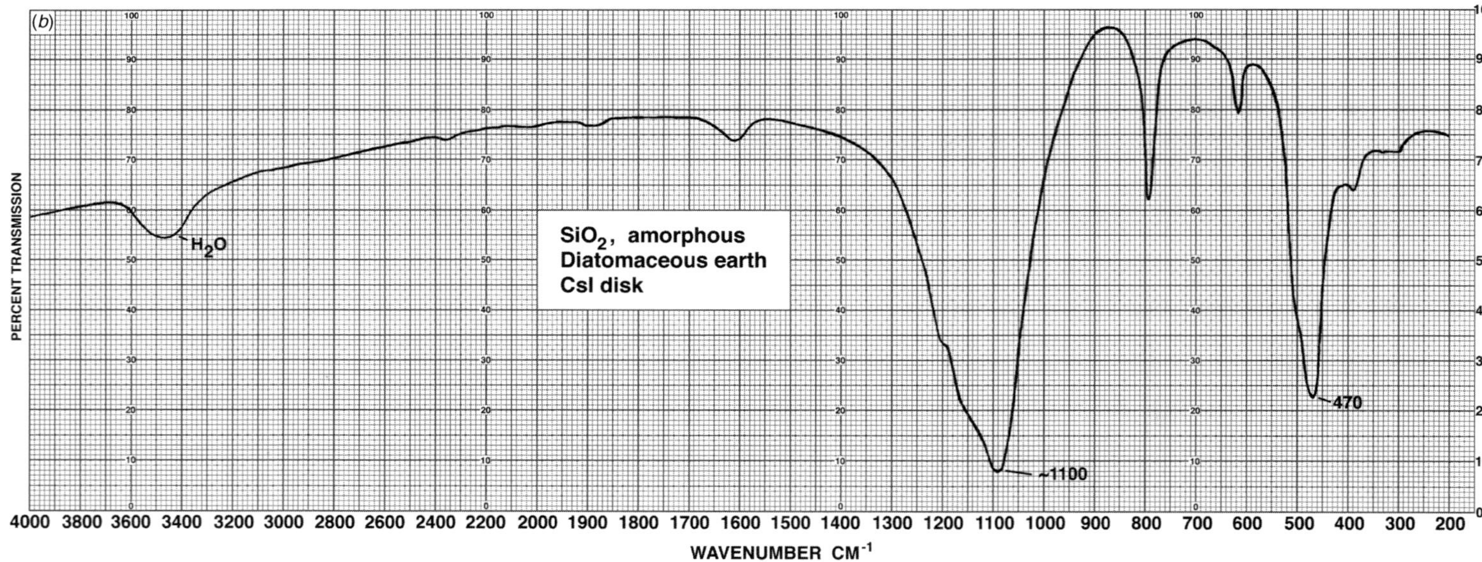

Fig. 11.27 *(Continued)*

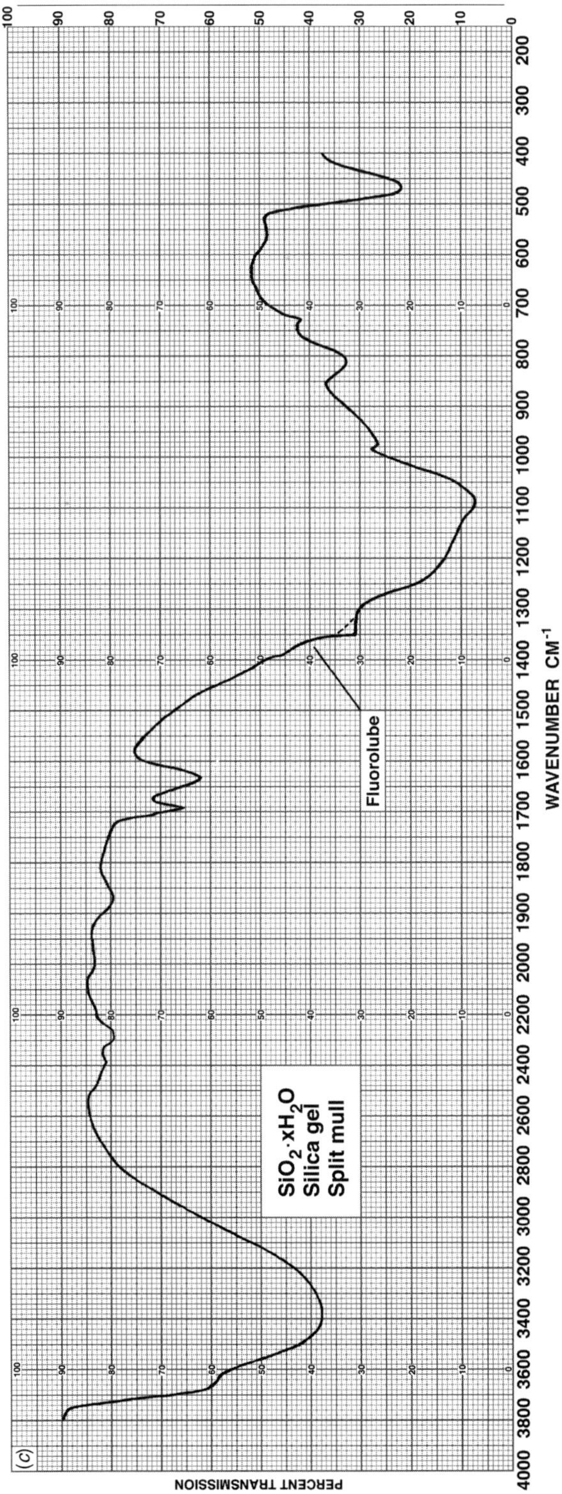

Fig. 11.27 *(Continued)*

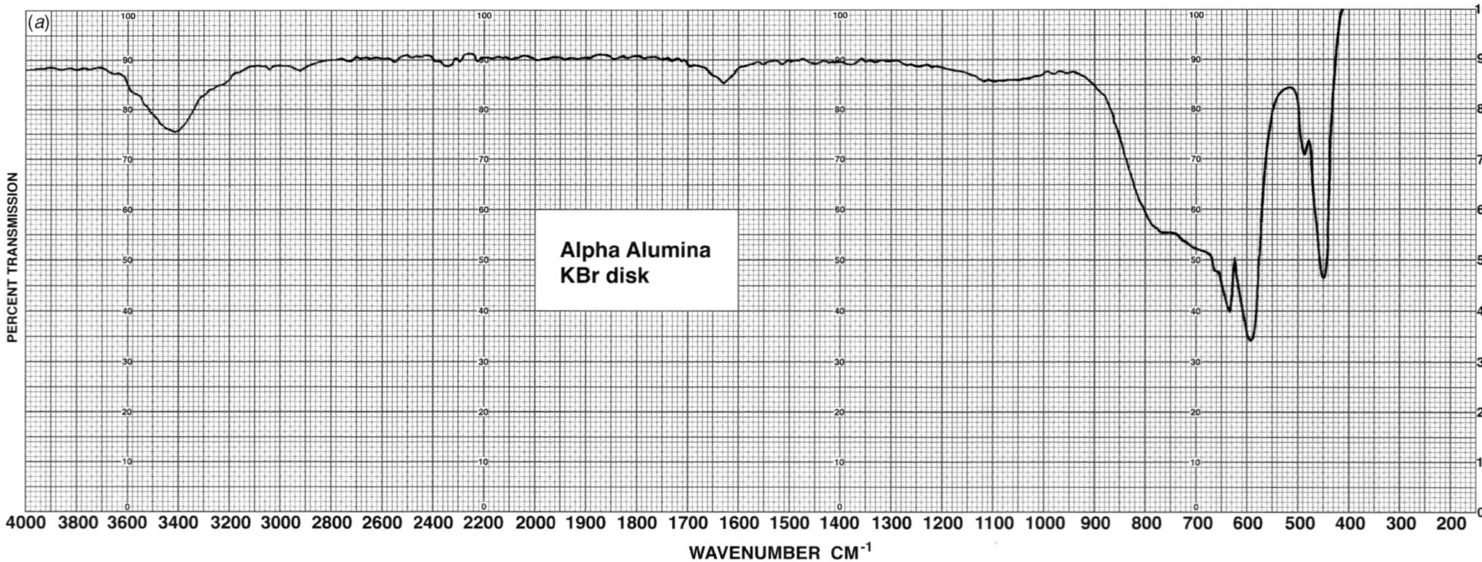

Fig. 11.28 Infrared spectra of α-, γ-, and ι-alumina.

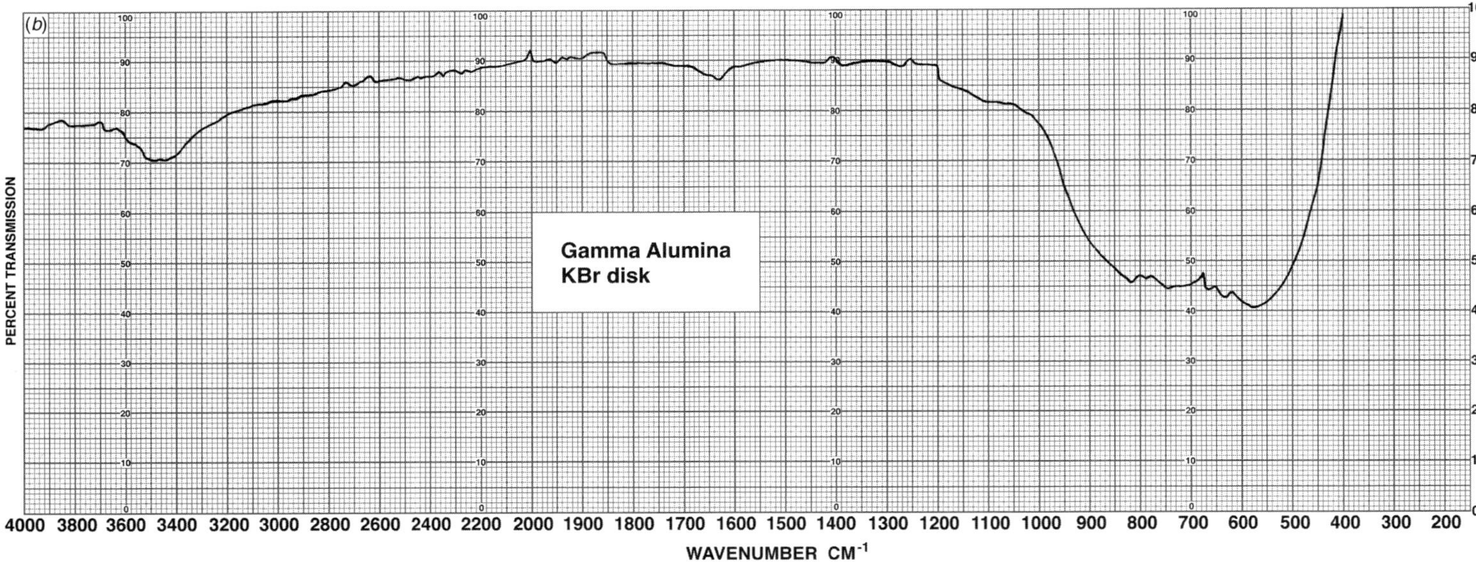

Fig. 11.28 (*Continued*)

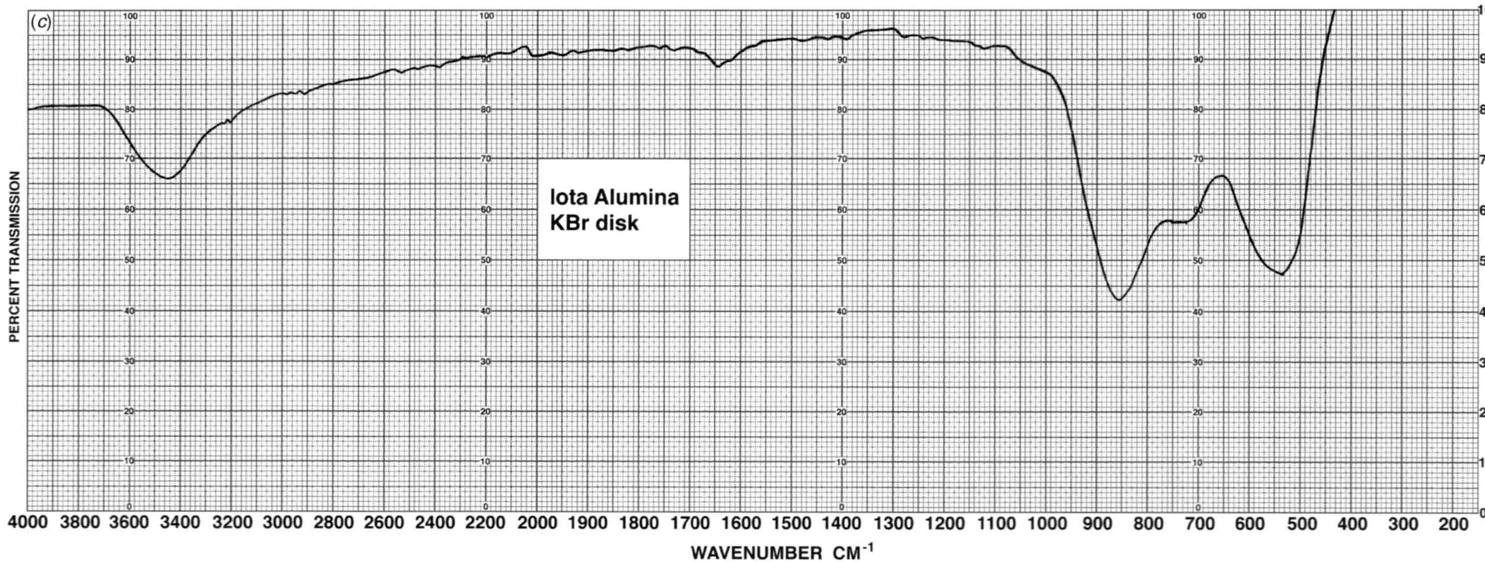

Fig. 11.28 *(Continued)*

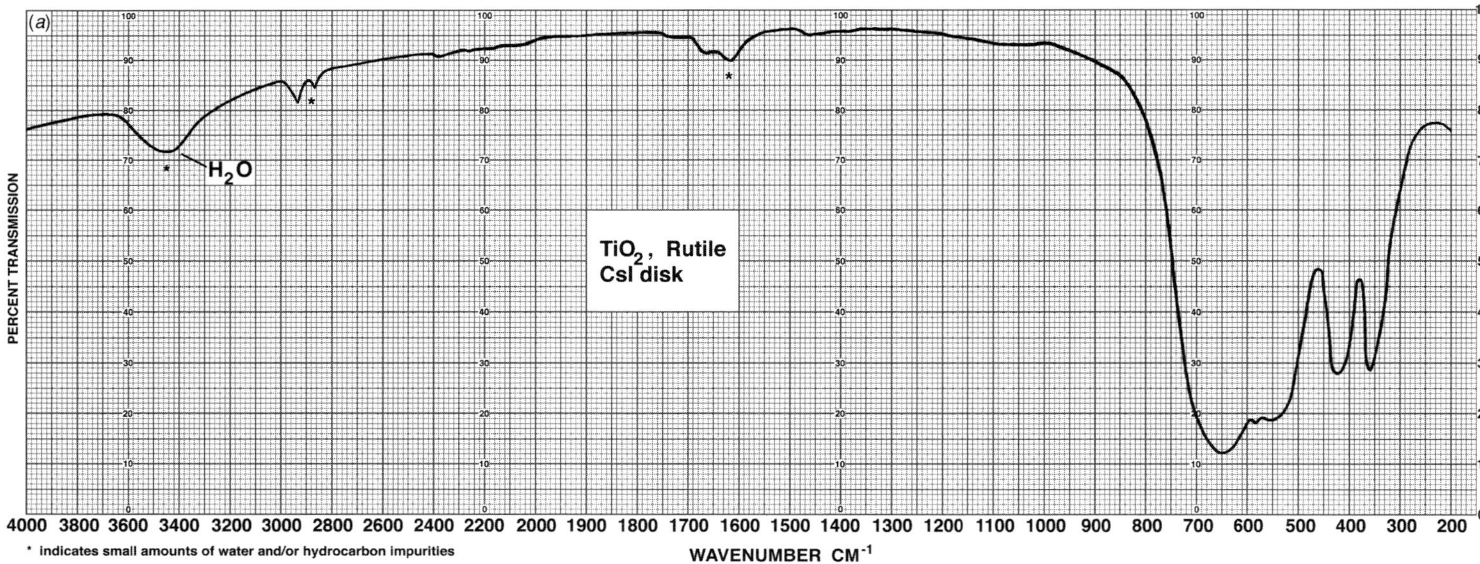

Fig. 11.29 Infrared spectra of TiO_2: (*a*) rutile; (*b*) anatase.

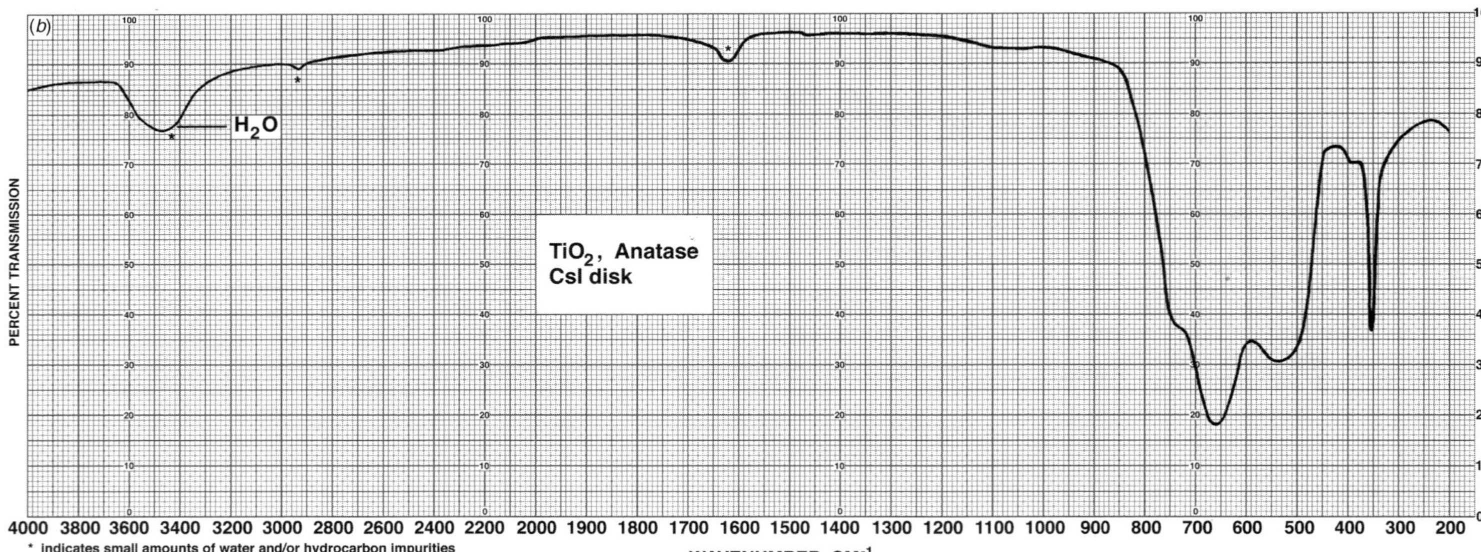

Fig. 11.29 *(Continued)*

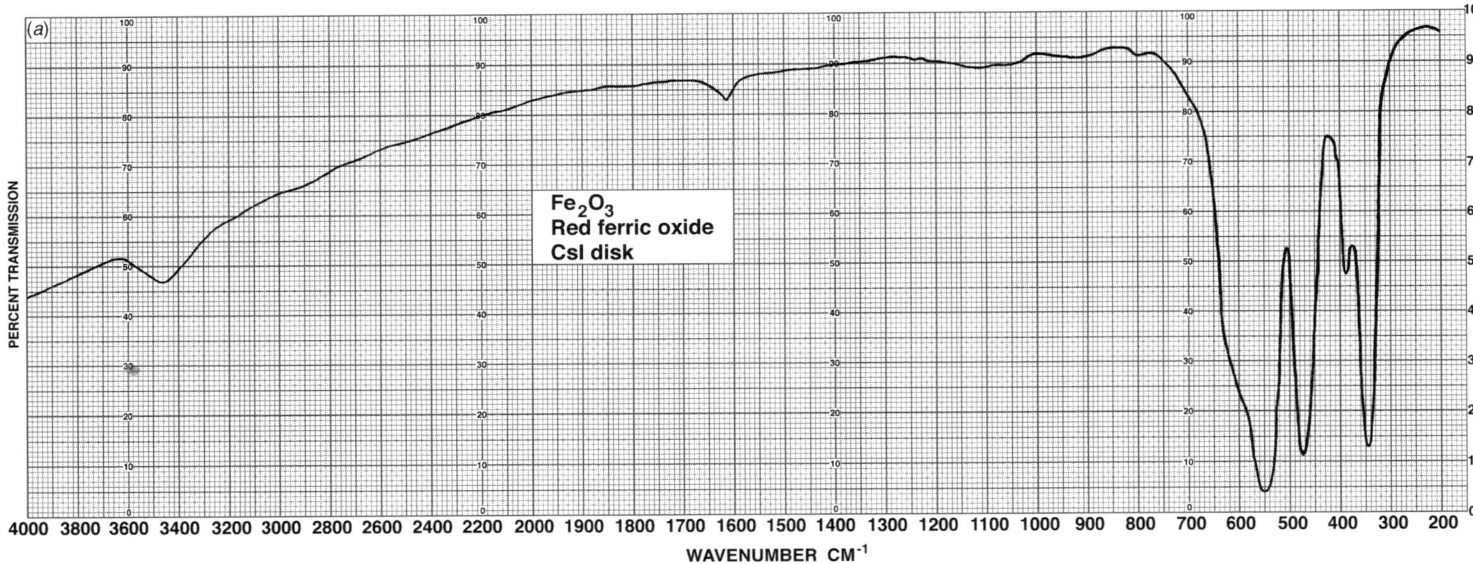

Fig. 11.30 Infrared spectra of (*a*) red ferric oxide (Fe_2O_3) and (*b*) yellow ferric oxide ($Fe_2O_3 \cdot H_2O$).

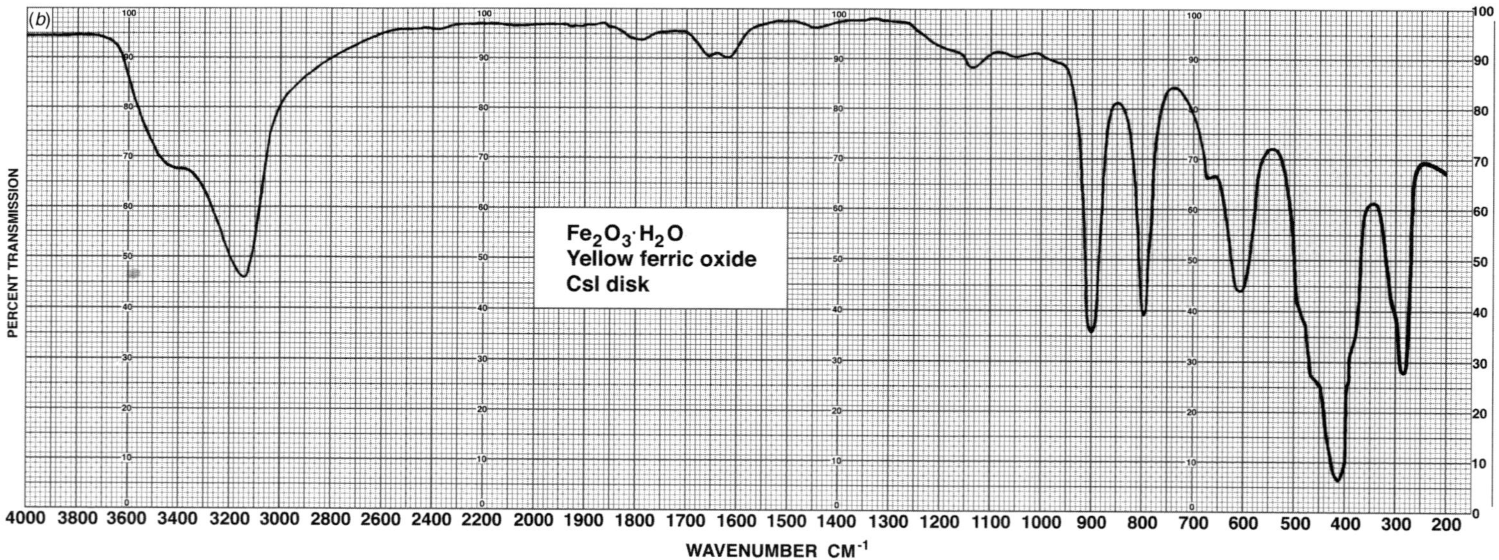

Fig. 11.30 *(Continued)*

C. Oxides often contain some water or M—OH groups. If so, there will be bands due to them. See the earlier spectrum of silica gel, Figure 11.27*c*.

D. Good reference on oxides: Nyquist and Kagel, *Infrared Spectra of Inorganic Compounds*. Pages 206–233 show the spectra of 55 oxides and p. 9 gives a correlation chart for 48 of them.

XI. Minerals

A. These too are used as fillers, extenders, pigments, or opacifiers, and sometimes there is need to identify them. *Examples*:

1. Fillers and extenders

$CaCO_3$	Talc	Kaolin	Diatomite (silica)
$BaSO_4$	Mica	Silica	

2. Pigments

White	TiO_2, ZnO
Black	Carbon black. Not a mineral, but included here because of its importance. Has absorption across the entire mid-IR, with no characteristic bands.
Red	Red ferric oxide
Yellow	Yellow ferric oxide, lead chromate

B. Minerals have characteristic infrared and Raman spectra.

1. Infrared—Powders are the best samples. It is difficult to obtain spectra of single crystals because they can seldom be cut thin enough to give acceptable transmission spectra. Reflection spectra can be obtained from single crystals, but they have poor contrast.
2. Raman—Thickness is not a problem. Well suited to studying either single crystals or powders. The preferred technique.

C. *Examples*:

1. Clays
   a. There are many types of clay. Their identification by IR is discussed in books by V. C. Farmer and by van der Marel and Beutelspacher.
   b. *Example*: China clay. See Figure 11.31.
      1) The sharp bands show that it is a well-defined substance.
      2) Note the two O—H stretches: how high (3700 and 3600 cm^{-1}) and how sharp they are. Because these bands are removed by heating, they must be due to H_2O rather than $[O—H]^-$. But how can water exist in this material without being hydrogen bonded?
2. Fluorapatite (often called just apatite). See Figure 11.32.
   This is $Ca_5(PO_4)_3F$. It has the strong P—O absorption of phosphate near 1050 cm^{-1}.
3. Hydroxyapatite. See Figure 11.33.
   a. This is $Ca_5(PO_4)_3(OH)$. The F^- ion of fluoroapatite has been replaced by $[O—H]^-$. It too has the very strong P—O band of phosphate near 1050 cm^{-1}.

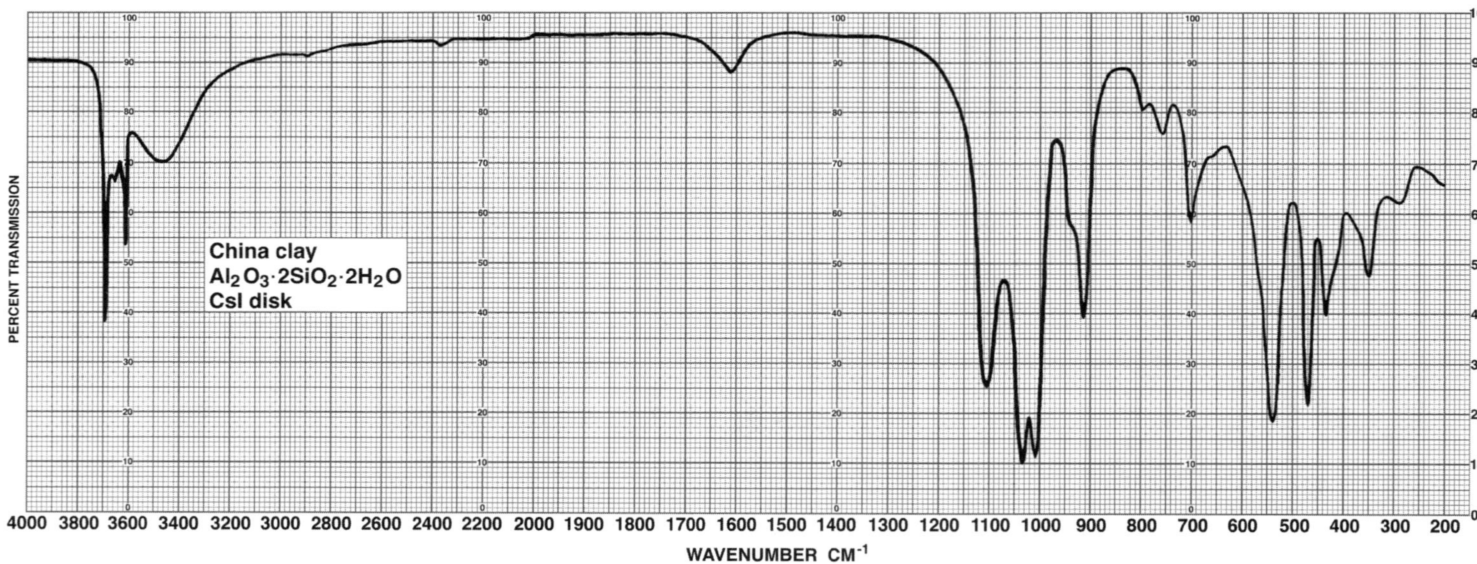

Fig. 11.31 Infrared spectrum of china clay, $Al_2O_3 \cdot 2SiO_2 \cdot 2H_2O$.

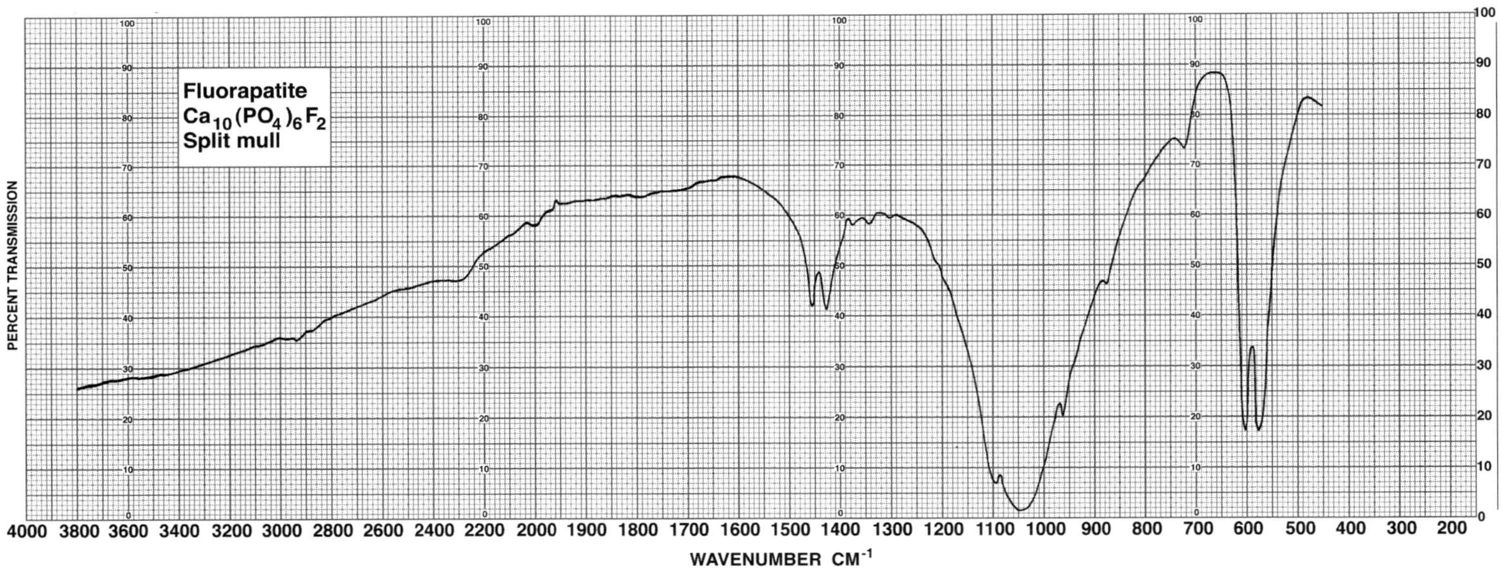

Fig. 11.32 Infrared spectrum of fluorapatite, $Ca_5(PO_4)_3F$.

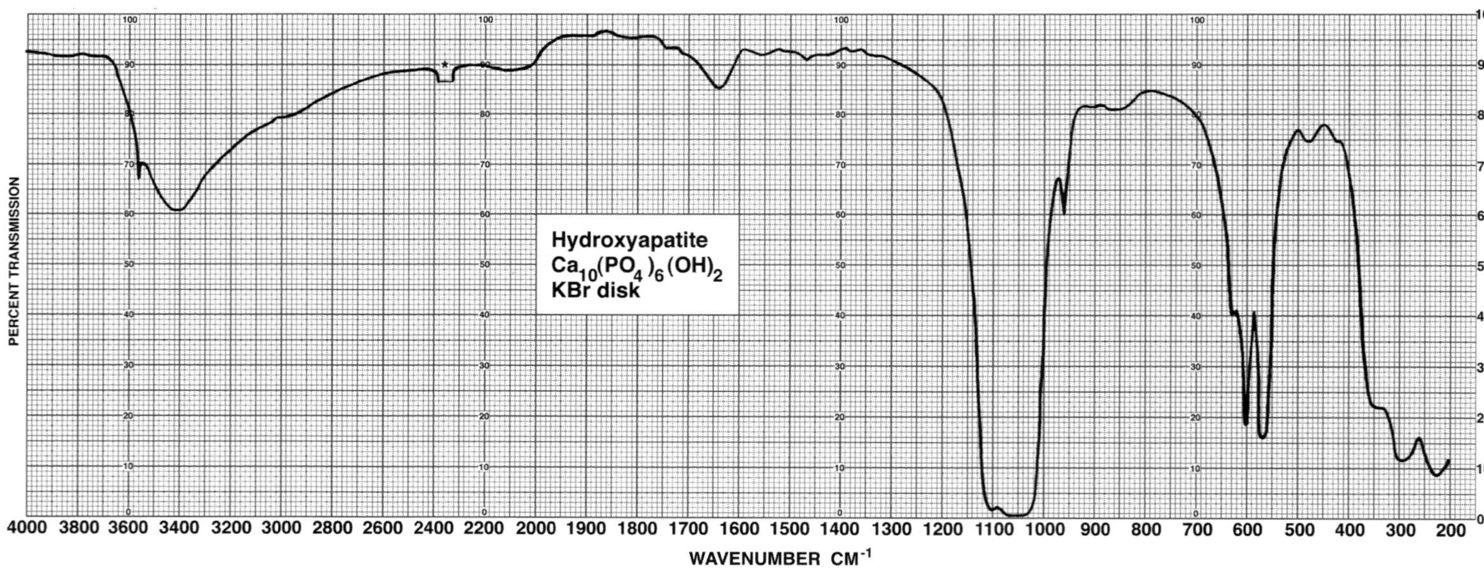

Fig. 11.33 Infrared spectrum of hydroxyapatite, $Ca_5(PO_4)_3(OH)$.

   b. Hydroxyapatite occurs in teeth. Fluoridation replaces some of the OH^- ions with F^- (cf. fluorapatite).
   c. Bone is a porous matrix of hydroxyapatite interlaced with collagen fibers and blood vessels.
4. Dolomite. See Figure 11.34.
   This is $CaCO_3 \cdot MgCO_3$. The carbonate bands are obvious.
5. Talc. See Figure 11.35.
   a. This has the empirical formula $3MgO \cdot 4SiO_2 \cdot H_2O$.
   b. It gives a nice spectrum
   c. There is a sharp band near 3520 cm^{-1} due to O—H stretch. It is sometimes destroyed when making a KBr disk.
6. Diamond
   a. You may wonder why one would ever need to know the IR spectrum of a diamond. *Reasons*:
      1) They are used as the anvils in diamond high-pressure cells, which are very useful sampling devices. Infrared radiation must pass through them, so their transmission is important.
      2) Diamonds are the windows on Golay cells when they are used as detectors in the far IR.
   b. Diamond is exceptional among hard crystals in having relatively little IR absorption. The reason is the high symmetry, making selection rules very restrictive.
   c. Most of the bands that are observed are due to impurities, except for lattice modes near 2000 cm^{-1}.
   d. Diamonds are divided into two classes on the basis of several physical properties: luminescence, IR spectra, UV transmission, and others.
      1) *Type I.* These have IR bands at 1400–1100 cm^{-1} of variable intensity. They are due to nitrogen impurity. About 99% of all diamonds are of this type.
      2) *Type II.* These are transparent from 43,500 cm^{-1} (in the UV) to 10 cm^{-1} (in the far IR), except for lattice modes near 2000 cm^{-1}. They therefore make good IR windows. Only about 1% of natural diamonds are of this type.
   e. Spectrum of a type II diamond several millimeters thick. See Figure 11.36.
7. FeS_2 (iron pyrites, or fool's gold)
   Has *no* bands in the mid-IR! This indicates heavy atoms and high symmetry.
8. Crystalline sulfur
   No IR bands above 500 cm^{-1}. Has Raman bands at 476, 437, 218, 194, and 152 cm^{-1} and lower.

XII. *References for IR spectra of inorganic materials.* See the Bibliography.
   A. The book by V. C. Farmer discusses the IR spectra of many other inorganic materials, such as Portland cement and inorganic glasses.

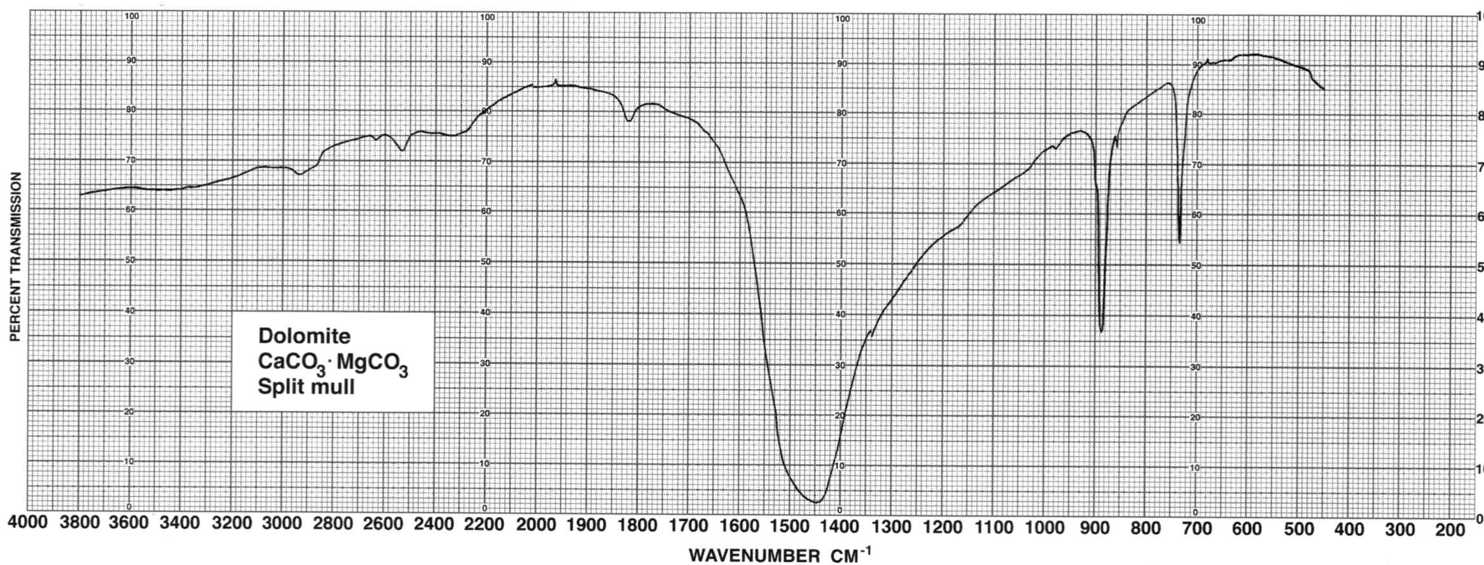

Fig. 11.34 Infrared spectrum of dolomite, $CaCO_3 \cdot MgCO_3$.

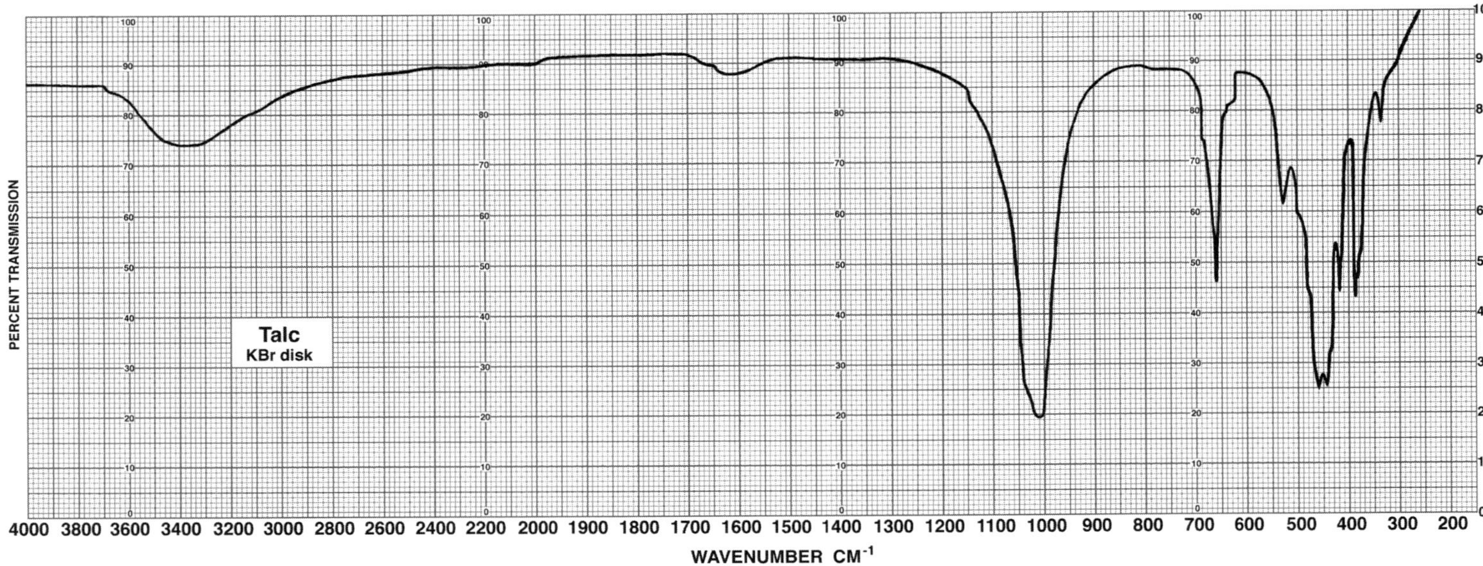

Fig. 11.35 Infrared spectrum of talc, $3MgO \cdot 4SiO_2 \cdot H_2O$.

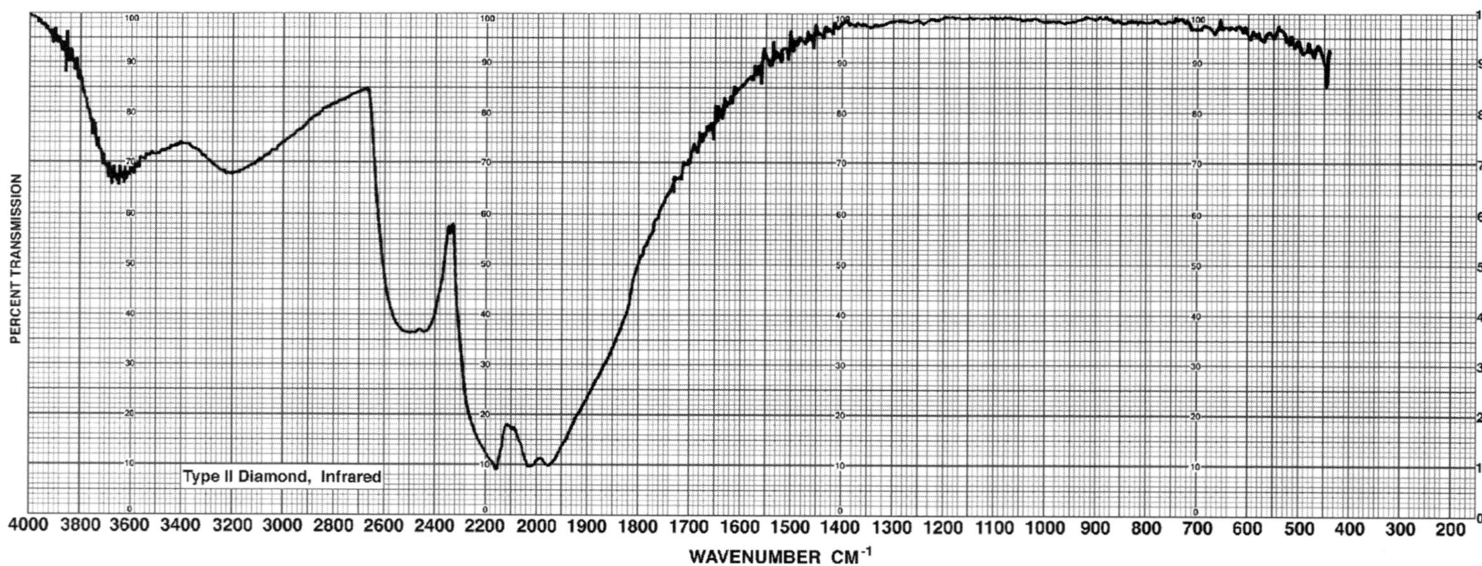

Fig. 11.36 Infrared spectrum of a natural uncut 0.38-carat type IIa diamond about 2 mm thick. (Type IIa diamonds show no impurity atoms detectable by IR.) Resolution 4 cm^{-1}. The fine structure near 1500 and 3650 cm^{-1} is due to water vapor.

B. *The two best collections of inorganic spectra are:*
   1. R. A. Nyquist, R. O. Kagel, C. L. Putzig, and M. A. Luegers, *The Handbook of Infrared and Raman Spectra of Inorganic Compounds and Organic Salts*, Academic, New York, 1996, 4 vols. *The best collection available as of* December 1998.
   2. R. A. Nyquist and R. O. Kagel, *Infrared Spectra of Inorganic Compounds (3800-45 cm^{-1})*, Academic, New York, 1971. About 900 spectra of high quality. *An excellent collection, but IR only.*

C. *Many* other references are given in the Bibliography.

XIII. Lattice vibrations (phonons)

A. It was said at the beginning of the chapter that these occur only in crystals and that they are usually below 300 cm^{-1}. They will now be described a little more fully.

B. Consider an NaCl crystal.
   1. It is made up of Na^+ and Cl^- ions. There are no molecules or polyatomic ions and therefore no internal vibrations.
   2. The Na^+ and Cl^- ions can oscillate around their equilibrium positions. There are $3n - 6$ such degrees of freedom, where n is the total number of ions in the piece of crystal. This is an enormous number. Most of these oscillations do not give IR absorption, however. The IR wavelength extends over about 100,000 unit cells. If the motions in one unit cell produce a change in the dipole moment but that change is canceled by contrary motions in another unit cell, there is no net change in the dipole moment and no resulting IR absorption. The chance of such cancellation among the 100,000 unit cells is very good.
   3. *Conclusion*: To have a net change in the dipole moment, all unit cells must be vibrating in the same manner and in phase (to a high approximation). Then there will be reinforcement and not cancellation.
      a. The physicist says the k vector must be zero.
      b. The resulting modes are *lattice vibrations*.
      c. This effects an enormous simplification. One need consider only the possible vibrations of one unit cell and find how many of these are IR active. Therefore the number of IR-active lattice modes is typically one to six, not millions. For each of these active modes, all the unit cells are vibrating in the same manner and in phase with one another.
   4. The NaCl crystal is cubic. Let us call the axes X, Y, and Z. There is one IR-active mode in which all the Na^+ ions move toward $+X$ while all the Cl^- ions move toward $-X$. Since NaCl is cubic and isotropic, the same thing occurs along the Y and Z axes. Thus this vibration is triply degenerate.
      a. This is the only IR-active lattice mode in NaCl. It occurs at 162 cm^{-1}.
      b. It is enormously intense. Because the particles that are vibrating are charged, the vibration produces a very large change in the dipole moment. This absorption is responsible for the low-frequency

cutoff in the transmission of NaCl. It is the reason NaCl windows are not used below about 650 cm^{-1}.

5. NaCl has no internal modes because the structural units are monoatomic ions.

C. If the crystal contains molecules or polyatomic ions, there are both internal and external modes.
   1. *Example 1*: neutral molecules (ice, naphthalene, benzoic acid, etc.)
      a. The *internal vibrations* are those of the molecule. They are essentially the same in the crystal as in the melt, a solution, or the gas, although there certainly will be small changes.
      b. The *external, or lattice, vibrations* are those in which the molecules or ions move relative to each other.
         1) They may be subdivided into:
            a) *Translatory modes*—relative motions of the centers of gravity of the molecules.
            b) *Rotatory modes*—hindered rotations of molecules or polyatomic ions. Also called *librations*. These do not occur for atoms or monoatomic ions.
         2) Lattice vibrations are also classified as optical branch and acoustical branch modes or as transverse optical (TO) and longitudinal optical (LO) modes. These are not important to us.
         3) Lattice vibrations disappear if the crystal is destroyed by any means—e.g., by melting or solution. They are a cooperative phenomenon of a highly ordered system.
         4) They are low in frequency—almost always <300 cm^{-1}. The low frequencies are due to the large masses and weak restoring forces. In exceptional cases lattice modes are higher. A hindered rotation of H_2O in $BaCl_2 \cdot 2H_2O$ is at 700 cm^{-1}. (See its spectrum in Figure 11.3.) In diamond the lattice modes are in the 2000-cm^{-1} region because covalent bonds are involved.
         5) Lattice frequencies are unique for a given crystalline substance and can be used for identification if this wavenumber range is available.
         6) They are more likely to be available from Raman spectra than from IR.
         7) Raman and IR lattice modes are usually different because they have separate selection rules.
   2. *Example 2*: K_2CO_3. This has the internal vibrations of CO_3^{2-} and the lattice vibrations of the K^+ and CO_3^{2-} ions.
   3. In general, *infrared* lattice modes are more intense for salts than for neutral molecules because charged particles are involved and the change in dipole moment is therefore larger.

D. *Recapitulation*: Lattice vibrations are associated only with the crystalline state. Their frequencies are usually low (<300 cm^{-1}), so they will seldom be seen in conventional IR spectra (above 400 cm^{-1}). They are often seen in Raman spectra of crystals, including crystalline powders.

12 Survey of Infrared and Raman Group Frequencies

DANA W. MAYO

I. Introduction

As described in Chapter 1, vibrational transitions may be observed through measurements of IR absorption or by Raman scattering experiments. The two effects involve different selection rules which often fortuitously give rise to highly complementary intensity observations (bands that are strong in one effect are weak in the other). This chapter will explore the complementary nature of IR and Raman spectra primarily within the context of *group frequencies*. Two important aspects central to these discussions that have been touched on in the earlier chapters and which are throughly explored here are (1) the range of Raman group frequencies and (2) the diagnostic value of applying data from both effects to the elucidation of a particular molecular structure or to the identification of an unknown material.

The collection of Raman-scattered radiation at about 90° to the exciting radiation has been described on page 17. The collection angle used for the majority of the Raman spectra shown in this chapter, which were recorded at an earlier date, utilized 180° back-scattered Raman radiation and employed a Cary model 81 dispersive spectrometer with either He–Ne 6328 Å, or Ar ion 5145 or 4880 Å CW laser excitation. The optical diagram (Figure 12.1) of how these observations were made shows the exciting laser radiation focused on a small triangular prism which is glued to the back of a hemispherical lens. The prism directs the exciting radiation to a focal point down the axis of a small quartz capillary tube which has a sealed and flattened end coupled to the front face of the lens with optical coupling compound. The sample (a few microliters) is placed in the capillary and the Raman scattering that is reflected off the walls of the capillary back to the hemispherical lens is

Course Notes on the Interpretation of Infrared and Raman Spectra, by Dana W. Mayo, Foil A. Miller, and Robert W. Hannah.
ISBN 0-471-24823-1

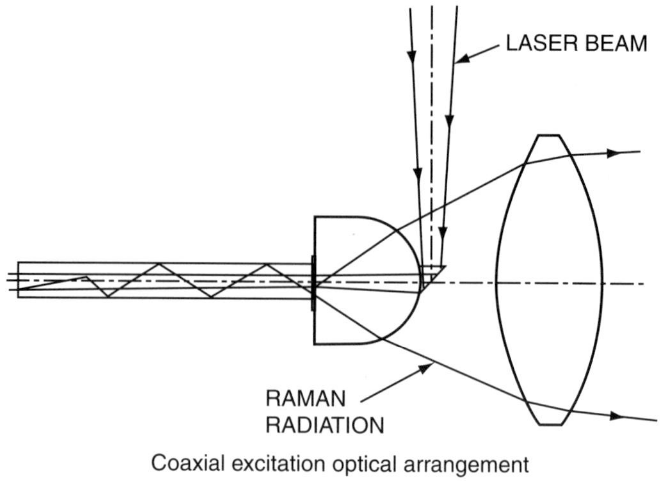

Fig. 12.1 Sampling optics of Cary 81 Raman spectrometer.

collected and focused on to a second lens which directs the radiation into the monochromator.

Now let us consider the 16 important partial structural units (PSUs) listed in Table 12.1. Twelve of these can be classified as possessing complementary intensities in the IR and Raman effects while the remaining 4 PSUs exhibit significant intensity in both effects.

Note that those groups which are weak in the IR and strong in the Raman are systems that are *not* highly polarized and possess relatively *high* bond order and/or lone-pair electron densities. Thus a relatively large change in polarizability and conversely a low change of dipole moment would be expected during vibrational displacement (mostly stretching modes). As scattering intensity is proportional to the square of the change in polarizability, these systems might be expected to give rise to strong Raman bands.

TABLE 12.1 IR Versus Raman Group Frequencies

Weak in IR, Strong in Raman	Strong in IR, Weak in Raman	Strong in Both
C≡C stretch	C—F stretch	C≡N stretch
C=C stretch	O—H stretch	C=O stretch
N=N stretch	N—H stretch	C—Cl stretch
S—H stretch	C—H out-of-plane bend	$-NO_2$ stretch
C=S stretch		
C—S stretch		
S—S stretch		
$-CH_2-$ twist and wag		

Conversely, the small change of dipole moment results in inherently weak IR absorption. *Eight examples are listed.*

Those bonds which are highly polarized (a large electronegativity difference between bonded atoms that should result in a relatively large dipole moment) with electron pairs tightly bound would be expected to exhibit the reverse effect (strong IR and weak Raman bands). *Four examples are listed.*

Finally, a few examples can be identified in which both high densities of polarizable electrons and significant dipole moments exist in the same system. As expected, these groups give rise to both strong Raman scattering and intense IR absorption bands. *Four examples are given.* In the case of the nitro group only the in-phase stretch is routinely found to be strong in both effects.

II. Survey

A survey of representative spectra follows. The IR absorption spectra have increasing intensity of absorption going down the vertical axis while Raman scattering intensity increases going up the vertical axis. The Raman frequencies are expressed on a wavenumber scale identical to that used for the IR spectra. It should be remembered that these values are, in fact, wavenumber shifts from the exciting line (6328 or 4880 Å).

III. Hydrocarbons

The IR and Raman spectra of *n*-hexane (Figure 12.2) reveal several important points:

A. The C—H stretching (3000–2800 cm^{-1}) and methyl antisymmetric (1460 cm^{-1}) bending modes have both modest and similar intensities in both effects (they give rise to very intense bands, but this is the result of the presence of many repeating groups in the overall structure). As we will see, the intensity of Raman scattering by these vibrational modes appears to be less influenced by adjacent inductive and resonance effects than is the case in IR spectra. Thus, the methyl and methylene stretching modes are generally observed in both effects (see also Chapter 2).

B. The first complementary group frequency in the hydrocarbons is the symmetric methyl bending mode at 1383 cm^{-1} which gives rise to a medium-to-strong band in the IR but is nearly absent in the Raman effect. It is therefore somewhat surprising that a weak overtone of this fundamental is observed near 2740 cm^{-1} on the low-wavenumber side of the Raman-active C—H stretching fundamentals. Why this transition is observed at all will be described later along with a few examples of systems where the fundamental is also observed in Raman spectra.

C. The next most prominent band in the IR occurs near 720 cm^{-1}. This absorption, which is assigned to the all-in-phase methylene rocking vibration, has little or no intensity in the Raman effect. In the IR this band has been identified in those systems possessing four or more methylene groups in a row (in systems with fewer than four methylene

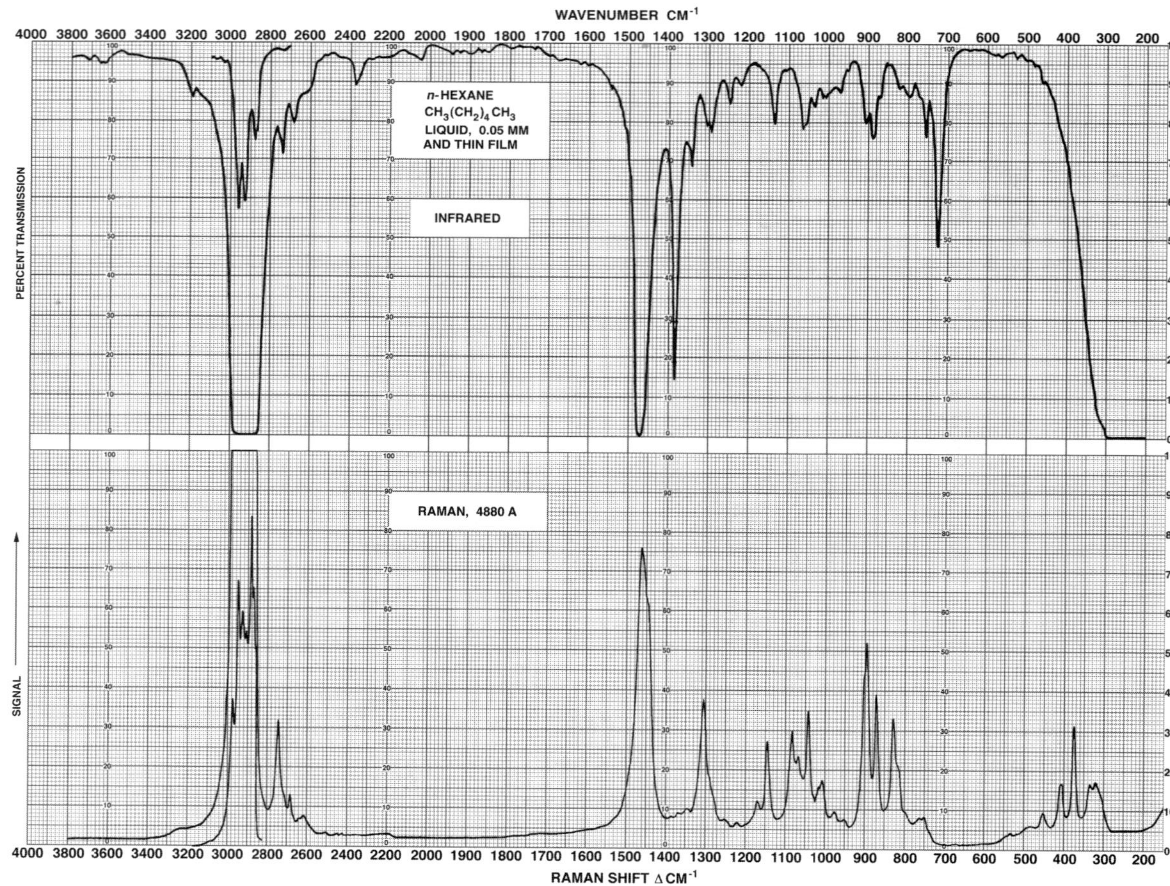

Fig. 12.2 Infrared and Raman spectra of *n*-hexane.

groups the wavenumber values drift upward and the intensity is weak to very weak). Positive identification of this weak-to-medium intensity band is difficult as it falls deep in the fingerprint region. If Raman data are available, a companion band with similar behavior can be found just above 1300 cm^{-1}. This scattering has been assigned to the all-in-phase methylene twisting mode (*active in the Raman but very weak in the IR*). Thus, both bands must be observed in order to assign the IR 720 cm^{-1} absorption to the methylene rocking mode (see also discussions in Chapter 2).

IV. Substitution by polar atoms and X—H groups

The *Raman* spectrum of di-*n*-hexyl ether (Figure 12.3) is nearly identical to that of *n*-hexane because the polar C—O stretching and bending modes make little contribution to the fingerprint region. The only clue that this material has incorporated a polar system is the strong band near 1120 cm^{-1} in the IR spectrum. The two spectra when taken together, however, strongly indicate the presence of an ether. If the hexane chain is substituted with polar hydroxyl or primary amine groups (see Figures 12.4 and 12.5), significant changes are observed in the IR *while* the Raman spectra continue to resemble the unsubstituted hexane. Although the single O—H stretching mode in hexanol is the most intense band in the IR, it is barely observed as a weak broad band in the Raman effect. In hexyl amine the N—H stretching modes are weaker in the IR and stronger in the Raman relative to the hydroxyl system, as might be expected because the electronegativity differences are considerably less in the latter case. In fact, the symmetric NH_2 stretches are easily observed in the Raman effect. While the O—H and $N<H_2$ bending modes and the C—O stretching mode can be easily assigned to strong or medium bands in the IR, they are poor Raman scatterers and are not identified in the spectra.

In the case of 1-chlorohexane (see Figure 12.6) both the IR and Raman spectra are similar to the hydrocarbon spectrum with the exception of the C—Cl stretching and bending regions where new strong-to-medium bands are observed in both spectra (as noted in Table 12.1). With the 1-substituted chlorine derivatives the Raman spectrum is a significant aid to making confident assignments of the C—Cl stretching mode because the presence of rotational isomers results in a doubling of the band. While one leg of the doublet can be assigned to a new strong band near 660 cm^{-1} in the IR and to the gauche isomer, the trans conformer which would be expected to occur at slightly higher frequencies is nearly accidentally degenerate with the methylene rocking vibration at 725 cm^{-1} (a strong band is observed in the IR just above 730 cm^{-1}). Because this bending vibration makes no contribution to this region of the Raman spectrum, the C—Cl stretching mode is easily observed there as a new band of medium intensity in the Raman effect at 735 cm^{-1} and is assigned to the trans isomer.

The same conformational isomerism has been found to occur in all of the 1-halogenated normal alkanes starting with the propanes and gives rise to a

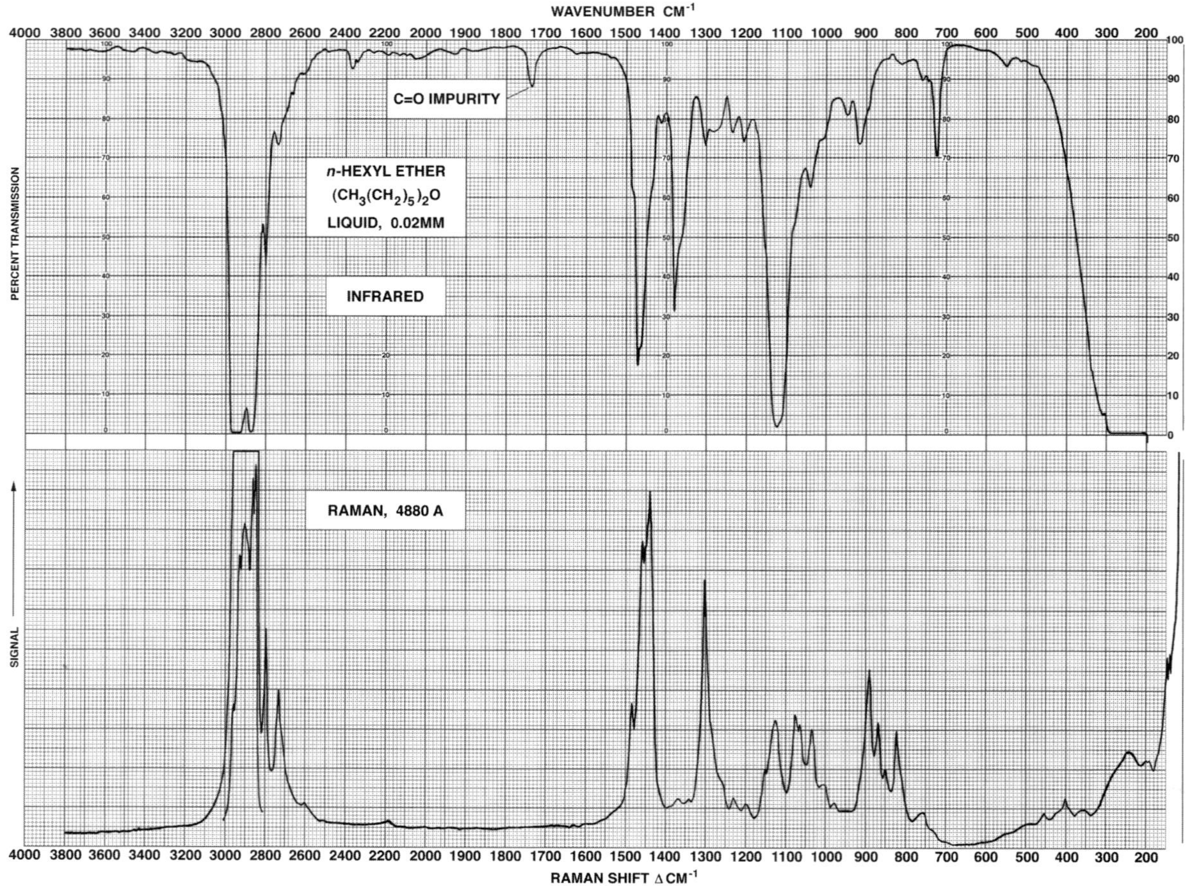

Fig. 12.3 Infrared and Raman spectra of di-n-hexyl ether.

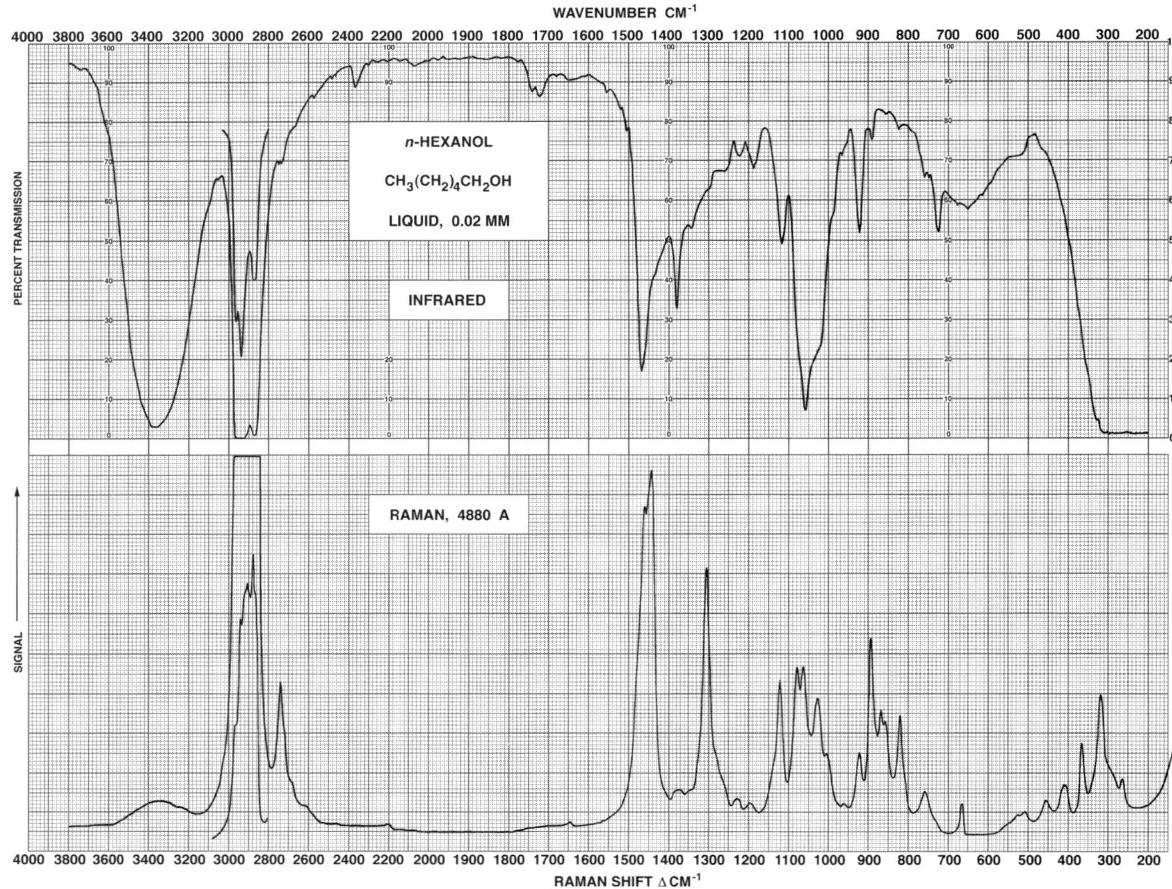

Fig. 12.4 Infrared and Raman spectra of *n*-hexanol.

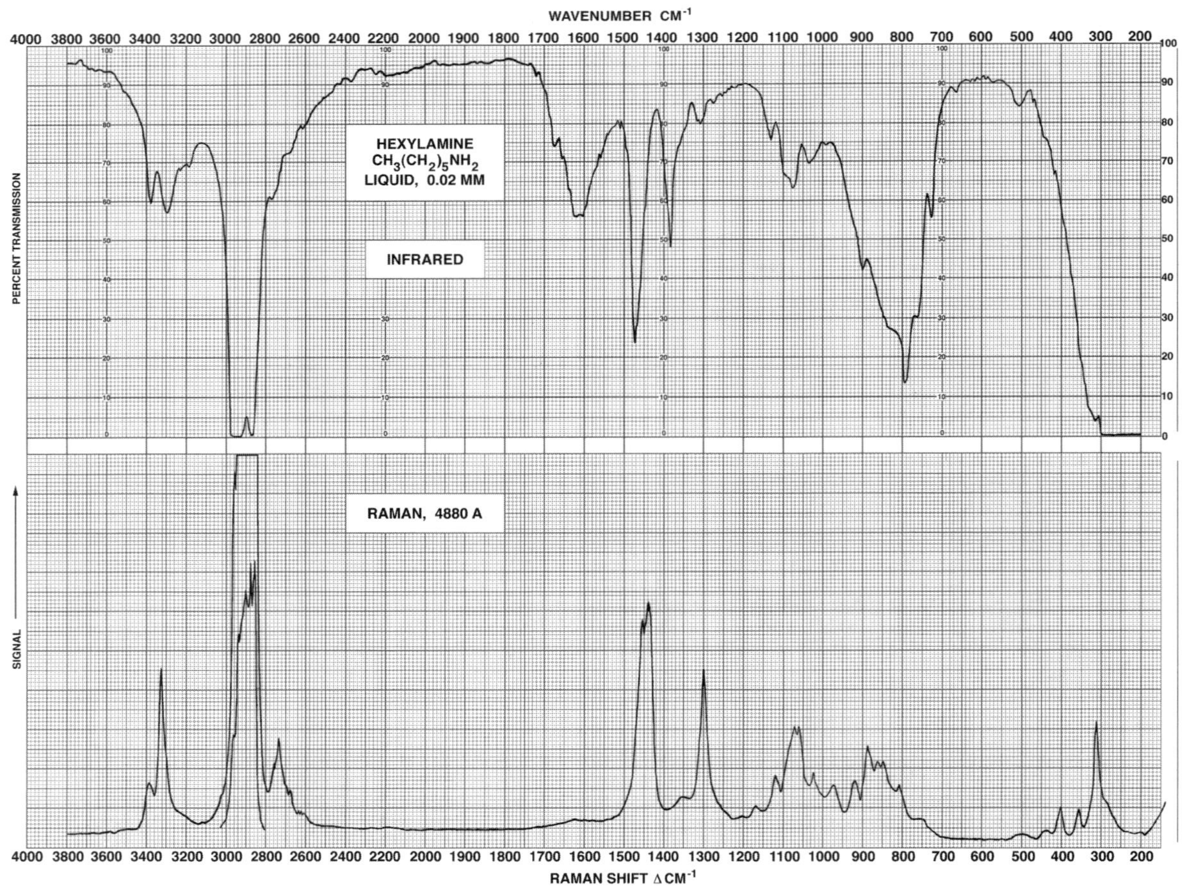

Fig. 12.5 Infrared and Raman spectra of *n*-hexylamine.

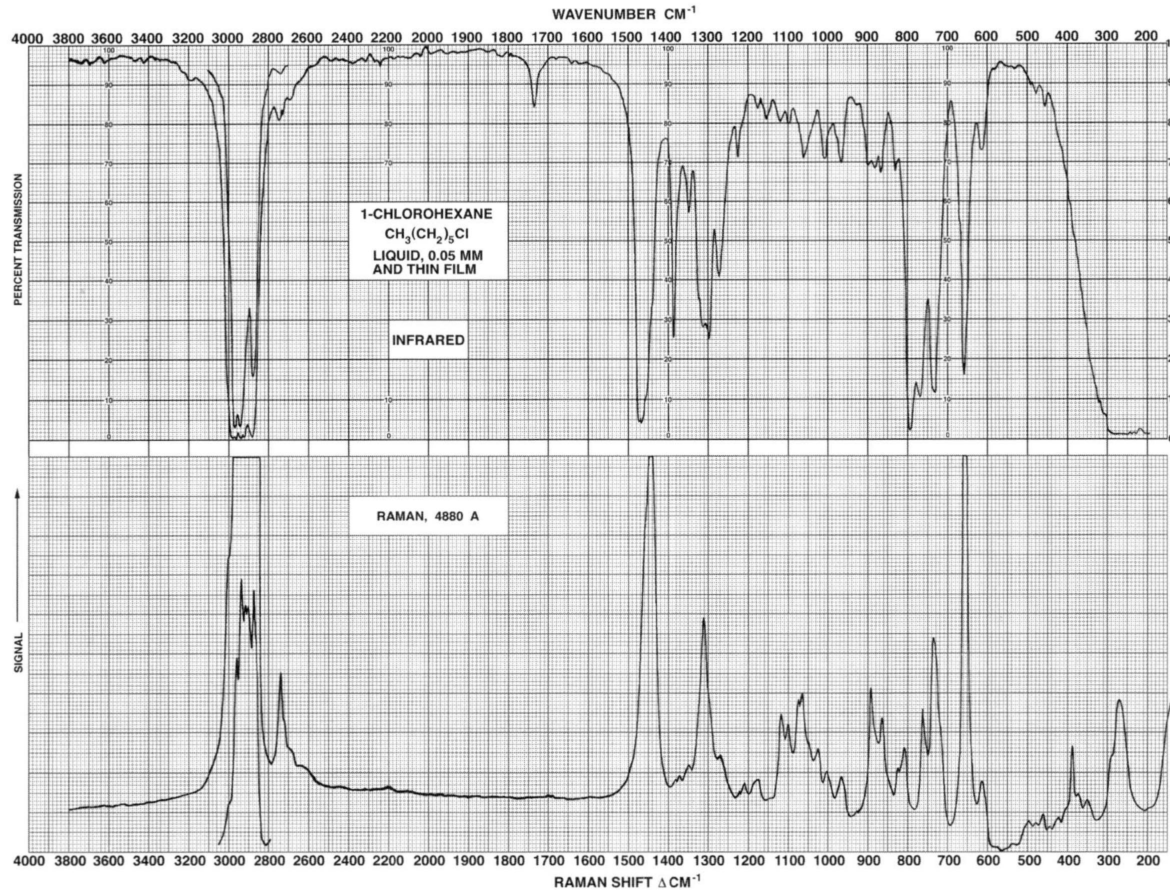

Fig. 12.6 Infrared and Raman spectra of 1-chlorohexane.

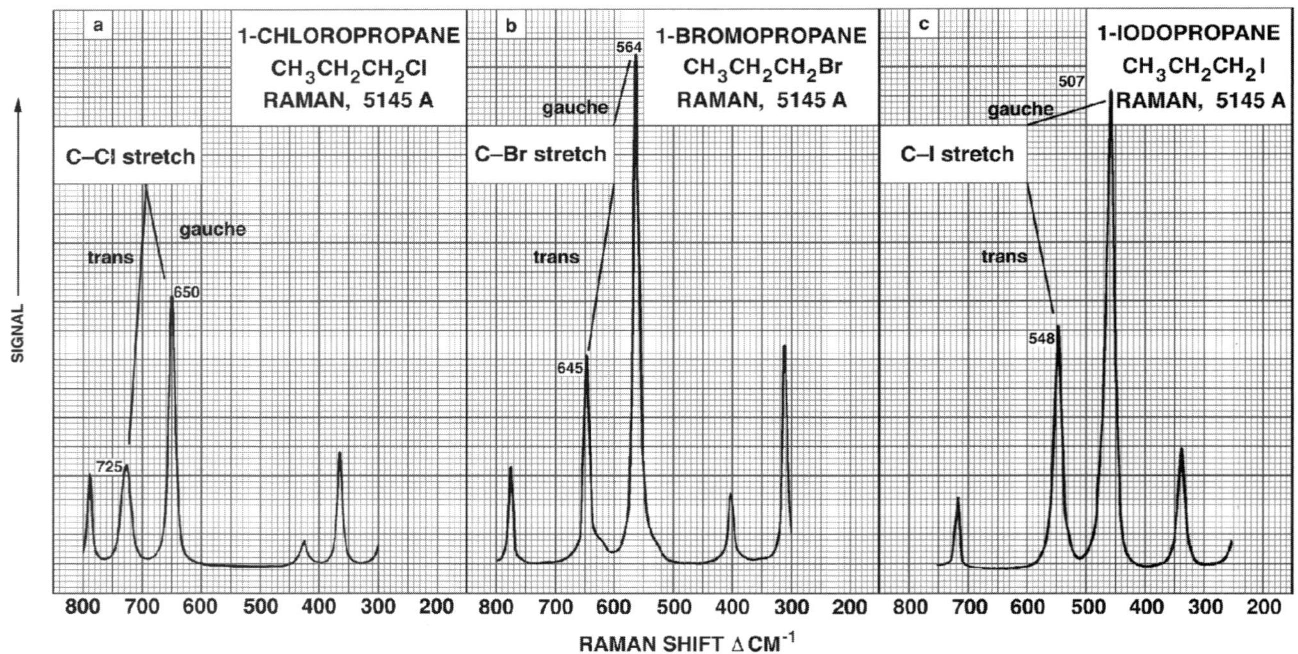

Fig. 12.7 Infrared and Raman spectra of 1-halopropanes.

pair of bands in the Raman that can be easily assigned to the C—X stretching mode. As expected, the wavenumber values decrease as the mass of the halogen increases and as the C—X bonds become weaker. This is shown in the case of the 1-halopropanes in Figures 12.7*a–c*.

V. Sulfur-containing groups

Four of the PSU groups listed in Table 12.1 are sulfur-containing systems.

A. The mercapto group

If the hexane chain is substituted with an —S—H group (see Figure 12.8), both the S—H (~2580 cm^{-1}) and C—S (660 cm^{-1}) stretching modes give rise to strong Raman bands. Note the very weak IR bands assigned to these two modes.

B. The disulfide group

L-Cystine (Figure 12.9) contains the S—S group, which gives rise to a very intense band in the Raman spectrum near 500 cm^{-1}. This mode has essentially no intensity in the IR. Because this structure contains highly polar groups, there is a rich collection of intense absorption bands in this region of the IR spectrum while the Raman spectrum is dominated by the single intense band from the nonpolar sulfur section of the molecule.

VI. Nitro group substitution

The nitro group is another example of a system which has strong bands in both effects, but the nitro group is a special case as it is also complementary in nature. The modes involved are the antisymmetric and symmetric N⋯O stretching fundamentals. The symmetric mode near 1350 cm^{-1} gives rise to intense bands in both effects while the antisymmetric mode near 1520 cm^{-1} is strong in the IR but weak in the Raman. A good example of this system is shown in the spectra of nitrobenzene (Figure 12.10) where the symmetric stretch of the nitro group is the strongest band in both the IR and Raman spectra while the antisymmetric stretch is a relatively weak band in the Raman effect. Another example of the enormous scattering power of the nitro group is in *p*-nitrotoluene (Figure 12.11). Again the nitro group clearly dominates the entire spectrum in both effects. This is particularly impressive because the benzene ring system itself is an efficient Raman scatterer.

VII. Aromatic substances

Aromatic ring stretching and bending modes are complementary in a number of cases. In the case of the parent system, benzene itself, the ring possesses a center of symmetry. Thus, the *rule of mutual exclusion* rigorously applies and only accidental degeneracies would be expected to be found in this highly symmetric system (Figure 12.12). Substitution on the ring often lowers the symmetry sufficiently that modes previously forbidden in one or both effects are now active in both. In the case of nitro substituents many of the details of the ring modes in the IR are masked, however, by the very intense nitro bands which fall in the same general region (see Figures 12.10 and 12.11).

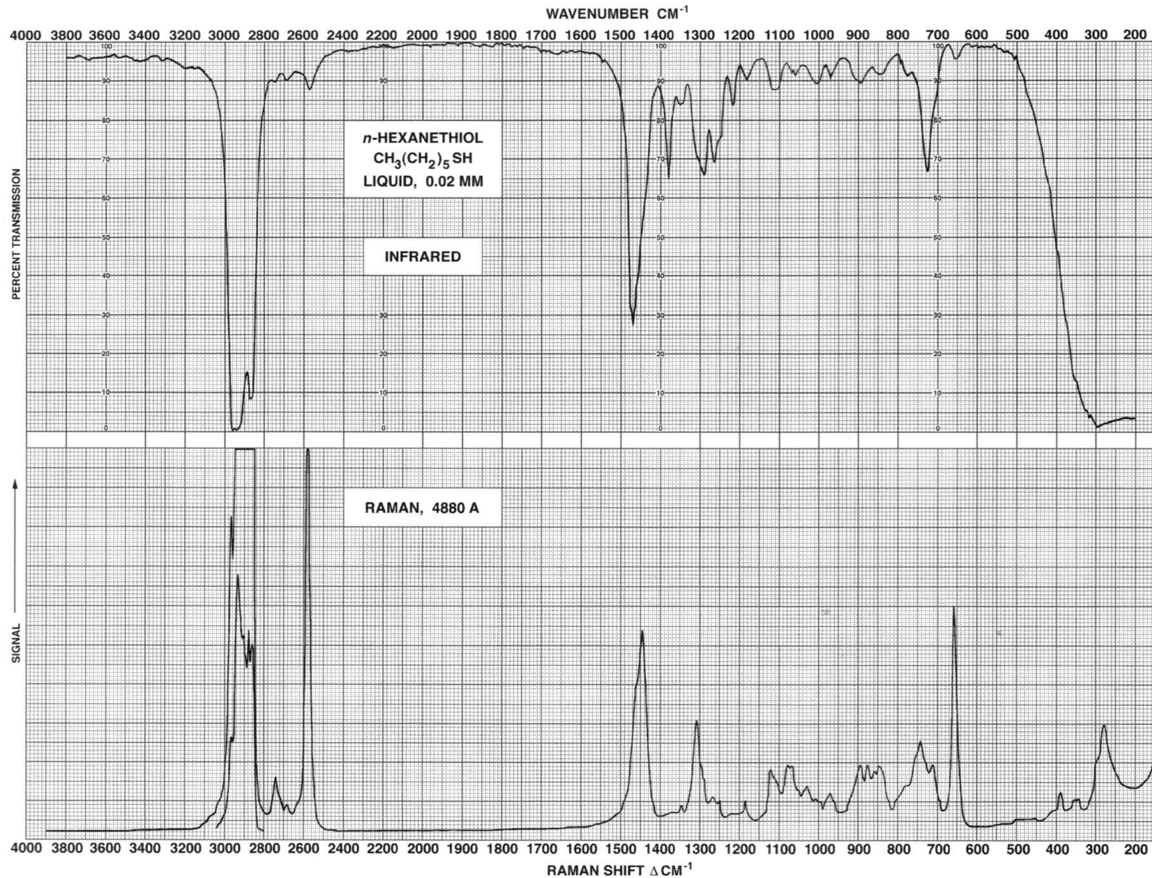

Fig. 12.8 Infrared and Raman spectra of 1-hexanethiol.

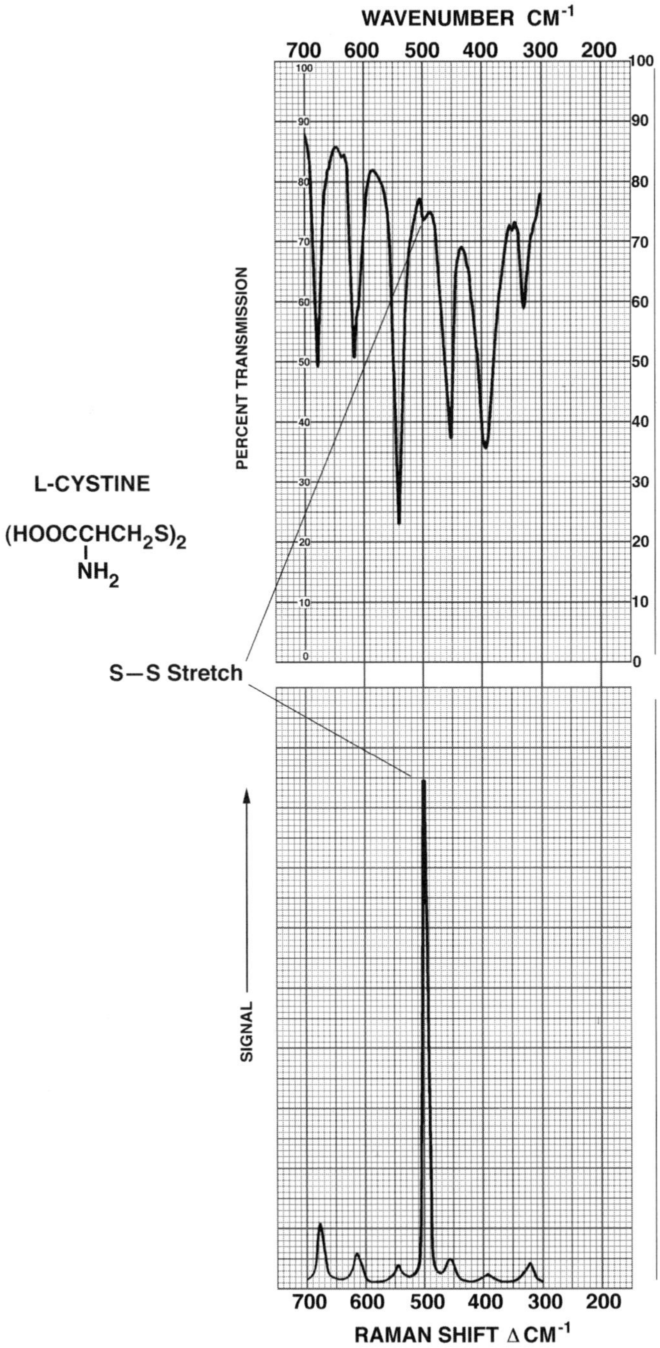

Fig. 12.9 Infrared and Raman spectra of L-cystine.

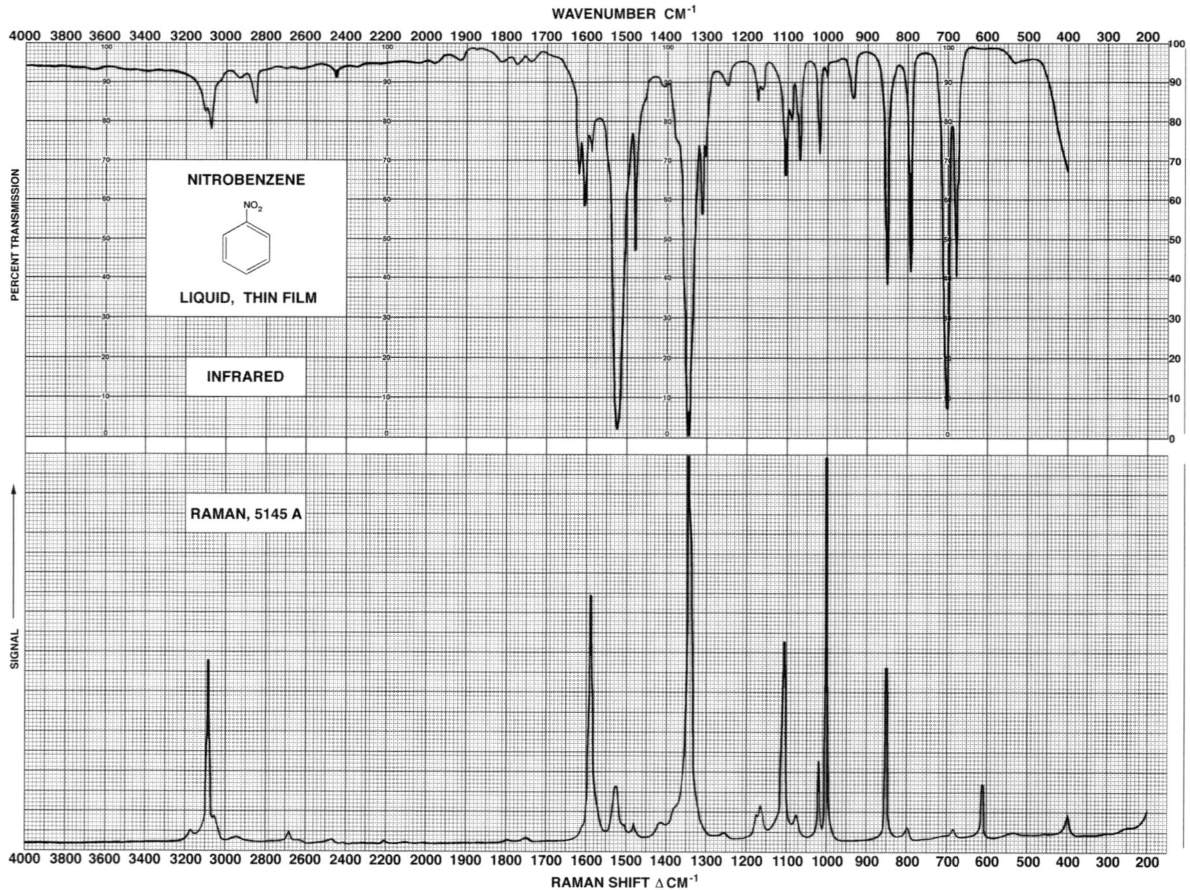

Fig. 12.10 Infrared and Raman spectra of nitrobenzene.

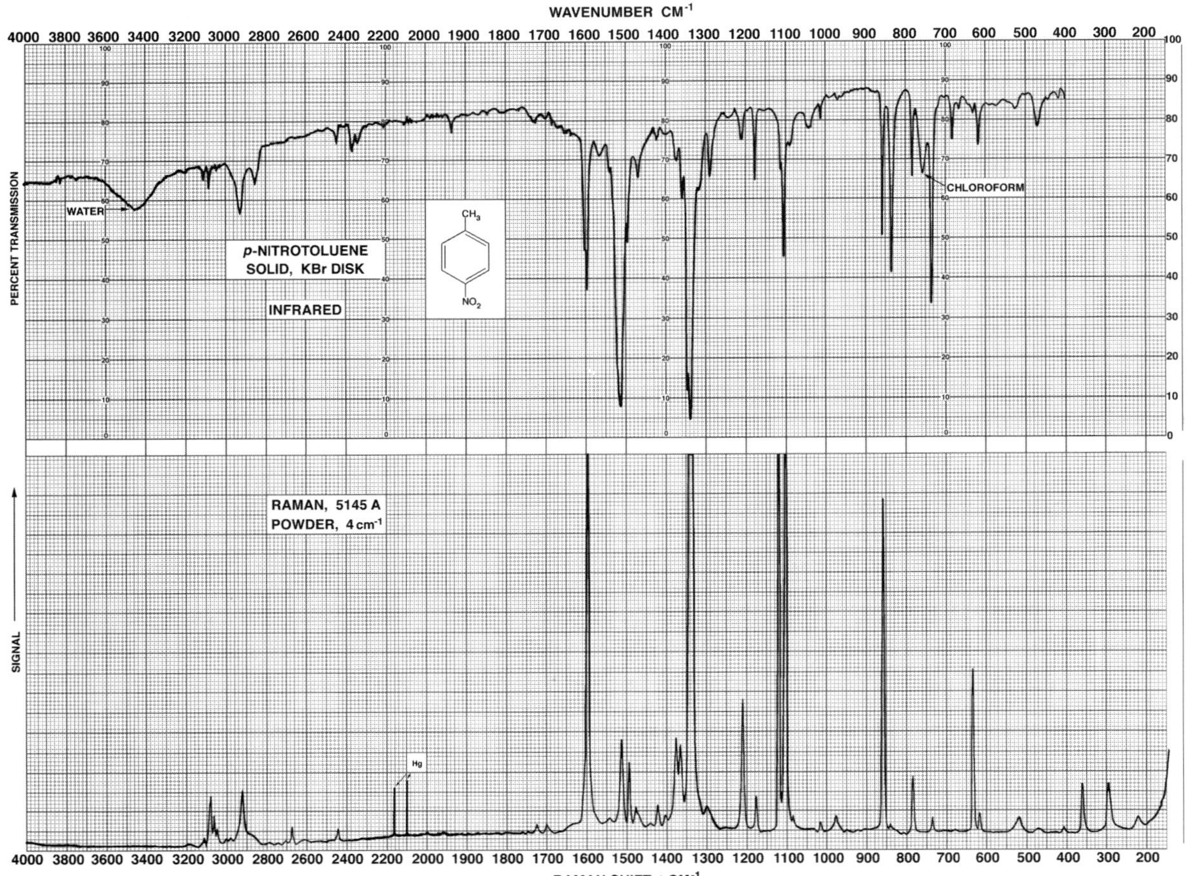

Fig. 12.11 Infrared and Raman spectra of *p*-nitrotoluene.

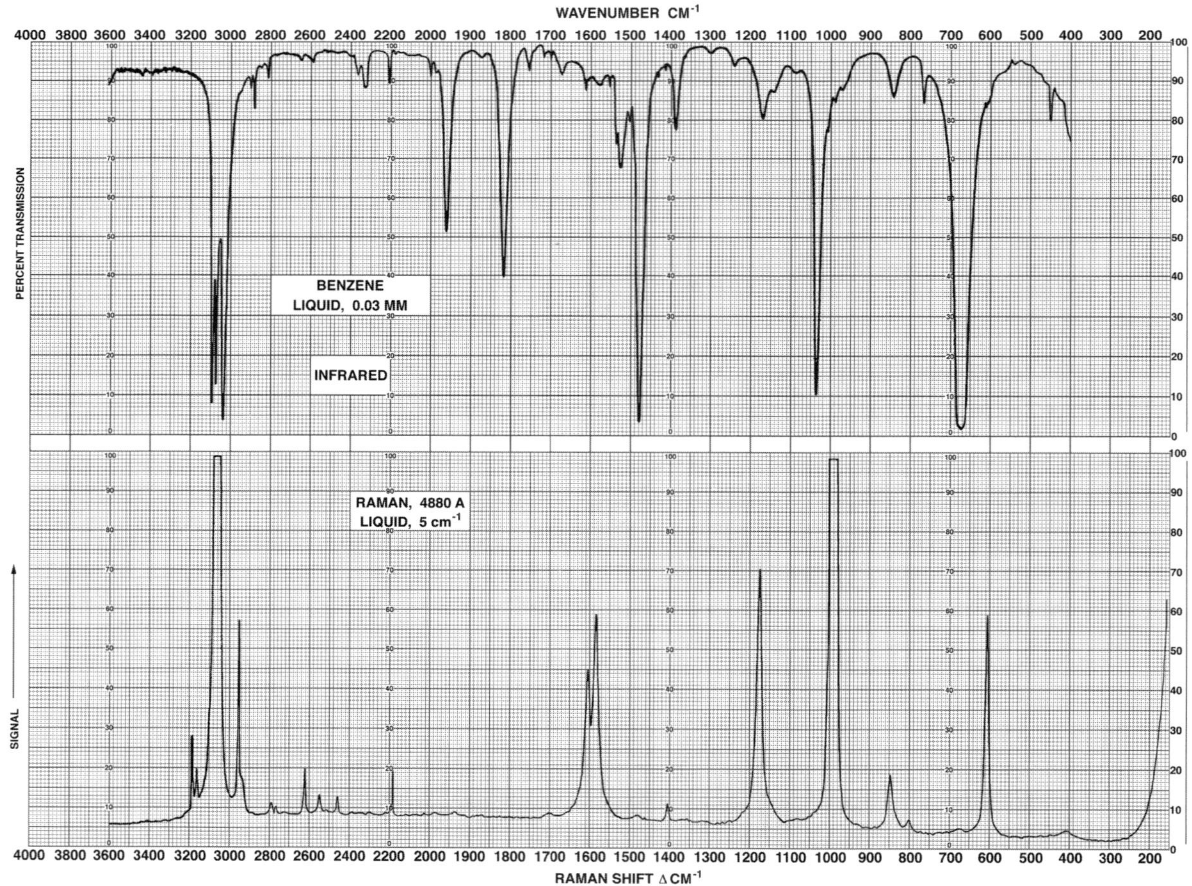

Fig. 12.12 Infrared and Raman spectra of benzene.

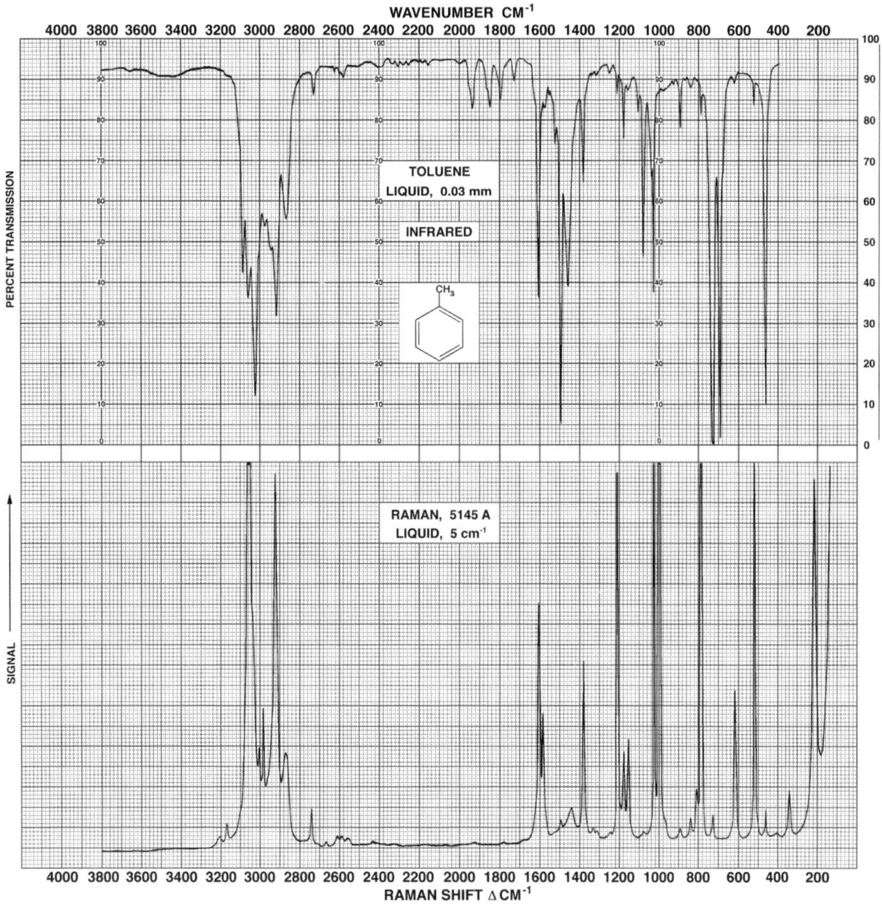

Fig. 12.13 Infrared and Raman spectra of toluene.

Alkyl-substituted rings, on the other hand, often clearly exhibit the ring vibrations, although a number of the modes may be vanishingly weak in one effect or the other. This is the case in toluene (Figure 12.13), where the analogs of classical benzene group frequency ring vibrations ν_{8a} (1605 cm^{-1}), ν_{8b} (1580 cm^{-1}), ν_{19a} (1494 cm^{-1}), ν_{19b} ($\sim$1460 cm^{-1}), ν_{12} (1001 cm^{-1}), ν_{6b} (618 cm^{-1}), and ν_4 (692 cm^{-1}) are all observed in one effect or the other but with only ν_{8a} occurring in both effects with significant intensity (see also earlier discussions in Chapter 5 for complete descriptions of these vibrations). Note that ν_{12} at 1010 ± 10 cm^{-1} is also easily recognized in the Raman spectra of *m*-xylene (Figure 12.14) and mesitylene (Figure 12.15) but is absent in the spectra of *o*- and *p*-xylene (Figures 12.16 and 12.17). It is also observed in the Raman spectra of nitrobenzene (Figure 12.10) unperturbed by the strongly electron-withdrawing substituent.

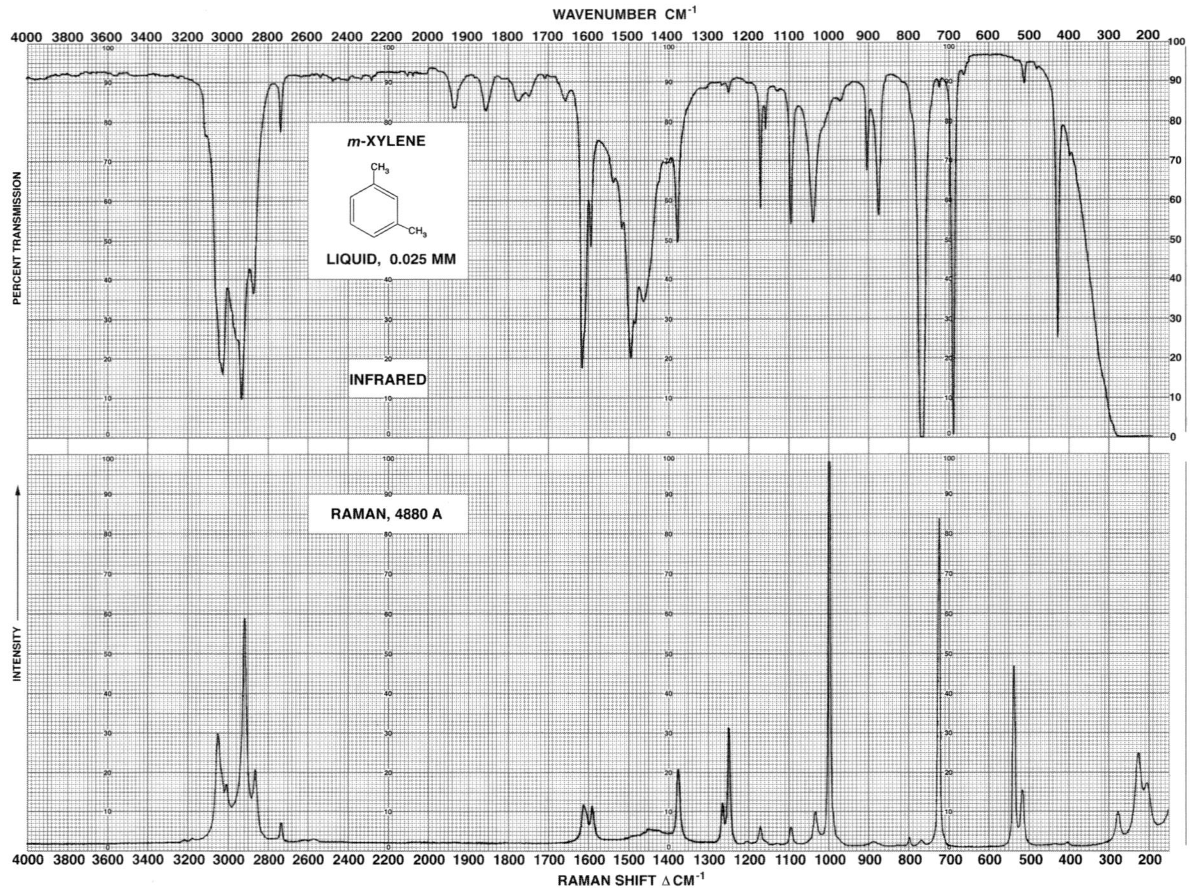

Fig. 12.14 Infrared and Raman spectra of *m*-xylene.

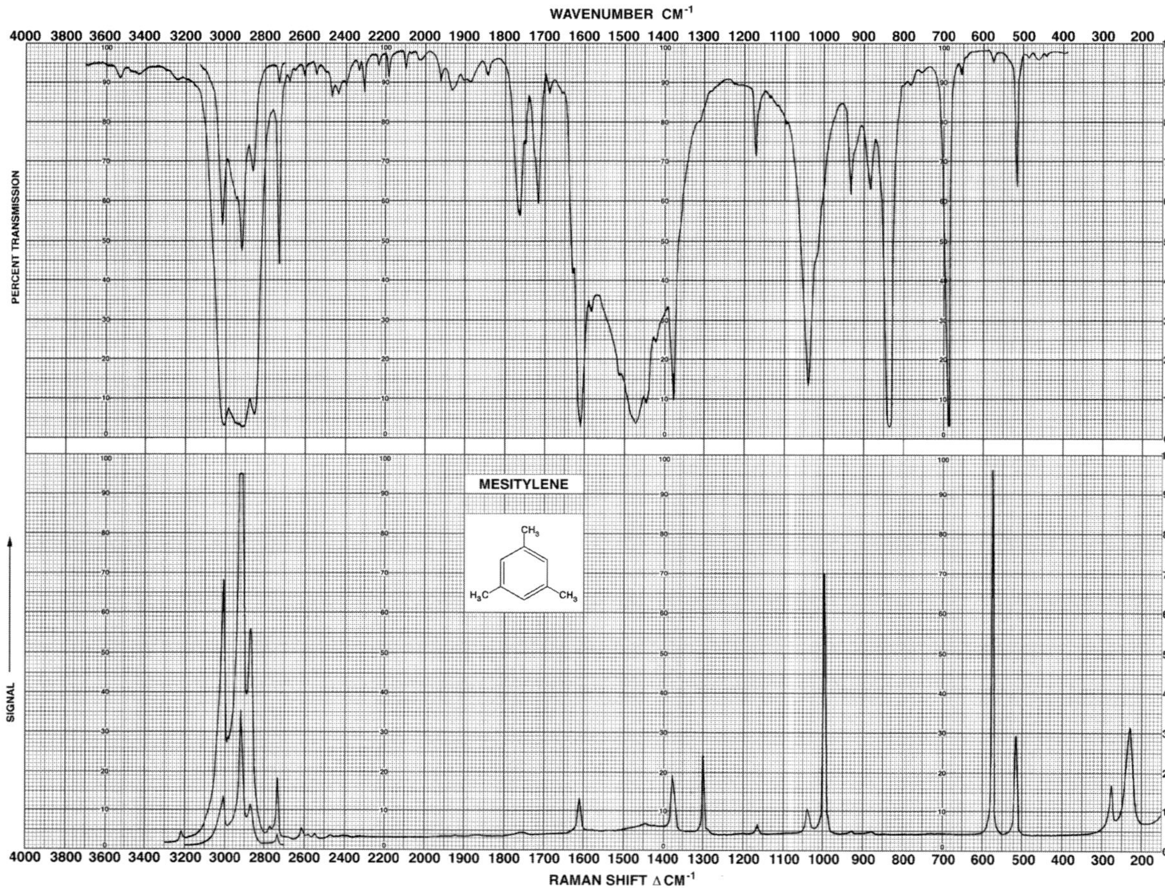

Fig. 12.15 Infrared and Raman spectra of mesitylene.

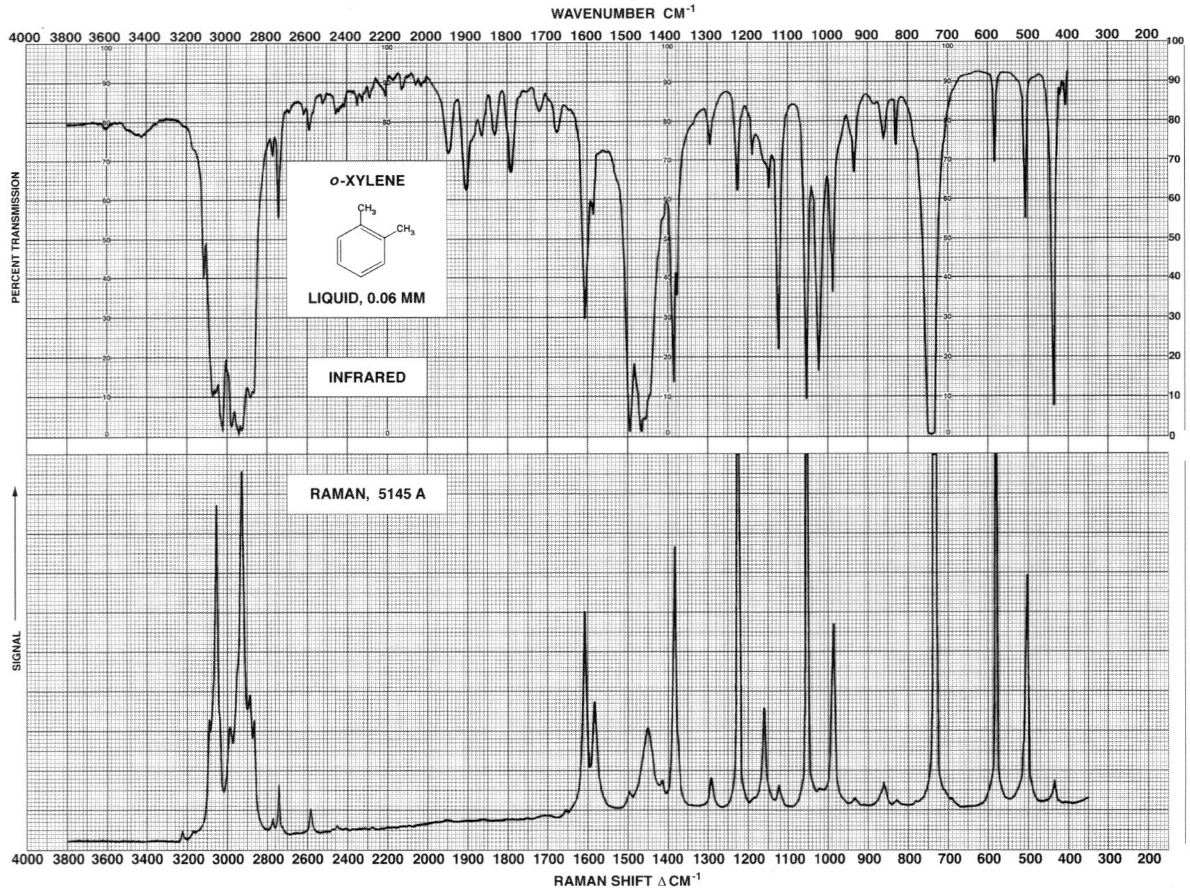

Fig. 12.16 Infrared and Raman spectra of *o*-xylene.

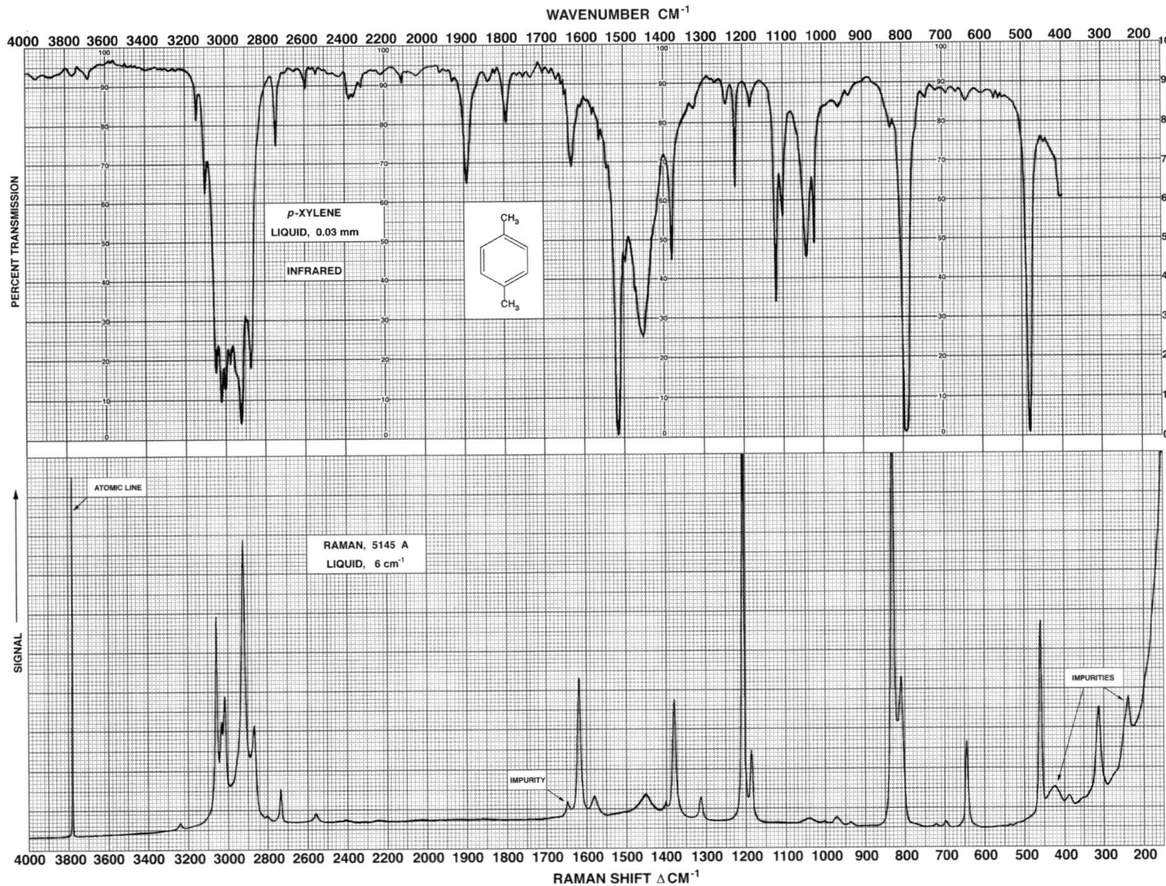

Fig. 12.17 Infrared and Raman spectra of *p*-xylene.

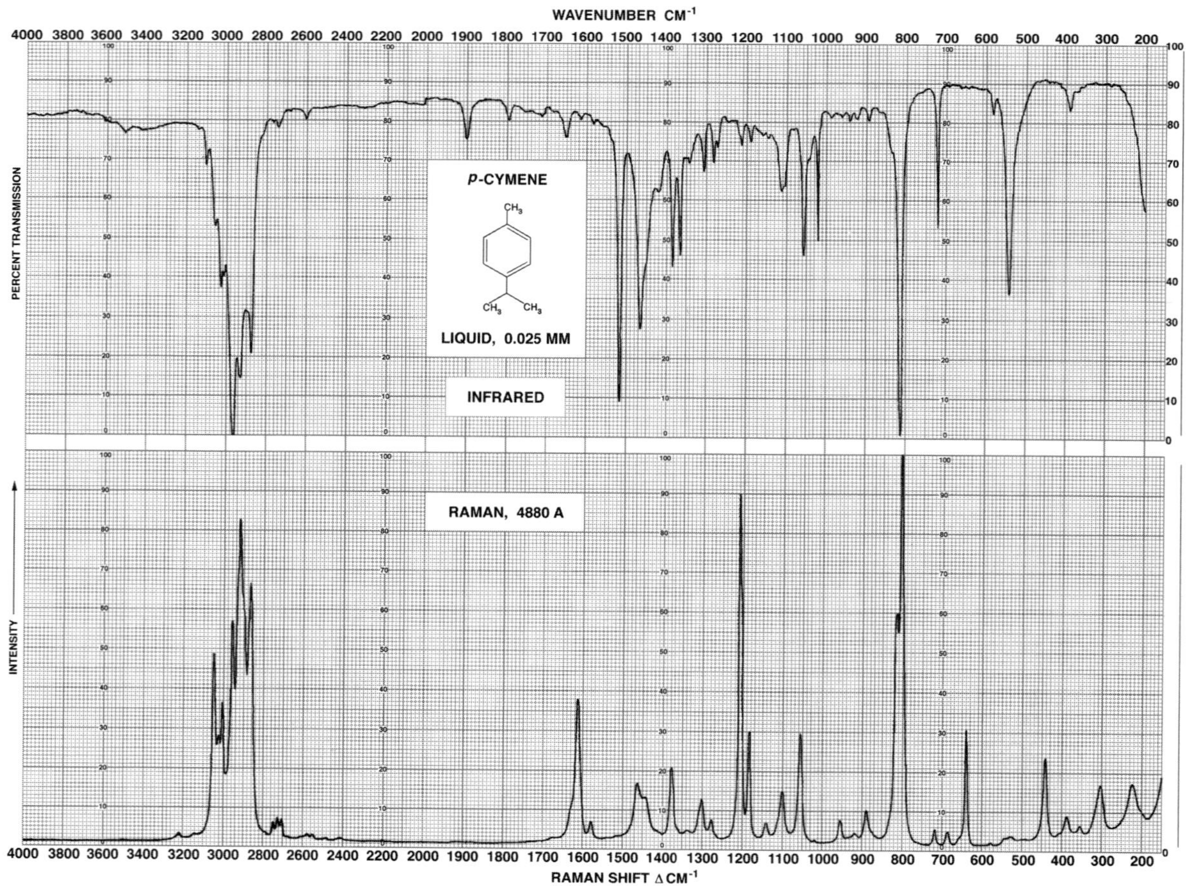

Fig. 12.18 Infrared and Raman spectra of *p*-cymene.

Note that while ν_8 (1600–1580 cm^{-1} doublet) is IR forbidden in symmetrically substituted para derivatives (such as *p*-xylene; Figure 12.17), it is also not observed in the IR spectra of those systems which possess pseudo centers of symmetry such as *p*-cymene (Figure 12.18). This ring stretch is allowed and observed, however, in unsymmetrically substituted para systems such as *p*-nitrotoluene (Figure 12.11).

In the Raman spectrum of the heterocyclic molecule pyridine (Figure 12.19), ν_{12} occurs close to 1030 cm^{-1} hardly perturbed from its location in the carbocyclic series at 1010 ± 10 cm^{-1}. Note also that ν_1, the Raman-active symmetric ring breathing vibration of benzene, also is observed in pyridine very close to its carbocyclic location of 992 cm^{-1}.

VIII. Nitrile substitutents

The nitrile functional group which possesses a triple bond and a high concentration of polarizable electrons is a highly efficient Raman scatterer and gives rise to intense bands which fall in an isolated region of the spectrum. It is an excellent group frequency in the Raman spectrum. The group possesses a highly polarized bond and permanent dipole moment which results in intense absorption in the IR. This normally vacant region of the vibrational spectrum allows the nitrile group (near 2255 cm^{-1}) to be easily observed in both effects (as listed in Table 12.1). A good example is the case of 1-hexanenitrile shown in Figure 12.20.

IX. Carbonyl groups

The substitution of conjugating carbonyl substituents on the ring does not shift ν_{12} (Raman 1010 ± 10 cm^{-1}) significantly, as shown by benzaldehyde (Figure 12.21). The single C—H group attached to the polar carbonyl is considerably more intense than the five hydrogens attached to the ring in the IR with the reverse occurring in the Raman. Conjugation of the carbonyl results, however, in an intensity jump for it in both the Raman and IR. The carbonyl stretch is the most intense band in the IR by far while it is the third most intense band in the Raman (also see Table 12.1).

The survey will now review the spectra of a dozen carbonyl-containing compounds.

A. Benzaldehyde (Figure 12.21)

As noted, the Raman and IR both exhibit strong bands in the carbonyl stretch region of benzaldehyde.

B. Hexanal (Figure 12.22)

Next consider the spectra of an aliphatic aldehyde, hexanal (Figure 12.22), in comparison to the aromatic aldehyde benzaldehyde. In the case of the aliphatic aldehyde note that in the IR spectrum, while the carbonyl intensity remains more intense than the aldehyde C—H stretch (one leg of the doublet is resolved), the aliphatic C—H stretch (11 C—H groups) is more intense than the single carbonyl oscillator. In the Raman spectrum, however, the unconjugated carbonyl intensity has clearly dropped somewhat relative to the aldehyde C—H stretch (2740 cm^{-1})

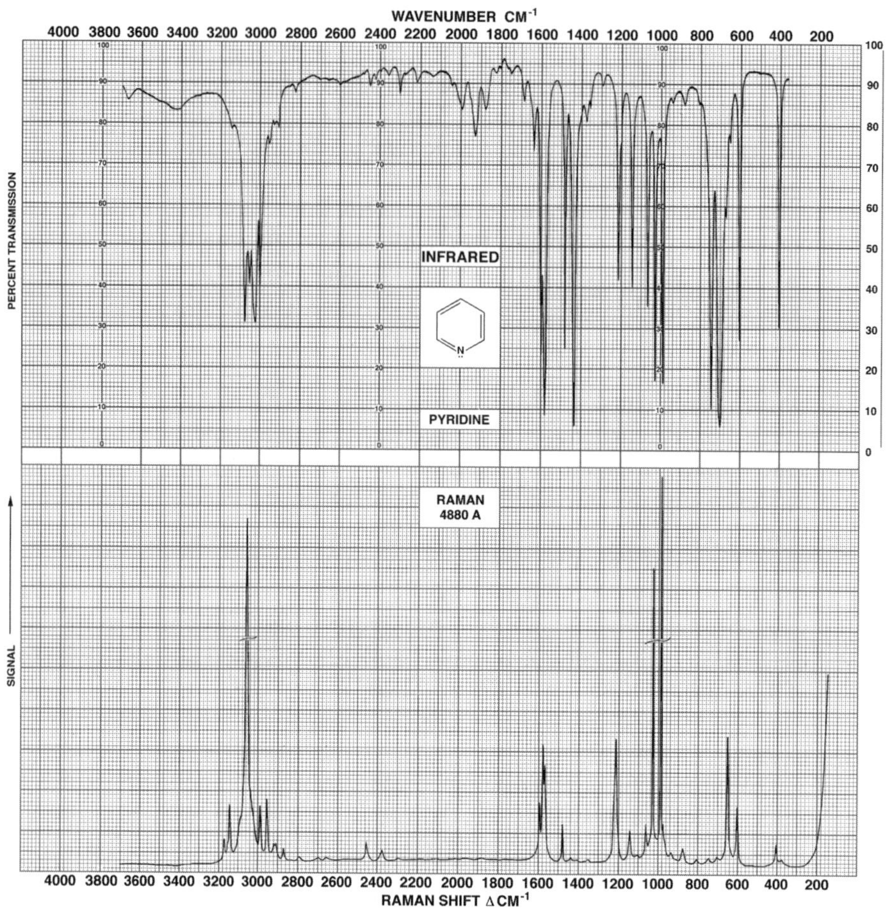

Fig. 12.19 Infrared and Raman spectra of pyridine.

and the aliphatic C—H bands are much more intense. In the Raman spectrum the carbonyl remains a band of reasonable intensity, but it is certainly not one of dominant strength in this aliphatic compound compared to its intensity in the aromatic series.

C. 3-Heptanone (Figure 12.23) and acetone (2-propanone) (Figure 12.24)

It is interesting to compare C_7 and C_3 aliphatic ketones. It might be expected that the carbonyls in these two cases would possess similar extinction coefficients. We might expect, therefore, the carbonyl band of acetone to be more intense relative to the region of the 6 C—H stretching modes as compared to the carbonyl stretch in 3-heptanone when that band is compared to the region of the 14 C—H stretches and that is exactly what is observed. In the Raman spectrum the carbonyl is clearly less intense than the C—H stretching regions. In acetone the carbonyl is the most intense band in the IR spectrum. Also, it should be noted that the

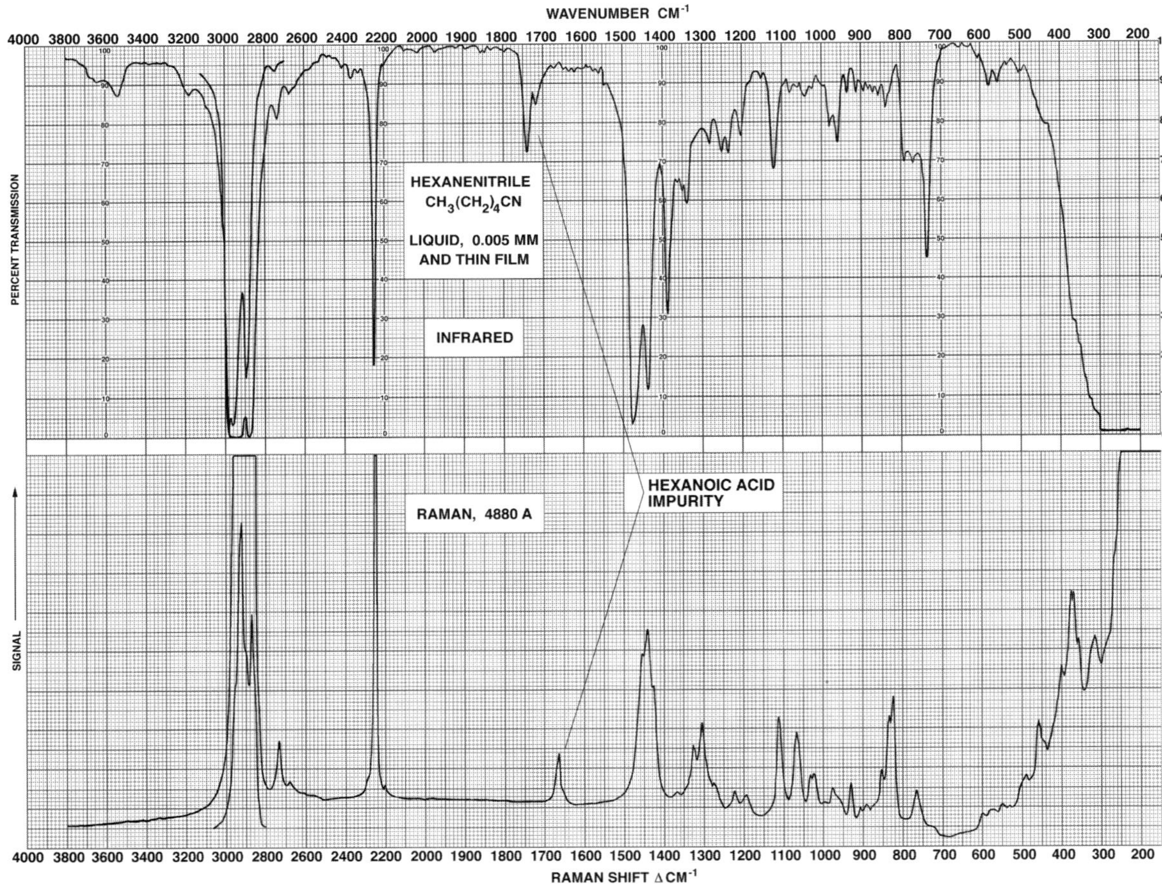

Fig. 12.20 Infrared and Raman spectra of 1-hexanenitrile.

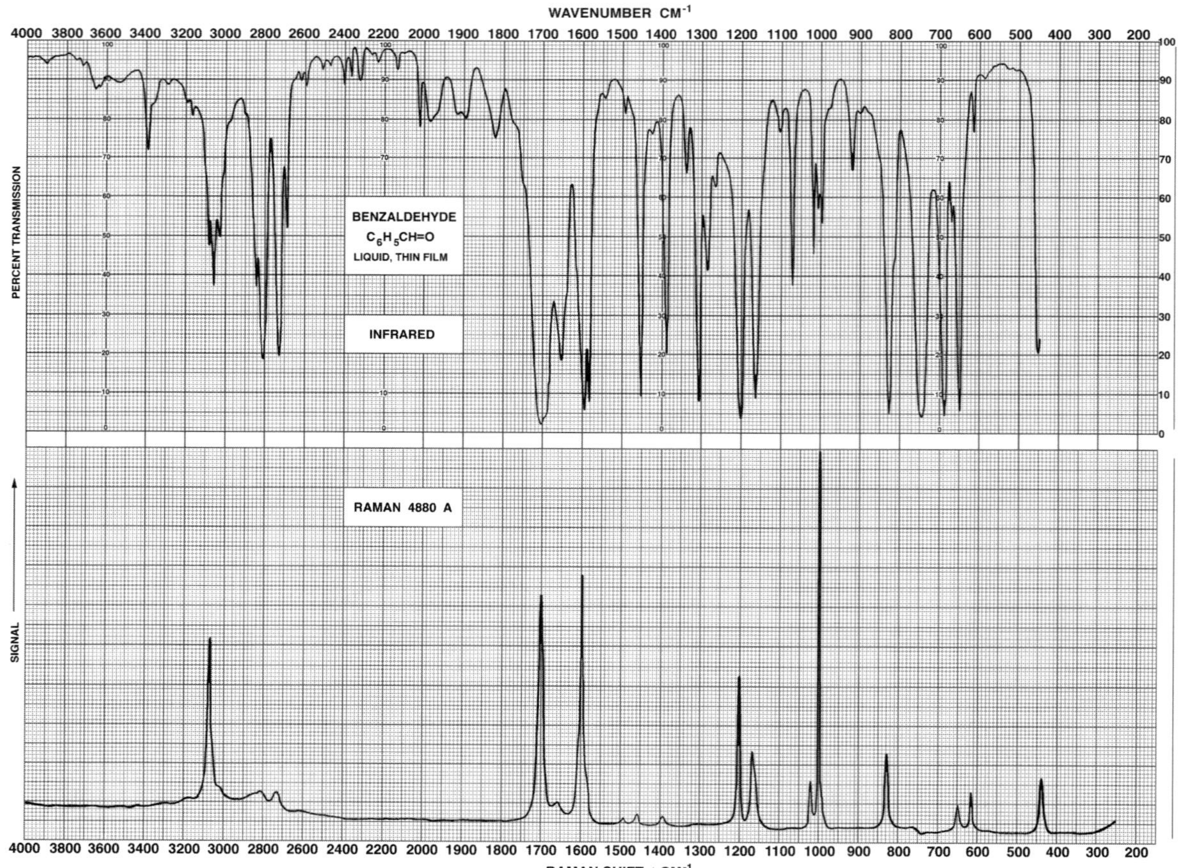

Fig. 12.21 Infrared and Raman spectra of benzaldehyde.

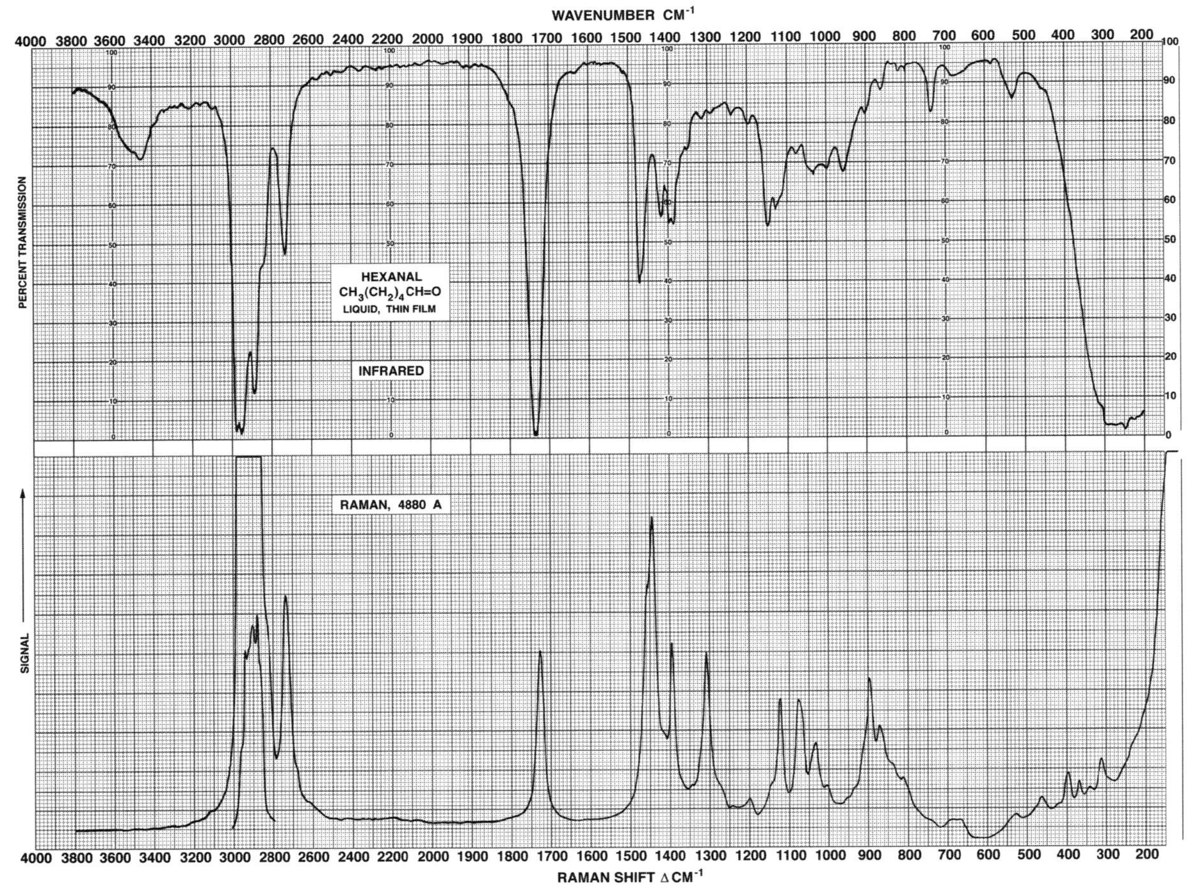

Fig. 12.22 Infrared and Raman spectra of hexanal.

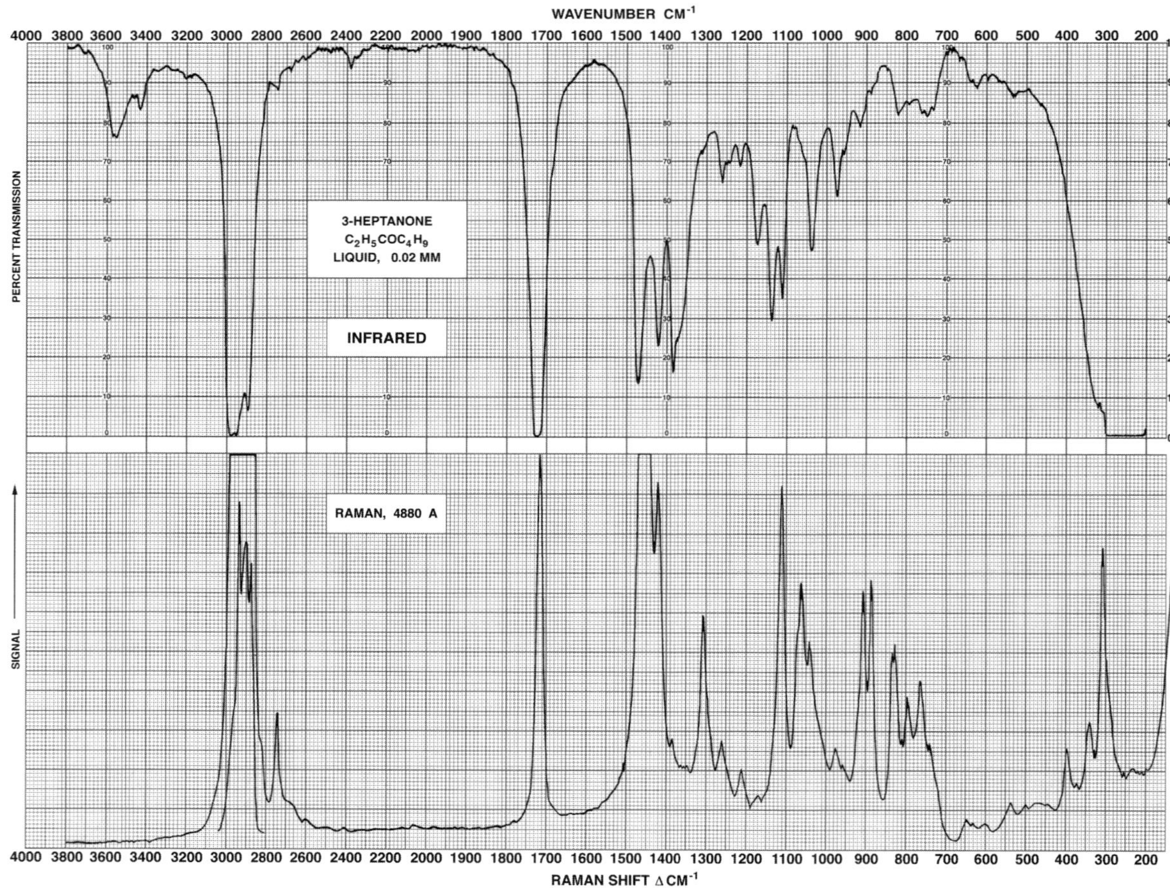

Fig. 12.23 Infrared and Raman spectra of 3-heptanone.

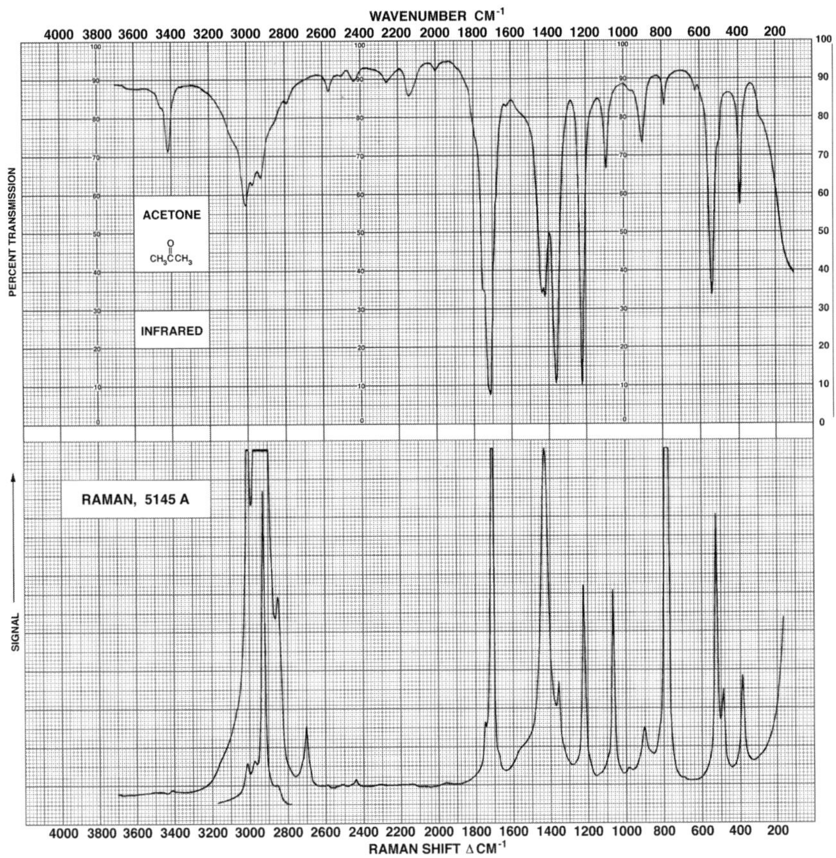

Fig. 12.24 Infrared and Raman spectra of acetone.

relative intensity of the overtone of the carbonyl (which falls near 3400 cm^{-1}) relative to the C—H stretch is greater in the C_3 compound than in the C_7 one.

D. Hexanoyl chloride (Figure 12.25)

It is observed that as the electronegativity of the groups substituted on the carbonyl increases, the frequency of the carbonyl stretch also increases. The intensity of the carbonyl band in the IR also increases rather significantly because the overall group has become very polar. As expected, the carbonyl group decreases in intensity in the Raman effect and becomes a weak- to medium-strength band and the same fate apparently occurs to the C—Cl stretch, which is assigned to strong bands in the IR near 680 and 570 cm^{-1} but now has weak counterparts in the Raman effect.

E. Hexyl acetate (Figure 12.26)

The intensity of the aliphatic ester carbonyl stretch is similar to that of the carbonyl of hexanal. This could have been predicted as the polarity

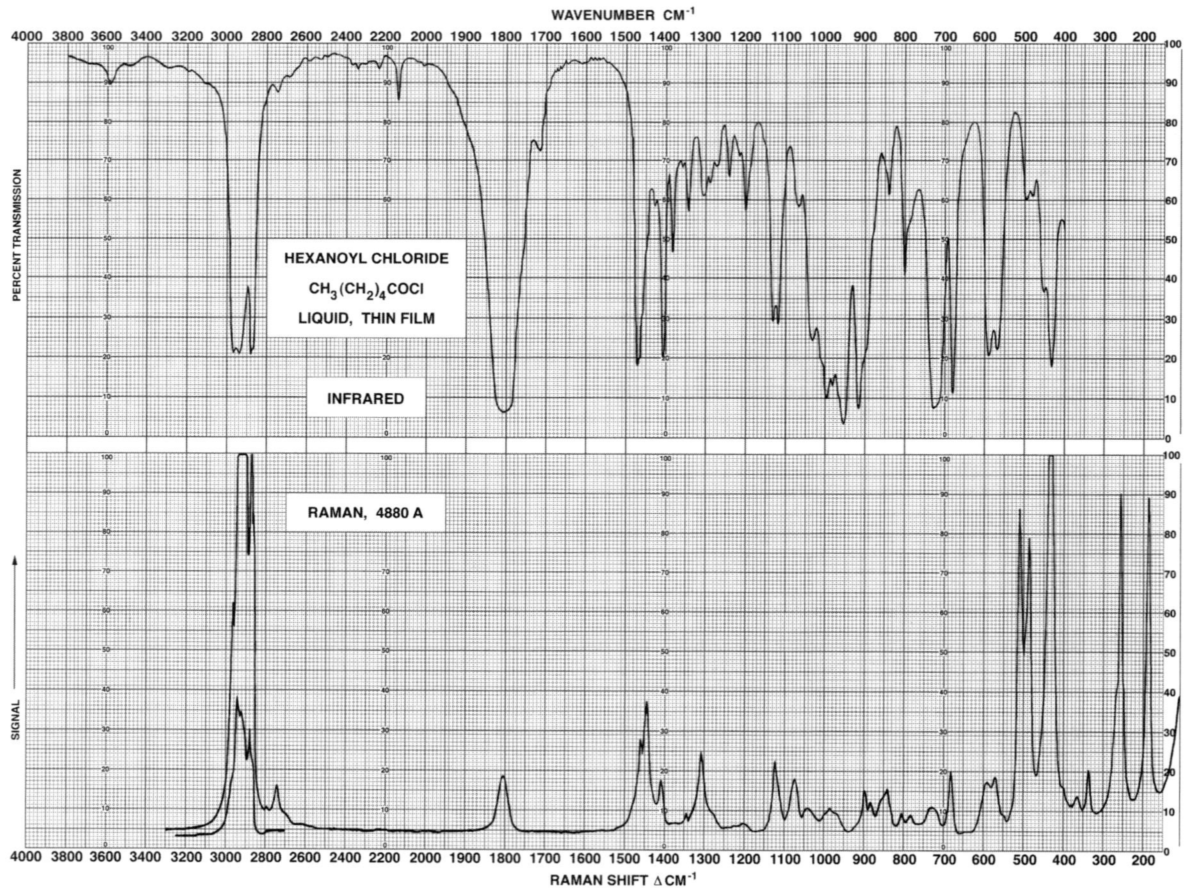

Fig. 12.25 Infrared and Raman spectra of hexanoyl chloride.

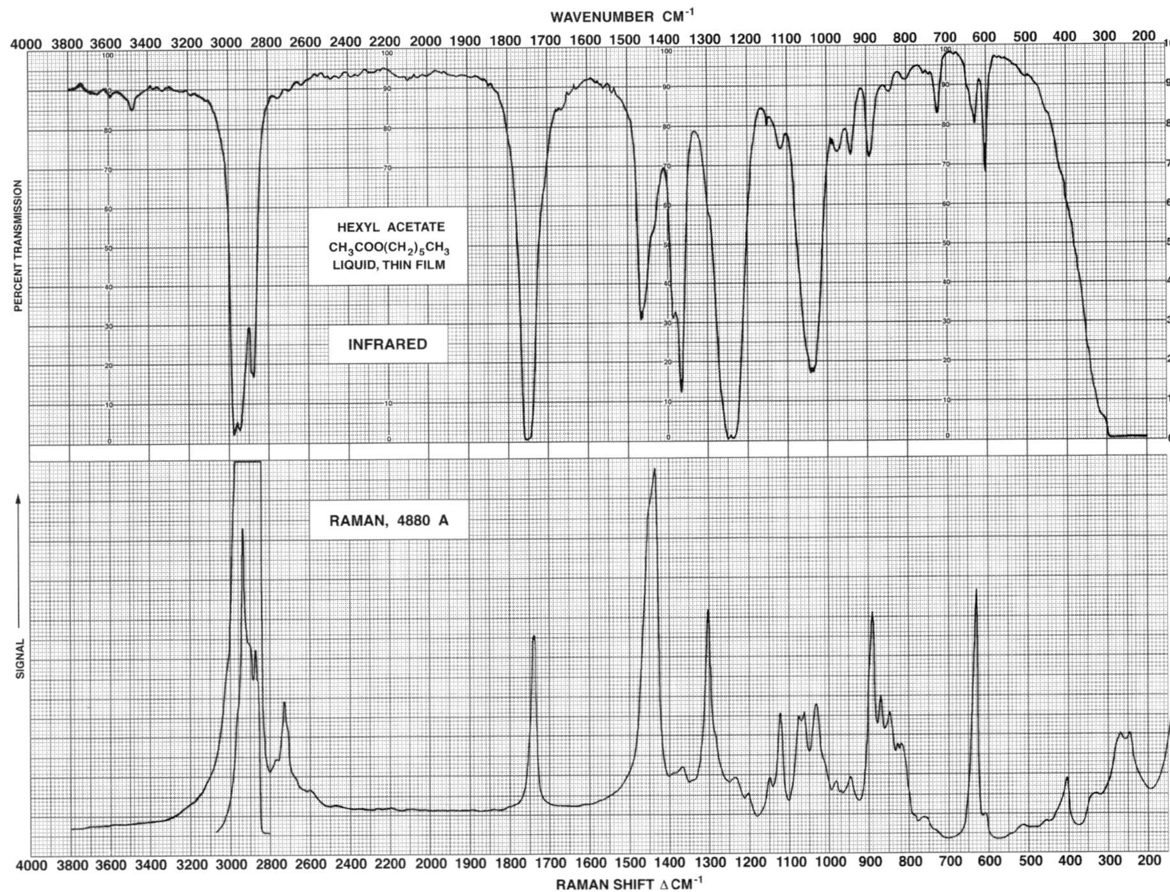

Fig. 12.26 Infrared and Raman spectra of 1-hexyl acetate.

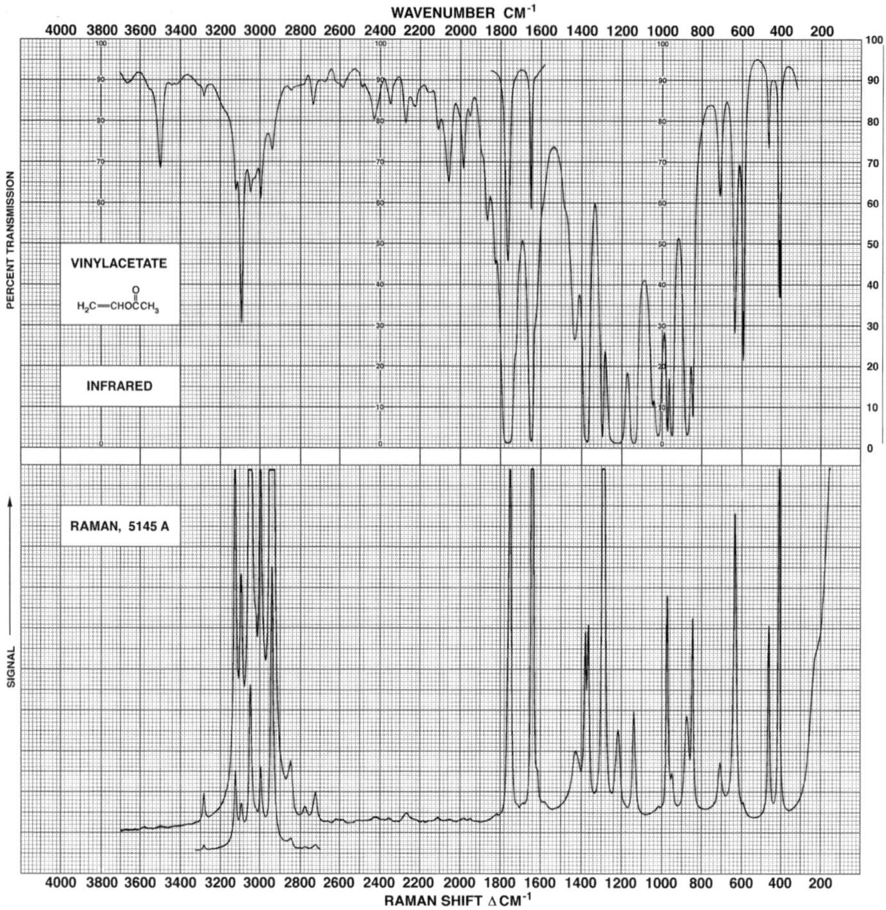

Fig. 12.27 Infrared and Raman spectra of vinyl acetate.

effects which govern the position of the carbonyl stretching frequency place aliphatic esters and aldehydes nearly in the same region.

F. Vinyl acetate (Figure 12.27)

In the case of vinyl acetate the carbonyl intensity in the Raman effect is difficult to relate to the aliphatic series, but the band is significantly weaker than the double-bond stretching frequency near 1655 cm^{-1} and the reverse of that found in the IR spectrum. The Raman band is in the medium to strong range.

G. Hexanamide (Figure 12.28)

This is a most impressive example of suppression of the carbonyl intensity in the Raman effect. The extremely polar primary amide group leaves the carbonyl with little or no intensity in the Raman spectrum and an enormously intense band in the IR. The carbonyl IR band often

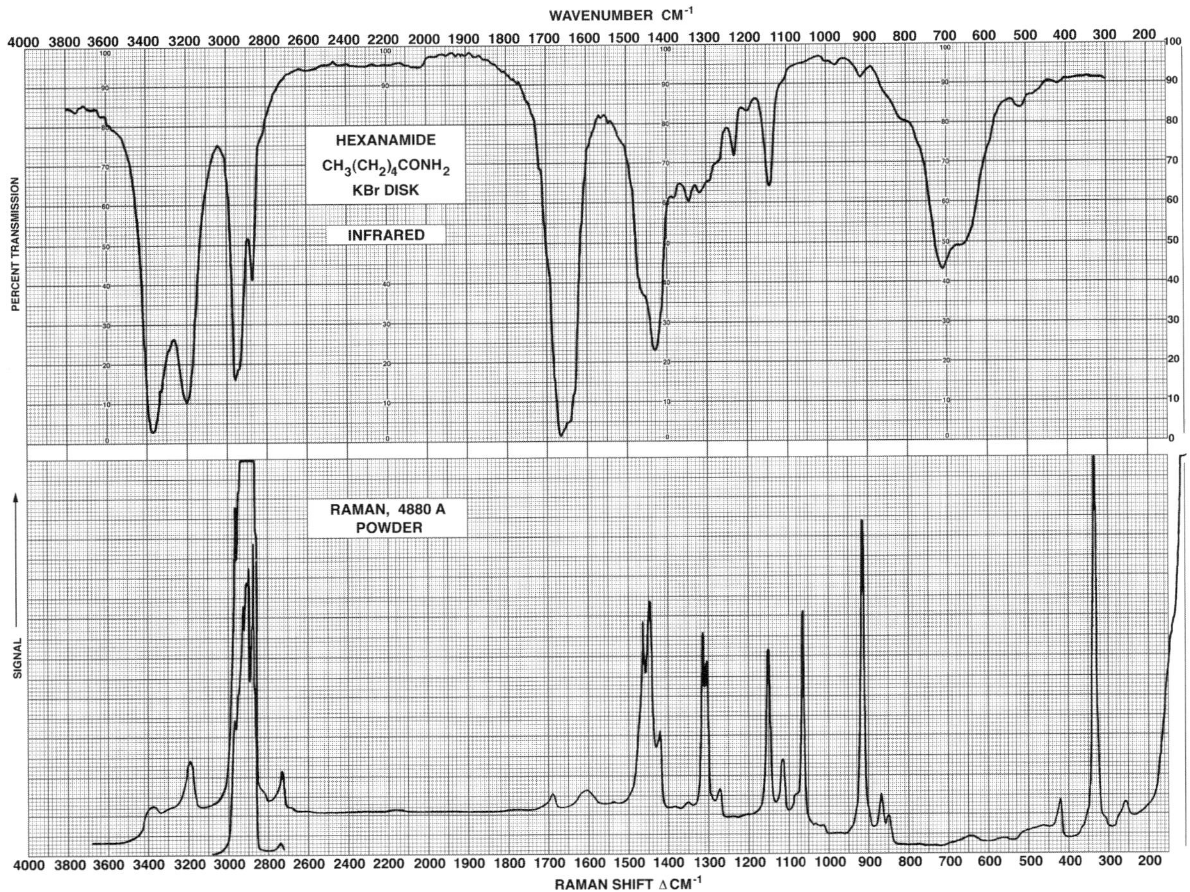

Fig. 12.28 Infrared and Raman spectra of hexanamide.

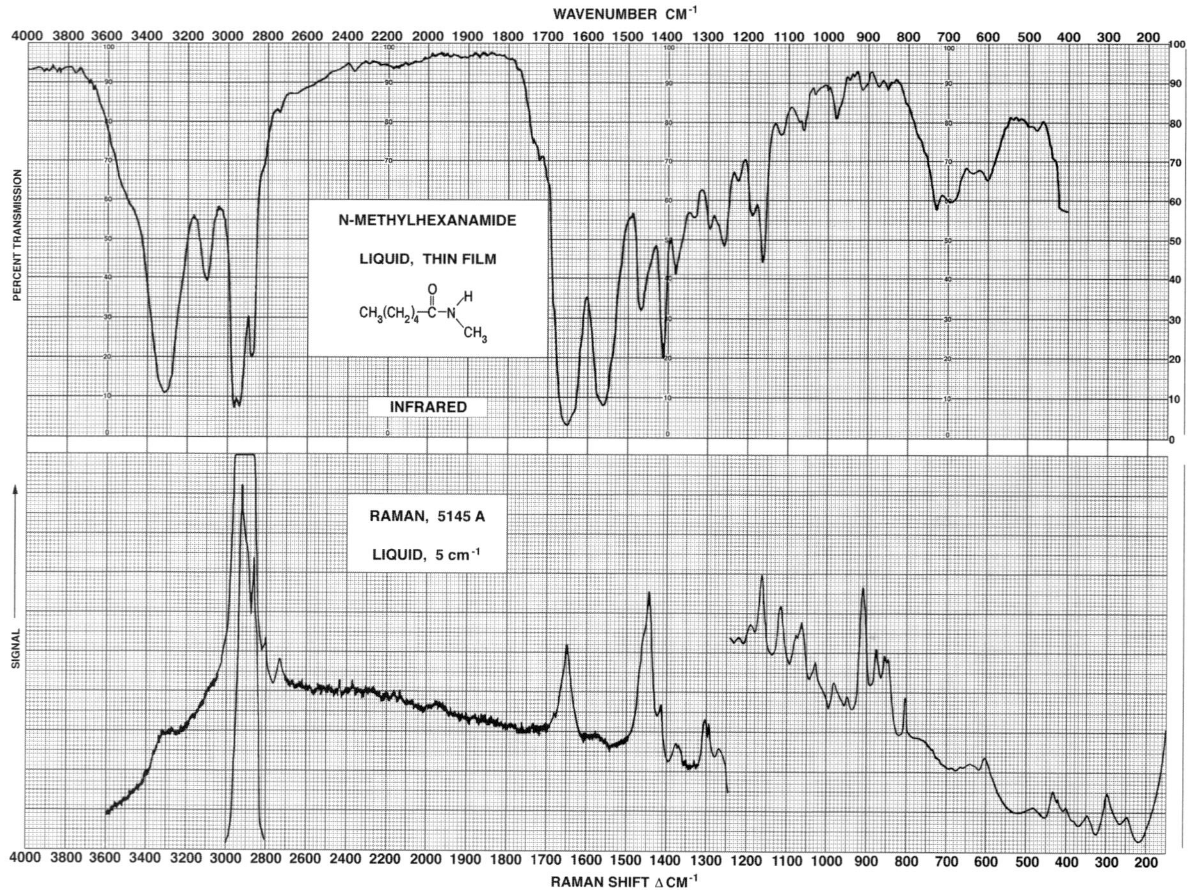

Fig. 12.29 Infrared and Raman spectra of *N*-methylhexanamide.

overlaps to such an extent with the NH_2 scissoring mode that the two band systems are unresolved (as is the case here). The two very weak bands in the Raman effect identify the band centers of the fundamentals. Note also the very weak scattering of the $N<H_2$ stretching modes in the Raman and the greatly enhanced $N<H_2$ stretch in the IR compared, for example, to that in 1-hexylamine (Figure 12.5).

H. *N*-Methyl hexanamide (Figure 12.29)

This is another example of polar influences on carbonyl intensity in the Raman spectrum, as shown above for hexanamide. Notice the influence of the nonpolar methyl group. The broad scattering background is indicative of the presence of a fluorescent impurity. Also note the vanishing intensity of both the N—H stretching band and strong 1560 cm^{-1} IR band in the Raman effect.

I. Hexanoic anhydride (Figure 12.30)

In the case of hexanoic anhydride the two carbonyls are coupled through the central oxygen atom and locked in a planar delocalized system. In open-chain anhydrides this positions the carbonyls in a nearly coparallel arrangement. Coupling splits the carbonyl band system into in-phase (the higher wavenumber mode) and out-of-phase (the lower wavenumber mode) vibrations. The splitting is usually $\sim$70 cm^{-1}. The upper member of the leg is more intense in the IR because the dipoles are oscillating in phase, and the same argument holds for the change of polarizability in the Raman spectrum. Note that the Raman doublet is weak to medium because the anhydride group has significant polarization.

J. Maleic anhydride (Figure 12.31)

In cyclic anhydrides the ring system can perturb the preferred parallel arrangement of the two carbonyls in open-chain systems, and in extreme cases such as in the five—membered ring the two oscillators can be drawn nearly back to back so that the dipoles are opposing each other. In this case the in-phase mode would be expected to be quite weak in the IR whereas the out-of-phase mode should be expected to be strong, which is what is observed. In the Raman effect the in-phase mode would be expected to be stronger then the out-of-phase one, which should be weak, and this is again what is observed. Of interest is the observation that in this conjugated and doubled carbonyl system the in-phase stretch is now the strongest band in the entire Raman spectrum (the only case where this occurs in any of the examples presented here). Note that the C=C double bond near 1590 cm^{-1} has very weak intensity in the IR while it is a medium band in the Raman effect.

K. Hexanoic acid (Figure 12.32)

This system exists as a centrosymmetric cyclic H-bonded dimer in neat samples. The carbonyls are coupled through the planar ring system and split to give two bands with a $\Delta\tilde{\nu}$ of approximately 50–60 cm^{-1}. One is IR active and the other Raman active (rule of mutual exclusion). The

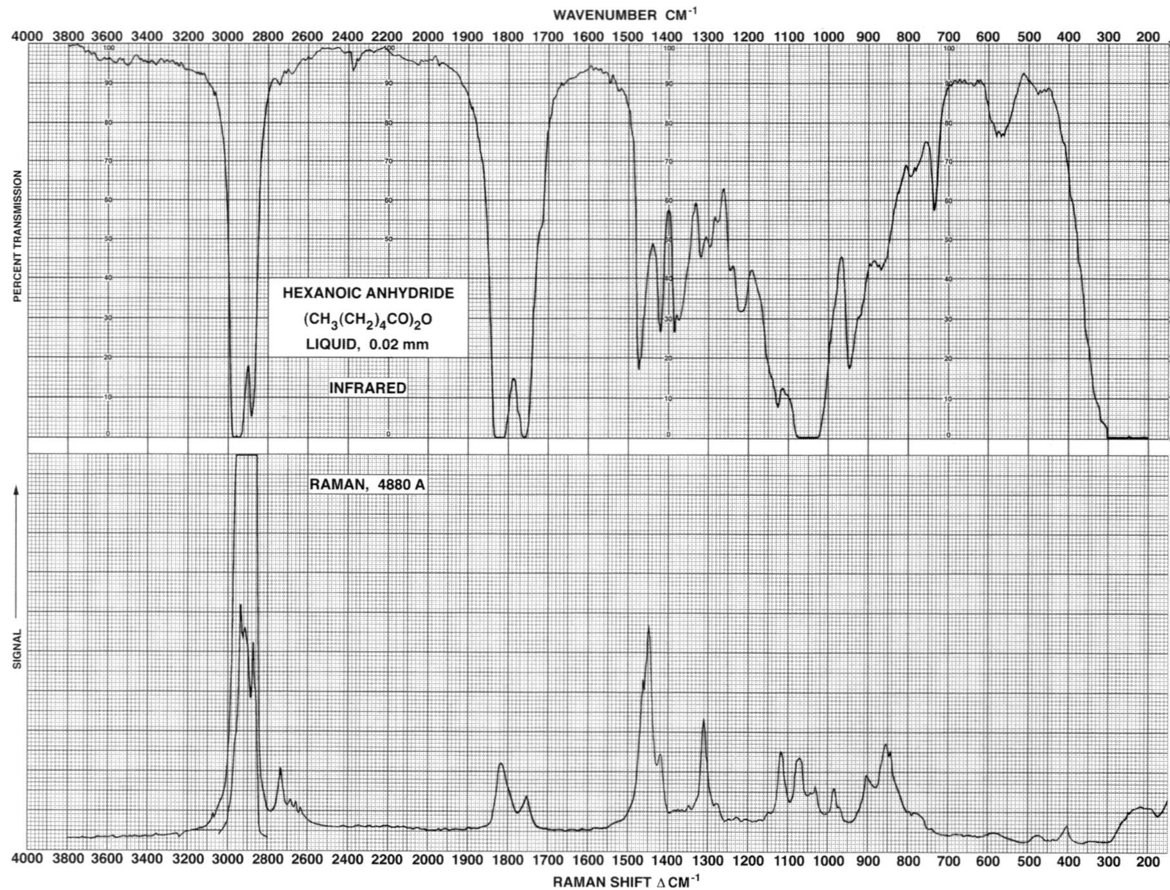

Fig. 12.30 Infrared and Raman spectra of hexanoic anhydride.

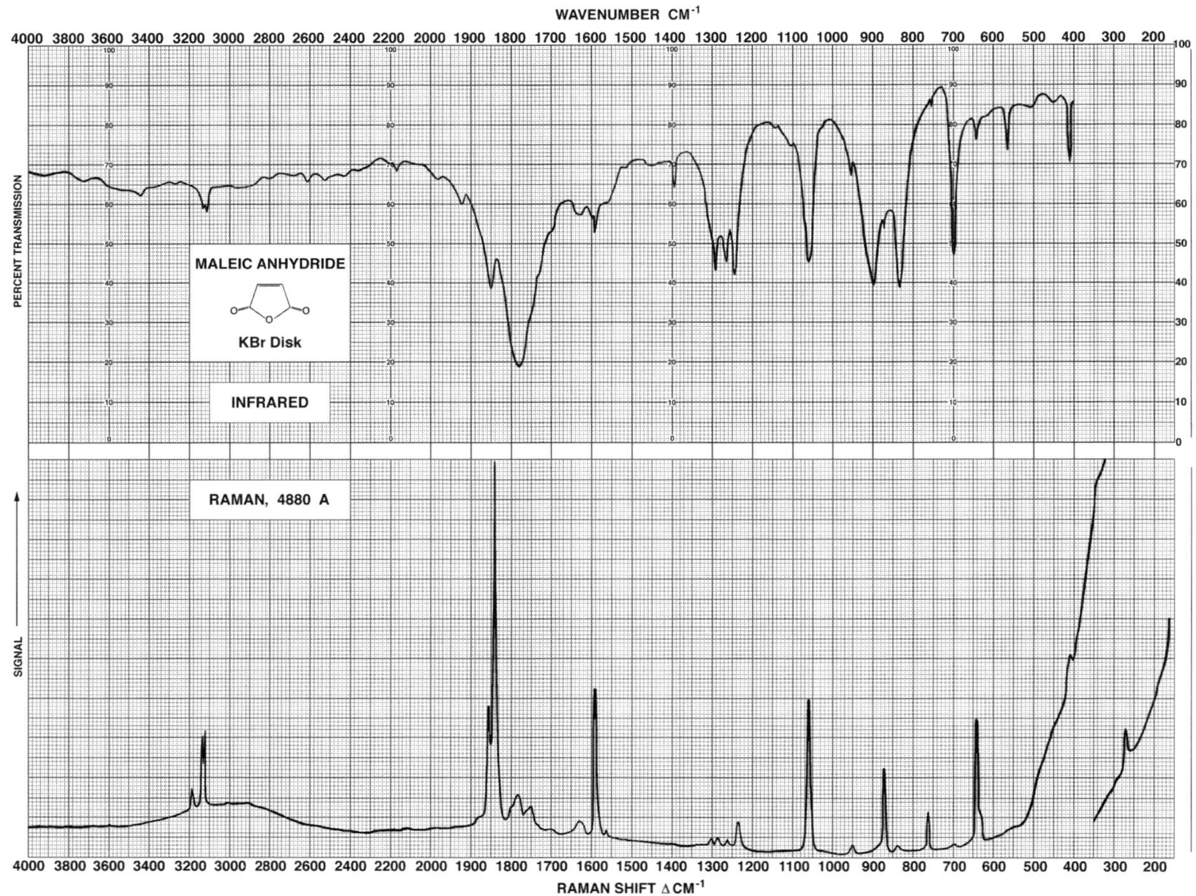

Fig. 12.31 Infrared and Raman spectra of maleic anhydride.

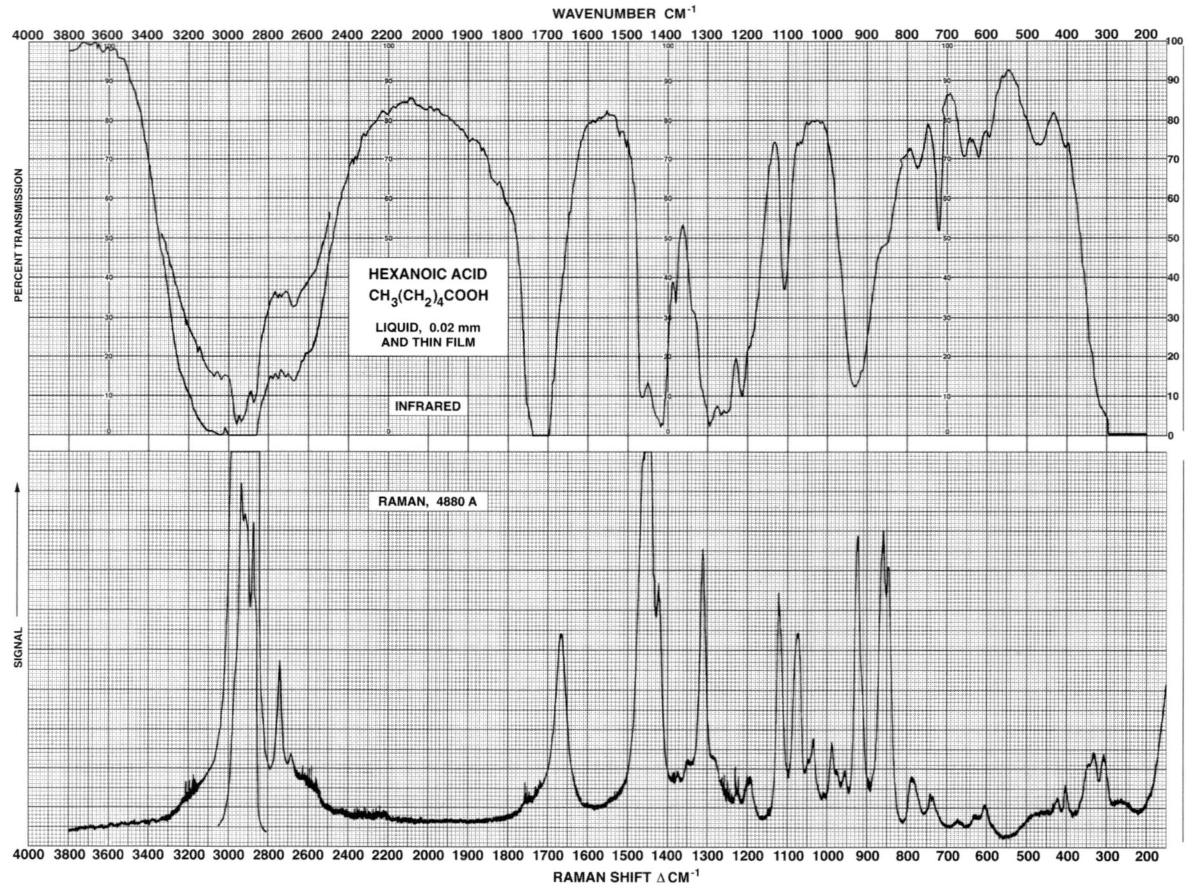

Fig. 12.32 Infrared and Raman spectra of hexanoic acid.

out-of-phase IR-active band occurs near 1715 cm^{-1} while the in-phase Raman-active mode is found near 1660 cm^{-1}. The IR band is broad and very intense while the Raman band is of weak to medium intensity.

X. Next, three C=C systems are examined (see Table 12.1).

A. The vinyl system, 1-hexene (Figure 12.33)

This olefin has a strong band in both the IR and Raman spectra near 1645 cm^{-1} assigned to the C=C stretching mode. It is one the few unconjugated simple olefins which possesses a strong band in the IR. The Raman band is the strongest band except for the C—H stretching region.

B. *trans*-2-Hexene (Figure 12.34)

The pseudo center of symmetry present in this system operates to greatly weaken the intensity of the olefinic C=C stretching mode in the IR, where it occurs as a very very weak band. In the Raman spectrum, however, it is again the strongest band beyond the C—H stretching region. The Raman data allow the assignment of the very very weak peak in the IR with some confidence. Note that the olefinic C—H out-of-plane bending mode close to 970 cm^{-1}, which is very strong in the IR, vanishes in the Raman effect. Particularly interesting is the observation of a sharp weak band near 1380 cm^{-1} assigned to the symmetrical methyl deformation in the Raman effect. As noted earlier, when methyl groups are bonded to sp^3-hybridized carbons, this bending frequency has vanishing intensity in the Raman effect. When the carbon is sp^2 or *sp* hybridized, weak sharp bands are observed in the Raman.

C. Styrene (Figure 12.35)

When a vinyl group is conjugated, as in styrene, the double bond gains considerable intensity in the IR. In styrene it gives rise to a medium–strong band (1633 cm^{-1}) while in the Raman spectrum it is the strongest band in the entire spectrum (the main competition comes from ν_{12} near 1000 cm^{-1}). Note that the three very strong IR-active C—H out-of-plane bending modes (vinyl 992 and 910 cm^{-1}, ring 772 cm^{-1}) are scarcely seen in the Raman effect. (The Raman band near 1000 cm^{-1} is ν_{12}, whereas the IR band at 992 cm^{-1} is the vinyl C—H out-of-plane bend.)

XI. Applications

This survey of the complementarity of IR and Raman spectra is completed with three examples where the comparison of both the IR and Raman spectra can lead to a deeper understanding of the nature of the sample.

A. Because the saturated carbon skeleton of organic compounds scatters Raman radiation more effectively than it absorbs IR radiation, the fingerprint region in the Raman spectrum has the potential of being somewhat more diagnostic for the type of skeleton present in a substance. For example, it is easy to differentiate between an open-chain monoterpene like ocimenol and the bicyclic monoterpene Δ^3-carene. In the open-chain case many rotational conformers are present that have similar but not

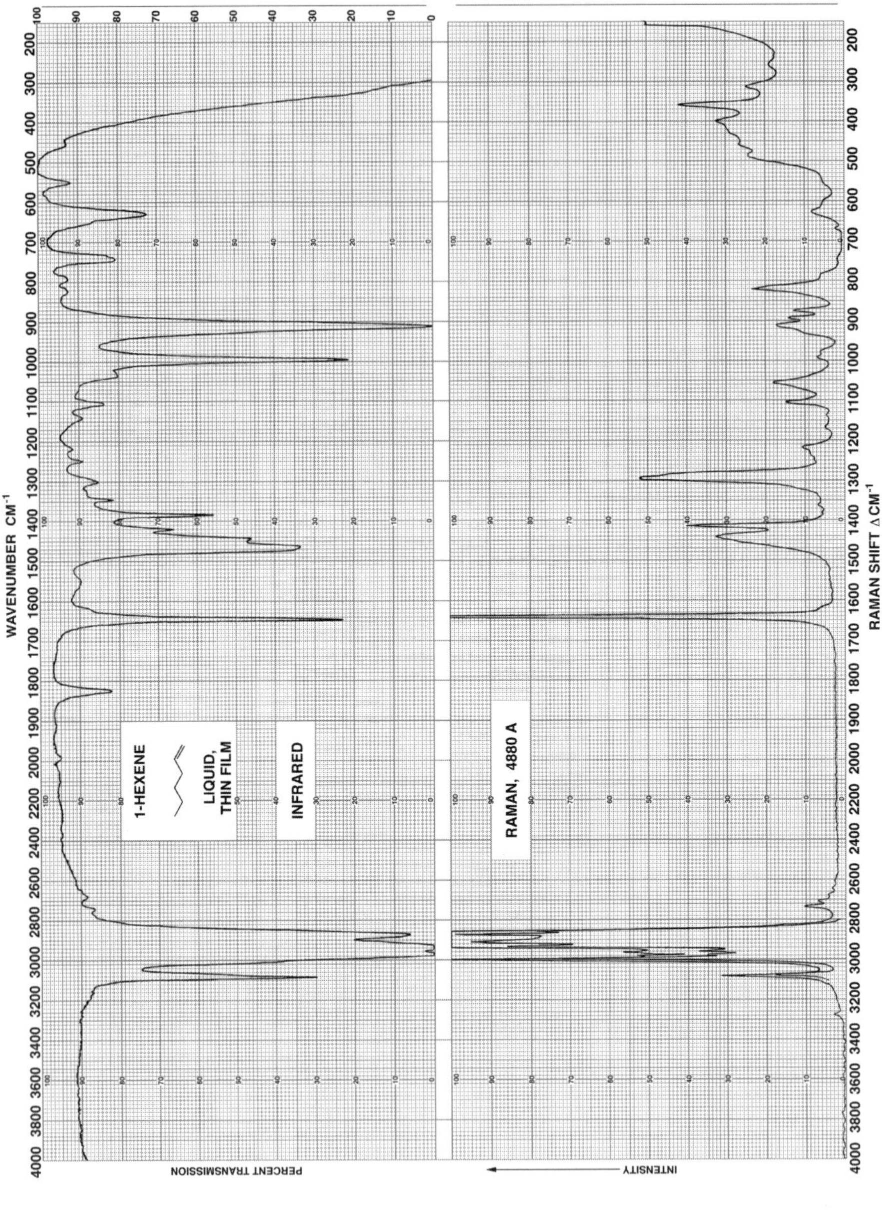

Fig. 12.33 Infrared and Raman spectra of 1-hexene.

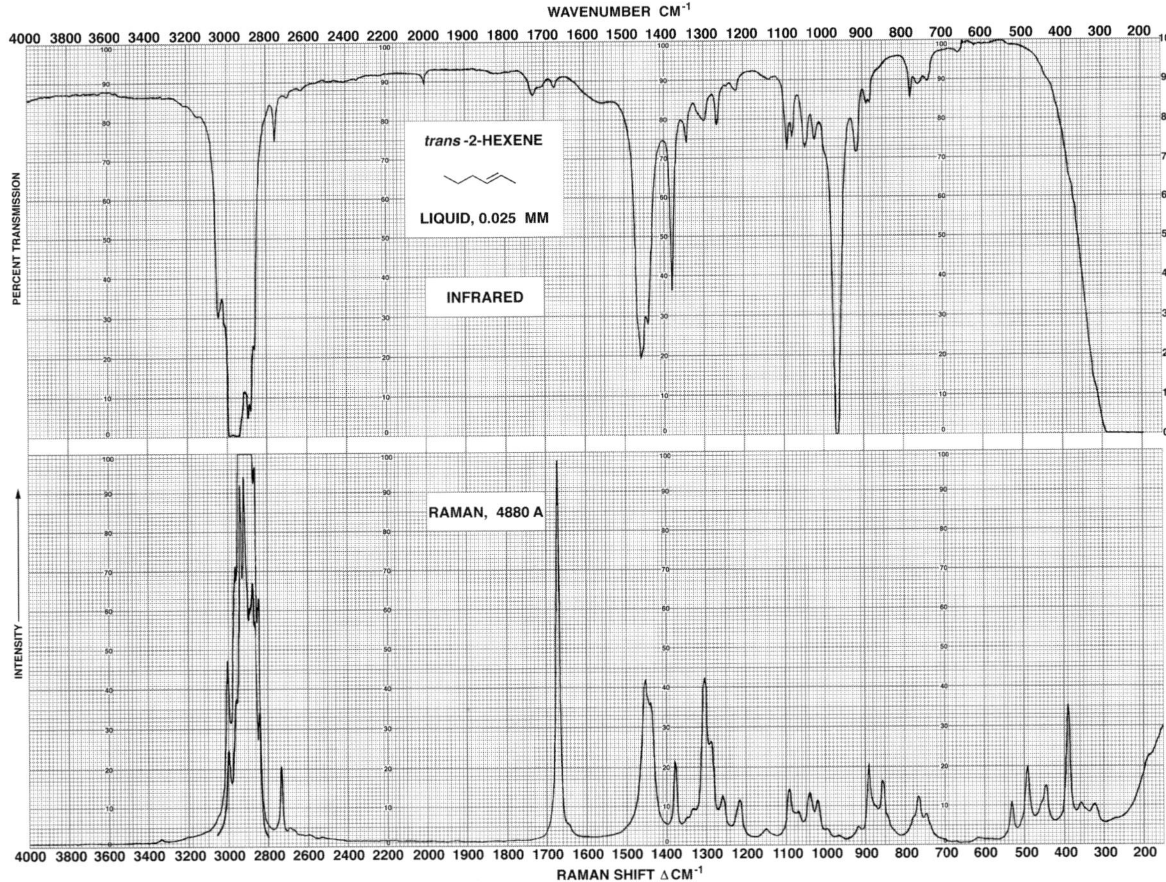

Fig. 12.34 Infrared and Raman spectra of *trans*-2-hexene.

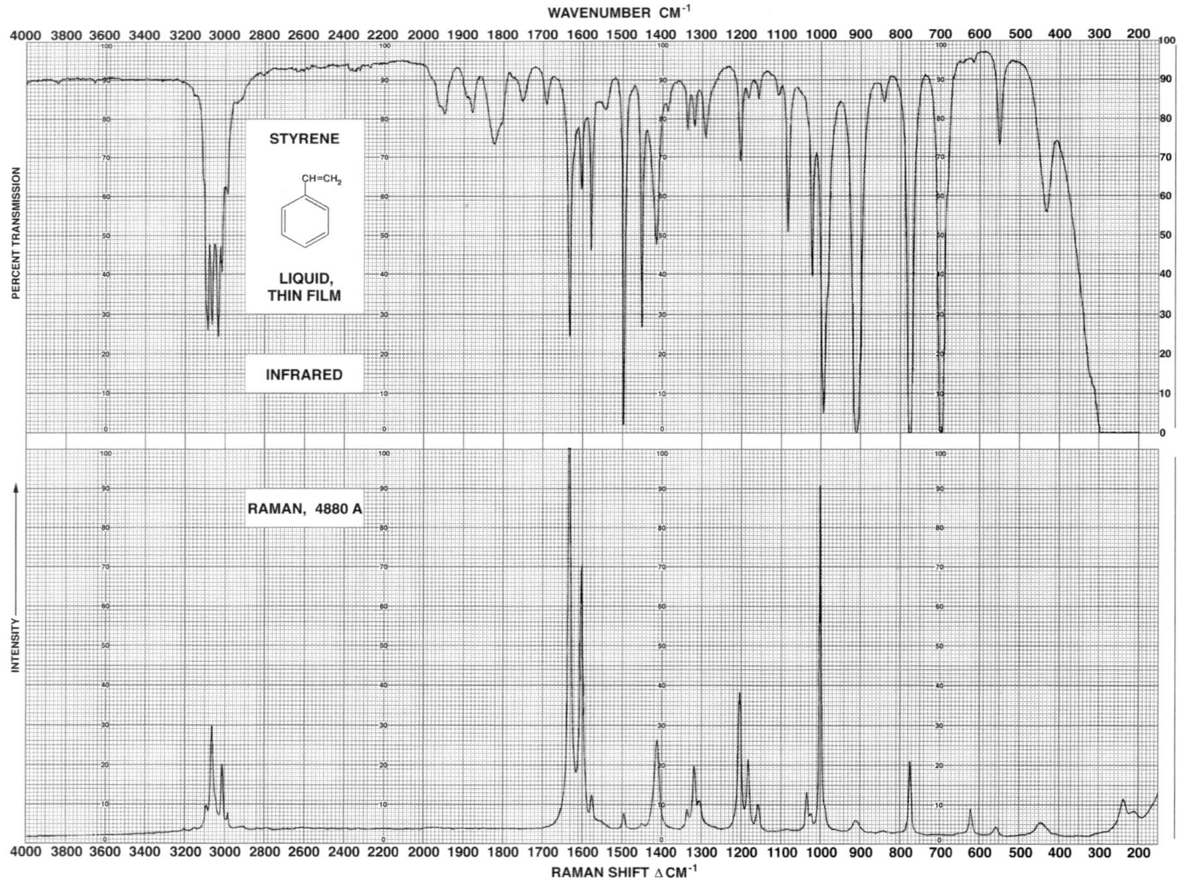

Fig. 12.35 Infrared and Raman spectra of styrene.

identical spectra. The result is a significant band broadening and a weak diffuse spectrum in the fingerprint region. On the other hand, the bicyclic compound has only two possible conformers. As a result the fingerprint region of the Raman spectrum of the latter compound contains a rich collection of sharp bands. Thus, it possible to say with some confidence that one sample is likely to be an open-chain system and the other a ring system. In some isolated instances it is, in fact, possible to identify a particular skeleton from its fingerprint pattern. The IR adds little to the ascertaining the type of skeleton present.

B. A number of cases where the *rule of mutual exclusion* applies have been noted throughout the chapter. In certain instances the rule can be powerfully applied to structure determination. For example, Levin has shown that in the three pairs of photodimers given below the correct structure can be assigned to each member of the pair based on the number of coincidences between the IR and Raman spectra. This is possible because one of the products possesses a center of symmetry and the other product does not. The system with the center of symmetry should have a greatly reduced number of coincidences because only accidental degeneracies may occur (see Table 12.2).

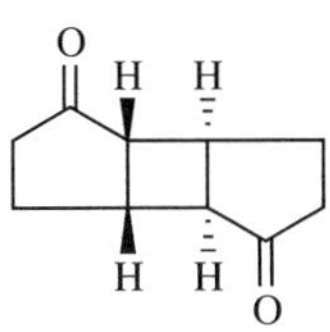

1

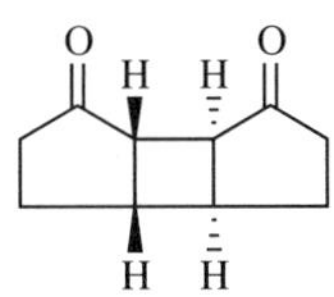

2

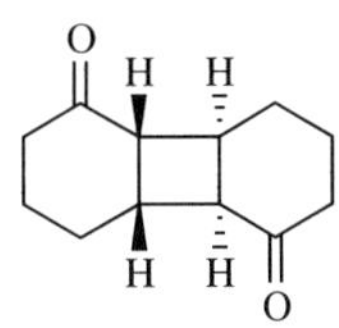

3

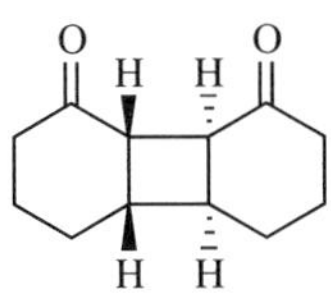

4

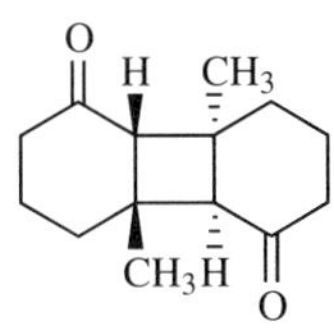

5

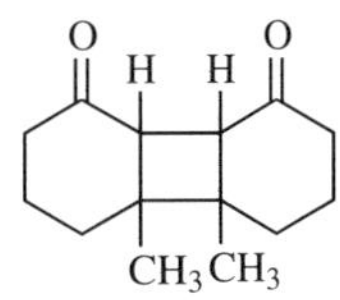

6

TABLE 12.2 Number of observed infrared and Raman bands for compounds 1 to 6, and number of coincidences between them.

	Head-to-Tail Photodimer				Head-to-Head Photodimer		
Compound	R	IR	C	Compound	R	IR	C
1	24	24	6	**2**	31	36	26
3	24	46	10	**4**	33	42	24
5	34	34	7	**6**	41	46	28

Note: R, IR, and C denote observed numbers of Raman lines, infrared peaks, and coincidences, respectively.

C. Because the methyl symmetric deformation mode is not normally observed in the Raman effect when the methyl group is bonded to an sp^3 carbon, it is surprising that, in fact, the overtone of this fundamental can be located as a weak band in the 2700 cm^{-1} region. This is an unusual case where an overtone is more intense than the fundamental. This reversal of intensification is the result of a second-order coupling (Fermi resonance) between the bending mode and the symmetric methyl stretching vibration. The overtone robs intensity from the fundamental and becomes more intense than its own normal mode. If two methyl groups are attached to the same carbon, a spatial coupling occurs and the band doubles. While this can be inferred in the IR where the splitting is smaller, ~10–25 cm^{-1}, and the bands tend to overlap, by doubling the separation, the Raman overtones clearly show the interaction. The resulting splitting has even been shown to occur in systems in which the methyl groups are held on adjacent (vicinal) carbons, as in *o*-xylene (Figure 12.16). The *o*-xylene case has been further substantiated by isotopic substitution of a single $–CD_3$ group which uncouples the interaction and collapses the doublet.

XII. Conclusion

It is possible to make far more confident vibrational assignments and sample identifications when both Raman and IR data are available. It is clear that with the significant improvement in ease of sampling with spectrometers employing FT systems and near-IR lasers Raman data will become much more readily available.

Exercise Section III

EXERCISE 11

Spectra Index Nos. 3-A **and** 5-C

Description of Samples: Both IR and Raman spectra are consistent only with gas-phase spectra (see experimental conditions). The IR and Raman spectra are of pure organic compounds of simple structure. Note how these gas-phase IR spectra differ from the solid- and liquid-phase spectra we have seen heretofore. The remarkable fine structure, seen especially on the 950 cm^{-1} band, tells us that the molecule has at least one very low moment of inertia. In other words, it is an exceptionally light molecule.

Objective: Identify the compounds with the help of their characteristic frequencies and the correlation charts. Once the spectrum has been established by comparison to catalog data such as the *Coblentz Society Desk Book* by C. D. Craver (Part 1 in the Bibliography) or the *Aldrich Gas Phase Handbook*, refer to the Answers to the Exercises for a detailed discussion.

Proposed Structures

3-A?
5-C?

Comments

Course Notes on the Interpretation of Infrared and Raman Spectra, by Dana W. Mayo, Foil A. Miller, and Robert W. Hannah.
ISBN 0-471-24823-1

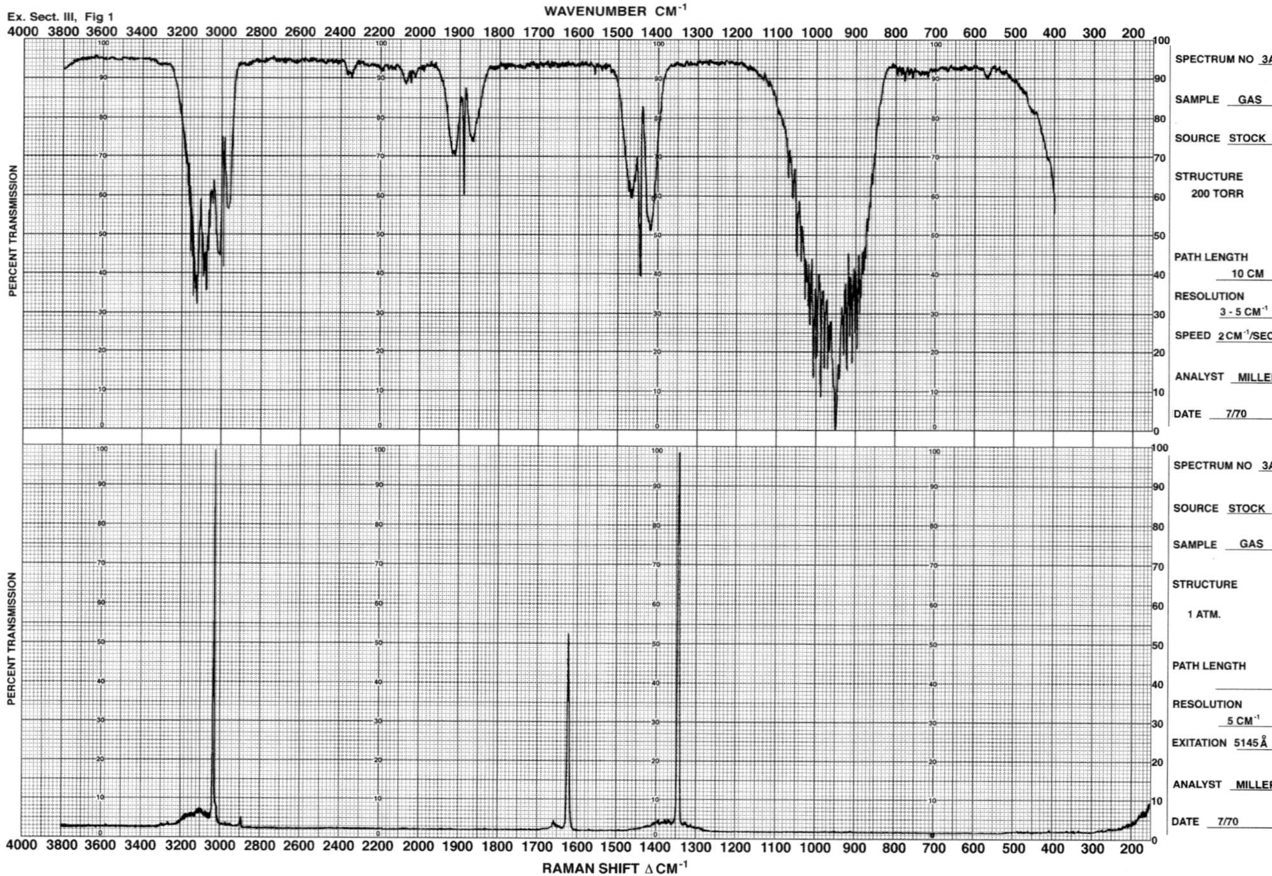

Fig. 1 Unknown 3-A.

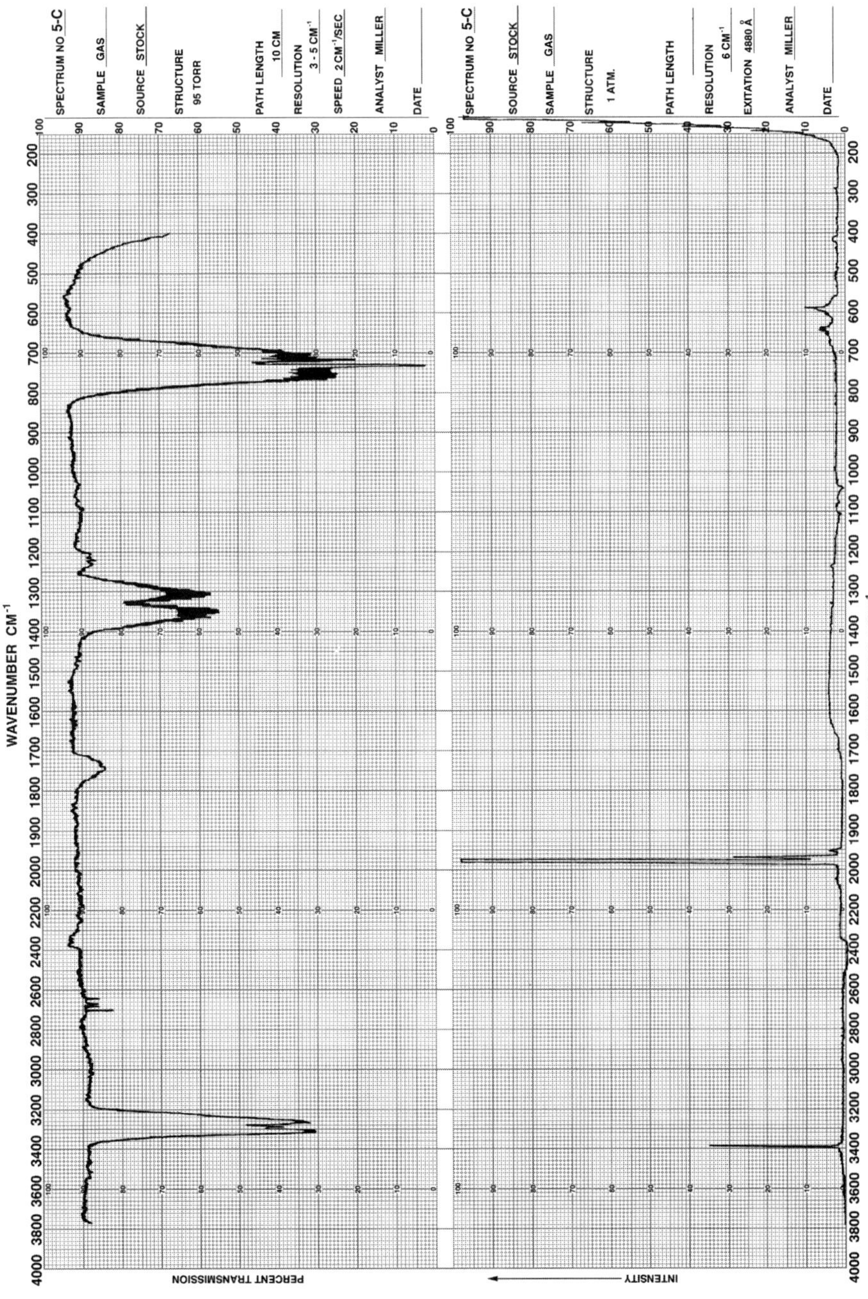

Fig. 2 Unknown 5-C.

EXERCISE 12

Spectra Index Nos. 1-A **and** 20-C

Description of Samples: Both sets of spectra are those of pure organic compounds of simple structure. Both compounds exhibit H bonding. Compound 20-C was shown from mass spectral data to have a molecular weight under 115.

Objective: Identify the compounds with the help of their characteristic frequencies and the correlation charts. Once the spectrum has been established by comparison to catalog data such as the *Coblentz Society Desk Book* or the *Aldrich Handbook*, refer to the Answers to the Exercises for a detailed discussion of the answer.

Proposed Structures

1-A?
20-C?

Comments

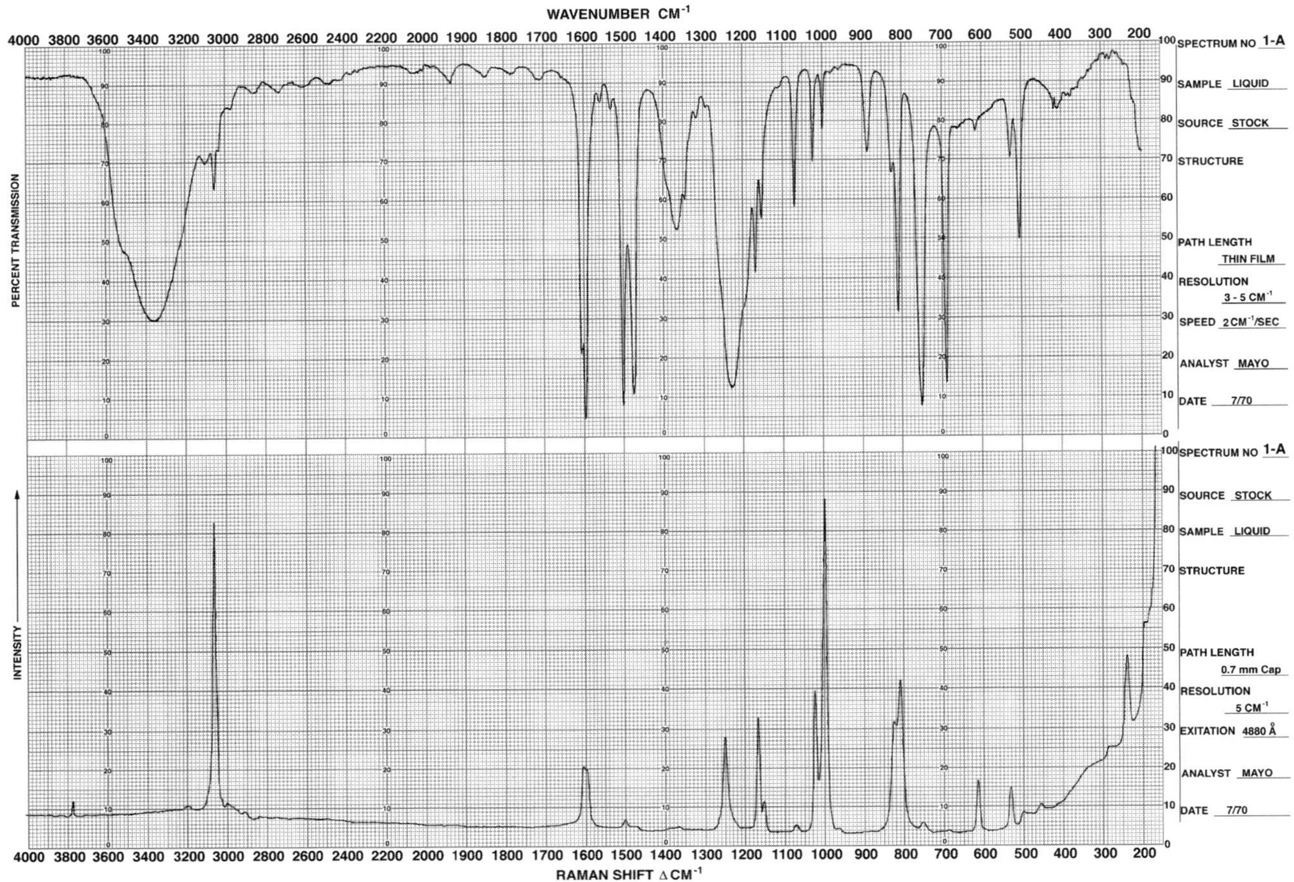

Fig. 3 Unknown 1-A.

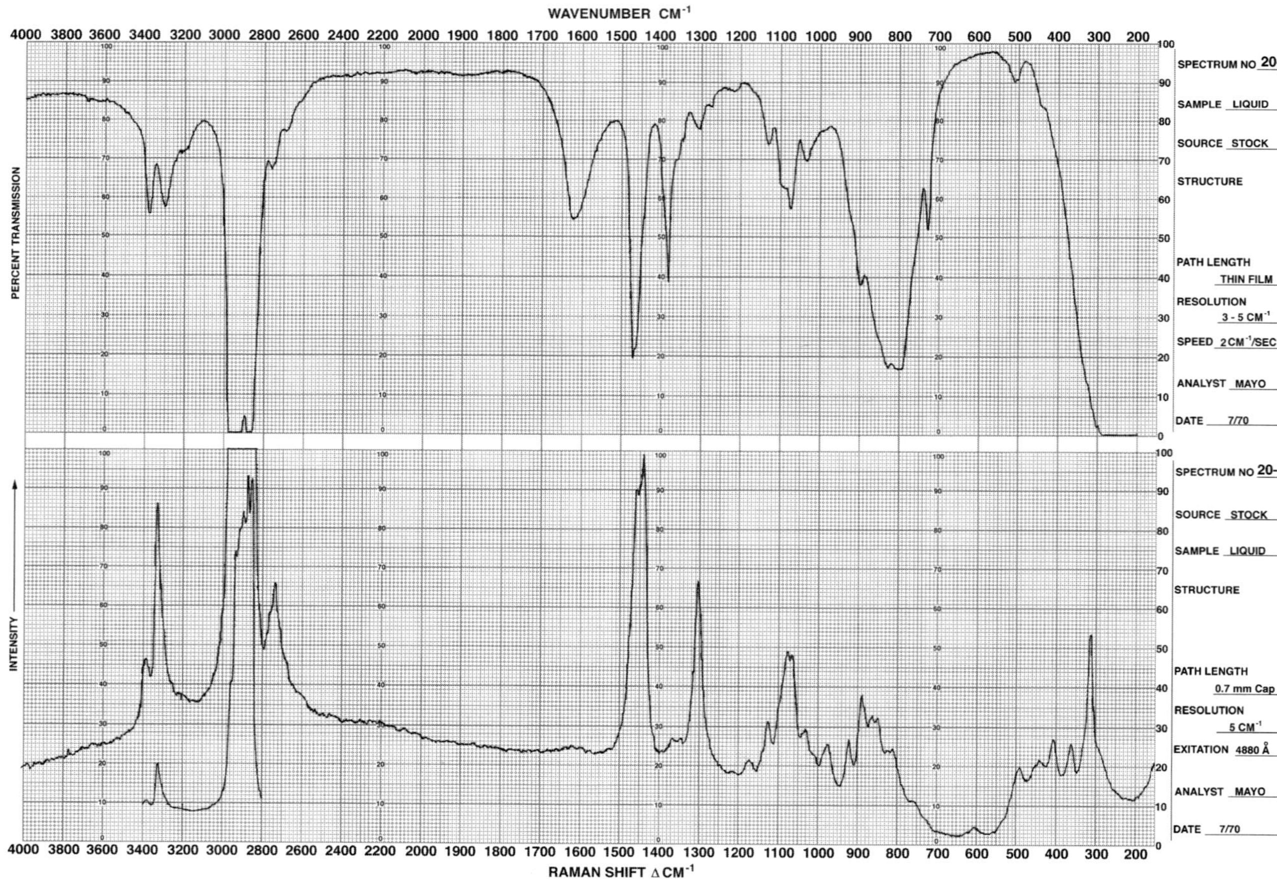

Fig. 4 Unknown 20-C.

EXERCISE 13

Spectra Index Nos.: None Used

Description of Samples: Two liquid samples (IR spectra only).

Objective: To assess the accuracy of the labeling and/or purity of the two samples.

I. First sample labeled "benzene"
After examining the spectrum for information concerning the accuracy of the label and the sample purity, refer to Answers to the Exercises for a detailed discussion.

"Benzene" spectrum?

Comment on Spectrum See Figure 5*a*.

II. Second sample labeled "di-*n*-butylamine" by the manufacturer.
After examining the spectrum for information concerning the accuracy of the label and the sample purity, refer to Answers to the Exercises for a detailed discussion.

"Di-*n*-butylamine" spectrum?

Comment on Spectrum See Figure 5*b*.

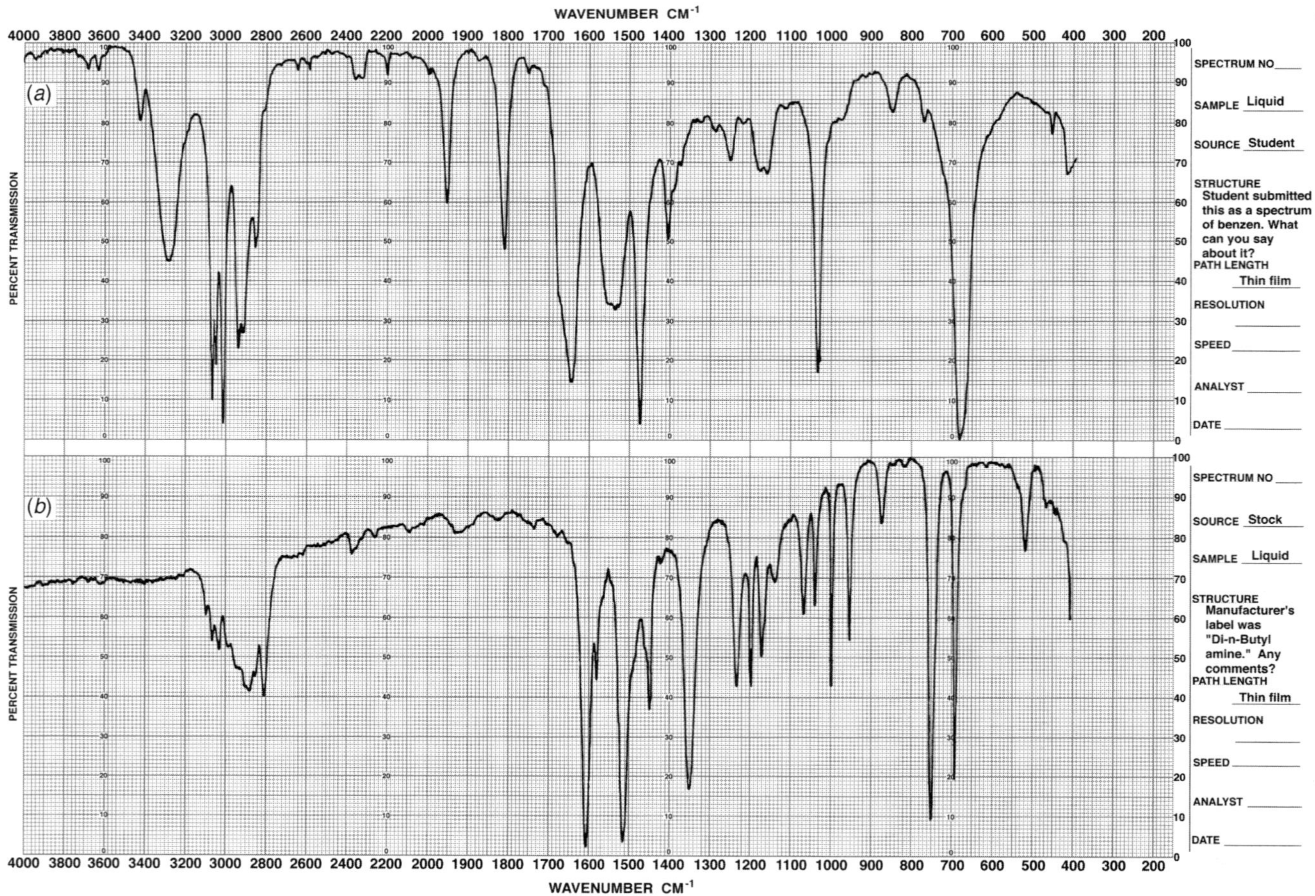

Fig. 5 (*a*) Unknown "benzene" and (*b*) unknown "di-*n*-butylamine."

EXERCISE 14

Spectra Index Nos. 6-B **and** 19-A

Description of Samples: Both unknowns are pure samples that have spectra representative of the more difficult, but still interpretable, spectra of complex molecules.

Objective: Identify the compounds with the help of their characteristic frequencies and the correlation charts. Once a spectrum has been established by comparison to catalog data such as the *Coblentz Society Desk Book* or the *Aldrich Handbook*, refer to the Answers to the Exercises for a detailed discussion of the answer.

I. Unknown 6-B on sodium fusion gave a precipitate of AgCl, a positive test for the presence of halogen, and from the nature of the precipitate the halogen most likely is chlorine. The IR and Raman spectra of 6-B were obtained with two different sets of instruments on several different samples. *The IR spectrum of the higher concentration sample (6-Ba) contains the most important clue to the structure of the unknown.*

Proposed Structure

6-B?
6-Ba?

Comments See Figures 6 and 7.

II. Spectrum 19-A belongs to a compound employed extensively as a monomer in the plastic-polymer industry. The IR spectrum of a second sample of this material (19-Aa) recorded on a different instrument is also shown. Spectrum 19-Aa was obtained at an intermediate path length compared to the two sample path lengths used in spectrum 19-A.

Proposed Structure

19-A and 19-Aa?

Comments See Figures 8 and 9.

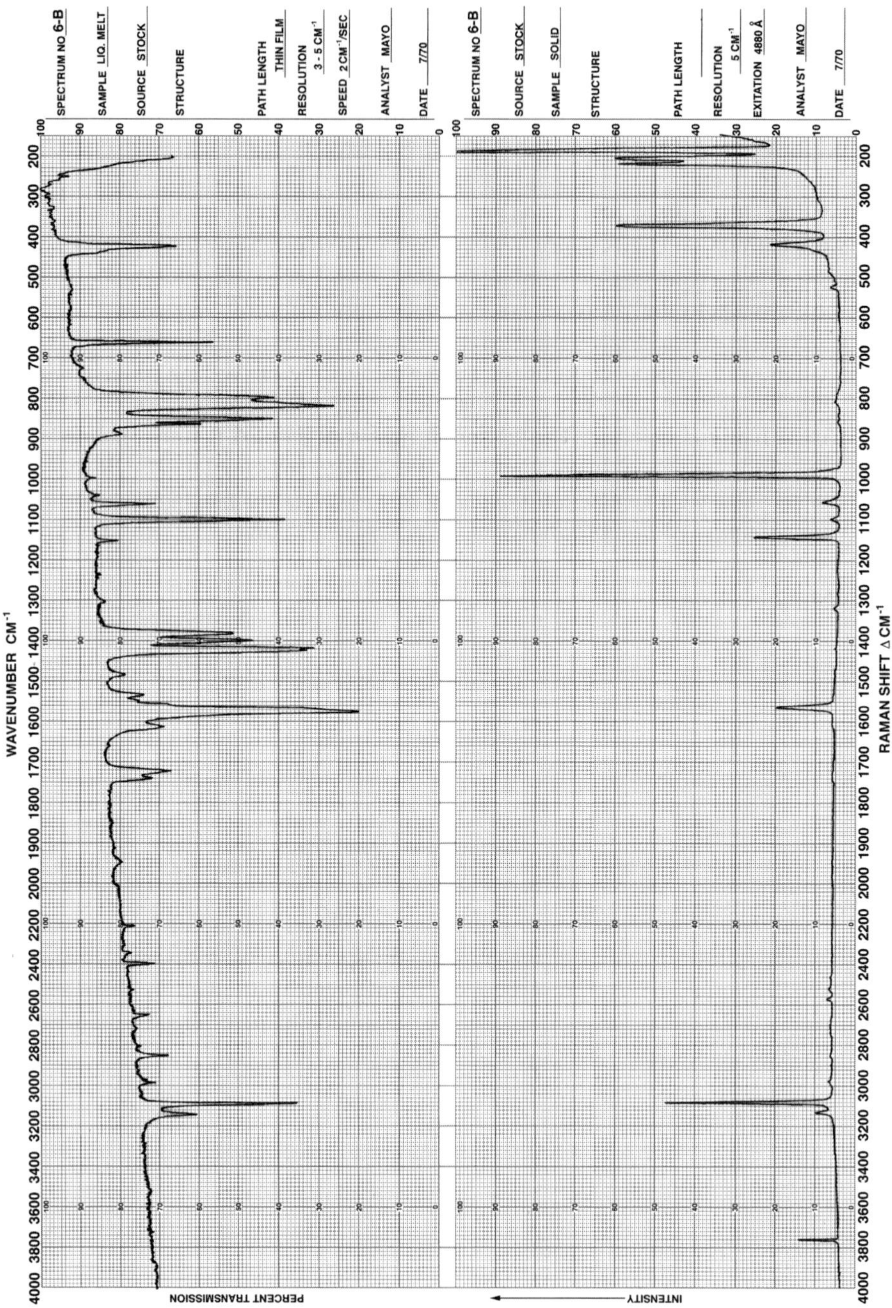

Fig. 6 Unknown 6-B.

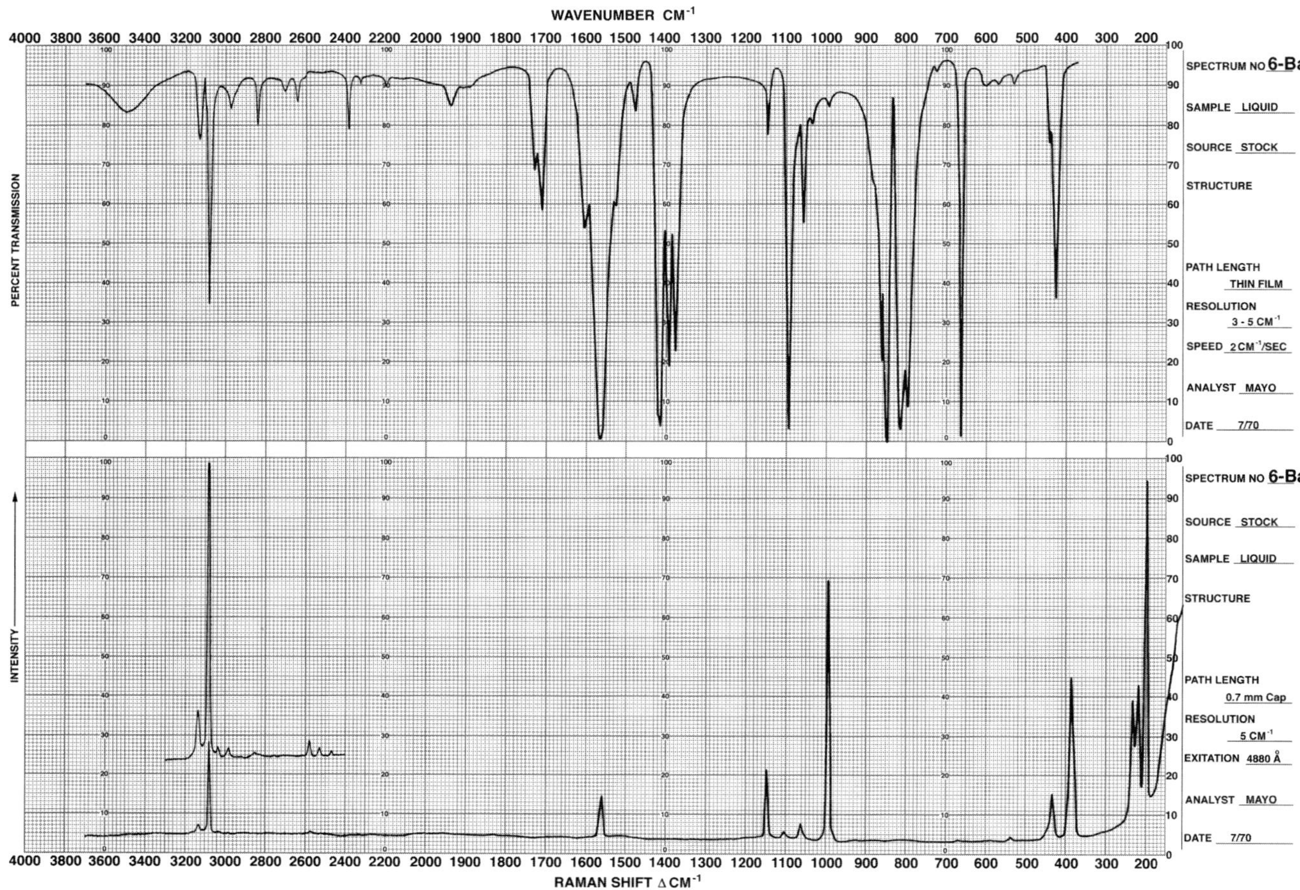

Fig. 7 Unknown 6-Ba.

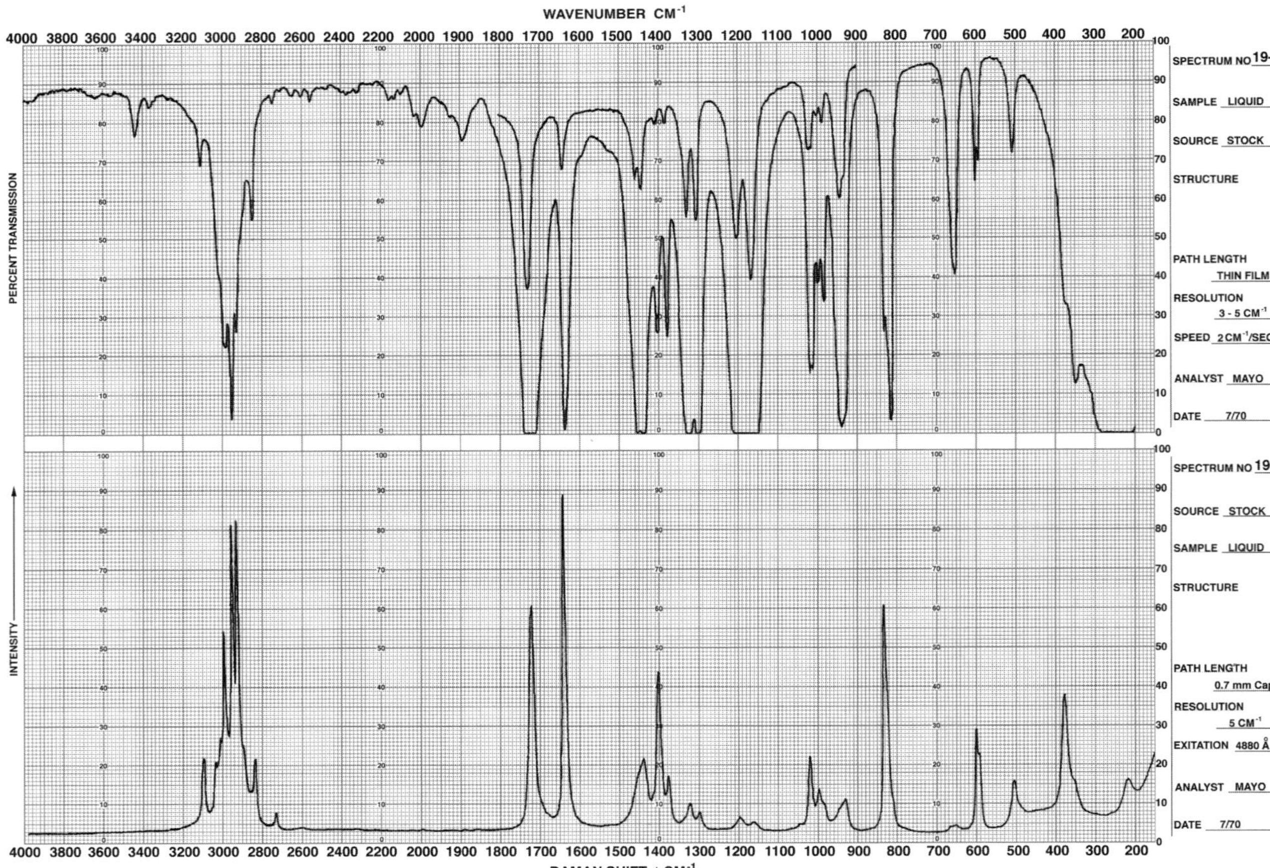

Fig. 8 Unknown 19-A.

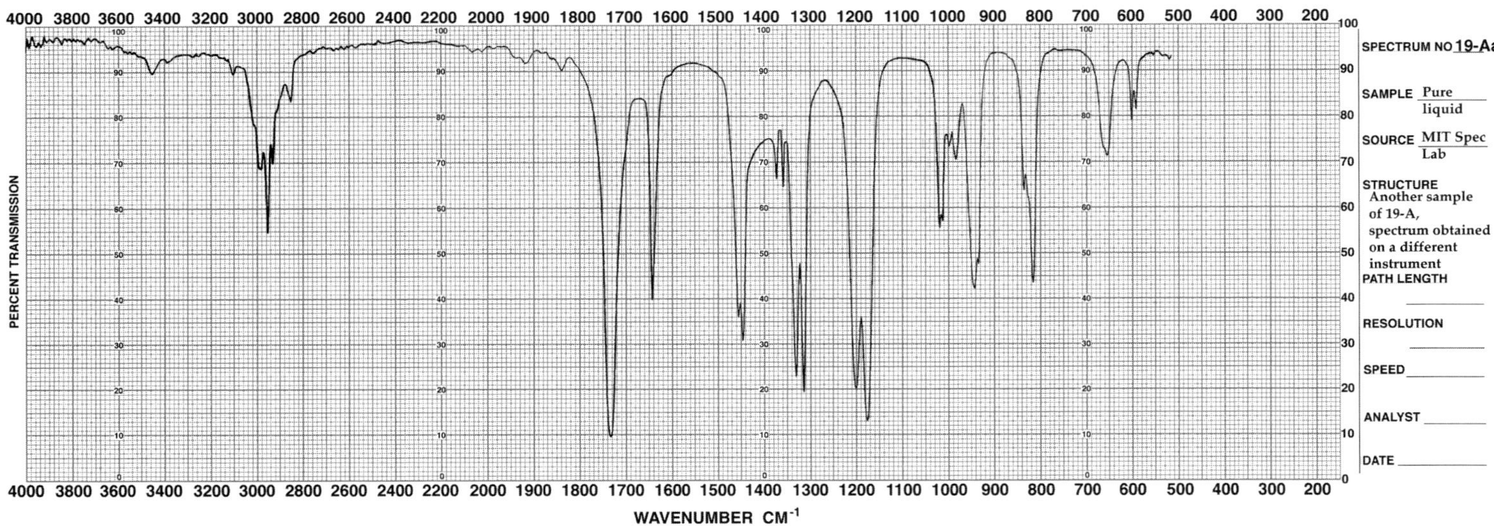

Fig. 9 Unknown 19-Aa.

EXERCISE 15

Spectrum Index No. 18-A

Description of Sample: This spectrum is that of a pure organic compound of simple structure. Following sodium fusion, compound 18-A gave a positive sodium nitroprusside (violet color) test, which indicates the presence of sulfur in the sample.

Objective: Identify the compound with the help of its characteristic frequencies and the correlation charts. Once a spectrum has been established by comparison to catalog data such as the *Coblentz Society Desk Book* or the *Aldrich Handbook*, refer to the Answers to the Exercises for a detailed discussion.

Proposed Structure

18-A?

Comments

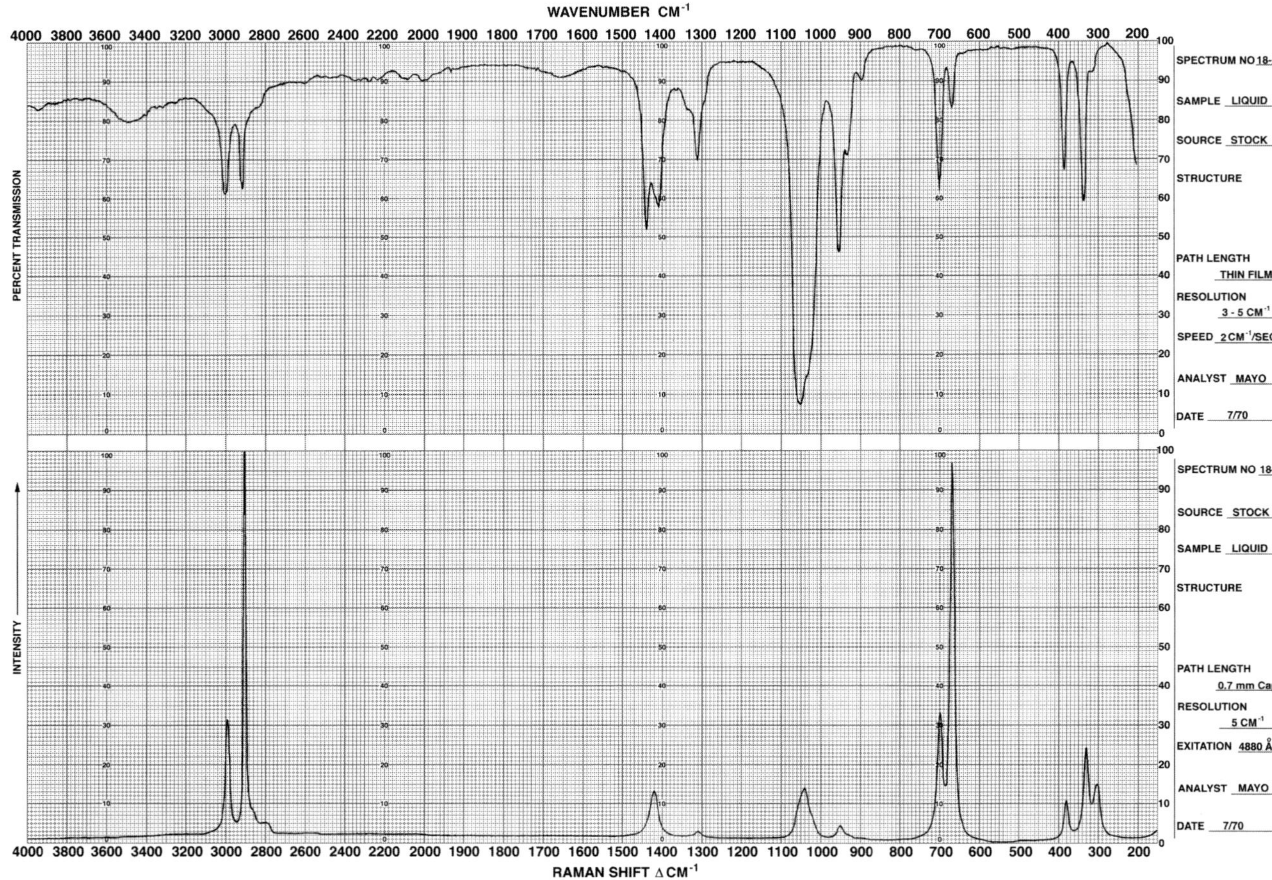

Fig. 10 Unknown 18-A.

EXERCISE 16

Spectra Index Nos. 21-A **and** 21-B **(IR Spectra Only)**

Description of Samples: The compounds of curve index nos. 21-A and 21-B are organic esters of inorganic acids.

Objective: Attempt an identification of the class of esters from the characteristic group frequencies in their spectra. Reference compounds similar to compound 21-A are found in the *Aldrich Handbook of Spectra* and can be used for comparison, while 21-B is found in both the *Coblentz Society Desk Book* and the *Aldrich Handbook*. Once a spectrum has been established, refer to the Answers to the Exercises for a detailed discussion.

Proposed Structures

21-A?

21-B?

Comments

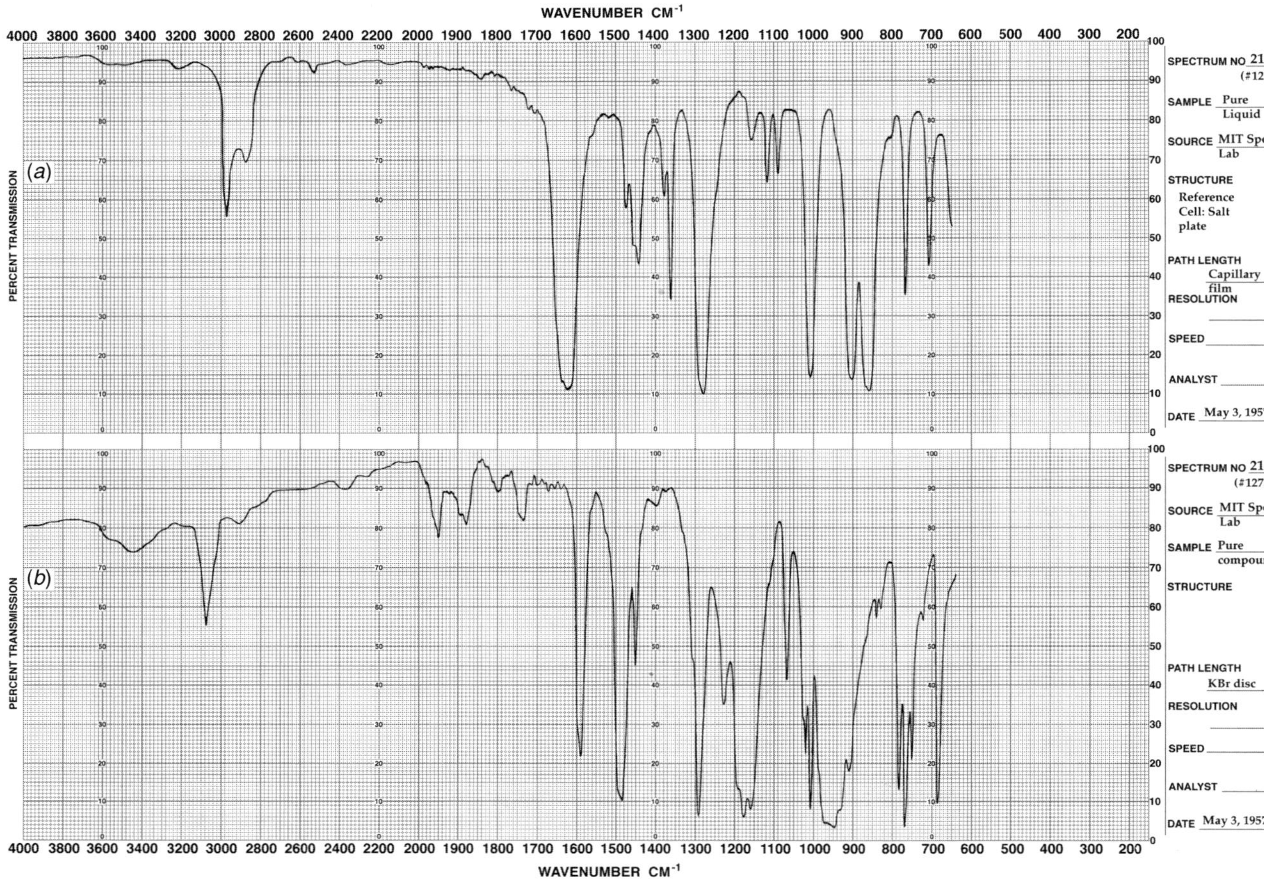

Fig. 11 (*a*) Unknown 21-A and (*b*) unknown 21-B.

EXERCISE 17

Spectra Index Nos. 7-A **and** 7-B **(IR Spectra Only)**

Description of Samples: Both spectra were obtained with solid samples, compound 7-A as a suspension in Nujol and 7-B as a KBr pressed disc.

Objective: Identify the materials. Since the simplicity of the spectra and, in the case of 7-A, the broad nature of the principal band suggest inorganic ions, try identification first with the help of the spectra given in Chapter 11, Aldrich collection, or an article by Miller and Wilkins, *Anal. Chem.* **24**, 1253 (1952). Remember that curve index no. 7-A contains Nujol bands. After preliminary identification of the ions (mainly the anions), check the individual spectra for final identification. Once a spectrum has been established, refer to the Answers to the Exercises for a detailed discussion.

Proposed Structures

7-A? M&W curve no. —

7-B? M&W curve no. —

Comment on Any Observed Differences

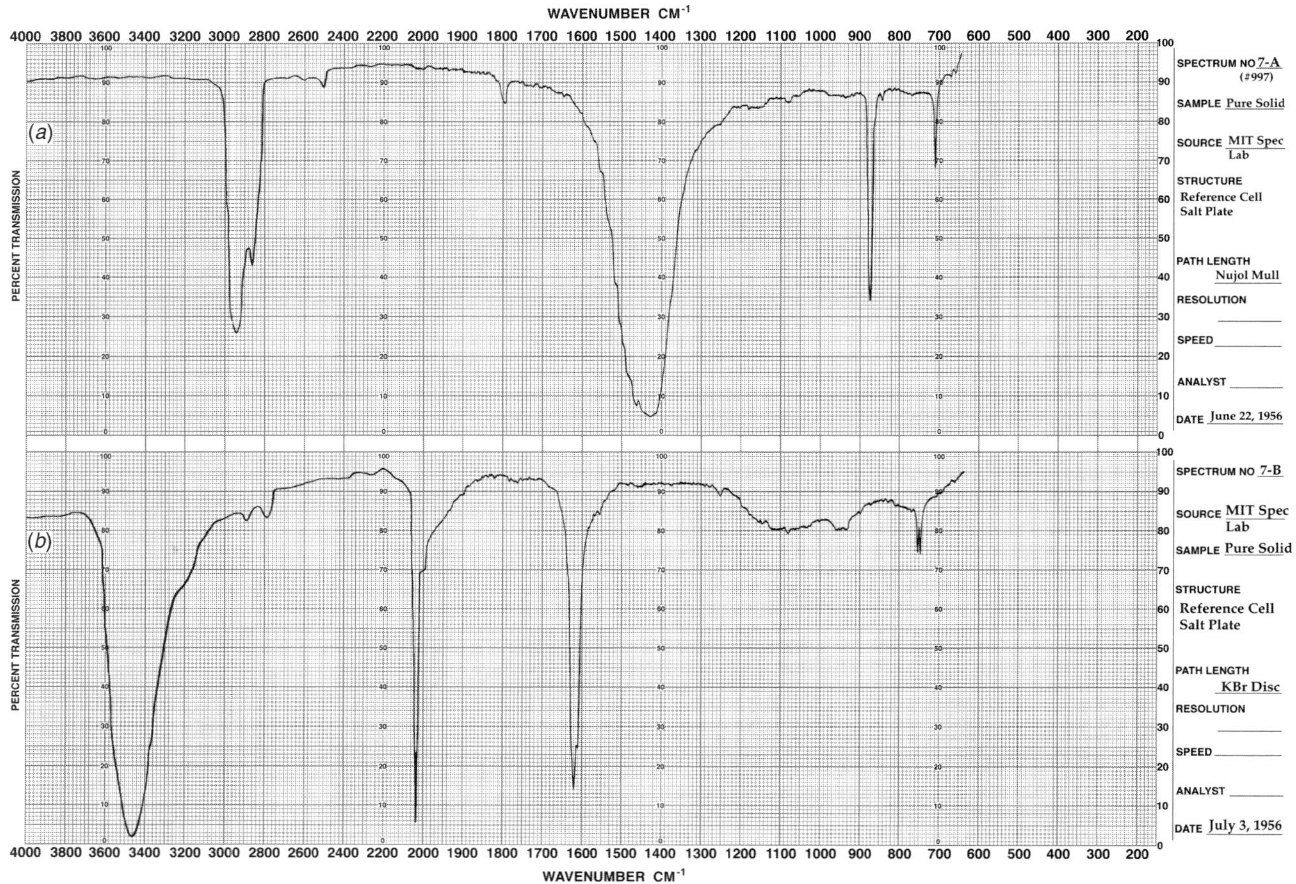

Fig. 12 (*a*) Unknown 7-A and (*b*) unknown 7-B.

EXERCISE 18

Spectra Index Nos. 15-C **&** 17-C, 16-C **&** 16-D, **and** 17-A & 10-C **(IR Spectra Only)**

Description of Samples: This set of unknowns contains four different polymer samples plus a polymer prototype molecule (17-A).

The polymer found in problem 15-C is derived from the monomer identified as unknown 19-A.

The polymer spectra shown in 16-C & 16-D were obtained by two different sampling techniques. The latter spectrum (16-D) is introduced to illustrate the advantage gained in polymer sampling from multiple internal reflection techniques.

Samples 17-C and 10-C represent further examples of relatively common high-molecular-weight polymers.

Objective: Identify the structural groups in the polymer film from its spectrum and infer the monomeric units from which the polymer was formed.

After a monomer has been proposed, it is possible to check the tentative structure by referring, for example, to Hummel's *Atlas of Polymer and Plastics Analysis*, Vol. I, if it is available to you. Once the spectrum has been established by comparison, refer to the Answers to the Exercises for a detailed discussion.

Proposed Structures

I. Curves 15-C and 17-C are those of high-molecular-weight polymers formed from single monomers.

15-C?
17-C?

Comments See Figure 13.

II. Curve 16-C is the spectrum of a sample which clearly is too thick to be accurately employed in automatic retrieval approaches to identification. Curve 16-D, however, is of the *same sample* obtained via ATR sampling.

16-C?
16-D?

Comments See Figure 14.

III. Curve 17-A is a low-molecular-weight example of a polymer prototype. Identify the structure of 17-A.

17-A?
Curve 10–C is that of a high-molecular-weight polymer formed from a single monomer. Identify the monomer used in forming the polymer 10-C.

10-C?

Comments See Figure 15.

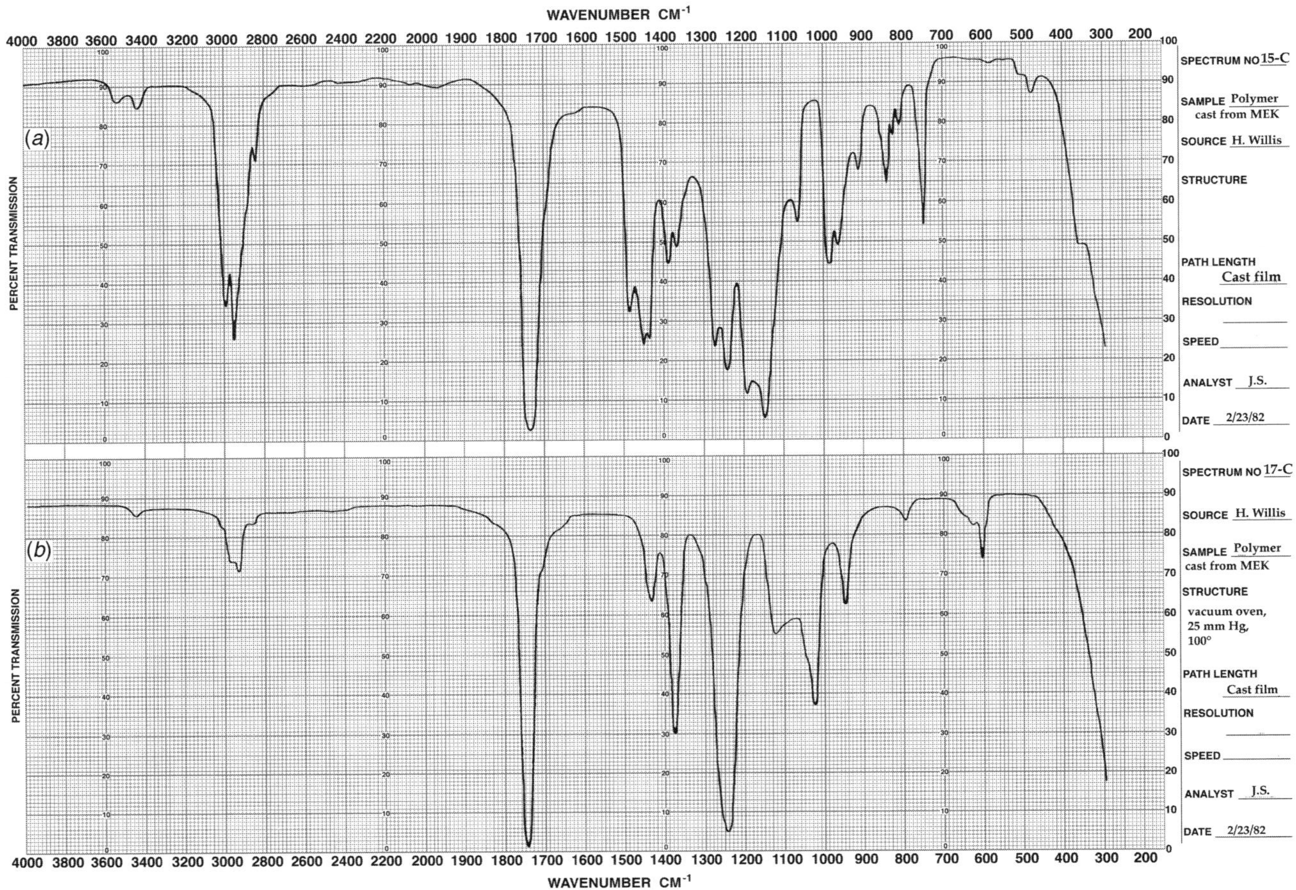

Fig. 13 (*a*) Unknown 15-C and (*b*) unknown 17-C.

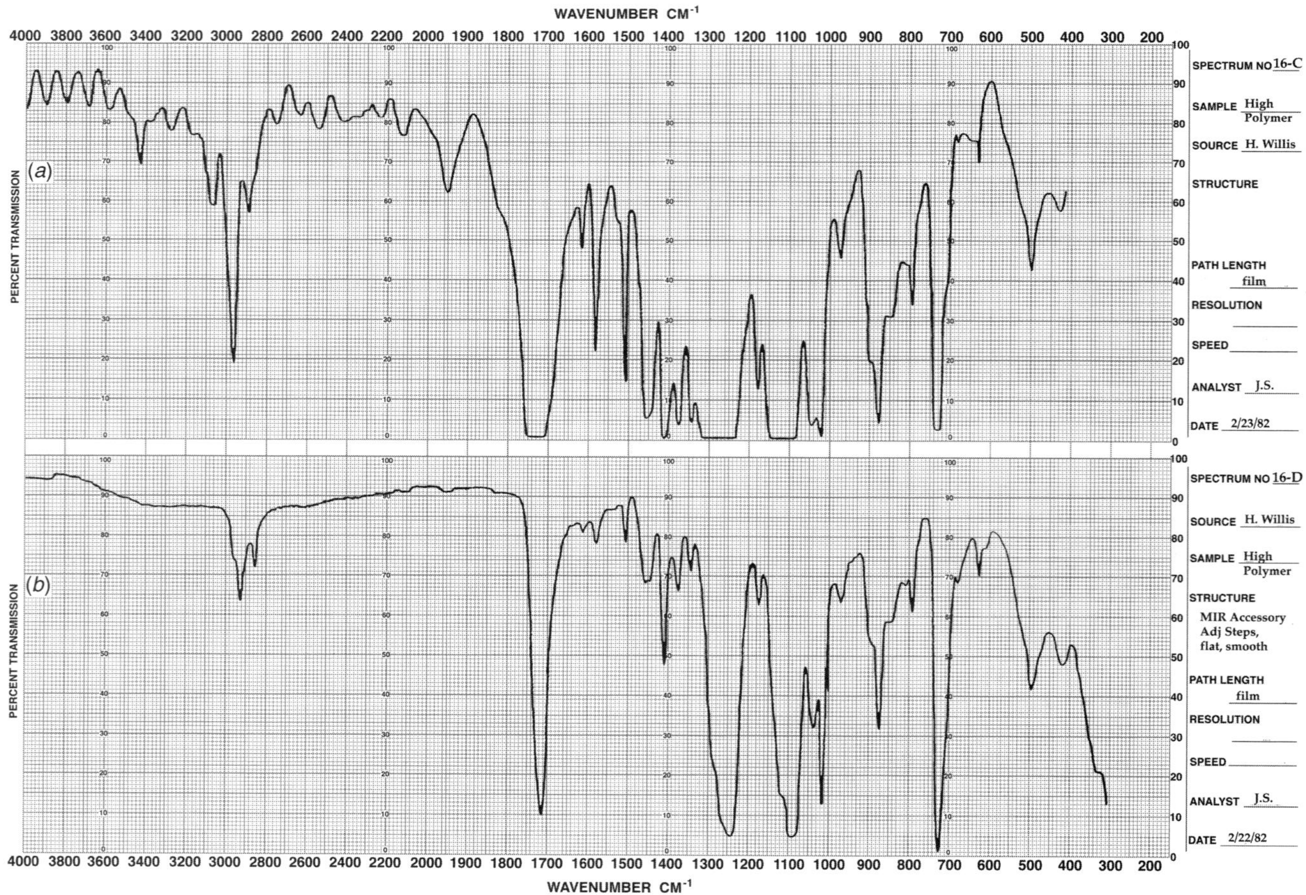

Fig. 14 (*a*) Unknown 16-C and (*b*) unknown 16-D.

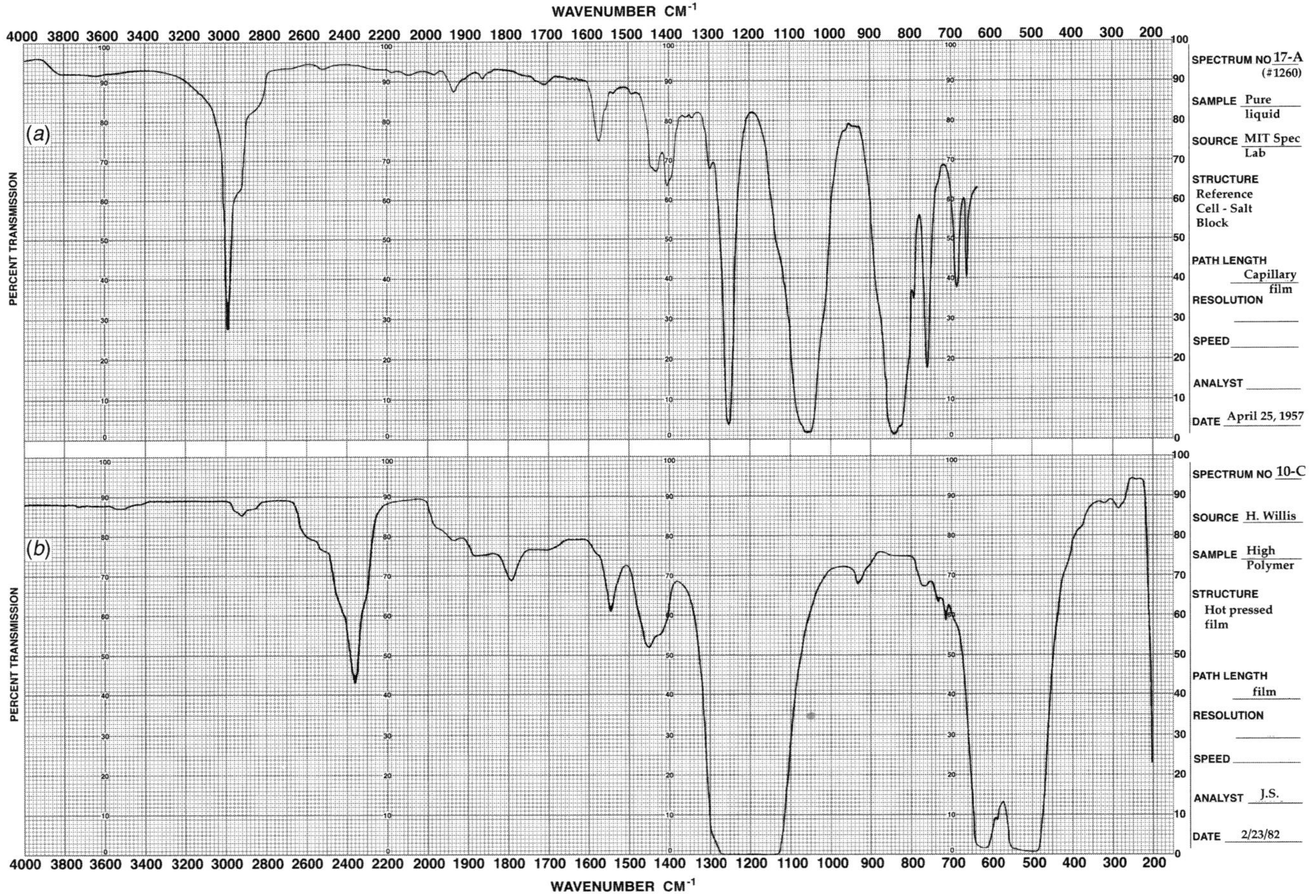

Fig. 15 (*a*) Unknown 17-A and (*b*) unknown 10-C.

EXERCISE 19

Spectra Index Nos. 34E **and** 35E

Description of Sample: The IR and Raman spectra of two relatively common polymers are shown. One of the samples contains an element other than C, H, O, and N. Note the value of having both the Raman and IR data available for identification problems.

Objective: Identify the class of polymer and the monomer or monomers used in each polymerization. After a monomer has been proposed, it is possible to check the tentative structure by referring, for example, to Hummel's *Atlas of Polymer and Plastics Analysis*, Vol. I, if it is available to you. Once the spectrum has been established by comparison, refer to the Answers to the Exercises for a detailed discussion.

Proposed Structures

34E?

35E?

Comments

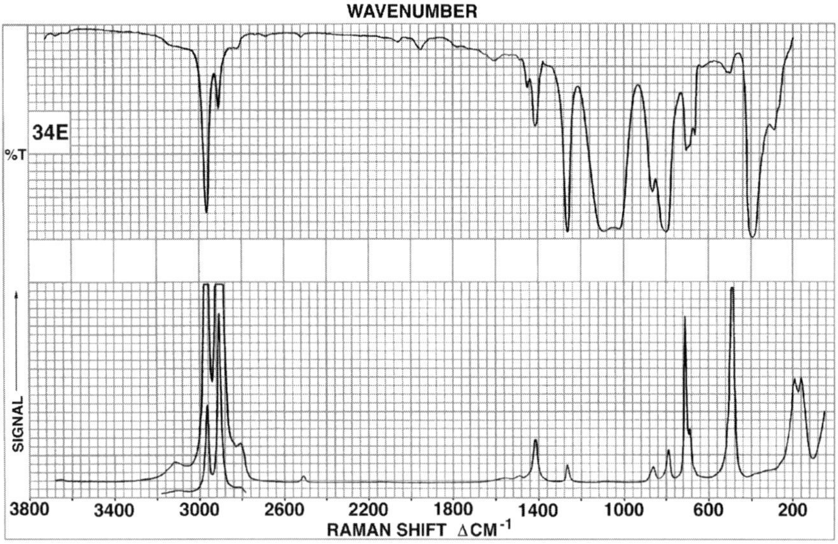

Fig. 16 Unknown 34E.

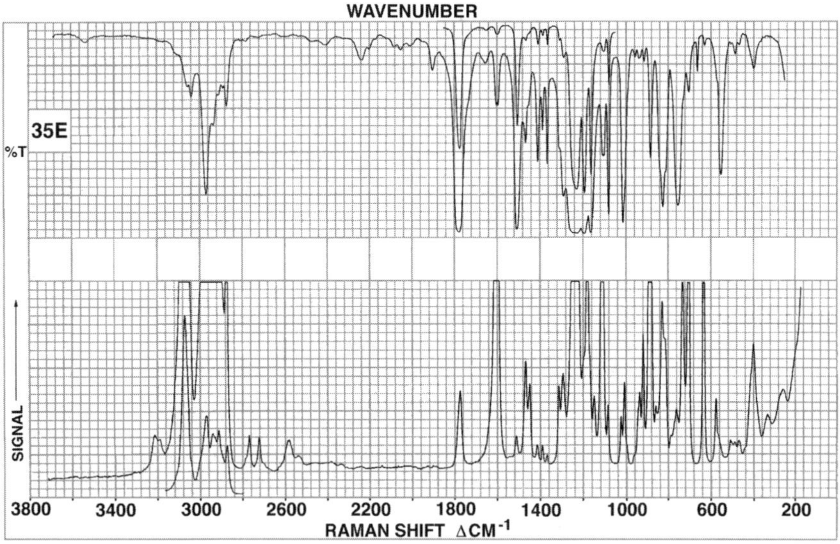

Fig. 17 Unknown 35E.

13 Sample-Handling Techniques

ROBERT W. HANNAH

This chapter is concerned with some fundamental aspects of sample preparation. Our IR instruments are very good. They are certainly better than what was available a decade ago, maybe even what was available two or three years ago. The software available makes them even more powerful. These improvements demand that you pay more attention to your sampling rather than less. It is probable that too many of us overlook this in our laboratories. Computers will not correct sloppy work—rather they will emphasize the bad aspects of a badly prepared sample. So a sample properly prepared and properly presented to the IR instrument is important, especially since the sample may and generally does interact with the instrument to change the optical properties of the latter. Your goal is to reduce that interaction to at least a common level for the same kinds of samples. Therefore, it is instructive to discuss some of the things which must be considered in the handling of samples, specifically the handling of liquids and solids. The handling of sealed and demountable cells, the KBr disc technique, and mulling and reflection techniques will be discussed in detail. These are the most common techniques used in the laboratory and the ones which, because of their simplicity, are very often not applied optimally. Specialized techniques such as IR microscopy will not be discussed. In all cases the choice of sampling technique will be dictated by the questions being asked about the sample and the experience of the analyst. A suggested procedure for selecting a sampling technique for a specific problem is given at end of the chapter.

To begin, consider the sampling facilities available for handling liquids (Table 13.1).

First, consider running the liquid sample in a solution if it happens to be soluble in an IR-transparent solvent. Second, if it is insoluble, run the material as a capillary film—meaning, taking a pair of IR windows, placing a drop of fluid on one of them, and placing the other crystal on top. If the second or top window is touched down at one edge at an angle and closed as one would close a book, the capillary film moves

Course Notes on the Interpretation of Infrared and Raman Spectra, by Dana W. Mayo, Foil A. Miller, and Robert W. Hannah.
ISBN 0-471-24823-1

TABLE 13.1 Liquid Sampling

1. Solution, if soluble
2. If insoluble, as a capillary film
3. In a thin sealed cell
4. Internal reflection

ahead of the closure and produces a continuous film which is free from bubbles. Third, run the same liquid in a thin-sealed cell. Fourth, run the liquid by internal reflection. The latter is a particularly good technique for looking at strong IR absorption bands and can yield good quantitative data as well.

Dealing with a solid is summarized in Table 13.2.

First, treat it as a solution in an appropriate solvent; Second, the sample might be prepared as a thin film. For polymers, thin films may be scanned as they are manufactured or as films you cast yourself. There are several useful techniques for casting films. The IR window is placed on a warm (not hot) hot plate. Paper tissue may be placed underneath the window to prevent direct contact with the hot plate to reduce the chance of fracture. Several drops of the solution are then placed on the window, covering the entire surface. As it warms, the solvent evaporates, leaving behind a thin film. However, the film tends to be thicker around the edges of the window and thinner in the middle, which, unfortunately, is the area the instrument sees. In some cases, this nonuniformity of thickness may be avoided by placing a beaker over the window allowing the solvent to evaporate in its own vapors. This very often produces a nice uniform film across the window. There are many other techniques for casting films, which may be found in the literature.

Polymers may also be sampled by placing them between two Teflon-coated steel blocks, both of which are heated to the softening point of the polymer. These are mounted in a hydraulic press and a thin film is pressed. Fringes can be reduced by sandblasting the surface of the steel blocks prior to coating with Teflon spray.

Third, solids can be handled as a mull or a KBr disk, which will be discussed later in detail. Fourth, still another way to handle solids is by pyrolysis. Fifth, reflection techniques such as specular reflectance, transflectance, internal reflection, and diffuse reflectance may also be used for solids.

TABLE 13.2 Solid Sampling

1. Solution, if soluble
2. Thin film
3. Mull
4. KBr disk
5. Pyrolyzate
6. Specular reflection, transflectance, or internal reflection
7. Diffuse reflection

GENERAL COMMENTS

A wide variety of window materials can be used for handling liquids, films, or solids. Figure 13.1 shows spectra of several window materials, probably the most common windows in the IR laboratory (sodium chloride, potassium bromide, and cesium iodide). If it is only necessary to obtain spectra out to 600 cm^{-1}, sodium chloride is satisfactory, except that over the last 50 cm^{-1} the transmission falls off. Potassium bromide goes out to 400 cm^{-1}, cesium iodide to 200 cm^{-1}. There is an interesting correlation that exists: As the transmission frequency limit moves to lower values, the price goes up.

For those samples that may contain water, other windows may be used (Figure 13.2).

Calcium fluoride is transparent to 1000 cm^{-1}, crystal clear, glasslike in its behavior, and difficult to damage. Barium fluoride is an insoluble window material that is transparent to about 700 cm^{-1}. However, it is somewhat brittle and may fracture if heated or cooled too quickly. In addition, it will react with sulfates and ammonium salts. Silver bromide is transparent to 300 cm^{-1}, but it is very soft and should not be used for sealed cells because it distorts easily. It is a better choice than silver chloride because it does not react nearly as fast to UV radiation from fluorescent lights. The windows given in the lower part of the figure are chemically resistant. Once again, their price increases as the low-frequency limit decreases. The transmission tends to be less for these latter windows because the refractive index is higher with subsequently higher reflection losses from the surfaces. They may also show interference fringes when used (as sealed cells) even when filled. Their refractive indices are around 2.2 while the refractive index for many organic compounds is around 1.5. Thus, there is a significant difference between the window and the sample resulting in interference fringes from the cell even if it is filled. On the other hand, they are extremely resistant to thermal shock. Zinc selenide is probably the material of choice today for IR reflection elements.

HANDLING OF SEALED CELLS

The structure of a sealed cell (Figure 13.3) determines how it should be used. There are two metal plates (back and front) and two windows separated by a spacer. One of the windows is drilled with holes that are used during filling of the cell. The windows are mounted together in the metal frame. Polymeric spacers allow the cell to be easily disassembled and reconstructed with a different thickness of spacer. The normal sealed cell uses an amalgamated metal spacer and is difficult for many of us to take apart and rebuild. One should therefore be careful when handling a sealed cell, and the cell is usually put together and left together. Normally, sealed cells are assembled with either silver or lead spacers with a mercury amalgam on the surface of the spacer. Therefore, the seal between the spacers and the windows is a wet seal. Wet seals must be handled properly. How should we fill such a cell? If the cell is thicker than 0.1 mm, a syringe filled with the liquid sample may be used.

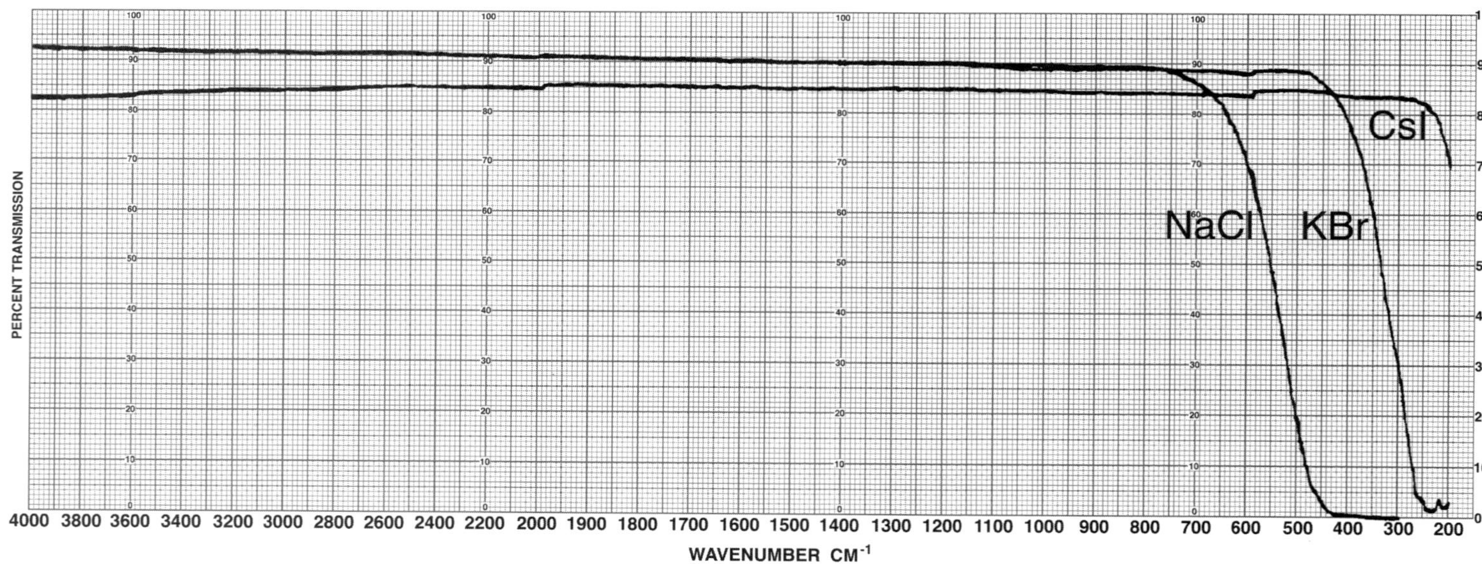

Fig. 13.1 Infrared spectra of water-soluble window materials.

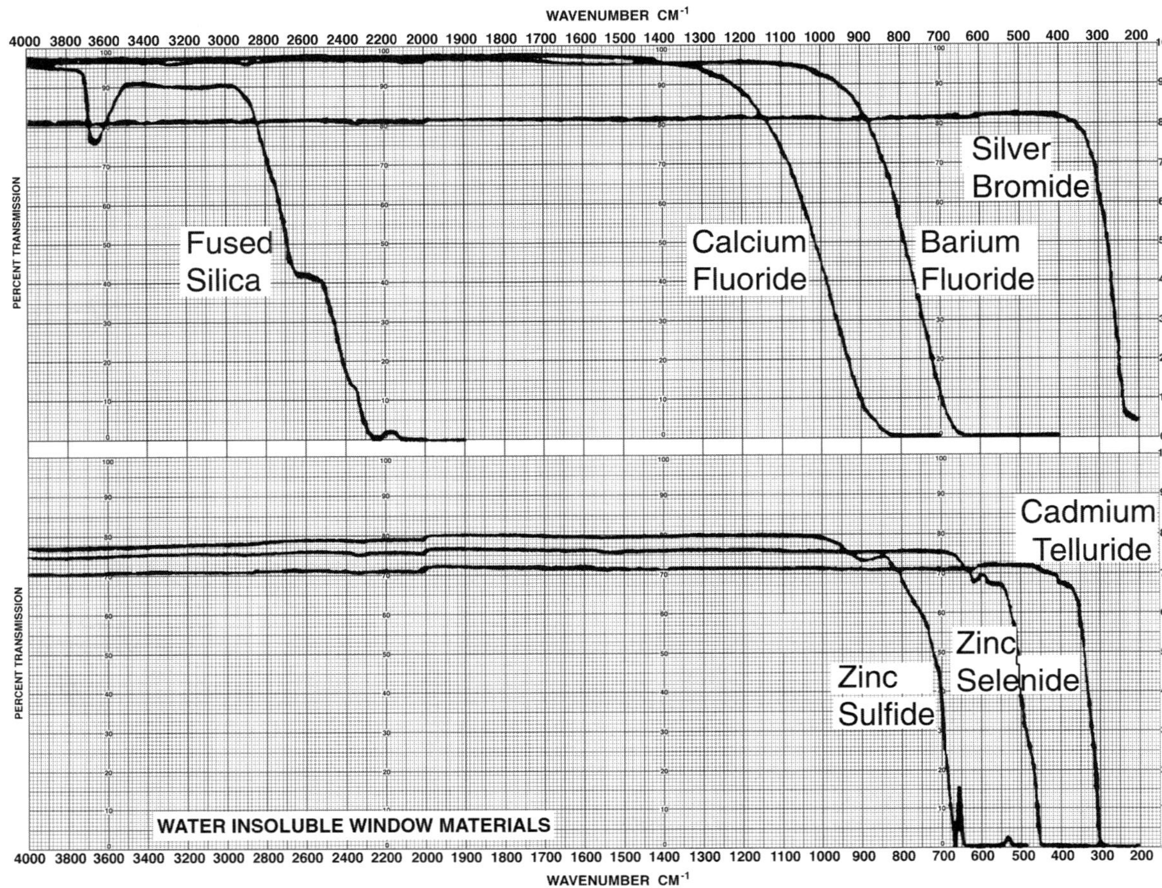

Fig. 13.2 Infrared spectra of water-insoluble window materials.

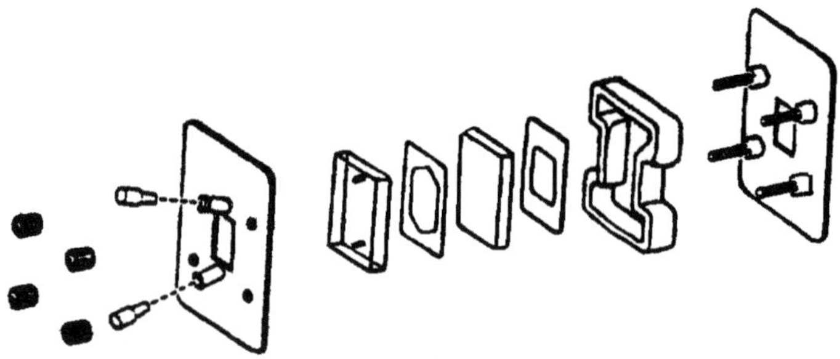

Fig. 13.3 Schematic structure of sealed cell.

Hold the cell at a slightly inclined angle and press gently on the syringe to fill the cell. If the cell is less than 0.1 mm, however, it should be filled in the following manner. Place an empty syringe on the upper port of the cell, as shown in Figure 13.4, and the syringe containing the sample in the lower port.

Fill the cell *not* by pushing on the liquid filled syringe but rather by withdrawing on the empty syringe. This produces a small vacuum in the cell causing liquid to flow from the liquid-filled syringe. This procedure reduces the potential for breaking the seal in these thin cells. There is even some art in the process of sealing the cell with the plugs once it has been filled. If the plugs are just shoved into the ports, they will put hydraulic pressure on the liquid and the spacers and may force the seal open. Therefore, excess liquid should be removed from the filling ports. Finally, the plugs should be placed in the ports and rotated to aid in producing a tight seal between the plug and the cell. Cleaning of the cell is done in the same way as that used for filling the cell. Dry air should be used to evaporate any residual solvent. In quantitative situations, the cell and sample should be allowed to come to temperature equilibrium with the instrument before measurements are made.

The two most common solvents for use in the IR are carbon tetrachloride and carbon disulfide (Figure 13.5).

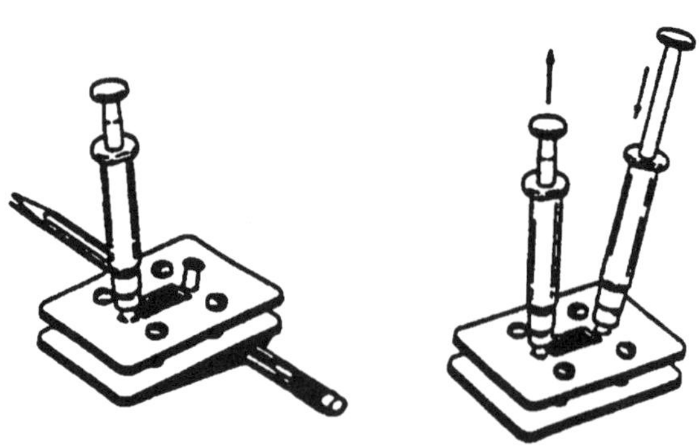

Fig. 13.4 Procedure for filling sealed cells.

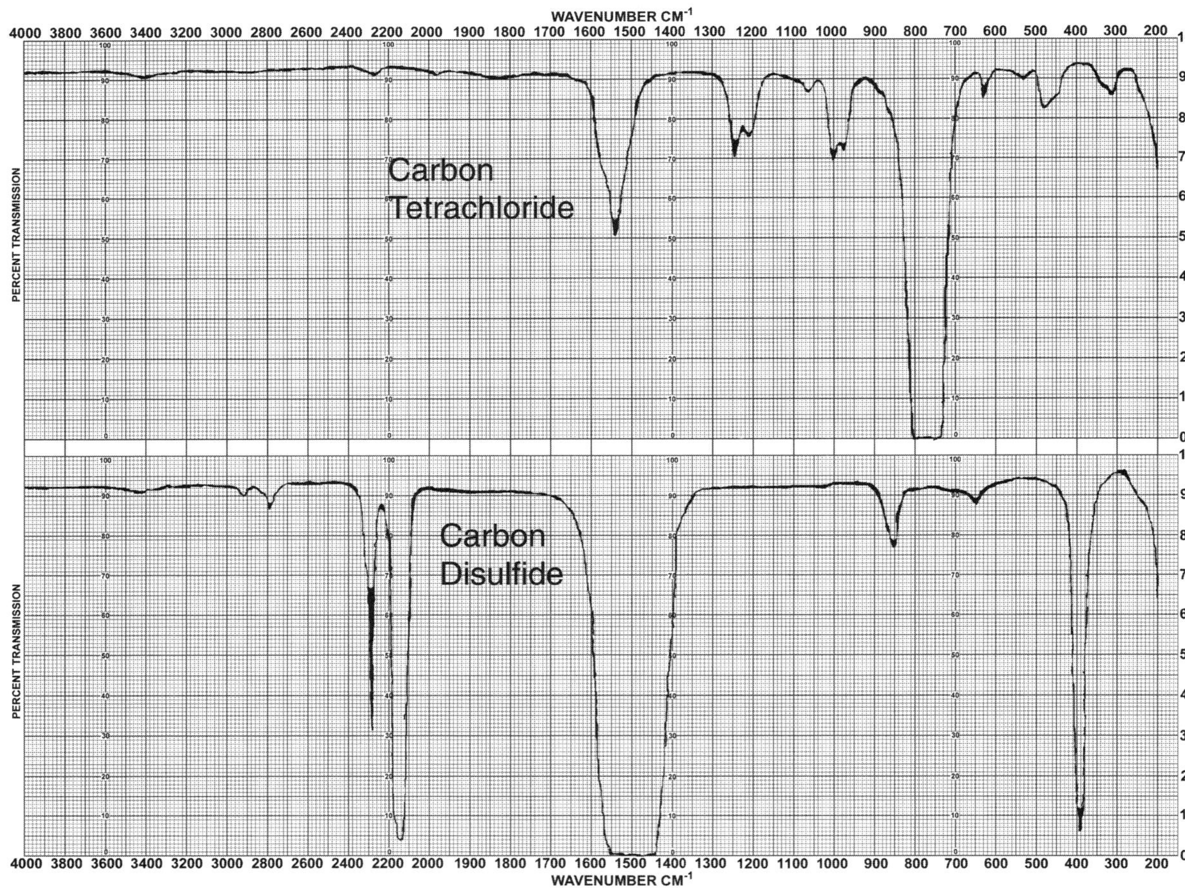

Fig. 13.5 Spectra of carbon tetrachloride and carbon disulfide.

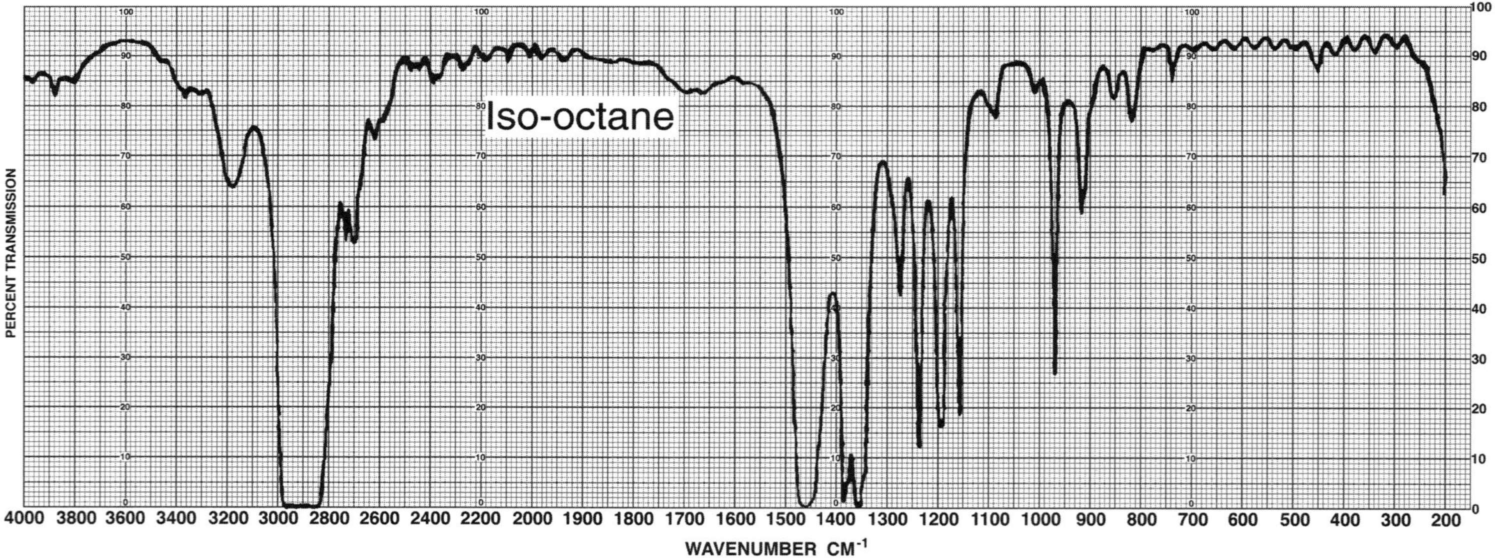

Fig. 13.6 Spectrum of isooctane.

Carbon tetrachloride from 4000 to 1350 cm^{-1} and carbon disulfide from 1350 to 400 cm^{-1} are often used as a solvent pair. Unfortunately, both have some toxic properties and should be handled only under well-ventilated conditions. In addition, carbon disulfide has a very low flash point and any source of heat may lead to an explosion. Isooctane (Figure 13.6) and hexane are also useful solvents, although they are not very polar and therefore not very good solvents.

Chloroform is more polar and can be used in some circumstances, as can acetone, dioxane, and even water. In general, as the solvent polarity increases, its IR spectrum becomes more intense and often more complex, contributing therefore to significant interferences with absorption bands of the sample.

It is sometimes desirable, for reasons discussed earlier, to obtain spectra of materials in solution. Absorption bands due to the solvent generally will obscure some of the solute absorption bands, and in these cases the spectrum of the pure solvent is obtained in the same absorption cell and subtracted (in absorbance) from the solution spectrum. Some scaling of the pure solvent spectrum will be required since there are fewer solvent molecules in the optical path of the cell for the solution. For example, a 10% solution will require a multiplication of the solvent spectrum in absorbance by about 0.9 before calculating the difference spectrum. Even when this factor is accurately chosen, there may be residual spectral distortions in the calculated solute spectrum due to solvent–solute molecular interactions that may effect band position, intensity, and shape.

In regions where the solvent totally absorbs, it is not possible to calculate the difference since the mathematical subtraction involves subtracting infinite absorbance from infinite absorbance. In these regions of total absorption the result may be full-scale noise or with some software packages a constant absorbance will be calculated. While the latter is aesthetically more pleasing, in either case the results in these regions of total absorption are meaningless. It is always advisable to know the spectrum of the solvent to take into account the effects of total absorption.

What thickness should be used, in general? (See Table 13.3.) For nonpolar materials that are normally weak absorbers, the thickness should be 0.1–0.2 mm, sometimes 0.5 mm or even thicker, to have detail in the spectra; for polar materials, it should be 0.015–0.050 mm, and for solutions, 0.1–2 mm. If the problem is a quantitative one, 0.1–2-mm cell paths are more commonly used. For quantitative applications the cell should be allowed to come to thermal equilibrium in the instrument, particularly if high precision is desired. That can take as long as one-half hour, sometimes longer, over which time the absorption will be changing as the temperature of the cell changes in the beam of the instrument.

TABLE 13.3 Choice of Cell Thickness

Nonpolar materials	0.1–0.2 mm (sometimes 0.5 mm or more)
Polar materials	0.05–0.15 mm
Solutions	0.1–1 mm (sometimes several centimeters)

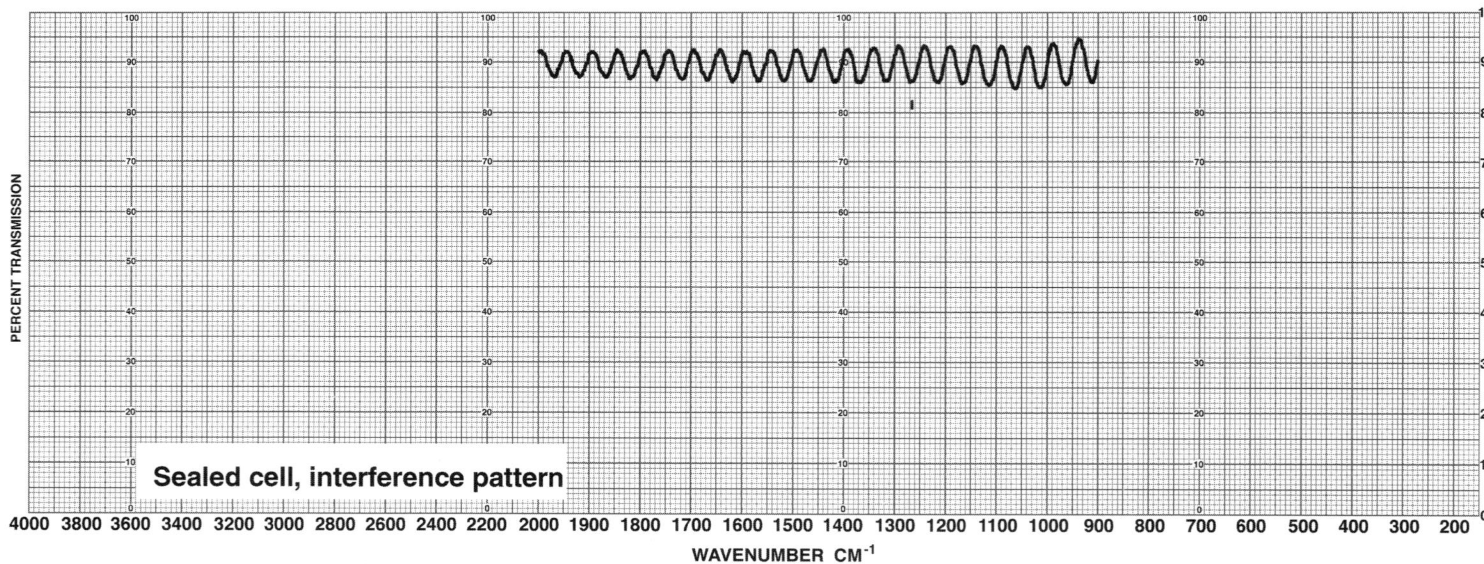

Fig. 13.7 Interference fringes from empty sealed cell for thickness calculation.

There is a general criterion that defines what constitutes a proper choice of cell thickness. If one is dogmatic about it, then the cell thickness should be chosen such that the strongest band in the spectrum has a transmission between 1 and 5%. That is a reasonable rule, particularly for reference spectra, but there are exceptions to it. A far better generalization is to use the thickness that gives you the information you want from the sample. On the other hand, if quantitative work is to be done, the thickness should be chosen such that the transmission of the analytical band is roughly between 20 and 80%. Thus, there are several factors that dictate cell thickness and the decision should be a pragmatic one—use a thickness that corresponds to the analytical problem.

Generally, the supplier indicates the thickness of a sealed cell. However, for quantitative measurements, it is necessary to know the cell thickness accurately. The thickness can be calculated from the interference fringe pattern produced by obtaining the spectrum of the empty cell. Interference fringes arise as a result of constructive and destructive light interference at the interfaces of the cell windows. Resolvable fringes will be produced for cells having a thickness in the range of 0.015 to about 0.5 mm. The relevant equation is

$$t\,(\text{cm}) = \frac{\Delta M}{2(\nu_2 - \nu_1)}$$

where ΔM is the number of full fringes between ν_2 and ν_1. In Figure 13.7, a typical interference fringe pattern is shown. Based on the fringe spacing, the thickness can be calculated as

$$t = \frac{21}{2(1968 - 913)} = 0.0995\,\text{mm}$$

If the cell is thicker than about 0.5 mm, the thickness can be determined by applying Lambert's law, $A = ab$, where A is the absorbance of a peak for which the absorption coefficient a is known and b is the thickness of the cell.

The spectrum shown in Figure 13.8 is one that everyone should look at. It is a spectrum of a silicone. It will appear many times as an impurity in samples, so it should be committed to memory.

HANDLING OF SOLID SAMPLES

Mulling Technique

To obtain the spectrum of a solid powder, it is generally necessary to reduce the particle size to something less than the wavelength of the light being used for the measurement so as to reduce light-scattering effects. A mortar and pestle are preferable, although mechanical grinders will suffice in many cases. The grinding

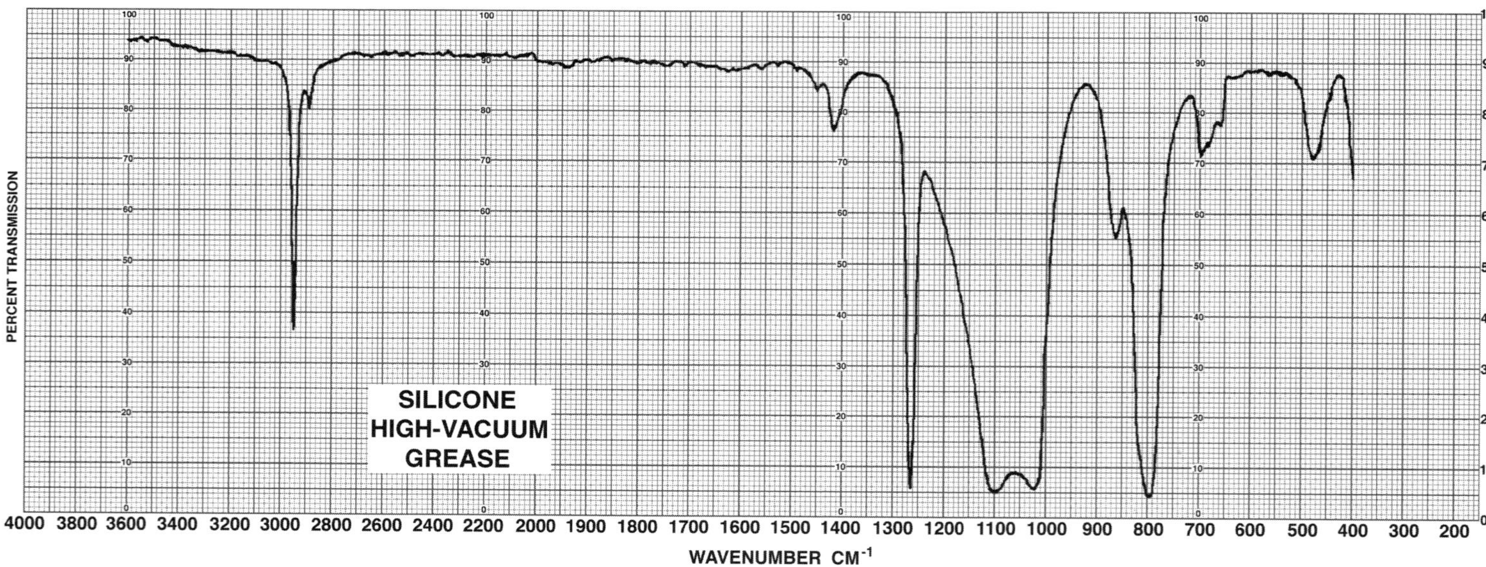

Fig. 13.8 Spectrum of a silicone, the most common impurity in the IR laboratory.

process can be controlled better and the analyst can see what is happening at any time. A sufficiently large mortar should be used along with a large enough pestle to put enough pressure to bear on the sample to grind it adequately. In the IR this means reducing the particle size to something less than a few micrometers. That is not an easy job for many materials. The process of grinding involves placing a small amount of material in the mortar, perhaps 25 mg, and the more sample there is, the longer it will take to grind. After crushing the sample, it is ground with a circular motion until it cakes on the side of the mortar and forms a glossy surface reflecting visible light. The particle size will then be less than IR wavelengths, and light scattering in the IR from the sample will be appreciably reduced. After the sample has been ground as described, a mulling fluid such as mineral oil or a perhalogenated hydrocarbon is added (see Figure 13.9).

Perfluorohydrocarbon from 4000 to 1350 cm^{-1} and mineral oil from 1350 to 400 cm^{-1} are sometimes used as a pair. The mixing is done by adding a small drop of the mulling fluid to the mortar, mixing the sample, and mulling fluid with a linear motion until a paste of about the same consistency as Vaseline is produced. At that point, take a clean rubber policeman (a device available from most chemical supply houses), scrape the paste out of the mortar, and spackle it on an IR window. Cover with a second window and press it into a thin film, moving the two windows back and forth to produce the film. The spectrum may then be scanned. The grinding process may take on the order of a few seconds for a soft powder to perhaps as long as 3–10 minutes for hard crystalline materials. In any event, the grinding must be done correctly prior to adding the mulling fluid because the particle size will not be reduced any further once the mulling fluid has been added.

A mull pair of stearic acid is shown in Figure 13.10, and it can easily be seen how the mulling fluids interfere with observation of the regions of the spectrum of stearic acid.

What are the advantages and disadvantages of the mulling technique? See Table 13.4.

KBr Disk Technique

Once again, the sample must be ground to reduce the particle size below the wavelength of the light before mixing the sample with KBr powder. The grinding techniques described for mulling may be applied here also. In addition, one of the

TABLE 13.4 Mulling Technique

Advantages	Disadvantages
1. Rapid and simple 2. Does not require expensive equipment such as dies and presses 3. Lower probability of reaction with the mulling fluid	1. Interference from absorption bands of the mulling agent 2. Sample not usually recoverable

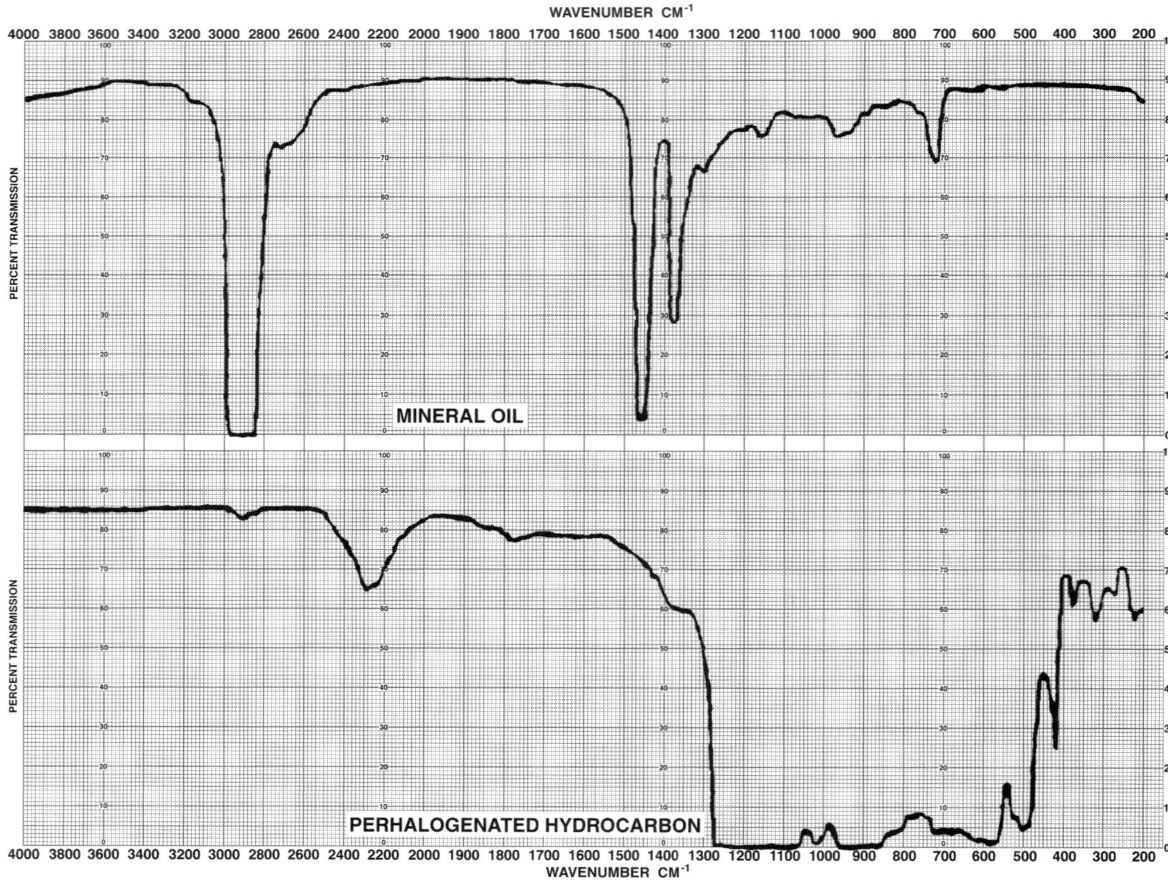

Fig. 13.9 Capillary film spectra of fluids used for mulling.

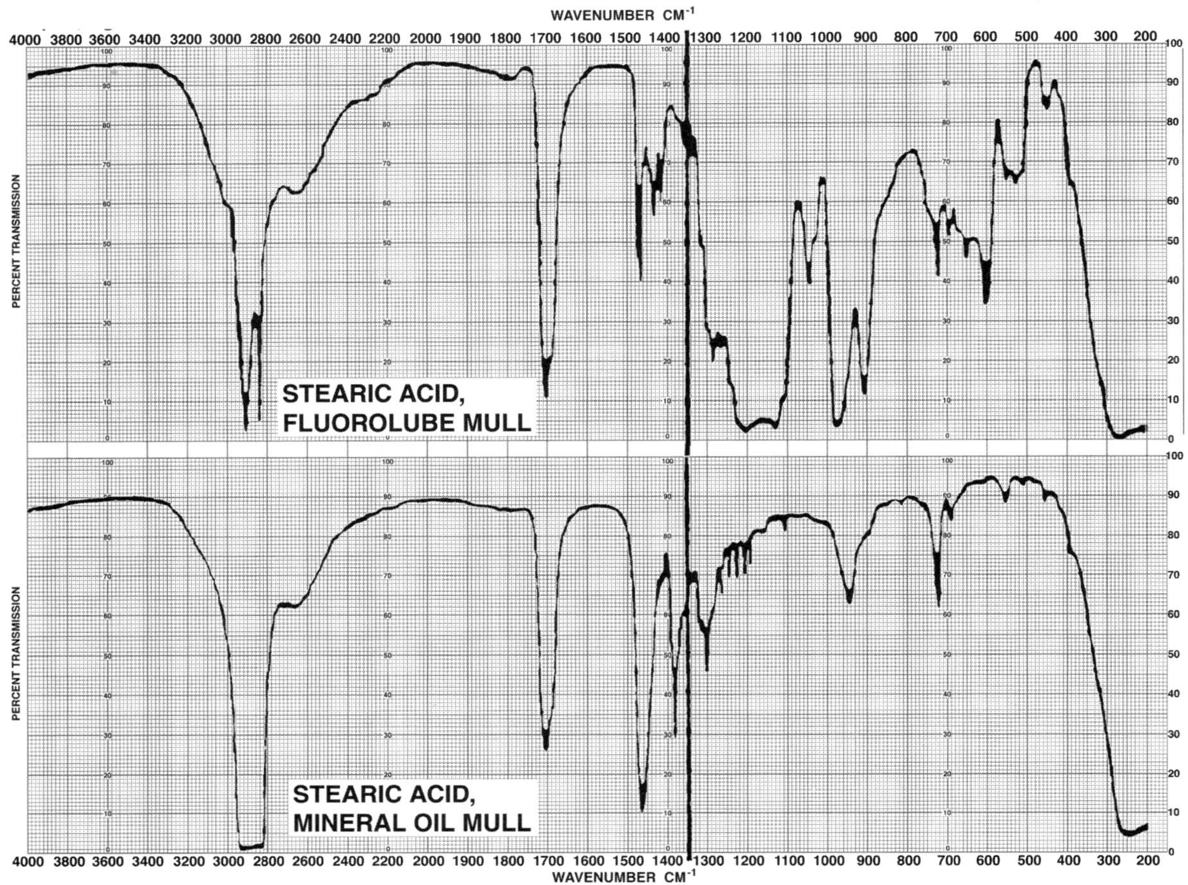

Fig. 13.10 Mull pair of stearic acid.

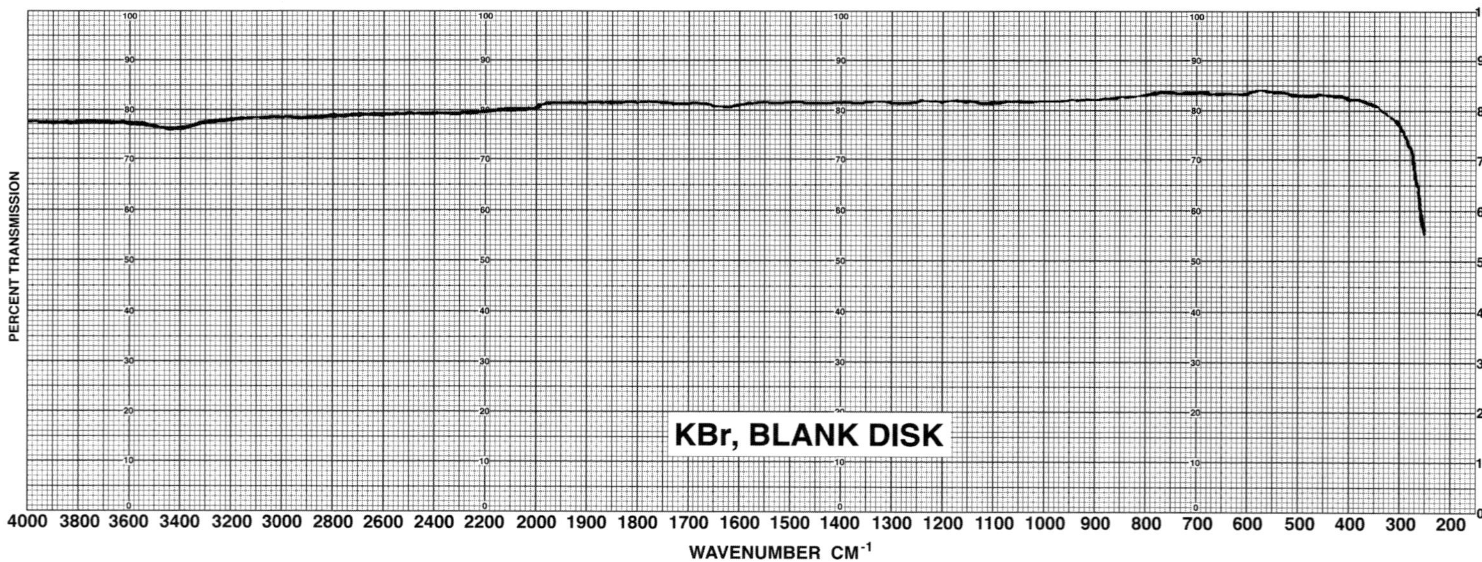

Fig. 13.11 Spectrum of a good blank KBr disk.

techniques particularly useful for reducing the particle size of minerals and inorganics is to put about 10 mg of the sample in the mortar along with about 10 drops of ethanol and to grind the material to dryness. This works quite well. Regardless of the grinding technique, a portion of the ground sample, about 3 mg, is mixed gently with 300–350 mg of spectral-quality potassium bromide powder for a 13-mm disk. Under no circumstances should the KBr powder be ground further. If it is, the net result is to create reactive, new crystal faces that take up water from the atmosphere to produce water bands in the spectrum near 3400 and 1640 cm^{-1}.

It is important to start with good potassium bromide; it must be dry and of spectral quality. A blank should be obtained for each new bottle of KBr powder. The KBr should be put in a drying oven at 110°C roughly once a week and dried overnight. A vacuum oven is even better.

Mixing to produce a homogeneous mixture is a critical step. A procedure that has been found useful is to place the weighed, preground sample in the mortar, add an equal volume (roughly) of KBr powder, and mix gently with the sample. Now add a volume of KBr roughly equal to the new volume in the mortar and mix. Continue doing this until all of the KBr powder has been added and mixed.

The sample mixed with the KBr is now poured into a die and the top ram is inserted and rotated a few times to distribute the powder evenly in the die. A vacuum is applied for 1 or 2 minutes and the powder is then pressed in a hydraulic press at the recommended pressure for the die. The spectrum may then be obtained. Since potassium bromide is being used, the lower frequency limit will be about 300 cm^{-1}. If the pellet has white areas, it is likely that the sample–KBr mixture was not leveled properly after loading the die. If it is required to go below 300 cm^{-1}, use cesium iodide, which is pressed slightly differently. For cesium iodide, apply a vacuum for 1–2 minutes after filling the die, apply pressure for 1–2 minutes, release the pressure, allow it to sit for 15–20 seconds and apply pressure again, release the pressure, and repeat the process one more time. Cesium iodide apparently requires time to relieve the strains induced during pressing in order to get a clear pellet. Spectra of KBr powder and of several samples are shown in Figures 13.11–13.15. Figure 13.11 is the spectrum of a blank KBr disk and shows weak water bands near

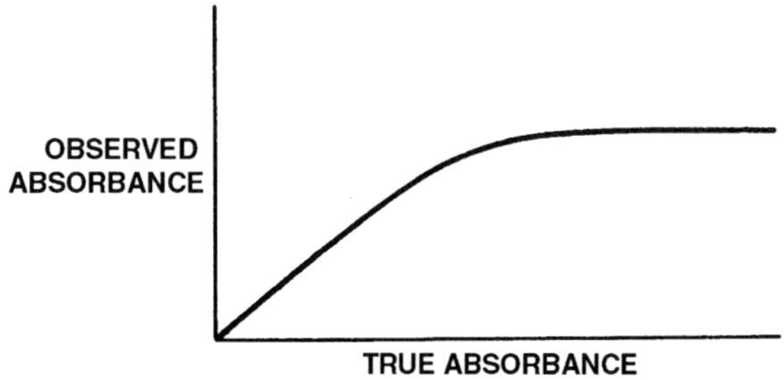

Fig. 13.12 Effect of sample inhomogeniety on measured absorbance.

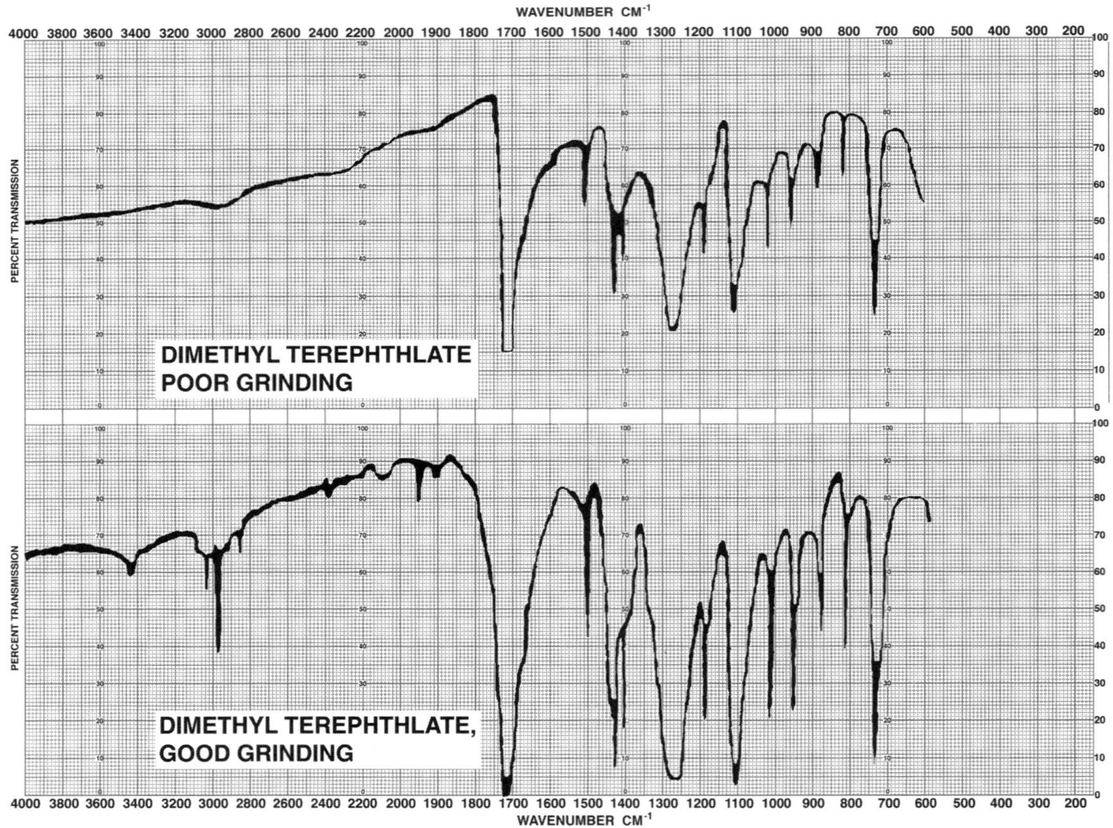

Fig. 13.13 KBr disk spectra of dimethyl terephthalate: upper spectrum, an improperly ground sample; lower spectrum, a properly ground sample.

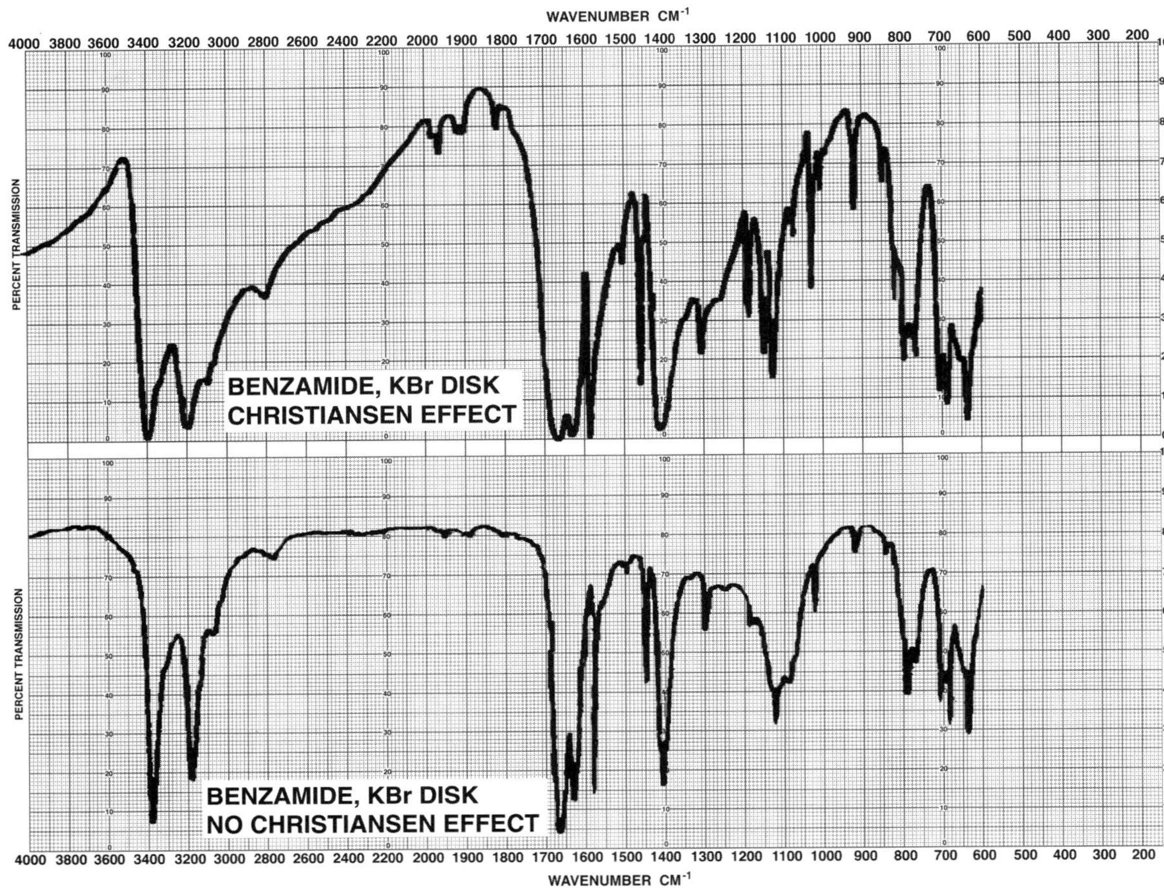

Fig. 13.14 Spectra of benzamide showing the Christiansen effect.

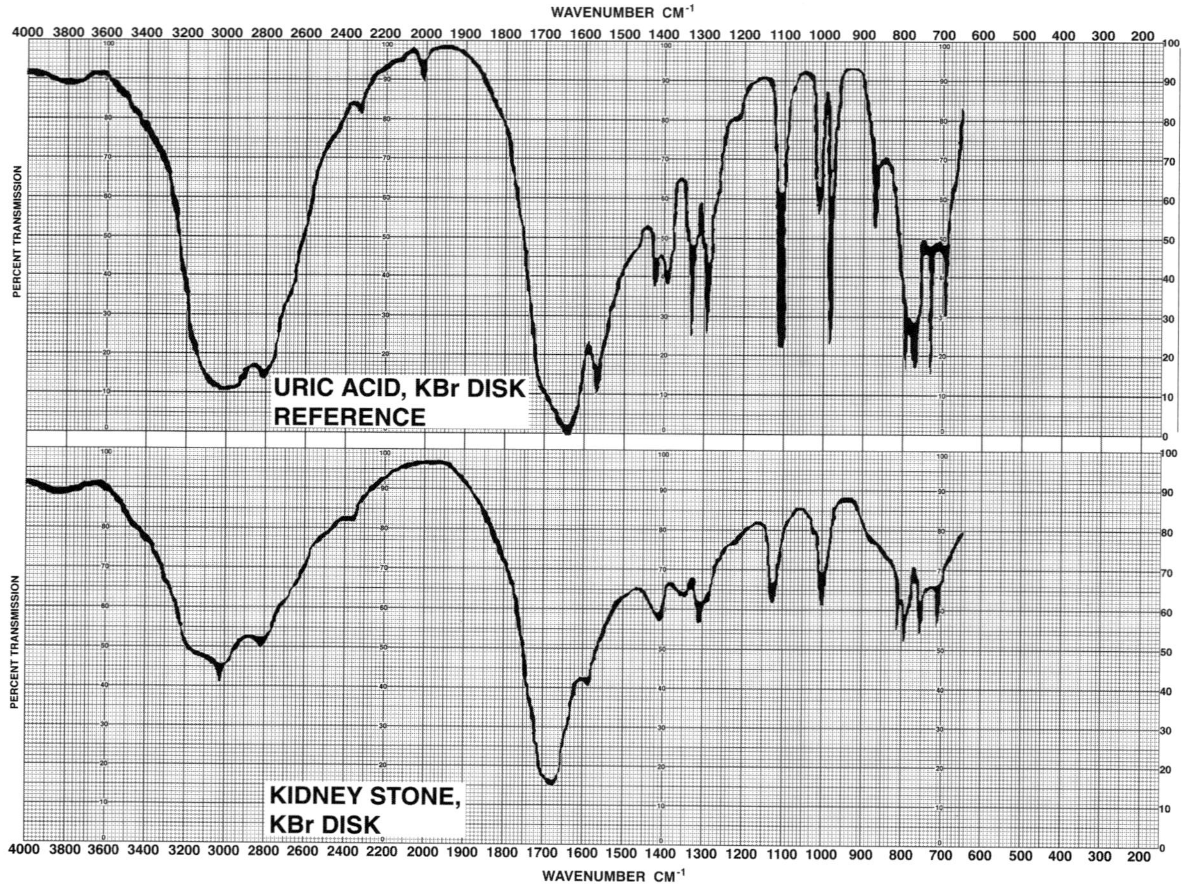

Fig. 13.15 KBr disk spectra of uric acid (upper) and kidney stone (lower).

3400 and 1640 cm^{-1} due to adsorption of water vapor after grinding. These bands would be much stronger if the KBr powder had been ground or not dried properly. As noted before, it is important both for the mulling technique and the KBr technique that a uniform dispersion is obtained. If this is not the case, then the measured absorbance will be less than the true absorbance due to the fact some of the radiation reaching the detector does not pass through sample particles (Figure 13.12). Figure 13.13 shows two spectra of dimethyl terephthalate and indicates the effects of improper grinding of the sample. The band contours in the upper spectrum show distortions due to scattering, the so-called Christiansen scattering, and refraction from larger particles, while the lower spectrum shows symmetrical band shapes.

A second example of Christiansen scattering is shown in Figure 13.14. The sample is benzamide. When the upper and lower spectra are compared, it is very easy to see the symmetric band shapes in the lower spectrum and the asymmetric shapes in the upper curve resulting from scattering from particles that are too large.

Figure 13.15 shows the spectrum of uric acid for reference and the lower curve is that of a kidney stone which is seen to be uric acid by comparison with the reference spectrum. This application of the KBr technique has been applied in clinical situations.

Figure 13.16 shows the spectra of α-alumina trihydrate (gibbsite) in the upper curve and α-alumina (corundum) in the lower curve. Both samples were prepared by grinding with ethanol as described previously. Notice the sharp symmetrical bands even for the bands near 3500 cm^{-1} in the upper curve. For the lower curve, some Christiansen scattering is present, but it should be remembered that corundum has a hardness of about 9 and is extremely difficult to grind.

Some analysts have recommended that the KBr disk containing the sample be heated several hours at 110°C to remove water. Any water in the pressed disk will not be removed by this technique, and furthermore there are likely to be changes in the sample during heating due to reaction with the KBr or to polymorphic changes. This technique is *not* recommended. Figure 13.17 shows the result of this treatment on a sample of secobarbital.

Advantages and disadvantages of the KBr disk technique are given in Table 13.5.

TABLE 13.5 KBr Pellet Technique

Advantages	Disadvantages
1. No interfering absorption bands	1. Requires accessories such as dies and presses
2. Smaller sample sizes than for mulls	2. KBr susceptible to H_2O adsorption
3. Quantitative analysis easier	3. Polymorphic changes possible due to pressure
4. Long-term storability	4. Possible ion exchange between KBr and sample

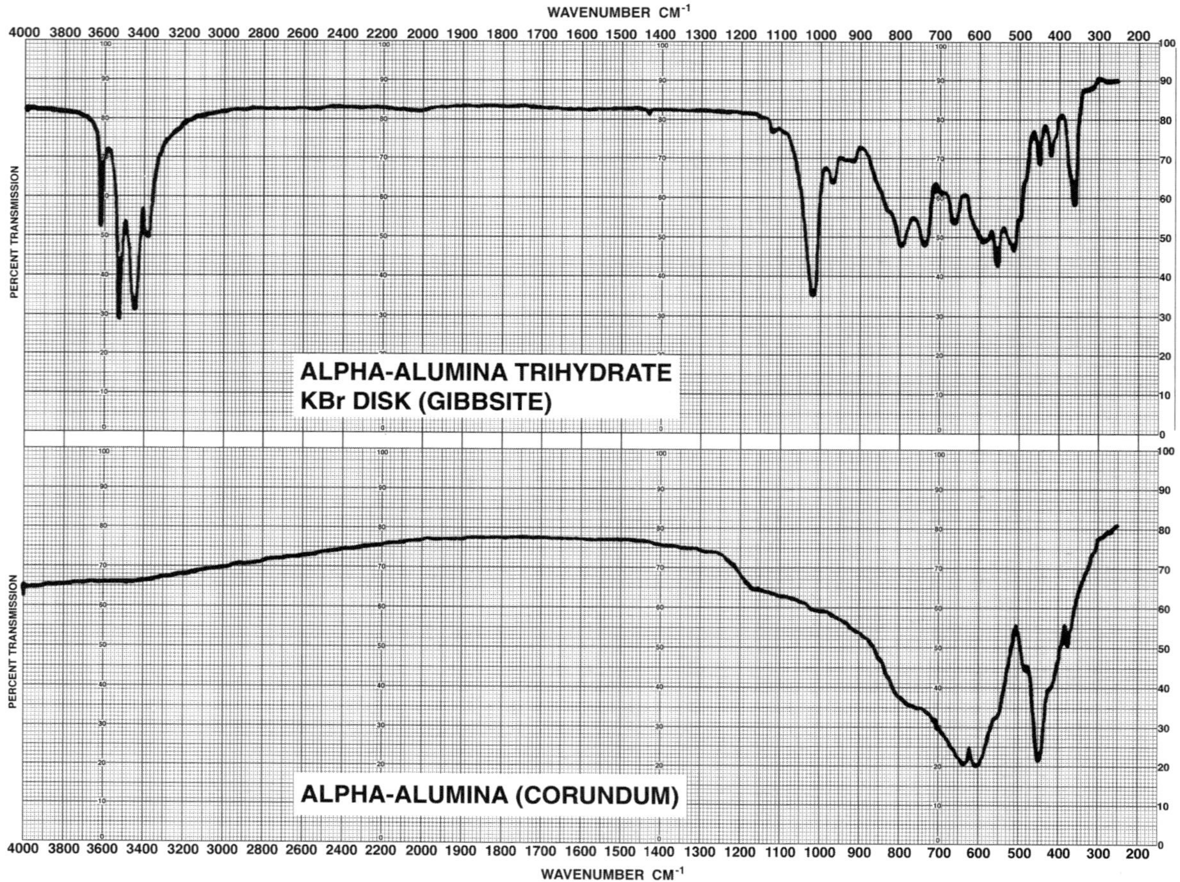

Fig. 13.16 KBr disk spectra of α-alumina trihydrate (upper) and α-alumina (lower).

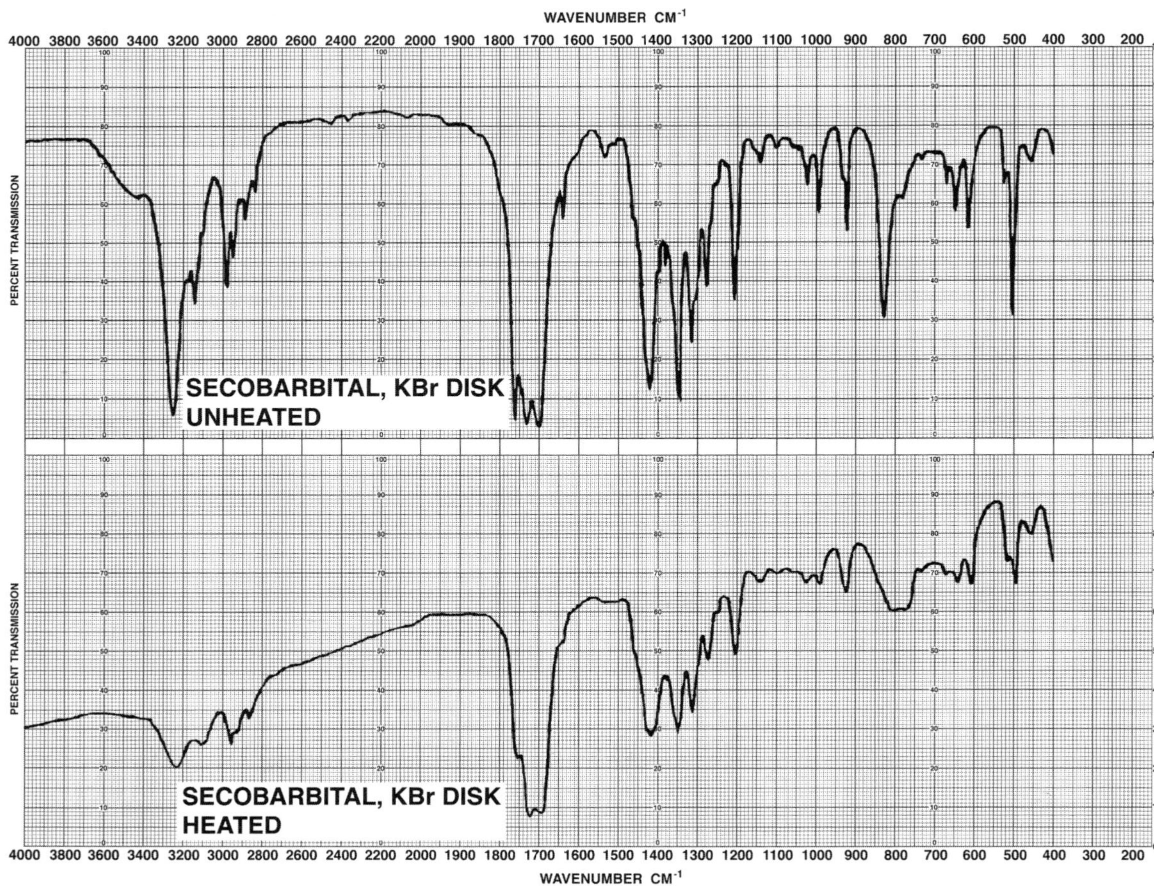

Fig. 13.17 Secobarbital in KBr disk showing effect of heating.

REFLECTION TECHNIQUES: TRANSFLECTANCE, SPECULAR, INTERNAL, AND DIFFUSE

Transmission–reflection or transflectance (reflection from the front surface is usually termed specular reflection) may be applied to samples that are pieces of metal with coatings on their surfaces. Radiation from the instrument is passed through an optical accessory (Figure 13.18), which directs the light to the surface of the sample where it passes through the coating, is reflected from the metal underneath, passes back through the coating again, and is collected and passed into the instrument. The spectrum that is recorded is essentially a transmission spectrum (e.g., Figure 13.19). This is the spectrum of a two-layer coating on the inside of a soft-drink can composed of polyvinyl chloride and epoxy layers. Artifacts may appear in the vicinity of very strong absorption bands due to specular reflection from the front surface of the coating.

Internal Reflection

Internal reflection is an important technique but one that may be misused. When it first became available in the early 1960s, it was thought to be a panacea for every kind of sampling problem. Hours were spent preparing a sample for internal reflection when it would have taken minutes for a KBr disk or a mull. Figure 13.20 shows a schematic for an internal reflection crystal.

The entrance and exit faces are cut at an angle. For crystals like zinc selenide, which is the most common one, the angle is generally 45°. The radiation is brought

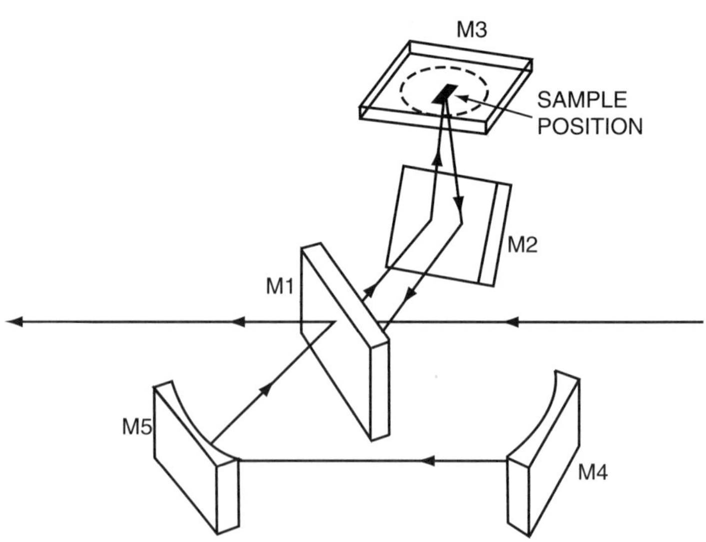

Fig. 13.18 Schematic of transflectance and specular reflectance accessory.

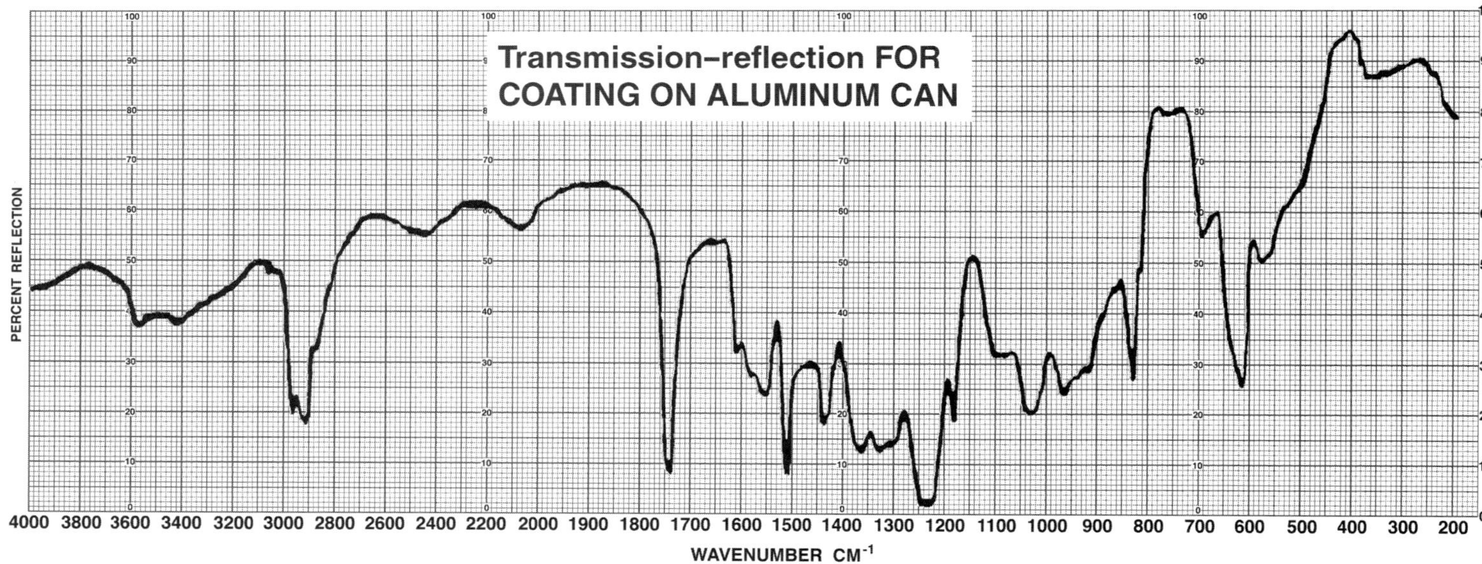

Fig. 13.19 Transmission–reflection spectrum of coating on inside of aluminum can.

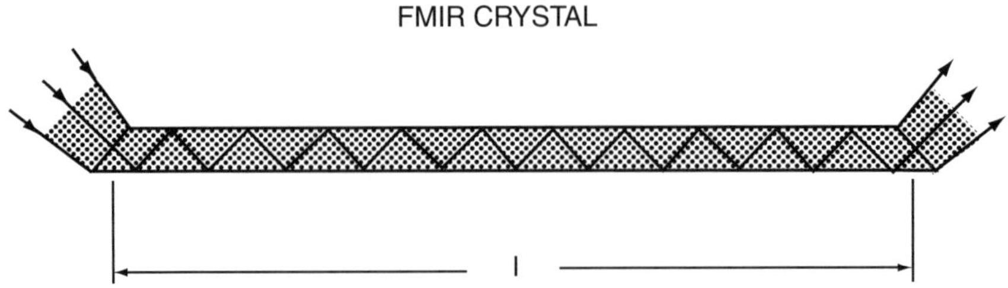

Fig. 13.20 Schematic of internal reflection crystal.

into the front face of the crystal. At an angle above the so-called critical angle, total internal reflection occurs within the crystal and the radiation just flows along, staying within the crystal like a waveguide, and comes out at the other end. The sample is placed tightly against the surface of the crystal. Some of the radiation penetrates beyond the crystal–sample interface, interaction occurs between the radiation and the sample, and a spectrum is produced. The radiation that penetrates beyond the crystal interface is referred to as the evanescent wave. Figures 13.21 and 13.22 describe the penetration depth, critical angle, and evanescent wave.

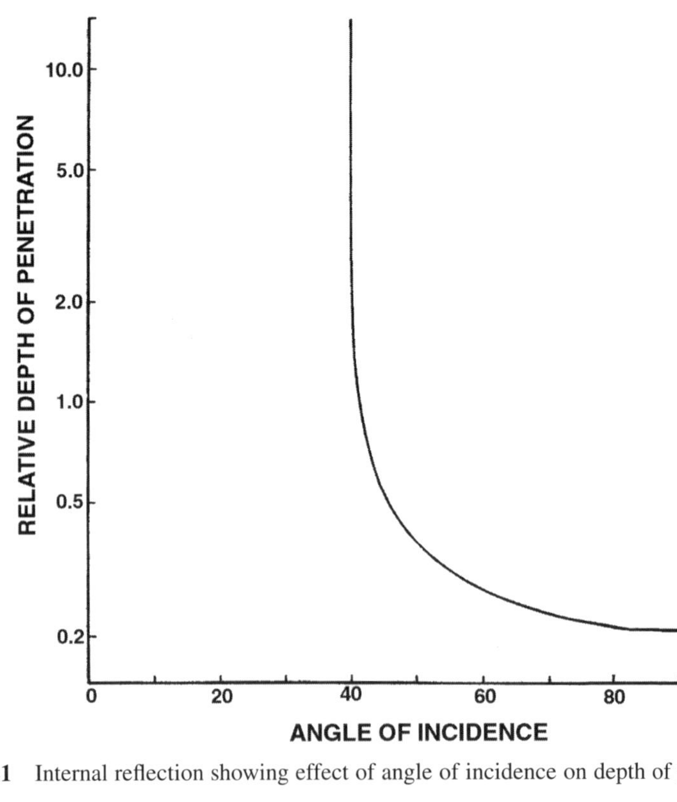

Fig. 13.21 Internal reflection showing effect of angle of incidence on depth of penetration.

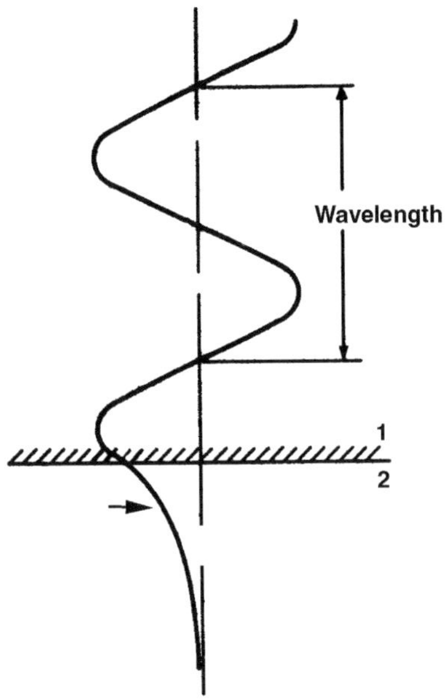

Fig. 13.22 Internal reflection, evanescent wave.

From Figure 13.21, it should be observed that as the angle of incidence decreases from 90°, the depth of penetration into the sample of the evanescent wave and, therefore, the effective sample thickness increase. The slope becomes infinite at the critical angle; all the radiation passes into the sample and is lost. At angles above the critical angle, the effective thickness is about 0.2 of a wavelength for a ZnSe internal crystal.

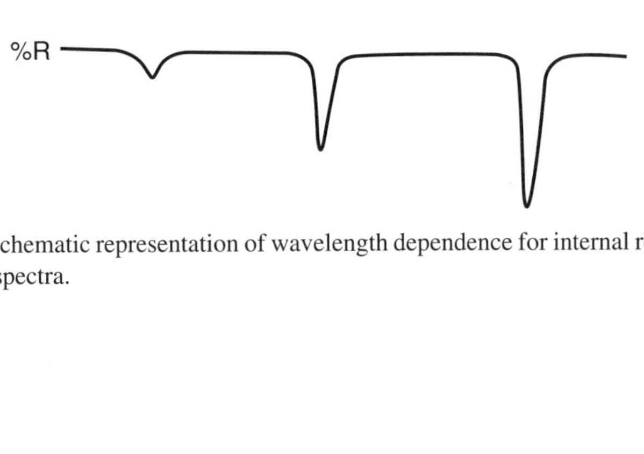

Fig. 13.23 Schematic representation of wavelength dependence for internal reflection versus transmission spectra.

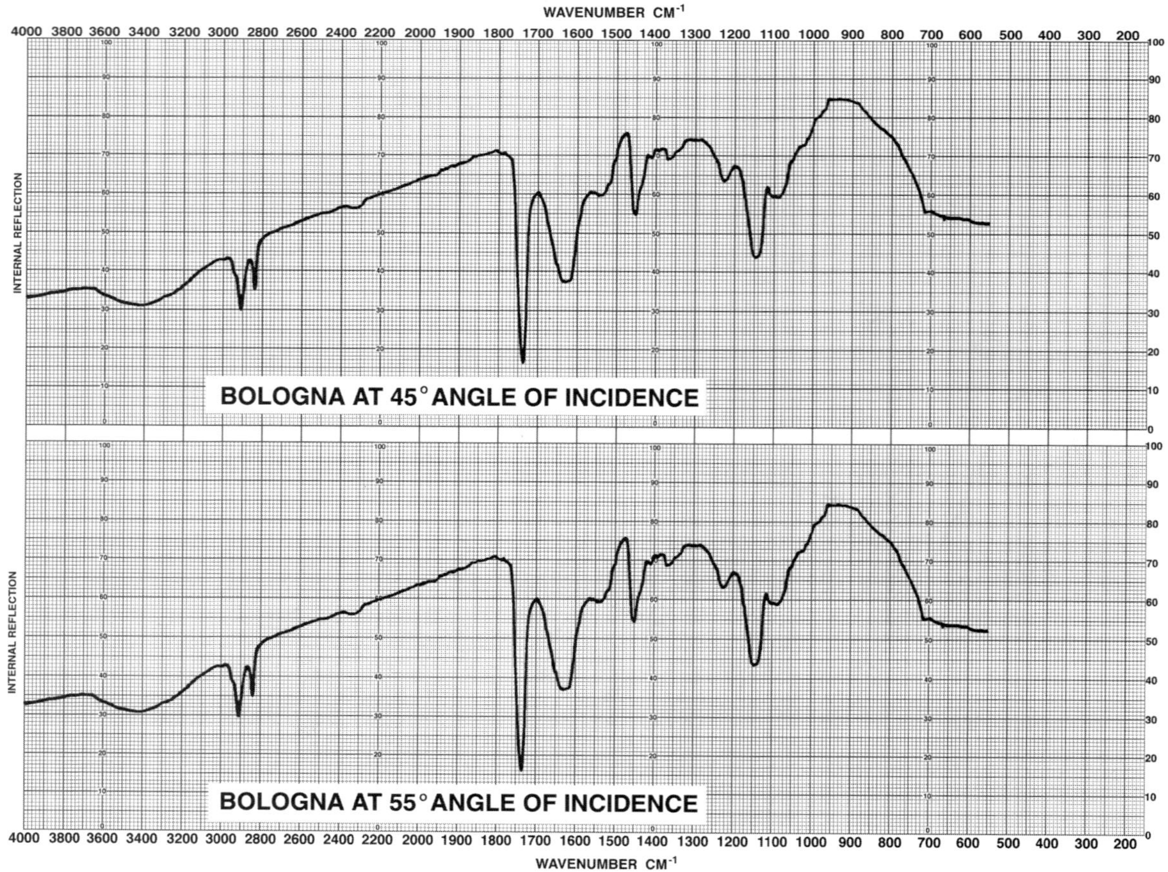

Fig. 13.24 Internal reflection spectra of bologna at two different angles of incidence.

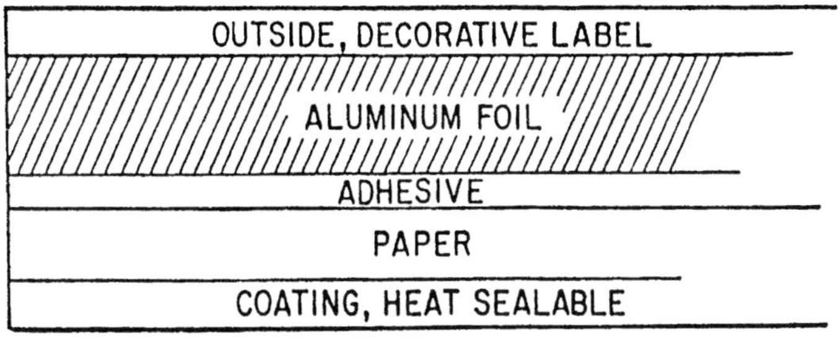

Fig. 13.25 Schematic diagram of soap overwrap.

The evanescent wave, shown schematically in Figure 13.22, decreases exponentially in amplitude with distance away from the interface between the crystal and the sample. Since the intensity varies as the square of the amplitude, its value falls off much more rapidly, thus leading to the very small effective sample thickness.

The so-called depth of penetration d_p is represented in Figure 13.21 and is the distance at which the amplitude of the radiant electric field reaches $1/e$ of the initial amplitude at the interface:

$$d_p = \frac{\lambda}{[2\pi(\sin^2\theta - n_{21}^2)^{1/2}]}$$

where λ is the wavelength, θ is the angle of incidence, and n_{21} is the ratio of the refractive index of the sample and the internal reflection crystal.

From the equation defining the depth of penetration, it can be seen that the internal reflection sample thickness varies with wavelength; longer wavelength bands appear stronger relative to shorter wavelength bands when compared with transmission spectra band intensities. This wavelength dependence is shown schematically in Figure 13.23.

Figures 13.24–13.27 show several examples of applications. The internal spectra of bologna in Figure 13.24 are instructive in two senses. First, the broad band due to water is clearly seen near 3450 cm^{-1}. Bands due to protein are seen at 1650 and 1550 cm^{-1}, and those due to fat are seen at 1745 and 1150 cm^{-1}. These bands may be used to quantitatively determine the amount of protein, fat, and water in a food sample. Second, these two spectra were obtained at two different angles of incidence and therefore represent two different thicknesses, albeit very thin. This offers scientific proof that no matter how thin you slice it, it is still bologna.

Figure 13.25 shows, schematically, the structure of a wrapper for a bar of soap. Internal reflection spectra of the organic components can be obtained easily. For example, in Figure 13.26, the spectra of the outside coating of the overwrap and the printed adhesive on the end flaps are shown. These are identified as ethylene vinyl acetate copolymer.

Figure 13.27 demonstrates a problem and a benefit of internal reflection. This is the spectrum of a residue left on the crystal after obtaining a spectrum of one of the

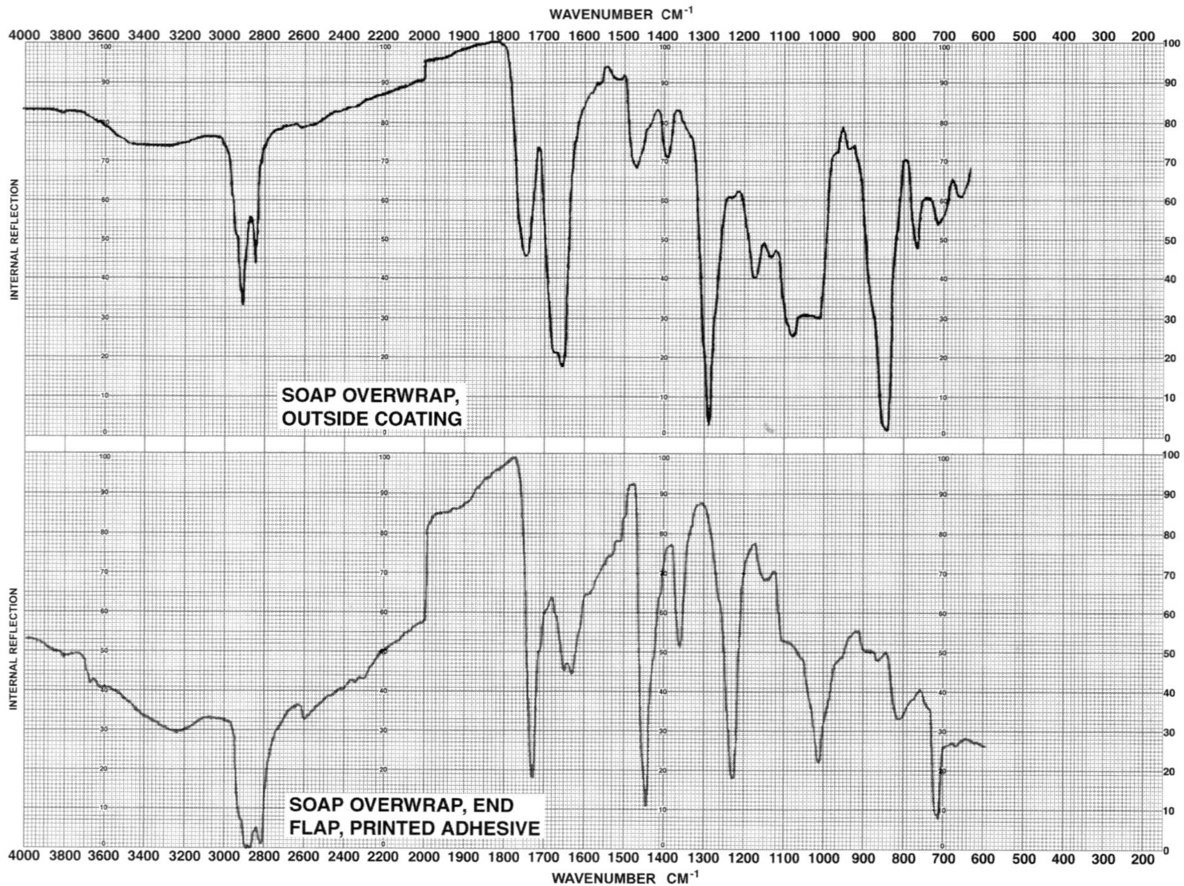

Fig. 13.26 Upper spectrum, outside coating; lower spectrum, printed adhesive on end flaps.

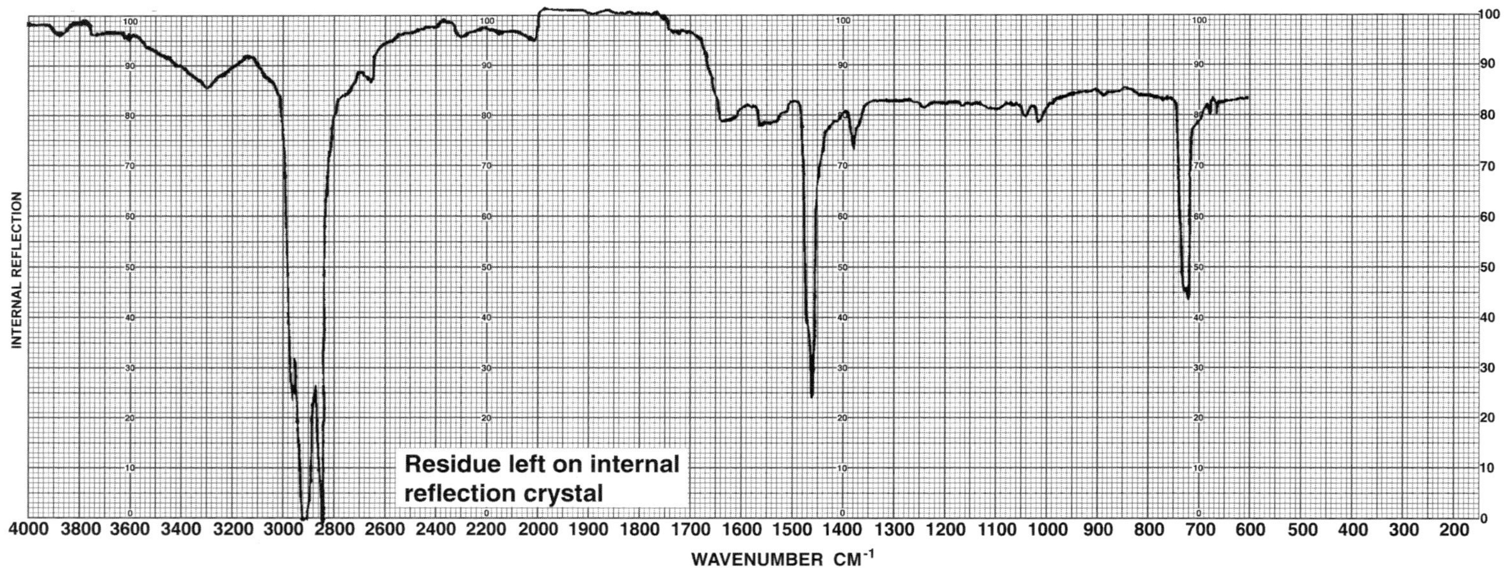

Fig. 13.27 Residue left on internal reflection crystal after running spectrum of soap overwrap.

layers. It is identified as being the spectrum of a hydrocarbon wax. Without running the blank crystal between samples, the presence of the wax would not have been detected. Furthermore, it would have contributed to subsequent spectra as an impurity. Cleaning of the internal crystal between samples is often necessary for this reason.

Other applications of internal reflection include detection of environmentally induced changes such as sun damage in paint coatings long before they are detectable by visual examination. Furthermore, the chemistry of the changes can be characterized from the IR spectrum. Materials having strong bands can be studied by internal reflection since the sample thickness is so small; water solutions would be examples of such samples.

Diffuse Reflection

Diffuse reflectance occurs when light is randomly scattered from the sample (Figure 13.28). Samples should normally be prepared as dilute (0.5–2%) mixtures in IR quality KBr or KCl. Grinding the sample prior to mixing with KBr or KCl will generally yield better quality spectra. Homogeneous mixtures can be prepared as described for the KBr technique or by using a vibrating mill. The sample mixture is placed in the sample cup of the accessory and excess powder may be tapped off or the surface may be scraped with a spatula held at about a 45° angle; the former procedure is preferred. The object is to leave a scattering surface. The sample cup is then placed in the diffuse reflectance accessory and the spectrum is obtained. This is a relatively low energy measurement, and more signal averaging may be

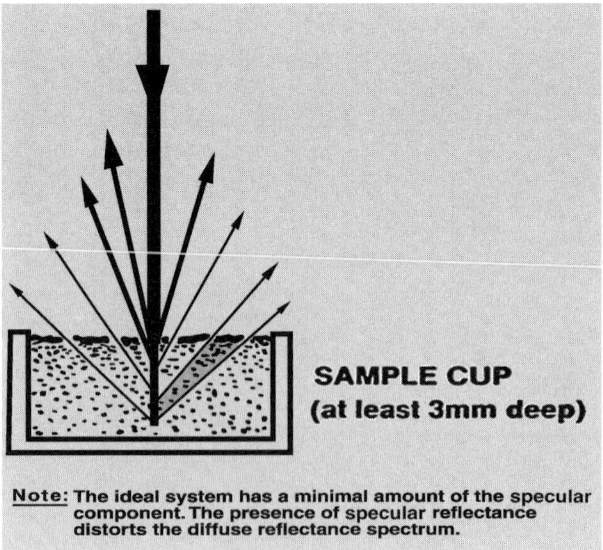

Fig. 13.28 Diffuse reflectance experiment.

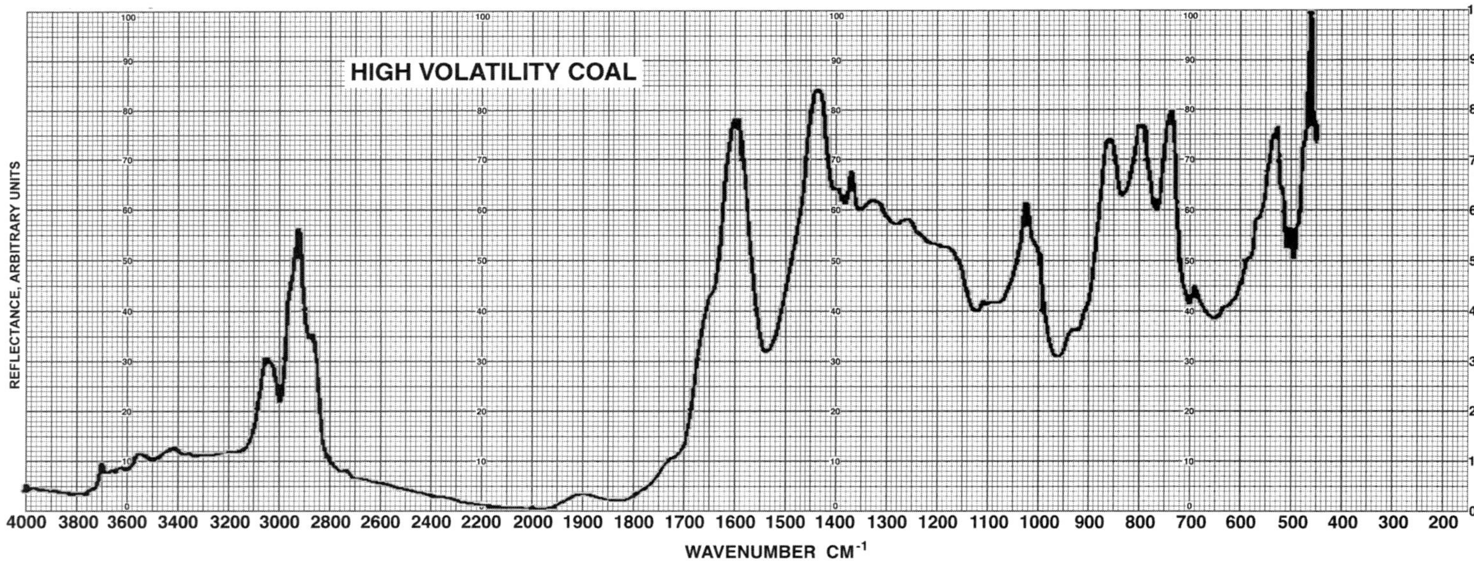

Fig. 13.29 Diffuse reflectance spectrum of high-volatility coal.

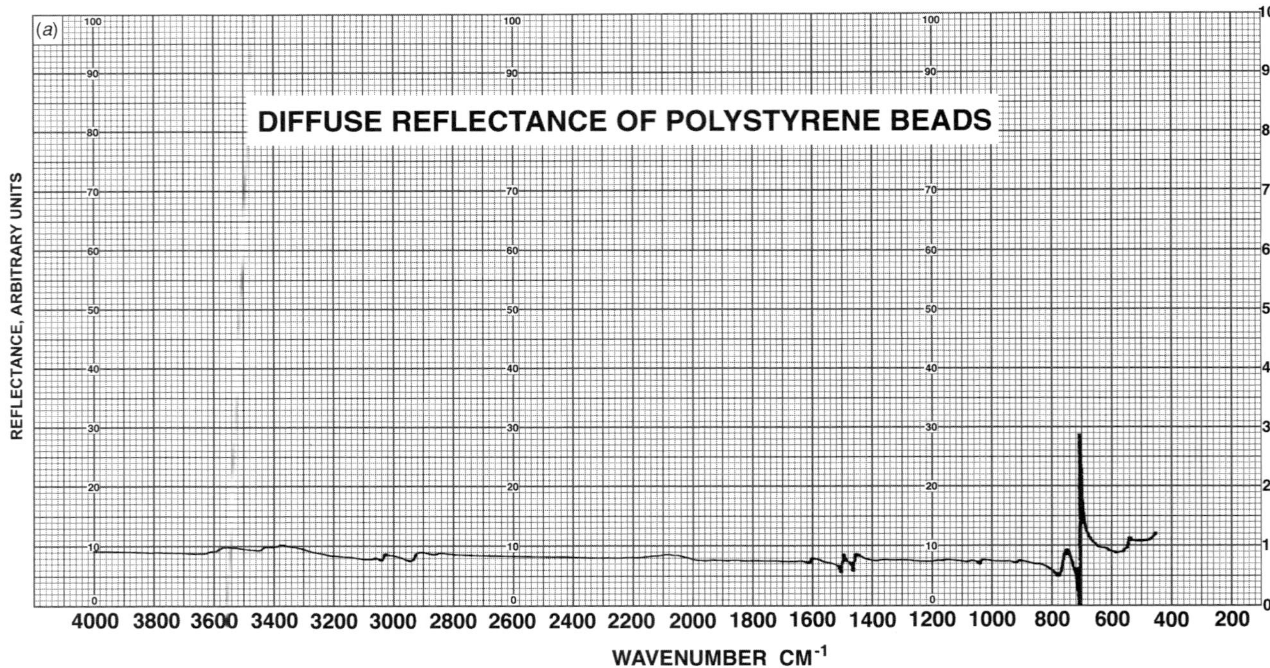

Fig. 13.30a Diffuse reflectance spectra of polystyrene beads: (*a*) direct; (*b*) by silicon carbide abrasion technique.

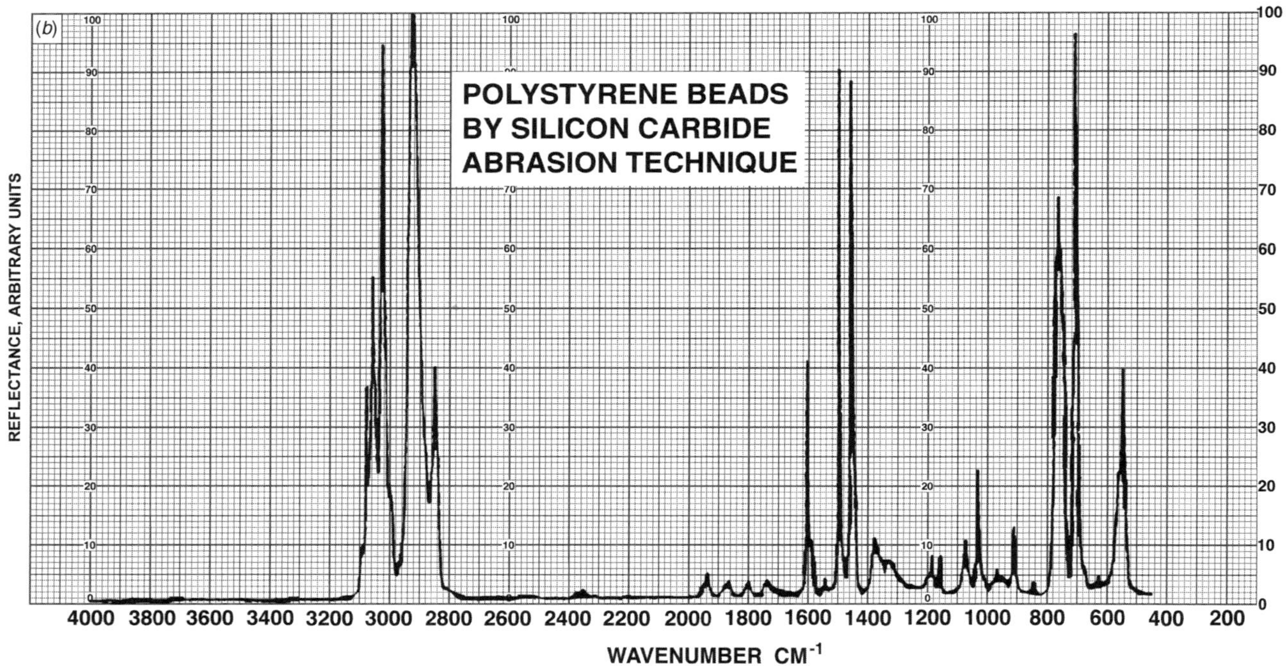

Fig. 13.30b (*Continued*)

TABLE 13.6 Diffuse Reflectance

Advantages	Disadvantages
1. High sensitivity for microsampling 2. Sample preparation minimal although grinding will improve spectra 3. With care can be used quantitatively 4. Wide variety of solid samples 5. Possibility of ion exchange minimized (compared to KBr disk technique)	1. Intensities not always consistent with transmission spectra intensities, making use of search routines difficult 2. Grinding can result in polymorphic changes for some samples

required. The background spectrum should be that of the pure matrix, i.e., KBr or KCl powder. Because the scattered light is a complex mixture of light scattered from surfaces of the powder in the cup as well as light that has passed through absorbing sample particles, the Kubelka–Munk algorithm is often applied to produce spectra that are more nearly like transmission spectra. This is particularly true in quantitative problems.

The example shown in Figure 13.29 is the spectrum of a high-volatility coal. The sample was prepared by grinding as described under the discussion of mulling at a concentration of about 10% with KBr powder. Coal is a difficult sample to prepare for an IR spectrum, and diffuse reflectance provides a relatively easier way to produce good data. Quantitative analysis has been performed on coal samples using this technique (see Fredericks et al.).

Figure 13.30 shows spectra of polystyrene beads. The upper spectrum was obtained by placing beads in the diffuse reflectance cup with KBr powder. The lower curve was obtained using an abrasion obtained using a technique developed by Spragg. In this procedure, a background spectrum is obtained by placing a piece of 400-grit silicon carbide paper on the top of the diffuse reflectance cup. The sample is prepared by lightly abrading one of the beads with the silicon carbide paper to produce a white deposit on the paper. The silicon carbide paper is then cut to fit and placed on top of the diffuse reflectance cup. The spectrum is substantially better than the upper curve and can be easily recognized as polystyrene.

Table 13.6 summarizes the advantages and disadvantages of diffuse reflectance.

CONCLUSION

The sampling techniques described in this chapter provide opportunities to handle all kinds of samples under a variety of conditions. The wide variety of sampling choices is one of the strong attributes of IR spectroscopy, which makes it so useful in the analytical laboratory. Perhaps the most critical part of making a choice is to have the best possible definition of the analytical problem.

14 Infrared Spectra of Mixtures

ROBERT W. HANNAH

I. Introduction

Many samples received in the analytical laboratory are not nice pure materials whose spectra lend themselves to the kind of interpretation procedures that have been described in this text. Rather, they are mixtures of two to many components. In many cases, it may be possible to characterize the general types of compounds present, but the severe overlap of bands which occurs, particularly in the fingerprint region, often makes the interpretation difficult or virtually impossible. There are several steps which may be defined in the processing of such a mixture:

A. Obtain a definition of the analytical problem. This may require several conversations before agreement on a real and complete definition understood by the client and the analyst is reached.
B. Obtain a history of the sample. This may lead to an understanding of sources of contamination and a further clarification of the problem definition.
C. Are any other physical or chemical properties known? Has gas or liquid chromatography been performed, thus indicating the number of components? Has a mass spectrum been obtained and, therefore, molecular weights of the components? Does a UV spectrum exist? And so on.
D. Run the IR spectrum.
E. Identify as many as possible of the functional groups present. Use the interpretive skills developed in this text to characterize the major functional groups.
F. On the basis of information obtained in A and E above, apply other instrumental techniques and separation procedures. Knowing which major functional groups are present should aid in defining a separation procedure.
G. Characterize and identify the separated components by group frequency analysis and by direct spectral comparison with reference data. Use other

Course Notes on the Interpretation of Infrared and Raman Spectra, by Dana W. Mayo, Foil A. Miller, and Robert W. Hannah.
ISBN 0-471-24823-1

instrumental and wet chemical methods as needed. Do not forget melting and boiling points.

H. Make sure the sum of the spectra of the components is the spectrum of the mixture. Components can be lost during the separation processes, and contaminants may be added. Processing blanks through the same procedures as the unknown is a necessary part of good analytical practice.

II. Separation techniques

In the following sections, a series of case studies will be described which demonstrate many of the separation procedures outlined in this section. In each case the separation procedure will be described and the results summarized.

A. *Column chromatography.* Glass columns 1 cm diameter by 20 cm long packed with silica or alumina can be used to separate materials based on their polarity and solubility in solvents and/or solvent mixtures. The fractions collected are evaporated and spectra obtained. Blanks for the solvents must be obtained. All operations must be carried out in a good hood.

B. *Thin-layer chromatography (TLC).* Equipment for this is available from laboratory supply houses. Small amounts of mixtures are separated and observed by UV irradiation or by spraying a parallel set of spots with sulfuric acid. The components of interest are extracted with solvents and spectra obtained.

C. *TLC and Wick-Stick technique.* This will be described in detail later in this chapter. The Wick-Stick allows the soluble components to be separated from the chromatographic media.

D. *Gas chromatography.* Combination of IR with gas chromatography to identify the components in a complex mixture has been a long desired goal. The high sensitivity and speed of Fourier transform spectroscopy made the combination feasible. This will not be discussed further in this chapter.

E. *Dialysis.* This is a separation technique based on molecular size. Smaller molecules pass through a membrane; larger molecules remain behind. An example is given.

F. *Pyrolysis.* A destructive distillation of the sample is performed and the liquid and/or vapor products examined as a means of identifying the original material. It is particularly useful for highly crosslinked insoluble materials. This will be discussed in detail.

G. *Internal reflection–extraction.* Small amounts of material extracted from samples such as fabrics or polymers may be sampled easily and produce good internal reflection spectra. An example will be given.

H. *Liquid chromatography.* As for gas chromatography, the combination with IR allows components in mixtures to be separated and spectra obtained. Combining with internal reflection techniques allows very small fractions to be processed and spectra obtained. This will not be discussed further.

I. *Computer manipulation.* Using the power of the software algorithms available, it is often possible to carry out separations or to distinguish differences without application of any physical or chemical separation processes. The results may be easier to obtain if it is possible to partially separate components, e.g., partially extract the plasticizer from a polymer. Subsequent application of absorbance subtraction routines allows the spectra of the plasticizer and the polymer to be obtained. Several applications will be described.

J. *Infrared microscopy.* The IR microscope provides sufficient spatial resolution in many cases to allow isolation of components in a mixture based on crystalline form or other optical properties. An application will be discussed.

III. Examples

A. Lubricating grease

The problem in this case study was to identify the major constituents in a sample of lubricating grease.

1. Run a spectrum of the grease (Figure 14.1). The sample was prepared by smearing a small amount of the grease on a KBr window and squeezing into a thin film with another KBr window.
2. Extract with methyl isobutyl ketone (MIBK). Put some of the grease in a beaker, add MIBK, and stir thoroughly. Perform this operation in a well-ventilated hood.
3. Filter the insolubles and wash the precipitate carefully with several small washings of MIBK. Evaporate the filtrate in a hood until most of the MIBK is gone. Submit part of the precipitate for elemental analysis.
4. Prepare fractions for IR examination. The solid precipitate was prepared as a KBr disk (Figure 14.2). Several drops of the filtrate (after being nearly evaporated to dryness) were prepared as a cast film (Figure 14.3).
5. Convert the thickening agent, a soap, to the free acid by treatment with dilute HCl and extract with chloroform (see Figure 14.4). This step was performed because large sets of reference spectra of carboxylic acid salts were not available, while good collections of reference spectra of carboxylic acids were available.
6. Interpretation
   a. The solid portion yielded a spectrum (Figure 14.2) of a carboxylic acid salt. This is the thickening agent. Elemental analysis showed the major cation present to be lithium.
   b. Conversion of the salt to the free acid (Figure 14.4) allowed it to be identified as 12-hydroxy stearic acid. Therefore, the thickening agent is lithium 12-hydroxy stearate.
   c. The filtrate yielded the spectrum (Figure 14.3) of a hydrocarbon oil, and this is the basic lubricating ingredient in the grease.
7. Remember to carry a blank through these procedures in order to observe any contaminants from the solvents or glassware.

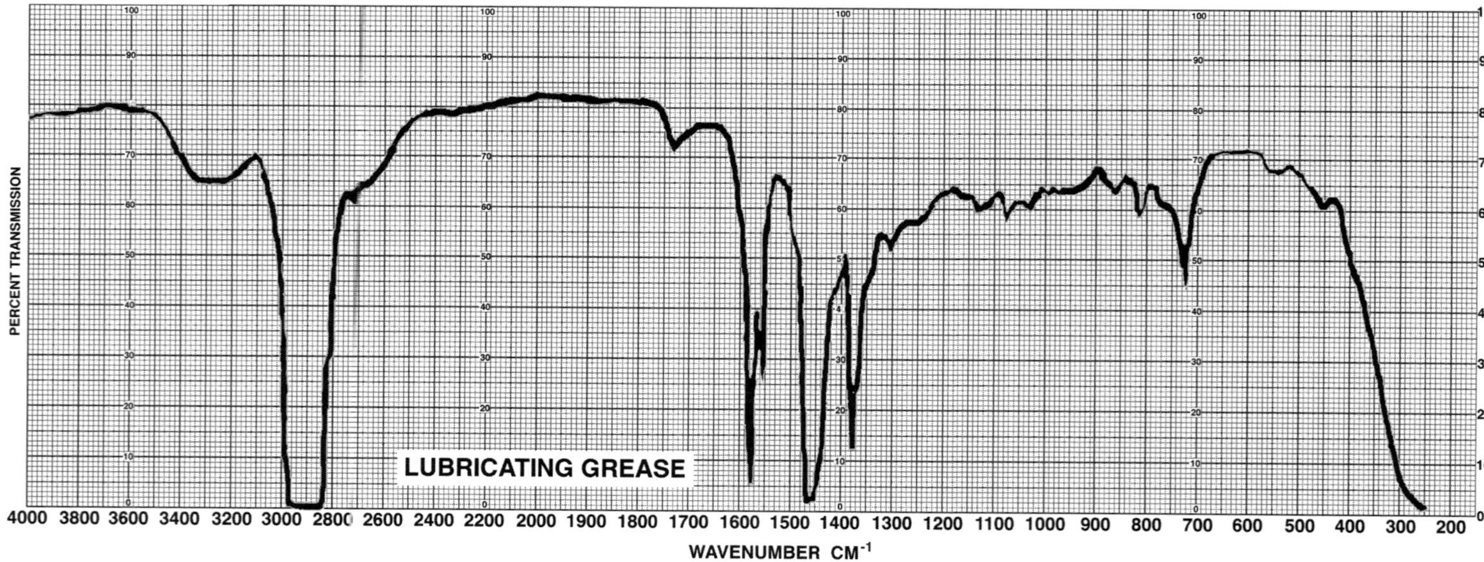

Fig. 14.1 Infrared spectra of lubricating grease.

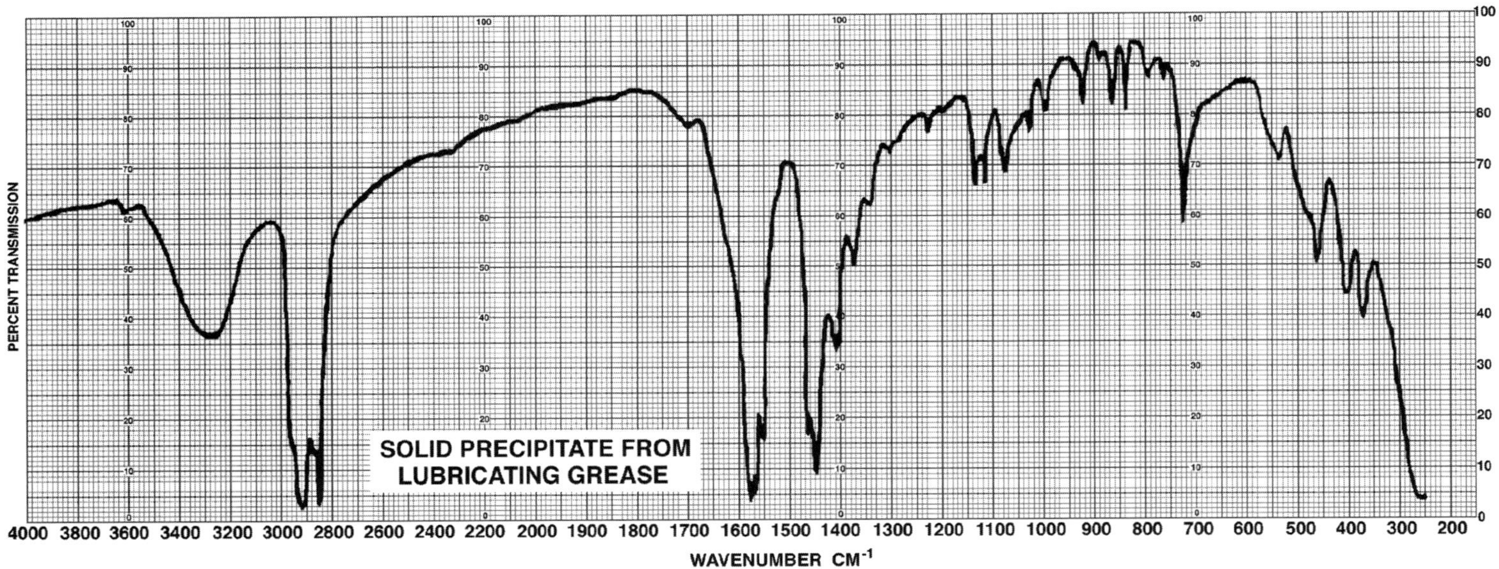

Fig. 14.2 Spectrum of solid precipitate from lubricating grease.

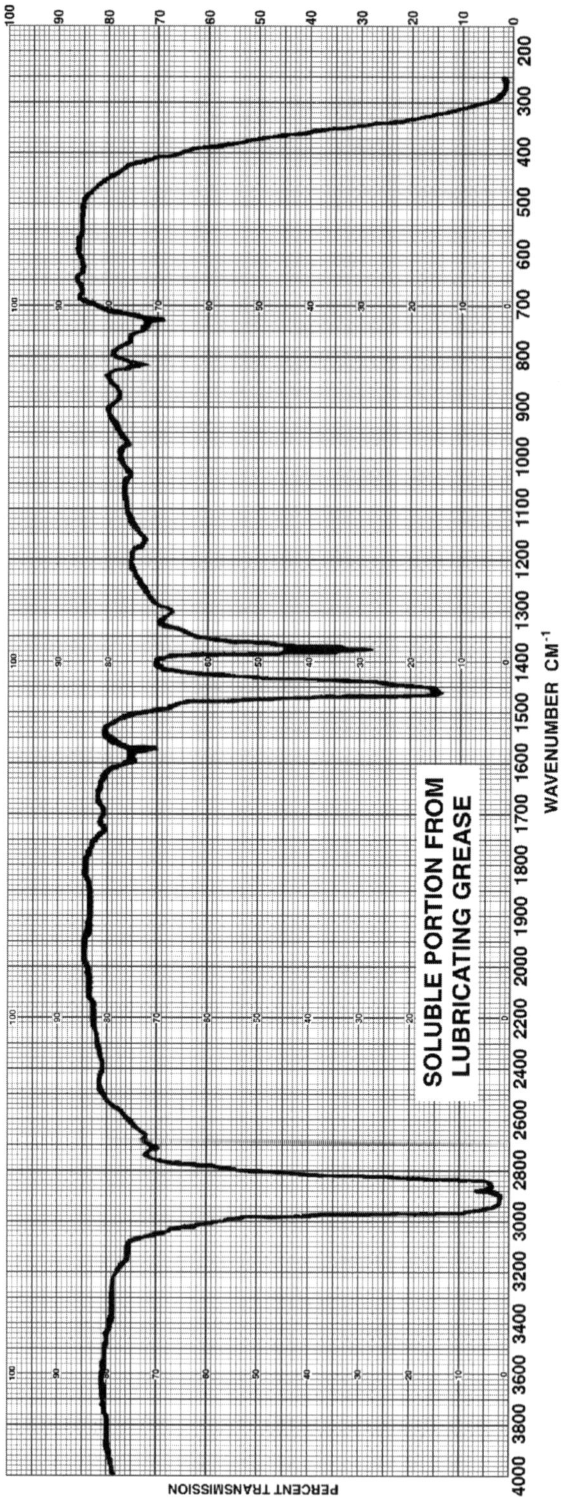

Fig. 14.3 Spectrum of soluble portion from lubricating grease.

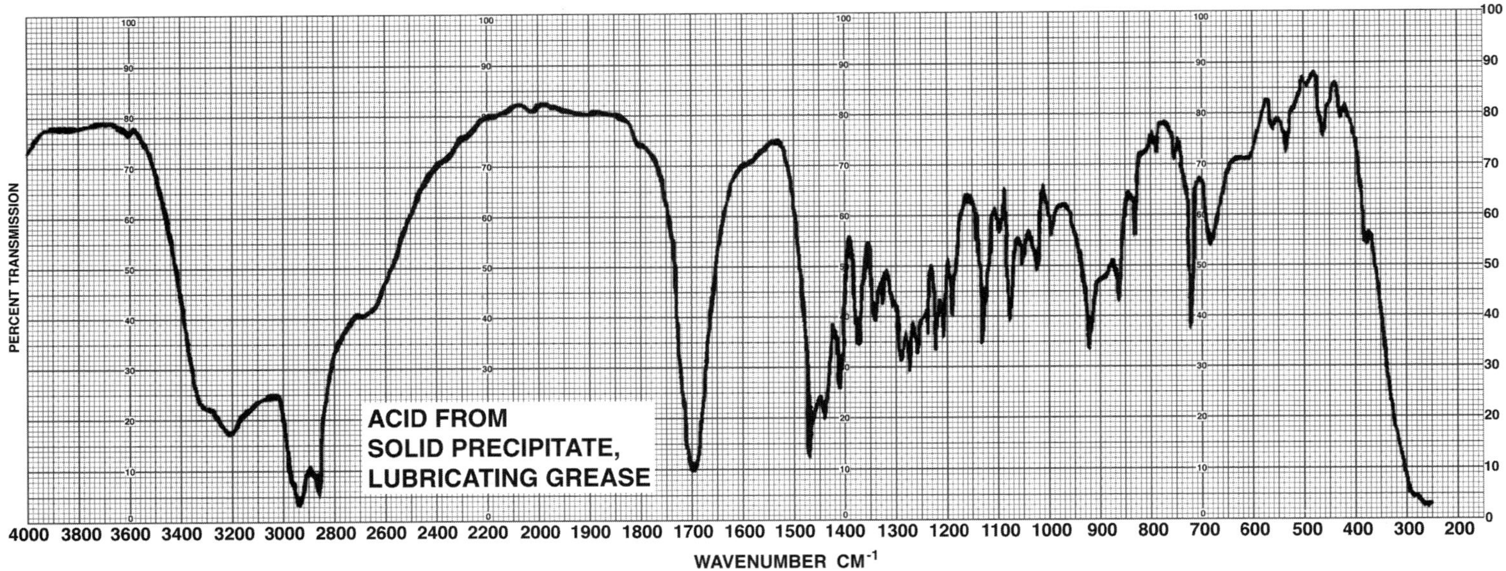

Fig. 14.4 Spectrum of acid from thickening agent in lubricating grease.

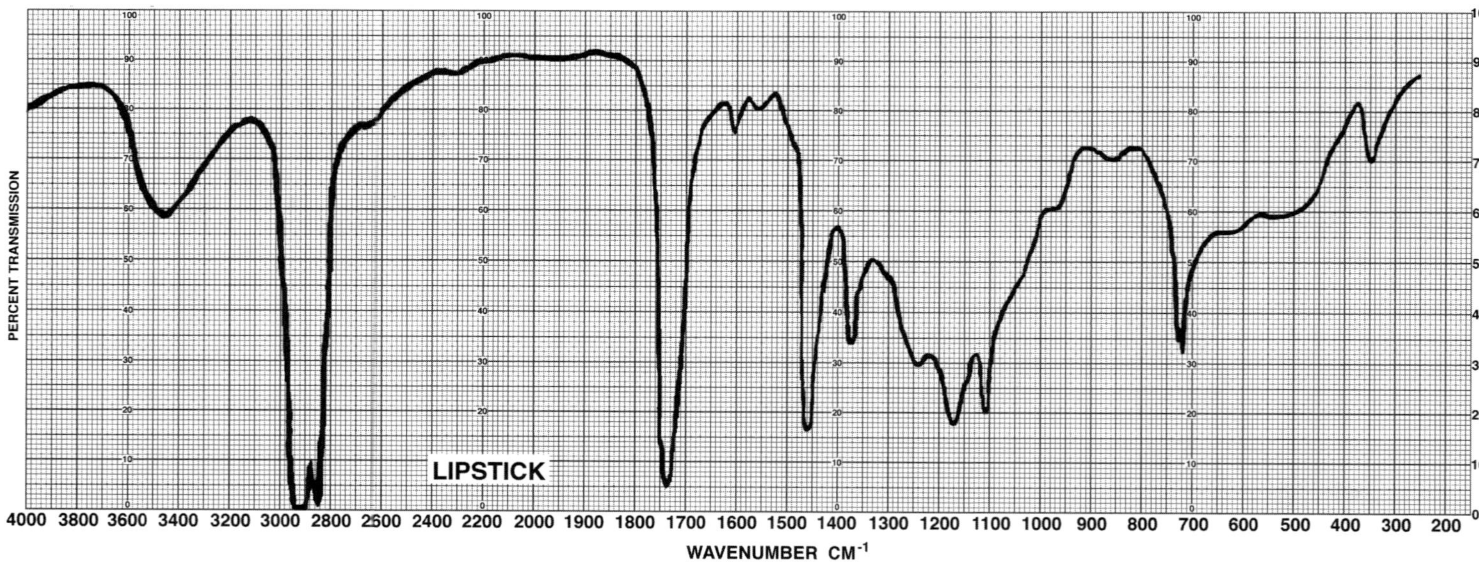

Fig. 14.5 Infrared spectrum of sample of lipstick.

8. Make sure spectra of the components (Figures 14.2 and 14.3), when added together, account for all the bands seen in the original spectrum (Figure 14.1).

B. Lipstick analysis

1. Obtain a spectrum of the lipstick sample. (See Figure 14.5.) This is done simply by smearing the lipstick on a CsI window.
2. Extract a small portion of lipstick with hexane, evaporate the hexane, and wash the hexane-soluble portion with hot ethanol, cool, and filter to isolate the wax. Prepare as a cast film (Figure 14.6).
3. Extract hexane insolubles with ethanol. Evaporate ethanol solution to dryness and prepare a KBr disk of the dyes (Figure 14.7*a*).
4. Evaporate the hexane-soluble portion and ethanol-soluble portion (step 2) on a KBr window to obtain a spectrum of an oil (Figure 14.7*b*).
5. Ash a portion and prepare the ash as a KBr disk to obtain a spectrum of the inorganic fillers (Figure 14.8). This is an excellent means to separate inorganics from organics.
6. Interpretation
   a. Figure 14.5, the spectrum of the lipstick, indicates a medium-intensity OH stretch, aliphatic CH stretches, an ester carbonyl, and in the 600–250 cm^{-1} region some inorganics. Based on these facts and some knowledge of what a lipstick might contain, the separation procedure was outlined as described.
   b. The wax (Figure 14.6) was identified as most likely beeswax, a common component in lipstick at the time of this analysis.
   c. The dye (Figure 14.7*a*) was identified as D&C Orange #17.
   d. The oil (Figure 14.7*b*) was identified as an oil similar to castor oil. Exact identification is not possible from the IR alone since many of these oils have very similar spectra. Nevertheless, castor oil was known to be used in lipstick.
   e. The inorganics (Figure 14.8) were identified as barium sulfate and the rutile form of titanium dioxide.
   f. These components account for all the bands seen in the original lipstick spectrum.

C. Thin-layer chromatography

The problem was to identify the individual barbiturates in a mixture of barbiturates. While barbiturates all have very distinctive spectra, they have similar IR spectra, and in a mixture of even two barbiturates, generally it is virtually impossible to identify the individual components. Therefore, a separation procedure, such as TLC, is advised.

1. Spot the sample on a silica gel TLC plate.
2. Develop the chromatogram with a chloroform–acetone solvent mixture.
3. Detect the spots with a UV lamp.
4. Scrape off the spots and extract with chloroform. It may be necessary to filter to remove any silica gel from the TLC plate. Alternatively, the

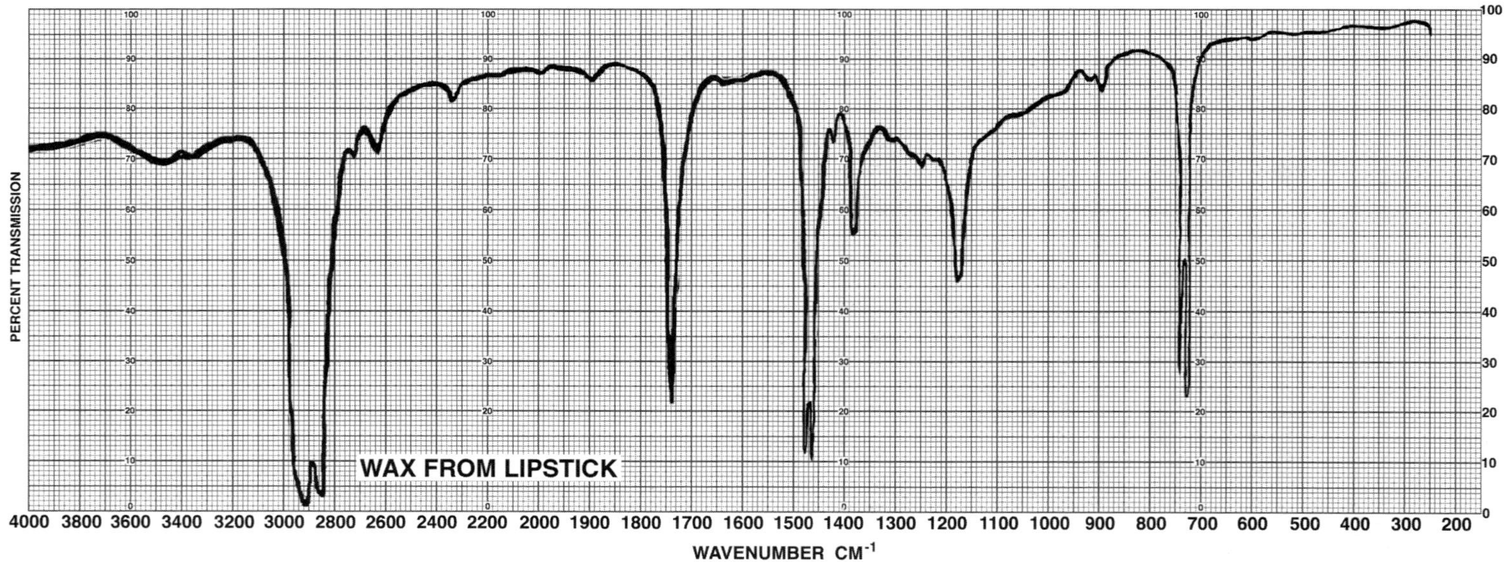

Fig. 14.6 Spectrum of wax from sample of lipstick.

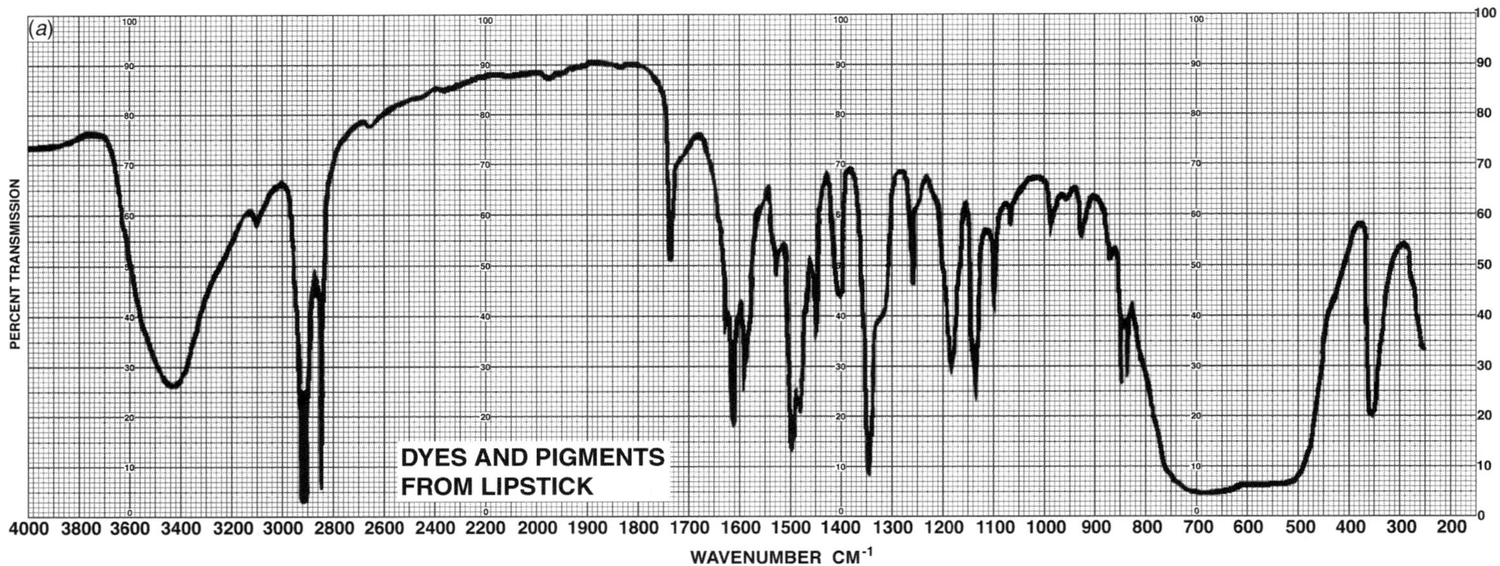

Fig. 14.7a Spectra of (*a*) dyes from lipstick and (*b*) oil from lipstick.

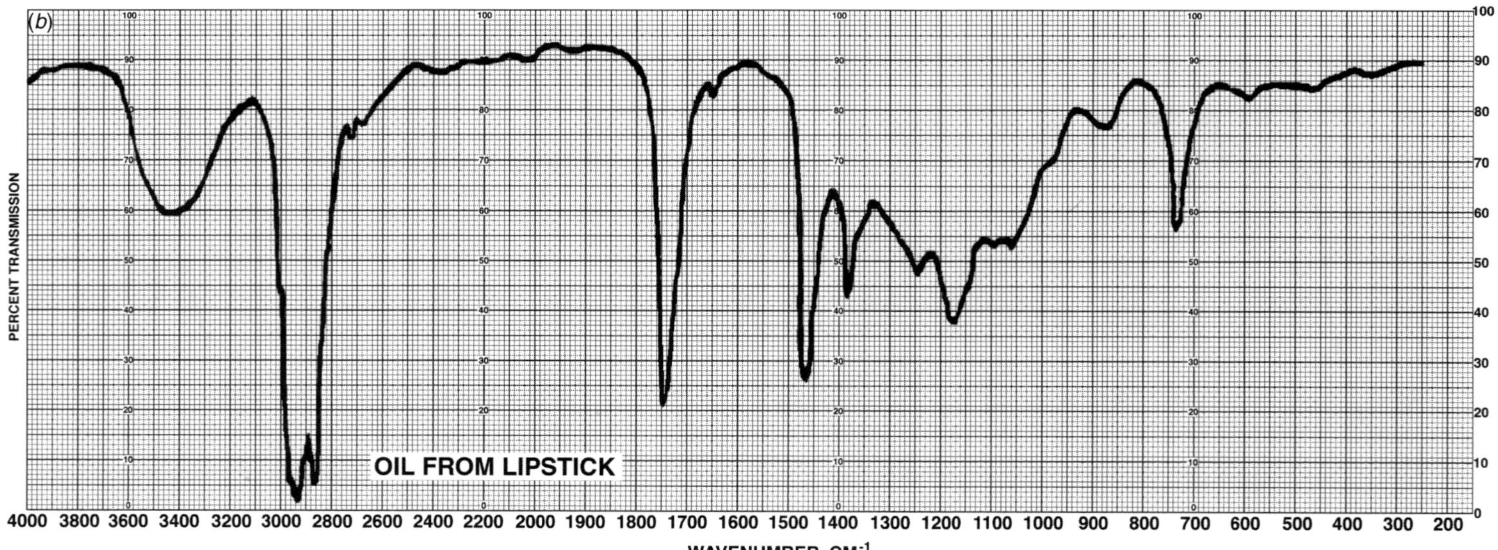

Figure 14.7b *(Continued)*

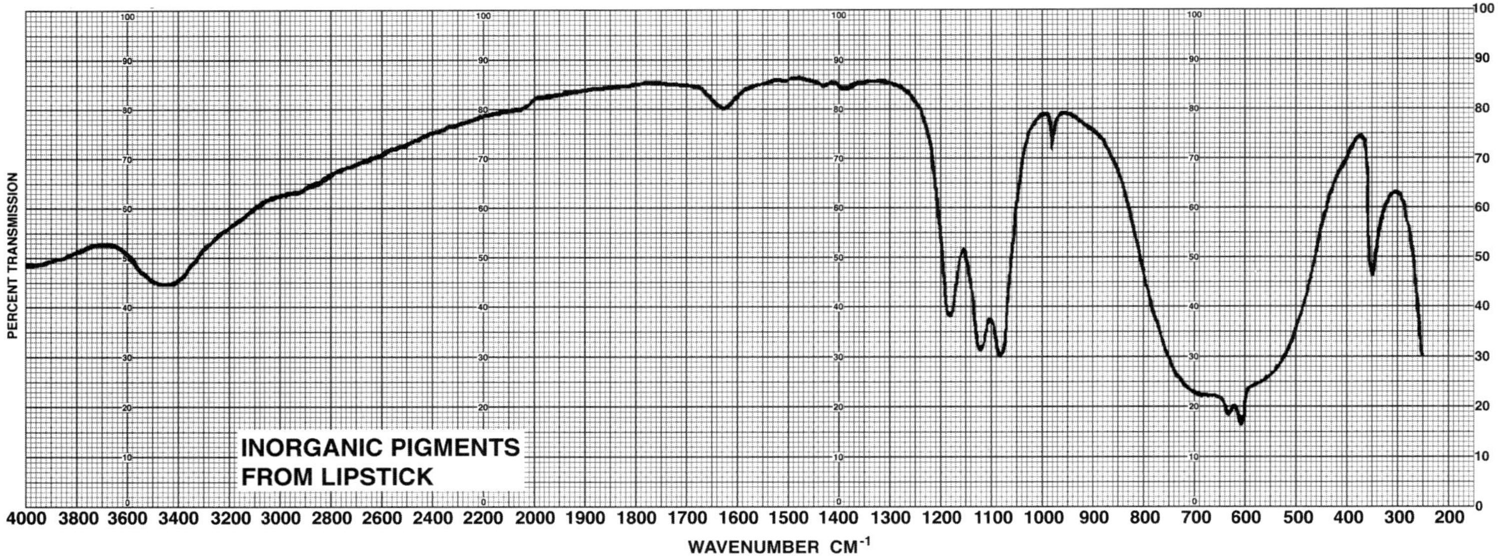

Fig. 14.8 KBr disk spectrum of inorganics in lipstick.

Wick-Stick technique or a derivative of the technique may be used to remove silica gel (see 7 below).

5. Evaporate the chloroform and prepare as micro-KBr disks (Figures 14.9 and 14.10).
6. Interpretation
   a. Two of the spots yielded excellent IR spectra of barbital and amobarbital (Figures 14.9*a,b*).
   b. The third spot was contaminated with silica gel that caused considerable light-scattering effects in the 4000–2500 cm^{-1} region and contributed a strong underlying band near 1000 cm^{-1} (Figure 14.10*a*). The silica gel was removed by the Wick-Stick technique, and the resulting spectrum was identified as that of hexobarbital (Figure 14.10*b*).
7. Description of the Wick-Stick technique

   The Wick-Stick is a triangular wedge of compressed KBr powder that is porous. It is about 8 mm wide at the base, 25 mm high, and 1.25 mm thick. About 1–2 mL of solution is placed in a vial; the vial is approximately 5 cm high and 1.25 cm in diameter. The Wick-Stick is inserted in a metal holder and dropped into the vial. A cap with a hole in it is placed on top. The solution rises to the tip of the KBr wedge and evaporates from the tip. Soluble material rises to the tip along with the solvent and is concentrated there as the solvent evaporates. Insoluble material is left behind. The tip is broken off and a micro-KBr disk is prepared. Slight warming and a slow stream of nitrogen across the hole speed the evaporation. *It is extremely important to run blanks on all solvents used in the procedure. All evaporations should be done in a well-ventilated hood.*

   In the event a Wick-Stick is not available, the following has been found useful as a substitute. Fill a 3-mm-thick-wall capillary [about 1.5 mm inside diameter (i.d.) by 3.0 cm long] with KBr powder and pack it firmly. Some experimentation will be necessary to determine the proper filling and packing process. At the top end create a cone of KBr from which the solvent can evaporate. This is placed in the vial as described above. After evaporation, the conical tip is used to make the micro-KBr disk.

D. Extraction, precipitation, and column chromatography.

   This example involved the identification of the major components in an adhesive. *All of the steps should be carried out in a well-ventilated hood.*

   1. Obtain the spectrum of the adhesive as a thin film on a KBr window (Figure 14.11, IR).
   2. Dissolve the adhesive in toluene. Filter fillers and pigments and examine separately. These will not be discussed further.
   3. Precipitate the polymer by adding methanol to the toluene solution. Filter and wash the precipitate with several milliliters of toluene.

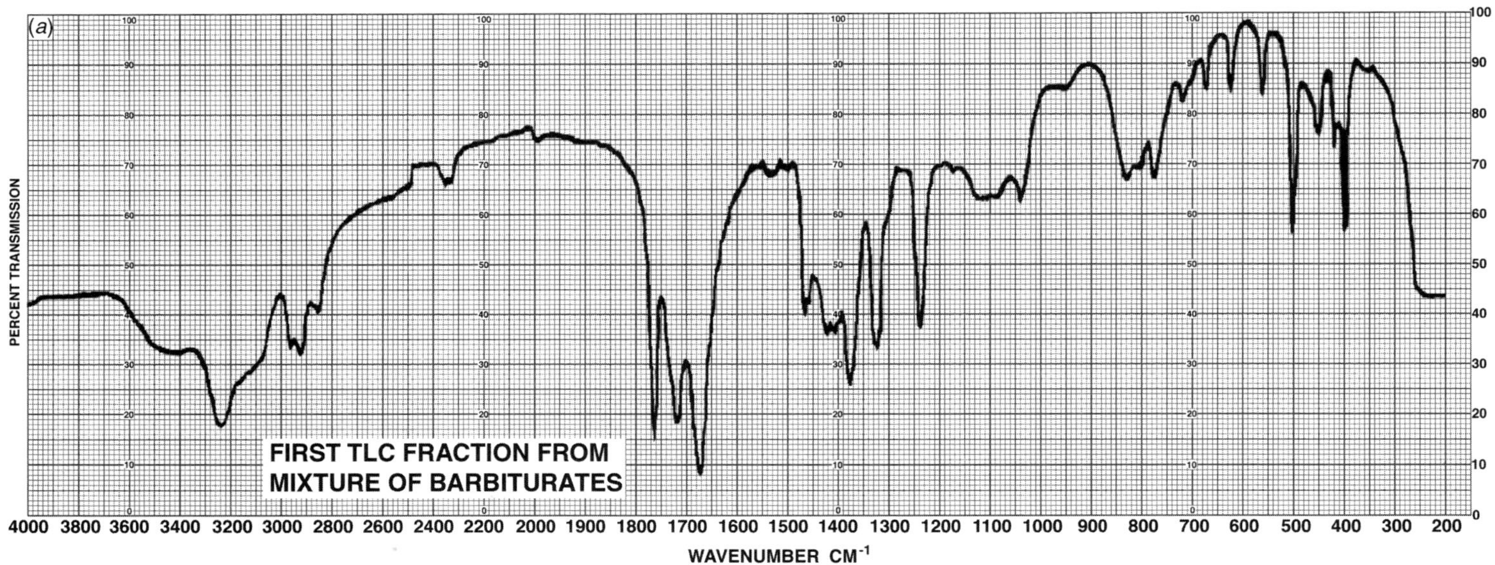

Fig. 14.9a Infrared spectra of (*a*) barbital from TLC separation and (*b*) amobarbital from TLC separation.

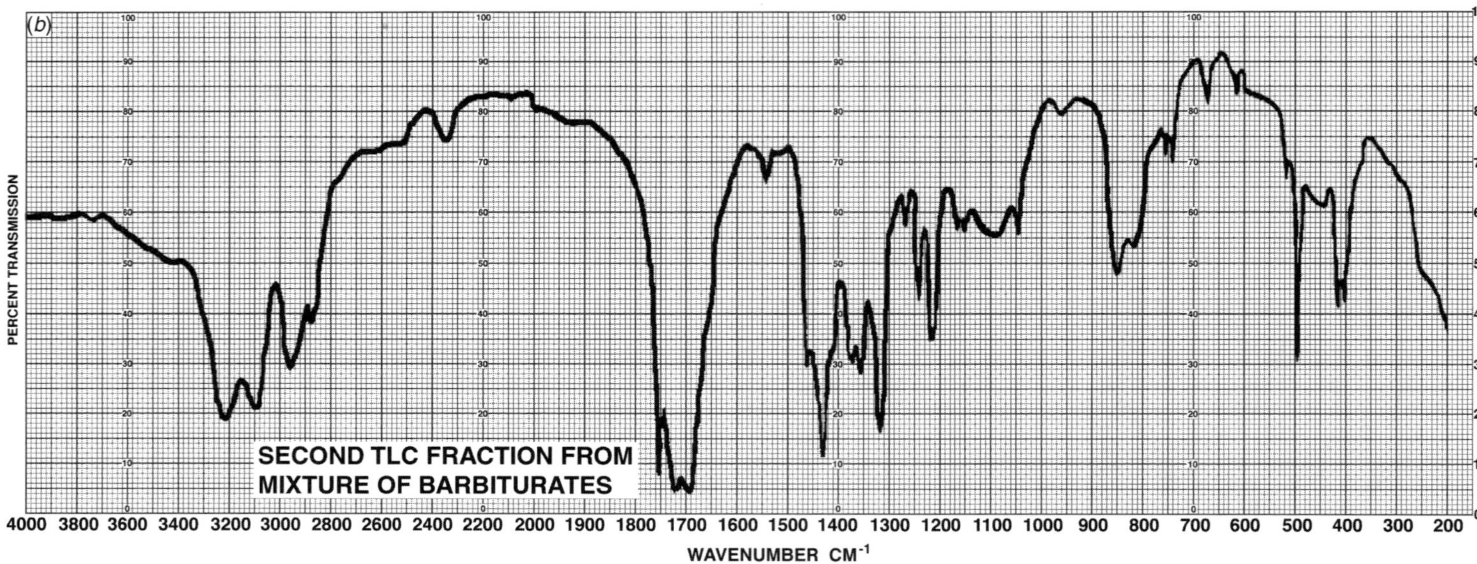

Figure 14.9b (*Continued*)

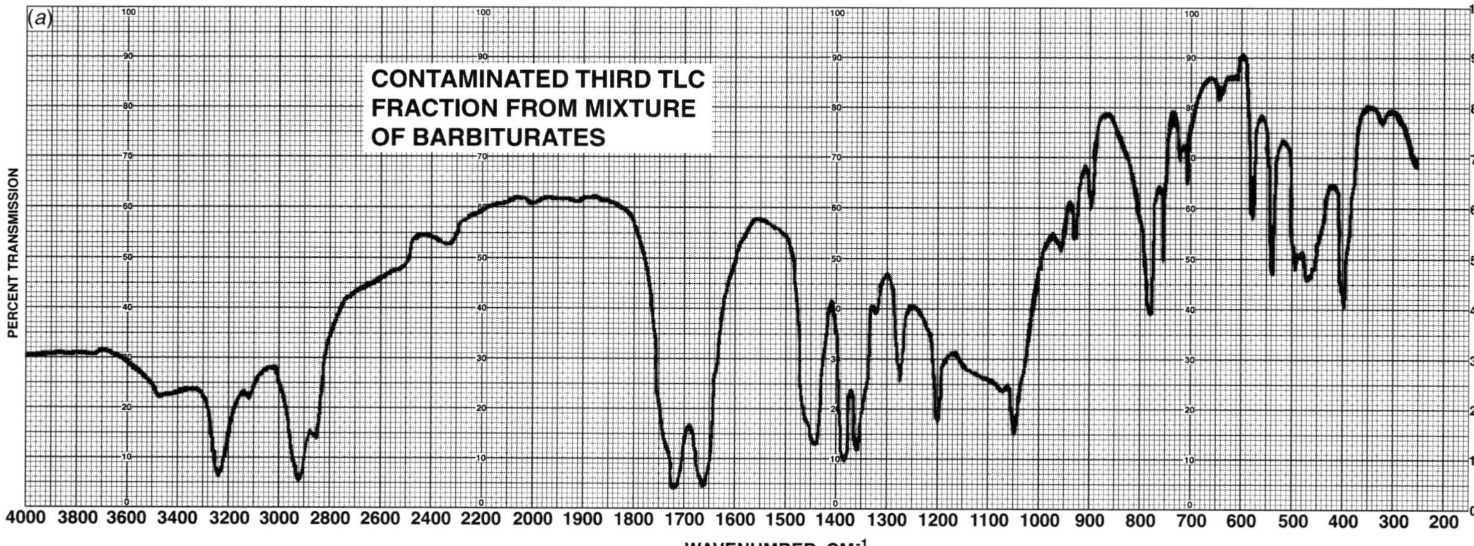

Fig. 14.10a Infrared spectra of (*a*) hexobarbital from TLC separation contaminated with silica gel and (*b*) hexobarbital after removing silica gel.

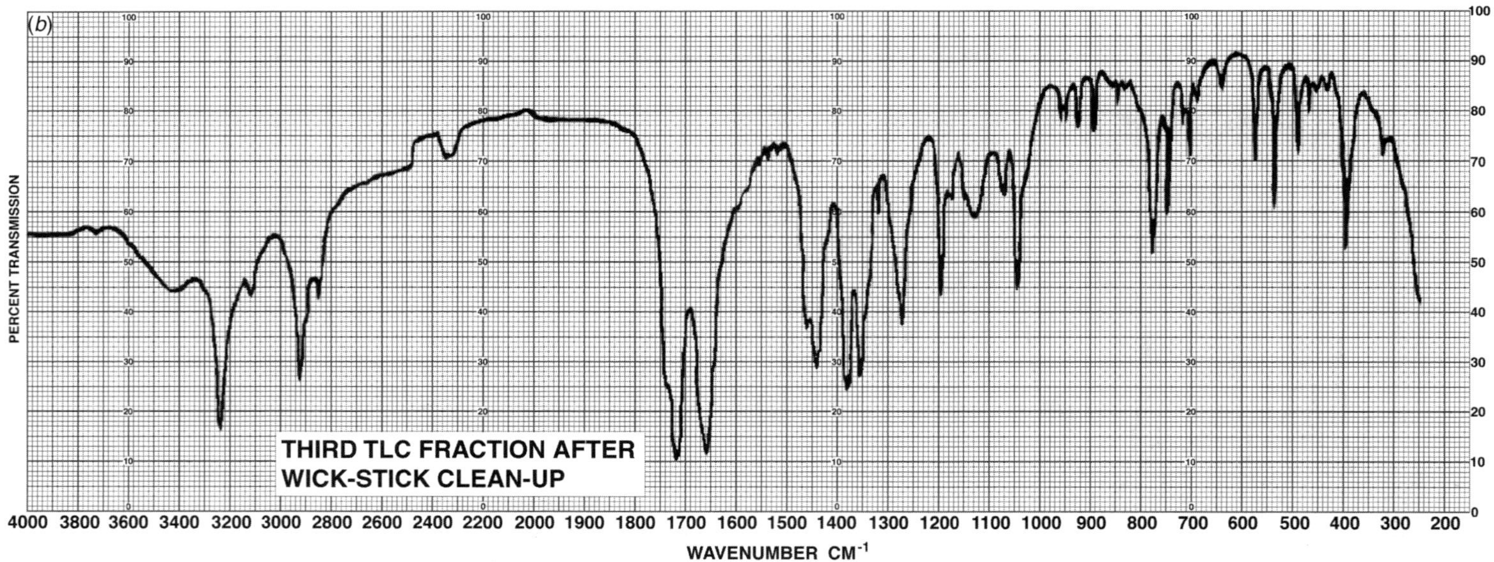

Figure 14.10b *(Continued)*

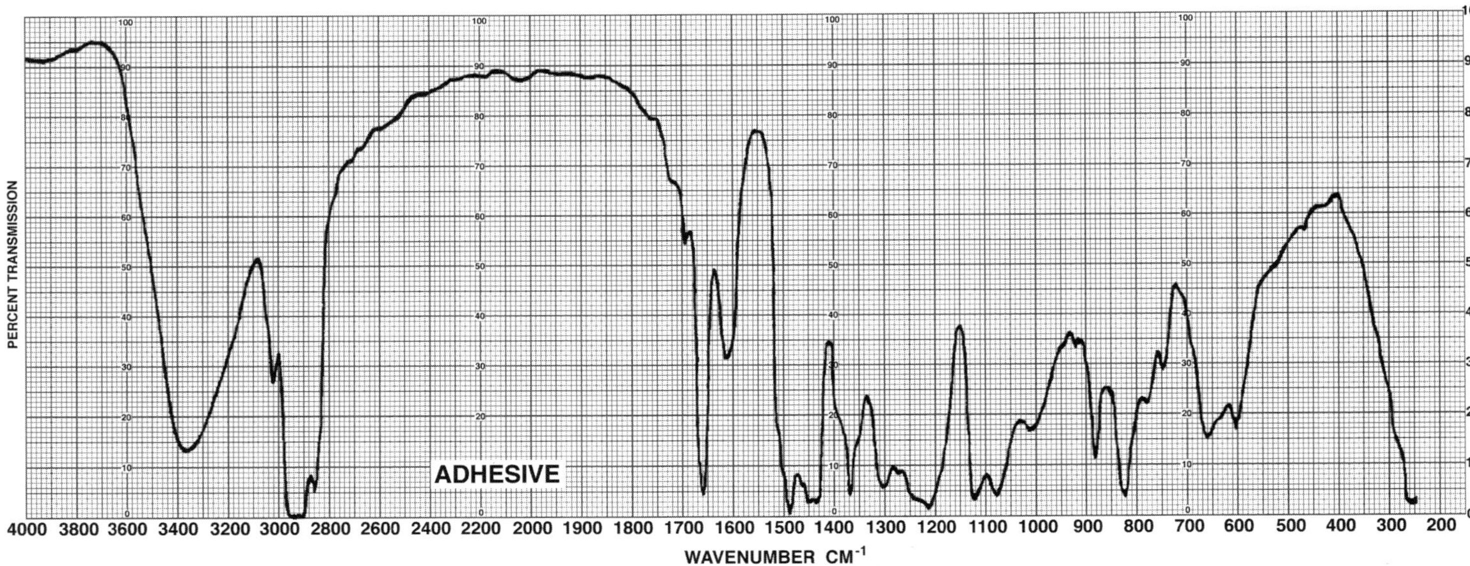

Fig. 14.11 Infrared spectrum of adhesive.

Transfer the precipitate to a KBr window and press into a thin film to obtain the spectrum (Figure 14.12*a*).

4. Evaporate the remaining toluene solution and dissolve the residue in hexane. Place the hexane solution on a 25-cm-long, 1-cm-i.d. aluminum oxide column. Elute successively with hexane, hexane–methyl ethyl ketone (MEK), and ethanol and collect each fraction.
5. Evaporate the solvents from each fraction in a hood and obtain the spectra of the residues (Figures 14.12*b* and 14.13).
6. Interpretation
   a. The precipitate was identified as neoprene (Figure 14.12*a*).
   b. This first residue was identified as a phenolic resin (Figure 14.12*b*).
   c. The next residue was a solid and was prepared as a KBr disk (Figure 14.13). It was identified as *N*-phenyl-β-napthylamine, an antirust agent.
7. The original spectrum of the adhesive is largely the sum of the spectra of the phenolic resin and neoprene. The last fraction was obviously at a low concentration, and its presence would not have been detected without carrying out the separation procedure described.
8. Blanks of all solvents were obtained in order to eliminate them as possible sources of any contamination.

E. Extraction and internal reflection

The problem definition was to identify a fire retardant on a polyester fabric and to trace its behavior during laundering. In this example, the high sensitivity of a multiple internal reflection device was used to obtain IR spectra of very thin films.

1. The spectrum of the fabric was obtained by pressing the fabric against a 25-reflection KRS-5 internal reflection crystal (Figure 14.14*a*). This is the spectrum of a polyester. There is no evidence for the presence of a fire retardant in this spectrum.
2. Using a clean KRS-5 internal reflection crystal, a background was obtained. A piece of the fabric was placed on the surface of the crystal while it was lying horizontally in a well-ventilated hood. The fabric was then saturated with chloroform. After approximately 30 seconds the fabric was removed and the chloroform was allowed to evaporate from the KRS-5 crystal. After evaporation, a hazy appearance was noted on the crystal face. This produced the spectrum in Figure 14.14*b*. The internal reflection spectrum of a sample of tris(2,3-dibromopropyl)phosphate (TDBPP) is given in Figure 14.14*c* for reference. Clearly, the fire retardant on the fabric is TDBPP. Weak bands in the 1720-cm^{-1} region are probably due to a trace of polyester fiber or possibly residual lubricant on the fiber.
3. A new sample of fabric was laundered and the extraction procedure repeated. After 10 washings it was no longer possible to detect the fire retardant by the technique described, although the fabric retained its fire-resistant properties.

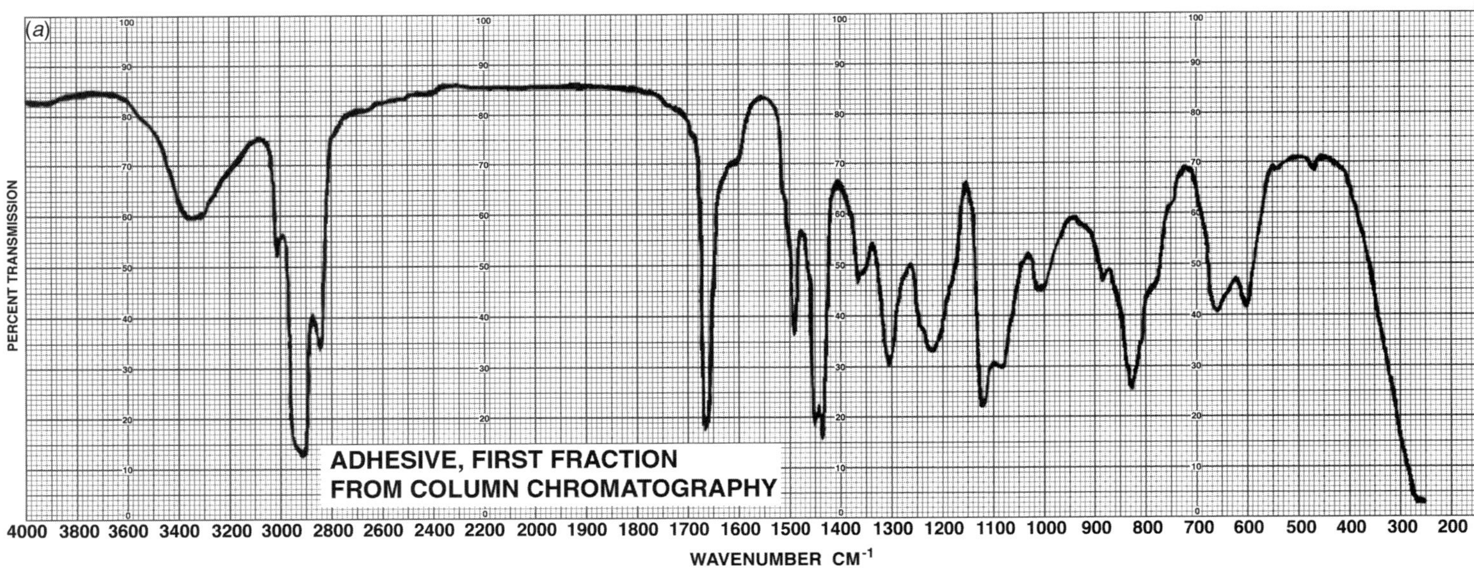

Fig. 14.12a Infrared spectra of fraction collected from column chromatography separation of adhesive: (*a*) first fraction; (*b*) second fraction.

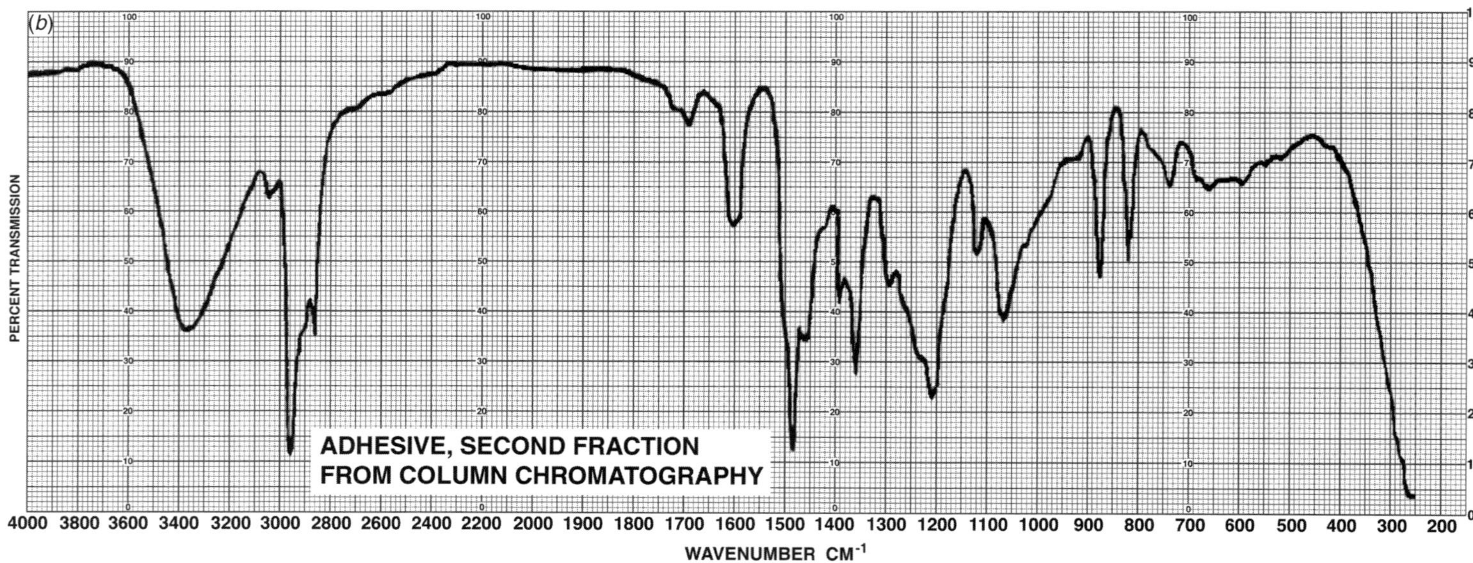

Figure 14.12b *(Continued)*

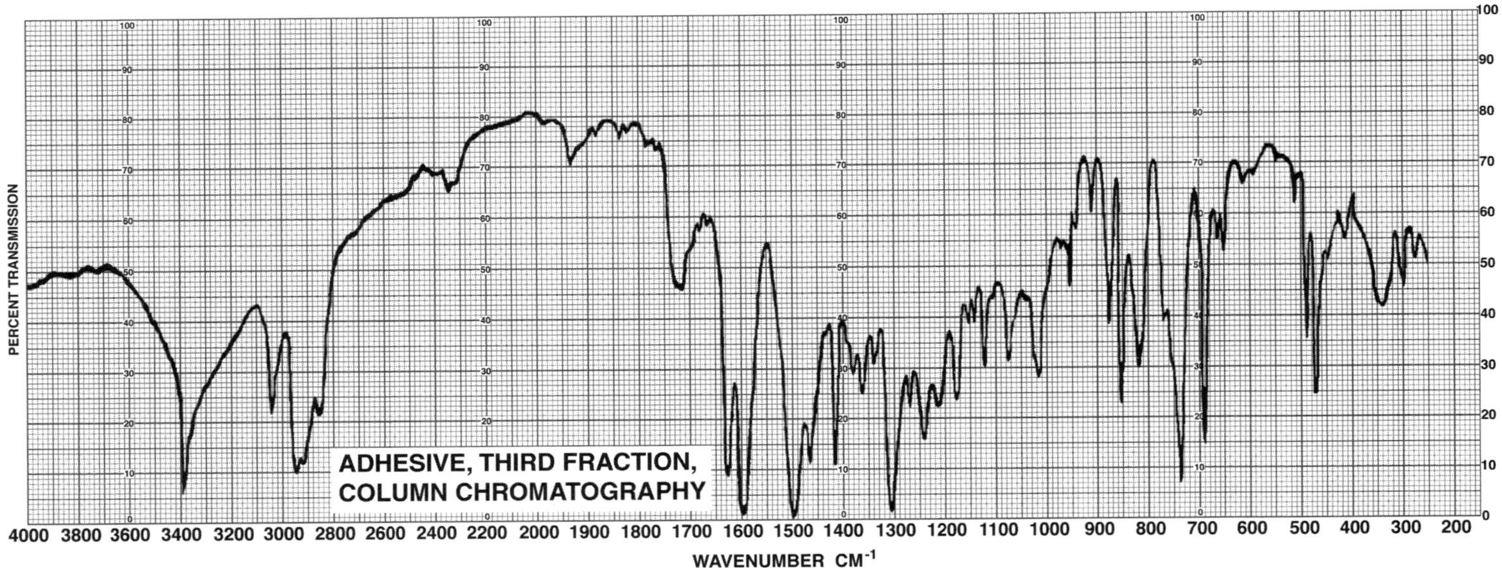

Fig. 14.13 Infrared spectrum of KBr disk preparation of third fraction.

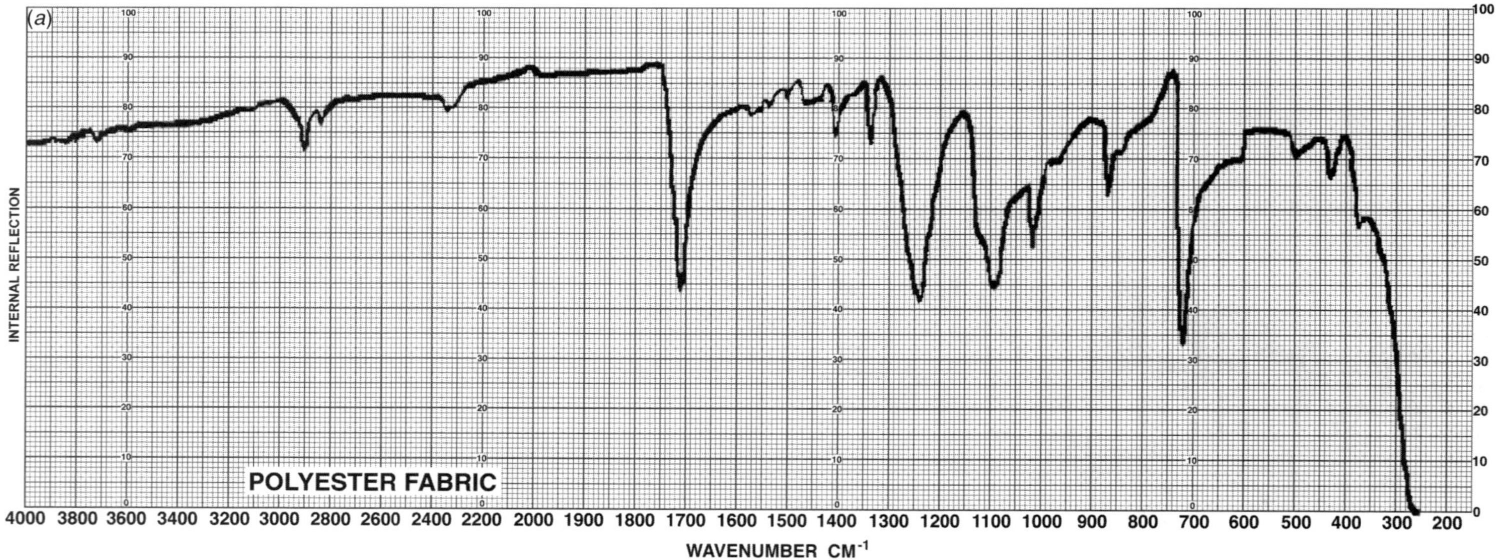

Fig. 14.14a (*a*) Internal reflection spectrum of polyester fabric. (*b*) Spectrum of extracted fire retardant from polyester fabric. (*c*) Reference spectrum of fire retardant.

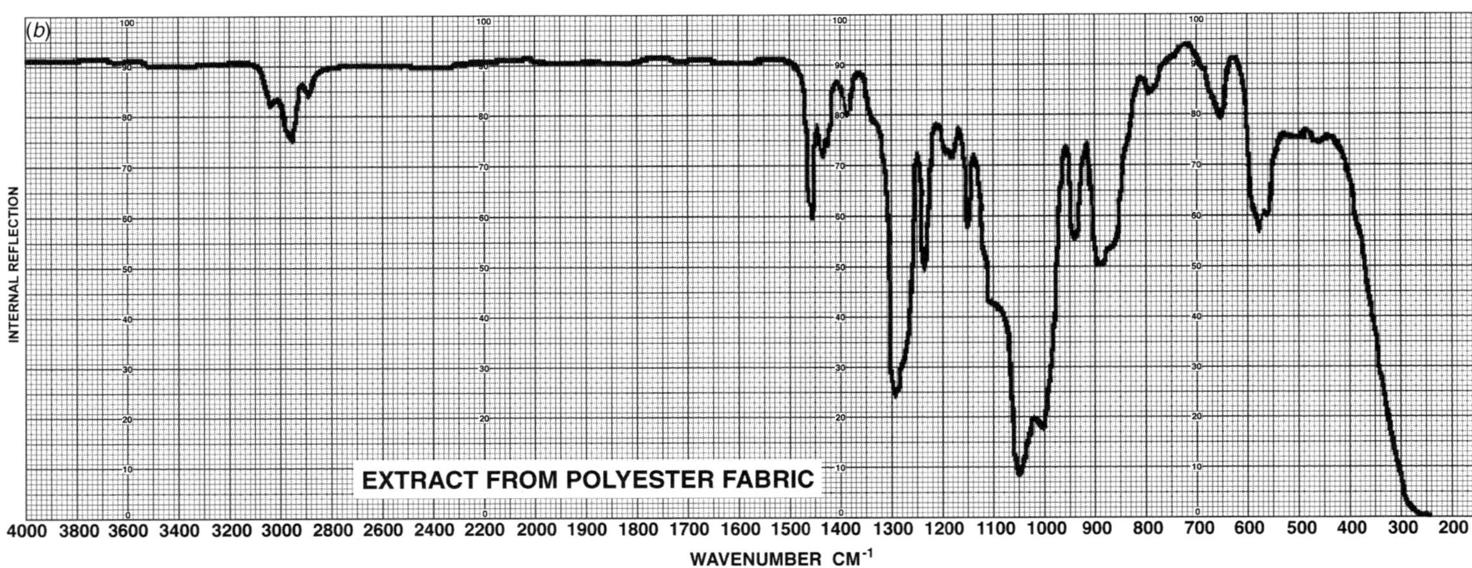

Figure 14.14b (*Continued*)

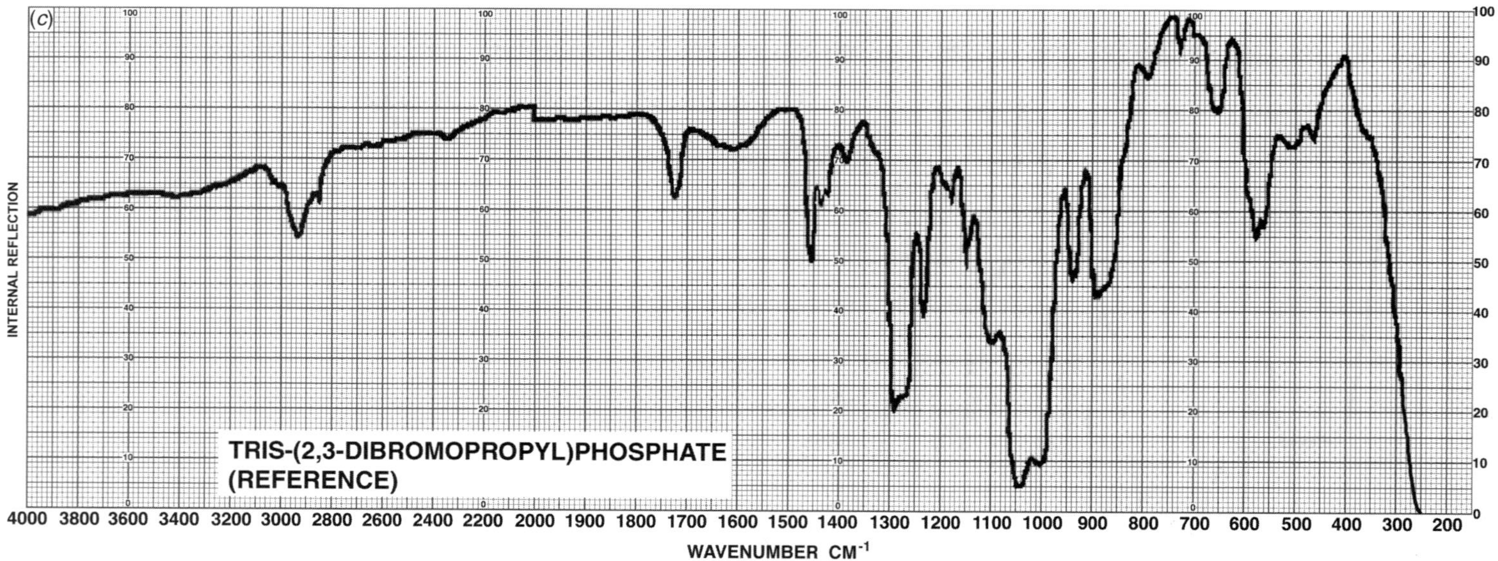

Figure 14.14c (*Continued*)

4. It is possible to think of other applications of this simple technique: for example, extraction of plasticizer from polymers or solvent-soluble materials in fertilizers or in pharmaceuticals.

F. Pyrolysis

1. Definition

   Pyrolysis is a technique that involves destructive distillation of a material yielding liquid and/or gaseous products that produce IR spectra characteristic of the original material. It was first described by Harms in 1953 and has been described in detail by Wake. Sadtler Research has published a large collection of reference spectra of pyrolyzates.

2. Applications

   Applications of pyrolysis include crosslinked polymers that are very difficult to sample by other techniques. Other types of problems might include identifying the elastomer in carbon-filled rubbers or the binder in a grinding wheel. As Wake has indicated, care must be taken to remove additives by extraction prior to pyrolysis.

3. The procedure is as follows: A small piece of the sample is extracted with solvents to remove additives. The piece is then placed in a pyrex tube and heated with the tube held horizontally in a Bunsen flame. As the sample decomposes, a tan-colored liquid will condense on the sides of the tube. This is the pyrolyzate. It can be removed from the tube with a clean rubber policeman, transferred to an IR window, and scanned as a capillary film.

   There are several variations of the technique. Commercial devices have been offered which provide for evacuation and pyrolysis of the sample heated on a platinum ribbon. The liquid pyrolyzate is condensed on an internal reflection crystal mounted directly in the pyrolysis accessory. Spectra of any vapor-phase species may also be obtained with this accessory. A similar device constructed from available laboratory equipment will be described in a later example.

4. Possibly the most useful variation involves a micropyrolysis. A glass capillary approximately 1 mm in diameter is sealed at one end in a Bunsen flame. A very small piece of the sample is placed in the open end of the capillary and is tapped down to the bottom of the tube. Pyrolysis is carried out by heating with a paper match. The pyrolyzate will be seen as a tan-colored liquid.

   After the capillary has cooled, the tube near the sealed end is scratched with a file and the end broken off below the tan pyrolyzate. The tube is then touched down to a few drops of chloroform, which enter the tube by capillary action. The tube is then tilted back and forth to dissolve the pyrolyzate.

   A small piece of KBr window is prepared as follows: An area about 3 mm by 3 mm in the center is scratched with a fine point. The crystallites of KBr from the scratching are left behind to trap the pyrolyzate during the evaporation step (Figure 14.15).

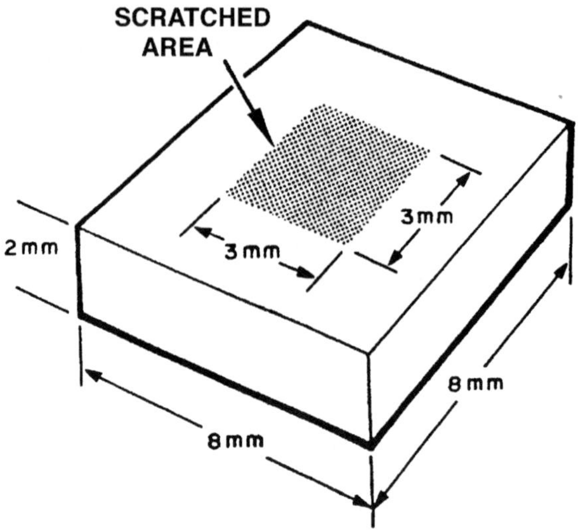

Fig. 14.15 Example IR window prepared for cast films of small amount of sample.

The capillary containing the chloroform solution is then touched down and lifted quickly from the scratched area. If contact is maintained too long, the solution will flood the entire surface of the crystal and not just the scratched area. The object is to concentrate the pyrolyzate as much as possible. After the chloroform has evaporated after the quick contact, the process is repeated until all of the pyrolyzate has been transferred to the crystal.

In general, since IR beam sizes at the sampling point are larger than the scratched area, it is advisable to prepare a mask from aluminum foil that has an aperture the same size as the scratched area. This is placed on the window and a second window is placed on top to trap the liquid. The spectrum may then be obtained.

5. Two examples will be described.
   a. A tiny chip of a polymer had been found in an electrical switch that had failed. The problem was to identify the polymer. Micropyrolysis was applied and the spectrum shown in Figure 14.16 was obtained. This is the spectrum of the pyrolyzate of a phenolic resin. Knowing which component had failed allowed design changes to prevent subsequent failures.
   b. The sample was a grinding wheel and the problem was to identify the organic binder. Attempts to dissolve the binder were unsuccessful, and pyrolysis was performed on a small piece broken from the grinding wheel. The resultant spectrum (Figure 14.17) was slightly

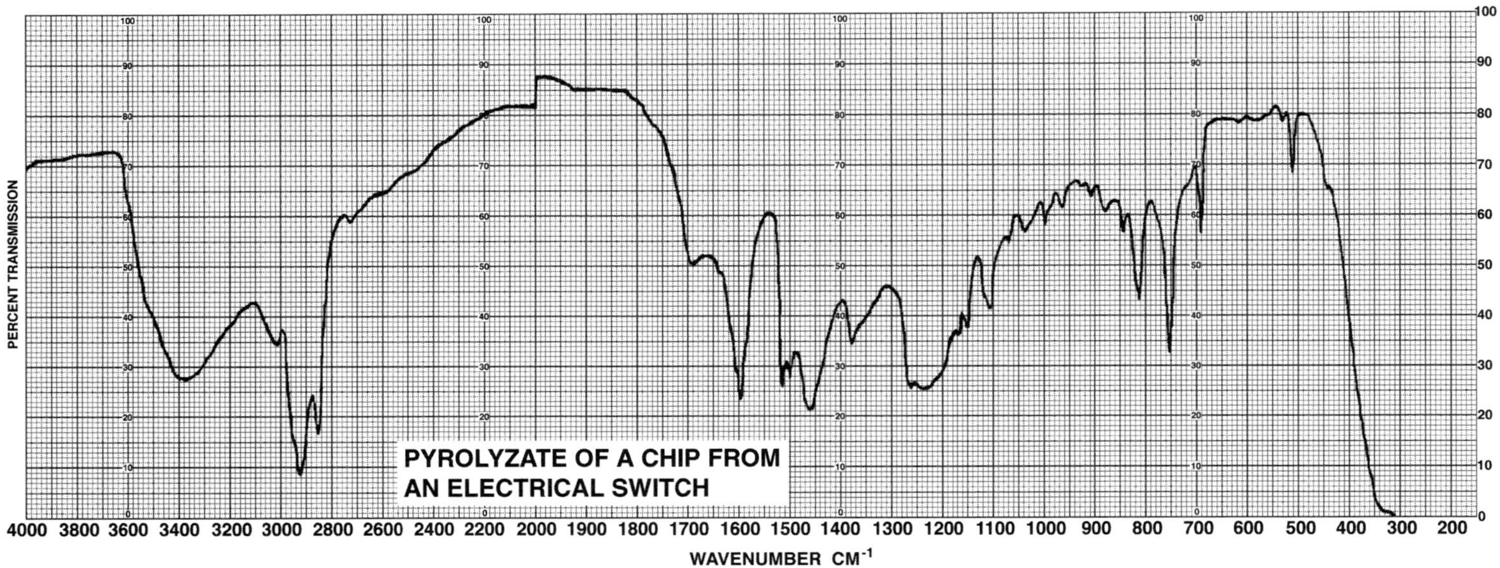

Fig. 14.16 Infrared spectrum of pyrolyzate of chip from electrical switch.

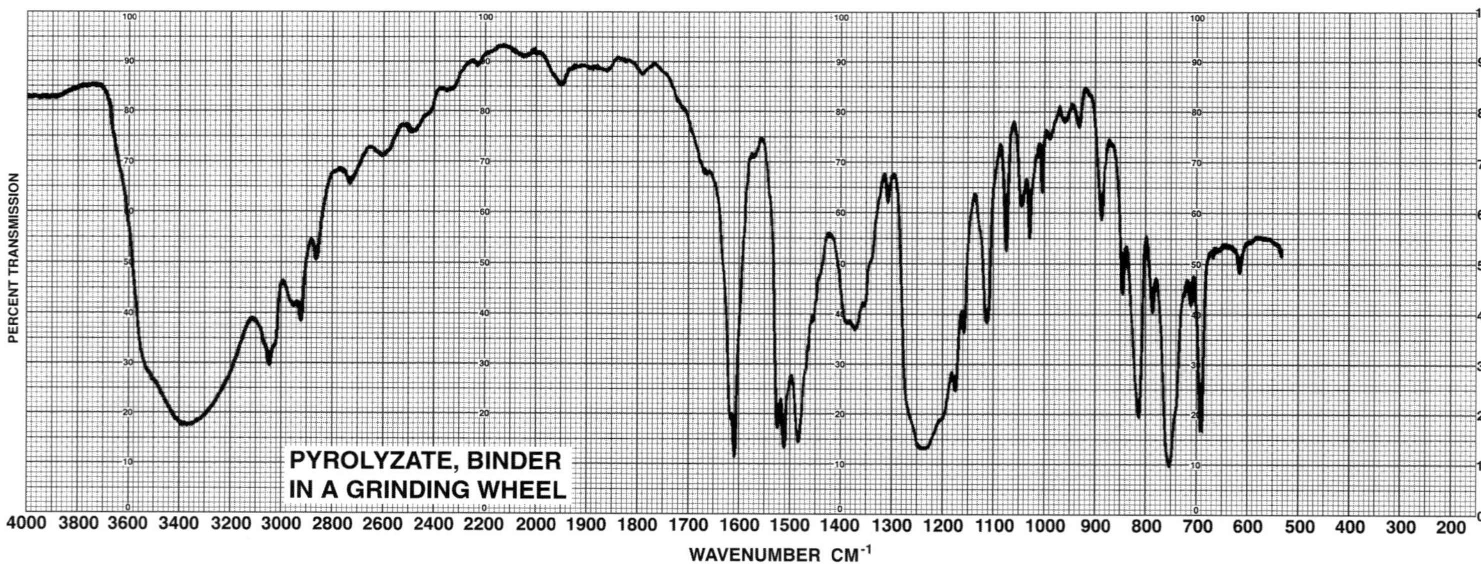

Fig. 14.17 Infrared spectrum of pyrolyzate from binder in grinding wheel.

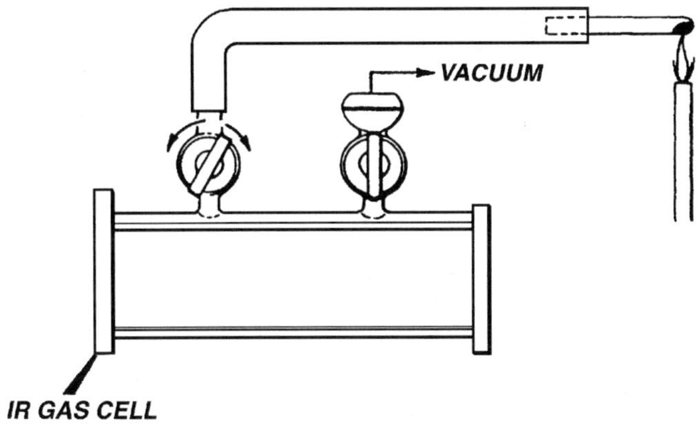

Fig. 14.18 Apparatus for collecting liquid- and vapor-phase pyrolyzates.

different from Figure 14.16 and was identified as arising from an epoxy resin.

6. Vapor- and liquid-phase pyrolyzates were prepared using the apparatus shown in Figure 14.18. For some materials only vapor-phase products are produced. In some cases, the liquid pyrolyzates are very similar while the vapor-phase products are very different. The windows on the 10-cm cell were of a nonhydroscopic material since water is generally produced during pyolysis. After assembling the apparatus, the entire system was evacuated, and the valve to the vacuum pump was closed. The sample was pyrolyzed in a Bunsen flame. Liquid collected on the wall of the tube and the vapor-phase pyrolyzate was trapped in the gas cell.
   a. As an example of the results, the pyrolysis products of a polybutadiene and a chlorinated polymer were examined. The liquid pyrolyzates are shown in Figure 14.19. There are small differences, but these could be due to slight differences in the conditions during pyrolysis, e.g., temperature differences.

      On the other hand, the vapor-phase pyrolyzates are very different (see Figure 14.20). The sharp series of bands near 2800 cm^{-1} is due to HCl and clearly come from the decomposition of the chlorinated polymer.
   b. A sample of plasticized polyvinyl chloride was pyrolyzed and the liquid and vapor products examined (Figure 14.21). The liquid-phase spectrum is the spectrum of a phthalic anhydride from the plasticizer, and the vapor phase shows spectra of several compounds, including vinyl chloride. It should be noted that polyvinyl chloride does not yield any liquid pyrolyzate; rather it yields only gas-phase species.

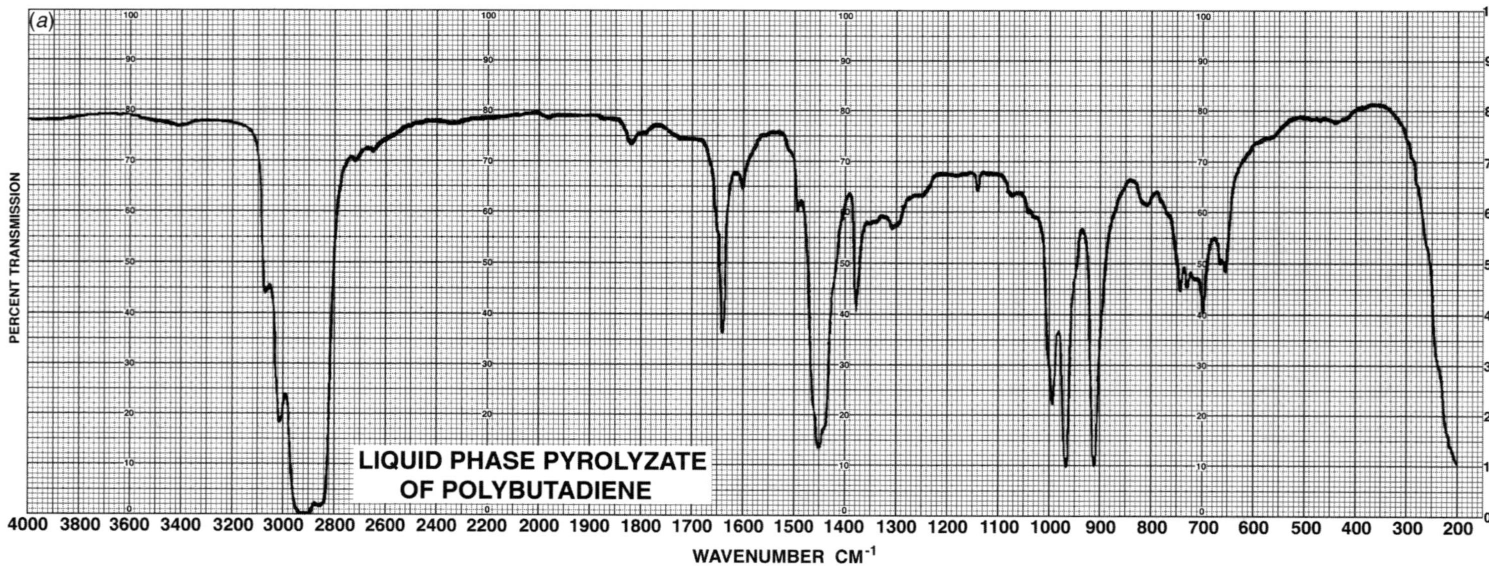

Fig. 14.19a Infrared spectra of (*a*) liquid-phase pyrolyzates of polybutadiene and (*b*) chlorinated polymer.

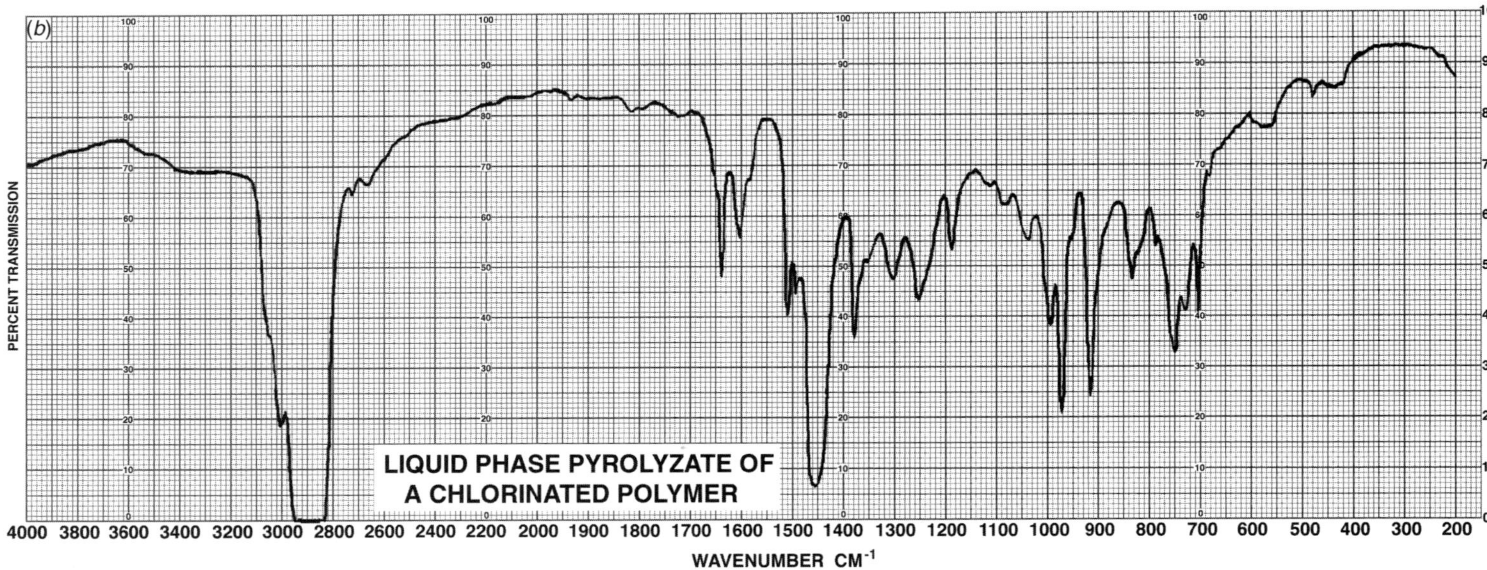

Figure 14.19b (*Continued*)

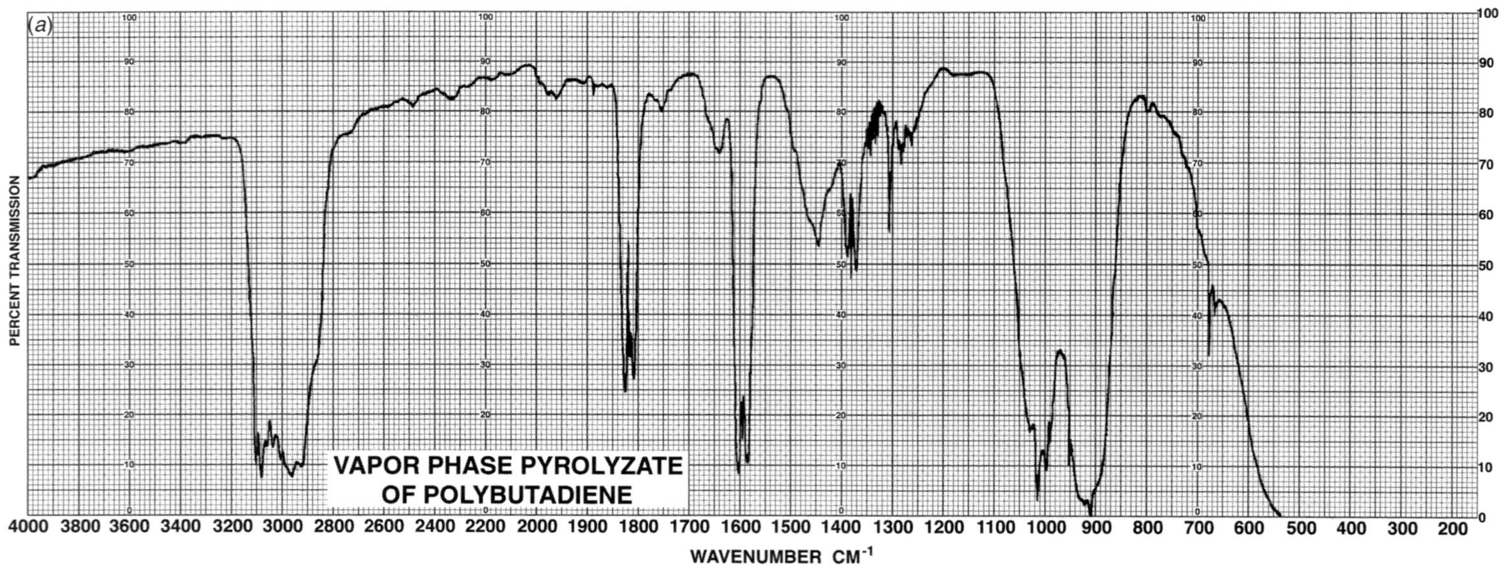

Fig. 14.20a Infrared spectra of (*a*) vapor-phase pyrolyzates of polybutadiene and (*b*) chlorinated polymer.

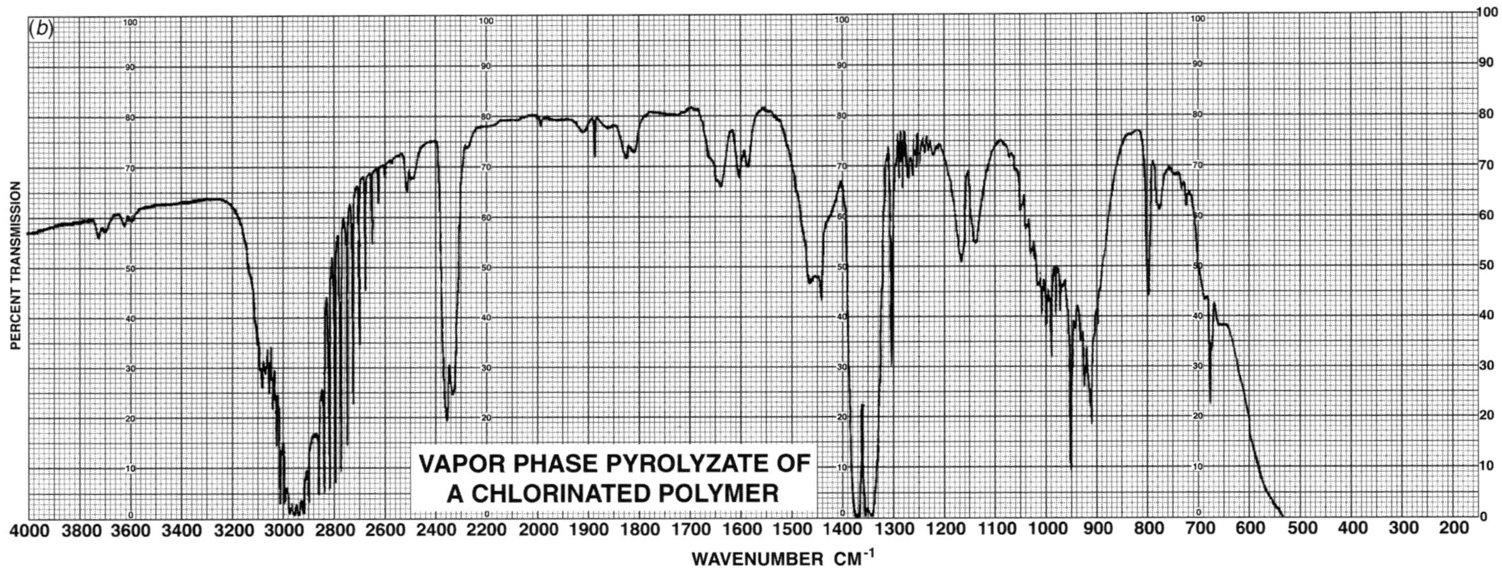

Figure 14.20b *(Continued)*

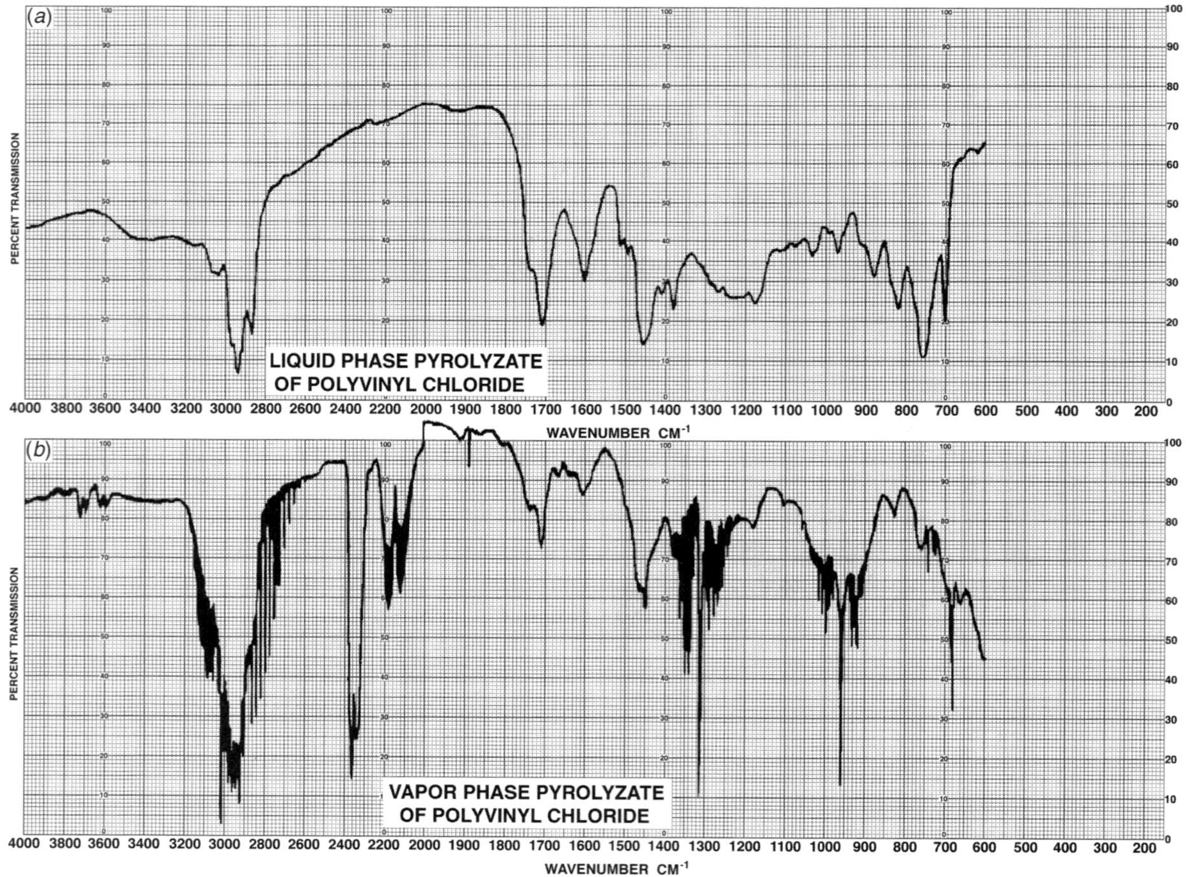

Fig. 14.21 Infrared spectra of pyrolyzates of polyvinyl chloride: (*a*) liquid and (*b*) vapor phases.

TABLE 14.1 Pyrolysis Products from Selected Fabrics

Fabric	Temperature, °C	Gases Evolved
Dynel	800	HCN, HCl, C_2H_4
SEF	800	HCN, HCl, C_2H_4
Kynol	1000	H_2O
Nomex	1000	HCN
Nylon	800	C_2H_4
Kohjin/cotton	800	C_2H_4, C_6H_6
TP-cotton	800	H_2O, C_2H_4
FR polyester	1000	C_2H_4, C_2H_2
PFR rayon	600	C_3H_6, C_2H_2, C_2H_4
FR acetate	800	H_2O

Table 14.1 summarizes the pyrolysis products from a number of fabrics. The temperature at which pyrolysis was performed is given in column 2.

G. Carbon-filled materials

While pyrolysis may be applied to these kinds of samples with good success, internal reflection may yield a sufficiently useful spectrum to allow identification of the major constituent. Another technique developed by L. Kalervo (private communication) involves elimination of the carbon as an interference.

It is well known that carbon black is a strong, very broad IR absorber. The technique developed by Kalervo involves dissolution of the sample and removal of the carbon filler.

1. Procedure

   A 3–4 mm diameter piece of the sample is placed in a reflux condensing system with *m*-nitrotoluene in a well-ventilated hood. After refluxing for several hours, most samples will dissolve. Several drops of the solution are transferred to a separatory funnel and 3 mL of methyl isobutyl ketone is added. Water is added a drop at a time and the separatory funnel shaken after each drop. It is important that this be done dropwise. The dropwise addition of water and shaking are continued until a layer separation occurs. At this point the carbon black will either be suspended in the water layer or lie at the interface between the layers. The lower layer can be taken from the separatory funnel and a cast film prepared.

2. The example shown is that of a carbon-filled butyl rubber. Direct internal reflection application yielded the spectrum in Figure 14.22*a*. The effect of the carbon black can be seen and only weak butyl rubber bands are observed. Following elimination of the carbon, Figure 14.22*b* was obtained. Certainly this is a much easier spectrum to work with for identification. Other types of carbon-filled rubbers and polyethylene have been successfully treated with this technique.

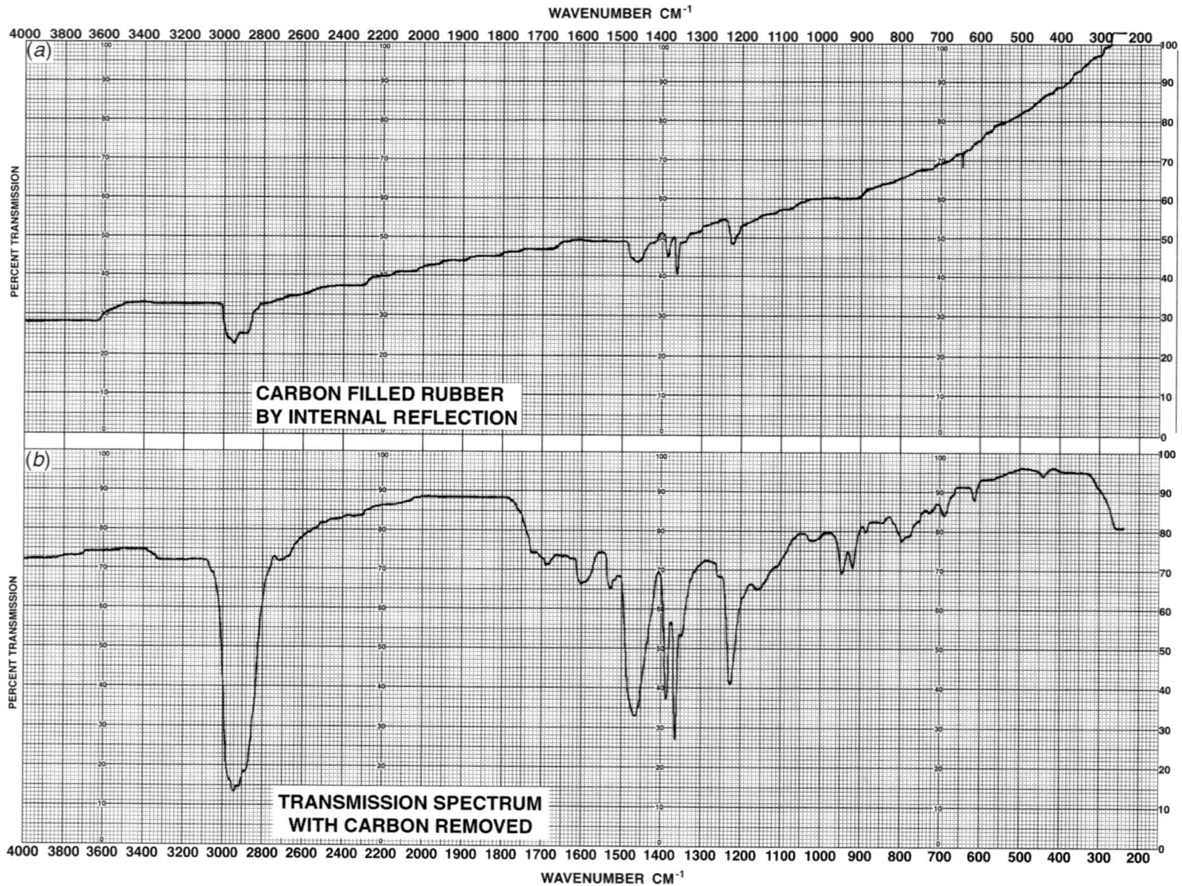

Fig. 14.22 (*a*) Infrared spectrum of carbon-filled butyl rubber by internal reflection (*b*) transmission spectrum of cast film of same sample after removing carbon black.

H. Dialysis

This is a separation technique based on differences in molecular size. Small molecules pass through a membrane leaving larger molecules behind. The example chosen was that of obtaining the spectrum of an additive in gasoline.

1. Procedure

   The volatile components in the gasoline were evaporated on a steam table under a well-ventilated hood. This took approximately 10 hours. About 2–4 mL of the residue was transferred to a rubber finger cot contained in a Soxhlet thimble and the sample was refluxed in a Soxhlet extractor with toluene. The resultant toluene solution was evaporated and a capillary film prepared from the final few drops.
2. The spectrum of the separated additive is shown in Figure 14.23*a*. A reference spectrum of a phosphate-type gasoline additive is shown for reference (Figure 14.23*b*).

I. Computer applications

It is often possible to obtain information about a complex mixture without doing physical separations but rather by applying computer subtraction.

1. The example chosen was described by Coates and was concerned with the spectra of gasoline samples. At the change of season, the composition of gasoline is also changed to provide better engine performance in cold versus warm weather.
   a. The procedure is to obtain the spectra of the gasoline samples, convert the spectra to absorbance, and subtract one from the other.
   b. The spectra are shown in Figure 14.24*a*. They each had the same octane number. Differences can be seen directly. After computer subtraction the differences were emphasized substantially (Figure 14.24*b*). Absorption bands going upward and near 1650, 910, and 990 cm^{-1} indicate a higher concentration of olefins in one sample, while bands going downward and near 1600, 1500, and 730 cm^{-1} indicate a higher concentration of aromatics in the other sample.
2. A second example is that of two hydrocarbon fuels, one a heating oil, the other a diesel fuel. Spectra of the two samples are shown in Figure 14.25*a*. They are nearly identical. Following conversion to absorbance, one spectrum was subtracted from the other to yield Figure 14.25*b*. Once again bands in Figure 14.25*b* in the upper direction are indicative of higher concentrations of a carbonyl-containing additive (1720 cm^{-1}) in the fuel oil as well as higher concentrations of saturated hydrocarbons [bands near 1460, 1380, and 720 cm^{-1} for CH_2 bend, CH_3 symmetric deformation, and $(CH_2)_{n>4}$ rock, respectively]. On the other hand, downward-going bands near 1600 and 1500 cm^{-1} and sharp bands near 810, 790, and 740 cm^{-1} are indicative of higher aromatic content in the diesel oil.

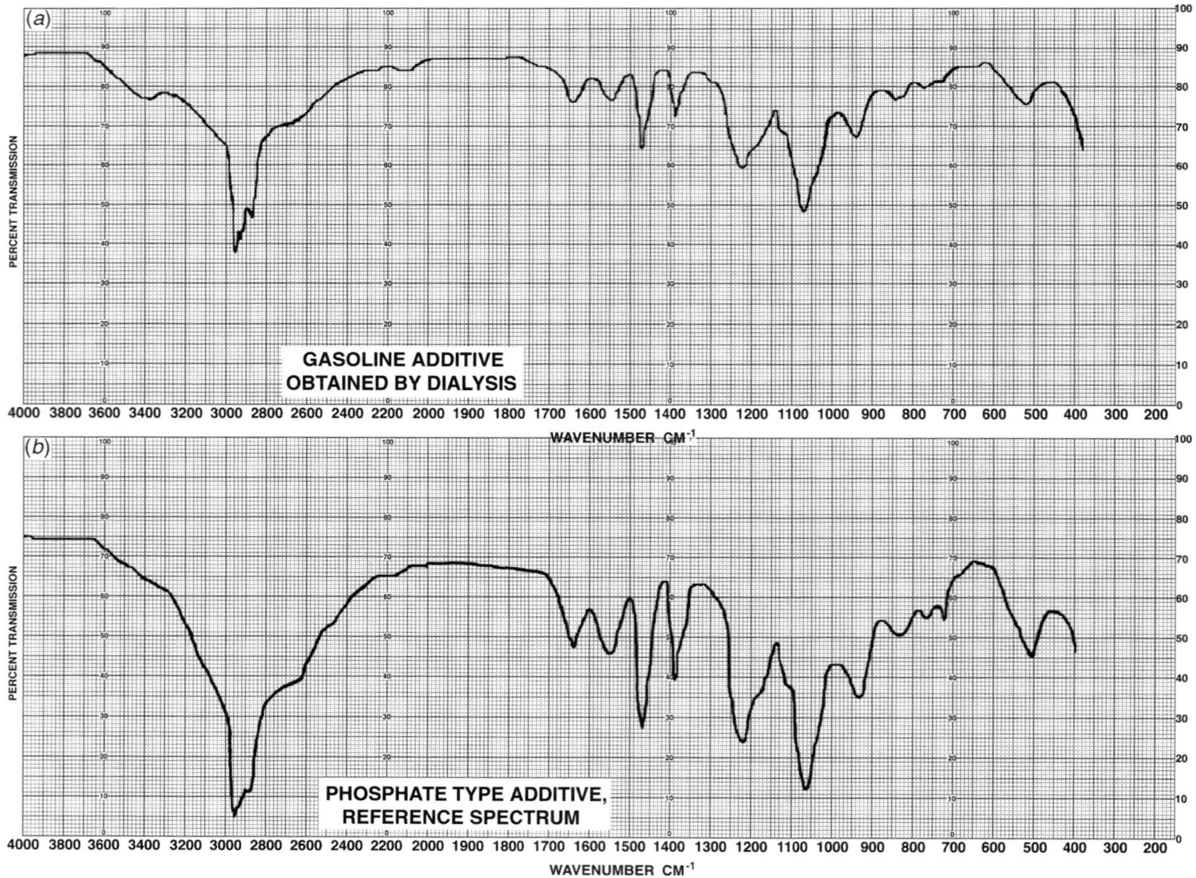

Fig. 14.23 (*a*) Infrared spectrum of separated additive. (*b*) Infrared reference spectrum of additive.

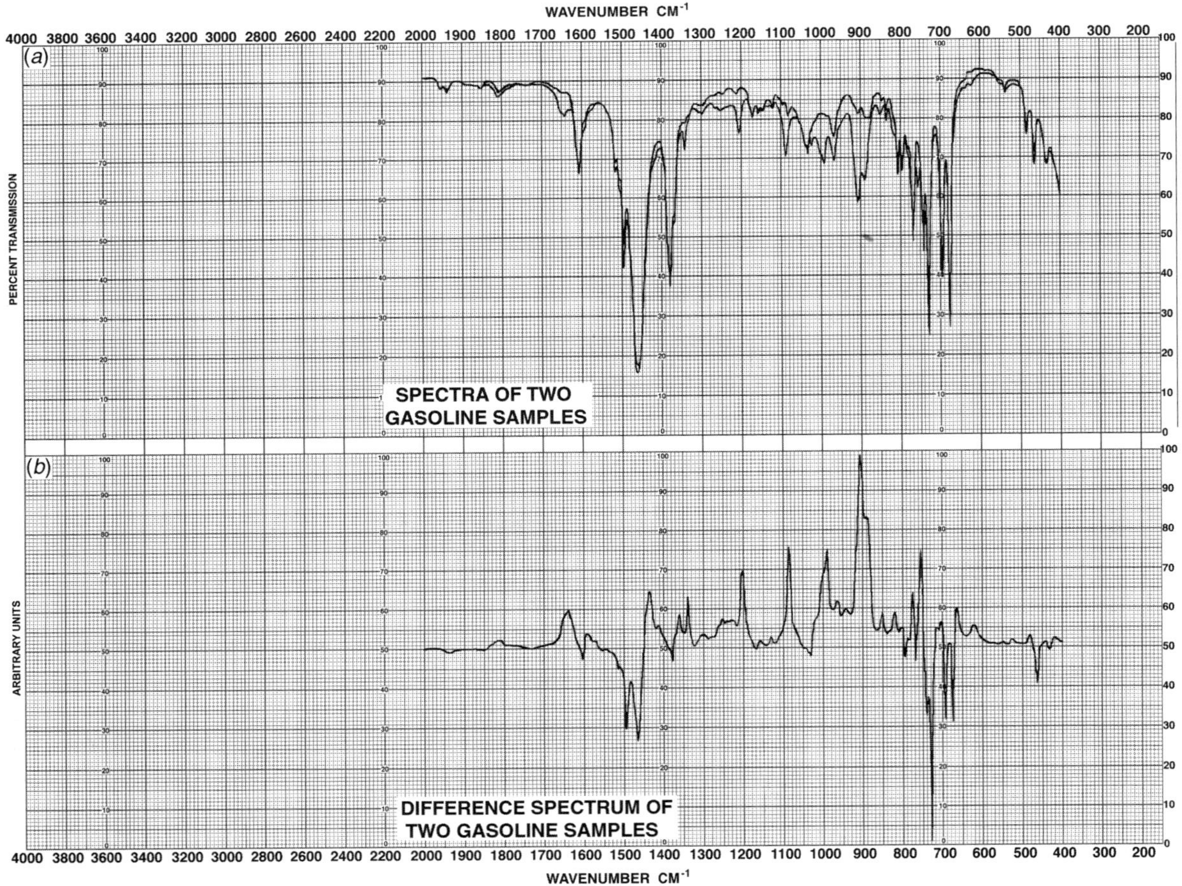

Fig. 14.24 (*a*) Infrared spectra of two gasoline samples. (*b*) Infrared spectra of difference between gasolines.

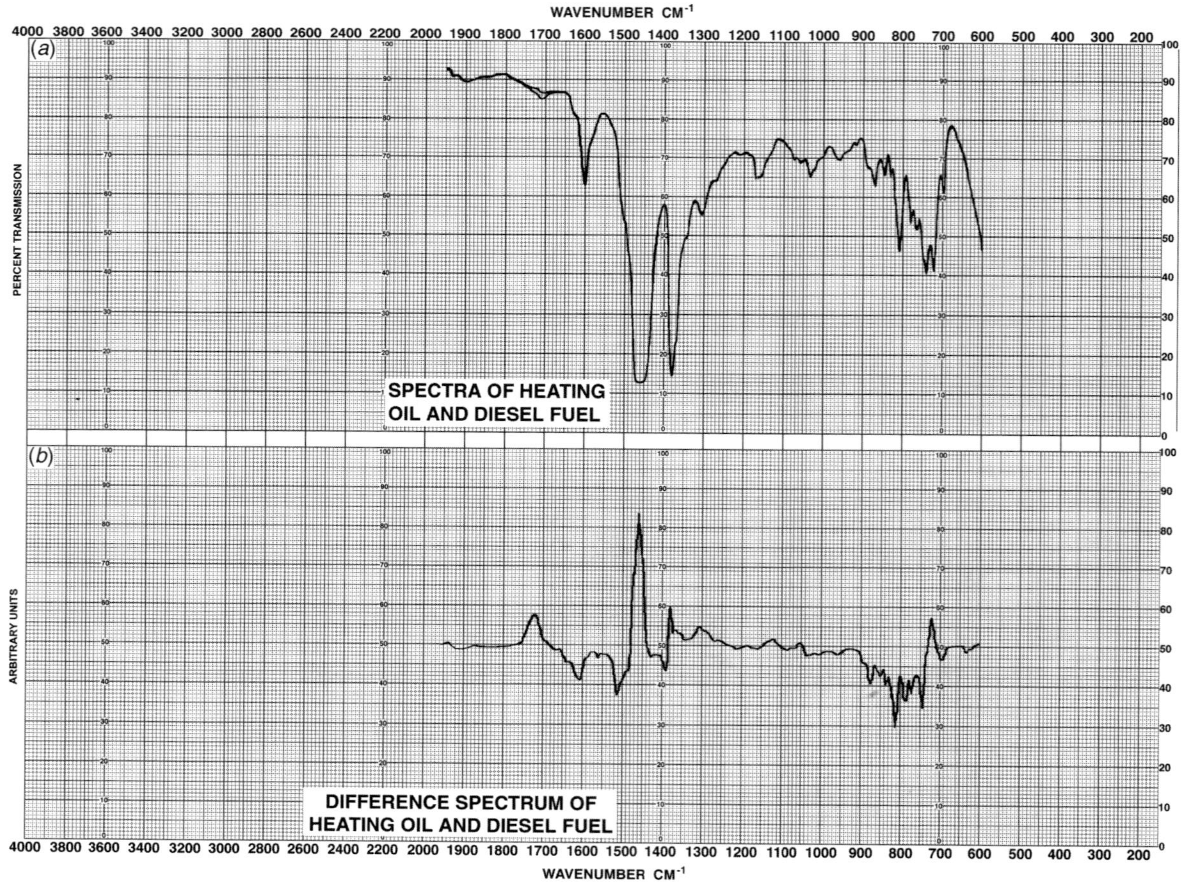

Fig. 14.25 (*a*) Infrared spectra of two hydrocarbon fuels. (*b*) Infrared difference spectrum between two fuel oils.

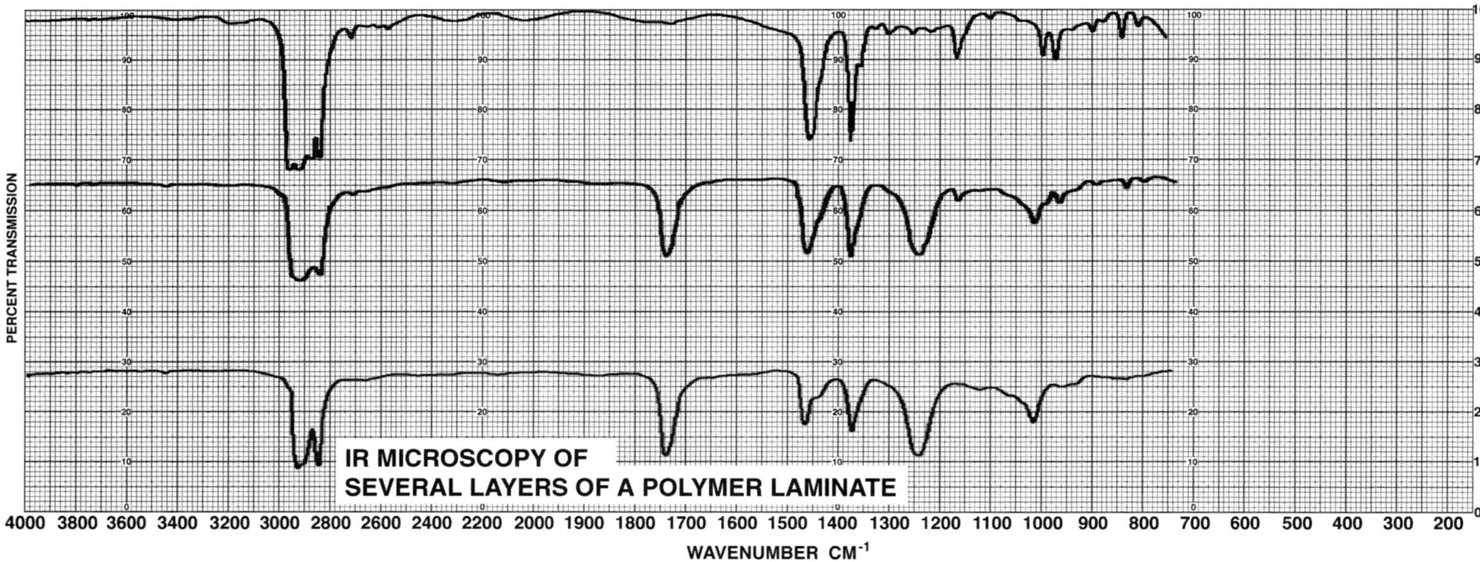

Fig. 14.26 Infrared spectra of several layers from polymer laminate by IR microscopy.

J. The IR microscope

It is possible to use the optical resolution of IR microscopy to isolate spectra of components in a solid mixture without any additional separation. For example, when examined under a microscope, mixtures of powders may show different crystalline forms for different components. It is possible to obtain the spectra of the different crystals by setting the microscope apertures to see only one crystal type at a time.

A different example involves a multilayer polymer film. A thin section can be microtomed from an edge and examined by looking downward through the layers, adjusting the apertures of the microscope to isolate each layer in turn.

Spectra of several layers of such a laminate are shown in Figure 14.26. This kind of sample would be extremely difficult to sample by any other technique.

IV. It should be obvious that many other applications could be described. It is useful to note that the techniques described in this chapter can be performed by the analyst without relying on other busy analysts in the laboratory. Scheduling is therefore much more straightforward. Application of separation techniques to other problems requires only the imagination of the analyst.

15 Answer to Chapter 5 Figure 5.30

UNKNOWN

The challenge here is the interpretation of the infrared and Raman spectra of an unknown that is described only as a simple monomer. Because Chapter 5 has been focused on the spectra of aromatic compounds it would not be unreasonable, however, for the reader to make the assumption that at least some aromatic component may be present in the structure of the unknown. The approach to the solution is to proceed exactly as was illustrated in the case of stryrene and polystyrene in the preceding section of Chapter 5 (VI, A).

When both the infrared and Raman spectra are available and presented on the same scale it is helpful to examine the same two spectral regions simultaneously. This approach can take advantage of the complementary nature of the two effects.

Thus, starting at the high wavenumber end of the two spectra the absence of any scattering or absorption in the 4000–3100 cm^{-1} region in both the infrared and Raman means that it is unlikely that any O—H or N—H systems are present. However, C—H groups clearly are expected and are shown to be present because the region where absorption and scattering are first observed is found to be between 3100–2850 cm^{-1}. Finding bands above and below the 3000 cm^{-1} implies that the molecule contains both sp^2 (above 3000 cm^{-1}) and sp^3 (below 3000 cm^{-1}) C—H groups. The relative intensities of the two regions are roughly similar in the infrared which would imply that there are similar amounts of unsaturated and saturated C—H present, or perhaps there might be slightly more unsaturated C—H present as the intensities of sp^2 C—H oscillators are slightly weaker than their aliphatic counterpart. The simple nature and low intensity of the region below 3000 cm^{-1} in the Raman (two weak bands above 2900 cm^{-1}) suggests that the aliphatic component is likely to be methyl and probably only one or two are present.

The next spectral region with absorption is 2000–1700 cm^{-1}. The bands found here are weak in the infrared and exhibit no scattering in the Raman. The infrared absorption pattern does show the characteristics which are associated with the

Course Notes on the Interpretation of Infrared and Raman Spectra, by Dana W. Mayo, Foil A. Miller, and Robert W. Hannah.
ISBN 0-471-24823-1

substitution on aromatic ring systems. In this case the pattern looks somewhat similar to the one found in styrene and as in that case this region is left for further evaluation till more information on the potential structure has been developed.

The 1600 cm^{-1} region in the infrared also appears quite similar that of styrene. Thus, there is a strong band near 1630 cm^{-1} that would appear to be the >C=C< stretch of an unsymmetrically substituted and conjugated olefin. In addition the in-plane ring stretches ν_{8a} and ν_{8b} are found at 1600 and 1575 cm^{-1} with relative intensities that support the presence of conjugation of the ring with a substituent as in styrene. The assignment of the 1630 cm^{-1} band to an olefinic double bond stretch is strongly supported by the Raman spectrum because a strong band is also found near 1630 cm^{-1} in this effect. The Raman spectrum also exhibits bands that can be assigned to ν_{8a} and ν_{8b}.

The presence of an aromatic ring is confirmed by the assignment of ν_{19a} and ν_{19b} near 1495 and 1445 cm^{-1} in the infrared. The broadening of the 1445 cm^{-1} component supports the presence of either aliphatic methylene or methyl groups or both. The strong band near 1378 cm^{-1} can be assigned with reasonable confidence to the symmetric methyl bending mode and the observation of this band supports the presence of at least one methyl group. All three of these modes are known to have little or no intensity in the Raman effect.

Both the Raman and infrared possess only weak to medium bands between 1300–1000 cm^{-1}. As in the case of styrene the bands observed in this region are of little value as group frequencies.

Just below 1000 cm^{-1} in the Raman spectrum at 995 cm^{-1} is the most intense band in the latter effect. This is a highly reliable Raman band and can be confidently assigned to ν_{12}, a symmetric in-plane ring-bending mode. This scattering is highly diagnostic for mono-, meta- or 1,3,5-substitution of the benzene ring.

The unknown appears to involve an olefin substituted on an aromatic ring. This conclusion is based on evidence from: (a) the relative intensities of ν_{8a} and ν_{8b}, and (b) the low frequency value for the >C=C< stretch in the 1600 cm^{-1} region. Thus it becomes important to examine the region between 1000–800 cm^{-1} because this region can help to assign the substitution pattern of the olefin. The presence of a strong band at 895 cm^{-1} is very strong evidence for the presence of a 1,1-disubstituted (>C=CR_2) system. Clearly, one of the R groups is likely to be a phenyl group. Thus, a partial structure is arrived at this stage.

With the substitution pattern established for the olefin attention is now focused on determining the substitution pattern of the aromatic ring.

That the ring is mono-substituted and therefore that the methyl or an R-group is not directly attached to the ring is supported by the following evidence.

First, there are two very strong bands in the infrared near 765 and 705 cm^{-1}. These bands can be assigned to the C—H all-in-phase out-of-plane bend of the ring plus the ν_4 ring puckering vibration. These two bands strongly argue for either mono- or meta-substitution of the ring.

Second, the Raman spectra has a band near 620 cm^{-1} which can be assigned to a ring mode (ν_{6b}) which supports either mono- or para-substitution.

Third, it appears that the overtone of the very strong 890 cm^{-1} band assigned to the C—H out-of-plane bend of the vinylidene group falls very near 1800 cm^{-1} and gives rise to the anomalous intensity of the mono-substituted ring pattern similar to the case which occurred in styrene.

Thus, it is established that the structure of the unknown is a 1,1-disubstituted olefin with one of the R-groups an unsubstituted phenyl group. The second R-group on the double bond can be tentatively assigned as a methyl group based on the following logic: (a) two bands in Raman C—H stretching region above 2900 cm^{-1}, (b) the infrared band at 1378 cm^{-1} assigned the methyl symmetric deformation, and (c) the relative intensities of the C—H stretch in the infrared. The elimination of the possibility of the R-group as an ethyl group while likely is not certain at this point. The Raman spectra, however, adds one further piece of vital information that helps to solidify the assignment of the methyl group. The Raman spectrum has a weak band near 1378 cm^{-1} that corresponds to the infrared band assigned to the methyl deformation. Normally this band has no intensity in the Raman spectrum, however, if the methyl group is substituted on an sp^2 hybridized carbon the bending mode in the Raman is intensified and becomes observed. It is never very strong but it is present as a recognizable band. The Raman spectrum of the unknown has the band and therefore a confident assignment of the second R-group as methyl and not ethyl or propyl can be made.

Thus, it is now possible to propose the structure of the unknown as the simple monomer α-methylstyrene and that is the correct structure.

Answers to the Exercises

ROBERT W. HANNAH and DANA W. MAYO

EXERCISE SECTION I

Exercise 1

***Curves** 9C, 8B, 8C, 9B, **and** 3C.* The description for this exercise states that these are spectra of pure alkene hydrocarbons in the C_6 boiling range.

***Curve** 9C*

IR	Raman	Assignment
3084	3084	sp^2 C—H stretch, vinyl or vinylidene
—	3000	sp^2 C—H stretch, vinyl
2980–2850	2980–2850	sp^3 C—H stretch, CH_2 and CH_3
1825	—	overtone, 2 times 910
1647	1647	C=C stretch, vinyl, vinylidene, or cis
1463	—	CH_3 antisymmetric deformation
1443	—	CH_2 bend
1385	—	CH_3 symmetric deformation
994	—	trans wag, out of plane, vinyl
910	—	terminal wag, out of plane, vinyl

The legend on the curve indicates the sample was prepared as a thin film and is therefore a liquid. Starting from 4000 cm^{-1}, the first significant band appears near 3084 cm^{-1}. Given the introductory information, the shape, position, and relative intensity, this band may be assigned to an olefinic (sp^2) C—H stretch. Further, this high frequency may be assigned to a terminal $=CH_2$ group for a vinyl or vinylidene group. There should be a second band near 3000 cm^{-1} for a vinyl group, but this is often overlapped by the strong aliphatic CH stretching bands and

Course Notes on the Interpretation of Infrared and Raman Spectra, by Dana W. Mayo, Foil A. Miller, and Robert W. Hannah.
ISBN 0-471-24823-1

may not be resolved. The band at 3000 cm^{-1} is clearly seen in the Raman, however, and it may be concluded that a terminal vinyl group is most likely. The C=C double-bond stretch region nearby should be examined next. A band is found near 1647 cm^{-1} in the IR that is in the range where vinyl, vinylidene, and cis structures would be expected to have a band. It should be noted that a very strong band appears at 1647 cm^{-1} in the Raman. This is a very reliable group frequency for the C=C stretch in the Raman for all alkenes. For olefins the next region to be examined in the IR is between 600 and 1000 cm^{-1}, where the out-of-plane bends for olefinic C—H appear. The two strongest bands in this region are at 910 and 994 cm^{-1}. This pair of bands is assigned to the out-of-plane $=CH_2$ wag and the out-of-plane trans C—H wag, respectively, for the vinyl group.

The rest of the major bands in the spectrum are due to vibrations assigned to sp^3-hybridized carbon. Bands between 3000 and 2850 cm^{-1} in both the IR and Raman may be assigned to CH_3 and CH_2 stretching vibrations. Bands near 1463 and 1443 cm^{-1} are assigned to the CH_3 antisymmetric deformation and the CH_2 bend (scissors), respectively. The band near 1385 cm^{-1} is assigned to the CH_3 symmetric deformation (umbrella). It is a singlet, thus eliminating *gem*-dimethyl, isopropyl, and tertiary butyl groups. Given these assignments, there are only two possible structures for a C_6 olefin with a terminal vinyl group.

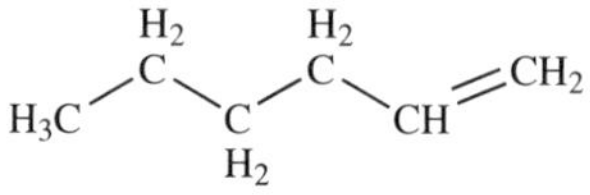

Structure 1

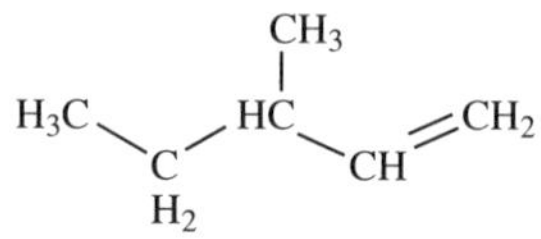

Structure 2

Distinguishing between these two structures is best done by comparison with reference spectra. When this is done, the answer is found to be structure 1, **1-hexene**.

Curve 8B

IR	Raman	Assignment
3030	3005	sp^2 CH stretch for cis-, trans-, and trisubstituted olefin
2990–2850	2990–2850	sp^3 CH stretch for $—CH_2—$ and $—CH_3$ groups
—	1665	C=C stretch for trans-, tri-, and tetrasubstituted groups
1460	1460	CH_3 antisymmetric deformation
1440	1440	CH_2 symmetric bend (scissors)
1380	1380	CH_3 symmetric deformation (umbrella)
968	—	out-of-plane CH bend for trans-substituted olefin

The conclusion is that the structure is a trans-substituted olefin. There are three possible structures:

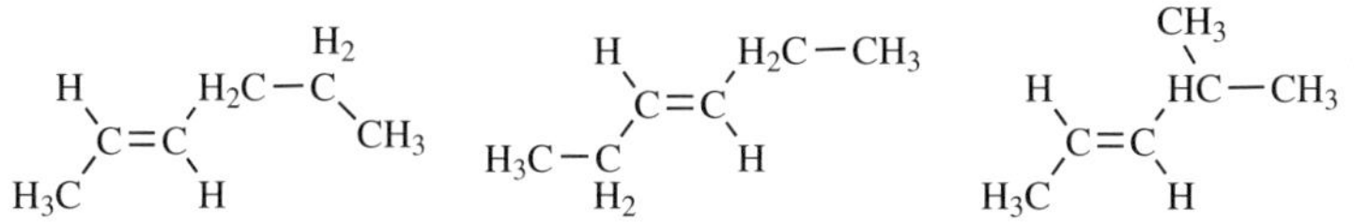

Structure 1 Structure 2 Structure 3

If structure 1 or 3 is correct, there should be a weak band in the Raman near 1380 cm^{-1}. A band appears at that frequency and it may be concluded that there is a $-CH_3$ group on the double bond. This eliminates structure 2. If structure 3 is correct, then the symmetric CH_3 deformation near 1375 cm^{-1} should be a doublet with bands near 1380 and 1360 cm^{-1} and about 20 cm^{-1} apart. The band is a singlet and therefore structure 3 is eliminated.

In the absence of a Raman spectrum, it is only necessary to examine the reference spectra for structures 1 and 2. These are *trans*-2-hexene and *trans*-3-hexene. The answer is structure 1, ***trans*-2-hexene**.

***Curve** 8C*

IR	Raman	Assignment
3040	2995	sp^2 C—H stretch, trans, cis, or tri
2980–2880	2980–2880	sp^3 C—H stretch
—	1670	C=C stretch, trans, tri, or tetra
1472	—	CH_3 antisymmetric deformation
—	1450	CH_3 deformation
1387, 1368	—	CH_3 symmetric deformation for *gem*-dimethyl
—	1377	CH_3 symmetric deformation, on double bond
972	—	out-of-plane trans C—H bend

The sp^2 C—H stretch at 3040 cm^{-1} is in the low end of the range and is therefore assigned to the trans-, cis-, or trisubstituted double bond. Similarly, the C=C double-bond stretch at 1670 cm^{-1} is in the high range and strong in the Raman and may be assigned to trans-, tri-, or tetrasubstituted double bond. (Since there is an sp^2 C—H stretch for hydrogen on the double bond, the tetrasubstituted form may be eliminated.) The absence of the band in the IR at 1670 cm^{-1} is highly indicative of a symmetrically substituted double bond with a center of symmetry or perhaps a pseudo center of symmetry. The suggestion is to look in the 1000–650 cm^{-1} region for the strongest band to determine the substitution on the double bond. There is a band at 972 cm^{-1} that is assigned to the trans out-of-plane C—H bend.

The bands in the region from 2970 to 2870 cm^{-1} are assigned to sp^3 C—H stretches. Looking in the 1450 cm^{-1} region, bands are found for the CH_3 and/or CH_2 bending modes. The 1375 cm^{-1} region in the IR shows two bands in the IR at 1387 and 1368 cm^{-1} ($\Delta\tilde{\nu} = 19$ cm^{-1}), and this indicates a *gem*-dimethyl or isopropyl group. The Raman also has a band at 1378 cm^{-1}, and this will only occur if there is a CH_3 attached directly to an sp^2-hybridized carbon, i.e., a C=C double bond in this case. Given these assignments, there is only one structure that can be written, that of **4-methyl-*trans*-pentene-2**. Even if the Raman data were not available, there is still only one structure that can be written for a C_6 olefinic hydrocarbon.

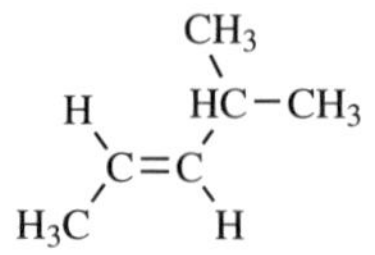

Curve 9B

IR	Raman	Assignment
3080	—	sp^2 C—H stretch, $=CH_2$
—	3000	sp^2 C—H stretch
2980–2880	2980–2880	sp^3 C—H stretches
1782 (vw)	—	overtone of 890 band
1650	1650	C=C stretch, vinyl, vinylidene, or cis
1463	—	CH_3 antisymmetric deformation
1450	1450	CH_2 bend
1380	1382(?)	CH_3 symmetric deformation
1364(?)	—	Possible CH_3 symmetric deformation, *gem*-dimethyl
890	—	out-of-plane C—H bend for vinylidene

The high frequency for the sp^2 C—H stretch and the lower frequency for the C=C stretch are consistent with vinyl or vinylidene but not with cis substitution. Looking in the region 1000–650 cm^{-1} for the out-of-plane bends for C=C substitution, there is a band at 890 cm^{-1} that may be assigned to the vinylidene structure. For a C_6 hydrocarbon there are three possible structures, 2,3-dimethyl-butene-1, 2-ethyl-butene-1, or 2-methyl-pentene-1.

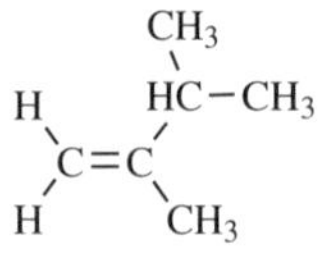

Structure 1

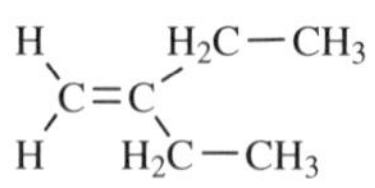

Structure 2

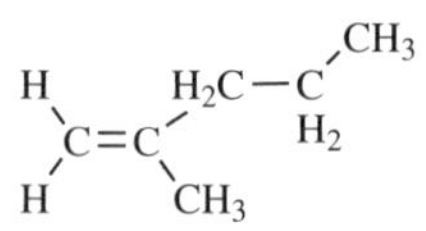

Structure 3

The sp^3 C—H stretching and deformation regions confirm the presence of sp^3 C—H groups, and since this region is fairly complex, it may be concluded that both $—CH_2—$ and $—CH_3$ groups are present.

If the 1375-cm^{-1} region is examined, a band is found in the IR at 1380 cm^{-1} that is assigned to the symmetric CH_3 deformation. The question arises whether the weak shoulder at 1364 cm^{-1} is another CH_3 deformation that is part of a pair of bands (along with the 1380 cm^{-1} band) characteristic of the *gem*-dimethyl or isopropyl groups? There is also a weak shoulder in the Raman at 1382 cm^{-1}, and if it is concluded that this is really present, then a CH_3 must be attached directly to the double bond. However, this assignment must be considered as tentative. At this point it is recommended that the three spectra corresponding to the three structures be compared to reference spectra. When this is done, the answer is found to be structure 1, **2,3-dimethyl-butene-1**.

A few words of explanation are needed to understand the 1380 cm^{-1} bands. Normally the isopropyl doublet has about equal intensities for the two bands in the doublet. From the structure given as the answer, it is known that there is another $—CH_3$ group attached directly to the double bond and this $—CH_3$ also has a contribution to the intensity of the 1380 cm^{-1} band in the IR. This contribution to the intensity of the 1380 cm^{-1} band makes the 1380 cm^{-1} higher part of the doublet appear stronger than it normally is for the isopropyl group and also causes the lower part of the doublet to be partially overwhelmed by the 1380 cm^{-1} band. Also, in the Raman the 1380 cm^{-1} band is partially obscured by the somewhat stronger band near 1400 cm^{-1}, thus making the presence and assignment of this band uncertain. In this case comparison with reference spectra totally resolves the question.

Curve 3C

IR	Raman	Assignment
3020	3005	sp^2 C—H stretch, trans, cis, or tri
2980–2870	2980–2870	sp^3 C—H stretch for CH_3 and CH_2
1658	1656	C=C stretch, vinyl, cis or vinylidene
1470	1450	sp^3 CH_3 antisymmetric and/or CH_2 bends
1378	—	CH_3 symmetric deformation
720 (m, br)	—	out-of-plane C—H bend, cis

The low frequency for the sp^2 stretch at 3020 cm^{-1} in the IR indicates trans-, cis-, or trisubstitution. This combined with the C=C stretch at 1658 (vinyl, cis, or vinylidene) strongly suggests cis-substitution. Looking in the region from 1000 to 650 cm^{-1} a broad, medium-to-weak band should be found for the cis configuration and a band is found at 720 cm^{-1}. This is too broad and strong to be assigned to the $(CH_2)_{n>4}$ rocking mode for a CH_2 sequence. It is assigned therefore to the out-of-plane C—H bend for cis-substitution. The 2980–2870 cm^{-1} region is assigned to CH_3 and CH_2 stretches, and the bands at 1470 and 1378 cm^{-1} in the IR and at 1450 cm^{-1} in the Raman confirm the presence of these groups. The 1378 cm^{-1} band is a singlet in the IR and any branching leading to an isopropyl or *gem*-dimethyl group is eliminated. This leaves two possible structures for a C_6 hydrocarbon, *cis*-hexene-3 and *cis*-hexene-2.

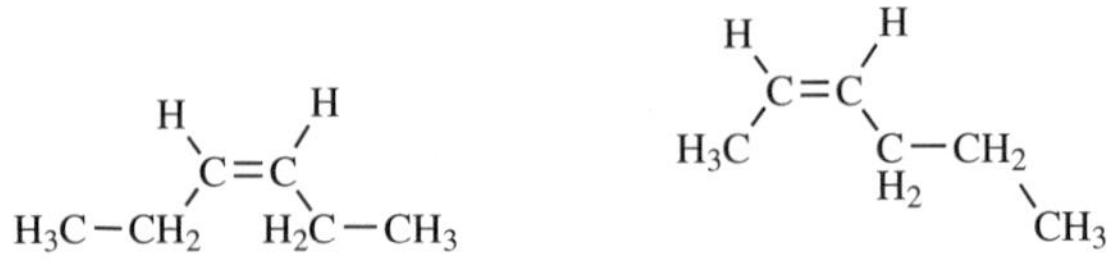

Structure 1 Structure 2

There is the possibility of a very weak band at 1375 cm^{-1} in the Raman that would suggest a CH_3 bound directly on the double bond, but this is too weak to be given much importance. (If it were taken as being present, then the hexene-2 isomer would be favored.) In the absence of any conclusive evidence to indicate which isomer is correct, reference spectra should be consulted. When this is done, the answer is found to be structure 1, ***cis*-hexene-3**.

Exercise 2

This set of spectra is for pure hydrocarbon liquids. The first thing one should notice is the significantly larger number of bands that are narrower than in the preceding set of curves. This fact marks the set as having aromatic or other structures that do not allow for rotational isomers that contribute to band broadening.

Curve 6-A

IR	Raman	Assignment
3025	3050	sp^2 C—H stretch, olefin or aromatic
2925	2920	sp^3 C—H antisymmetric stretch for CH_3
2870	2865	sp^3 C—H symmetric stretch for CH_2
1940–1740	—	sum tone region for aromatic substitution
1616, 1592	1615, 1592	ν_{8a} and ν_{8b}, aromatic ring stretch
1493, 1460	—	ν_{19a} and ν_{19b}, aromatic ring stretch
1378	1378	CH_3 symmetric deformation
—	998	mono-, meta-, or 1,3,5-trisubstituted
770	—	out-of-plane aromatic C—H bend
698	—	out-of-plane ring puckering

The band at 3025 cm^{-1} is assigned to an sp^2 hybridized C—H stretch, either olefinic or aromatic. The absence of a band in the IR and more importantly in the Raman in the 1680–1620 cm^{-1} region effectively eliminates consideration of an olefinic double bond. This suggests that confirmation for the aromatic structure be considered. The pair of bands near 1600 cm^{-1} and the pair near 1500 and 1450 cm^{-1} are assigned to the ν_{8a}, ν_{8b} and ν_{19a}, ν_{19b} modes, respectively, for the aromatic ring. The band at 770 cm^{-1} is assigned to the out-of-plane C—H bending for mono- or meta-substitution. The band at 688 cm^{-1} is assigned to the out-of-plane ring bending in which alternate carbons are moving in phase in and out of the plane of the ring. This band only occurs for mono-, meta-, or 1,3,5-substitution. The sum tone region from 1940 to 1740 cm^{-1} shows a pattern that is characteristic of meta-substitution. It is important that the pattern of weak bands, the number of them, and the relative intensities and appearance of shoulders in this region be considered, not the actual band positions. The Raman band at 998 cm^{-1} is also assigned to mono-, meta-, or 1,3,5-substitution and is confirmation of the IR conclusion. Finally, if the ring were monosubstituted, there should be a weak band in the 650–600 cm^{-1} region in the Raman; this band is not present and thus the structure is probably not monosubstituted.

The remaining question is what are the substituents on the ring? There are bands at 2925 and 2870 cm^{-1} in the IR and at 2920 and 2865 cm^{-1} in the Raman. These are in the right region for sp^3 hybridized C—H stretches and are assigned to these vibrations. There is appreciable broadening for the 1450 cm^{-1} band, and this is attributed to underlying sp^3 C—H bending. There is a singlet at 1378 cm^{-1} in both the IR and Raman, and this is assigned to the symmetric bend of the $-CH_3$ group. From the presence of this band in the Raman it is concluded that at least one of the $-CH_3$ groups must be attached directly to the ring in order to explain its presence in the Raman spectrum. The other position could be another $-CH_3$ group, but from the evidence an ethyl, propyl, or perhaps a slightly larger group cannot be ruled out. Reference spectra should be consulted starting with 1,3-dimethylbenzene and then 1-methyl-3-ethyl benzene, etc. When this is done, the answer is found to be **1,3-dimethyl benzene**, or ***meta*-xylene**.

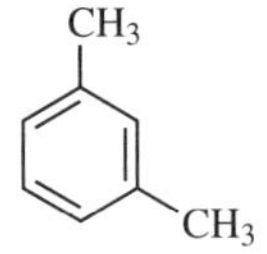

***Curve** 18-B*

IR	Raman	Assignment
3080–3020	3080–3020	sp^2 C—H stretch
1950–1690	—	sum tone region for aromatic substitution
1822	—	overtone for 910 sp^2 out-of-plane C—H bend
1633	1630	C=C stretch
1603, 1580	1602, 1586	ν_{8a}, ν_{8b} aromatic ring stretch
1508, 1452	—	ν_{19a}, ν_{19b} aromatic ring stretch,
—	1002	mono-, meta-, or 1,3,5-trisubstituted
992, 910	—	vinyl group, $-CH=CH_2$
780	—	out-of-plane aromatic C–H bend
698	—	out-of-plane ring puckering
—	622	ν_{6b} in-plane ring deformation

The preponderance of sharp bands in this spectrum strongly indicates very little or no contribution from rotational isomers, and the presence of olefinic or aromatic groups musts be considered.

The bands in the 3050 cm^{-1} region are assigned to either olefinic or aromatic C—H stretches or both. There are no bands in the sp^3 C—H stretch region from 2950 to 2820 cm^{-1}, so aliphatic substitution is not present. The next strong band in the spectrum is at 1633 cm^{-1} in the IR and 1630 cm^{-1} in the Raman. (Small differences in frequency such as these are probably due to slight errors in registration on the recorder paper.) This band is assigned to the C=C stretch and therefore an olefinic group is present in the molecule. It is in the low end of the region for vinyl, vinylidene, and cis-substitution. Since the frequency is in the low end, it may indicate that the olefinic double bond is conjugated. The pair of bands at 1603 and 1585 cm^{-1} are assigned to the ν_{8a} and ν_{8b}, respectively, in-plane aromatic ring stretch. Note that the 1585 cm^{-1} band is stronger than the 1603 cm^{-1} band, indicating that the substituent on the ring is conjugated to the ring. The 1508 and 1452 cm^{-1} bands are sharp and are assigned to the ν_{19a} and ν_{19b} in-plane aromatic ring stretches.

The next strong bands in the IR are at 992 and 910 cm^{-1} and are assigned to the out-of-plane olefinic C—H bend for a vinyl group. Also in the IR the bands at 780 and 698 cm^{-1} are assigned to the out-of-plane C—H bend for mono- or meta-substitution and the out-of-plane ring puckering for mono-, meta-, or 1,3,5-tri-substitution. If the sum tone region from 1950 to 1690 cm^{-1} is examined, the pattern would seem to be most like that for monosubstitution except for the anomalous intensity of the band at 1822 cm^{-1}. The reason for this anomaly may be understood by remembering that the overtone of the 910 cm^{-1} band for the vinyl group will be near 2×910 (or slightly higher), or just above 1820 cm^{-1}. This overtone is overlapping the normal intensity for the sum tone pattern for aromatic ring substitution, causing the relative intensities to appear anomalous.

The Raman band at 1002 cm^{-1} is assigned to mono-, meta-, or 1,3,5-trisubstitution, and the IR conclusion that monosubstitution is present is confirmed by the appearance of this band. In addition, there is a very weak band at 622 cm^{-1} in the Raman, and this may be assigned to the ν_{6b} in-plane ring deformation for monosubstitution. This assignment must be considered as weak evidence for monosubstitution.

The spectral evidence indicates that the unknown is an aromatic with a conjugated substituent. The substituent contains a vinyl group that has its double bond conjugated to the ring. There are no other substituents present based on the absence of bands in the spectrum to indicate their presence. The only structure that satisfies all of the evidence is **vinyl benzene**, or **styrene**.

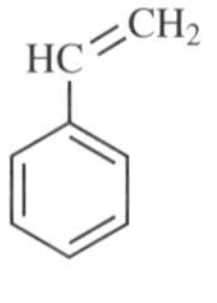

Curve 6-C

IR	Raman	Assignment
3100–3020	3050, 3005	sp^2 C—H stretch
2980–2870	2970–2870	sp^3 C—H stretch
1900–1650	—	sum tone region for aromatic substitution
—	1612, 1578	ν_{8a} and ν_{8b} aromatic ring stretch
1516, 1464	—	ν_{19a} and ν_{19b} aromatic ring stretch
1384, 1368	—	symmetric deformation for *gem*-dimethyl
—	1378	symmetric deformation for CH_3 on sp^2 C
813	—	out-of-plane C—H bend for para-substitution
—	805	fingerprint band
—	640	ν_{6b} in-plane ring puckering for mono- or para-substituted phenyl

The bands in the 3100–3005 cm^{-1} region are assigned to sp^2 C—H stretches and may be due to either olefinic or aromatic structures or both. On examining the 1650 cm^{-1} region for the C=C stretch, a band is not observed (in particular, in the Raman where the group frequency is a highly reliable one) and the presence of an olefinic group must be ruled out. The other possibility is that an aromatic ring is present. If the 1600 and 1500 cm^{-1} regions are examined for the ring stretching vibrations for the aromatic ring, then some confusion arises. There are no bands in the IR near 1600 cm^{-1}, but bands do appear at 1612 and 1578 cm^{-1} in the Raman. This would occur if the ring has a center of symmetry or at least a pseudo center of symmetry, for example, was para-substituted with identical or similar groups. If this is assumed to be the case, then the bands at 1516 and 1464 cm^{-1} can be assigned to the ν_{19a} and ν_{19b} ring stretches, respectively. The substitution on the ring may be determined from the strong bands in the 900–650 cm^{-1} region and from the sum tone pattern in the 1950–1650 cm^{-1} region. The band near 813 cm^{-1} may be assigned to para-substitution. The sum tone pattern confirms this assignment.

The remaining question is what are the substituents. The bands in the 2980–2870 cm^{-1} region are indicative of sp^3 C—H stretches, and it is possible, based on the complexity of this region, that both $—CH_3$ and $—CH_2—$ groups are present. The 1450 cm^{-1} band is overlapped by the aromatic ring stretch but is broader than might be expected if only the aromatic stretch is contributing. It may be assumed that some contribution to the band comes from an aliphatic C—H bend, either CH_3 or CH_2 or both. The CH_3 symmetric deformation bands near 1380 cm^{-1} in the IR appear as a doublet. The band pair is assigned to a *gem*-dimethyl or isopropyl group. The Raman also shows a band at 1378 cm^{-1}. This is assigned to the symmetric deformation of a $—CH_3$ group on an sp^2 carbon, in this case on the aromatic ring.

The conclusions are that there is a para-substituted aromatic ring with identical or nearly identical substituents. One of the substituents is a $—CH_3$ group based on the appearance of a band at 1378 cm^{-1} in the Raman. In the absence of the Raman spectrum this conclusion cannot be reached, and one or more —CH_2— groups may separate the —CH_3 group from the ring. While this makes the final identification somewhat more difficult, it only adds one or two additional structures to be considered when reference spectra are consulted. The other substituent is most probably an isopropyl group, but one or more —CH_2— groups may separate it from the ring. Once again, looking at reference spectra starting with *para*-isopropyl toluene, *para*-2-methyl-propyl toluene, *para*-isopropyl ethyl benzene, etc., the former, ***para*-isopropyl toluene**, is found to be correct. The common name is ***p*-cymene**. It should be noted that *para*-isopropyl ethylbenzene is eliminated by the presence of the 1378 cm^{-1} band in the Raman that is assigned to the CH_3 symmetric deformation for a methyl group attached to an sp^2 carbon. If Raman data were not available, then this possibility needs to be considered.

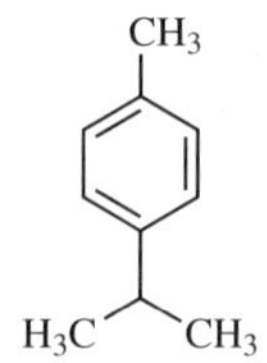

Exercise 3

***Curve** 16-A.* This is the spectrum of a hydrocarbon with a molecular weight of 70.

IR	Raman	Assignment
3025	3022	sp^2 C—H stretch
2980–2860	2980–2860	sp^3 C—H stretches
1678	1677	C=C stretch
1455	1454	CH_2 and/or CH_3 bends
1380	1382	CH_3 symmetric deformation
804	—	out-of-plane olefinic C—H bend

The band at 3024 cm^{-1} is assigned to the C—H stretch for an sp^2-substituted carbon and may be associated with either an olefinic or aromatic group. The absence of band pairs at 1600, 1580 cm^{-1} and 1500, 1450 cm^{-1} eliminates aromatics from consideration. The band at 1678 cm^{-1} in both the IR and the Raman is assigned to the C=C stretch. Looking in the 1000–650 cm^{-1} region for the out-of-plane olefinic bends shows a band at 804 cm^{-1}. This is assigned to a trisubstituted structure. The 1378 cm^{-1} band in the IR is a single band indicating no *gem*-dimethyl groups are present. If all of the substituents on the double bond are —CH_3 groups, then the molecular weight is the sum of 5 carbons, or 60 plus 10 hydrogens, or 10, making the sum equal to 70. This is the only possible arrangement that will yield a molecular weight of 70. Note that this also explains the presence of the 1378 cm^{-1} band in the Raman since all of the $-CH_3$ groups are attached to the double bond. The answer is **2-methyl-butene-2**.

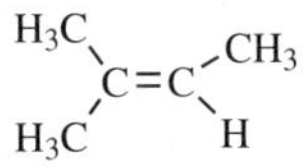

Curve 14-B. This is a spectrum of a well-known high-molecular-weight polymer.

IR	Raman	Assignment
2920, 2850	2880, 2830	sp^3 CH_2 stretches
1475	1475	sp^3 CH_2 bend, scissors
1463	—	sp^3 CH_2 bend, scissors
—	1440	?
1372 (vw)	—	CH_3 symmetric deformation, umbrella
730, 720	—	$(CH_2)_n$ in-phase rock, $n > 4$

The spectrum is a very simple one with only a few bands. There are only bands that may be assigned to $-CH_2-$ groups except for the very weak band at 1372 cm^{-1}. This band is assigned to the symmetric deformation of the $-CH_3$ group, but it has very weak intensity, and the number of $-CH_3$ groups compared to the number of $-CH_2-$ groups must be small. The only high-molecular-weight polymer that satisfies the evidence is polyethylene. The band at 1378 cm^{-1} is the CH_3 symmetric deformation, and its presence is due to methyl end groups and chain branches. The split band near 725 cm^{-1} is assigned to the $(CH_2)_{n>4}$ in-phase rock, and it is split because there are two $-CH_2-$ chains in the unit cell leading to chain–chain interaction. This splitting is due to the presence of crystalline structure in polyethylene, but it must be noted that it should not be used quantitatively because there is a band due to amorphous polyethylene that underlies the 725 cm^{-1} doublet and contributes significantly to the intensity. The chain–chain interaction also causes the 1475 cm^{-1} band to appear as a doublet in this spectrum.

The presence of the 1372 cm^{-1} band indicates some chain branching. Side chains will cause less crystalline and more amorphous structures to occur. The weak band near 1300 cm^{-1} is characteristic of amorphous content, while the weak band near 1895 cm^{-1} is a better indication of crystalline content.

The answer is **polyethylene**. The monomer is **ethylene**.

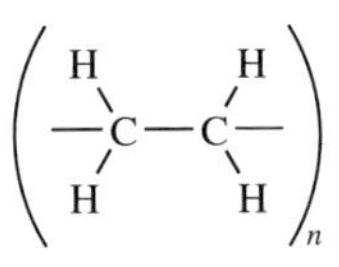

Polyethylene

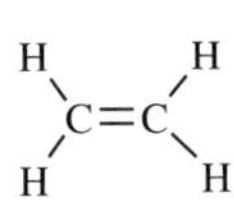

Ethylene

Exercise 4

Curves Index, Fractions A and B

Fraction A	Fraction B	Assignment
—	3090 (m)	sp^2 C—H stretch, terminal CH_2
3054 (w)	—	(?) possibly sp^2 C—H stretch, nonterminal
—	2249 (s)	C≡N stretch, not conjugated
2219 (vs)	—	C≡N stretch, conjugated
—	1661 (s)	C=C stretch
1635 (s)	—	C=C stretch
—	1416 (m)	$=CH_2$ in-plane bend
—	904 (vs)	$=CH_2$ out-of-plane bend for vinylidene
800 (s)	—	=CH out-of-plane bend for trisubstitution

This problem can be solved by looking at the position of the C≡N stretching bands in the 2250 cm^{-1} region. The unconjugated position is 2250 ± 10 cm^{-1} and fraction B has the band at this position. In structure **I** the nitrile group is not conjugated and, therefore, fraction B must be structure **I**.

The remainder of the two spectra also supports this conclusion. The 3090 cm^{-1} band for fraction B is assigned to the terminal $=CH_2$ stretch for an sp^2-hybridized carbon, either vinyl or vinylidene. The 904 cm^{-1} band is assigned to the out-of-plane $=CH_2$ bend for a vinylidene group. (Note that if the structure had a vinyl group, then a strong band would have to appear in addition to the 904 cm^{-1} band near 990 cm^{-1}.) The weak band near 1820 cm^{-1} is further conformation of the vinylidene structure since the overtone of the 904 cm^{-1} band should appear but at slightly higher frequency than twice the fundamental. The 800 cm^{-1} band for fraction A is assigned to the out-of-plane bend for the trisubstituted double bond. This is consistent with the earlier conclusion also that fraction B must be structure **I**.

The positions and intensities of the C=C stretch may be better understood now. In fraction A (structure **II**), conjugation has lowered the frequencies of the C=C stretch and the C≡N stretch from their unconjugated positions in the spectrum of fraction B (structure **I**).

Thus **fraction A is structure II** and **fraction B is structure I**.

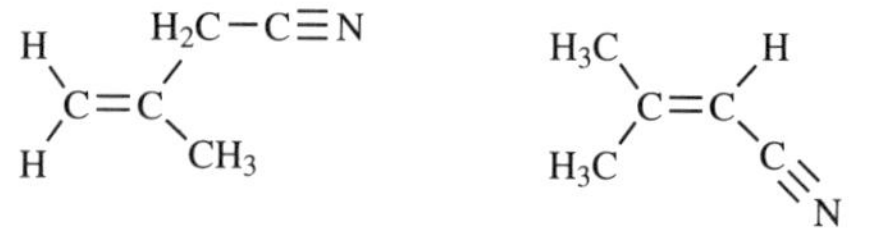

Structure I, Fraction B Structure II, Fraction A

***Curve** 14-C.* This spectrum is that of a high-molecular-weight polymer.

IR	Assignment
3632, 3539	(?) (See discussion)
2938, 2875	sp^3 C—H stretches, probably CH_2 stretches
2248	C≡N stretch
1453	sp^3 CH bend, CH_3 and/or CH_2

The most important band in the spectrum is the 2248 cm^{-1} band that is assigned to the unconjugated nitrile stretch. The most likely simple polymer containing a nitrile group is polyacrylonitrile, and by referring to reference spectra, this identification is confirmed. The bands at 3632 and 3539 cm^{-1} may present some confusion unless it is known that this polymer was prepared by an emulsion polymerization process. The assignment of these bands is to the presence of some residual water and a trace of an antioxidant. The answer is **polyacrylonitrile**.

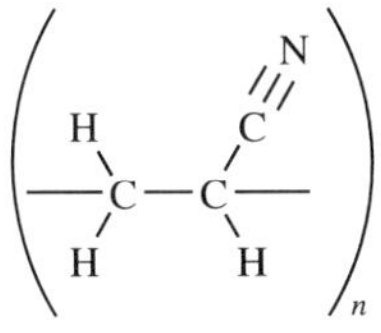

Polyacrylonitrile

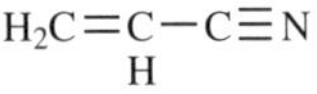

Acrylonitrile (monomer)

EXERCISE SECTION II

Exercise 5

***Curve** 2-B.* This spectrum is that of a pure organic compound of a simple structure. It is known to contain nitrogen based on a positive Prussian blue test.

IR	Raman	Assignment
3074	3072	sp^2 C–H stretch
2232	2235	C≡N stretch
1975–1775	—	sum tone pattern for monosubstituted ring
1600, 1580	1600	ν_{8a}, ν_{8b} in-plane aromatic ring deformation
1492, 1448	—	ν_{19a}, ν_{19b} in-plane aromatic ring deformation
—	1001	mono-, meta-, or 1,3,5-trisubstituted
748	—	out-of-plane CH bend, mono or meta
690	—	ν_4 out-of-plane ring bend, mono-, meta-, or 1,3,5-trisubstituted
—	620	ν_{6b} in-plane ring deformation, mono- or para-substituted

The first thing that should be noted about this spectrum is the preponderance of sharp, needlelike bands. This is characteristic of molecules that have rigid structures that do not allow for rotational isomers to be present. In particular, this is characteristic of aromatic molecules, and this must be considered in the interpretation.

The band near 3073 cm^{-1} may be assigned to either the terminal CH_2 stretch for an olefin or the C–H stretch for hydrogens on an aromatic ring. There are no bands in the 2990–2850 cm^{-1} region for the sp^3 C–H stretching vibrations for an aliphatic, thereby eliminating any aliphatic segments from consideration. If the unknown is an olefin, then a band should appear in the region from 1680 to 1630 cm^{-1}, in particular in the Raman, where it is expected to be very strong. A band does not appear in this region in either the IR or the Raman. Looking in the IR in the 1600, 1580 cm^{-1} and 1500, 1450 cm^{-1} regions for the aromatic ring deformations ν_{8a}, ν_{8b} and ν_{19a}, ν_{19b}, respectively, these two pairs of bands are observed, and it may be concluded that the unknown is an aromatic. The band at 748 cm^{-1} in the IR is assigned to the out-of-plane C—H bends for mono- or meta-substitution. The 690 cm^{-1} band, ν_4, is assigned to the out-of-plane ring bend involving the 1,3,5 and 2,4,6 carbons moving in phase but in opposite directions. This band appears for mono-, meta-, and 1,3,5-substitution. There is a band in the Raman at 1001 cm^{-1} that is assigned to the in-plane C—H bend for mono-, meta-, or 1,3,5-trisubstitution. In addition, the Raman has a weak band at 622 cm^{-1} that may be assigned to the ν_{6b} in-plane ring deformation for a mono- or para-substituted aromatic. The sum tone pattern between 1975 and 1700 cm^{-1} shows a series of four bands that is very similar to that shown by monosubstitution. It may be concluded from this evidence that a monosubstituted aromatic ring is present in the unknown.

There is a strong band near 2233 cm^{-1} in both the IR and Raman, and this is assigned to the conjugated C≡N stretch. There is no evidence for any other groups being present. It may be concluded from this that the unknown is **benzonitrile**.

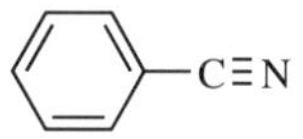

Exercise 6

***Curve** 12-C.* This is the spectrum of a well-known high-molecular-weight polymer.

IR	Raman	Assignment
3080–3020	3070	sp^2 C—H stretch
2920, 2850	2920, 2850	sp^3 C—H stretch, CH_2 group
1950–1740	—	sum tones for monosubstitution
1603, 1583	1602, 1580	ν_{8a}, ν_{8b} in-plane aromatic ring deformation
1494, 1454	—	ν_{19a}, ν_{19b} in-plane aromatic ring deformation
—	1000	phenyl ring for mono-, meta-, or 1,3,5-trisubstitution
760	—	out-of-plane C—H bend for mono- or meta-substitution
702	—	out-of-plane ring bending, mono-, meta-, or 1,3,5-trisubstitution
—	620	ν_{6b} in-plane ring deformation, mono- or para-substitution

Once again, it should be noted that the spectrum contains many sharp, needlelike bands. However, in this spectrum there are several bands that are appreciably broader than in the preceding example, curve 2-B. This broadening arises due to rotational isomers being present for the singly bonded carbons for the saturated, sp^3-hybridized carbons in the molecule. It may be concluded that the unknown has both aromatic and aliphatic groups.

The bands in the 3080–3020 cm^{-1} region are assigned to sp^2 C—H stretches. Looking in the 1680–1620 cm^{-1} region, there are no bands, particularly in the Raman, where a strong band should appear for an olefin. However, looking in the 1600, 1580 cm^{-1} and 1500, 1450 cm^{-1} regions, bands are found in the IR that are assigned to ν_{8a}, ν_{8b} and ν_{19a}, ν_{19b}, respectively, aromatic ring deformations. The band at 760 cm^{-1} in the IR is assigned to the out-of-plane C—H bend for mono- or meta-substitution. The band at 702 cm^{-1} is assigned to the out-of-plane ring bend for mono-, meta-, or 1,3,5-trisubstitution. There is a band in the Raman at 1000 cm^{-1} that is characteristic of mono-, meta-, and trisubstitution also. Finally, the band at 620 cm^{-1} in the Raman indicates mono- or para-substitution as well. The sumtone pattern in the IR also agrees with the monosubstitution conclusion. The remaining bands that must be explained are those at 2920 and 2850 cm^{-1}. These are the normal position for the antisymmetric and symmetric stretches for sp^3 —CH_2— groups. Note that there are no —CH_3 groups because of the lack of a sharp band in the IR at 1378 ± 5 cm^{-1}.

The spectral evidence suggests that the aromatic ring is monosubstituted with —CH_2— groups only. The only well-known high-molecular-weight polymer that satisfies this conclusion is **polystyrene**.

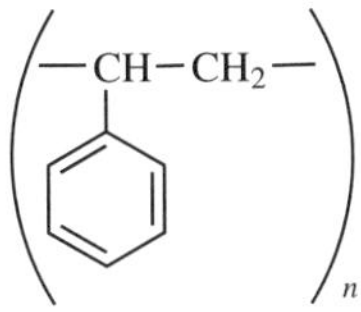

Exercise 7

Curve 1-B. The unknown is a pure organic liquid of simple structure. This compound gave a positive Prussian blue test for nitrogen.

IR	Raman	Assignment
3320	3305	N—H stretch
3112	—	overtone of the 1565 cm^{-1} amide II band
2960	2930	sp^3 C—H stretch
1650	1650	carbonyl stretch, amide I, secondary amide
1565	—	amide II of secondary amide
1450	—	CH_3 antisymmetric deformation or CH_2 bend
1418	—	symmetric deformation of CH_3 on nitrogen
1372	1372	CH_3 symmetric deformation
1300	1300	amide III of secondary amide (unreliable)
725	—	N—H out-of-plane bend

This curve is a classical secondary-amide spectrum. The band at 3320 cm^{-1} is assigned to the N—H stretch. There is only one band in this region, as opposed to the expected two bands if the unknown were a primary amide having two N—H bonds. The band at 1650 cm^{-1} is the C=O stretch for a secondary amide shifted to lower frequencies by hydrogen bonding and by conjugation with the lone pair of electrons on the nitrogen. It is referred to as the amide I band. The band at 1565 cm^{-1} is assigned to a mix of the C—N stretch and the N—H in-plane bend. This is the amide II band, and it occurs only for the trans secondary-amide structure. There are actually two vibrations associated with the mixing of the vibrations of these two groups. The second band is not a good group frequency and normally appears between 1300 and 1220 cm^{-1} for secondary amides. It is referred to as amide III. The weak-to-medium intensity band at 3112 cm^{-1} is the overtone of the 1565 cm^{-1} amide II group frequency. It has gained substantially in intensity by Fermi interaction with the N—H stretch at 3320 cm^{-1}. The band at 725 cm^{-1} is assigned to the out-of-plane N—H bend for a secondary amide and is also not a good group frequency. It is useful to confirm the structure once the amide I and amide II bands have been assigned. Note that it is quite broad due to hydrogen bonding.

What are the substituents on the carbonyl and on the secondary nitrogen? From the Raman where a band appears at 1371 cm^{-1} that is assigned to a —CH_3 group on an sp^2 carbon, it must be concluded that the substituent on the carbonyl is a methyl group. The band at 1418 cm^{-1} (see Table 2.4 for CH_3 attached to elements other than carbon) is assigned to a —CH_3 group on nitrogen. In the absence of the Raman spectrum reference spectra should be consulted. The analyst should start with the simplest structure that has methyl groups on both the carbonyl and nitrogen and add methylene groups until the answer is found. It is clear that there is no *gem*-dimethyl or tertiary butyl groups since the 1372 cm^{-1} band in the IR is not split. Furthermore the C—H stretching region (2960–2930 cm^{-1}) is very simple and the hydrocarbon substituents must likewise be simple.

Based on the available evidence, it may be concluded that the answer is ***N*-methyl acetamide**.

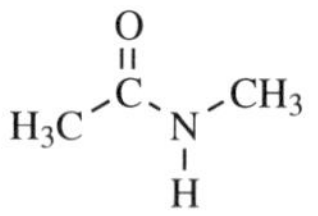

Curve 19-B. This compound gave a positive Beilstein test for halogens. It is extremely reactive when in contact with moist air.

IR	Raman	Assignment
3600		? Overtone (2X1810 C=O stretch)
3024, 2938 (vw)		? sp^2 and/or sp^2 C—H stretch
1810	1803	C=O stretch
1363	1358	CH_3 symmetric deformation
1102		?
595	592	C—Cl stretch
440	440	C—Cl bend

All of the bands in this spectrum are broad. The strongest band is at 1810 cm^{-1} in the IR and at 1803 cm^{-1} in the Raman spectrum. This is unusually high for a carbonyl frequency but given its high intensity and its position in the carbonyl region it must be assigned to a carbonyl. The reason it is so high is that there is a substantial inductive effect due to the presence of a highly electronegative chlorine on the carbonyl group. In this case, a chlorine is attached to the carbonyl and the band is assigned to an acid-chloride type of compound. The band at 595 cm^{-1} is assigned the C—Cl stretch while the band at 440 cm^{-1} is assigned to the C—Cl bend. Neither of these are good group frequencies but given their presence they serve as evidence for the halogen to be chlorine. The band at 1363 cm^{-1} in the IR (1358 cm^{-1} in the Raman) is assigned to the symmetric deformation of the CH_3 group when the CH_3 is attached to a carbon. It appears in the Raman as a result of being attached to an sp^2 hybridized carbon, that is the carbonyl carbon. The position of this band at a lower frequency than normal is consistent with the fact that is attached to sp^2 hybridized carbon (see the note at the bottom of the table for CH_3 groups attached to elements other than an sp^3 carbon).

Why does the CH_3 deformation intensity appear so much stronger than the CH_3 stretching frequency? Normally, the CH_3 stretching is much stronger than the bending modes. Also, why is the CH_3 anti-symmetric stretch shifted above 3000 cm^{-1}? The answer to both of these questions is associated with the fact the CH_3 group is attached to the carbonyl group. This would appear to be a general rule and if the CH_3 stretch and bending modes have these relative intensities then the analyst should expect the CH_3 group to be attached directly to a carbonyl group. The reason the CH_3 stretch is shifted above 3000 cm^{-1} is attributed to the presence of the electronegative chlorine on the carbonyl.

The band at 1100 cm^{-1} lies in the fingerprint region and even though it is quite strong it must be assigned to a fingerprint band. It must be taken into account when comparing with reference spectra to confirm the identification.

The answer is **acetyl chloride**.

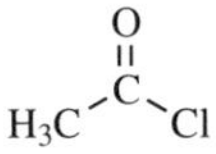

Acetyl Chloride

Exercise 8

***Curve** 3-**B**.* This unknown is a pure organic liquid. Halogens are not present. However, the compound is highly reactive when in contact with moist air or protic solvents.

IR	Raman	Assignment
3036, 2950	3030, 2950	sp^3 CH_3 stretch
1830, 1758	1820, 1755	C=O stretch ($\Delta v = 70$ cm^{-1}, consistent with anhydride)
1372	1375	CH_3 symmetric deformation
1130	—	antisymmetric C—O—C stretch

All of the bands in this spectrum are broad, indicating the presence of rotational isomers and/or hydrogen bonding.

The strongest bands appear in the C=O region at 1830 and 1758 cm^{-1}. The appearance of two carbonyl bands in the high end of the frequency range for carbonyls indicates an open-chain anhydride. This conclusion is consistent with the fact that the compound is highly reactive in moist air or in contact with protic solvents.

There are no bands for aromatic or olefinic groups. Once again the C—H stretching bands at 3036 and 2950 cm^{-1} may be assigned to $-CH_3$ groups attached directly to a carbonyl. These two bands are weaker than expected compared with the intensity of the symmetric deformation for CH_3 at 1372 cm^{-1}. There is a band in the Raman spectrum at 1375 cm^{-1} assigned to the $-CH_3$ group, and this only occurs if the $-CH_3$ group is attached directly to an sp^2-hybridized carbon, in this case on the carbonyl group. Since it is more likely that an anhydride will have the same substituent on both carbonyls, the unknown spectrum should be compared with a spectrum of acetic anhydride and then with the next one or two anhydrides in the homologous series. When this is done, the answer is found to be **acetic anhydride.**

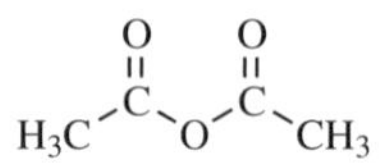

The prism spectrum that accompanies this problem shows an additional strong band at 5.72 μm (1703 cm^{-1}). There is a very significant broadening under the C—H stretching bands near 3.3 μm (3000 cm^{-1}). This evidence suggests an acidic —COOH group is present. The assignment of the 5.72 μm band to an acid carbonyl and the 11.2 μm band to the out-of-plane deformation of the acid dimer would be consistent with an assignment of a broad acidic O—H band under the C—H stretching band.

It is known that the compound is extremely reactive in the presence of water. It must be concluded that this "reference compound" has reacted with water or water vapor to produce some acetic acid. The answer to the second question is **acetic anhydride contaminated with acetic acid.**

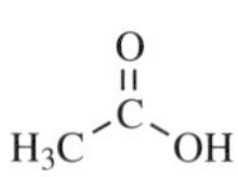

Acetic Acid

Exercise 9

Curve *2-A*. This is a pure compound of simple structure.

IR	Raman	Assignment
2980, 2860	2980, 2880	sp^3 C—H stretches for CH_3
2940	2940	sp^3 C—H stretch for CH_2
1453	1458	CH_2 bend and/or CH_3 antisymmetric deformation
1387	—	CH_3 symmetric deformation
1354	—	?
1130	—	C—O—C antisymmetric stretch for saturated ether

There are no bands that may be assigned to aromatic, olefinic, OH, or carbonyl groups. The C—H stretching region is characteristic of a molecule that contains both $-CH_3$ and $-CH_2-$ groups except the frequencies have been shifted slightly to higher values. In addition, the intensity of the 2860 cm^{-1} band is greater than normal. Remember that the positions quoted as normal are for normal hydrocarbons, and the bands for the stretches of the $-CH_3$ and $-CH_2-$ groups may shift if attached to atoms other than carbon, particularly electronegative groups.

The band at 1453 cm^{-1} is assigned to the antisymmetric deformation of the $-CH_3$ group and/or the bend of the $-CH_2-$ group. The band at 1387 cm^{-1} is assigned to the symmetric deformation of the $-CH_3$ group, but it is slightly higher than normal. The band at 1355 cm^{-1} is too low to be assigned as part of a CH_3 doublet for a *gem*-dimethyl or tertiary butyl group and is considered a fingerprint band.

The strongest band in the spectrum is at 1130 cm^{-1}. This is assigned to the antisymmetric stretch of a saturated ether group, C—O—C.

Reference spectra should be consulted starting with diethyl ether. (A symmetric ether is much more likely.) Dimethyl ether is eliminated from consideration since it is a gas at room temperature. When reference spectra are consulted, the answer is found to be **diethyl ether**.

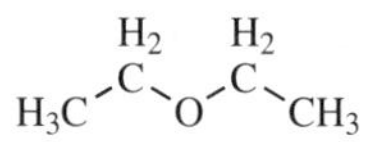

Exercise 10

***Curve** M-12.* The sample yielding curve M-12 is a solid. The spectrum is linear in micrometers.

What spectral features should be expected for each of the three structures?

1. a. Acidic OH stretch near 3000 cm^{-1} (3.3 μm) underlying the C—H stretch
   b. A cumulated double bond stretch near 1975 cm^{-1} (4.9 μm)
   c. An acidic C=O stretch near 1718 cm^{-1} (5.82 μm)
   d. The out-of-plane deformation of the acid dimer near 950 cm^{-1} (10.5 μm)
2. a. All of the acid-related bands cited for structure 1, i.e., 3.3, 5.82, and 10.5 μm (3000, 1718, and 950 cm^{-1})
   b. A weak C≡C stretch near 2120 cm^{-1} (4.7 μm). This may be so weak that it is not observed at all.
3. a. All of the acid-related bands noted for structure 1, i.e., 3.3, 5.72, and 10.5 μm
   b. A terminal C—H stretch for an acetylenic group near 3300 cm^{-1} (3 μm). This band is generally sharper than bands for other groups appearing near here, such as OH and NH stretches.
   c. A weak C≡C stretch near 2130 cm^{-1}

The spectrum has acid-related bands near 3.0 (vs), 5.9, and 11.5 μm. It also has a fairly sharp band at 3.0 μm (3320 cm^{-1}, half width of 30–40 cm^{-1}). This is the acetylenic C—H stretch. Note that the abscissa scale, the wavelength scale, is linear in micrometers. This compresses the higher frequency bands and expands the widths of the lower frequency bands. Thus, bands in the lower wavelength (higher wavenumber) region are compressed, while those bands in the higher wavelength (lower wavenumber) region appear broader.

Given the presence of the terminal C—H stretch for an acetylenic compound, the answer is **structure 3**.

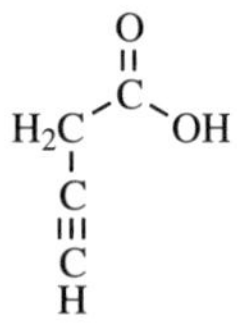

***Curve** M-40.* This sample resulted from a reaction that should have yielded one of three structures.

Once again, the question arises of which bands should each structure have that could be used to distinguish among the three structures.

Structure A. There should be an OH stretch near 3.0 μm and no carbonyl near 5.7 μm.

Structure B. There should be an broad OH stretch and a carbonyl band near 5.9 μm (1700 cm^{-1}).

Structure C. There should be two carbonyl bands near 5.7 μm (1720 cm^{-1}) and 5.95 μm (1680 cm^{-1}) but no OH band.

The spectrum has an OH stretch at 3.08 μm (3250 cm^{-1}). It appears narrower than expected because of the linear wavelength abscissa and the subsequent compression of the shorter wavelengths. It actually has a half width of about 200 cm^{-1}. This is typical of a hydrogen-bonded OH group. The spectrum also exhibits a carbonyl band at 5.86 μm (1707 cm^{-1}). Only **compound B** has both groups, and this must be the answer.

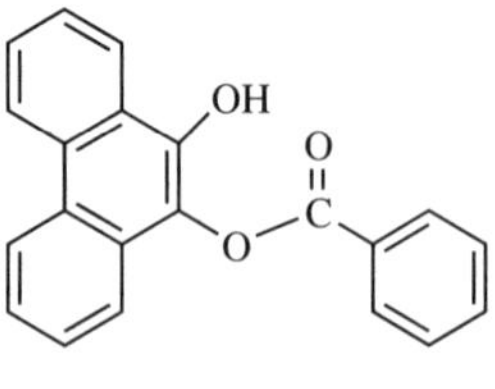

EXERCISE SECTION III

Exercise 11

***Curve** 3-A.* This is a gas phase spectrum.

IR	Raman	Assignment
3120, 3078	—	sp^2 C—H stretch (band center, 3100)
—	3020	sp^2 C—H stretch
2990	—	sp^3 C—H stretch (?)
1890	—	sum tone
—	1620	C=C stretch
1443	—	CH_2 scissors, out of phase
—	1340	CH_2 scissors, in phase
950	—	CH out-of-plane bend, all hydrogens in phase

A molecule has three principal moments of inertia for rotation about its three principal axes. If at least one of these moments is very small, the IR spectrum in the gas phase shows a great deal of fine structure on both sides of its vibrational bands. The 950 cm^{-1} band in this spectrum is a good example. If the moments of inertia are a little larger or the instrumental resolution is a little poorer, only the envelope of this fine structure is seen. One then observes an apparent doublet (at 3120 and 3078 cm^{-1}) or triplet (e.g., the bands centered at 2990, 1890, and 1443 cm^{-1}). If the moments of inertia are still larger, even this detail is lost, and one observes the usual bell-shaped curve, as in liquids. For both doublets and triplets, the center frequency is the pure vibrational transition and is the wavenumber that should be used in assignments. From the fine structure observed in this spectrum, one knows immediately that the sample is a very light molecule and therefore a small one. Even propane and benzene are too large.

The other characteristic of the IR and Raman spectra is the total lack of any coincidences in the observed bands between the two spectra. This is very strong evidence for a center-of-symmetry in the molecule.

The band centered at 3099 cm^{-1} in the IR (3120 and 3078 cm^{-1}) is assigned to the sp^2 C—H antisymmetric stretch. The 3020 cm^{-1} band in the Raman is the sp^2 C—H symmetric stretch. The 1620 cm^{-1} band in the Raman is assigned to the C=C stretch. There is no band in the IR at this frequency, and this fact agrees with the earlier conclusion that the molecule is very symmetrical. It is concluded that the molecule is an olefin and, given this, the region between 1000 and 650 cm^{-1} should be examined for the strongest band that could be assigned to the out-of-plane C—H bend. There is a band at 950 cm^{-1} in the IR and this is assigned to the out-of-plane bend.

The band at 1880 cm^{-1} is a sum tone of the mode at 950 cm^{-1} in the IR and an extremely weak Raman band at 943 cm^{-1} (not seen in this Raman spectrum). (It is not the overtone of the 950 cm^{-1} band.) The 1443 cm^{-1} band in the IR and the 1340 cm^{-1} band in the Raman are assigned to the two CH_2 scissorings, out of phase and in phase, respectively.

Given the presence of a center of symmetry, the low molecular weight of the molecule, and a C=C double-bond stretch in the Raman, the analyst should consider the smallest olefin molecule satisfying these requirements. A reference spectrum of **ethylene** reveals this is the answer.

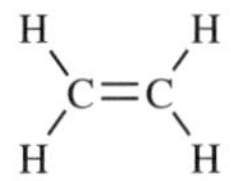

***Curve** 5-C.* This is a pure organic compound of simple structure. The spectrum was obtained in the gas phase.

IR	Raman	Assignment
—	3380	*sp* C—H stretch, in phase
3280	—	*sp* C—H stretch, out of phase (center of the doublet)
—	1975	C≡C stretch
1330	—	sum tone
732	—	≡C—H bend

Careful examination of the IR and Raman spectra reveals that there are no coincidences between the IR and Raman spectra. Thus, there is almost certainly a center of symmetry.

Splitting of the IR bands into many sharp lines is due to rotational splitting of the vibrational energy levels and indicates a low-molecular-weight molecule.

The band at 3280 cm^{-1} in the IR is in the region for the (out-of-phase) *sp* C—H stretch for a hydrogen on a carbon–carbon triple bond. N—H and O—H stretches may be eliminated since they usually occur at somewhat higher frequencies in the gas phase (3400 cm^{-1} for the N—H stretch, 3600 cm^{-1} for the O—H stretch). The Raman band at 3380 cm^{-1} is the (in-phase) *sp* C—H stretch.

The strong band in the Raman at 1925 cm^{-1} is in the cumulated double-bond range. If this is correct, the IR should be strong and the Raman very weak. This is obviously not true so another assignment must be considered.

If the unknown were acetylene, the first member of the homologous series, the C≡C stretch should be below the usual frequency (2120 cm^{-1}) for the monosubstituted triple bond. The 732 cm^{-1} band can be assigned to the C—H bend for a hydrogen bonded to a carbon–carbon triple bond.

Given the center of symmetry and the statement that this is a simple molecule, a reference spectrum of acetylene should be consulted. When this is done, the answer is found to be **acetylene**.

HC≡CH

The band at 1330 cm^{-1} is a sum tone with contributions from the 732 cm^{-1} band in the IR and a very weak band at 590 cm^{-1} in the Raman.

Exercise 12

Curve 1-A. This is pure organic compound exhibiting hydrogen bonding.

IR	Raman	Assignment
3350	—	O—H stretch, broad due to hydrogen bonding
3046	3058	sp^2 C—H stretch
1950–1700	—	sum tone pattern, monosubstitution
1606, 1598	1608	ν_{8a}, ν_{8b} aromatic ring deformation
1502, 1477	—	ν_{19a}, ν_{19b} aromatic ring deformation
1367 (broad)	—	O—H bend and C–O stretch
1230 (broad)	—	C–O stretch and O—H bend
—	1000	mono-, meta-, or 1,3,5-trisubstituted phenyl
752	—	mono- or meta-substitution
690	—	mono-, meta-, or 1,3,5-tri-substitution
—	620	ν_{6b} in-plane deformation of mono- or para-substituted phenyl
~660 (vvb)	—	O—H out-of-plane bend

The band at 3350 cm^{-1} is strong and broad. It is assigned tentatively to a hydrogen-bonded O—H stretch. It could be an N—H stretch, but this would require the presence of a broad band in the N—H bending region. The Raman spectrum offers important clues to the assignment of this band. An N—H stretch would be weak in the Raman but probably would appear, while the O—H stretch would be extremely weak. Since the latter is the case, the assignment to an O—H stretch is more likely. Confirmation of this assignment will be considered in a later paragraph.

The bands at 3046 and 3058 cm^{-1} are assigned to sp^2 C—H stretching vibrations. Note that there is no significant IR absorption at 3000–2800 cm^{-1} and therefore no sp^3 C—H. A strong band does not appear in the 1680–1620 cm^{-1} region, and it may be concluded that an olefin is not present. There are pairs of bands at 1606 and 1597 cm^{-1} and at 1502 and 1477 cm^{-1} in the IR that are assigned to the aromatic ring deformations. Note that the 1597 cm^{-1} band is stronger than the 1606 cm^{-1} band. This occurs when the aromatic ring is conjugated, or a heteroatom with a lone pair of electrons is attached directly to the ring.

It is appropriate to seek confirmation for the presence of O—H groups by looking in the O—H bending region. Strong bands occur at 1367 and 1230 cm^{-1}. These are in the region for group frequencies assigned to aromatic alcohols, that is, phenols. The 1367 cm^{-1} band is due to a mixed mode of vibration composed mostly of an O—H bend but containing significant C—O stretching as well. The band at 1230 cm^{-1} is mostly due to C—O stretching but also has contributions from O—H bending.

The band at 752 cm^{-1} is assigned to the out-of-plane C—H bending for a mono- or meta substitution, while the 690 cm^{-1} band is characteristic of mono-, meta-, or 1,3,5-trisubstitution. The sum tone pattern is consistent with monosubstitution. The Raman spectrum serves to confirm this conclusion since a band appears at 1000 cm^{-1} due to an in-plane vibration for a mono-, meta-, or 1,3,5-tri-substitution. The 620 cm^{-1} band is assigned to the ν_{6b} vibration of a monosubstituted aromatic.

There is no evidence for any other substituents. The conclusion is that the unknown is a monosubstituted aromatic and the substituent is an OH group. The answer is **phenol.**

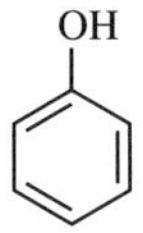

Note that the melting point of phenol is 41°C. This is just above room temperature. Any melting due to warming in the beam of an IR instrument would lead to additional band broadening and perhaps band shifts. While this is not likely to happen in an FT instrument, it may have occurred for older reference spectra obtained on a dispersive instrument where appreciable heating may occur from the source radiation.

***Curve** 20-C.* This is a spectrum of a pure organic compound of simple structure exhibiting hydrogen bonding. It has a molecular weight of less than 115.

IR	Raman	Assignment
3376	3380	N—H antisymmetric stretch
3300	3320	N—H symmetric stretch
2970–2850	2970–2850	sp^3 C—H stretch, CH_2 and/or CH_3
1620	—	NH_2 scissors
1470	1440	CH_2 scissors and CH_3 antisymmetric deformation
1379	—	CH_3 symmetric deformation
—	1300	$(CH_2)_n$, $n \geq 4$, CH_2 twist
800	—	NH_2 wag
728	—	$(CH_2)_n$, $n \geq 4$, CH_2 rock

The bands at 3378 and 3300 cm^{-1} in the IR and Raman are assigned to the N—H stretching vibrations of a primary amine NH_2 group. This is confirmed by the presence of the NH_2 bending mode at 1620 cm^{-1} and the NH_2 wag at 800 cm^{-1}. (The latter is not a good group frequency, but it is useful for confirmation when the NH stretching and bending modes are present.) Note that all of the NH-related bands are broad due to hydrogen bonding.

Bands in the 2970–2850 cm^{-1} region are assigned to sp^3 C—H stretching and indicate the presence of $-CH_3$ and $-CH_2-$ groups. This is confirmed by the presence of bands in the 1450 cm^{-1} region due to the CH_2 scissors and the CH_3 antisymmetric deformation. The band at 1379 cm^{-1} is assigned to the symmetric deformation for the $-CH_3$ group. It is not split and the *gem*-dimethyl or tertiary butyl group may be eliminated.

There is a weak band at 728 cm^{-1} in the IR, suggesting a long sequence of $-CH_2-$ groups with at least four groups in the sequence. This is confirmed by the presence of a band in the Raman at 1300 cm^{-1}.

It is concluded that the unknown is a primary amine with at least four $-CH_2-$ groups in a sequence. The molecular weight was stated to be less than 115. Subtracting 16 for the $-NH_2$ group and 15 for a terminal $-CH_3$ group leaves 84. This could be due to six $-CH_2-$ groups, but this would yield a molecular weight of exactly 115. This would be too large since the molecular weight must be less than 115. Comparison with a reference spectrum of ***n*-hexyl amine** with a molecular weight of 101 shows this to be the answer.

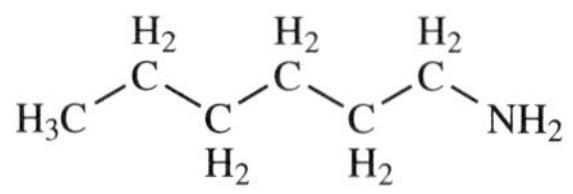

Exercise 13

There are two problems in this exercise. The objective is to assess the accuracy of the labeling and/or purity of the two samples.

1. Sample 1 is labeled benzene. Comment on the spectrum.
2. Sample 2 is labeled di-*n*-butylamine by the manufacturer. Comment on the spectrum.

Sample 1

IR	Assignment
3285	N—H stretch
3066, 3030, and 3008	benzene bands
2940–2850	sp^3 C—H stretch, CH_2 and/or CH_3
1954, 1810	benzene sum tones
1645	amide I
1540	amide II
1474, 1405	benzene bands
1025, 680	benzene bands

Benzene is present in the sample as shown by comparison with a reference spectrum of benzene. When the benzene bands are labeled on the spectrum, the remaining strong bands are at 3285, 1645, and 1540 cm^{-1}. This is the classic set of bands for a secondary amide. The 3285 cm^{-1} band is due to the N—H stretch, while the 1645 and 1540 cm^{-1} bands are the amide I and amide II bands. Examination of the instrument log revealed that the preceding sample in the cell was *n*-methyl hexamide.

The unknown spectrum is that of a sample of **benzene contaminated by *N*-methyl hexamide**.

Benzene

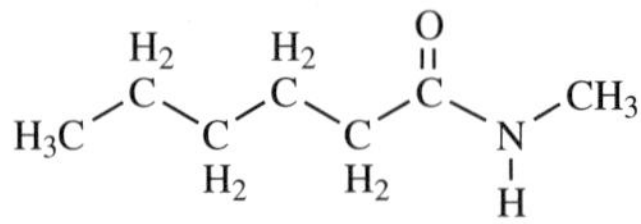

N-methyl hexamide

Sample 2. Labeled as di-*n*-butylamine.

IR	Assignment
3100–3030	sp^2 C—H stretch
2950–2805	sp^3 C—H stretch
1608, 1582	ν_{8a}, ν_{8b} aromatic ring deformations
1515, 1448	ν_{19a}, ν_{19b} aromatic ring deformations
1350	?
750	mono- or meta-substitution
692	mono-, meta-, or 1,3,5-trisubstitution
1940–1700	sum tone pattern, uncertain

The sample shows bands due to sp^2 and sp^3 C—H stretches. Therefore, it cannot be an aliphatic amine or it must be highly contaminated with an aromatic or an olefin. The presence of the 1608, 1582 cm^{-1} pair and the 1515, 1448 cm^{-1} pair for an aromatic compound indicates an aromatic is present. The 750- and 692 cm^{-1} bands are assigned to mono or meta ring substitution. The sum tone pattern is weak and not useful. As a result, the ring substitution cannot be specified exactly but it must be either mono or meta.

Note that the sp^3 C—H stretching region is somewhat anomalous since the strongest band in the region is seen at 2805 cm^{-1}. It should be remembered that this would occur when the —CH_3 group is attached to a heteroatom having a lone pair of electrons. Bohlman studied this carefully, and the 2805 cm^{-1} band is assigned to a CH_3 bonded to nitrogen. Bonding to oxygen is eliminated since strong bands do not occur in the C—O stretching region, 1300–1200 cm^{-1}. There are neither N—H stretching nor bending bands, and it must be concluded that the unknown is an aromatic tertiary amine. The answer is found by comparison with reference spectra to be ***N,N*-dimethylaniline**. The sample was mislabeled.

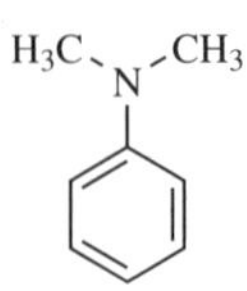

Exercise 14

***Curves 6B and** 6-Ba.* This is a pure sample that contains chlorine. Curve 6Ba is a thicker sample.

IR	Raman	Assignment
3140, 3084	3140, 3084	sp^2 C—H stretch
1950, 1740, 1720	—	sum tones
1612, 1576	—	ν_{8a}, ν_{8b} aromatic ring deformation
—	—	ν_{19a}, ν_{19b} aromatic ring deformation
—	992	1,3,5-trisubstituted aromatic
850	—	1,3,5-trisubstituted C—H out of plane
818	—	some C—Cl stretch and some ring deformation
661	—	1, 1,3, or 1,3,5 ring out of plane
430	430	C—Cl stretch (?)

The preponderance of sharp bands is the first thing to be noted. The bands at 3140 and 3084 cm^{-1} are assigned to sp^2 C—H stretches. The absence of a strong band in the Raman in the 1680–1620 cm^{-1} region virtually eliminates an olefinic double bond as being present. If aromatics are considered, then the 1612 and 1576 cm^{-1} bands may be assigned to the ν_{8a}, ν_{8b} aromatic ring deformation. Where, then, is the 1500, 1450 cm^{-1} pair for the other ring deformation? Using normal coordinate methods Nyquist and others have calculated that the 1420 cm^{-1} band may be assigned to the ν_{19a} part of the doublet and it is shifted to this significantly lower frequency by the chlorine.

How is the aromatic ring substituted? The band at 850 cm^{-1} may be assigned tentatively to 1,3,5-trisubstituted or 1,2,4,5-tetrasubstituted aromatic since it falls in the right range. The Raman band at 992 cm^{-1} favors the trisubstituted ring. If the sum tone pattern is considered, particularly in the spectrum for the thicker sample, then the 1950, 1740, and 1720 cm^{-1} pattern with the relative intensities as shown would seem to agree with 1,3,5-trisubstitution. Further, there are no bands in the 605–630 cm^{-1} region in the Raman where the ν_{6b} band for mono- and para-substituted aromatics should appear, thus eliminating these substitutions. It should be noted that the 850 and 661 cm^{-1} bands are shifted slightly from their normal positions. Again, according to Nyquist, this may be due to some in-plane bends overlapping the usual aromatic bands in the region, leading to the intensity changes and frequency shifts.

There are no bands that suggest aliphatic or other substituents. Only chlorine should be considered as a ring substituent. When a reference spectrum for 1,3,5-trichlorobenzene is examined, it will be found to be the answer.

Where are the C—Cl stretches? This is not clear, but it is likely that the band at 430 cm^{-1} is probably mostly C—Cl stretch. The 818 cm^{-1} band has been calculated to have some in-plane ring deformation and a C—Cl stretching component as well.

The conclusion is that the unknown is a 1,3,5-trisubstituted aromatic with chlorine being the only substituent. The answer is **1,3,5-trichlorobenzene**.

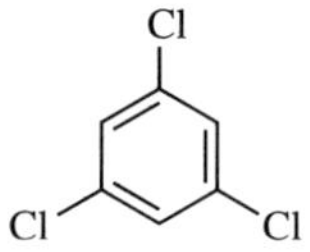

***Curve** 19-A.* This is a pure compound used as a monomer in the polymer industry.

IR	Raman	Assignment
3106	3094	sp^2 C—H stretch
—	2998	sp^2 C—H stretch
2980–2850	2960–2840	sp^3 C—H stretch
1890	—	overtone of 940 band
1730	—	C=O stretch
1638	1642	C=C stretch
1450	1440	sp^3 CH_2 and/or antisymmetric CH_3 bend
1403	1403	=CH_2 in-plane bend
1380	1380	CH_3 symmetric deformation
1325, 1300	—	C—O—C stretch, ester
1205, 1169	—	C—O—C stretch, ester
940	—	out-of-plane =CH_2 bend

The interpretation of this spectrum brings together many of the considerations that have been applied in interpreting other spectra in this and the preceding sets of spectra.

The bands at 3106 cm^{-1} in the IR and at 3094, 2998 cm^{-1} in the Raman are assigned to sp^2 C—H stretching vibrations. There are no bands at 1600, 1580 cm^{-1} and 1500, 1450 cm^{-1} in the IR for the aromatic ring deformations, and this eliminates aromatics as a consideration. The bands at 3106 cm^{-1} in the IR and at 3094 and 2998 cm^{-1} in the Raman are therefore assigned to an olefinic C—H stretch. Furthermore, since there are two =C—H stretches and one is near 3100 cm^{-1}, there appears to be a terminal =CH_2 group. There is a band at about 1640 cm^{-1} in both the IR and Raman (much stronger relative intensity in the Raman), and this must be assigned to the C=C stretch. It is in the lower end of the range for the C=C stretch, and this suggests vinyl, vinylidene, or cis configurations, possibly lowered somewhat by conjugation.

The bands at 2980 and 2850 cm^{-1} are assigned to sp^3 C—H stretches. This is confirmed by the presence of bands at 1450 cm^{-1} for the CH_2 bend and/or the CH_3 antisymmetric deformation as well as the 1380 cm^{-1} band assigned to the CH_3 symmetric deformation. The 1380 cm^{-1} band also occurs in the Raman, and this can only appear if the CH_3 group is attached to an sp^2 carbon, i.e., a doubly bonded carbon.

There is a very strong band in the IR at 1730 cm^{-1} that is assigned to the C=O stretch. It is low for an ester, but the presence of strong bands at 1325 and 1300 cm^{-1} suggests that the carbonyl is an ester carbonyl. The bands at 1205 and 1169 cm^{-1} lend support for this conclusion, although taken alone they are less reliable as a group frequency. The fact that the C=O frequency is somewhat low for an ester indicates it is conjugated. This must mean it is conjugated to the olefinic double bond since this is the only other sp^2 carbon available. Conjugation would also lower the C=C frequency, and it has already been noted that the C=C band frequency is quite low.

It is now appropriate to ask where the out-of-plane C–H bending vibration appears? There is a strong band at 940 cm^{-1}, and this is close to where the trans configuration out-of-plane mode should occur. However, the presence of two bands in the sp^2 stretching region and the strength of the C=C stretch in the IR tend to discredit the presence of a trans configuration.

What then is the assignment of the 940 cm^{-1} band? The usual position for the out-of-plane bend for a vinylidene $=CH_2$ is about 890 cm^{-1}. If the double bond is conjugated, as has been concluded, to the carbonyl, then the out-of-plane bend is raised by about 40 cm^{-1} by resonance effects. Thus it should appear at about 940–930 cm^{-1}, and it does. There should be an overtone of this band at a frequency somewhat higher than twice the fundamental. A weak band does appear at 1890 cm^{-1} that is about 10 cm^{-1} higher than twice the fundamental frequency. This strongly supports the suggested assignment.

The evidence contained in the Raman spectrum for the presence of a CH_3 on a double bond points toward a CH_3 attached either directly to the C=C bond or to the carbonyl. The latter cannot be true since one side of the C=O is conjugated to a C=C group and the other side bonded to the ester oxygen. This leaves the possibilities fairly limited, and a methacrylate is one of the possible structures satisfying the evidence. The R group on the ester singly bonded oxygen is uncertain, but the simplicity in the sp^3 C–H stretching region suggests a $-CH_3$ group in that position. In addition, it was stated in the definition of the problem that the unknown is a monomer used in the polymer industry to make a common polymer. Methyl methacrylate is used to make polymethylmethacrylate, a very common polymer. The answer is **methyl methacrylate**.

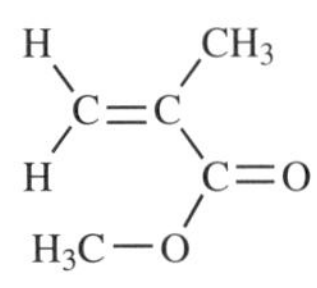

It is interesting to consider why the $=CH_2$ terminal wag increases in frequency to 940 cm^{-1} from the normal position at 890 cm^{-1}. Resonance is responsible for the decrease in the carbonyl frequency from the normal position of 1750–1730 cm^{-1}.

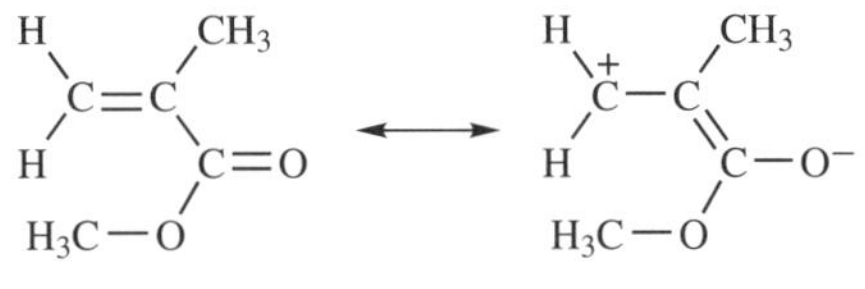

If the electron distribution shown on the right is contributing more to the actual electron distribution, then this explains the drop in frequency of the carbonyl since there is more single-bond character in the C=O bond.

Exercise 15

***Curve** 18-A.* This is a pure compound which gave a positive test for the presence of sulfur. Both IR and Raman spectra are available.

IR	Raman	Assignment
3000	2996	sp^3 C—H stretch
2916	2908	sp^3 C—H stretch
1440	1420	sp^3 CH_2 and/or antisymmetric CH_3 bend
1315	1310 (w)	S—CH_3 symmetric deformation
1052	1040 (w)	S=O stretch, sulfoxides
700, 670 (w)	700, 670 (s)	C—S—C stretch

A very simple doublet centered near 2960 cm^{-1} with weak absorption in the IR and a strong lower leg in the Raman indicates the presence of a very small alkyl group, probably methyl because of the center of the doublet at 2960 cm^{-1}.

The lack of a 1375 cm^{-1} band (corresponding to the symmetric methyl deformation) in the IR is not consistent with the C—H stretching region unless the potential methyl group(s) is (are) not on carbon. With the presence of sulfur, a S—CH_3 could be a possibility. The band at 1315 cm^{-1} fits for methyl on sulfur and supports the assignments in the C—H stretching region. A very strong band in the IR near 1050 cm^{-1} with sulfur present indicates the likely presence of a sulfoxide group (the S=O stretch). The 700, 670 cm^{-1} doublet, which is strong in the Raman and weak to medium in the IR, is consistent with C—S stretch. The most likely candidate for this unknown therefore would be **dimethyl sulfoxide**. A check of reference spectra should confirm the assignment. The 700, 670 cm^{-1} doublet appears to be the antisymmetric and symmetric C—S—C stretches with the lower leg of the doublet assigned to the symmetric stretching mode because of the intensity of the Raman band.

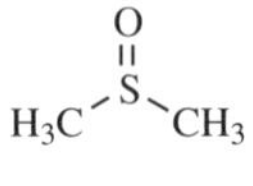

Exercise 16

***Part I: Curve** 21-A.* This is a pure compound which is an organic ester of an inorganic acid. The latter information would imply that we are looking for something like dimethyl sulfate, but as sulfur is not indicated to be present, that particular ester of an inorganic acid is probably unlikely. Only IR data are available.

IR	Assignment
2970, 2870	sp^3 C—H stretch
1642 (s)	O—NO_2 antisymmetric stretch
1460–1445	sp^3 CH_2 and/or antisymmetric CH_3 bend
1380	CH_3 symmetric deformation
1360	too low for CH_3 symmetric deformation
1280 (s)	O—NO_2 symmetric stretch

The weak doublet in the 3000–2850 cm^{-1} region indicates the molecule has a small aliphatic component which likely will be the R group on the inorganic ester. The weak band at 1380 cm^{-1} indicates, however, that R is not simply methyl as there has to be methyl on carbon. The two very strong bands near 1640 and 1280 cm^{-1} strongly indicate the presence of a nitrate ester. The R group must be simple (no branching) but not just a methyl group. There is a band near 1360 cm^{-1}, which is too low to be the lower leg of a methyl doublet (gem or tertiary substitution). Thus, the best choice would be a simple straight-chain R group on a nitrate. Ethyl nitrate, **$CH_3CH_2ONO_2$**, is the answer.

Part II: Curve 21-B. This is a pure compound and is another organic ester of an inorganic acid. The information supplied implies that we are looking for something like dimethyl sulfate, but as sulfur is not indicated to be present, in this case the particular ester of an inorganic acid is probably unlikely to be a sulfate. Only IR data are available.

IR	Assignment
3070	sp^2 C—H stretch
1950, 1890, 1800, 1735	sum tone pattern for monosubstitution
1595, 1590	aromatic, ν_{8a}, ν_{8b} ring stretch
1490, 1450	aromatic, ν_{19a}, ν_{19b} ring stretch
1170 (s)	P=O stretch of phosphate esters
960 (s)	P—O—C aromatic stretch.
770	C—H ring out-of-plane bend (mono, meta)
685	ring puckering (mono, meta)

No aliphatic C—H stretching. Only C—H absorption is above 3000 cm^{-1}, which indicates that this is likely to be an aromatic ester of an inorganic acid. The sum tone pattern is a good match for monosubstitution and is backed up by strong bands in the 770 and 685 cm^{-1} regions. Thus, a phenyl ester is a possibility. The very strong bands near 1170 and 960 cm^{-1} strongly suggest that the material is an aromatic phosphate ester. Thus, triphenylphosphate becomes a prime candidate. A check of reference spectra confirms that the unknown is **triphenylphosphate**.

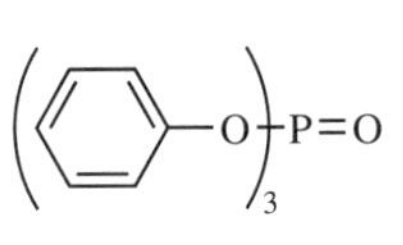

Exercise 17

Curve *7-A.* The spectra of both unknowns 7-A and 7-B have the appearance of inorganic material. This decision is mainly based on the highly simplified spectrum that is observed. Thus, the use of inorganic reference material will be helpful in identifying these unknowns.

In the case of 7-A, the sample was prepared as a Nujol mull so that aliphatic C—H bands will be present in the spectrum. They are identified by an asterisk in the table. No Raman data were available.

IR	Assignment
2980–2850*	sp^3 C—H stretch
1790	overtone or combination band
1430	C=O stretch of carbonate ion
875	C=O bend of carbonate
720*	sp^3 CH_2 rock

The spectrum of the unknown has only two bands. One is a very very intense band centered near 1430 cm^{-1} and arises from the C—O stretch of the carbonate ion. When this band occurs together with a band near 875 cm^{-1}, the sample is highly likely to be a carbonate. The identification of the counterion may be difficult unless additional information is available.

In this case the sample is $\mathbf{CaCO_3}$.

Curve *7-B.* The spectra of both unknowns 7-A and 7-B have the appearance of inorganic material. This decision is mainly based on the highly simplified spectrum that is observed. Thus, the use of inorganic reference material will be helpful in identifying these unknowns.

In the case of 7-B, the sample was prepared as a KBr disk and was shown to be a dihydrate. Thus, we would expect to see the presence of water in the spectrum. No Raman data were available.

IR	Assignment
3470 (br)	O—H stretch, H bonded
2045	SC≡N stretch
1620	H_2O bend
755, 745 (doublet)	C—S bend (split by crystal effects) (or possibly a water lattice mode)

The spectrum of the unknown has only one or two bands. One is a very intense band centered near 2045 cm^{-1} and arises from the C≡N stretch of the cyano group. When this band occurs together with a band near 750 cm^{-1}, the sample is likely to be barium thiocyanate. The identification of the counterion may be easier in this case as there are relatively few hydrated thiocyanates.

The sample is $\mathbf{Ba(SCN)_2 \cdot 2H_2O}$.

Exercise 18

***Curve** 15-C.* This is a polymer derived from the monomer identified as unknown 19-A. It is a polymer made from a single monomer.

IR	Assignment
3430	overtone of C=O at 1735
2990, 2950	*sp*^{3} C–H stretch
1735	C=O stretch
1462	CH_3 symmetric deformation, CH_3 on oxygen
1450	*sp*^{3} CH_2 scissors and/or CH_3 antisymmetric deformation
1383	CH_3 symmetric deformation
1270–1240	O=<u>C–O</u>–C stretch
1190–1150	O=C–<u>O–C</u> stretch

There are no significant bands above 3000 cm^{-1} except the weak 3430 cm^{-1} band that will be assigned later. The bands at 2990 and 2950 cm^{-1} are assigned to *sp*^{3} C–H stretching modes and the 1450 cm^{-1} band is assigned to the *sp*^{3} CH_3 antisymmetric deformation and/or CH_2 bending. The band at 1385 cm^{-1} is due to the symmetric CH_3 symmetric deformation.

The strongest bands in the spectrum are at 1735 cm^{-1}, the carbonyl stretching vibration, and the doublets centered at 1250 and 1170 cm^{-1}. These latter bands are assigned to the ester C–O stretching modes with the higher frequency band attributed to the C–O attached to the C=O and the 1170 cm^{-1} band to the C–O on the other side of the "ether"oxygen. The weak band at 3430 cm^{-1} is the overtone of the C=O stretch. The band at 1462 cm^{-1} can be assigned to the CH_3 symmetric deformation for a CH_3 attached to an oxygen.

Since it is known this is the polymer formed from the monomer giving curve 19-A that was identified as methyl methacrylate, the polymer must be **polymethyl-methacrylate**.

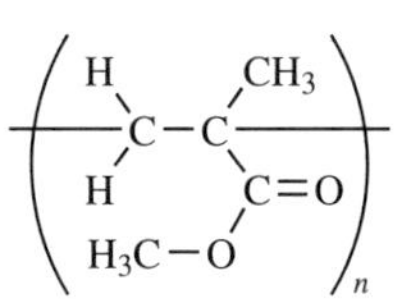

The doublets shown at 1250 and 1170 cm^{-1} are very characteristic of long sequences of methacrylate groups in the polymer. If a random copolymer is formed with, for example, styrene, then these two bands shift to 1210 and 1130 cm^{-1} and appear as broad, single bands.

***Curve** 17-C.* This is the spectrum of a fairly common high-molecular-weight polymer made from a single monomer.

IR	Assignment
3450	overtone of carbonyl at 1743
2980–2930	sp^3 C–H stretch
1743	C=O stretch
1435	sp^3 CH_3 antisymmetric deformation and/or CH_2 bend
1377	symmetric CH_3 deformation
1245	O=C–O–C stretch
1025	O=C–O–C stretch

There are no bands above 3000 cm^{-1} except a weak band at 3450 cm^{-1}. This will be assigned later. The bands between 2980 and 2930 cm^{-1} are assigned to sp^3 C–H stretching vibrations. There is a band at 1435 cm^{-1} that is due to the sp^3 CH_3 antisymmetric deformation and/or the CH_2 bend. The 1377 cm^{-1} band is assigned to the CH_3 symmetric deformation.

The strongest band in the spectrum occurs at 1743 cm^{-1}, and this must be the C=O stretch. The weak band at 3450 cm^{-1} is now assigned to the overtone of the carbonyl. The strong band at 1245 cm^{-1} and the somewhat weaker band at 1025 cm^{-1} are due to thc C–O–C stretching vibrations. The former band is assigned to the C–O, where the oxygen is attached to the carbonyl carbon. The band at 1025 cm^{-1} is due to the other C–O bond vibration. This latter band is a less reliable group frequency, but its presence provides good conformational evidence for the presence of the ester group. It should be noted that the 1245 cm^{-1} band while in the general region for the C–O stretch for esters is somewhat higher in the region, and this position is characteristic of the acetate group.

The CH_3 symmetric deformation at 1377 cm^{-1} is stronger than expected; indeed, it is stronger than the C–H stretching bands. This will happen when the $-CH_3$ group is attached to an sp^2 carbon, in this case to the carbonyl carbon.

The evidence indicates the unknown to have an acetate group. The most common acetate polymer consistent with the evidence is **polyvinylacetate**, and comparison with reference spectra will prove this is the answer.

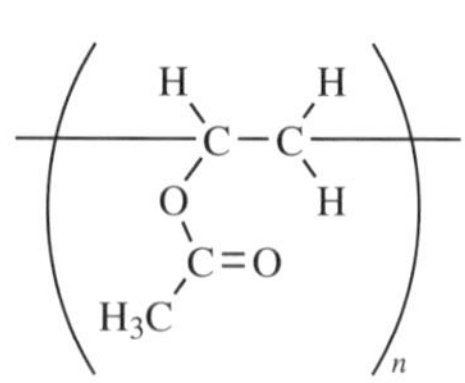

***Curves** 16-C **and** 16-D.* Curve 16-C is a spectrum of a sample that is much too thick to obtain accurate band frequencies for the strong bands.

Curve 16-D was obtained by internal reflection that allows the band frequencies to be read.

IR	Assignment
3070	sp^2 C—H stretch
2960	sp^3 C—H stretch
1712	C=O stretch
1615, 1580	ν_{8a}, ν_{8b} aromatic ring deformation
1510, 1450	ν_{19a}, ν_{19b} aromatic ring deformation
1240	O=$\underline{\text{C—O}}$—C stretch
1090	O=C—$\underline{\text{O—C}}$ stretch
870	C—H out-of-plane bend, para-substitution
722	?

The band at 3070 cm^{-1} is assigned to the sp^2 C—H stretch. There is no evidence for C=C in the 1680–1620 cm^{-1} region, although that does not eliminate the presence of one. The bands at 1615 and 1580 cm^{-1} due to the ν_{8a} and ν_{8b} ring deformations confirm an aromatic group, and the doublet at 1510 and 1450 cm^{-1} due to the ν_{19a} and ν_{19b} ring deformations supports this conclusion. The 1580 cm^{-1} band is stronger than the 1615 cm^{-1} band, and this occurs when the aromatic ring is conjugated.

The band at 2960 cm^{-1} is assigned to the sp^3 C—H stretch. When the 1450 cm^{-1} band is examined, it is broader than would be expected if it were due only to the ν_{19b} aromatic deformation. The width is increased by the presence of the underlying CH_2 bend and/or the CH_3 antisymmetric deformation. There is no band in the 1380 cm^{-1} symmetric deformation region for the CH_3 group, and the presence of CH_3 groups can be eliminated as a result of the absence of this band.

The strongest band in the spectrum is at 1712 cm^{-1}, and this is the C=O stretch. The presence of bands near 1240 and 1090 cm^{-1} assigned to C—O—C stretching modes provides evidence that the 1712 cm^{-1} C=O stretch is due to an ester carbonyl. The C=O band is somewhat low for an ester carbonyl, but if it were conjugated to the ring, the frequency would be expected to be lower due to resonance effects. This would also lead to an increase in the intensity of the 1580 cm^{-1} band compared to the 1615 cm^{-1} band as noted previously.

The next question is how the aromatic ring is substituted. There is a band at 870 cm^{-1} that is in the range for para-substitution. The sum tone region is not useful due to the overlap of the strong C=O stretching band at 1712 cm^{-1}. There are also interference fringes in the transmission spectrum (not seen in the internal reflection spectrum) related to reflections within the thin polymer film.

The best approach to determining the answer to the problem is to examine reference spectra of para-substituted aromatic-ester-type polymers such as alkyds or polyesters. When this is done, the answer is found to be **polyethylene terephthlate**.

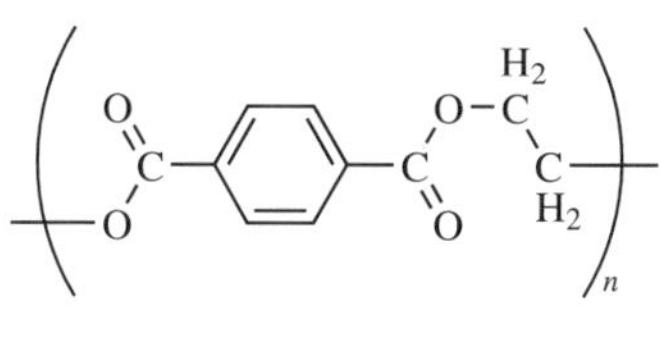

There are several questions that must be considered regarding this spectrum and the accompanying internal reflection spectrum. The band at 725 cm^{-1} is stronger than the 870 cm^{-1} band and it has not been assigned. It is probably a fingerprint band. This indicates the unreliability of trying to assign all bands in a given spectrum, even some of the strong bands.

The C—H stretching bands in the internal reflection spectrum would seem to indicate only sp^3 C—H stretching modes. This band, however, is due to an impurity on the internal reflection crystal possibly occurring as a result of handling the crystal. It is very difficult to eliminate such impurity films on internal reflection crystals, and a spectrum of the crystal should be obtained without sample before obtaining the sample spectrum.

There are interference fringes in the transmission spectrum that are related to the thickness of the polymer film. The equation that applies is

$$T = \frac{\Delta n}{2\eta(\nu_1 - \nu_2)}$$

where η is the refractive index of the film and v_1 and v_2 are the wavenumber values for the highest and lowest wavenumber fringes, respectively. The term Δn is the number of maxima (or minima) between v_1 and v_2. When values for this film are put in the equation, the thickness may be calculated as follows:

$$T = \frac{3}{2 \times 1.5(3900 - 3580)} = 0.003\,\text{cm}$$

***Curve** 17-A.* This is a low-molecular-weight example of a polymer prototype often incorporated in higher molecular weight liquids and resins.

IR	Assignment
2995	sp^3 C—H stretch, CH_3
1255	symmetric CH_3 deformation when attached to Si
1060	Si—O—Si stretch
840, 760	Si—$(CH_3)_3$

The absence of bands in the N—H and O—H stretching regions as well as the absence of a band in the carbonyl region is important. The band at 2995 cm^{-1} is assigned to the sp^3 C—H stretch for a —CH_3 group. It is somewhat higher than normal, but it must be remembered that the CH_3 stretching frequencies that have been noted to occur within a few wavenumbers of 2960 and 2870 cm^{-1} are for —CH_3 groups attached to other sp^3-hybridized carbons.

The 1255 cm^{-1} band is extremely characteristic of the symmetric methyl deformation when the methyl is attached directly to a silicon. The broad, very strong band at 1060 cm^{-1} may be assigned to the Si—O—Si stretch. This falls in the ether C—O stretching region also, but examination of spectra of ethers reveals a much different C—H stretching region and stronger intensities for the C—H bending bands near 1450 and 1380 cm^{-1}. While this is not sufficient evidence to totally exclude ethers, it should raise a warning flag that something else is going on. The relative strength of the 1060 cm^{-1} band must also be taken into consideration. The Si—O—Si stretching band is one of the strongest absorption bands observed in the IR.

The bands at 840 and 760 cm^{-1} are characteristic as a pair for a Si—$(CH_3)_3$ group. If fewer than three CH_3 groups are attached to the silicon, only one strong band appears near 800 cm^{-1}.

The sample is stated to be a polymer prototype molecule. The spectral evidence indicates that both Si—O—Si groups and Si—$(CH_3)_3$ groups are present. Referring to reference spectra shows the unknown to be **hexamethyl disiloxane**.

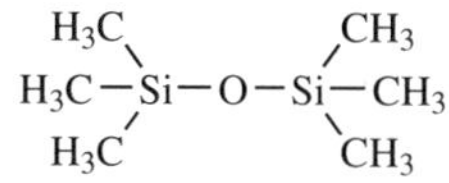

Curve 10-C. This sample is a high-molecular-weight polymer formed from a single monomer. Identify the polymer and the monomer used to form the polymer.

IR	Assignment
2360	sum tone, 2 × 1200
1200	C—F stretch
620, 520	CF_2 scissors

The spectrum is unique in having virtually no C—H stretching bands. Also, no OH, NH, C=C, aromatic, and C=O bands are observed. The strongest band in the spectrum occurs near 1200 cm^{-1}, and this can be assigned to the C—F stretching vibration. For a —CF_2— group, two bands should be expected in the 1400–1000 cm^{-1} region, but the 1200 cm^{-1} band is so strong it is not seen as two bands. The bands at 620 and 520 cm^{-1} are due to the CF_2 bending modes. The only other band of any significance is the 2360 cm^{-1} band, and this is due to the overtone of one of the unresolved components of the 1200 cm^{-1} band.

The answer is **polytetrafluoroethylene**.

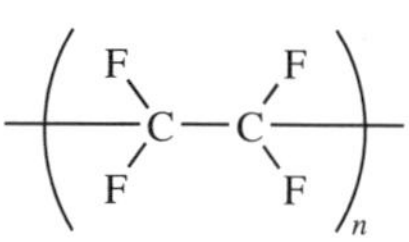

It is extremely difficult to produce a film thin enough to observe the bands in the 1200 cm^{-1} region. An internal reflection spectrum should be considered for very strong absorbers such as this sample because the effective thickness determined by the physics of internal reflection is only a fraction of a wavelength.

Exercise 19

***Part I: Curve** 34E.* The spectra of both unknowns 34E and 35E are spectra of two relatively common polymers. We were told that one of the samples contains an element other than C, H, O, and N, which would help in a preliminary glance at the spectra.

In the case of 34E the sample has a relatively simple spectrum with bands in the following regions. Raman data were available.

IR	Raman	Assignment
2960, 2910	2960, 2910	sp^3 C—H stretch, CH_2 and/or CH_3 (this range is high for $-CH_2-$)
1260	—	Si—CH_3 symmetric bend
1060	—	Si—O—Si stretch
800	—	Si—C stretch

2960 and 2910 cm^{-1}, aliphatic C—H stretch (no alkene or aromatic C—H): Aliphatic symmetric methyl bend is shifted if methyl is present; this indicates a possible heteroatom methyl system and the 1260 cm^{-1} band would fit for Si substitution.

Very strong doublet band centered near 1060 cm^{-1} (no intensity in Raman spectrum indicates that a halogen is unlikely, also a simple ether is unlikely as methyl bend is shifted if present): Strong bands at 1260 and 1060 cm^{-1} are consistent with Si—CH_3 (symmetric bend) and Si—O—Si (stretch $\geq$4 units for doublet).

The answer is a **polydimethylsiloxane**, **$Me_3SiO(Me_2SiO)_xSiMe_3$**, polymer.

***Part II: Curve** 35E.* The spectra of both unknowns 34E and 35E are spectra of two relatively common polymers. We were told that one of the samples contains an element other than C, H, O, and N, which would help in a preliminary glance at the spectra. As Part I was found to have another element, we would expect the 35E should only be composed of C, H, O, or N.

In the case of 35E the sample has a relatively complex spectrum with many sharp bands. Raman data are available.

IR	Raman	Assignment
—	3071	sp^2 C—H stretch, ring
2970	2970	sp^3 C—H stretch, CH_2 and/or CH_3
1773	1773	(this range is very high for C=O)
1601	1601	ν_{8a}, ν_{8b}, aromatic ring
1500–1450		ν_{19a}, ν_{19b}, aromatic ring
1232	1232	C—O stretch
828	828	C—H out-of-plane bend, para-substitution
—	640	ring bend, para

In the 3100–2800 cm^{-1} region, aliphatic and either aromatic or alkene C—H stretch is present. Therefore, this unknown will likely be a mixed aliphatic–alkene or aliphatic–aromatic substance.

(Doublet in the Raman 2800–2700 cm^{-1} region indicates a potential *gem*-methyl system is present.)

The strong carbonyl near 1780 cm^{-1} (high-wavenumber value) coupled with a very strong band near 1230 cm^{-1} suggests a possible ester polymer. The high-wavenumber value for the ester carbonyl with identical values in the IR and Raman spectra strongly suggests a polycarbonate.

The presence of an aromatic ring follows from the strong 1605 cm^{-1} band in the Raman with a weaker counterpart pair (1605, 1595 cm^{-1}) in the IR. Also, the ν_{19a} and ν_{19b} bands near 1500, 1450 cm^{-1} are observed in the IR. Strong bands near 830 cm^{-1} (IR) and 640 cm^{-1} (Raman) support para-substitution of the ring.

The symmetric methyl deformation mode near 1375 cm^{-1} is split, which further supports the presence of a branched methyl system (this band is very weak in the Raman effect). Here is a case where the Raman fundamental is weaker than the observed overtone. The relative intensities of the methyl doublet in the IR suggest that another absorption band is masked by the lower leg of the doublet. This shift of intensities could be misleading and result in the erroneous assignment of the presence of *t*-butyl groups, however the unmasked Raman overtones show the expected intensity relationship at 2770 and 2730 cm^{-1} ($\Delta\tilde{\nu} \sim 40\,cm^{-1}$).

The answer is a **polycarbonate**.

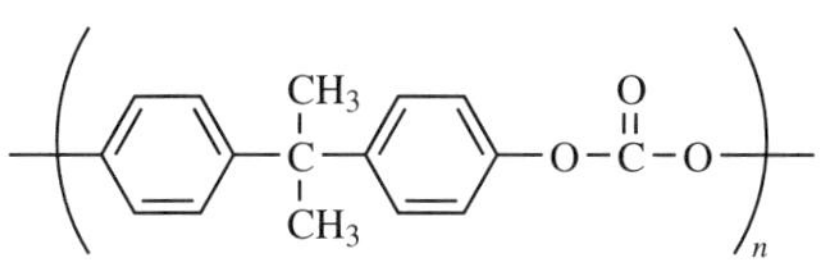

Bibliography

FOIL A. MILLER

The books and articles listed here deal with applications of vibrational spectroscopy to chemical problems. Entries are alphabetical, usually by author. *Those marked with an asterisk are especially good.* The list is not intended to be complete—it would be far too long. A more extensive compilation is given in the *Coblentz Society Desk Book* edited by C. D. Craver and given in Part 1. Those references dealing primarily with Raman spectroscopy are gathered at the end. The material is arranged as follows:

1. Basics: a core collection on infrared spectroscopy
2. General texts
3. Libraries of infrared reference spectra
4. Infrared group frequencies
5. Gases and vapors
6. Polymers and coatings
7. Inorganics and organometallics
8. Biomedical, drug, and forensic applications
9. Other classes of compounds
   A. Essential oils
   B. Isotopically labeled compounds
   C. Silicones
   D. Organophosphorus compounds
   E. Pesticides
   F. Propellants and explosives
   G. Simpler molecules, fundamental frequencies for
10. Surface studies
11. Near-infrared spectra
12. Far-infrared spectra
13. Instrumentation and techniques
14. Miscellaneous infrared topics
15. Raman spectroscopy

Course Notes on the Interpretation of Infrared and Raman Spectra, by Dana W. Mayo, Foil A. Miller, and Robert W. Hannah.
ISBN 0-471-24823-1

1. Basics: a core collection on infrared spectroscopy
   *A. L. J. Bellamy, *The Infrared Spectra of Complex Molecules*, 3rd ed., Vol. 1, Chapman and Hall, 1975. The best collection of infrared group frequencies.
   *B. L. J. Bellamy, *The Infrared Spectra of Complex Molecules*, Vol. II, *Advances in Infrared Group Frequencies*, 2nd ed., Chapman and Hall, 1980. An extensive discussion of the factors that affect group frequencies.
   *C. C. D. Craver (Ed.), *Coblentz Society Desk Book of Infrared Spectra*, 2nd ed., Coblentz Society, Kirkwood, MO. 540 pp. Useful introduction, with extensive bibliography. 900 high-quality reference spectra. Highly recommended.
   *D. A. Lee Smith, *Applied Infrared Spectroscopy*, Wiley-Interscience, New York, 1979. 322 pp. Excellent.
   *E. D. Lin-Vien, N. B. Colthup, W. G. Fateley, and J. G. Grasselli, *Handbook of Infrared and Raman Characteristic Group Frequencies*, Academic, New York, 1991. 503 pp. Excellent discussion of the two kinds of group frequencies. Also contains IR and Raman spectra of 111 compounds reproduced from several sources.
   *F. W. J. Potts, *Chemical Infrared Spectroscopy*, Vol. 1, *Techniques*, Wiley, New York, 1963. 322 pp. Excellent.

2. General texts
   A. American Society for Testing and Materials (ASTM), Committee E-13, *Manual on Practices in Molecular Spectroscopy*, 4th ed., ASTM Philadelphia, PA, 1979. 162 pp.
   *B. R. P. Bauman, *Absorption Spectroscopy*, Wiley, New York, 1962. 611 pp. Both theory and experiment.
   C. E. G. Brame and J. G. Grasselli (Eds.), *Infrared and Raman Spectroscopy* (3 parts), Decker, 1977. A series, with chapters on applications to textiles, foods, petroleum products, etc.
   *D. N. B. Colthup, L. H. Daly, and S. E. Wiberly, *Introduction to Infrared and Raman Spectra*, 3rd ed., Academic, New York, 1990. 560 pp. General coverage, but particularly good for group frequencies and their physical interpretation.
   E. W. O. George and P. McIntyre, *Infrared Spectroscopy*, Wiley, New York, 1987. 560 pp.
   *F. R. N. Jones and C. Sandorfy, Applications of Infrared and Raman Spectra in W. West (Ed.), *Chemical Applications of Spectroscopy*, Vol. IX in Weissburger's series *Technique of Organic Chemistry*, Interscience, New York, 1956. 334 pp.
   G. D. N. Kendall (Ed.), *Applied Infrared Spectroscopy*, Reinhold, 1966. 560 pp. Chapters by specialists.

3. Libraries of infrared reference spectra
   *A. *Coblentz Society Spectra*, Coblentz Society, Kirkwood, MO. An extensive collection of critically evaluated spectra.

B. D. Dolphin and A. E. Wick, *Tabulation of Infrared Spectral Data*, Wiley, New York, 1977. Tabulated band positions. Useful for unusual molecules.

*C. J. G. Grasselli and W. M. Ritchey (Eds.), *Atlas of Spectral Data and Physical Constants for Organic Compounds*, 2nd ed., CRC Press, Boca Raton, FL, 1975.

*D. D. Hummel. See Polymers and Coatings, Part 6.

*E. R. J. Keller, *The Sigma Library of FT-IR Spectra*, Edition I, Vols. 1 and 2, Sigma Chemical Co., St. Louis, MO, 1986. Over 10,000 spectra of compounds of biochemical interest, 4 per page. An extension of the Aldrich collection.

F. R. W. A. Oliver and B. Marsden, A Bibliography of Published Collections of [Infrared] Spectral Data, *Eur. Spectrosc. News*, **33** 33–37 (1980). A very extensive bibliography, with a good index by chemical class. It is unfortunate that this journal is not widely available.

*G. Pachler, Matlok, and Gremlich, *Merck FT-IR Atlas*, VCH Publishers, New York, 1988. In German and English. 3050 spectra, three per page; well indexed.

*H. C. J. Pouchert (Ed.), *Aldrich Library of Infrared Spectra*, 3rd ed., Aldrich Chemical Co., Milwaukee, WI, 1981. 12,000 spectra in one volume, eight per page.

*I. *Sadtler Research Laboratories Spectra*, Sadtler Research Laboratories, Philadelphia, PA. The largest published collection.

*J. B. Schrader, *Raman/Infrared Atlas of Organic Compounds*, 2nd ed., VCH Publishers, New York, 1989. 1118 pp. Raman and IR spectra plotted on the same scale for about 1000 compounds. The spectra are large; they are vertical on the page, and each page has one pair of spectra. Very good.

*K. *Sprouse Collection of Infrared Spectra*, Elsevier Science, Journal Information Center, New York.

1) Book I, *Polymers.* 415 spectra.
2) Book II, *Solvents by Cylindrical Internal Reflectance*, 1987. 776 pp. 350 spectra of a wide variety of liquids, well presented. Good indices. Very useful.
3) Book III, *Surface Active Agents.*

4. Infrared group frequencies

*A. Bellamy, Vols. I and II. See Part 1.

B. F. F. Bentley, L. D. Smithson, and A. L. Rozek, *Infrared Spectra and Characteristic Frequencies, 700–300* cm^{-1}, Interscience, New York, 1968.

C. N. B. Colthup, Spectra-Structure Correlations in the Infra-Red Region, *J. Opt. Soc. Am.* **40**, 397–400 (June 1950). The original Colthup chart.

*D. Colthup, Daly, and Wiberly. See Part 2.

*E. Jones and Sandorfy. See Part 2.

*F. Lin-Vien, Colthup, Fateley, and Grasselli. See Part 1.

G. R. C. Lord and F. A. Miller, Factors Influencing Characteristic Vibrational Frequencies of Molecules: Intramolecular Effects, *Appl. Spectrosc.* **10**, 115–123 (1956).

H. K. Nakanishi, *Infrared Absorption Spectroscopy—Practical*, 2nd ed., Holden-Day, 1977. Contains a set of problems useful for self-instruction.

5. Gases and vapors

*A. D. G. Erley and B. H. Blake, *Infrared Spectra of Gases and Vapors*, Vol. II, *Grating Spectra*, Chemical Physics Research Laboratory, Dow Chemical Company, Midland, MI, 1965. 132 spectra.

B. D. G. Murcray and A. Goldman (Eds.), *CRC Handbook of High Resolution Infrared Library Spectra of Atmospheric Interest*, CRC Press, Boca Raton, FL, 1981.

C. R. A. Nyquist, *The Interpretation of Vapor Phase Infrared Spectra*, Sadtler Research Laboratories. Vol. 1, *Interpretation.* Vol. 2, *Spectra.*

D. R. H. Pierson, A. N. Fletcher, and E. S. C. Gantz, Catalog of IR Spectra for Qualitative Analysis of Gases, *Anal. Chem.* **28**, 1218 (1956). 66 spectra.

E. Welti, *Infrared Vapour Spectra*, Heyden, 1970. 222 pp, 300 spectra.

F. The Coblentz Society and Sadtler Research Labs. have collections of gas spectra.

6. Polymers and coatings

A. D. I. Bower and W. F. Maddams, *The Vibrational Spectroscopy of Polymers*, Cambridge University Press, 1992.

*B. Federation of Societies for Coatings Technology (FSCT), *An Infrared Spectroscopy Atlas*, FSCT, Philadelphia, PA, 1982, 896 pp. Text, 1400 spectra, 1500 references. Excellent.

*C. D. L. Harms, Identification [of Polymers] by Infrared Spectra of Their Pyrolysis Products, *Anal. Chem.* **25**, 1140–1155 (1953). 30 spectra.

*D. J. Haslam, H. A. Willis, and D. C. M. Squirrel, *Identification and Analysis of Plastics*, 2nd ed., Heyden, 1980. 748 pp, 296 spectra.

E. J. C. Henniker, *Infrared Spectrometry of Industrial Polymers*, Academic, New York, 1967. 229 pp. Text only; no reference spectra.

F. D. O. Hummel, *Infrared Spectra of Polymers in the Medium and Long Wavelength Regions*, Interscience, New York, 1966. 207 pp, 192 spectra.

*G. D. O. Hummel, *Atlas of Polymer and Plastics Analysis*, 3rd ed., Vol. 1, *Defined Polymers*, Carl Hauser Verlag, Munich. 1000 pp, 2809 spectra at 3 spectra/page. In English and German.

*H. D. O. Hummel and F. Scholl, *Atlas of Polymer and Plastics Analysis*, 2nd ed., Verlag Chemie International, New York, 1979. Three volumes: 1) *Polymers: Structures and Spectra.*

2) *Plastics, Fibres, Rubber, Resins.*
3) *Additives and Processing Aids.*
Very good.

I. W. J. Irwin, *Analytical Pyrolysis. A Comprehensive Guide*, Dekker, 1982.

J. R. E. Kagarise and L. A. Weinberger, *Infrared Spectra of Plastics and Resins*, NRL Report 4369, Naval Research Laboratory, Washington, DC, 1954. PB 111438. Obtainable from National Technical Information Service, Dept. of Commerce, Springfield, VA. 57 spectra.

K. S. S. Stimler and R. E. Kagarise, same title, *Part 2. Materials Developed Since 1954*, NRL Report 6392, Naval Research Laboratory, Washington, DC, 1962. AD 634427. 46 spectra.

L. D. S. Cain and S. S. Stimler, same title, *Part 3. Related Polymeric Materials (Elastomers)*, NRL Report 6503, Naval Research Laboratory, Washington, DC, 1967. AD 649004. 47 spectra.

M. D. N. Kendall, chapter on plastics. See Part 2.

*N. J. L. Koenig, *Spectroscopy of Polymers*, American Chemical Society, Washington, DC, 1991. 328 pp. Includes IR, Raman, and NMR spectra.

O. R. A. Nyquist, *Infrared Spectra of Plastics and Resins*, Chemical Physics Research Laboratory, Dow Chemical Co., Midland, MI, 1961. 125 spectra.

P. P. C. Painter, M. M. Coleman, and J. L. Koenig, *The Theory of Vibrational Spectroscopy and Its Application to Polymeric Materials*, Wiley-Interscience, New York, 1982.

Q. H. W. Siesler and K. Holland-Moritz, *Infrared and Raman Spectroscopy of Polymers*, Dekker, 1980. 400 pp.

R. W. C. Wake, *Analysis of Rubber and Rubber-Like Polymers*, 1st ed., Wiley-Interscience, New York, 1958. Discusses pyrolysis. In 2nd ed. (1969) reference spectra are not given.

S. R. Zbinden, *Infrared Spectroscopy of High Polymers*, Academic, New York, 1964.

T. M. V. Zeller and S. C. Pattacini, *The Infrared Grating Spectra of Polymers*, Perkin-Elmer Corp., Norwalk, CT, 1975. Application Note #13. 29 spectra.

U. The Coblentz Society and Sadtler Research Labs. have extensive collections of polymer spectra. See Part 3.

7. Inorganics and organometallics

A. D. M. Adams, *Metal-Ligand and Related Vibrations*, Edward Arnold, London, 1967. 379 pp.

B. L. C. Afremow and J. J. Vandeberg, High Resolution Spectra of Inorganic Pigments and Extenders in the Mid-Infrared Region from 1500 to 200 cm^{-1}, *J. Paint Technol.* **38**, 169–202 (1966). 78 spectra.

C. V. C. Farmer, *Infrared Spectra of Minerals*, Mineralogical Society, London, 1974. 590 spectra.

D. J. R. Ferraro, *Low Frequency Vibrations of Inorganic and Coordination Compounds*, Plenum, New York, 1971. ca. 300 pp.

E. J. R. Ferraro (Ed.), *The Sadtler Infrared Handbook of Minerals and Clays*, Sadtler Research Laboratories, Philadelphia, PA, 1982.

F. J. A. Gadsden, *Infrared Spectra of Minerals and Related Inorganic Compounds*, Butterworths, 1975. 277 pp.

G. N. N. Greenwood and E. J. F. Ross, *Index of Vibrational Spectra of Inorganic and Organometallic Compounds*, Butterworths.
1) Vol. 1, 1972. Covers 1935–1960. 762 pp.
2) Vol. 2, 1975. Covers 1961–1963. 916 pp.
3) Vol. 3, 1977. 1616 pp.

H. F. R. Haba and C. L. Wilson, *Talanta* **11**, 21–26 (1964). A scheme for the IR identification of polyatomic negative ions. Uses a preliminary separation.

I. J. M. Hunt, M. P. Wisherd, and L. C. Bonham, *Anal. Chem.* **22**, 1478 (1950). Spectra of about 60 minerals. Be careful using these. Their procedure to sort by particle size sometimes fractionated the samples.

J. C. Karr, Jr. (Ed.), *Infrared and Raman Spectroscopy of Lunar and Terrestrial Materials*, Academic Press, New York, 1975. Text and references; very few spectra.

K. J. R. Lehr, et al., *Crystallographic Properties of Fertilizer Compounds*, National Fertilizer Development Center, Muscle Shoals, AL, Chemical Engineering Bulletin No. 6, May 1967. 207 infrared spectra, 4000–400 cm^{-1}, of well-characterized nitrates, phosphates, carbonates, etc.

L. E. Maslowsky, *Vibrational Spectra of Organometallic Compounds*, Wiley, New York, 1977. 1380 tabulated spectra. 124 plotted.

M. F. A. Miller, G. L. Carlson, F. F. Bentley, and W. H. Jones, Infrared Spectra of Inorganic Ions in the Cesium Bromide Region (700–300 cm^{-1}), *Spectrochim. Acta* **16**, 135–235 (1960). An extension of the next paper.

*N. F. A. Miller and C. H. Wilkins, Infrared Spectra and Characteristic Frequencies of Inorganic Ions, *Anal. Chem.* **24**, 1253–1294 (1952). 160 reference spectra, linear in micrometers 4000–650 cm^{-1}.

O. H. Moenke, *Mineralspektren*, Akademie-Verlag, Berlin, 1962. 343 spectra.

P. K. Nakamoto, *Infrared and Raman Spectroscopy of Inorganic and Coordination Compounds*, 4th ed., Wiley, New York, 1986. 484 pp.

*Q. R. A. Nyquist and R. O. Kagel, *Infrared Spectra of Inorganic Compounds (3800–45 cm^{-1})*, Academic, New York, 1971. 495 pp. About 900 spectra of high quality. An excellent collection, IR only.

*R. R. A. Nyquist, R. O. Kagel, C. L. Putzig, and M. A. Luegers, *The Handbook of Infrared and Raman Spectra of Inorganic Compounds and Organic Salts*, 4 vols., Academic, New York, 1996. The best collection available as of 12/98. This is available on CD-ROM.

S. S. D. Ross, *Inorganic Infrared and Raman Spectra*, McGraw-Hill, New York, 1972. 414 pp.

T. J. W. Salisbury, L. S. Walter, N. Vergo, and D. M. D'Aria, *Infrared (2.1–25 μm) Spectra of Minerals*, Johns Hopkins University Press, Baltimore, MD, 1991. Spectra in transmittance (KBr disks), specular reflection, and diffuse reflection for 130 carefully characterized minerals. Typically 4 spectra/mineral, but linear in micrometers. Numerical data on a CD-ROM included.

U. H. W. van der Marel and H. Beutelspacher, *Atlas of Infrared Spectroscopy of Clay Minerals and Their Admixtures*, Elsevier, 1976. 1180 spectra.

8. Biochemical, drug, and forensic applications
   A. Association of Official Analytical Chemists (AOAC), *Infrared and Ultraviolet Spectra of Some Compounds of Pharmaceutical Interest*, AOAC, Washington, DC, 1972.
   B. R. W. Hannah and S. C. Pattacini, *The Identification of Drugs from Their Infrared Spectra*, Applications Study #11, Perkin-Elmer Corp., Norwalk, CT, 1972.
   C. J. Holubeck and O. Strouf, *Spectral Data and Physical Constants of Alkaloids*, Heyden, 1965.
   D. *Infrared Absorption Spectra of Steroids, An Atlas*, Interscience, New York.
      1) Vol. I, Dobriner, Katzenellenbogen, and Jones, 1953.
      2) Vol. II, Roberts, Gallagher, and Jones, 1958.
   E. H. H. Mantsch and Dennis Chapman (Eds.), *Infrared Spectroscopy of Biomolecules*, Wiley-Liss, New York, 1996. 359 pp.
   F. M. J. Maunder, *Practical Hints on Infrared Spectrometry from a Forensic Analyst*, Adam Hilger, London, 1972.
   G. T. Mills III and J. Conrad Roberson, *Instrumental Data for Drug Analysis*, CRC Press, Boca Raton, FL, 1993. 7 volumes covering IR, NMR, UV, and other data. Especially Vols. 1–4 for IR spectra.
   H. A. C. Moffat (Senior Ed.), *Clarke's Isolation and Identification of Drugs*, Pharmaceutical Press, London, England. In the U. S.: Rittenhouse Book Distributors, King of Prussia, PA, 1986. About 500 IR spectra of drugs.
   I. F. S. Parker, *Applications of Infrared Spectroscopy in Biochemistry, Biology, and Medicine*, Plenum, New York, 1971.

9. Other classes of compounds
   A. Essential oils
      1) J. Bellanato and A. Hidalgo, *Infrared Analysis of Essential Oils*, Heyden, 1971. 164 pp., 214 spectra.
      2) W. W. Morris, High Resolution Infrared Spectra of Fragrance and Flavor Compounds, *J. Assoc. Off. Anal. Chem.* **56**, 1027 (1973).
      3) J. A. Wenninger, R. L. Yates, et al., High Resolution Infrared Spectra of Some Naturally Occurring Sesquiterpene Hydrocarbons.

   - a) I. *J. Assoc. Off. Anal. Chem.* **50**, 1313–1335 (1967).
   - b) II. Ibid., **53**, 949 (1970).
4) See also: a) Kendall, pp. 285–311. See Part 2.
   b) Sadtler collection.

B. Isotopically labeled compounds
   1) S. Pinchas and I. Laulicht, *Infrared Spectra of Labelled Compounds*, Academic, New York, 1971. 371 pp.
   2) Numerous papers by K. Nakamoto on compounds with isotopes of Fe, Mn, and other metals

C. Silicones
   1) E. D. Lipp and A. Lee Smith, Silicones: Infrared, Raman, Near-Infrared, and Ultraviolet Spectra, in A. L. Smith (Ed.), *The Analytical Chemistry of Silicones*, Wiley-Interscience, New York, 1991, Chapter 11, pp. 305–345. Has several labeled IR spectra.

D. Organophosphorus compounds
   1) R. A. Nyquist and W. J. Potts, Jr., Analytical Chemistry of Phosphorus Compounds, in M. Halman (Ed.), *Chemical Analysis Series*, Wiley-Interscience, New York, 1972, pp. 169–293, Chapter 5.
   2) J. V. Pustinger, Jr., W. T. Cave, and M. L. Nielsen, Infrared Spectra of Inorganic Phosphorus Compounds, *Spectrochim. Acta* **15**, 909–925 (1959). 57 spectra.
   3) L. C. Thomas, *Interpretation of the Infrared Spectra of Organophosphorus Compounds*, Heyden, ca. 1975. 257 pp.

E. Pesticides
   1) R. C. Gore, R. W. Hannah, S. C. Pattacini, and T. J. Porro, Pesticide Residues: Infrared and Ultraviolet Spectra of 76 Pesticides, *J. Assoc. Off. Agric. Chem.* **54**, 1040–1082 (1971).
   2) W. M. Morris, Jr., and E. O. Haenni, Infrared Spectra of Pesticides, *J. Assoc. Off. Agric. Chem.* **46**, 964–992 (1963). 40 spectra.
   3) T. Visser, *Infrared Spectra of Pesticides*, Marcel Dekker, New York, 1993. 440 pp, over 400 spectra.

F. Propellants and explosives
   1) F. Pristera and W. Fredericks, *Compilation of Infrared Spectra of Ingredients of Propellants and Explosives*, U.S. Army Munitions Command, Picatinny Arsenal, Dover, NJ, 1965.
   2) F. Pristera et al., *Anal. Chem.* **32**, 495–508 (1960). 68 spectra.

G. Simpler molecules, fundamental frequencies for
   *1) T. Shimanouchi, Tables of Molecular Vibrational Frequencies.
      - a) Consolidated Volume I, SD Catalog No. C13, 48:39, U.S. Government Printing Office, Washington, DC, 1972.
      - b) Part 5. *J. Phys. Chem. Ref. Data* **1**, 189–216 (1972). Reprint No. 5.
      - c) Part 6. Ibid. **2**, 121–162 (1973). Reprint No. 21.
      - d) Part 7. Ibid. **2**, 225–256 (1973). Reprint No. 25.
      - e) Part 8. Ibid. **3**, 269–306 (1974). Reprint No. 49.

10. Surface studies
   A. M. L. Hair, *Infrared Spectroscopy in Surface Chemistry*, Dekker, 1967. 315 pp.
   B. L. H. Little, *Infrared Spectra of Adsorbed Species*, Academic, New York, 1966. 428 pp.
   C. W. Suëtaka, *Surface Infrared and Raman Spectroscopy*, Plenum, New York, 1995. 270 pp.
   D. J. T. Yates, Jr., and T. E. Madey (Eds.), *Vibrational Spectroscopy of Molecules on Surfaces*, Plenum, New York, 1987. 484 pp.
   E. There are many papers on this subject.

11. Near-infrared spectra
   A. W. Kaye, Near Infrared Spectroscopy. I. Spectral Identification and Analytical Applications, *Spectrochim. Acta* **6**, 257–287 (1954).
   B. W. Kaye, Near Infrared Spectroscopy. II. Instrumentation and Technique, *Spectrochim. Acta* **7**, 181–204 (1955).
   C. O. H. Wheeler, Near Infrared Spectra of Organic Compounds, *Chem. Rev.* **59**, 629–666 (1959).
   D. There are many papers since 1980 on quantitative analysis by near IR spectroscopy.

12. Far-infrared spectra
   A. A. Finch, P. N. Gates, K. Radcliffe, F. N. Dickson, and F. F. Bentley, *Chemical Applications of Far Infrared Spectroscopy*, Academic, New York, 1970. 277 pp.
   *B. K. D. Möller and W. G. Rothschild, *Far Infrared Spectroscopy*, Wiley-Interscience, New York, 1971. 796 pp.

13. Instrumentation and techniques
   *A. A. R. H. Cole (for IUPAC), *Table of Wavenumbers for the Calibration of Infrared Spectrometers*, 2nd ed., Pergamon, New York, 1977.
   B. P. B. Coleman et al., *Practical Sampling Techniques for Infrared Analysis*, CRC Press, Boca Raton, FL, 1993. 301 pp.
   C. J. R. Ferraro and K. Krishnan, *Practical Fourier Transform Infrared Spectroscopoy*, Academic, New York, 1990. 534 pp.
   *D. N. J. Harrick, *Internal Reflection Spectroscopy*, Wiley-Interscience, New York, 1967. Obtainable from Harrick Scientific Corp., Croton Dam Road, Box 867, Ossining, NY 10562.
   *E. P. R. Griffiths and J. A. deHaseth, *Fourier Transform Infrared Spectrometry*, Wiley, New York, 1986. 656 pp.
   F. H. J. Humecki (Ed.), *Practical Guide to Infrared Microspectroscopy*, Marcel Dekker, 1995. 488 pp. 12 chapters by various authors.
   G. G. W. J. Irwin, *Analytical Pyrolysis. A Comprehensive Guide*, Dekker, 1982.
   H. R. G. Messerschmidt and M. A. Harthcock, *Infrared Microspectroscopy*, Dekker, 1988. 312 pp.
   I. Potts. See Part 1.

J. W. J. Potts, Jr., and A. L. Smith, Optimizing the Operating Parameters of Infrared Spectrometers, *Appl. Opt.* **6**, 257–265 (1967).

*K. Smith. See Part 1.

*L. A. L. Smith, Trace Analysis by Infrared Spectroscopy, in G. D. Christian and J. B. Callis (Eds.), *Trace Analysis: Spectroscopic Methods for Molecules*, Wiley, 1986, pp. 175–284. Excellent.

M. R. A. Spragg, A Rapid Sample Preparation Technique for Diffuse Reflectance Measurements, *Appl. Spectrosc.* **38**, 604–605 (1984). Describes his method of rubbing a hard polymer on SiC paper.

14. Miscellaneous infrared topics

A. Coblentz Society Specifications for Evaluation of Infrared Spectra.

1) *Anal. Chem.* **38**, 27A (No. 9, 1966).

2) *Anal. Chem.* **47**, 945A (Sept. 1975). Class II spectra.

B. E. R. Lippincott, The Limitations and Advantages of Infrared Spectroscopy in Patent Problems, *J. Patent Office Soc.* **45**, 380 (1963).

15. Raman spectroscopy

*A. F. R. Dollish, W. G. Fateley, and F. F. Bentley, *Characteristic Raman Frequencies of Organic Molecules*, Wiley, New York, 1974. 443 pp, 108 spectra. The best book to date on this subject.

B. S. K. Freeman, *Applications of Laser Raman Spectroscopy*, Wiley, New York, 1974. Applications to organic chemistry only.

C. J. G. Grasselli and B. J. Bulkin (Eds.), *Analytical Raman Spectroscopy*, Wiley, New York, 1991. 400 pp. Chapters by various authors.

*D. J. G. Grasselli, M. K. Snavely, and B. J. Bulkin, *Chemical Applications of Raman Spectroscopy*, Wiley-Interscience, New York, 1981. 198 pp. Excellent.

E. P. Hendra, C. Jones, and G. Warnes, *Fourier Transform Raman Spectroscopy. Instrumentation and Chemical Applications*, Ellis Horwood, New York, 1991. 311 pp.

*F. Schrader. See Part 3.

G. D. P. Strommen and K. Nakamoto, *Laboratory Raman Spectroscopy*, Wiley, New York, 1984. 138 pp.

H. H. Szymanski (Ed.), *Raman Spectroscopy. Theory and Practice*, Plenum, New York.

1) Vol. 1, 1967. 255 pp. Contains some excellent chapters, especially the introductory one by L. A. Woodward.

2) Vol. 2, 1970. 221 pp.

I. See also Part 2.

J. There are many more references on Raman spectroscopy.

INDEX

Absorption mechanism, of IR radiation, 13–17
Accuracy, of labeling and/or purity, of two liquid samples, exercises, 405–406, 533–534
Acetylene-d_2, isotopic substitution effects on group frequencies, 59
Acetylene frequencies, 85–91
  3-bromopropyne, 90–91
  carbon-carbon stretch, 86–90
  carbon-hydrogen bend, 86
  carbon-hydrogen stretch, 85
  hexyne, 90
  isotopic substitution effects on group frequencies, 59
  mono- and dissubstituted, 85
Addition reactions, polymers formation, 276–279
Adhesive example, mixtures, extraction, precipitation, and column chromatography, 474, 479–480, 481–483
Adsorbed water, inorganic materials, 300
Alcohols
  C-O single bonds, 213
  X-H systems spectra, 169
Aliphatic halocompounds, halogen group frequencies, 246
Alkanes, 33–72
  branched-chain hydrocarbon spectra, 50–52
    carbon-hydrogen bending vibrations, 50–52
    carbon-hydrogen stretching vibrations, 50
  complex molecular structure analysis, 69–71
  cyclic alkane carbon-hydrogen stretching and bending modes, 52–56
    hybridization effects on bending modes, 54–56
    hybridization effects on stretching modes, 52–54
  general appearance, 33–36
    mineral oil (Nujol), 33–36
    saturated hydrocarbons, 33
  heteroatom substitution effects
    generally, 60–67
    on methyl symmetric bending mode, 67–69
  hyperconjugation effects, 69
  isotopic substitution effect on group frequencies, 56–60
  normal hydrocarbon spectra, 36–50
    carbon-carbon stretching and bending vibrations, 48–50
    carbon-hydrogen bending vibrations, 42–47
    carbon-hydrogen oscillator factors leading to good group frequencies, 39–41
    carbon-hydrogen stretching modes, 38–39
    carbon-hydrogen stretching vibration, 36–38
    coupling in carbon-hydrogen stretching vibrations, 41–42
    intensity of carbon-hydrogen bending modes, 48
    intensity of carbon-hydrogen stretching modes, 42
Alkenes (olefins), 73–84
  carbon-carbon stretch, 74–79
  carbon-hydrogen stretch, 73–74

Course Notes on the Interpretation of Infrared and Raman Spectra, by Dana W. Mayo, Foil A. Miller, and Robert W. Hannah.
ISBN 0-471-24823-1

Alkenes (olefins) (*Continued*)
  exercises, 145–150, 509–514
  hydrogen out-of-plane bending modes, 79–84
  procedures recommended for, 84
Allene, isotopic substitution effects on group frequencies, 59
Allene-d_4, isotopic substitution effects on group frequencies, 59
Amides, 205–209
  "amide I" band, 205–206
  "amide II" band, 206–207
  characteristic frequencies, 205
  examples, 207–209
  N-H stretch, 205
Amines, N-H group frequencies, X-H systems spectra, 173
Amine salts, N-H group frequencies, X-H systems spectra, 173, 178
Ammonium ion, inorganic salts, 305
Ancient collection, prism spectra from, exercises, 259–260, 527–528
Anhydrides, C-O single bonds, 213
Apatite (fluorapatite), 345
Aromatic compounds, 101–140
  appearance, 101
  carbon-carbon ring stretching modes, 108–121
    606 cm^{-1} degenerate in-plane bend, 119–121
    1010 cm^{-1} (ν_{12}) Raman band, 117–119
    1500 and 1450 cm^{-1} degenerate pair, 115–117
    1600 cm^{-1} degenerate pair, 108–115
    symmetric ring breathing mode (ν_1), 119
  carbon-hydrogen bending vibrations, 121–129
    in-plane C-H bending, 121–123
    out-of-plane C-H and C-C bending, 123–126
    sum tones in 2000–1650 cm^{-1} region, 126–129
  carbon-hydrogen stretching vibrations, 105–108
  classification of, 103–104
  good group frequencies, 103
  group frequencies, 365, 368–377
  heterocyclic spectra, 129–133
    carbon-hydrogen out-of-plane bending vibrations, 131–133
    carbon-hydrogen stretching vibrations, 129
    ring C-C stretching vibrations, 129–131
    sum tones in 2100–1650 cm^{-1} region, 133
  interpretation, 133–140
  vibrational mode, 101–103

Band intensity criteria, 7–8
Bandwidth criteria, 8
Barbiturates example, mixtures, thin-layer chromatography (TLC), 469, 474, 475–478
Bending vibrations, carbon-hydrogen, normal hydrocarbon spectra, alkanes, 42–47
Benzaldehyde, carbonyl group frequencies, 377
Bicarbonate, inorganic salts, 317
Bisulfate, inorganic salts, 306, 311
Branched-chain hydrocarbon spectra, alkanes, 50–52
  carbon-hydrogen bending vibrations, 50–52
  carbon-hydrogen stretching vibrations, 50
3-Bromopropyne, acetylenes, characteristic frequencies, 90–91
Butyl rubber (polybutadiene), 261–262, 265

Carbonate, inorganic salts, 311, 317
Carbon-carbon ring stretching modes (aromatic compounds), 108–121
  606 cm^{-1} degenerate in-plane bend, 119–121
  1010 cm^{-1} (ν_{12}) Raman band, 117–119
  1500- and 1450 cm^{-1} degenerate pair, 115–117
  1600 cm^{-1} degenerate pair, 108–115
  symmetric ring breathing mode (ν_1), 119
Carbon-carbon stretch
  acetylenes, characteristic frequencies, 86–90
  alkenes (olefins), 74–79
Carbon-carbon stretching and bending vibrations, normal hydrocarbon spectra, alkanes, 48–50
Carbon-filled materials example, mixtures, 497–498
Carbon-hydrogen bending
  acetylenes, characteristic frequencies, 86
  aromatic compounds, 121–129
    in-plane C-H bending, 121–123
    out-of-plane C-H and C-C bending, 123–126

sum tones in 2000–1650 cm^{-1} region, 126–129
branched-chain hydrocarbon spectra, alkanes, 50–52
intensity of, normal hydrocarbon spectra, alkanes, 48
normal hydrocarbon spectra, alkanes, 42–47
Carbon-hydrogen oscillator factors, leading to good group frequencies, normal hydrocarbon spectra, alkanes, 39–41
Carbon-hydrogen out-of-plane bending vibrations, aromatic compounds, heterocyclic spectra, 131–133
Carbon-hydrogen stretch
acetylenes, characteristic frequencies, 85
alkenes (olefins), 73–74
intensity of, normal hydrocarbon spectra, alkanes, 42
normal hydrocarbon spectra, alkanes, 38–39
Carbon-hydrogen stretching vibrations
aromatic compounds, 105–108
heterocyclic spectra, 129
branched-chain hydrocarbon spectra, alkanes, 50
coupling in, normal hydrocarbon spectra, alkanes, 41–42
normal hydrocarbon spectra, alkanes, 36–38
Carbonyl compounds, 179–204
electronic effects, 184–189
example, 199–201
geometric effects, 183–184
interaction effects, 189–199
intermolecular, 198–199
intramolecular, 189–198
mass effects, 180–183
overview, 179–180
perturbing factors summary, 201–204
Carbonyl group frequencies, 377–393
3-heptanone, 378, 382–383
benzaldehyde, 377
hexanal, 377–378
hexanamide, 386–389
hexanoic acid, 389, 392–393
hexanoic anhydride, 389
hexanoyl chloride, 383
hexyl acetate, 383, 385–386
maleic anhydride, 389
N-Methyl hexanamide, 389
vinyl acetate, 386
Carboxylate ion, 210
Carboxylic acids, X-H systems spectra, 169–172
C=C systems, group frequencies, 393
Chromate, inorganic salts, 311
Clays, 345
Collision model, for scattering, Raman spectroscopy, 18–19
Colthup-type chart, inorganic salts, 303
Column chromatography, mixtures, 462, 474, 479–480
Complex molecular structure analysis, alkanes, 69–71
Complex molecules, identification of unknown pure samples, exercises, 407–411, 535–538
Computer manipulation, mixtures, 463, 499–503
Condensation reaction, polymers formation, 273–276
Conformational isomerism, polymers, 288–290
Coordinated water, inorganic materials, 300
Copolymers, formation of, 279–286
C-O single bonds, 210–215
alcohols, 213
anhydrides, 213
esters, 213
ethers, 210–213
peroxides, 213
substances containing, 210
Coupling
aromatic compounds, carbon-hydrogen stretching vibrations, 106
in carbon-hydrogen stretching vibrations, normal hydrocarbon spectra, alkanes, 41–42
infrared and Raman spectra, 26–30
Covalent bonds, Inorganic materials, 297
Crosslinked polymers, formation of, 286–287
Crystalline sulfur, 349
Crystallinity, polymers, 290–291
Crystals, lattice vibrations (phonons), 353–354
Cumulated double-bond systems, 96–99. *See also* Triple bond molecule frequencies
linear systems, 96–98
nonlinear systems, 99
Cyanide, inorganic salts, 322, 327
Cyclic alkane carbon-hydrogen stretching and bending modes, alkanes, 52–56

Cyclic alkane carbon-hydrogen stretching and bending modes, alkanes (*Continued*)
  hybridization effects on bending modes, 54–56
  hybridization effects on stretching modes, 52–54

Dialysis, mixtures, 462, 499
Diamond, 349
Dibasic orthophosphate, inorganic salts, 311
Diffuse reflection, sample-handling techniques, 456–460
Dolomite, 349

Electronic effects, carbonyl compounds, 184–189
Energy levels, in scattering, Raman spectroscopy, 19–20
Esters, C-O single bonds, 213
Ethers, C-O single bonds, 210–213
Ethylene, isotopic substitution effects on group frequencies, 58
Ethylene-d_4, isotopic substitution effects on group frequencies, 58
Exercises
  alkenes (olefins), 145–150, 509–514
  assignment of, 141–144
  complex molecules, identification of unknown pure samples, 407–411, 535–538
  liquid samples, accuracy of labeling and/or purity of two samples, 405–406, 533–534
  methyl-*n*-propyl isoamyl amine oxide pyrolysis, 155–156, 518–519
  organic compound, pure
    of simple structure, 257–258, 526–527
    unknown, 247–248, 521–522
  organic esters of inorganic acids, identification of, 414–415, 538–539
  organic liquids, 151–154, 514–518
    pure, of simple structure, 251–256, 523–526
  polymer, high-molecular-weight, 249–250, 522–523
  polymer film, identification of functional groups in, 157–158, 161, 518–519, 520–521
  polymers
    class identification of, 422–423, 546–547
    polymer prototype molecule and, identification of, 418–421, 541–545
  prism spectra from ancient collection, 259–260, 527–528
  pure organic compounds of simple structure, identification of, 399–404, 529–533
    sodium fusion, 412–413, 538
  reaction products, liquid mixture of, distillation, 159–160, 520–521
  solid samples, identification of, 416, 540
Extraction chromatography, adhesive example, mixtures, 474, 479–480

Field effects, carbonyl compounds, 197–198
Fingerprint frequencies
  below 1500 cm^{-1}, 2–3
  vibrational frequencies, 1
Fire retardant example, mixtures, 480, 484–487
First-order coupling effects, carbonyl compounds, 189–191
Flow charts, polymers, 291–295
Fluorapatite (apatite), 345

Gas chromatography, mixtures, 462
Geometric effects, carbonyl compounds, 183–184
Good group frequencies
  aromatic compounds, 103
  carbon-hydrogen oscillator factors leading to, alkanes, 39–41
Group frequencies, 3–7. *See also* Infrared and Raman group frequencies
  absence of, 3
  defined, 3
  desirable qualities, 3
  empirical determination, 3
  examples, 4–7
  procedures for use of, 3–4
  vibrational frequencies, 1

Halogen group frequencies, 241–246
  aliphatic halocompounds, 246
  C-Cl, C-Br, and C-I, 241–246
  C-F groups, 241
  generally, 217–218
Halosilanes, silicon compounds, 229
Heavy atoms, salts, 303
3-Heptanone, carbonyl group frequencies, 378, 382–383
Heteroatom substitution effects, alkanes
  generally, 60–67

on methyl symmetric bending mode, 67–69
Heterocyclic spectra (aromatic compounds), 129–133
carbon-hydrogen out-of-plane bending vibrations, 131–133
carbon-hydrogen stretching vibrations, 129
ring C-C stretching vibrations, 129–131
sum tones in 2100–1650 cm^{-1} region, 133
Hexanal, carbonyl group frequencies, 377–378
Hexanamide, carbonyl group frequencies, 386–389
Hexanoic acid, carbonyl group frequencies, 389, 392–393
Hexanoic anhydride, carbonyl group frequencies, 389
Hexanoyl chloride, carbonyl group frequencies, 383
trans-2-Hexene, group frequencies, 393
Hexyl acetate, carbonyl group frequencies, 383, 385–386
Hexyne, acetylenes, characteristic frequencies, 90
High-molecular-weight polymer, exercises, 249–250, 522–523
Hybridization effects
on bending modes, cyclic alkane carbon-hydrogen stretching and bending modes, alkanes, 54–56
on stretching modes, cyclic alkane carbon-hydrogen stretching and bending modes, alkanes, 52–54
Hydrocarbons, group frequencies, 357–359
Hydrogen bonding, X-H systems spectra, 164–165
Hydrogen out-of-plane bending modes, alkenes (olefins), 79–84
Hydroxyapatite, 345, 349
Hyperconjugation effects, alkanes, 69

Inductive effects, carbonyl compounds, 184–189
Infrared and Raman group frequencies, 355–398. *See also* Group frequencies
applications, 393, 397–398
aromatic substances, 365, 368–377
carbonyl groups, 377–393
3-heptanone, 378, 382–383
benzaldehyde, 377
hexanal, 377–378
hexanamide, 386–389
hexanoic acid, 389, 392–393
hexanoic anhydride, 389
hexanoyl chloride, 383
hexyl acetate, 383, 385–386
maleic anhydride, 389
N-Methyl hexanamide, 389
vinyl acetate, 386
C=C systems, 393
hydrocarbons, 357–359
nitrile substitutents, 377
nitro group substitution, 365
overview, 355–357
substitution by polar atoms and X-H groups, 359–365
sulfur-containing groups, 365
survey, 357
Infrared and Raman spectra, 1–32
absorption of IR radiation mechanism, 13–17
band intensity criteria, 7–8
bands expected in, 8–13
bandwidth criteria, 8
comparisons of, 20–22
coupling of vibrations, 26–30
dividing point (1500 cm^{-1}), 2–3
above, 2
below (fingerprint region), 2–3
group characteristics tabulation, 31–32
group frequency, 3–7
absence of, 3
defined, 3
desirable qualities, 3
empirical determination, 3
examples, 4–7
procedures for use of, 3–4
infrared region names, 30–31
intensities, 22–25
infrared, 22–24
Raman, 24–25
Raman group frequencies, 25–26
Raman spectroscopy, 17–22
vibrational frequencies, 1–2
fingerprint frequencies, 1
group frequencies, 1
historical perspective, 2
Infrared microscopy, mixtures, 463, 504
Infrared radiation, absorption mechanism of, 13–17
Inorganic acids, organic esters of, identification of, exercises, 414–415, 538–539
Inorganic materials, 297–354
KBr pressed disk technique, 333

Inorganic materials (*Continued*)
lattice vibrations (phonons), 353–354
lower frequency IR bands, 327
minerals, 345–349
overview, 297
oxides, 333–345
salts, 303–327
ammonium ion, 305
bicarbonate, 317
bisulfate, 306, 311
carbonate, 311, 317
chromate, 311
Colthup-type chart, 303
cyanide, 322, 327
dibasic orthophosphate, 311
heavy atoms, 303
nitrate, 317, 322
nitrite ion, 322
orthophosphate, 311
planar trigonal ions, 311
sodium sulfate, 303
sulfate ion, 305–306
tetrahedral ions, 303–305
thiocyanate, 327
simple spectrum from solid sample, 327, 333
split mull, 297–300
ugly spectra, 327
water, 300–302
weak band, 333
In-plane C-H bending, aromatic compounds, carbon-hydrogen bending vibrations, 121–123
In-plane O-H bend, X-H systems spectra, 169
Intensities
of carbon-hydrogen bending modes, normal hydrocarbon spectra, alkanes, 48
of carbon-hydrogen stretching modes, normal hydrocarbon spectra, alkanes, 42
infrared, 22–24
Raman, 24–25
Interaction effects, carbonyl compounds, 189–199
intramolecular, 189–199
Internal reflection, sample-handling techniques, 448–456
Internal reflection-extraction, mixtures, 462, 497–498
Intramolecular H bonding, carbonyl compounds, 193–195
Intramolecular interaction effects, carbonyl compounds, 189–198
Iron pyrite, 349
Isotopic substitution effects, on group frequencies, alkanes, 56–60

KBr disk technique
inorganic materials, 333
solids, sample-handling techniques, 437–447

Lattice vibrations (phonons), inorganic materials, 353–354
Lattice water, inorganic materials, 300
Lipstick example, mixtures, 468, 469, 470–473
Liquid chromatography, mixtures, 462
Liquid mixture, of reaction products, exercises, 159–160, 520–521
Liquid samples
accuracy of labeling and/or purity of two, exercises, 405–406, 533–534
sample-handling techniques, 425–426
Liquid water, inorganic materials, 300
Lower frequency IR bands, inorganic materials, 327
Lubricating grease example, mixtures, 463–469

Maleic anhydride, carbonyl group frequencies, 389
Mass effects, carbonyl compounds, 180–183
N-Methyl hexanamide, carbonyl group frequencies, 389
Methyl-*n*-propyl isoamyl amine oxide pyrolysis, exercises, 155–156, 518–519
Methyl symmetric bending mode, hetero-atom substitution effects on, alkanes, 67–69
Mineral oil (Nujol), *n*-hexane compared with, alkanes, 33–36
Minerals, inorganic materials, 345–349
MIT-Bowdoin College Summer Infrared course, history of, xvii–xxi
Mixtures, 461–504
examples
adhesive, extraction, precipitation, and column chromatography, 474, 479–480, 481–483
barbiturates, thin-layer chromatography (TLC), 469, 474, 475–478
carbon-filled materials, 497–498

computer manipulation, 499–503
dialysis, 499
fire retardant, 480, 484–487
infrared microscopy, 504
lipstick, 468, 469, 470–473
lubricating grease, 463–469
pyrolysis, 487–497
overview, 461–462
separation techniques, 462–463
Mulling technique, solids, sample-handling techniques, 435–437

N-H group frequencies, X-H systems spectra, 172–178
amine salts, 173, 178
generally, 172
N-H stretches, 172–173
primary amines, 173
secondary amines, 173
N-H stretch
amides, 205
N-H group frequencies, X-H systems spectra, 172–173
Nitrate
inorganic salts, 317, 322
nitrogen-oxygen compounds, n bonded to two oxygen, 222–224
Nitriles, characteristic frequencies, 91–96
frequency of stretch, 91–93
intensity of stretch, 93
Nitrile substitutents, group frequencies, 377
Nitrite ion, inorganic salts, 322
Nitrites, nitrogen-oxygen compounds, n doubly bonded to one oxygen, 219–220
Nitro, nitrogen-oxygen compounds, n bonded to two oxygen, 220–222
Nitrogen-oxygen compounds, 217–224
behavior, 218
generally, 217–218
N bonded to two oxygen, 220–224
nitrate, 222–224
nitro, 220–222
symmetric stretches, 220
N doubly bonded to one oxygen, 218–220
nitrites, 219–220
nitroso, 218–219
nitrosoamines, 220
normal frequency, 218
Nitro group substitution, group frequencies, 365
Nitroso, nitrogen-oxygen compounds, N doubly bonded to one oxygen, 218–219
Nitrosoamines, nitrogen-oxygen compounds, n doubly bonded to one oxygen, 220
N-Methyl hexanamide, carbonyl group frequencies, 389
Normal hydrocarbon spectra (alkanes), 36–50
carbon-carbon stretching and bending vibrations, 48–50
carbon-hydrogen bending vibrations, 42–47
carbon-hydrogen oscillator factors leading to good group frequencies, 39–41
carbon-hydrogen stretching modes, 38–39
carbon-hydrogen stretching vibration, 36–38
coupling in carbon-hydrogen stretching vibrations, 41–42
intensity of carbon-hydrogen bending modes, 48
intensity of carbon-hydrogen stretching modes, 42

O-H group frequencies, X-H systems spectra, 165–169
O-H in-plane bend, X-H systems spectra, 169
Olefins. *See* Alkenes (olefins)
Organic compound, pure
of simple structure, exercises, 257–258, 526–527
unknown, exercises, 247–248, 521–522
Organic compounds, pure, of simple structure, identification of
exercises, 399–404, 529–533
sodium fusion, exercises, 412–413, 538
Organic esters, of inorganic acids, identification of, exercises, 414–415, 538–539
Organic liquids
exercises, 151–154, 514–518
pure, of simple structure, exercises, 251–256, 523–526
Orthophosphate, inorganic salts, 311
Out-of-plane bending vibrations, carbon-hydrogen, aromatic compounds, heterocyclic spectra, 131–133
Out-of-plane C-H and C-C bending, aromatic compounds, carbon-hydrogen bending vibrations, 123–126
Out-of-plane C-O-H bend, X-H systems spectra, 169
Oxides, inorganic materials, 333–345

Peroxides, C-O single bonds, 213
Phonons (lattice vibrations), inorganic materials, 353–354
Phosphorous compounds, 231–233
  example, 233
  generally, 217–218, 231
  PH and PH_2 bends and P-CH_3, 231–232
  P-H stretch, 231
  P-O-C stretches, 233
  P-O-P stretch, 233
  P=O stretch, 232–233
Planar trigonal ions, inorganic salts, 311
Polar atoms, X-H groups and, substitution by, group frequencies, 359–365
Polybutadiene (butyl rubber), 261–262, 263, 265
Polymer(s), 261–296
  butyl rubber (polybutadiene), 261–262, 265
  class identification of, exercises, 422–423, 546–547
  conformational isomerism, 288–290
  crystallinity, 290–291
  flow charts, 291–295
  formation of, 266–268, 273–279
    addition reactions, 276–279
    condensation reaction, 273–276
    copolymers, 279–286
    crosslinked polymers, 286–287
    reaction conditions, 279
    tacticity, 287–288
  high-molecular-weight, exercises, 249–250, 522–523
  overview, 261
  polybutadiene, 261, 263, 265
  polymer prototype molecule and, identification of, exercises, 418–421, 541–545
  polystyrene, 261, 264, 265–266
  problems of, 268–273
Polymer film, identification of functional groups in, exercises, 157–158, 161, 518–519, 520–521
Polystyrene, polymers, 261, 264, 265–266
Polyvinyl chloride (PVC), 266–268
Precipitation chromatography, adhesive example, mixtures, 474, 479–480
Primary amines, N-H group frequencies, X-H systems spectra, 173
Prism spectra, from ancient collection, exercises, 259–260, 527–528
Pure organic compound
  of simple structure, exercises, 257–258, 526–527
  unknown, exercises, 247–248, 521–522
Pure organic compound of simple structure, identification of
  exercises, 399–404, 529–533
  sodium fusion, exercises, 412–413, 538
Pure organic liquids, of simple structure, exercises, 251–256, 523–526
Pyrolysis, mixtures, 462, 487–497

Raman spectra, sulfur compounds, 241
Raman spectroscopy, 17–22
Reaction products, liquid mixture of, exercises, 159–160, 520–521
Reflection techniques, 448–460
  diffuse reflection, 456–460
  internal reflection, 448–456
Resonance effects, carbonyl compounds, 184–189
Ring C-C stretching vibrations, aromatic compounds, heterocyclic spectra, 129–131

Salts (inorganic materials), 303–327
  ammonium ion, 305
  bicarbonate, 317
  bisulfate, 306, 311
  carbonate, 311, 317
  chromate, 311
  Colthup-type chart, 303
  cyanide, 322, 327
  dibasic orthophosphate, 311
  heavy atoms, 303
  nitrate, 317, 322
  nitrite ion, 322
  orthophosphate, 311
  planar trigonal ions, 311
  sodium sulfate, 303
  sulfate ion, 305–306
  tetrahedral ions, 303–305
  thiocyanate, 327
Sample-handling techniques, 425–460
  liquids, 425–426
  overview, 425
  reflection techniques, 448–460
    diffuse reflection, 456–460
    internal reflection, 448–456
  sealed cells, 427, 430–435
  solids, 426, 435–447
    KBr disk technique, 437–447
    mulling technique, 435–437
  window materials, 427

Saturated hydrocarbons, vibrational modes in, alkanes, 33
Sealed cells, sample-handling techniques, 427, 430–435
Secondary amines, N-H group frequencies, X-H systems spectra, 173
Second-order coupling effects, carbonyl compounds, 195–197
Silicon compounds, 224–231
  generally, 217–218, 224
  halosilanes, 229
  Si-CH_3 deformations, 225–229
  Si-H bands, 224–225
  siloxanes, 229
Siloxanes, silicon compounds, 229
Simple spectrum, from solid sample, inorganic materials, 327, 333
Sodium fusion, pure organic compounds of simple structure, identification of, exercises, 412–413, 538
Sodium sulfate, spectra of, 303
Solid samples
  handling techniques, 426, 435–447
    KBr disk technique, 437–447
    mulling technique, 435–437
  identification of, exercises, 416, 540
  simple spectrum from, inorganic materials, 327, 333
Split mull, inorganic materials, 297–300
Styrene, group frequencies, 393
Sulfate ion, inorganic salts, 305–306
Sulfur compounds, 233–241
  C-S and S-S stretches, 235
  C-S single-bond stretch, 237–241
  C-S stretch, 237
  generally, 217–218
  Raman spectra, 241
  S-H stretch, 233–235
  S=O stretch, 235–237
Sulfur-containing groups, group frequencies, 365
Sum tones in 2000–1650 cm^{-1} region, aromatic compounds, carbon-hydrogen bending vibrations, 126–129
Sum tones in 2100–1650 cm^{-1} region, aromatic compounds, heterocyclic spectra, 133

Tacticity, polymer formation, 287–288
Talc, 349
Tetrahedral ions, inorganic salts, 303–305
Thin-layer chromatography (TLC), mixtures, 462, 469, 474
Thiocyanate, inorganic salts, 327
trans-2-Hexene, group frequencies, 393
Transannular interactions, carbonyl compounds, 198
Triple bond molecule frequencies, 85–99. *See also* Cumulated double-bond systems
  acetylenes, 85–91
    3-bromopropyne, 90–91
    carbon-carbon stretch, 86–90
    carbon-hydrogen bend, 86
    carbon-hydrogen stretch, 85
    hexyne, 90
    mono- and dissubstituted, 85
  cumulated double-bond systems, 96–99
  nitriles, 91–96
    frequency of stretch, 91–93
    intensity of stretch, 93
    related groups, 93–96

Ugly spectra, inorganic materials, 327

Vibrational frequencies, 1–2
  fingerprint frequencies, 1
  group frequencies, 1
  historical perspective, 2
Vinyl acetate, carbonyl group frequencies, 386
Vinyl system, group frequencies, 393

Water, inorganic materials, 300–302
Wick-Stick technique, mixtures, 462, 474
Window materials, sample-handling techniques, 427

X-H groups, polar atoms and, substitution by, group frequencies, 359–365
X-H systems spectra, 163–178
  alcohols, 169
  carboxylic acids, 169–172
  hydrogen bonding, 164–165
  N-H group frequencies, 172–178
    amine salts, 173, 178
    generally, 172
    N-H stretches, 172–173
    primary amines, 173
    secondary amines, 173
  O-H group frequencies, 165–169
  stretches, generally, 163–164